FE/EIT EXAM PRE

FUNDAMENTALS OF ENGINEERING

Seventeenth Edition

Donald G. Newnan, PhD, PE Civil Eng., Editor

David R. Arterburn, PhD, New Mexico Institute of Mining and Technology

E. Vernon Ballou, PhD, Senior Chemical Engineer

Gary R. Crossman, PE, Old Dominion University

Fidelis O. Eke, PhD, University of California, Davis

Brian Flinn, PhD, PE, University of Washington

James R. Hutchinson, PhD, University of California, Davis

Lincoln D. Jones, PE, San Jose State University

Sharad Laxpati, PhD, University of Illinois, Chicago

Robert F. Michel, PE, Old Dominion University

Charles E. Smith, PhD, Oregon State University

Lawrence H. Van Vlack, PhD, PE, University of Michigan

President: Roy Lipner
Vice President of Product Development and Publishing: Evan M. Butterfield
Editorial Project Manager: Laurie McGuire
Director of Production: Daniel Frey
Production Editor: Caitlin Ostrow
Creative Director: Lucy Jenkins

Published by Kaplan AEC Education
30 South Wacker Drive
Chicago, IL 60606-7481
(312) 836-4400
www.kaplanaecengineering.com

Printed in the United States of America.

07 08 09 10 9 8 7 6 5 4 3 2 1

CONTENTS

PERMISSIONS

"NCEES Model Rules of Professional Conduct," Chapter 6, reprinted by permission of NCEES
Source: *NCEES Model Rules*, National Council of Examiners for Engineering and Surveying. 2003.

Figure 10.7 reprinted by permission of ASME.
Source: Moody, L., F, *Transactions of the ASME*, Volume 66: pp. 671–684. 1944.

Figure 11.5 courtesy of DuPont.
Source: *Thermodynamic Properties of HFC-134a*, DuPont Company.

Figures 13.4a-b, 13.5a-b, 13.6 a, 13.9, 13.10, Chapter 13 Exhibit 2, 13.18, 13.21, 13.22 and 13.23 used by permission of John Wiley & Sons, Inc.
Source: Callister, William D., Jr. *Materials Science and Engineering: An Introduction, 6/e*. J. Wiley & Sons. 2003.

Figures 13.4c, 13.5c, and 13.6b reprinted by permission of the estate of William G. Moffatt.
Source: Moffatt, William G. *The Structure and Property of Materials,* Volume 1. J. Wiley & Sons. 1964.

Tables 13.4 and 13.5 reprinted by permission of McGraw-Hill Companies.
Source: Fontana, M., *Corrosion Engineering*. McGraw-Hill Companies.

Figure 13.13 used by permission of ASM International.
Source: Mason, Clyde W., *Introductory Physical Metallurgy*: p. 33. 1947.

Figure 13.26 used by permission of ASM International.
Source: Rinebolt, J.A., and W. J. Harris, Jr., "Effect of Alloying Elements on Notch Toughness of Pearlitic Steels." *Transactions of ASM*, Volume 43: pp. 1175–1201. 1951

CHAPTER 1

Introduction

Donald G. Newnan

OUTLINE

HOW TO USE THIS BOOK

Fundamentals of Engineering: Exam Preparation for the FE is designed to help you prepare for the Fundamentals of Engineering/Engineer-in-Training exam. The book covers the full breadth and depth of topics covered by the morning section of the exam, as well as the general discipline afternoon exam. If you plan to take one of the six discipline-specific afternoon exams, you should consider obtaining an additional license review book that focuses on that discipline. Engineering Press offers several.

Each chapter of this book covers one major topic on the exam, reviewing important terms, equations, concepts, analysis methods, and typical problems. Solved examples are provided throughout each chapter to help you apply the concepts and to model problems you may see on the exam. After reviewing the topic, you can work the end-of-chapter problems to test your understanding. The problems are typical of what you will see on the exam, and complete solutions are provided so that you can check your work and further refine your solution methodology.

After reviewing individual topics, you should take the Sample Exam at the end of the book. The morning portion of the Sample Exam is the same length as the actual exam. To fully simulate the exam experience, you should answer these 120 questions in an uninterrupted four-hour period, without referring back to any content in the rest of the book. The afternoon portion of the Sample Exam models the 60-question afternoon general discipline exam. Answer these questions if you plan to take the general option on the actual afternoon exam, or if you simply want more practice on morning topics. Like the morning portion, the afternoon portion of the exam will have a four-hour time limit.

When you've completed the Sample Exam, check the provided solutions to determine your correct and incorrect answers. This should give you a good sense of topics you may want to spend more time reviewing. Complete solution methods are shown, so you can see how to adjust your approach to problems as needed.

The following sections provide you with additional details on the process of becoming a licensed professional engineer and on what to expect at the exam.

BECOMING A PROFESSIONAL ENGINEER

To achieve registration as a Professional Engineer, there are four distinct steps: (1) education, (2) the Fundamentals of Engineering/Engineer-in-Training (FE/EIT) exam, (3) professional experience, and (4) the professional engineer (PE) exam. These steps are described in the following sections.

Education

Generally, no college degree is required to be eligible to take the FE/EIT exam. The exact rules vary, but all states allow engineering students to take the FE/EIT exam before they graduate, usually in their senior year. Some states, in fact, have no education requirement at all. One merely need apply and pay the application fee. Perhaps the best time to take the exam is immediately following completion of related coursework. For most engineering students, this will be the end of the senior year.

Fundamentals of Engineering/ Engineer-In-Training Examination

This eight-hour, multiple-choice examination is known by a variety of names—Fundamentals of Engineering, Engineer-in-Training (EIT), and Intern Engineer—but no matter what it is called, the exam is the same in all states. It is prepared and graded by the National Council of Examiners for Engineering and Surveying (NCEES).

Experience

States that allow engineering seniors to take the FE/EIT exam have no experience requirement. These same states, however, generally will allow other applicants to substitute acceptable experience for coursework. Still other states may allow a candidate to take the FE/EIT exam without any education or experience requirements.

Typically, four years of acceptable experience is required before one can take the Professional Engineer exam, but the requirement may vary from state to state.

Professional Engineer Examination

The second national exam is called Principles and Practice of Engineering by NCEES, but many refer to it as the Professional Engineer exam or PE exam. All states, plus Guam, the District of Columbia, and Puerto Rico, use the same NCEES exam. Review materials for this exam are found in other engineering license review books.

FUNDAMENTALS OF ENGINEERING/ ENGINEER-IN-TRAINING EXAMINATION

Laws have been passed that regulate the practice of engineering in order to protect the public from incompetent practitioners. Beginning in 1907 the individual states began passing *title* acts regulating who could call themselves engineers and offer services to the public. As the laws were strengthened, the practice of engineering was limited to those who were registered engineers, or to those working under the supervision of a registered engineer. Originally the laws were limited to civil engineering, but over time they have evolved so that the titles, and some-times the practice, of most branches of engineering are included.

There is no national licensure law; licensure is based on individual state laws and is administered by boards of registration in each state. Table 1.1 is a listing of the state, boards of registration and territory.

Table 1.1 State Boards of Registration for Engineers

State	Web site	Telephone
AL	www.bels.alabama.gov	334-242-5568
AK	www.commerce.state.ak.us/occ/pael.cfm	907-465-1676
AZ	www.btr.state.az.us	602-364-4930
AR	www.arkansas.gov/pels	501-682-2824
CA	Dca.ca.gov/pels/contacts/htm	916-263-2230
CO	www.dora.state.co.us/aes	303-894-7788
CT	State.ct.us/dcp	860-713-6145
DE	www.dape.org	302-368-6708
DC	www.asisvcs.com/indhome _fs.asp?cpcat=en09statereg	202-442-4320
FL	www.fbpe.org	850-521-0500
GA	www.sos.state.ga.us/plb/pels/	478-207-1450
GU	www.guam-peals.org	671-646-3138 or 3115
HI	www.Hawaii.gov/dcca/pbl	808-586-2702
ID	www.ipels.idaho.gov	208-373-7210
IL	www.idfpr.com	217-524-3211
IN	www.in.gov/pla/bandc/engineers	317-234-3022
IA	www.state.ia.us/engls	515-281-7360
KS	www.kansas.gov/ksbtp	785-296-3054
KY	www.kyboels.ky.gov	502-573-2680
LA	www.lapels.com	225-925-6291
ME	www.maine.gov/professionalengineers/	207-287-3236
MD	www.dllr.state.md/us	410-230-6322
MA	www.mass.gov/dpl/boards/en/	617-727-9957
MI	www.michigan.gov/engineers	517-241-9253
MN	www.aelslagid.state.mn.us	651-296-2388
MS	www.pepls.state.ms.us	601-359-6160
MO	www.pr.mo.gov/apelsla.asp	573-751-0047
MP		(011) 670-664-4809
MT	www.engineer.mt.gov	406-841-2367
NE	www.ea.state.ne.us	402-471-2021

(*Continued*)

Table 1.1 State Boards of Registration for Engineers *(Continued)*

State	Web site	Telephone
NV	www.boe.state.nv.us	775-688-1231
NH	www.state.nh.us/jtboard/home.htm	603-271-2219
NJ	www.state.nj.us/lps/ca/nonmedical/pels.htm	973-504-6460
NM	www.state.nm.us pepsboard	505-827-7561
NY	www.op.nysed.gov	518-474-3817 x140
NC	www.ncbels.org	919-791-2000
ND	www.ndpelsboard.org/	701-258-0786
OH	www.ohiopeps.org	614-466-3651
OK	www.pels.state.ok.us/	405-521-2874
OR	www.osbeels.org	503-362-2666
PA	www.dos.state.pa.us/eng	717-783-7049
PR	www.estado.gobierno.pr/ingenieros.htm	787-722-2122 x232
RI	www.bdp.state.ri.us	401-222-2565
SC	www.llr.state.sc.us/POL/Engineers	803-896-4422
SD	www.state.sd.us/dol/boards/engineer	605-394-2510
TN	www.state.tn.us/commerce/boards/ae/	615-741-3221
TX	www.tbpe.state.tx.us	512-440-7723
UT	www.dopl.utah.gov	801-530-6396
VT	vtprofessionals.org	802-828-2191
VI	www.dlca.gov.vi/pro-aels.html	340-773-2226
VA	www.dpor.virginia.gov	804-367-8512
WA	www.dol.wa.gov/engineers/engfront.htm	360-664-1575
WV	www.wvpebd.org	304-558-3554
WI	www.drl.state.wi.us	608-266-2112
WY	engineersandsurveyors.wy.gov	307-777-6155

Examination Development

Initially, the states wrote their own examinations, but beginning in 1966 the NCEES took over the task for some of the states. Now the NCEES exams are used by all states. Thus it is easy for engineers who move from one state to another to achieve licensure in the new state. About 50,000 engineers take the FE/EIT exam annually. This represents about 65% of the engineers graduated in the United States each year.

The development of the FE/EIT exam is the responsibility of the NCEES Committee on Examination for Professional Engineers. The committee is composed of people from industry, consulting, and education, all of whom are subject-matter experts. The test is intended to evaluate an individual's understanding of mathematics, basic sciences, and engineering sciences obtained in an accredited bachelor degree of engineering. Every five years or so, NCEES conducts an engineering task analysis survey. People in education are surveyed periodically to ensure the FE/EIT exam specifications reflect what is being taught.

The exam questions are prepared by the NCEES committee members, subject matter experts, and other volunteers. All people participating must hold professional licensure. When the questions have been written, they are circulated for review in workshop meetings and by mail. Currently, the exam is written in SI units, although some problems may also be solved using engineering units. All problems are four-way multiple choice.

Examination Structure

The FE/EIT exam is divided into a morning four-hour section and an afternoon four-hour section. There are 120 questions in the morning section and 60 in the afternoon.

The morning exam covers the topics that make up roughly the first 2½ years of a typical engineering undergraduate program. Table 1.2 summarizes these topics and the percentage of exam problems you can expect to see on each. All examinees take the same morning exam.

Seven different exams are in the afternoon test booklet, one for each of the following six branches: civil, mechanical, electrical, chemical, industrial, environmental. A general exam is included for those examinees not covered by the six engineering branches. Each of the six branch exams consists of 60 problems covering coursework in the specific branch of engineering. The general exam, also 60 problems, has topics that are similar to the morning topics. Thus the afternoon exam may benefit those specializing in one of the six engineering branches. Most of the test's topics cover the third and fourth year of college courses. These are the courses you will use for the balance of your engineering career, so the test becomes focused to your own needs.

Table 1.2 Morning Topics for the FE/EIT Exam

Topic	Percentage of Exam Problems	Subtopics
Chemistry	9	Acids and bases; equilibrium; equations; metals and nonmetals; nomenclature; oxidation and reduction; periodic table; states of matter; stoichiometry.
Computers	7	Spreadsheets; terminology; structured programming.
Electricity and magnetism	9	Charge, energy, current, voltage, and power; work done in moving a charge in an electric field; force between charges; current and voltage laws; equivalent circuits; capacitance and inductance; reactance and impedance, susceptance and admittance; AC circuits; basic complex algebra.
Engineering economics	8	Discounted cash flow; cost; analyses; uncertainty.
Engineering mechanics	10	Resultants of force systems; centroid of area; concurrent force systems; equilibrium of rigid bodies; frames and trusses; friction; area moments of inertia; linear motion; angular motion; mass moments of inertia; friction; impulse and momentum; work, energy and power.
Engineering probability and statistics	7	Measures of central tendencies and dispersions; probability distributions; conditional probabilities; estimation; regression and curve fitting; expected value; hypothesis testing.
Ethics and business practices	7	Code of ethics; agreements and contracts; ethical vs. legal; professional liability; public protection issues.
Fluid mechanics	7	Flow measurement; fluid properties; fluid statics; energy, impulse and momentum equations; pipe and other internal flow.
Material properties	7	Properties (chemical, electrical, physical, mechanical); corrosion mechanisms and control; materials (engineered, ferrous, non-ferrous).
Mathematics	15	Analytic geometry; differential equations; differential calculus; integral calculus; matrix operations; roots of equations; vector analysis.
Strength of materials	7	Shear and moment diagrams; stress types; stress strain caused by axial loads, bending loads, torsion, shear; combined stresses; columns; indeterminant analysis; plastic vs. elastic deformation.
Thermodynamics	7	Thermodynamic laws (1st, 2nd); availability and reversibility; cycles; energy, heat and work; ideal gases; mixture of gases; phase changes; properties (enthalpy, entropy); heat transfer.

Graduate engineers will find the afternoon branch test to their advantage, as the broad fundamentals test usually causes them to do a good deal of review of their earliest classwork.

We recommend that civil, mechanical, electrical, chemical, enviromental, and industrial engineers take their branch exam. All others should take the general examination. Analysis of pass rates on previous exams shows that examinees taking the discipline-specific exam in their branch of engineering typically do better than those taking the general exam!

At the beginning of the afternoon test period, examinees will mark the answer sheet as to which branch exam they are taking. You could quickly scan the test, judge the degree of difficulty of the general versus the branch exam, then choose the test to answer. We do not recommend this practice, as you would waste time in determining which test to write. Further, you could lose confidence during this indecisive period.

Table 1.3 summarizes the major subjects for the six exams, including the percentage of problems you can expect to see on each one.

Taking the Examination

The National Council of Examiners for Engineering and Surveying (NCEES) prepares FE/EIT exams for use on a Saturday in April and October each year. Some state boards administer the exam twice a year; others offer the exam only once a year. The scheduled exam dates are

	April	**October**
2007	21	27
2008	12	25
2009	25	24
2010	17	30

Those wishing to take the exam must apply to their state board several months before the exam date.

Examination Procedure

Before the morning four-hour session begins, the proctors pass out exam booklets and a scoring sheet to each examinee. Space is provided on each page of the examination booklet for scratchwork. The scratchwork will *not* be considered in the scoring. Proctors will also provide each examinee with a mechanical pencil for use in recording answers; this is the only writing instrument allowed. Do not bring your own lead or eraser. If you need an additional pencil during the exam, a proctor will supply one.

The examination is closed book. You may not bring any reference materials with you to the exam. To replace your own materials, NCEES has prepared a *Fundamentals of Engineering (FE) Supplied-Reference Handbook.* The handbook contains engineering, scientific, and mathematical formulas and tables for use in the examination. Examinees will receive the handbook from their state registration board prior to the examination. The *FE Supplied-Reference Handbook* is also included in the exam materials distributed at the beginning of each four-hour exam period.

Table 1.3 FE Afternoon Exams

Branch Exam	Topics	Percentage of Problems
Chemical	Chemical reaction engineering	10
	Chemical engineering thermodynamics	10
	Chemistry	10
	Computer usage in chemical engineering	5
	Fluid dynamics	10
	Heat transfer	10
	Mass transfer	10
	Material/energy balances	15
	Process control	5
	Process design and economic optimization	10
	Safety, health, and environmental	5
Civil	Construction management	10
	Environmental engineering	12
	Hydraulics and hydrologic systems	12
	Materials	8
	Soil mechanics and foundations	15
	Structural analysis	10
	Structural design	10
	Surveying	11
	Transportation	12
Electrical	Circuits	16
	Communications	9
	Computer systems	10
	Control systems	10
	Digital systems	12
	Electromagnetics	7
	Electronics	15
	Power	13
	Signal processing	8
Environmental	Air quality engineering	15
	Environmental science and management	15
	Solid and hazardous waste engineering	15
	Water and wastewater engineering	30
	Water resources	25
General	Advanced engineering mathematics	10
	Application of engineering mechanics	13
	Biology	5
	Electricity and magnetism	12
	Engineering economics	10
	Engineering of materials	11
	Engineering probability and statistics	9
	Fluids	15
	Thermodynamics and heat transfer	15

(Continued)

Table 1.3 FE Afternoon Exams *(Continued)*

Branch Exam	Topics	Percentage of Problems
Industrial	Engineering economics	15
	Facilities and logistics	12
	Human factors, productivity, ergonomics, and work design	12
	Industrial management	10
	Manufacturing and production systems	13
	Modeling and computation	12
	Quality	11
Mechanical	Fluid mechanics and fluid machinery	15
	Heat transfer	10
	Kinematics, dynamics, and vibrations	15
	Materials and processing	10
	Measurement, instrumentation, and controls	10
	Mechanical design and analysis	15
	Refrigeration and HVAC	10
	Thermodynamics and energy conversion processes	15

There are three versions (A, B, and C) of the exam. These have the major subjects presented in a different order to reduce the possibility of examinees copying from one another. The first subject on your exam, for example, might be fluid mechanics, while the exam of the person next to you may have electrical circuits as the first subject.

The afternoon session begins following a one-hour lunch break. The afternoon exam booklets will be distributed along with a scoring sheet. There will be 60 multiple choice questions, each of which carries twice the grading weight of the morning exam questions.

If you answer all questions more than 15 minutes early, you may turn in the exam materials and leave. If you finish in the last 15 minutes, however, you must remain to the end of the exam period to ensure a quiet environment for all those still working, and to ensure an orderly collection of materials.

Examination-Taking Suggestions

Those familiar with the psychology of examinations have several suggestions for examinees:

1. There are really two skills that examinees can develop and sharpen. One is the skill of illustrating one's knowledge. The other is the skill of familiarization with examination structure and procedure. The first can be enhanced by a systematic review of the subject matter. The second, exam-taking skills, can be improved by practice with sample problems—that is, problems that are presented in the exam format with similar content and level of difficulty.

2. Examinees should answer every problem, even if it is necessary to guess. There is no penalty for guessing. The best approach to guessing is to first eliminate the one or two obviously incorrect answers among the four alternatives. If this can be done, the chance of selecting a correct answer obviously improves from 1 in 4 to 1 in 2 or 3.
3. Plan ahead with a strategy and a time allocation. There are 120 morning problems in 12 subject areas. Compute how much time you will allow for each of the 12 subject areas. You might allocate a little less time per problem for the areas in which you are most proficient, leaving a little more time in subjects that are more difficult for you. Your time plan should include a reserve block for especially difficult problems, for checking your scoring sheet, and finally for making last-minute guesses on problems you did not work. Your strategy might also include time allotments for two passes through the exam—the first to work all problems for which answers are obvious to you, the second to return to the more complex, time-consuming problems and the ones at which you might need to guess.
4. Read all four multiple-choice answers options before making a selection. All distractors (wrong answers) are designed to be plausible. Only one option will be the best answer.
5. Do not change an answer unless you are absolutely certain you have made a mistake. Your first reaction is likely to be correct.
6. If time permits, check your work.
7. Do not sit next to a friend, a window, or other potential distraction.

License Review Books

To prepare for the FE/EIT exam you need two or three review books.

1. This book, to provide a review of the common four-hour morning examination.
2. An afternoon supplement book (for example, *Civil Engineering: Exam Preparation for the FE*) if you are going to take one of the six branch exams. If you plan to take the general afternoon exam, this book covers those topics, so no supplement book is needed.
3. *Fundamentals of Engineering (FE) Supplied-Reference Handbook.* At some point this NCEES-prepared book will be provided to applicants by their State Registration Board. You may want to obtain a copy sooner so you will have ample time to study it before the exam. You must, however, pay close attention to the *FE Supplied-Reference Handbook* and the notation used in it, because it is the only book you will have at the exam.

Textbooks

If you still have your university textbooks, they can be useful in preparing for the exam, unless they are out of date. To a great extent the books will be like old friends with familiar notation. You probably need both textbooks and license review books for efficient study and review.

Examination Day Preparations

The exam day will be a stressful and tiring one. You should take steps to eliminate the possibility of unpleasant surprises. If at all possible, visit the examination site ahead of time. Try too determine such items as

1. How much time should I allow for travel to the exam on that day? Plan to arrive about 15 minutes early. That way you will have ample time, but not too much time. Arriving too early, and mingling with others who are also anxious, can increase your anxiety and nervousness.
2. Where will I park?
3. How does the exam site look? Will I have ample workspace? Will it be overly bright (sunglasses), or cold (sweater), or noisy (earplugs)? Would a cushion make the chair more comfortable?
4. Where are the drinking fountain and lavatory facilities?
5. What about food? Most states do not allow food in the test room (exceptions for ADA). Should I take something along for energy in the exam? A light bag lunch during the break makes sense.

Items to Take to the Examination

Although you may not bring books to the exam, you should bring the following:

- *Calculator*—Beginning with the April 2004 exam, NCEES has implemented a more stringent policy regarding permitted calculators. In brief, you may bring a battery-operated, silent, nonprinting, noncommunicating calculator. For more details, see the NCEES Web site (www.ncees.org), which includes a list of permitted calculators. You also need to determine whether your state permits pre-programmed calculators. Bring extra batteries for your calculator just in case, and many people feel that bringing a second calculator is also a very good idea.
- *Clock*—You must have a time plan and a clock or wristwatch.
- *Exam Assignment Paperwork*—Take along the letter assigning you to the exam at the specified location to prove that you are the registered person. Also bring something with your name and picture (driver's license or identification card).
- *Items Suggested by Your Advance Visit*—If you visit the exam site, it will probably suggest an item or two that you need to add to your list.
- *Clothes*—Plan to wear comfortable clothes. You probably will do better if you are slightly cool, so it is wise to wear layered clothing.

Special Medical Condition

If you have a medical situation that may require special accommodation, you need to notify the licensing board well in advance of exam day.

Examination Scoring

The questions are machine-scored by scanning. The answer sheets are checked for errors by computer. Marking two answers to a question, for example, will be detected and no credit will be given.

CHAPTER 2

Precalculus and Analytic Geometry

David R. Arterburn

OUTLINE

ALGEBRA

Logarithms

Definition. If b is a finite positive number, other than 1, and $b^x = N$, then x is the logarithm of N to the base b, or $\log_b N = x$. If $\log_b N = x$, then $b^x = N$.

Properties of Logarithms

$$\log_b b = 1; \quad \log_b 1 = 0; \quad \log_b 0 = \begin{cases} +\infty, & \text{when } b \text{ lies between 0 and 1} \\ -\infty, & \text{when } b \text{ lies between 1 and } \infty \end{cases}$$

$$\log_b M \bullet N = \log_b M + \log_b N \quad \log_b M/N = \log_b M - \log_b N$$

$$\log_b N^p = p \log_b N \quad \log_b \sqrt[r]{N^p} = \frac{p}{r} \log_b N$$

$$\log_b N = \log_a N / \log_a b \quad \log_b b^N = N; \quad b^{\log_b N} = N$$

Systems of Logarithms

Common (Briggsian)—base 10.

Natural (Napierian or hyperbolic)—base 2.7183 (designated by e or ε).

The abbreviation of *common logarithm* is log, and the abbreviation of *natural logarithm* is ln.

The Solution of Algebraic Equations

The Quadratic Equation

If $ax^2 + bx + c = 0$, then

$$x = \frac{-b \pm \sqrt{b^2 - 4ac}}{2a}$$

If $b^2 - 4ac > 0$, the roots are real and unequal; if $b^2 - 4ac = 0$, the roots are real and equal; if $b^2 - 4ac < 0$, the roots are imaginary.

Progressions

Arithmetic Progression

An arithmetic progression is $a, a + d, a + 2d, a + 3d, \ldots$, where d = common difference.

The nth term is $t_n = a + (n - 1)d$

The sum of n terms is $S_n = \frac{n}{2}[2a + (n-1)d] = \frac{n}{2}(a + t_n)$

Geometric Progression

A geometric progression is $a, ar, ar^2, ar^3, \ldots$, where r = common ratio.

The nth term is $t_n = ar^{n-1}$

The sum of n terms is $S_n = a\left(\frac{1-r^n}{1-r}\right)$

If $r^2 < 1$, S_n approaches a definite limit as n increases indefinitely, and

$$S_\infty = \frac{a}{1-r}$$

COMPLEX QUANTITIES

Definition and Representation of a Complex Quantity (Fig. 2.1)

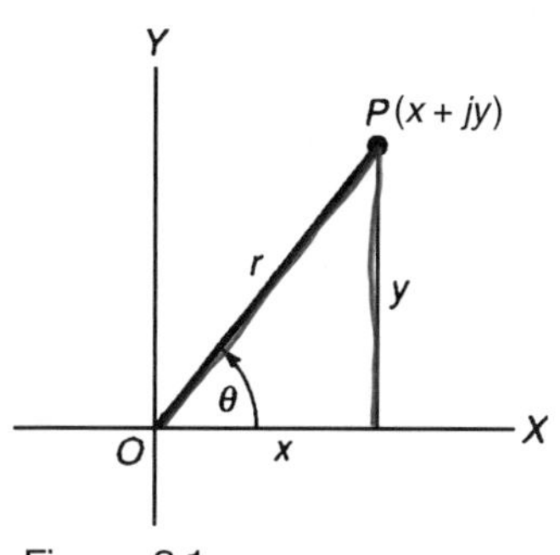

Figure 2.1

If $z = x + jy$, where $j = \sqrt{-1}$ and x and y are real, z is called a complex quantity and is completely determined by x and y.

If $P(x, y)$ is a point in the plane (Fig. 2.1), then the segment OP in magnitude and direction is said to represent the complex quantity $z = x + jy$.

If θ is the angle from OX to OP and r is the length of OP, then $z = x + jy = r(\cos\theta + j\sin\theta) = re^{j\theta}$, where $\theta = \tan^{-1} y/x$, $r = +\sqrt{x^2 + y^2}$, and e is the base of natural logarithms. The pair $x + jy$ and $x - jy$ are called complex conjugate quantities.

Properties of Complex Quantities

Let z, z_1, and z_2 represent complex quantities; then

Sum or Difference: $z_1 \pm z_2 = (x_1 \pm x_2) + j(y_1 \pm y_2)$.

Equation: If $z_1 = z_2$, then $x_1 = x_2$ and $y_1 = y_2$.

Periodicity: $z = r(\cos\theta + j\sin\theta) = r[\cos(\theta + 2k\pi) + j\sin(\theta + 2k\pi)]$, or $z = re^{j\theta} = re^{j(\theta + 2k\pi)}$ and $e^{j2k\pi} = 1$, where k is any integer.

Exponential-Trigonometric Relations: $e^{jz} = \cos z + j\sin z$, $e^{-jz} = \cos z - j\sin z$,

$\cos z = \frac{1}{2}(e^{jz} + e^{-jz})$, $\sin z = \frac{1}{2j}(e^{jz} - e^{-jz})$.

TRIGONOMETRY

Definition of an Angle

An angle is the amount of rotation (in a fixed plane) by which a straight line may be changed from one direction to any other direction. If the rotation is counterclockwise, the angle is said to be positive; if clockwise, negative.

Measure of an Angle

A degree is $\frac{1}{360}$ of the plane angle about a point, and a radian is the angle subtended at the center of a circle by an arc equal in length to the radius.

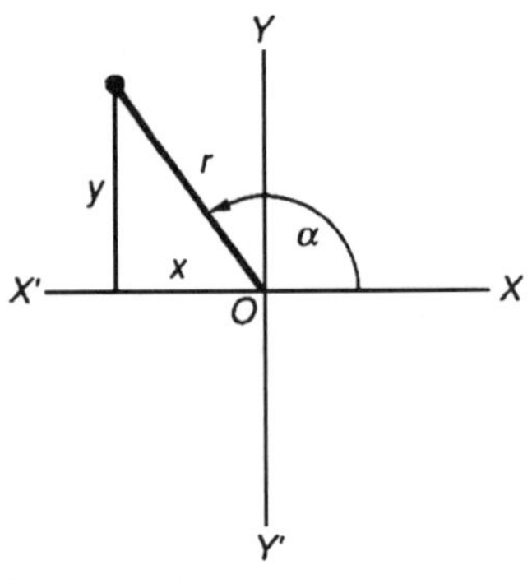

Figure 2.2

Trigonometric Functions of an Angle (Fig. 2.2)

sine (sin) $\alpha = y/r$ cotangent (cot) $\alpha = x/y$
cosine (cos) $\alpha = x/r$ secant (sec) $\alpha = r/x$
tangent (tan) $\alpha = y/x$ cosecant (csc) $\alpha = r/y$

The variable x is positive when measured along OX and negative along OX'. Similarly, y is positive when measured parallel to OY, and negative parallel to OY'.

Fundamental Relations among the Functions

$$\sin\alpha = \frac{1}{\csc\alpha};\quad \cos\alpha = \frac{1}{\sec\alpha};\quad \tan\alpha = \frac{1}{\cot\alpha} = \frac{\sin\alpha}{\cos\alpha}$$

$$\csc\alpha = \frac{1}{\sin\alpha};\quad \sec\alpha = \frac{1}{\cos\alpha};\quad \cot\alpha = \frac{1}{\tan\alpha} = \frac{\cos\alpha}{\sin\alpha}$$

$$\sin^2\alpha + \cos^2\alpha = 1;\quad \sec^2\alpha - \tan^2\alpha = 1;\quad \csc^2\alpha - \cot^2\alpha = 1$$

Functions of Multiple Angles

$$\sin 2\alpha = 2\sin\alpha\cos\alpha$$

$$\cos 2\alpha = 2\cos^2\alpha - 1 = 1 - 2\sin^2\alpha = \cos^2\alpha - \sin^2\alpha$$

$$\tan 2\alpha = (2\tan\alpha)/(1 - \tan^2\alpha)$$

$$\cot 2\alpha = (\cot^2\alpha - 1)/(2\cot\alpha)$$

Functions of Half Angles

$$\sin\frac{1}{2}\alpha = \sqrt{\frac{1-\cos\alpha}{2}};\quad \cos\frac{1}{2}\alpha = \sqrt{\frac{1+\cos\alpha}{2}}$$

$$\tan\frac{1}{2}\alpha = \frac{1-\cos\alpha}{\sin\alpha} = \frac{\sin\alpha}{1+\cos\alpha} = \sqrt{\frac{1-\cos\alpha}{1+\cos\alpha}}$$

Functions of Sum or Difference of Two Angles

$$\sin(\alpha \pm \beta) = \sin\alpha\cos\beta \pm \cos\alpha\sin\beta$$

$$\cos(\alpha \pm \beta) = \cos\alpha\cos\beta \mp \sin\alpha\sin\beta$$

$$\tan(\alpha \pm \beta) = \frac{\tan\alpha \pm \tan\beta}{1 \mp \tan\alpha\tan\beta}$$

Sums, Differences, and Products of Two Functions

$$\sin\alpha + \sin\beta = 2\sin\frac{1}{2}(\alpha+\beta)\cos\frac{1}{2}(\alpha-\beta)$$

$$\sin\alpha - \sin\beta = 2\cos\frac{1}{2}(\alpha+\beta)\sin\frac{1}{2}(\alpha-\beta)$$

$$\cos\alpha + \cos\beta = 2\cos\frac{1}{2}(\alpha+\beta)\cos\frac{1}{2}(\alpha-\beta)$$

$$\cos\alpha - \cos\beta = 2\sin\frac{1}{2}(\alpha+\beta)\sin\frac{1}{2}(\alpha-\beta)$$

$$\tan\alpha \pm \tan\beta = \frac{\sin(\alpha + \beta)}{\cos\alpha\cos\beta}$$

$$\sin^2\alpha - \sin^2\beta = \sin(\alpha+\beta)\sin(\alpha-\beta)$$

$$\cos^2\alpha - \cos^2\beta = \sin(\alpha+\beta)\sin(\alpha-\beta)$$

$$\cos^2\alpha - \sin^2\beta = \cos(\alpha+\beta)\cos(\alpha-\beta)$$

$$\sin\alpha\sin\beta = \frac{1}{2}\cos(\alpha-\beta) - \frac{1}{2}\cos(\alpha+\beta)$$

$$\cos\alpha\cos\beta = \frac{1}{2}\cos(\alpha-\beta) + \frac{1}{2}\cos(\alpha+\beta)$$

$$\sin\alpha\cos\beta = \frac{1}{2}\sin(\alpha+\beta) + \frac{1}{2}\sin(\alpha-\beta)$$

Properties of Plane Triangles (Fig. 2.3)

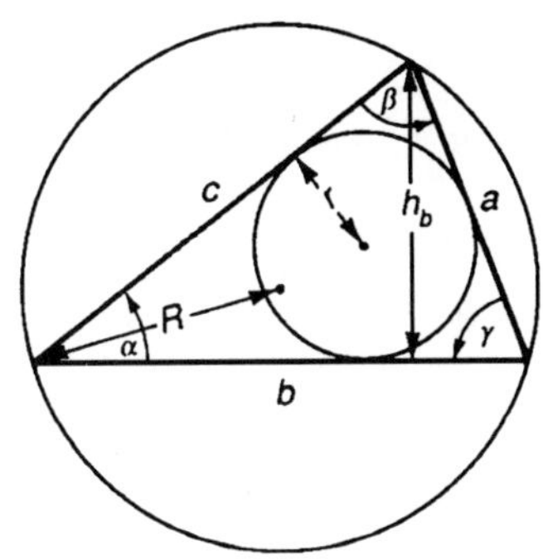

Figure 2.3

Notation. α, β, γ = angles; a, b, c = sides; A = area; h_b = altitude on b; $s = \frac{1}{2}(a + b + c)$; r = radius of inscribed circle; R = radius of circumscribed circle.

$$\alpha + \beta + \gamma = 180° = \pi \text{ radians}$$

$$\frac{a}{\sin\alpha} = \frac{b}{\sin\beta} = \frac{c}{\sin\gamma}$$

$$\frac{a+b}{a-b} = \frac{\tan\frac{1}{2}(\alpha+\beta)}{\tan\frac{1}{2}(\alpha-\beta)}$$

$$a^2 = b^2 + c^2 - 2bc\cos\alpha \qquad a = b\cos\gamma + c\cos\beta$$

$$\cos\alpha = \frac{b^2 + c^2 - a^2}{2bc} \qquad \sin\alpha = \frac{2}{bc}\sqrt{s(s-a)(s-b)(s-c)}$$

$$\sin\frac{\alpha}{2} = \sqrt{\frac{(s-b)(s-c)}{bc}} \qquad \cos\frac{\alpha}{2} = \sqrt{\frac{s(s-a)}{bc}}$$

$$\tan\frac{\alpha}{2} = \sqrt{\frac{(s-b)(s-c)}{s(s-a)}} = \frac{r}{s-a}$$

$$h_b = c\sin\alpha = a\sin\gamma = \frac{2}{b}\sqrt{s(s-a)(s-b)(s-c)}$$

$$r = \sqrt{\frac{(s-a)(s-b)(s-c)}{s}} = (s-a)\tan\frac{\alpha}{2}$$

$$R = \frac{a}{2\sin\alpha} = \frac{abc}{4A}$$

$$A = \frac{1}{2}bh_b = \frac{1}{2}ab\sin\gamma = \frac{a^2\sin\beta\sin\gamma}{2\sin\alpha} = \sqrt{s(s-a)(s-b)(s-c)} = rs$$

Solution of the Right Triangle (Fig. 2.4)

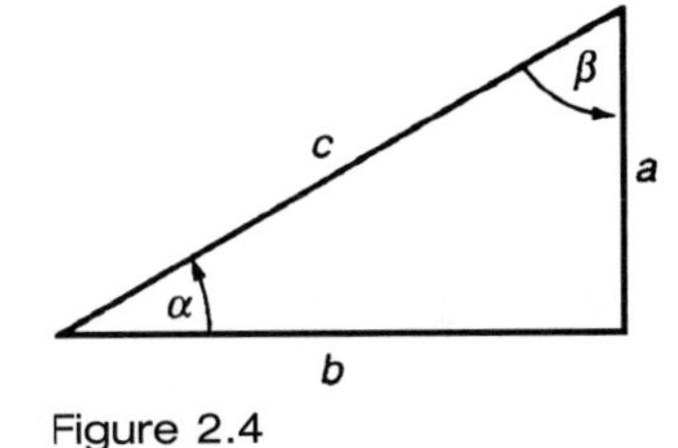

Figure 2.4

Given any two sides, or one side and any acute angle, α, use the following to find the remaining parts:

$$\sin\alpha = \frac{a}{c} \qquad \cos\alpha = \frac{b}{c} \qquad \tan\alpha = \frac{a}{b} \qquad \beta = 90° - \alpha$$

$$a = \sqrt{(c+b)(c-b)} = c\sin\alpha = b\tan\alpha$$

$$b = \sqrt{(c+a)(c-a)} = c\cos\alpha = \frac{a}{\tan\alpha}$$

$$c = \frac{a}{\sin\alpha} = \frac{b}{\cos\alpha} = \sqrt{a^2 + b^2}$$

$$A = \frac{1}{2}ab = \frac{a^2}{2\tan\alpha} = \frac{b^2\tan\alpha}{2} = \frac{c^2\sin 2\alpha}{4}$$

Solution of Oblique Triangles

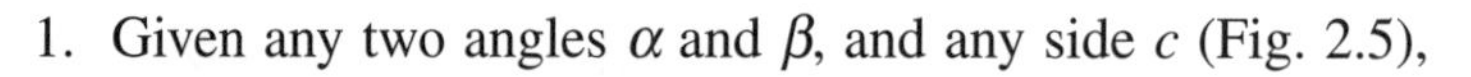

1. Given any two angles α and β, and any side c (Fig. 2.5),

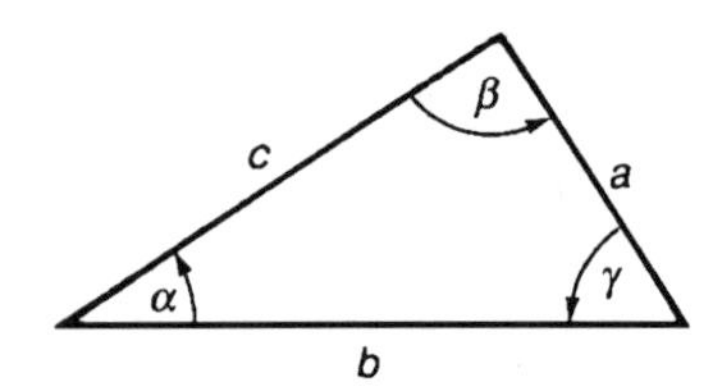

Figure 2.5

$$\gamma = 180° - (\alpha + \beta); \qquad a = \frac{c\sin\alpha}{\sin\gamma}; \qquad b = \frac{c\sin\beta}{\sin\gamma}$$

2. Given any two sides a and c, and an angle opposite one of these, say α (Fig. 2.6),

$$\sin\gamma = \frac{c\sin\alpha}{a}, \qquad \beta = 180° - (\alpha + \gamma), \qquad b = \frac{a\sin\beta}{\sin\alpha}$$

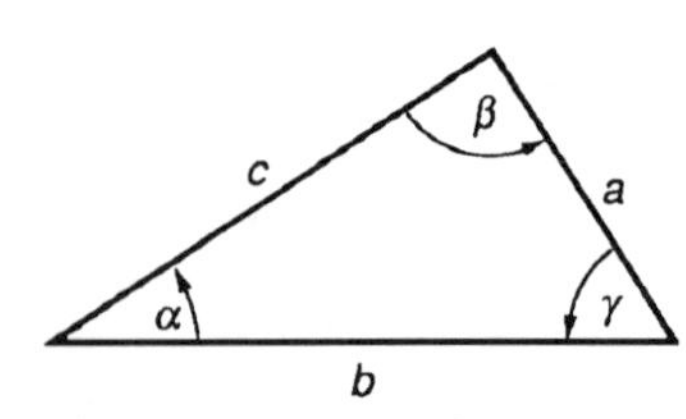

Figure 2.6

In both cases above, γ may have two values: $\gamma_1 < 90°$ and $\gamma_2 = 180° - \gamma_1 > 90°$. If $\alpha + \gamma_2 > 180°$, use only γ_1.

3. Given any two sides b and c and their included angle α, use any one of the following sets of formulas (Fig. 2.7):

$$\frac{1}{2}(\beta+\gamma) = 90° - \frac{1}{2}\alpha; \qquad \tan\frac{1}{2}(\beta-\gamma) = \frac{b-c}{b+c}\tan\frac{1}{2}(\beta+\gamma);$$
$$\beta = \frac{1}{2}(\beta+\gamma) = \frac{1}{2}(\beta-\gamma); \qquad \gamma = \frac{1}{2}(\beta+\gamma) - \frac{1}{2}(\beta-\gamma); \qquad a = \frac{b\sin\alpha}{\sin\beta} \tag{2.1}$$

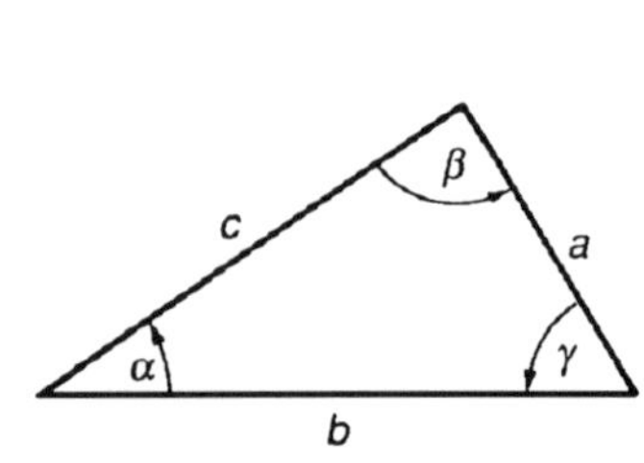

Figure 2.7

$$a = \sqrt{b^2 + c^2 - 2bc\cos\alpha}; \qquad \sin\beta = \frac{b\sin\alpha}{a}; \qquad \gamma = 180° - (\alpha + \beta) \tag{2.2}$$

$$\tan\gamma = \frac{c\sin\alpha}{b - c\cos\alpha}; \qquad \beta = 180° - (\alpha + \gamma); \qquad a = \frac{c\sin\alpha}{\sin\gamma} \tag{2.3}$$

4. Given the three sides a, b, and c, use the following sets of formulas (Fig. 2.8):

$$s = \frac{1}{2}(a+b+c); \qquad r = \sqrt{\frac{(s-a)(s-b)(s-c)}{s}}$$
$$\tan\frac{1}{2}\alpha = \frac{r}{s-a}; \qquad \tan\frac{1}{2}\beta = \frac{r}{s-b}; \qquad \tan\frac{1}{2}\gamma = \frac{r}{s-c} \tag{2.4}$$

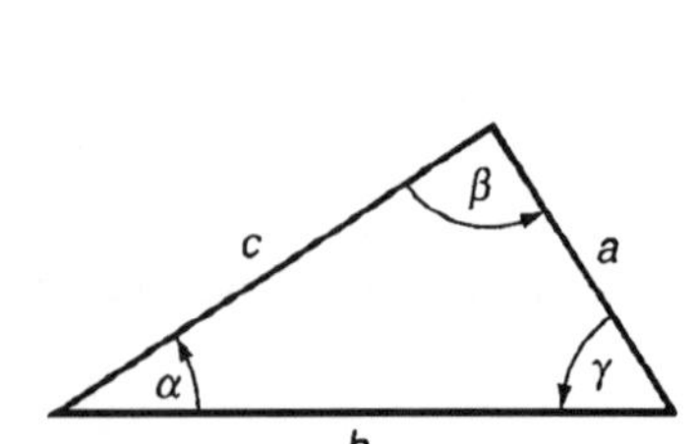

Figure 2.8

$$\cos\alpha = \frac{b^2 + c^2 - a^2}{2bc}; \qquad \cos\beta = \frac{c^2 + a^2 - b^2}{2ca}; \qquad \gamma = 180° - \beta(\alpha + \beta) \tag{2.5}$$

GEOMETRY AND GEOMETRIC PROPERTIES (MENSURATION)

Notation: a, b, c, d, and s denote lengths, A denotes area, V denotes volume.

Right Triangle (Fig. 2.9)

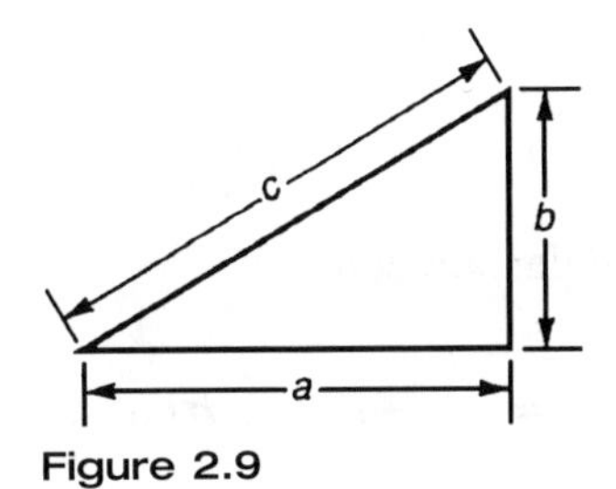

Figure 2.9

$$A = \frac{1}{2}ab$$

$$c = \sqrt{a^2 + b^2}, \quad a = \sqrt{c^2 - b^2}, \quad b = \sqrt{c^2 - a^2}$$

Oblique Triangle (Fig. 2.10)

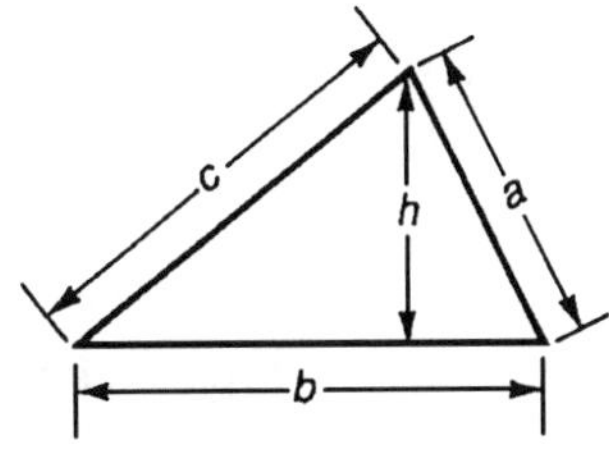

Figure 2.10

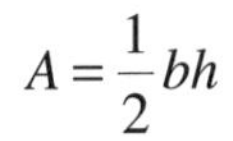

$$A = \frac{1}{2}bh$$

Equilateral Triangle (Fig. 2.11)

All sides and all angles are equal.

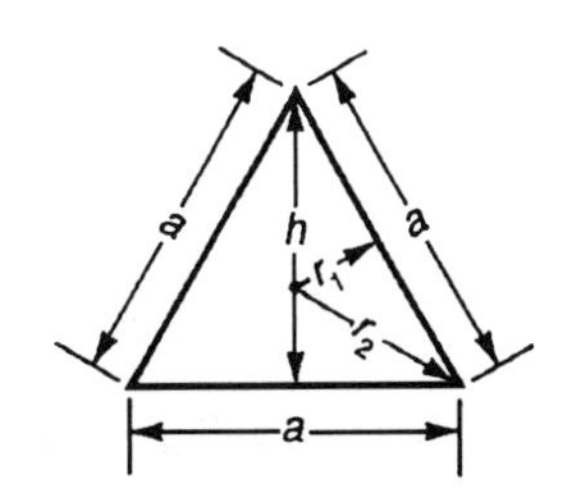

Figure 2.11

$$A = \frac{1}{2}ah = \frac{1}{4}a^2\sqrt{3}, \qquad h = \frac{1}{2}a\sqrt{3}, \qquad r_1 = \frac{a}{2\sqrt{3}}, \qquad r_2 = \frac{a}{\sqrt{3}}$$

Square (Fig. 2.12)

All sides are equal, and all angles are 90°.

$$A = a^2, \qquad d = a\sqrt{2}$$

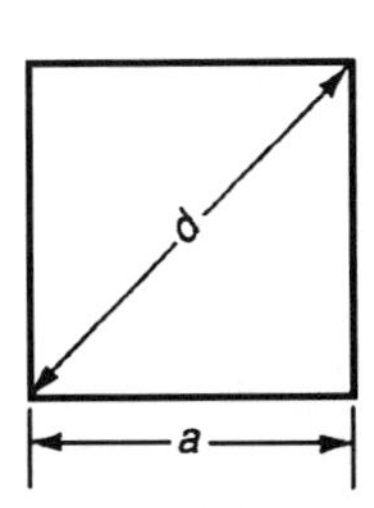

Figure 2.12

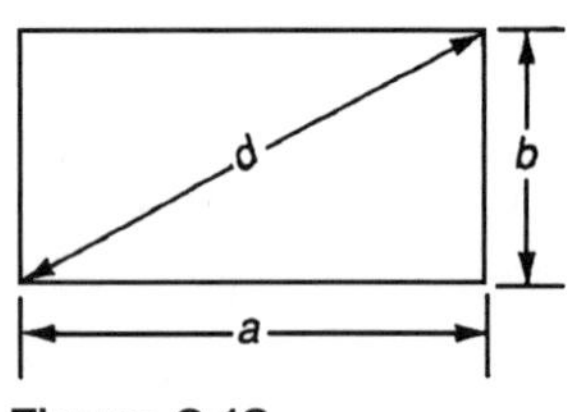

Figure 2.13

Rectangle (Fig. 2.13)

Opposite sides are equal and parallel, and all angles are 90°.

$$A = ab, \qquad d = \sqrt{a^2 + b^2}$$

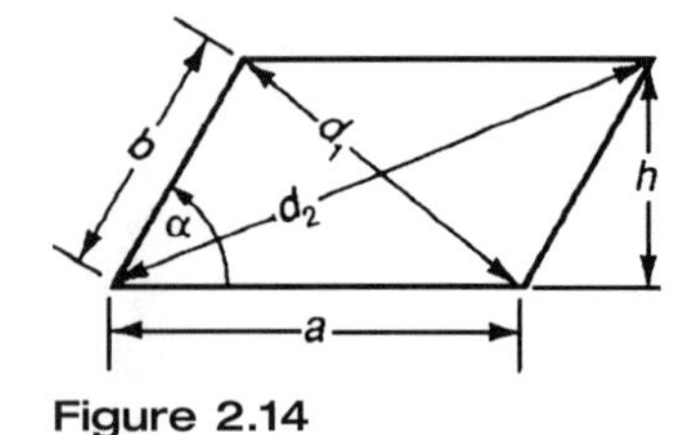

Figure 2.14

Parallelogram (Fig. 2.14)

Opposite sides are equal and parallel, and opposite angles are equal.

$$A = ah = ab\sin\alpha, \quad d_1 = \sqrt{a^2 + b^2 - 2ab\cos\alpha}, \quad d_2 = \sqrt{a^2 + b^2 + 2ab\cos\alpha}$$

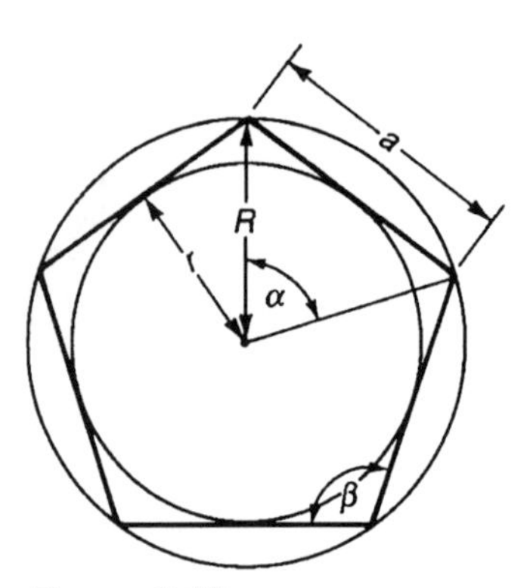

Figure 2.15

Regular Polygon of *n* Sides (Fig. 2.15)

All sides and all angles are equal.

$$\beta = \frac{n-2}{n}180^\circ = \frac{n-2}{n}\pi \text{ radians}, \qquad \alpha\frac{360^\circ}{n} = \frac{2\pi}{n} \text{ radians}, \qquad A = \frac{nar}{2}$$

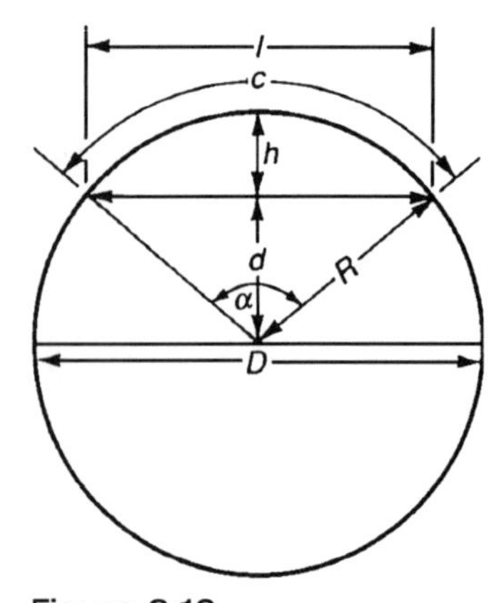

Figure 2.16

Circle (Fig. 2.16)

Notation. C = circumference, α = central angle in radians.

$$C = \pi D = 2\pi R$$

$$c = R\alpha = \frac{1}{2}D\alpha = D\cos^{-1}\frac{d}{R} = D\tan^{-1}\frac{1}{2d}$$

$$l = 2\sqrt{R^2 - d^2} = 2R\sin\frac{\alpha}{2} = 2d\tan\frac{\alpha}{2} = 2d\tan\frac{c}{D}$$

$$d = \frac{1}{2}\sqrt{4R^2 - l^2} = \frac{1}{2}\sqrt{D^2 - l^2} = R\cos\frac{\alpha}{2} = \frac{1}{2}l\cot\frac{c}{D}$$

$$h = r - d$$

$$\alpha = \frac{c}{R} = \frac{2c}{D} = 2\cos^{-1}\frac{d}{R} = 2\tan^{-1}\frac{l}{2d} = 2\sin^{-1}\frac{l}{D}$$

$$A_{(\text{circle})} = \pi R^2 = \frac{1}{4}\pi D^2 = \frac{1}{2}RC = \frac{1}{4}DC$$

$$A_{(\text{sector})} = \frac{1}{2}Rc = \frac{1}{2}R^2\alpha = \frac{1}{8}D^2\alpha$$

$$A_{(\text{segment})} = A_{(\text{sector})} - A_{(\text{triangle})} = \frac{1}{2}R^2(\alpha - \sin\alpha) = \frac{1}{2}R\left(c - R\sin\frac{c}{R}\right)$$

$$= R^2\sin^{-1}\frac{l}{2R} - \frac{1}{4}l\sqrt{4R^2 - l^2} = R^2\cos^{-1}\frac{d}{R} - d\sqrt{R^2 - d^2}$$

$$= R^2\cos^{-1}\frac{R-h}{R} - (R-h)\sqrt{2Rh - h^2}$$

Ellipse (Fig. 2.17)

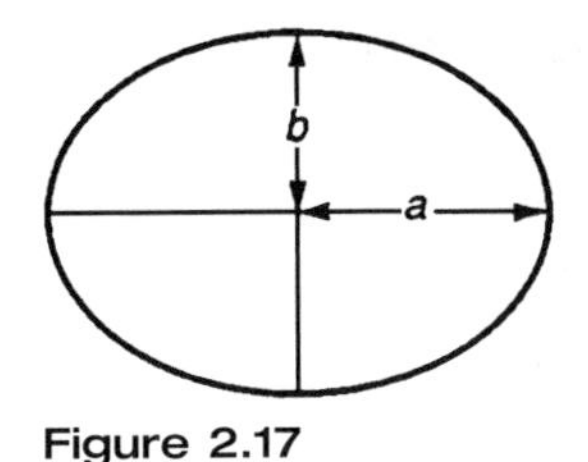

Figure 2.17

$$A = \pi ab$$

$$\text{Perimeter } (s) = \pi(a+b)\left[1 + \frac{1}{4}\left(\frac{a-b}{a+b}\right)^2 + \frac{1}{64}\left(\frac{a-b}{a+b}\right)^4 + \frac{1}{256}\left(\frac{a-b}{a+b}\right)^6 + \cdots\right]$$

$$\text{Perimeter } (s) \approx \pi\frac{a+b}{4}\left[3(1+\lambda) + \frac{1}{1-\lambda}\right], \quad \text{where } \lambda = \left[\frac{a-b}{2(a+b)}\right]^2$$

Parabola (Fig. 2.18)

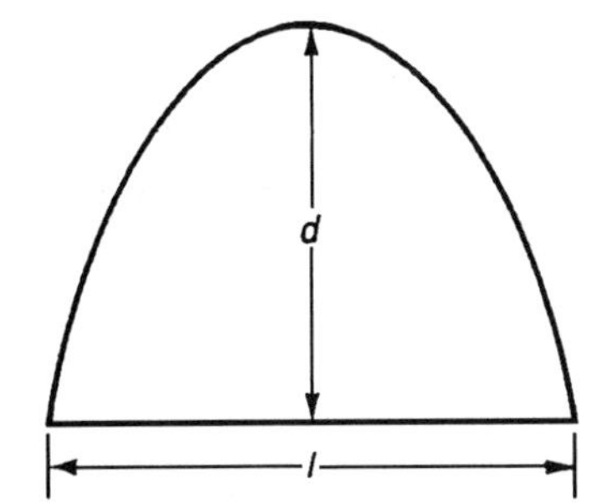

Figure 2.18

$$A = \frac{2}{3}ld$$

Cube (Fig. 2.19)

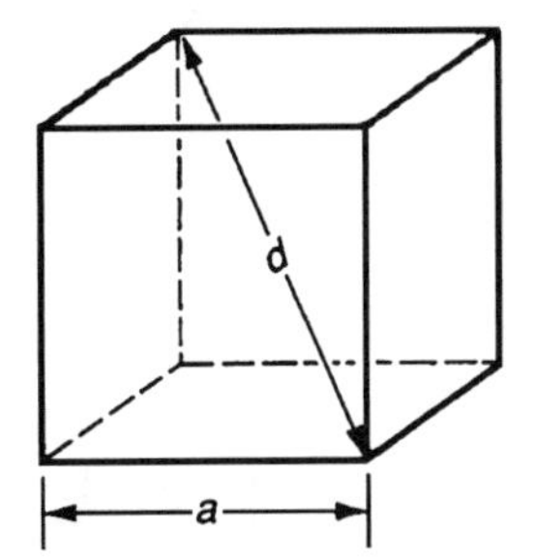

Figure 2.19

$$V = a^3 \qquad d = a\sqrt{3}$$

Total surface area $= 6a^2$

Prism or Cylinder (Fig. 2.20)

V = (area of base) × (altitude, h)

Lateral area = (perimeter of right section) × (lateral edge, e)

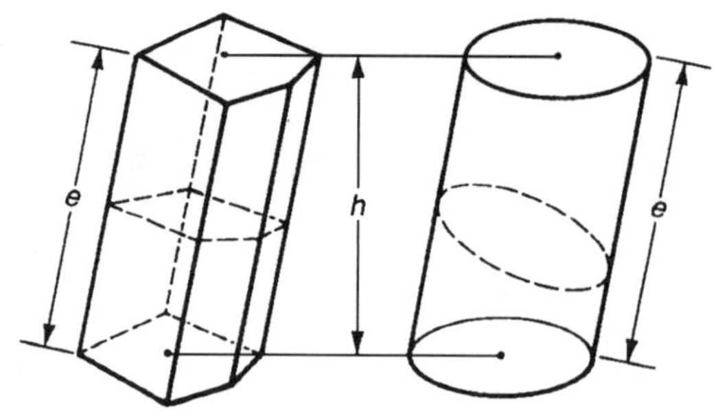

Figure 2.20

Pyramid or Cone (Fig. 2.21)

$V = \frac{1}{3}$ (area of base) × (altitude, h)

Lateral area of regular figure = $\frac{1}{2}$ (perimeter of base) × (slant height, s)

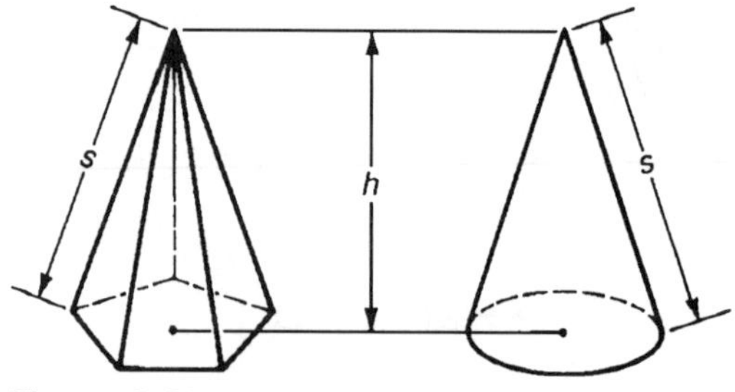

Figure 2.21

Sphere (Fig. 2.22)

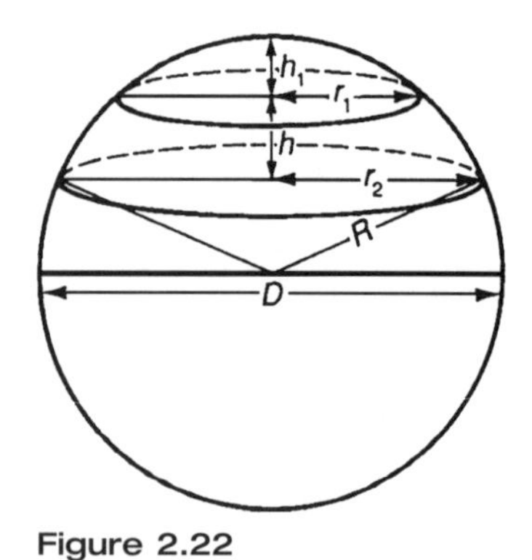

Figure 2.22

$$A_{(\text{sphere})} = 4\pi R^2 = \pi D^2$$

$$A_{(\text{zone})} = 2\pi Rh = \pi Dh$$

$$V_{(\text{sphere})} = \frac{4}{3}\pi R^3 = \frac{1}{6}\pi D^3$$

$$V_{(\text{spherical sector})} = \frac{2}{3}\pi R^2 h = \frac{1}{6}\pi D^2 h$$

$$V_{(\text{spherical segment of one base})} = \frac{1}{6}\pi h_1\left(3r_1^2 + h_1^2\right) = \frac{1}{3}\pi h_1^2(3R - h_1)$$

$$V_{(\text{spherical segment of two bases})} = \frac{1}{6}\pi h(3r_1^2 + 3r_2^2 h^2)$$

PLANE ANALYTIC GEOMETRY

Rectangular Coordinates (Fig. 2.23)

Let two perpendicular lines, $X'X$ (x-axis) and $Y'Y$ (y-axis) meet at a point O (origin). The position of any point $P(x, y)$ is fixed by the distances x (abscissa) and y (ordinate) from $Y'Y$ and $X'X$, respectively, to P. Values of x are positive to the right and negative to the left of $Y'Y$; values of y are positive above and negative below $X'X$.

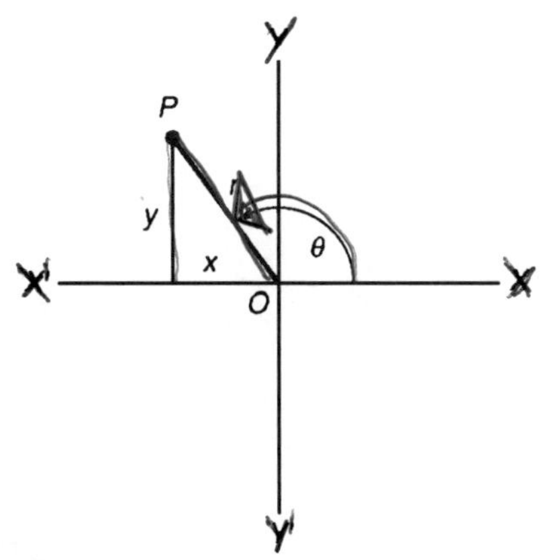

Figure 2.23

Polar Coordinates

Let O (origin or pole) be a point in the plane and OX (initial line) be any line through O. The position of any point P (r, θ) is fixed by the distance r (radius vector) from O to the point and the angle θ (vectorial angle) measured from OX to OP (Fig. 2.23).

A value for r is positive measured along the terminal side of θ; a value for θ is positive measured counterclockwise and negative measured clockwise.

Relations Connecting Rectangular and Polar Coordinates

$$x = r\cos\theta, \quad y = r\sin\theta$$

$$r = \sqrt{x^2 + y^2}, \quad \theta = \tan^{-1}\frac{y}{x}, \quad \sin\theta = \frac{y}{\sqrt{x^2+y^2}}, \quad \cos\theta = \frac{x}{\sqrt{x^2+y^2}}, \quad \tan\theta = \frac{y}{x}$$

Points and Slopes (Fig. 2.24)

Let $P_1(x_1, y_1)$ and $P_2(x_1, y_1)$ be any two points, and let α_1 be the angle from the x axis to P_1P_2, measured counterclockwise.

The length P_1P_2 is $d = \sqrt{(x_2 - x_1)^2 + (y_2 - y_1)^2}$.

The mid-point of P_1P_2 is $\left(\frac{x_1 + x_2}{2}, \frac{y_1 + y_2}{2}\right)$.

The point that divides P_1P_2 in the ratio n_1:n_2 is $\left(\frac{n_1x_2 + n_2x_1}{n_1 + n_2}, \frac{n_1y_2 + n_2y_1}{n_1 + n_2}\right)$.

The slope of P_1P_2 is $\tan\alpha = m = \frac{y_2 - y_1}{x_2 - x_1}$.

The angle between two lines of slopes m_1 and m_2 is $\beta = \tan^{-1}\frac{m_2 - m_1}{1 + m_1m_2}$.

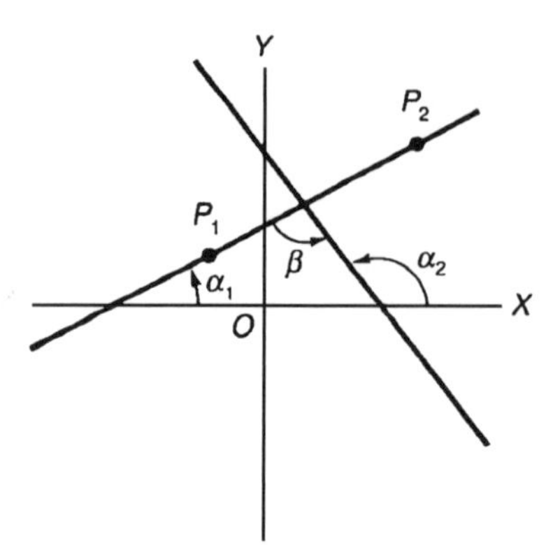

Figure 2.24

Two lines of slopes m_1 and m_2 are perpendicular if $m_2 = -\frac{1}{m_1}$.

Locus and Equation

The collection of all points that satisfy a given condition is called the **locus** of that condition; the condition expressed by means of the variable coordinates of any point on the locus is called the **equation of the locus**.

The locus may be represented by equations of three kinds: (1) a rectangular equation involves the rectangular coordinates (x, y); (2) a polar equation involves the polar coordinates (r, θ); (3) parametric equations express x and y or r and θ in terms of a third independent variable called a parameter.

The following equations are generally given in the system in which they are most simply expressed; sometimes several forms of the equation in one or more systems are given.

Straight Line (Fig. 2.25)

$Ax + By + C = 0$ $[-A/B = \text{slope}]$

$y = mx + b$ $[m = \text{slope}, b = \text{intercept on } OY]$

$y - y_1 = m(x - x_1)$ $[m = \text{slope}, P_1(x_1, y_1) \text{ is a known point on the line}]$

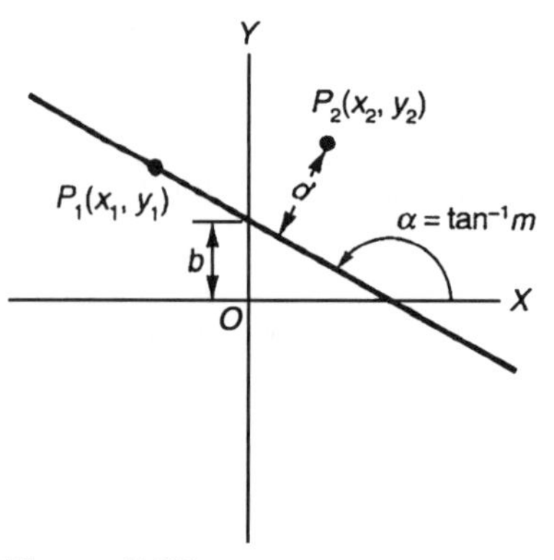

Figure 2.25

Circle

The locus of a point at a constant distance (radius) from a fixed point C (center) is a circle.

$(x-h)^2+(y-k)^2=a^2$ $C(h, k)$, radius $=a$
$r^2+b^2 \pm 2\, br\cos(\theta-\beta)=a^2$ $C(b, \beta)$, radius $=a$ [Fig. 2.26(a)]

$x^2+y^2=2ax$ $C(a, 0)$, radius $=a$
$r=2a\cos\theta$ $C(a, 0)$, radius $=a$ [Fig. 2.26(b)]

$x^2+y^2=2ay$ $C(0, a)$, radius $=a$
$r=2a\sin\theta$ $C(0, a)$, radius $=a$ [Fig. 2.26(c)]

$x^2+y^2=a^2$ $C(0, 0)$, radius $=a$
$r=a$ $C(0, 0)$, radius $=a$ [Fig. 2.26(d)]
$x=a\cos\phi$, $y=a\sin\phi$ $\phi=$ angle from OX to radius

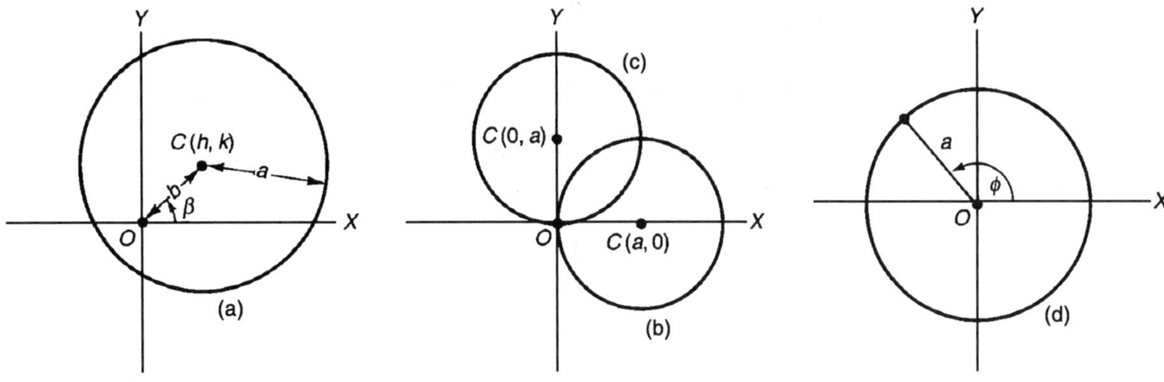

Figure 2.26

Conic (Fig. 2.27)

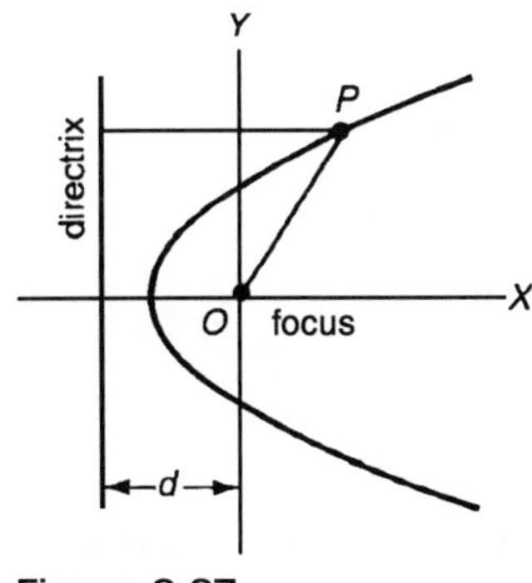

Figure 2.27

A **conic** is the locus of a point whose distance from a fixed point (focus) is in a constant ratio e, called the eccentricity, to its distance from a fixed straight line (directrix).

$$x^2+y^2=e^2(d+x)^2 \qquad d = \text{distance from focus to directrix}$$

$$r=\frac{de}{1-e\cos\theta}$$

The conic is called a parabola when $e = 1$, an ellipse when $e < 1$, and a hyperbola when $e > 1$.

Parabola

A parabola is a special case of a conic where $e = 1$.

$(y-k)^2=a(x-h)$ Vertex (h, k), axis $\parallel OX$
$y^2=ax$ Vertex $(0, 0)$, axis along OX [Fig. 2.28(a)]

$(x-h)^2=a(y-k)$ Vertex (h, k), axis $\parallel OY$
$x^2=ay$ Vertex $(0, 0)$, axis along OY [Fig. 2.28(b)]

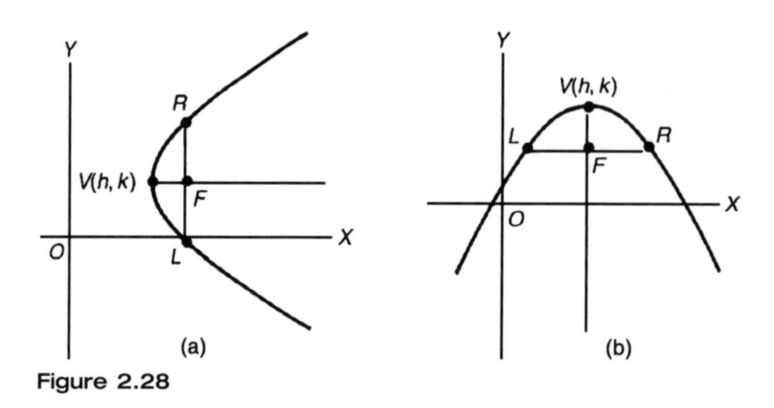

Figure 2.28

Distance from vertex to focus = $VF = \frac{1}{4}a$. Latus rectum = $LR = a$.

Ellipse

This is a special case of a conic where $e < 1$.

$$\frac{(x-h)^2}{a^2} + \frac{(y-k)^2}{b^2} = 1 \qquad \text{Center } (h, k), \text{ axes } \parallel OX, OY$$

$$\frac{x^2}{a^2} + \frac{y^2}{b^2} = 1 \qquad \text{Center } (0, 0), \text{ axes along } OX, OY$$

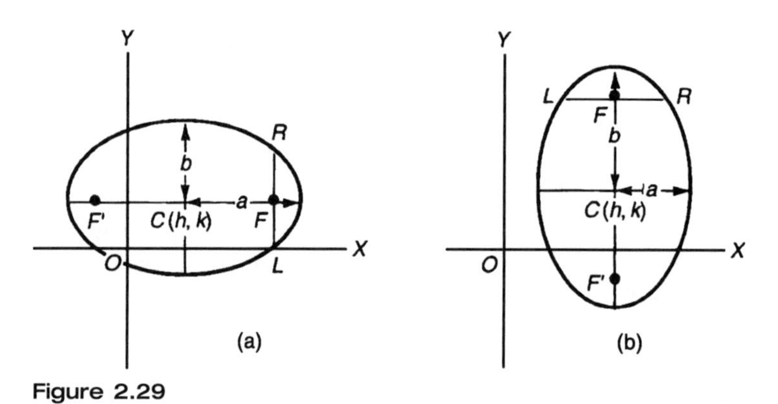

Figure 2.29

	$a > b$, Fig. 2.29(a)	$b > a$, Fig. 2.29(b)
Major axis	$2a$	$2b$
Minor axis	$2b$	$2a$
Distance from center to either focus	$\sqrt{a^2 - b^2}$	$\sqrt{b^2 - a^2}$
Latus rectum	$\frac{2b^2}{a}$	$\frac{2a^2}{b}$
Eccentricity, e	$\sqrt{\frac{a^2 - b^2}{a}}$	$\sqrt{\frac{b^2 - a^2}{b}}$
Sum of distances of any point P from the foci, $PF' + PF$	$2a$	$2b$

Hyperbola

This is a special case of a conic where $e > 1$.

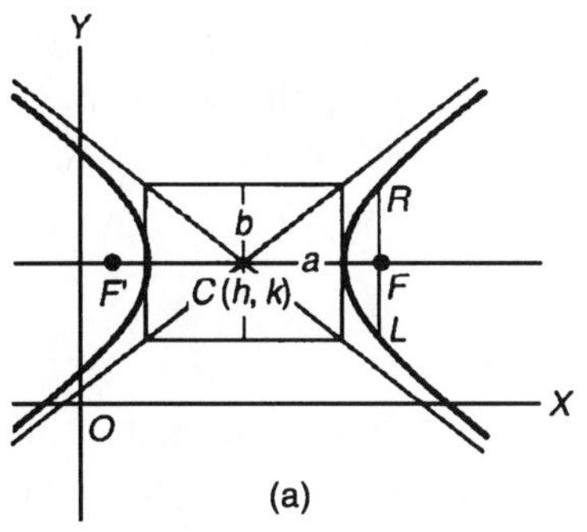

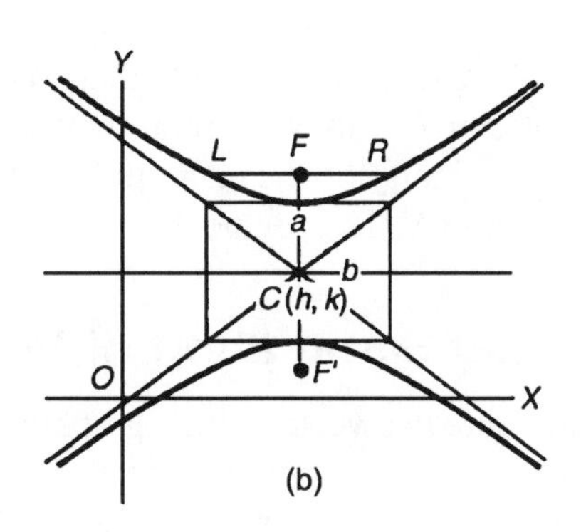

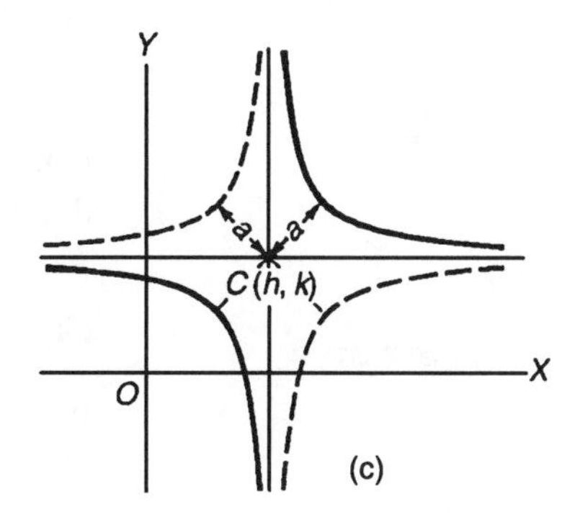

Figure 2.30

$$\frac{(x-h)^2}{a^2} - \frac{(y-k)^2}{b^2} = 1 \qquad C(h,\ k),\ \text{transverse axis} \parallel OX$$

$$\frac{x^2}{a^2} - \frac{y^2}{b^2} = 1 \qquad C(0,\ 0),\ \text{transverse axis along } OX$$

$$\frac{(y-k)^2}{a^2} - \frac{(x-h)^2}{b^2} = 1 \qquad C(h,\ k),\ \text{transverse axis} \parallel OY$$

$$\frac{y^2}{a^2} - \frac{x^2}{b^2} = 1 \qquad C(0,\ 0),\ \text{transverse along } OY$$

Transverse axis = $2a$; conjugate axis = $2b$

Distance from center to either focus = $\sqrt{a^2+b^2}$

Latus rectum = $\dfrac{2b^2}{a}$

Eccentricity, $e = \dfrac{\sqrt{a^2+b^2}}{a}$

Difference of distances of any point from the foci = $2a$.

The asymptotes are two lines through the center to which the branches of the hyperbola approach arbitrarily closely; their slopes are $\pm b/a$ [Fig. 2.30(a)] or $\pm a/b$ [Fig. 2.30(b)].

The rectangular (equilateral) hyperbola has $b = a$. The asymptotes are perpendicular to each other.

$$(x-h)(y-k) = \pm e = \sqrt{2} \qquad \text{Center } (h,\ k),\ \text{asymptotes} \parallel OX, OY$$

$$xy = \pm e = \sqrt{2} \qquad \text{Center } (0,\ 0),\ \text{asymptotes along } OX,\ OY$$

The + sign gives the solid curves in Fig. 2.30(c); the – sign gives the dotted curves in Fig. 2.30(c).

VECTORS

Definition and Graphical Representation of a Vector (Fig. 2.31)

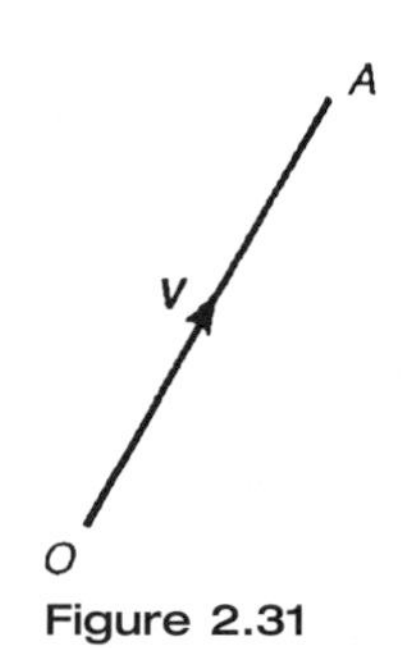

Figure 2.31

A vector (**V**) is a quantity that is completely specified by magnitude *and* a direction. A scalar (s) is a quantity that is completely specified by a magnitude *only*.

The vector (**V**) may be represented geometrically by the segment $\overrightarrow{OA}$, the length of OA signifying the magnitude of **V** and the arrow carried by OA signifying the direction of V. The segment $\overrightarrow{AO}$ represents the vector −**V**.

Graphical Summation of Vectors

If $\mathbf{V}_1$ and $\mathbf{V}_2$ are two vectors, their graphical sum $\mathbf{V} = \mathbf{V}_1 + \mathbf{V}_2$ is formed by drawing the vector $\mathbf{V}_1 = \overrightarrow{OA}$, from any point O, and the vector $\mathbf{V}_2 = \overrightarrow{AB}$ from the end of $\mathbf{V}_1$ and joining O and B; then $\mathrm{V} = \overrightarrow{OB}$. Also, $\mathbf{V}_1 + \mathbf{V}_2 = \mathbf{V}_2 + \mathbf{V}_1$ and $\mathbf{V}_1 + \mathbf{V}_2 - \mathbf{V} = 0$ (Fig. 2.32(a)).

Similarly, if $\mathbf{V}_1, \mathbf{V}_2, \mathbf{V}_3, \ldots, \mathbf{V}_n$ are any number of vectors drawn so that the initial point of one is the end point of the preceding one, then their graphical sum $\mathbf{V} = \mathbf{V}_1 + \mathbf{V}_2 + \cdots + \mathbf{V}_n$ is the vector joining the initial point of $\mathbf{V}_1$ with the end point of $\mathbf{V}_n$ (Fig. 2.32(b)).

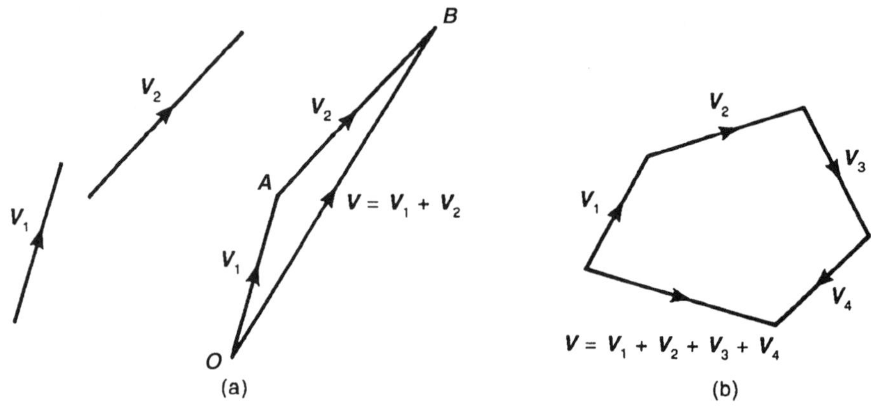

Figure 2.32

Analytic Representation of Vector Components

A vector **V** that is considered as lying in the x-y coordinate plane (Fig. 2.33(a)) is completely determined by its horizontal and vertical components x and y. If **i** and **j** represent vectors of unit magnitude along OX and OY, respectively, and a and b are the magnitude of x and y, then V may be represented by $\mathbf{V} = a\mathbf{i} + b\mathbf{j}$, its magnitude by $|\mathbf{V}| = +\sqrt{a^2 + b^2}$, and its direction by $\alpha = \tan^{-1} b/a$.

A vector **V** in three-dimensional in space is completely determined by its components x, y, and z along three mutually perpendicular lines OX, OY, and OZ, directed as shown in Fig. 2.33(b). If **i**, **j**, and **k** represent vectors of unit magnitude along OX, OY, OZ, respectively, and a, b, and c are the magnitudes of the components x, y, and z and respectively, then **V** may be represented by $\mathbf{V} = a\mathbf{i} + b\mathbf{j} + c\mathbf{k}$, its magnitude by, $|\mathbf{V}| = +\sqrt{a^2 + b^2 + c^2}$, and its direction by $\cos\alpha : \cos\beta : \cos\gamma = a : b : c$.

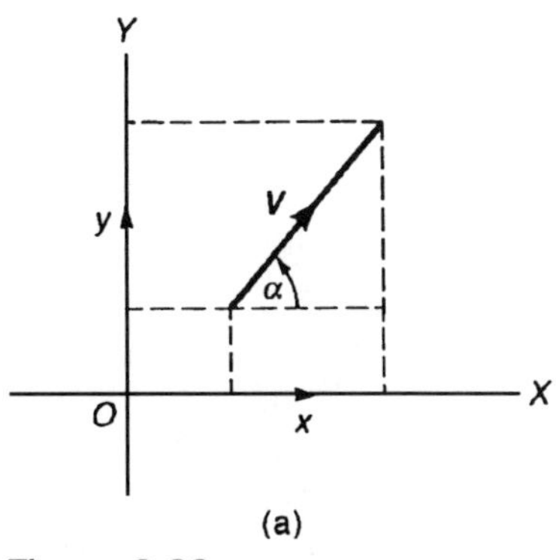

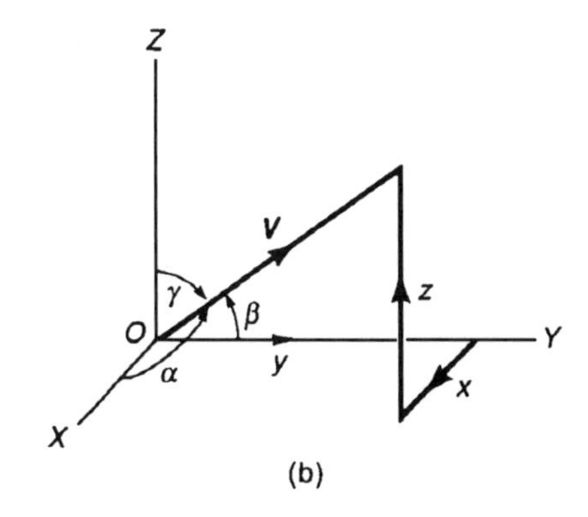

Figure 2.33

Properties of Vectors

$$\mathbf{V} = a\mathbf{i} + b\mathbf{j} \quad \text{or} \quad \mathbf{V} = a\mathbf{i} + b\mathbf{j} + c\mathbf{k}$$

Vector Sum V of any Number of Vectors, $V_1, V_2, V_3, \ldots$

$$\mathbf{V} = \mathbf{V}_1 + \mathbf{V}_2 + \mathbf{V}_3 + \cdots = (a_1 + a_2 + a_3 + \cdots)\,\mathbf{i} + (b_1 + b_2 + b_3 + \cdots)\,\mathbf{j} + (c_1 + c_2 + c_3 + \cdots)\,\mathbf{k}$$

Product of a Vector V and a Scalar *s*

The product $s\mathbf{V}$ has the same direction as $\mathbf{V}$, and its magnitude is s times the magnitude of $\mathbf{V}$.

$$s\mathbf{V} = (sa)\,\mathbf{i} + (sb)\,\mathbf{j} + (sc)\,\mathbf{k}$$
$$(s_1 + s_2)\,\mathbf{V} = s_1\mathbf{V} + s_2\mathbf{V} \qquad (\mathbf{V}_1 + \mathbf{V}_2)s = \mathbf{V}_1 s + \mathbf{V}_2 s$$

Scalar Product of Two Vectors: $V_1 \bullet V_2$ (Fig. 2.34)

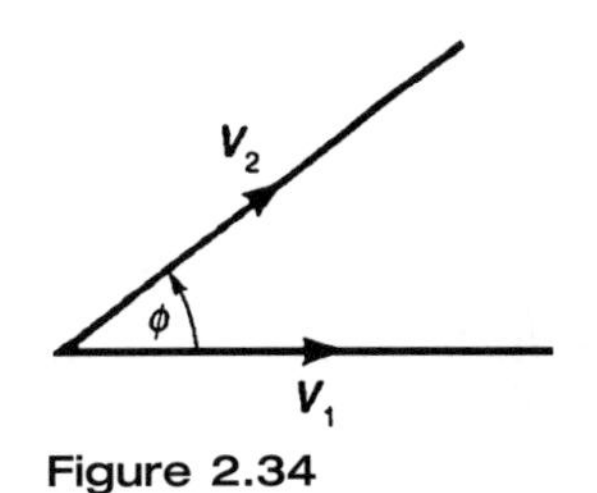

Figure 2.34

$\mathbf{V}_1 \bullet \mathbf{V}_2 = |\mathbf{V}_1||\mathbf{V}_2| \cos\phi$, where ϕ is the angle between $\mathbf{V}_1$ and $\mathbf{V}_2$

$\mathbf{V}_1 \bullet \mathbf{V}_2 = \mathbf{V}_2 \bullet \mathbf{V}_1$; $\mathbf{V}_1 \bullet \mathbf{V}_1 = |\mathbf{V}_1|^2$; $(\mathbf{V}_1 + \mathbf{V}_2) \bullet \mathbf{V}_3 = \mathbf{V}_1 \bullet \mathbf{V}_3 + \mathbf{V}_2 \bullet \mathbf{V}_3$

$(\mathbf{V}_1 + \mathbf{V}_2) \bullet (\mathbf{V}_3 + \mathbf{V}_4) = \mathbf{V}_1 \bullet \mathbf{V}_3 + \mathbf{V}_1 \bullet \mathbf{V}_4 + \mathbf{V}_2 \bullet \mathbf{V}_3 + \mathbf{V}_2 \bullet \mathbf{V}_4$

$\mathbf{i} \bullet \mathbf{i} = \mathbf{j} \bullet \mathbf{j} = \mathbf{k} \bullet \mathbf{k} = 1$; $\mathbf{i} \bullet \mathbf{j} = \mathbf{j} \bullet \mathbf{k} = \mathbf{k} \bullet \mathbf{i} = 0$

In a plane, $\mathbf{V}_1 \bullet \mathbf{V}_2 = a_1 a_2 + b_1 b_2$; in space, $\mathbf{V}_1 \bullet \mathbf{V}_2 = a_1 a_2 + b_1 b_2 + c_1 c_2$.

The scalar product of two vectors $\mathbf{V}_1 \bullet \mathbf{V}_2$ is a scalar quantity and may physically represent the work done by a constant force of magnitude $|\mathbf{V}_1|$ on a particle moving through a distance $|\mathbf{V}_2|$, where ϕ is the angle between the direction of the force and the direction of motion.

Vector Product of Two Vectors: $V_1 \times V_2$ (Fig. 2.35)

The vector product is $\mathbf{V}_1 \times \mathbf{V}_2 = \mathbf{l}\,|\mathbf{V}_1|\,\mathbf{V}_2|\sin\phi$, where ϕ is the angle from $\mathbf{V}_1$ to $\mathbf{V}_2$ and $\mathbf{l}$ is a unit vector perpendicular to the plane of the vectors $\mathbf{V}_1$ to $\mathbf{V}_2$ and so directed that a right-handed screw driven in the direction of $\mathbf{l}$ would carry $\mathbf{V}_1$ into $\mathbf{V}_2$.

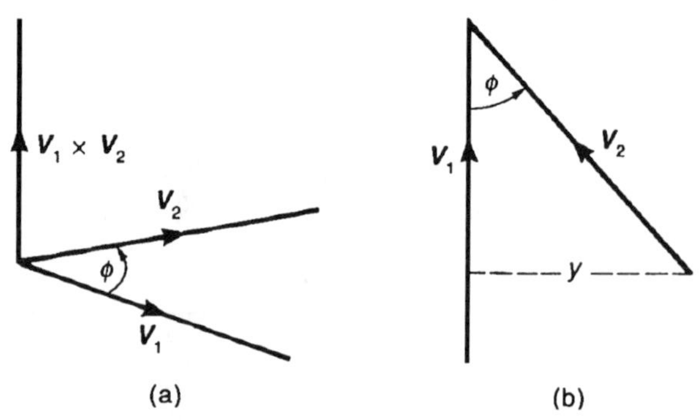

Figure 2.35

$$\mathbf{V}_1 \times \mathbf{V}_2 = -\mathbf{V}_2 \times \mathbf{V}_1; \quad \mathbf{V}_1 \times \mathbf{V}_1 = 0$$
$$(\mathbf{V}_1 + \mathbf{V}_2) \times \mathbf{V}_3 = \mathbf{V}_1 \times \mathbf{V}_3 + \mathbf{V}_2 \times \mathbf{V}_3$$
$$\mathbf{V}_1 \bullet (\mathbf{V}_2 \times \mathbf{V}_3) = \mathbf{V}_2 \bullet (\mathbf{V}_3 \times \mathbf{V}_1) = \mathbf{V}_3 \bullet (\mathbf{V}_1 \times \mathbf{V}_2)$$
$$\mathbf{i} \times \mathbf{i} = \mathbf{j} \times \mathbf{j} = \mathbf{k} \times \mathbf{k} = 0; \quad \mathbf{i} \times \mathbf{j} = \mathbf{k}; \quad \mathbf{j} \times \mathbf{k} = \mathbf{i}; \quad \mathbf{k} \times \mathbf{i} = \mathbf{j}$$

In the *x-y* plane, $\mathbf{V}_1 \times \mathbf{V}_2 = (a_1 b_2 - a_2 b_1)\mathbf{k}$.

In space, $\mathbf{V}_1 \times \mathbf{V}_2 = (b_2 c_3 - b_3 c_2)\,\mathbf{i} + (c_3 a_1 - c_1 a_3 \mathbf{j}) + (a_1 b_2 - a_2 b_1)\,\mathbf{k}$.

The vector product of two vectors is a vector quantity and may physically represent the moment of a force $\mathbf{V}_1$ about a point O placed so that the moment arm is $y = |\mathbf{V}_2| \sin \phi$ (see Fig. 2.35(b)).

LINEAR ALGEBRA

Matrix Operations

Matrices are rectangular arrays of real or complex numbers. Their great importance arises from the variety of operations that may be performed on them. Using the standard convention, the across-the-page lines are called **rows** and the up-and-down-the-page lines are **columns**. Entries in a matrix are **addressed** with double subscripts (always row first, then column). Thus the matrix

$$\mathbf{A} = \begin{bmatrix} 1 & 2 & 3 \\ 0 & 9 & -3 \end{bmatrix}$$

is 2 × 3, and the "9" is a_{22}. The "2" is a_{12} and the "0" is a_{21}. One also can refer to entries with square brackets, the "9" being $[A]_{22}$ and the "3" $[A]_{13}$.

If two matrices are the same size, they may be added: $[A + B]_{ij} = [A]_{ij} + [B]_{ij}$. Thus,

$$\begin{bmatrix} 1 & 2 & 3 \\ 0 & 9 & -3 \end{bmatrix} + \begin{bmatrix} 1 & 3 \\ 2 & 4 \end{bmatrix}$$

is not defined, but

$$\begin{bmatrix} 1 & 2 & 3 \\ 0 & 9 & -3 \end{bmatrix} + \begin{bmatrix} 1 & 3 & 5 \\ 2 & 4 & 6 \end{bmatrix} = \begin{bmatrix} 2 & 5 & 8 \\ 2 & 13 & 3 \end{bmatrix}$$

is proper.

Any matrix may be multiplied by a **scalar** (a number): $[c\mathbf{A}]_{ij} = c[\mathrm{A}]_{ij}$, so that

$$5 \bullet \begin{bmatrix} 1 & 5 \\ 0 & 6 \end{bmatrix} = \begin{bmatrix} 5 & 25 \\ 0 & 30 \end{bmatrix}$$

The most peculiar matrix operation (and the most useful) is matrix multiplication. If **A** is $m \times n$ and **B** is $n \times p$, then $\mathbf{A} \bullet \mathbf{B}$ (or **AB**) is of size $m \times p$, and

$$[AB]_{ij} = \sum_{k=1}^{n} a_{jk} \bullet b_{kj}$$

The **dot product** (scalar product) of the ith row of **A** with the jth column of **B**, as in

$$\begin{bmatrix} 1 & 2 & 3 \\ 0 & 9 & -3 \end{bmatrix} \bullet \begin{bmatrix} 1 & 5 \\ 0 & 6 \end{bmatrix}$$

is not defined (owing to the mismatch of row and column lengths), but

$$\begin{bmatrix} 1 & 2 & 3 \\ 0 & 9 & -3 \end{bmatrix} \bullet \begin{bmatrix} 1 & 5 \\ 0 & 6 \\ 7 & 8 \end{bmatrix} = \begin{bmatrix} 1 \bullet 1 + 2 \bullet 0 + 3 \bullet 7 & 1 \bullet 5 + 2 \bullet 6 + 3 \bullet 8 \\ 0 \bullet 1 + 9 \bullet 0 - 3 \bullet 7 & 0 \bullet 5 + 9 \bullet 6 - 3 \bullet 8 \end{bmatrix} = \begin{bmatrix} 22 & 41 \\ -21 & 30 \end{bmatrix}$$

is correct.

A matrix with only one row or one column is called a vector, so a matrix times a vector is a vector (if defined). Thus $\mathbf{A}\ (m \times n) \bullet \mathbf{X}\ (n \times 1) = \mathbf{Y}(m \times 1)$, so a matrix can be thought of as an **operator** that takes vectors to vectors.

Another useful way of working with matrices is transposition: If **A** is $m \times n$, $\mathbf{A}^t$ is $n \times m$ and is the result of interchanging rows and columns. Hence

$$\begin{bmatrix} 1 & 2 & 3 \\ 0 & 9 & -3 \end{bmatrix}^t = \begin{bmatrix} 1 & 0 \\ 2 & 9 \\ 3 & -3 \end{bmatrix}$$

These various operations interact in the usual pleasant ways (and one decidedly unpleasant way) the standard convention is that all of the following combination are defined:

$$\begin{aligned} \mathbf{A} + \mathbf{B} &= \mathbf{B} + \mathbf{A} \\ \mathbf{A} + (\mathbf{B} + \mathbf{C}) &= (\mathbf{A} + \mathbf{B}) + \mathbf{C} \end{aligned}$$
$$c(\mathbf{A} + \mathbf{B}) = c \bullet \mathbf{A} + c \bullet \mathbf{B}$$
$$(c + d) \bullet \mathbf{A} = c \bullet \mathbf{A} + d \bullet \mathbf{A}$$
$$(-1) \bullet \mathbf{A} + \mathbf{A} = (0 \bullet \mathbf{A})$$
$$\mathbf{A} \bullet \mathbf{B} \neq \mathbf{B} \bullet \mathbf{A} \text{ (in general)}$$
$$\mathbf{A} \bullet (\mathbf{B} \bullet \mathbf{C}) = (\mathbf{A} \bullet \mathbf{B}) \bullet \mathbf{C}$$
$$\mathbf{A} \bullet (\mathbf{B} + \mathbf{C}) = \mathbf{A} \bullet \mathbf{B} + \mathbf{A} \bullet \mathbf{C}$$
$$(\mathbf{A} + \mathbf{B}) \bullet \mathbf{C} = \mathbf{A} \bullet \mathbf{C} + \mathbf{B} \bullet \mathbf{C}$$
$$(\mathbf{A} + \mathbf{B})^t = \mathbf{A}^t + \mathbf{B}^t$$
$$(\mathbf{A} \bullet \mathbf{B})^t = \mathbf{B}^t \bullet \mathbf{A}^t$$

In addition, matrices **I**, which are $n \times n$ and whose entries are 1 on the diagonal $i = j$ and 0 elsewhere, are multiplicative identities: $\mathbf{A} \bullet \mathbf{I} = \mathbf{A}$ and $\mathbf{I} \bullet \mathbf{A} = \mathbf{A}$. Here the two **I** matrices may be different sizes; for example,

$$\begin{bmatrix} 1 & 2 & 3 \\ 0 & 9 & -3 \end{bmatrix} \bullet \begin{bmatrix} 1 & 0 & 0 \\ 0 & 1 & 0 \\ 0 & 0 & 1 \end{bmatrix} = \begin{bmatrix} 1 & 2 & 3 \\ 0 & 9 & -3 \end{bmatrix}$$

but

$$\begin{bmatrix} 1 & 0 \\ 0 & 1 \end{bmatrix} \bullet \begin{bmatrix} 1 & 2 & 3 \\ 0 & 9 & -3 \end{bmatrix} = \begin{bmatrix} 1 & 2 & 3 \\ 0 & 9 & -3 \end{bmatrix}$$

I is called the *identity* matrix, and the size is understood from context.

Example **2.1**

Verify that the transpose of **A** + **BC** is $\mathbf{C}^t\mathbf{B}^t + \mathbf{A}^t$ if

$$\mathbf{A} = \begin{bmatrix} 1 & 1 \\ 2 & 3 \end{bmatrix} \quad \mathbf{B} = \begin{bmatrix} 1 & 2 & 3 \\ 0 & 9 & -3 \end{bmatrix} \quad \mathbf{C} = \begin{bmatrix} 1 & 5 \\ 0 & 6 \\ 7 & 8 \end{bmatrix}$$

Solution

$\mathbf{BC} \begin{bmatrix} 22 & 41 \\ -21 & 30 \end{bmatrix}$, so $\mathbf{A} + \mathbf{BC} = \begin{bmatrix} 23 & 42 \\ -19 & 33 \end{bmatrix}$ and $[\mathbf{A} + \mathbf{BC}]^t = \begin{bmatrix} 23 & -19 \\ 42 & 33 \end{bmatrix}$. On the other hand, $\mathbf{C}^t\mathbf{B}^t = \begin{bmatrix} 1 & 0 & 7 \\ 5 & 6 & 8 \end{bmatrix} \bullet \begin{bmatrix} 1 & 0 \\ 2 & 9 \\ 3 & -3 \end{bmatrix} = \begin{bmatrix} 22 & -21 \\ 41 & 30 \end{bmatrix}$ and

$$\mathbf{A}^t = \begin{bmatrix} 1 & 2 \\ 1 & 3 \end{bmatrix}, \text{ so } \mathbf{A}^t + \mathbf{C}^t\mathbf{B}^t = \mathbf{C}^t\mathbf{B}^t + \mathbf{A}^t = \begin{bmatrix} 23 & -19 \\ 42 & 33 \end{bmatrix}.$$

Types of Matrices

Matrices are classified according to their appearance or the way the act. If **A** is square and $\mathbf{A}^t = \mathbf{A}$, then **A** is called symmetric. If $\mathbf{A}^t = -\mathbf{A}$, then it is skew-symmetric.

If **A** has complex entries, **A*** then is called the Hermitian adjoint of **A**. If $\mathbf{A}^* = \overline{\mathbf{A}}^t$ (complex conjugate), then

$$\begin{bmatrix} 1+i & i \\ 3 & 4-i \end{bmatrix}^* = \begin{bmatrix} 1-i & -i \\ 3 & 4+i \end{bmatrix}^t = \begin{bmatrix} 1-i & 3 \\ -i & 4+i \end{bmatrix}$$

If **A** = **A***, then A is called Hermitian. If **A*** = –**A**, the name is skew-Hermitian.

If **A** is square and $a_{ij} = 0$ unless $i = j$, **A** is called diagonal. If **A** is square and zero below the diagonal ($[\mathbf{A}]_{ij} = 0$ if $i > j$), **A** is called upper triangular. The transpose of such a matrix is called lower triangular.

If A is square and there is a matrix $\mathbf{A}^{-1}$ such that $\mathbf{A}^{-1} \bullet \mathbf{A} = \mathbf{A} \bullet \mathbf{A}^{-1} = \mathbf{I}$, **A** is nonsingular. Otherwise, it is singular. If **A** and **B** are both nonsingular $n \times n$ matrices, then **AB** is nonsingular and $(\mathbf{AB})^{-1} = \mathbf{B}^{-1}\mathbf{A}^{-1}$, because $(\mathbf{AB})(\mathbf{B}^{-1}\mathbf{A}^{-1}) = \mathbf{A}(\mathbf{B}\,\mathbf{B}^{-1})\mathbf{A}^{-1} = \mathbf{AIA}^{-1} = \mathbf{AA}^{-1} = \mathbf{I}$, as does $(\mathbf{B}^{-1}\mathbf{A}^{-1}) \bullet (\mathbf{AB})$.

If $\mathbf{A}^t\mathbf{A} = \mathbf{AA}^t = \mathbf{I}$ and **A** is real, it is called orthogonal (the reason will appear below). If $\mathbf{A}^*\mathbf{A} = \mathbf{AA}^* = \mathbf{I}$ (**A** complex), **A** is called unitary. If **A** commutes with $\mathbf{A}^*$, so that $\mathbf{AA}^* = \mathbf{A}^*\mathbf{A}$, then **A** is called normal.

Elementary Row and Column Operations

The most important tools used in dealing with matrices are the elementary operations: R for row, C for column. If **A** is given matrix, performing $R(i \leftrightarrow j)$ on A means interchanging Row i and Row j. $R_i(c)$ means multiplying Row i by the number c (except $c = 0$). $R_j + cR_i$ means multiply Row i by c and add this result into Row j ($i \neq j$). Thus, if

$$\mathbf{A} = \begin{bmatrix} 1 & 2 & 3 \\ 4 & 5 & 6 \\ 7 & 8 & 0 \end{bmatrix},$$

then

$$\mathrm{R}(2 \leftrightarrow 3)\,(\mathbf{A}) = \begin{bmatrix} 1 & 2 & 3 \\ 7 & 8 & 0 \\ 4 & 5 & 6 \end{bmatrix} \qquad C_1(2)(\mathbf{A}) = \begin{bmatrix} 2 & 2 & 3 \\ 8 & 5 & 6 \\ 14 & 8 & 0 \end{bmatrix}$$

$$R_1 - R_2(\mathbf{A}) = \begin{bmatrix} -3 & -3 & -3 \\ 4 & 5 & 6 \\ 7 & 8 & 0 \end{bmatrix}$$

These operations are used in reducing matrix problems to simpler ones, as well as for other reasons mentioned below.

Example 2.2

Solve $\mathbf{AX} = \mathbf{B}$ where

$$\mathbf{A} = \begin{bmatrix} 1 & 2 & 3 \\ 4 & 5 & 6 \\ 7 & 8 & 9 \end{bmatrix}, \quad \mathbf{X} = \begin{bmatrix} x \\ y \\ z \end{bmatrix}, \quad \mathbf{B} = \begin{bmatrix} 1 \\ 1 \\ 1 \end{bmatrix}$$

Solution

Form the "augmented" matrix

$$[\mathbf{A}|\mathbf{B}] = \begin{bmatrix} 1 & 2 & 3 & 1 \\ 4 & 5 & 6 & 1 \\ 7 & 8 & 9 & 1 \end{bmatrix}$$

and perform elementary row operations on this matrix until the solution is apparent:

$$\begin{bmatrix} 1 & 2 & 3 & 1 \\ 4 & 5 & 6 & 1 \\ 7 & 8 & 9 & 1 \end{bmatrix} \begin{matrix} R_2 - 4R_1 \\ R_3 - 7R_1 \end{matrix} \begin{bmatrix} 1 & 2 & 3 & 1 \\ 0 & -3 & -6 & -3 \\ 0 & -6 & -12 & -6 \end{bmatrix} \begin{matrix} R_2\left(-\frac{1}{3}\right) \\ R_3\left(-\frac{1}{6}\right) \end{matrix} \begin{bmatrix} 1 & 2 & 3 & 1 \\ 0 & 1 & 2 & 1 \\ 0 & 1 & 2 & 1 \end{bmatrix}$$

$$R_3 - R_2 \begin{bmatrix} 1 & 2 & 3 & 1 \\ 0 & 1 & 2 & 1 \\ 0 & 0 & 0 & 0 \end{bmatrix}$$

The answer is now apparent: $y + 2z = 1$ and $x + 2y + 3z = 1$, or, z arbitrary, $y = 1 - 2z$, $x = 1 - 2(1 - 2z) - 3z = -1 + z$. This system of equations has an infinite number of solutions.

Example 2.3

Solve the system of equations

$$\begin{aligned} x + y - z &= a \\ 2x - y + 3z &= 2 \\ 3x + 2y + z &= 1 \end{aligned}$$

for x, y, and z in terms of a.

Solution

Strip off the variables x, y and z:

$$\begin{bmatrix} 1 & 1 & -1 & a \\ 2 & -1 & 3 & 2 \\ 3 & 2 & 1 & 1 \end{bmatrix} \begin{matrix} R_2 - 2R_1 \\ R_3 - 3R_1 \end{matrix} \begin{bmatrix} 1 & 1 & -1 & a \\ 0 & -3 & 5 & 2-2a \\ 0 & -1 & 4 & 1-3a \end{bmatrix}$$

$$\begin{matrix} R_2(2 \leftrightarrow 3) \\ R_2(-1) \end{matrix} \begin{bmatrix} 1 & 1 & -1 & a \\ 0 & 1 & -4 & 3a-1 \\ 0 & -3 & 5 & 2-2a \end{bmatrix} \begin{matrix} R_1 - R_2 \\ R_3 + 3R_2 \end{matrix} \begin{bmatrix} 1 & 0 & 3 & 1-2a \\ 0 & 1 & -4 & 3a-1 \\ 0 & 0 & -7 & 7a-1 \end{bmatrix}$$

The solution is now clear:

$$\begin{aligned} z &= \frac{7a-1}{-7} = -a + \frac{1}{7} \\ y &= 3a - 1 + 4z = 3a - 1 - 4a + \frac{4}{7} = -a - \frac{3}{7} \\ x &= 1 - 2a - 3z = 1 - 2a + 3a - \frac{3}{7} = a + \frac{4}{7} \end{aligned}$$

Example 2.4

Find $\mathbf{A}^{-1}$ if

$$\mathbf{A} = \begin{bmatrix} 1 & 1 & -1 \\ 1 & 2 & 3 \\ 3 & 2 & 1 \end{bmatrix}$$

Solution

Since this amounts to solving $\mathbf{AX} = \mathbf{B}$ three times, with

$$\mathbf{B} = \begin{bmatrix} 1 \\ 0 \\ 0 \end{bmatrix} \quad \mathbf{B} = \begin{bmatrix} 0 \\ 1 \\ 0 \end{bmatrix} \quad \mathbf{B} = \begin{bmatrix} 0 \\ 0 \\ 1 \end{bmatrix}$$

form

$$[\mathbf{A}|\mathbf{I}] = \begin{bmatrix} 1 & 1 & -1 & 1 & 0 & 0 \\ 1 & 2 & 3 & 0 & 1 & 0 \\ 3 & 2 & 1 & 0 & 0 & 1 \end{bmatrix}$$

and perform row operations until a solution emerges.

$$[\mathbf{A}\,|\,\mathbf{I}] \begin{matrix} R_2 - R_1 \\ R_3 - 3R_1 \end{matrix} \begin{bmatrix} 1 & 1 & -1 & 1 & 0 & 0 \\ 0 & 1 & 4 & -1 & 1 & 0 \\ 0 & -1 & 4 & -3 & 0 & 1 \end{bmatrix}$$

$$\begin{matrix} R_1 - R_2 \\ R_3 + R_2 \end{matrix} \begin{bmatrix} 1 & 0 & -5 & 2 & -1 & 0 \\ 0 & 1 & 4 & -1 & 1 & 0 \\ 0 & 0 & 8 & -4 & 1 & 1 \end{bmatrix}$$

$$\begin{matrix} R_3\left(\frac{1}{8}\right) \\ R_2 - 4R_3 \\ R_1 + 5R_3 \end{matrix} \begin{bmatrix} 1 & 0 & 0 & -\frac{1}{2} & -\frac{3}{8} & \frac{5}{8} \\ 0 & 1 & 0 & 1 & \frac{1}{2} & -\frac{1}{2} \\ 0 & 0 & 1 & -\frac{1}{2} & \frac{1}{8} & \frac{1}{8} \end{bmatrix}$$

Thus,

$$\mathbf{A}^{-1} = \begin{bmatrix} -\frac{1}{2} & -\frac{3}{8} & \frac{5}{8} \\ 1 & \frac{1}{2} & -\frac{1}{2} \\ -\frac{1}{2} & \frac{1}{8} & \frac{1}{8} \end{bmatrix}$$

Example 2.5

Verify that $\mathbf{A}^{-1}\mathbf{A} = \mathbf{I}$ in Example 2.4.

Solution

$$\mathbf{8A}^{-1}\mathbf{A} = \begin{bmatrix} -4 & -3 & 5 \\ 8 & 4 & -4 \\ -4 & 1 & 1 \end{bmatrix} \begin{bmatrix} 1 & 1 & -1 \\ 1 & 2 & 3 \\ 3 & 2 & 1 \end{bmatrix}$$

$$= \begin{bmatrix} -4-3+15 & -4-6+10 & 4-9+5 \\ 8+4-12 & 8+8-8 & -8+12-4 \\ -4+1+3 & -4+2+2 & 4+3+1 \end{bmatrix} = 8 \begin{bmatrix} 1 & 0 & 0 \\ 0 & 1 & 0 \\ 0 & 0 & 1 \end{bmatrix} = 8\mathbf{I}$$

Example 2.6

Describe the set of solutions of $\mathbf{AX} = \mathbf{B}$.

Solution

If $\mathbf{AX}_0 = \mathbf{B}$ is one solution, and $\mathbf{AY} = 0$, then $\mathbf{A}(\mathbf{X}_0 + \mathbf{Y})$ is a solution, so all solutions are of the form $\mathbf{X} = \mathbf{X}_0 + \mathbf{Y}$ where $\mathbf{AY} = 0$. Thus, if $\mathbf{N} = \{\mathbf{Y} : \mathbf{AY} = 0\}$ is the null space of $\mathbf{A}$, the set of solutions to $\mathbf{AX} = \mathbf{B}$ is $\mathbf{X}_0 + \mathbf{N} = \{\mathbf{X}_0 + \mathbf{Y} : \mathbf{Y} \in \mathbf{N}\}$.

Determinants

The determinant of a square matrix is a scalar representing the *volume* of the matrix in some sense. Matrices that are not square do not have determinants.

The determinant is frequently indicated by vertical lines, viz. $|A|$. It is a complicated formula, and one way to find it is by induction. The determinant of a 1×1 matrix is $|a| = a$. The determinant of a 2×2 matrix is

$$\begin{vmatrix} a & b \\ c & d \end{vmatrix} = ad - bc.$$

The determinant of an $n \times n$ matrix is given in term of n determinants, each of size $(n-1) \times (n-1)$. If $\mathbf{A}$ is $n \times n$ and $\mathbf{M}_{ij}$ is the matrix obtained by removing the ith row and the jth column from $\mathbf{A}$, then

$$|A| = \sum_{j=1}^{n} (-1)^{1+j} a_{1j} \, |M_{1j}|$$

Example 2.7

Find the determinant

$$\begin{vmatrix} 1 & 2 & 3 \\ 4 & 0 & 6 \\ 7 & 8 & 9 \end{vmatrix}$$

Solution

$$
\begin{aligned}
|A| &= (-1)^{1+1}a_{11}|M_{11}| + (-1)^{1+2}a_{12}|M_{12}| + (-1)^{1+3}a_{13}|M_{13}| \\
&= 1\begin{vmatrix} 0 & 6 \\ 8 & 9 \end{vmatrix} - 2\begin{vmatrix} 4 & 6 \\ 7 & 9 \end{vmatrix} + 3\begin{vmatrix} 4 & 0 \\ 7 & 8 \end{vmatrix} \\
&= -48 - 2(36-42) + 3(32) = 60
\end{aligned}
$$

Example 2.8

Find the determinant

$$
\begin{vmatrix} 0 & 0 & 2 & 0 \\ 1 & 2 & 7 & 3 \\ 4 & 0 & 3 & 6 \\ 7 & 8 & -6 & 9 \end{vmatrix}
$$

Solution

$$
\begin{aligned}
|A| &= a_{11}|M_{11}| - a_{12}|M_{12}| + a_{13}|M_{13}| - a_{14}|M_{14}| \\
&= 0|M_{11}| - 0\ |M_{12}| + 2\ |M_{13}| - 0\ |M_{14}| = 2 \bullet 60 = 120
\end{aligned}
$$

The last example provides a clue to the evaluation of large determinants, but the use of the first row of **A** in the definition of a determinant was arbitrary. For any row or column (fix i or j),

$$
|A| = \sum_{j=1}^{n} (-1)^{1+j} a_{i1j} |M_{ij}|
$$

The interaction of the determinant with elementary row or column operations is simple: Interchanging two rows changes the sign of the determinant; multiplying a row by a constant multiplies the determinant by that constant.

Example 2.9

Evaluate the determinant

$$
\begin{vmatrix} 1 & 2 & 3 & 4 \\ 1 & 1 & 1 & 0 \\ 4 & 0 & 3 & 2 \\ 0 & 3 & 0 & 1 \end{vmatrix}
$$

Solution

Choose a row or column with many zeroes and introduce still more:

$$
|A| = C_2 - 3C_4 |A| = \begin{vmatrix} 1 & -10 & 3 & 4 \\ 1 & 1 & 1 & 0 \\ 4 & -6 & 3 & 2 \\ 0 & 0 & 0 & 1 \end{vmatrix} = (-1)^{4+4} a_{44} \begin{vmatrix} 1 & -10 & 3 \\ 1 & 1 & 1 \\ 4 & -6 & 3 \end{vmatrix}
$$

$$
\underset{R_3 - 4R_1}{\overset{R_2 - R_1}{=}} \begin{vmatrix} 1 & -10 & 3 \\ 0 & 11 & -2 \\ 0 & 34 & -9 \end{vmatrix} = (-1)^{1+1} a_{11} \begin{vmatrix} 11 & -2 \\ 34 & -9 \end{vmatrix} = -99 + 68 = -31
$$

If a matrix is upper (or lower) triangular, the determinant is simply the product of the diagonal entries.

Example 2.10

Find the determinant

$$\begin{vmatrix} 3 & 4 & 5 & 6 & 7 \\ 0 & 1 & 2 & -1 & 6 \\ 0 & 0 & 2 & 5 & 9 \\ 0 & 0 & 0 & 1 & 2 \\ 0 & 0 & 0 & 3 & 5 \end{vmatrix}$$

Solution

Although this matrix is not upper triangular, one can easily make it so by using $R_5 - 3R_4$. The diagonal product $3 \bullet 1 \bullet 2 \bullet 1$ (new a_{55}) $= 6 \bullet (-1) = -6$ is $|A|$.

Example 2.11

Find which values, if any, of the number c make $\mathbf{A}$ singular if

$$\mathbf{A} = \begin{vmatrix} 1 & 2 & c \\ 4 & 5 & 6 \\ 1 & 1 & 1 \end{vmatrix}$$

Solution

$|\mathbf{A}| = (-1)^2(5-6) + (-1)^3(2)(4-6) + (-1)^4 c(4-5) = -1 + 4 - c = 0$. Hence $\mathbf{A}$ is singular for only one value of c, $c = 3$.

If $\mathbf{A}$ is a square matrix, the classical adjoint (adj) of $\mathbf{A}$ is the transpose of the matrix of *signed minors*: $\text{adj}\,(A) = [(-1)^{i+j}\,|M_{ij}|]^t$.

Example 2.12

Find the adjoint of an arbitrary 2×2 matrix, and compute $\mathbf{A} \bullet \text{adj}(A)$.

Solution

$$\text{adj}\begin{bmatrix} a & b \\ c & d \end{bmatrix} = \begin{bmatrix} |M_{11}| & -|M_{12}| \\ -|M_{21}| & |M_{22}| \end{bmatrix}^t = \begin{bmatrix} d & c \\ -b & a \end{bmatrix}^t = \begin{bmatrix} d & -b \\ -c & a \end{bmatrix}$$

$$\mathbf{A} \bullet \text{adj}(A) = \begin{bmatrix} a & b \\ c & d \end{bmatrix} \bullet \begin{bmatrix} d & -b \\ -c & a \end{bmatrix} = \begin{bmatrix} ad-bc & 0 \\ 0 & ad-bc \end{bmatrix} = |A| \bullet I$$

The result of Example 12 is not an accident, and it is the reason one is interested in the classical adjoint. Always, $\mathbf{A} \bullet \text{adj}(A) = \text{adj}(A) \bullet A = |A| \bullet I$.

Example 2.13

Using the adjoint, find

$$\begin{bmatrix} 1 & 2 \\ -3 & 6 \end{bmatrix}^{-1}$$

Solution

$$A^{-1} = |A|^{-1}\,\text{adj}(A) = \frac{1}{12}\begin{bmatrix} 6 & -2 \\ 3 & 1 \end{bmatrix}$$

Example 2.14

Using the adjoint, invert

$$\begin{bmatrix} 1 & 2 & 1 \\ 3 & 4 & 5 \\ 1 & -1 & 2 \end{bmatrix}$$

Solution

$$\text{adj}(A) = \begin{bmatrix} \begin{vmatrix} 4 & 5 \\ -1 & 2 \end{vmatrix} & -\begin{vmatrix} 3 & 5 \\ 1 & 2 \end{vmatrix} & \begin{vmatrix} 3 & 4 \\ 1 & -1 \end{vmatrix} \\ -\begin{vmatrix} 2 & 1 \\ -1 & 2 \end{vmatrix} & \begin{vmatrix} 1 & 1 \\ 1 & 2 \end{vmatrix} & -\begin{vmatrix} 1 & 2 \\ 1 & -1 \end{vmatrix} \\ \begin{vmatrix} 2 & 1 \\ 4 & 5 \end{vmatrix} & -\begin{vmatrix} 1 & 1 \\ 3 & 5 \end{vmatrix} & \begin{vmatrix} 1 & 2 \\ 3 & 4 \end{vmatrix} \end{bmatrix}^t$$

$$= \begin{bmatrix} 13 & -1 & -7 \\ -5 & 1 & 3 \\ 6 & -2 & -2 \end{bmatrix}^t = \begin{bmatrix} 13 & -5 & 6 \\ -1 & 1 & -2 \\ -7 & 3 & -2 \end{bmatrix}$$

Multiply any row of **A** by the corresponding column of adj(A) to get $|A|$. That is,

$$\begin{bmatrix} 1 & 2 & 1 \end{bmatrix} \bullet \begin{bmatrix} 13 \\ -1 \\ -7 \end{bmatrix} = 4 = |A|$$

Thus,

$$\mathbf{A}^{-1} = \frac{1}{4}\begin{bmatrix} 13 & -5 & 6 \\ -1 & 1 & -2 \\ -7 & 3 & -2 \end{bmatrix}$$

Cramer's Rule is a consequence of adj(A): If **A** is nonsingular, the ith component of the solution of $\mathbf{A}X = \mathbf{B}$ is $x_i = \frac{|A_i|}{|A|}$, where A_i is the result of replacing the ith column of **A** by **B**.

Example 2.15

Find x_2 in $\mathbf{AX} = \mathbf{B}$ by Cramer's Rule if

$$\mathbf{A} = \begin{bmatrix} 1 & 2 & 1 & 1 \\ 3 & 4 & 5 & -2 \\ 6 & 7 & 1 & 5 \\ -1 & 0 & 2 & 0 \end{bmatrix} \quad \text{and} \quad \mathbf{B} = \begin{bmatrix} 1 \\ 2 \\ 3 \\ 4 \end{bmatrix}$$

Solution

First,

$$|A| = \begin{vmatrix} 1 & 2 & 3 & 1 \\ 3 & 4 & 11 & -2 \\ 6 & 7 & 13 & 5 \\ -1 & 0 & 0 & 0 \end{vmatrix} = (-1)^{4+1}(-1)\begin{vmatrix} 2 & 3 & 1 \\ 4 & 11 & -2 \\ 7 & 13 & 5 \end{vmatrix}$$

$$= \begin{vmatrix} 0 & 0 & 1 \\ 8 & 17 & -2 \\ -3 & -2 & 5 \end{vmatrix} = (-1)^{1+3}(1)\begin{vmatrix} 8 & 17 \\ -3 & -2 \end{vmatrix} = -16 + 51 = 35$$

Next, the numerator of x_2 is

$$\begin{vmatrix} 1 & 1 & 1 & 1 \\ 3 & 2 & 5 & -2 \\ 6 & 3 & 1 & 5 \\ -1 & 4 & 2 & 0 \end{vmatrix} = \begin{vmatrix} 1 & 0 & 0 & 0 \\ 3 & -1 & 2 & -5 \\ 6 & -3 & -5 & -1 \\ -1 & 5 & 3 & 1 \end{vmatrix} = \begin{vmatrix} -1 & 2 & -5 \\ -3 & -5 & -1 \\ 5 & 3 & 1 \end{vmatrix} = \begin{vmatrix} -1 & 2 & -5 \\ 0 & -11 & 14 \\ 0 & 13 & -24 \end{vmatrix}$$

$$= -\begin{vmatrix} -11 & 14 \\ 13 & -24 \end{vmatrix} = -\begin{vmatrix} -11 & 14 \\ 2 & -10 \end{vmatrix} = -(110 - 28) = -82, x_2 = -\frac{82}{35}$$

PROBLEMS

2.1 The simplest value of $\frac{[(n+1)!]^2}{n!(n-1)!}$ is

a. n^2
b. $n(n+1)$
c. $n+1$
d. $n(n+1)^2$

2.2 If $x^{3/4} = 8$, x equals

a. 6 c. –9
b. 9 d. 16

2.3 If $\log_a 10 = 0.250$, $\log_{10} a$ equals

a. 4 c. 2
b. 0.50 d. 0.25

2.4 A right circular cone, cut parallel with the axis of symmetry, reveals a

a. circle
b. hyperbola
c. eclipse
d. parabola

2.5 The expression $\frac{6!}{3!0!}$ is equal to

a. ∞
b. 120
c. 2!
d. 0

2.6 To find the angles of a triangle, given only the lengths of the sides, one would use

a. the law of cosines
b. the law of tangents
c. the law of sines
d. the inverse-square law

2.7 If $\sin\alpha = \frac{a}{\sqrt{a^2+b^2}}$, which of the following statements is true?

a. $\tan^{-1}\frac{b}{a} = \frac{\pi}{2} - \alpha$

b. $\tan^{-1}\frac{b}{a} = -\alpha$

c. $\cos^{-1}\frac{b}{\sqrt{a^2+b^2}} = \frac{\pi}{2} - \alpha$

d. $\cos^{-1}\frac{a}{\sqrt{a^2+b^2}} = \alpha$

2.8 The sine of 840° equals
a. –cos 30°
b. –cos 60°
c. sin 30°
d. sin 60°

2.9 One root of $x^3 - 8x - 3 = 0$ is
a. 2
b. 3
c. 4
d. 5

2.10 Napierian logarithms have a base of
a. 3.1416
b. 2.171828
c. 10
d. 2.71828

2.11 $(5.743)^{1/30}$ equals
a. 1.03
b. 1.04
c. 1.05
d. 1.06

2.12 The value of $\tan(A + B)$, where $\tan A = 1/3$ and $\tan B = 1/4$ (A and B are acute angles) is
a. 7/12
b. 1/11
c. 7/11
d. 7/13

2.13 To cut a right circular cone in such a way as to reveal a parabola, it must be cut
a. perpendicular to the axis of symmetry
b. at any acute angle to the axis of symmetry
c. at any obtuse angle to the axis of symmetry
d. none of these

2.14 The equation of the line perpendicular to $3y + 2x = 5$ and passing through (–2, 5) is
a. $2x = 3y$
b. $2y = 3x$
c. $2y = 3x + 16$
d. $3x = 2y + 8$

2.15 If $\mathbf{A} = \begin{bmatrix} 1 & 2 & 3 \\ 1 & 2 & 9 \end{bmatrix}$ and $\mathbf{B} = \begin{bmatrix} 5 & 1 \\ 6 & 0 \\ 4 & 7 \end{bmatrix}$, the (2,1) entry of AB is
a. 29
b. 53
c. 33
d. 64

2.16 The inverse of the matrix $\begin{bmatrix} 1 & 1 \\ 3 & 2 \end{bmatrix}$ is

(a) $\begin{bmatrix} 2 & -1 \\ -3 & 1 \end{bmatrix}$ (b) $\begin{bmatrix} 2 & 3 \\ 1 & 1 \end{bmatrix}$ (c) $\begin{bmatrix} 1 & 3 \\ 1 & 2 \end{bmatrix}$ (d) $\begin{bmatrix} -2 & 1 \\ 3 & -1 \end{bmatrix}$

2.17 The determinant of the matrix is $\begin{bmatrix} 1 & 2 & -1 \\ 3 & 0 & 2 \\ 2 & -2 & -1 \end{bmatrix}$ is

a. 4 c. 24
b. 16 d. –16

2.18 In the system of equations

$$\begin{aligned} 3x_1 + 2x_2 - x_3 &= 5 \\ x_2 - x_3 &= 2 \\ x_1 + 2x_2 - 3x_3 &= -1 \end{aligned}$$

the value of x_2 = is
a. 2 c. 4
b. –1 d. 6

2.19 What is the determinant of M?

$$M = \begin{bmatrix} 0 & 1 & 1 & 1 \\ 1 & 1 & 1 & 1 \\ 1 & 1 & 3 & 1 \\ 2 & 1 & 3 & 4 \end{bmatrix}$$

a. –6 c. 0
b. 6 d. 7

SOLUTIONS

2.1 d. The value $(n+1)!$ may be written as $(n+1)(n)\,[(n-1)!]$. It may be written also as $n!(n+1)$. Hence the given expression may be written as follows:

$$\frac{\{(n+1)(n)\,[(n-1)!]\}\,\{n\,!(n+1)\}}{n\,!(n-1)!} = (n+1)^2 n$$

2.2 d. Raise both sides of the equation to the 4/3 power:

$$[x^{3/4}]^{4/3} = 8^{4/3}$$

$$x = \sqrt[3]{8^4} = \sqrt[3]{(2^3)^4} = 2^{\frac{3\bullet4}{3}} = 2^4 = 16$$

2.3 a. $\log_a 10 = 0.250$ can be written as $10 = a^{0.250}$. Taking $\log_{10}$,

$$\log_{10} 10 = \log_{10} a^{0.250}$$
$$1 = 0.250 \log_{10} a$$

and

$$\log_{10} a = \frac{1}{0.250} = 4$$

2.4 b.

2.5 b.

$$\frac{6!}{3!0!} = \frac{6 \times 5 \times 4 \times 3!}{3! \times 1} = 120$$

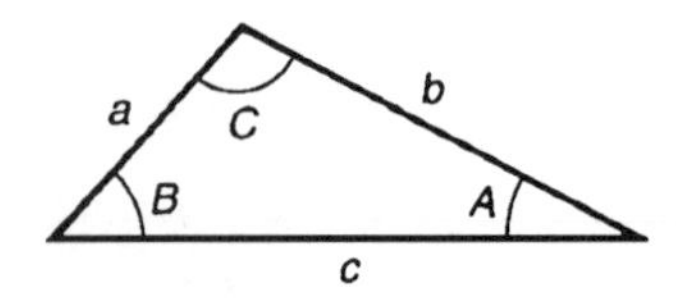

Exhibit 2.6

2.6 a. The law of cosines is $a^2 = b^2 + c^2 - 2bc \cos A$ for any plane triangle with angles A, B, C and sides a, b, c, respectively.

This law can be applied to solve for the angles, given three sides in a plane triangle (Exhibit 2.6).

Exhibit 2.7

2.7 a. The triangle appears in Exhibit 2.7.

$$\tan\left(\frac{\pi}{2} - \alpha\right) = \frac{b}{a}$$
$$\tan^{-1}\frac{b}{a} = \frac{\pi}{2} - \alpha$$

2.8 d.

$$840° = 2(360) + 120 = 2(2\pi) \text{ rad} + 120°$$
$$\sin\,[2(2\pi) \text{ rad} + 120°] = \sin 120° = \sin 60°$$

2.9 b. The solution is obtained by seeing which of the five answers satisfies the equation.

x	$x^3 - 8x - 3$
2	−11
3	0
4	29
5	82
6	165

2.10 d. Common logarithms have base 10. Napierian, or natural, logarithms have base $e = 2.71828$.

2.11 d.

$$\log(5.743)^{1/30} = \frac{1}{30}\log 5.743 = \frac{1}{30}(0.7592) = 0.0253$$

The antilogarithm of 0.0253 is 1.06.

2.12 c.

$$\sin(A + B) = (\sin A \cos B) + (\cos A \sin B)$$
$$\cos(A + B) = (\cos A \cos B) - (\sin A \sin B)$$
$$\tan(A+B) = \frac{\sin(A+B)}{\cos(A+B)} = \frac{(\sin A \cos B) + (\cos A \sin B)}{(\cos A \cos B) - (\sin A \sin B)}$$

Dividing by cos A cos B,

$$\tan(A+B) = \frac{\left(\frac{\sin A \cos B}{\cos A \cos B}\right) + \left(\frac{\cos A \sin B}{\cos A \cos B}\right)}{\left(\frac{\cos A \cos B}{\cos A \cos B}\right) - \left(\frac{\sin A \sin B}{\cos A \cos B}\right)} = \frac{\tan A + \tan B}{1 - \tan A \tan B}$$
$$= \frac{\frac{1}{3} + \frac{1}{4}}{1 - \frac{1}{3} \times \frac{1}{4}} = \frac{\frac{4}{12} + \frac{3}{12}}{1 - \frac{1}{12}} = \frac{\frac{7}{12}}{\frac{11}{12}} = \frac{7}{11}$$

The problem could also be solved by determining angle A (whose tangent is 1/3) and angle B (whose tangent is 1/4). Then we could find the tangent of $(A + B)$.

$$\tan^{-1}\frac{1}{3} = 18.435° \quad \tan^{-1}\frac{1}{4} = 14.036°$$
$$\tan(18.435 + 14.036)° = \tan(32.471°) = 0.6364 = \frac{7}{11}$$

2.13 d. To reveal a parabola, a right circular cone must be cut parallel to an element of the cone and intersecting the axis of symmetry.

2.14 c. Rewriting the given line, $y = \frac{5}{3} - \frac{2}{3}x$. This line has slope $-\frac{2}{3}$, so a perpendicular line must have slope $\frac{3}{2}$. Using the point-slope form, $\frac{y-5}{x+2} = \frac{3}{2}$. Simplifying, $y = \frac{3}{2}x + 8$.

2.15 b. To compute the (2,1) entry, take $[1 \quad 2 \quad 9] \bullet [5 \quad 6 \quad 4] = 5 + 12 + 36 = 53$.

2.16 d. To invert a 2 × 2 matrix,

$$\begin{bmatrix} a & b \\ c & d \end{bmatrix}^{-1} = \frac{1}{ad - bc}\begin{bmatrix} d & -b \\ -c & a \end{bmatrix} = \frac{1}{2-3}\begin{bmatrix} 2 & -1 \\ -3 & 1 \end{bmatrix}$$

2.17 c. This 3 × 3 determinant can be computed quickly be expanding it in minors, especially around the second column:

$$\begin{vmatrix} 1 & 2 & -1 \\ 3 & 0 & 2 \\ 2 & -2 & 1 \end{vmatrix} = (-1)^{1+2}(2)\begin{vmatrix} 3 & 2 \\ 2 & -1 \end{vmatrix} + (-1)^{2+2}(0)\begin{vmatrix} 1 & -1 \\ 2 & -1 \end{vmatrix} + (-1)^{3+2}(-2)\begin{vmatrix} 1 & -1 \\ 3 & 2 \end{vmatrix}$$

$$= -2(-3-4) + 0 + 2(2+3) = 24$$

2.18 d. By Cramer's Rule,

$$x_2 = \frac{\text{Det}\begin{bmatrix} 3 & 5 & -1 \\ 0 & 2 & -1 \\ 1 & -1 & -3 \end{bmatrix}}{\text{Det}\begin{bmatrix} 3 & 2 & -1 \\ 0 & 1 & -1 \\ 1 & 2 & -3 \end{bmatrix}} = \frac{3\begin{vmatrix} 2 & -1 \\ -1 & -3 \end{vmatrix} + 1\begin{vmatrix} 5 & -1 \\ 2 & -1 \end{vmatrix}}{3\begin{vmatrix} 1 & -1 \\ 2 & -3 \end{vmatrix} + 1\begin{vmatrix} 2 & -1 \\ 1 & -1 \end{vmatrix}} = \frac{3(-7)+(-3)=-24}{3(-1)+(-1)=-4} = 6$$

2.19 a. To evaluate a 4 × 4 matrix, one must do some row or column operations and expand by minors:

$$\begin{bmatrix} 0 & 1 & 1 & 1 \\ 1 & 1 & 1 & 1 \\ 1 & 1 & 3 & 1 \\ 2 & 1 & 3 & 4 \end{bmatrix} \sim \begin{bmatrix} 0 & 1 & 1 & 1 \\ 1 & 1 & 1 & 1 \\ 0 & 0 & 2 & 0 \\ 0 & -1 & 1 & 2 \end{bmatrix}$$

Taking minors of column 1,

$$\text{Det}(M) = 1 \bullet (-1)^{2+1}\text{Det}\begin{bmatrix} 1 & 1 & 1 \\ 0 & 2 & 0 \\ -1 & 1 & 2 \end{bmatrix} = -(4+2) = -6$$

CHAPTER 3

Calculus and Differential Equations

David R. Arterburn

OUTLINE

DIFFERENTIAL CALCULUS

Definition of a Function

Notation. A variable y is said to be a function of another variable x if, when x is given, y is determined.

The symbols $f(x)$, $F(x)$, etc., represent various functions of x.
The symbol $f(a)$ represents the value of $f(x)$ when $x = a$.

Definition of a Derivative

Let $y = f(x)$. If Δx is any increment (increase or decrease) given to x, and Δy is the corresponding increment in y, then the derivative of y with respect to x is the limit of the ratio of Δy to Δx as Δx approaches zero; that is,

$$\frac{dy}{dx} = \lim_{\Delta x \to 0} \frac{\Delta y}{\Delta x} = \lim_{\Delta x \to 0} \frac{f(x+\Delta x) - f(x)}{\Delta x} = f'(x)$$

Some Relations among Derivatives

If $x = f(y)$, then $\dfrac{dy}{dx} = 1 \div \dfrac{dx}{dy}$

If $x = f(t)$ and $y = F(t)$, then $\dfrac{dy}{dx} = \dfrac{dy}{dt} \div \dfrac{dx}{dt}$

If $y = f(u)$ and $u = F(x)$, then $\dfrac{dy}{dx} = \dfrac{dy}{du} \times \dfrac{du}{dx}$

Table of Derivatives

Functions of x are represented by u and v, and constants are represented by a, n, and e.

$$\frac{d}{dx}(x) = 1 \qquad \frac{d}{dx}(a) = 0$$

$$\frac{d}{dx}(u \pm v \pm \cdots) = \frac{du}{dx} \pm \frac{dv}{dx} \pm \cdots \qquad \frac{d}{dx}(au) = a\frac{du}{dx}$$

$$\frac{d}{dx}(uv) = u\frac{dv}{dx} + v\frac{du}{dx} \qquad \frac{d}{dx}\left(\frac{u}{v}\right) = \frac{v\frac{du}{dx} - u\frac{dv}{dx}}{v^2}$$

$$\frac{d}{dx}(u^n) = nu^{n-1}\frac{du}{dx} \qquad \frac{d}{dx}\log_a u = \frac{\log_a e}{u}\frac{du}{dx}$$

$$\frac{d}{dx}a^u = a^u \ln a \frac{du}{dx}$$

$$\frac{d}{dx}e^u = e^u\frac{du}{dx} \qquad \frac{d}{dx}u^v = vu^{v-1}\frac{du}{dx} + u^v \ln u \frac{dv}{dx}$$

$$\frac{d}{dx}\sin u = \cos u \frac{du}{dx} \qquad \frac{d}{dx}\cot u = -\csc^2 u \frac{du}{dx}$$

$$\frac{d}{dx}\cos u = -\sin u \frac{du}{dx} \qquad \frac{d}{dx}\sec u = \sec u \tan u \frac{du}{dx}$$

$$\frac{d}{dx}\tan u = \sec^2 u \frac{du}{dx} \qquad \frac{d}{dx}\csc u = -\csc u \cot u \frac{du}{dx}$$

$$\frac{d}{dx}\sin^{-1} u = \frac{1}{\sqrt{1-u^2}}\frac{du}{dx} \quad \text{where} \quad -\pi/2 \le \sin^{-1} u \ge \pi/2$$

$$\frac{d}{dx}\cos^{-1} u = -\frac{1}{\sqrt{1-u^2}}\frac{du}{dx} \quad \text{where} \quad 0 \le \cos^{-1} u \ge \pi$$

$$\frac{d}{dx}\tan^{-1} u = \frac{1}{1+u^2}\frac{du}{dx}$$

$$\frac{d}{dx}\cot^{-1} u = -\frac{1}{1+u^2}\frac{du}{dx}$$

$$\frac{d}{dx}\sec^{-1} u = \frac{1}{u\sqrt{u^2-1}}\frac{du}{dx} \quad \text{where} \quad 0 \le \sec^{-1} u \le \pi/2 \text{ and } -\pi \le \sec^{-1} u \le -\pi/2$$

$$\frac{d}{dx}\csc^{-1} u = -\frac{1}{u\sqrt{u^2-1}}\frac{du}{dx} \quad \text{where} \quad -\pi < \csc^{-1} u \le -\pi/2 \text{ and } 0 < \csc^{-1} u \le \pi/2$$

Slope of a Curve: Tangent and Normal (Fig. 3.1)

The slope of the curve (slope of the tangent line to the curve) whose equation is $y = f(x)$ is

$$\text{Slope} = m = \tan\phi = \frac{dy}{dx} = f'(x)$$

$$\text{Slope at } x_1 \text{ is } m_1 = f'(x_1)$$

The equation of a tangent line at $P_1(x_1, y_1)$ is $y - y_1 = m_1(x - x_1)$. The equation of a normal at $P_1(x_1, y_1)$ is

$$y - y_1 = -\frac{1}{m_1}(x - x_1)$$

The angle β of the intersection of two curves whose slopes at a common point are m_1 and m_2 is

$$\beta = \tan^{-1}\frac{m_2 - m_1}{1 + m_1 m_2}$$

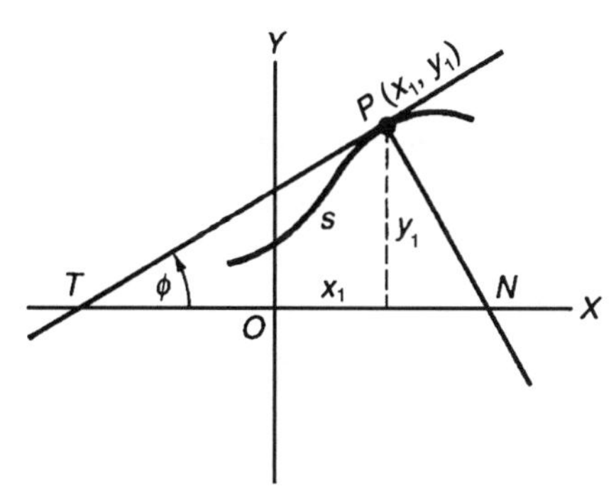

Figure 3.1

Maximum and Minimum Values of a Function (Fig. 3.2)

The maximum or minimum value of a function $f(x)$ in an interval from $x = a$ to $x = b$ is the value of the function which is larger or smaller, respectively, than the values of the function in its immediate vicinity. Thus, the values of the function at M_1 and M_2 in Fig. 3.2 are maxima, its values at m_1 and m_2 are minima.

Test for a maximum at $x = x_1$: $f'(x_1) = 0$ or, and $f''(x_1) < 0$

Test for minimum at $x = x_1$: $f'(x_1) = 0$ or ∞, and $f''(x_1) > 0$

If $f''(x_1) = 0$ or ∞, then for a maximum, $f'''(x_1) = 0$ or ∞ and $f^{IV}(x_1) < 0$; for a minimum, $f'''(x_1) = 0$ or ∞ and $f^{IV}(x_1) > 0$, and similarly if $f^{IV}(x_1) = 0$ or ∞, and so on, where f^{IV} represents the fourth derivative.

In a practical problem that suggests that the function $f(x)$ has a maximum or has a minimum in an interval from $x = a$ to $x = b$, simply equate $f'(x)$ to 0 and solve for the required value of x. To find the largest or smallest values of a function $f(x)$ in an interval from $x = a$ to $x = b$, find also the values $f(a)$ and $f(b)$. L and S

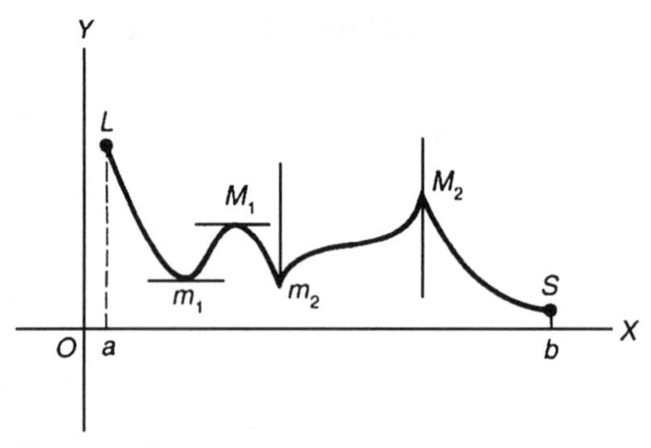

Figure 3.2

may be the largest and smallest values, although they are not maximum or minimum values (see Fig. 3.2).

Points of Inflection of a Curve (Fig. 3.3)

Wherever $f''(x) < 0$, the curve is concave down.

Wherever $f''(x) > 0$, the curve is concave up.

The curve is said to have a point of inflection at $x = x_1$ if $f''(x_1) = 0$ or ∞, and the curve is concave up on one side of $x = x_1$ and concave down on the other (see points I_1 and I_2 in Fig. 3.3).

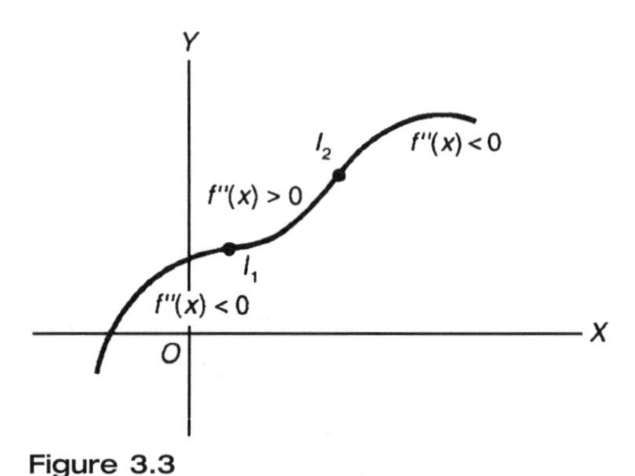

Figure 3.3

Taylor and Maclaurin Series

In general, any $f(x)$ may be expanded into a **Taylor series**:

$$f(x) = f(a) + f'(a)\frac{x-a}{1} + f''(a)\frac{(x-a)^2}{2!} + f'''(a)\frac{(x-q)^3}{3!} + \cdots$$
$$+ f^{(n-1)}(a)\frac{(x-a)^{n-1}}{(n-1)!} + R_n$$

where a is any quantity whatever, so chosen that none of the expressions $f(a)$, $f'(a)$, $f''(a), \ldots$ become infinite. If the series is to be used for the purpose of computing the approximate value of $f(x)$ for a given value of x, a should be chosen such that $(x-a)$ is numerically very small, and thus only a few terms of the series need be used. If $a = 0$, this series is called a Maclaurin series.

Evaluation of Indeterminate Forms

Let $f(x)$ and $F(x)$ be two functions of x, and let a be a value of x.

1. If $\dfrac{f(a)}{F(a)} = \dfrac{0}{0}$ or $\dfrac{\infty}{\infty}$, use $\dfrac{f'(a)}{F'(a)}$ for the value of this fraction.
   If $\dfrac{f'(a)}{F'(a)} = \dfrac{0}{0}$ or $\dfrac{\infty}{\infty}$, use $\dfrac{f''(a)}{F''(a)}$ for the value of this fraction, and so on.
2. If $f(a) \bullet F(a) = 0 \bullet \infty$ or if $f(a) - F(a) = \infty - \infty$, evaluate the expression by changing the product or difference to the form $\frac{0}{0}$ or $\frac{\infty}{\infty}$ and use the previous rule.
3. If $f(a)^{F(a)} = 0^0$ or ∞^0 or 1^∞, then form $e^{F(a)\bullet \ln f(a)}$, and the exponent, being of the form $0 \bullet \infty$, may be evaluated by rule 2.

Differential of a Function

If $y = f(x)$ and Δx is an increment in x, then the differential of x equals the increment of x, or $dx = \Delta x$; and the differential of y is the derivative of y multiplied by the differential of x; thus

$$dy = \frac{dy}{dx}dx = \frac{df(x)}{dx}dx = f'(x)\,dx \quad \text{and} \quad \frac{dy}{dx} = dy \div dx$$

If $x = f_1(t)$ and $y = f_2(t)$, then $dx = f_1'(t)\,dt$, and $dy = f_2'(t)\,dt$.

Every derivative formula has a corresponding differential formula; thus, from the Table of Derivatives above we have, for example,

$$d(uv) = u\,dv + v\,du; \quad d(\sin u) = \cos\, u\,du; \quad d(\tan^{-1} u) = \frac{du}{1+u^2}$$

Functions of Several Variables, Partial Derivatives, and Differentials

Let z be a function of two variables, $z = f(x, y)$; then its partial derivatives are

$$\frac{\partial z}{\partial x} = \frac{dz}{dx} \text{ when } y \text{ is kept constant} \qquad \frac{\partial z}{\partial y} = \frac{dz}{dy} \text{ when } x \text{ is kept constant}$$

Example 3.1

The cubic $y = x^3 + x^2 - 3$ has one point of inflection. Where does it occur?

Solution

The answer requires knowing where y' changes sign. Now $y' = 3x^2 + 2x$ and $y'' = 6x + 2$, which is 0 where $x = -\frac{1}{3}$. Thus the only inflection point is at $x = -\frac{1}{3}$.

Example 3.2

The function of Example 3.1 has one local maximum and one local minimum. Where are they?

Solution

Setting $y' = 0$ $(3x^2 + 2x = 0)$ yields $x = 0$ or $x = -\frac{2}{3}$. Since the second derivative is 2 at $x = 0$, this is the local minimum. At $x = -\frac{2}{3}$, $y'' = -2$, so $x = -\frac{2}{3}$ is the local maximum.

Example 3.3

Two automobiles are approaching the origin. The first one is traveling from the left on the x-axis at 30 mph. The second is traveling from the top on the y-axis at 45 mph. How fast is the distance between them changing when the first is at (–5, 0) and the second is at (0, 10)? (Both coordinates are in miles.)

Solution

If $x(t)$ is taken as the position of the first auto at time t and $y(t)$ as the position of the second auto at time t, then the distance between them at time t is $s(t) = \sqrt{[x(t)]^2 + [y(t)]^2}$. Using the chain rule,

$$s'(t) = \frac{ds}{dt} = \frac{1}{d_2 s(t)} \bullet \frac{d}{dt}\{[x(t)]^2 + [y(t)]^2\} = \frac{1}{2s(t)} \bullet [2x(t)x'(t) + 2y(t)y'(t)]$$

Now $x'(t) = 30$ and $y'(t) = -45$ for all t; and when $t = t_0$, $x(t_0) = -5$ and $y(t_0) = 10$. Therefore,

$$s'(t_0) = \frac{-2 \bullet 5 \bullet 30 - 2 \bullet 10 \bullet 45}{2\sqrt{(-5)^2 + (10)^2}} = \frac{-1200}{2\sqrt{125}} = \frac{-120}{\sqrt{5}} = 24\sqrt{5} \approx 54$$

Thus, the two automobiles are "closing" at about 54 mph.

Example 3.4

How close do the two automobiles in Example 3.3 get?

Solution

One wants to minimize $s(t)$ in Example 3.3, so set $s'(t) = 0$. Thus,

$$\frac{xx' + yy'}{s} = 0 \quad \text{or} \quad xx' + yy' = 0$$

Since $x' = 30$ and $y' = -45$, $30x = 45y$. However, since $x'(t) = 30$, $x(t) = 30t + x_0$, and similarly $y(t) = -45t + y_0$. If one takes $t_0 = 0$ when the problem starts, $x_0 = -5$ and $y_0 = 10$, so $30(30t - 5) = 45(-45t + 10)$ gives time of minimum distance. Solving for t, factor out 75 from both sides to get $2(6t - 1) = 3(-9t + 2)$, or $39t = 8$.

Thus the minimum distance occurs at 8/39 of an hour after the initial conditions of Example 3.3. At this time $x = -5 + 240/39$ and $y = 10 - 360/39$, so $x = 45/39$ and $y = 30/39$. The minimum distance is

$$s\left(\frac{8}{39}\right) = \frac{\sqrt{(45)^2 + (30)^2}}{39} = \frac{15}{39}\sqrt{9+4} = \frac{15\sqrt{13}}{39} \approx 1.4 \text{ miles}$$

Example **3.5**

In Example 3.3, which reaches the origin first, Car 1 or 2?

Solution

This is obvious if, in Example 3.4, one notices that x is positive and y is (still) positive. Alternatively, notice that Car 1 takes 5/30 of an hour to reach the origin and Car 2 takes 10/45 of an hour. The time $5/30 < 10/45$, so Car 1 gets there first.

INTEGRAL CALCULUS

Definition of an Integral

The function $F(x)$ is said to be the integral of $f(x)$ if the derivative of $F(x)$ is $f(x)$, or if the differential of $F(x)$ is $f(x)\,dx$. In symbols,

$$F(x) = \int f(x)\,dx \quad \text{if} \quad \frac{dF(x)}{dx} = f(x), \qquad \text{or} \quad dF(x) = f(x)dx$$

In general, $\int f(x)\,dx = F(x) + C$, where C is an arbitrary constant.

Fundamental Theorems on Integrals

$$\int df(x) = f(x) + C$$

$$\int df(x)\,dx = f(x)\,dx$$

$$\int [f_1(x) \pm f_2(x) \pm \cdots\,]\,dx = \int f_1(x)\,dx \pm \int f_2(x)\,dx \pm \cdots$$

$$\int af(x)\,dx = a\int f(x)\,dx, \text{ where } a \text{ is any constant}$$

$$\int u^n du = \frac{u^{n+1}}{n+1} + C \quad (n \neq -1), \text{ where } u \text{ is any function of } x$$

$$\int \frac{du}{u} = \ln u + C, \text{ where } u \text{ is any function of } x$$

$$\int u\,dv = uv - \int v\,du, \text{ where } u \text{ and } v \text{ are any functions of } x$$

$$\int [u(x) \pm v(x)]\,dx = \int u(x)\,dx \pm \int v(x)\,dx$$

$$\int \frac{dx}{ax+b} = \frac{1}{a}\ln|ax+b|$$

$$\int \frac{dx}{\sqrt{x}} = 2\sqrt{x}$$

$$\int a^x dx = \frac{a^x}{\ln a}$$

$$\int \sin x\,dx = -\cos x$$

$$\int \sin^2 x\,dx = \frac{x}{2} - \frac{\sin 2x}{4}$$

$$\int x \sin x \, dx = \sin x - x \cos x$$

$$\int \cos x \, dx = \sin x$$

$$\int \cos^2 x \, dx = \frac{x}{2} + \frac{\sin 2x}{4}$$

$$\int x \cos x \, dx = \cos x + \sin x$$

$$\int \sin x \cos x \, dx = (\sin^2 x)/2$$

$$\int \tan x \, dx = -\ln|\cos x| = \ln|\sec x|$$

$$\int \tan^2 x \, dx = \tan x - x$$

$$\int \cot x \, dx = -\ln|\csc x| = \ln|\sin x|$$

$$\int \cot^2 x \, dx = -\cot x - x$$

$$\int e^{ax} dx = (1/a)e^{ax}$$

$$\int \ln x \, dx = x[\ln(x) - 1] \qquad (x > 0)$$

Moment of Inertia

Moment of inertia J of a mass m:

About $OX: I_x = \int y^2 dm = \int r^2 \sin^2 \theta \, dm$

About $OY: I_y = \int x^2 dm = \int r^2 \cos^2 \theta \, dm$

About $O: J_0 = \int (x^2 + y^2) dm = \int r^2 \, dm$

Center of Gravity

Coordinates $(\bar{x}, \bar{y})$ of the center of gravity of a mass m:

$$\bar{x} = \frac{\int x \, dm}{\int dm}, \qquad \bar{y} = \frac{\int y \, dm}{\int dm}$$

The center of gravity of the differential element of area may be taken at its midpoint. In the above equations, x and y are the coordinates of the center of gravity of the element.

Work

The work W done in moving a particle from $s = a$ to $s = b$ against a force whose component in the direction of motion is F_s is

$$dW = F_s \, ds, \qquad W = \int_a^b F_s \, ds$$

where F_s must be expressed as a function of s.

Example 3.6

Consider the function $y = x^2 + 1$ between $x = 0$ and $x = 2$. What is the area between the curve and the x-axis?

Solution

$$A = \int_0^2 (x^2 + 1)\, dx = \left(\frac{x^3}{3} + x \right)\bigg|_0^2 = \frac{8}{3} + 2 = \frac{14}{3}$$

DIFFERENTIAL EQUATIONS

Definitions

A **differential equation** is an equation involving differentials or derivatives.

The **order** of a differential equation is the order of the derivative of highest order that it contains.

The **degree** of a differential equation is the power to which the derivative of highest order in the equation is raised, that derivative entering the equation free from radicals.

The **solution** of a differential equation is the relation involving only the variables (but not their derivatives) and arbitrary constants, consistent with the given differential equation.

The most **general solution** of a differential equation of the nth order contains n arbitrary constants. If particular values are assigned to these arbitrary constants, the solution is called a particular solution.

Notation

Symbol or Abbreviation	Definition
M, N	Functions of x and y
X	Function of x alone or a constant
Y	Function of y alone or a constant
C, c	Arbitrary constants of integration
$a, b, k, 1, m, n$	Given constants

Equations of First Order and First Degree: $M\,dx + N\,dy = 0$

Variables Separable: $X_1 Y_1\, dx + X_2 Y_2\, dy = 0$

Solution

$$\int \frac{X_1}{X_2}\, dx + \int \frac{Y_2}{Y_1}\, dy = 0$$

Linear Equation: $dy + (X_1 Y - X_2)\, dx = 0$

Solution

$$y = e^{-\int X_1 dx} \left(\int X_2 e^{\int X_1 dx}\, dx + C \right)$$

Second-Order Differential Equations

A second-order differential expression, $L(x, y, y', y'')$, is linear if

$$L(x, ay_1 + by_2, ay_1' + by_2', ay_1'' + by_2'') = aL(x, y_1, y_1', y_1'') + bL(x, y_2, y_2', y_2'')$$

or if it has the form

$$L(x, y_1, y_1', y_1'') = f(x)y + g(x)y' + h(x)y''$$

A second-order linear differential equation is

$$L(x, y, y', y'') = F(x)$$

If $F(x) \equiv 0$, it is homogeneous; if $F(x)$ is nonzero, it is inhomogeneous.

Constant Coefficients

If $L = ay'' + by' + cy$ where a, b, and c are constants with $a \neq 0$, the first step is to solve the associated homogenous equation $ay'' + by' + cy = 0$. By replacing y by $1, y'$ by r, and y'' by r^2, one obtains the characteristic equation $ar^2 + bc + c = 0$ with roots r_1 and r_2 obtained from factoring or from the quadratic formula. There are three cases to consider:

Case 1: $r_1 \neq r_2$, both real; $y = c_1 e^{r_1 x} + c_2 e^{e_2 x}$, where c_1 and c_2 are arbitrary constants.

Case 2: $r_1 = r_2$; $y = c_1 e^{r_1 x} + c_2 x e^{r_2 x}$.

Case 3: $r_1 = \alpha + j\beta$, $r_2 = \alpha - j\beta$, where α and β are real and $j^2 = -1$; $y = d_1 e^{r_1 x} + d_2 e^{r_2 x} = e^{\alpha x}(c_1 \sin \beta x + c_2 \cos \beta x)$. In particular, if $\alpha = 0$, $y = c_1 \sin \beta x + c_2 \cos \beta x$.

After finding the two solutions to the associated homogeneous equation (y_1 is the result of setting $c_1 = 1$ and $c_2 = 0$, whereas y_2 has $c_1 = 0$ and $c_2 = 1$), one proceeds in either of the following two ways.

Variation of Parameters

If $L(y) = F(x)$ in which the coefficient of y'' is 1, and $W(x) = y_1(x)y_2'(x) - y_1'(x)y_2(x)$ in which y_1 and y_2 are those solutions found above, and if

$$u_1' = \frac{-F(x)y_2(x)}{W(x)} \qquad \text{and} \qquad u_2' = \frac{F(x)y_1(x)}{W(x)}$$

one solution to the inhomogeneous equation is $y_p = u_1(x)y_1(x) + u_2(x)y_2(x)$.

Undetermined Coefficients

In this technique, one guesses y_p by using the following patterns.

One guesses that the solution may be of the same form as the $F(x)$ function but with coefficients to be determined. This method requires modification if $F(x)$ and $c_1y_1 + c_2y_2$ from the associated homogeneous equation interfere, but it is frequently easier than the integration required in the Variation of Parameters technique to construct u_1 and u_2 from their derivatives.

To guess, classify $F(x)$. If it is a polynomial of degree k, the guess will be a polynomial of degree k. However, if $r = 0$ occurs in the homogeneous equation, increase the degree of the polynomial by one. If $F(x)$ is a polynomial times e^{Ax}, so will be the guess. Once again, the degree may have to be increased by one. If $F(x)$ contains sines and cosines, so should the guess.

After making the guess, differentiate it twice and put it into the equation. The coefficients may be determined at this time.

When mixed forms of functions are present in $F(x)$—for example, $x^2 + 2 + 3\sin 2x$—treat the terms $x^2 + 2$ and $3\sin 2x$ independently. The principle of *superposition* then permits you to add the results.

Now, after finding the solution $c_1y_1 + c_2y_2$ to the associated homogeneous equation, and the particular solution y_p to the inhomogeneous equation, form the general solution $y = c_1y_1 + c_2y_2 + y_p$. If initial values are required, such as $y(1) = 2$ and $y'(1) = 3$, the final step is to determine the values of c_1 and c_2 that fit the initial conditions.

Example 3.7

Find the solution of $y'' + 2y' = x^2 + 2 + 3\sin 2x$ subject to $y(0) = 1, y'(0) = 0$.

Solution

Begin with $y'' + 2y' = 0$. The characteristic equation is $r^2 + 2r = 0$, which has roots 0 and -2. Thus, the two solutions to the associated homogeneous equation are $y_1 = 1$ and $y_2 = e^{-2x}$. To use the method of Undetermined Coefficients, guess $(ax^2 + bx + c) \bullet x$ for the $x^2 + 2$ term (the x is needed because $y_1 = 1$). Differentiate twice and insert in the equation: $(6ax + 2b) + 2(3ax^2 + 2bx + c)$ should be the same as $x^2 + 2$. Thus, $6a = 1, 4b + 6a = 0$, and $2b + 2c = 2$. Consequently, $a = \frac{1}{6}, b = -\frac{1}{4}$, and $c = \frac{5}{4}$. Next, guess $c\sin 2x + d\cos 2x$ for the other term. Then $y'' + 2y' = -4c\sin 2x - 4d\cos 2x + 2(2c\cos 2x - 2d\sin 2x)$ should match $3\sin 2x$, so $-4c-4d = 3$ and $-4d + 4c = 0$. Thus $c = d = -\frac{3}{8}$.

Putting all this together, one has the general solution

$$y = A + Be^{-2x} + \frac{1}{6}x^3 - \frac{1}{4}x^2 + \frac{5}{4}x - \frac{3}{8}\sin 2x - \frac{3}{8}\cos 2x$$

Now to fit the initial conditions,

$$y(0) = 1 = A + B - \frac{3}{8} \quad \text{and} \quad y'(0) = 0 = -2B + \frac{5}{4} - \frac{3}{4}$$

Hence, $B = \frac{1}{4}$ and $A = \frac{9}{8}$, so

$$y = \frac{9}{8} + \frac{1}{4}e^{-2x} + \frac{1}{6}x^3 - \frac{1}{4}x^2 + \frac{5}{4}x - \frac{3}{8}\sin 2x - \frac{3}{8}\cos 2x$$

Euler Equations

An equation of the form $x^2y'' + axy' + by = F(x)$, with a and b constants, may be solved as readily as the constant coefficient case. These are called **Euler equations**. Upon substituting $y = x^m$, one obtains the *indicial equation* $m(m - 1) + am + b = 0$. This quadratic equation has two roots, m_1 and m_2.

If $m_1 \neq m_2$, both real, then $y = c_1|x|^{m_1} + c_2|x|^{m_2}$.

If $m_1 = m_2$, then $y = |x|^{m_1}(c_1 + c_2\ln|x|)$.

If $m_1 = p + jq$ and $m_2 = p - jq$, then $y = |x|^p[c_1\cos(q\ln|x|) + c_2\sin(q\ln|x|)]$.

Once y is determined, y_p for the inhomogeneous equation may be found by Variation of Parameters. The method of Undetermined Coefficients is not recommended for Euler equations.

Higher-order linear equations with constant coefficients or of Euler form may be solved analogously.

Laplace Transform

The Laplace transform is an operation that converts functions of x on the half-line $[0, \infty]$ into functions of p on some half-line (a, ∞). The damping power of e^{-xp} is the basis for this useful technique. If $f(x)$ is a piecewise continuous function on $[0, \infty]$ that does not grow too fast, $L(f)$ is the function of p defined by

$$L[f(p)] = \int_0^\infty e^{-xp} f(x)\, dx$$

for the values of p for which the integral converges. For example, if $f(x) \equiv 1$,

$$L(f) = L(1) = \int_0^\infty e^{-xp} dx = \frac{1}{p} \quad (\text{for } p > 0)$$

As a further example,

$$L(f) = L(e^{ax}) = \int_0^\infty e^{-xp} e^{ax} dx = \int_0^\infty e^{-x(p-a)}\, dx = \frac{1}{p-a} \quad (\text{for } p > a)$$

The basic connection between the Laplace transform and differential equations is the following result achieved by integration by parts:

$$L[y'(p)] = \int_0^\infty e^{-xp} y'(x)\, dx = y(x)e^{-xp}\Big|_0^\infty + p\int_0^\infty e^{-xp} y(x)\, dx = -y(0) + pL[y(p)]$$

Consequently, the solution to $ay'' + by' + cy = F(x)$ may be obtained by transforming $L[ay'' + by' + cy] = L(F)$, so $a[p^2L(y) - py(0) - y'(0)] + b[pL(y) - y(0)] + cL(y) = L(F)$. Solving for $L(y)$,

$$L(y) = \frac{L(F) + apy(0) + ay'(0) + by(0)}{ap^2 + bp + c}$$

If one were able to "invert" this result,

$$y = L^{-1}\left[\frac{L(F) + apy(0) + ay'(0) + by(0)}{ap^2 + bp + c}\right]$$

the solution would appear, complete with initial values.The Laplace transform is invertible, and the process of finding $L^{-1}[f(p)]$ as a function of x is much like the process of integration.

Table 3.1 presents a tabulation of selected transforms, where the transform of $f(x)$ is called $F(p)$. Line 3 in Table 3.1 reveals that the operation of multiplying by x corresponds to the negative of the operation of differentiating with respect to p. Lines 2 and 8 have been discussed above. The δ in Line 1 is a *pseudo-function* with great utility defined by $\delta(x) = 0$ for all x except $x = 0$, and $\int_0^\infty \delta(x)\, dx = 1$, so $\delta(0) = +\infty$. The u in Line 7 is called the *Heaviside function*, and it

Table 3.1 Selected transforms

$f(x)$	$F(p)$
1. δ	1
2. 1	p^{-1}
3. $xf(x)$	$\dfrac{-dF}{dp}$
4. $e^{ax}f(x)$	$F(p-a)$
5. $\sin ax$	$\dfrac{a}{p^2+a^2}$
6. $\cos ax$	$\dfrac{p}{p^2+a^2}$
7. $u(x-c)f(x-c)$	$e^{-cp}F(p)$
8. $f'(x)$	$pF(p)-f(0)$
9. $\int_0^x f(u)g(x-u)\,du$	$F(p)\bullet G(p)$
10. $f(x)$, if f is periodic, of period L	$\dfrac{\int_0^L e^{-px}f(x)\,dx}{1-e^{-pL}}$

Figure 3.4

is *zero* until $x-c>0$. Thus $u(x-c)f(x-c)$ is $f(x)$ shifted right to the point $x=c$. For example, if $f(x)=x$ and $c=1$, $u(x-1)f(x-1)$ has the graph shown in Fig. 3.4, whereas f has the graph shown in Fig. 3.5. The operation in the left column of Line 9 is a new way to multiply functions, called **convolution**.

For an example of inversion of a transform, consider

$$F(p)=\frac{2p+3}{(p^2+1)(p-2)}$$

Begin by using partial fractions to write

$$F(p)=\frac{A}{p-2}+\frac{Bp+C}{p^2+1}$$

from which

$$F(p)=\frac{7/5}{p-2}+\frac{(-7/5)p-4/5}{p^2+1}$$

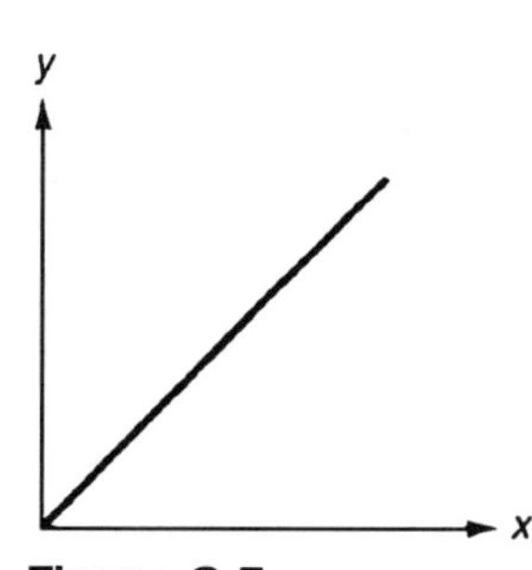

Figure 3.5

Looking at Table 3.1, one finds that $p/(p^2+1)$ is $L(\cos x)$, so

$$L^{-1}\left[\frac{(-7/5)p}{p^2+1}\right]$$

is $-(7/5)\cos x$, and similarly

$$L^{-1}\left(-\frac{4}{5}\bullet\frac{1}{p^2+1}\right)$$

is $-(4/5)\sin x$. Since $L(1)=1/p$, $L(e^{2x})$ is $1/(p-2)$ by Line 4, and combining these three terms yields

$$f(x)=\frac{7}{5}e^{2x}-\frac{7}{5}\cos x-\frac{4}{5}\sin x$$

The *Heaviside* operation in Line 7 leads to an easy solution of differential equations whose right-hand side is not continuous. For example, consider the response of $y''+y$ to a driving function $f(x)$ that is 1 for x between 0 and 2 and then becomes 0. Suppose $y(0) = 1$ and $y'(0) = 3$. By applying Line 8 in Table 3.1 twice,

$$L(y) = \frac{L(f)+p+3}{p^2+1}$$

and since $f(x) = 1 - u(x - 2)$,

$$L(f) = \frac{1}{p} - \frac{e^{-2p}}{p}$$

so

$$L(y) = \frac{p+3+(1-e^{-2p})/p}{p^2+1}$$

The e^{-2p} portion represents delay, so write

$$L(y) = \frac{p^2+3p+1}{p(p^2+1)} - \frac{e^{-2p}}{p(p^2+1)} = \frac{1}{p} + \frac{3}{p^2+1} - e^{-2p}\left(\frac{1}{p} - \frac{p}{p^2+1}\right)$$

from which $y = 1 + 3\sin x - u(x-2) + u(x-2)\cos(x-2)$.

PROBLEMS

3.1 $\int_{\pi/2}^{\pi} \sin 2x \, dx =$

a. 2 b. 1
c. 0 d. –1

3.2 $\int_0^2 x^2\sqrt{1+x^3}\, dx =$

a. 52/9 b. 0
c. 52/3 d. 26/3

3.3 If the first derivative of the equation of a curve is constant, the curve is a
a. circle
b. hyperbola
c. parabola
d. straight line

3.4 Which of the following is a characteristic of all trigonometric functions?
a. The values of all functions repeat themselves every 45 degrees.
b. All functions have units of length or angular measure.
c. The graphs of all functions are continuous.
d. All functions have dimensionless units.

3.5 For a given curve $y = f(x)$ that is continuous between $x = a$ and $x = b$, the average value of the curve between the ordinates at $x = a$ and $x = b$ is represented by

a. $\dfrac{\int_a^b x^2\, dy}{b-a}$ b. $\dfrac{\int_a^b y^2 dx}{b-a}$

c. $\dfrac{\int_a^b x\, dy}{a-b}$ d. $\dfrac{\int_a^b y\, dx}{b-a}$

3.6 If $y = \cos x$, $\dfrac{dy}{dx}$ is

a. $\sin x$ b. $-\tan x \cos x$

c. $\dfrac{1}{\sec x}$ d. $\sec x \sin x$

3.7 The derivative of $\cos^3 5x$ is
a. $3 \sin^2 5x$ c. $\cos^2 5x \sin x$
b. $15 \sin^2 5x$ d. $-15 \cos^2 5x \sin 5x$

3.8 If $x^3 + 3x^2y + y^3 = 4$ defines y implicitly, $dy/dx =$

a. $-\dfrac{x^2+2xy}{x^2+y^2}$ c. $-\dfrac{x^2+y^2}{x^2+2xy}$

b. $3x^2 + 3y^2$ d. $-\dfrac{x^2+2xy}{x^2+y^2}$

3.9 Estimate $\sqrt{34}$ using differentials. The answer is closest to

a. $6+\frac{1}{6}$ c. 6

b. $6-\frac{1}{6}$ d. $6-\frac{1}{3}$

3.10 The only relative maximum of $f(x)=x^4-\frac{4}{3}x^3-12x^2+1$ is

a. -1 c. 0

b. 1 d. -1

3.11 The area between $y=x^2$ and $y=2x+3$ is

a. 9 c. $6\frac{1}{3}$

b. 20 d. $10\frac{2}{3}$

3.12 The area enclosed by the curve $r=2(\sin\theta+\cos\theta)$ is

a. π c. 2π

b. $\frac{\pi}{2}$ d. $\pi\sqrt{2}$

3.13 The curve in Exhibit 3.13 has the equation $y=f(x)$. At point A, what are the values of $\frac{dy}{dx}$ and $\frac{d^2y}{dx^2}$?

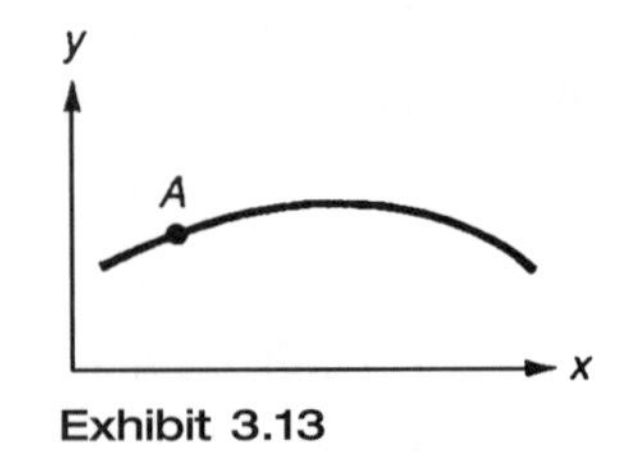

Exhibit 3.13

a. $\frac{dy}{dx}<0,\ \frac{d^2y}{dx^2}<0$ c. $\frac{dy}{dx}=0,\ \frac{d^2y}{dx^2}=0$

b. $\frac{dy}{dx}<0,\ \frac{d^2y}{dx^2}>0$ d. $\frac{dy}{dx}>0,\ \frac{d^2y}{dx^2}<0$

3.14 $\lim_{x\to 1}\frac{x^2-1}{x-1}=$

a. 2 c. 0

b. ∞ d. 1

3.15 The solution to $xy'+2y=e^{3x}$ is

a. $y=e^{3x}-\frac{e^{3x}}{x}+\frac{c}{x}$

b. $y=\frac{xe^{3x}-\frac{1}{3}e^{3x}+3c}{3x^2}$

c. $y=xe^{3x}-3e^{3x}+c$

d. $y+x=e^{3x}+c$

3.16 Solve $xy''-2(x+1)y'+(x+2)y=0$.

a. $y=Ae^x+Bx^3e^x$ c. $y=A\sin(x+1)+B\cos(x+2)$

b. $y=Ae^x+Be^{2x}$ d. $y=Ae^x+Be^{-x}$

3.17 Solve $y''+4y=8\sin x$.

a. $y=Ae^{2x}+Be^{-2x}$

b. $y=A\sin 2x+B\cos 2x$

c. $y=A\sin 2x+B\cos 2x+\sin x$

d. $y=A\sin 2x+B\cos 2x+\frac{8}{3}\sin x$

3.18 The family of trajectories orthogonal to the family $x^2 + y^2 = 2cy$ is

a. $x - y = c$ c. $x^2 + y^2 = c$
b. $x^2 - y^2 = cx$ d. $x^2 + y^2 = 2cx$

SOLUTIONS

3.1 d.

$$\int_{\pi/2}^{\pi} \sin 2x\, dx = -\frac{1}{2}\cos 2x = -\frac{1}{2}\cos 2\pi + \frac{1}{2}\cos \pi = -\frac{1}{2} - \frac{1}{2} = -1$$

3.2 a. Let $u = 1 + x^3$, so $du = 3x^2 dx$. The integral becomes $\frac{1}{3}\int_1^9 \sqrt{u}\, du = \frac{2}{9}u^{3/2} = \frac{2}{9}(27 - 1)$.

3.3 d. If $\frac{dy}{dx} = m$, $y = \int m\, dx = m\int dx = mx + b$, so $y = mx + b$ is a straight line.

3.4 d. All trigonometric functions are ratios of lengths, with the result that they are dimensionless.

3.5 d.

$$\text{Area} = \int_a^b y\, dx$$

$$\text{Average value} = \frac{\text{Area}}{\text{Base width}} = \frac{\int_a^b y\, dx}{b - a}$$

3.6 b. Since $\frac{dy}{dx} = -\sin x$ and $\tan x = \frac{\sin x}{\cos x}$, then $\sin x = \tan x \cos x$. Thus, the derivative is

$$\frac{dy}{dx} = -\tan x \cos x$$

3.7 d. Apply the chain rule. The "outside" function is u^3, so $y' = 3\cos^2 5x\,(\cos 5x)'$ $= 3\cos^2 5x\,(-\sin 5x) \bullet 5$.

3.8 a. Taking the derivative with respect to x, $3x^2 + 6xy + 3x^2y' + 3y^2y' = 0$, so $y' = -\frac{3x^2 + 6xy}{3x^2 + 3y^2}$.

3.9 b. Since $\sqrt{36} = 6$, take $x_0 = 36$ and $f(x) = \sqrt{x}$. In general,

$$\Delta y = f(x) - f(x_0) = f'(x_0)(x - x_0)$$

$$\sqrt{34} - \sqrt{36} = \frac{1}{2} \bullet \frac{1}{\sqrt{36}} \bullet (-2) = -\frac{1}{6}$$

3.10 b. Here, $f'(x) = 4x^3 - 4x^2 - 24x = x(4x^2 - 4x - 24) = 4x(x-3)(x+2) = 0$, so possible extrema are at 0, 3, and –2. Since $f''(0) = -24$, it is the maximum (3 and –2 are minima). Since $f(0) = 1$.

3.11 d. The line and the parabola intersect when $x^2 = 2x + 3$, or $x^2 - 2x + 1 = 4$, or $(x-1)^2 = 2^2$. The line is above the parabola, so

$$A = \int_{-1}^{3} (2x + 3 - x^2)\, dx = (x^2 + 3x - \tfrac{1}{3}x^3)\Big|_{-1}^{3} = 9 - \left(-\tfrac{5}{3}\right)$$

3.12 c. Multiply by r to obtain $x^2 + y^2 = 2y + 2x$, or $x^2 - 2x + y^2 - 2y = 0$, or $(x-1)^2 + (y-1)^2 = 2$ a circle centered at (1, 1) of radius $\sqrt{2}$. The area is $\pi(\sqrt{2})^2 = 2\pi$.

3.13 d. The first derivative $\frac{dy}{dx}$ is the slope of the curve. At point A the slope is positive. The second derivative $\frac{d^2y}{dx^2}$ gives the direction of bending. A negative value indicates the curve is concave downward.

3.14 a.

$$\lim_{x \to 1} \frac{x^2 - 1}{x - 1} = \lim_{x \to 1} \frac{(x-1)(x+1)}{x - 1} = \lim_{x \to 1}(x+1) = 2$$

3.15 b. This is linear equation, $y' + \frac{2}{x}y = \frac{1}{x}e^{3x}$. The integrating factor is $e^{\int \frac{2}{3} dx} = x^2$, so the equation becomes $d(x^2y) = xe^{3x}\, dx$. Integrating, $x^2y = \frac{1}{3}xe^{3x} - \frac{1}{9}e^{3x} + c$, or $y = \frac{1}{3x}e^{3x} - \frac{1}{9x^2}e^{3x} + \frac{c}{x^2}$.

3.16 a. By inspection, $y_1 = e^x$ is one solution. Use reduction of order to obtain

$$\left(\frac{y_2}{y_1}\right)' = \frac{e^{\int \frac{-2(x+1)}{x} dx}}{y_1^2} = \frac{e^{2x+2\ln x}}{e^{2x}} = x^2$$

Hence $\frac{y_2}{y_1} = \frac{x^3}{3}$, so $y_2 = \frac{x^3}{3}y_1$. Suppressing the $\frac{1}{3}$, $y_2 = x^3e^x$.

3.17 d. The associated homogeneous equation, $y'' + 4y = 0$, has the solution $y_h = A \sin 2x + B \cos 2x$. Using the method of undetermined coefficients,

$$y_p = a \sin x + b \cos x$$
$$y_p'' = -a \sin x - b \cos x$$
$$y_p'' + 4y_p = (-a + 4a) \sin x + (-b + 4b) \cos x = 8 \sin x$$

Thus, $b = 0$ and $a = \frac{8}{3}$.

3.18 d. Begin by eliminating c. Thus and $x^2 + y^2 = 2cy$ and $2x + 2yy' = 2cy'$, $c = \frac{x + yy'}{y'}$, $x^2 + y^2 = 2\frac{x + yy'}{y'}y$, $x^2y' + y^2y' = 2xy + 2y^2y'$, and $y' = \frac{2xy}{x^2 - y^2}$.

Now the orthogonal family will have $y'_{\text{new}} = -\frac{1}{y'_{\text{old}}}$, so $y'_{\text{old}} = \frac{y^2 - x^2}{2xy}$

Letting $u = \dfrac{y}{x}$, $xu' + u = \dfrac{u^2 - 1}{2u}$, $xu' = \dfrac{u^2 - 1 - 2u^2}{2u} = -\dfrac{1 + u^2}{2u}$, and $\dfrac{2u\,du}{1 + u^2} = -\dfrac{dx}{x}$. Integrating, $\ln(1 + u^2) = -\ln|x| + c$, and

$$\ln\left\{\left[1 + \left(\frac{y}{x}\right)^2\right]|x|\right\} = c$$

$$\left|x + \frac{y^2}{x}\right| = e^c = c_1 > 0, \quad \text{or} \quad x + \frac{y^2}{x} = c_2(=\pm c_1)$$

Thus $x^2 + y^2 = c_2 x$, or $x^2 + y^2 = 2cx$.

CHAPTER 4

Computers

Donald G. Newnan
with updates by Sharad Laxpati

OUTLINE

INTRODUCTION

The Fundamentals of Engineering exam contains seven questions concerning computers in the morning session and three questions in the afternoon session. These questions cover the topics of spreadsheets, algorithm flow charts, pseudocode, and data transmission and storage. Each of the branch-specific afternoon exams contain three questions on numerical methods related to that branch.

You should review the glossary of important computer related keywords that is presented at the end of this chapter to ensure that you have a basic understanding of this broad general topic. Should you find any term unfamiliar to you, a review of that topic would be warranted. The current exam does not include a programming language such as C, C++, Java, FORTRAN, or BASIC, but use of applications such as spreadsheets is included. You should be familiar with one of the popular spreadsheet programs such as Excel, Quattro Pro, or Lotus 1-2-3.

Number Systems

The number system most familiar to everyone is the decimal system based on the symbols 0 through 9. This base-10 system requires 10 different digits to create the representation of numbers.

A far simpler system is the binary number system. This base-2 system uses only the characters 0 and 1 to represent any number. A binary representation 110, for example, corresponds to the number.

$$1\times 2^2 + 1\times 2^1 + 0\times 2^0 = 4 + 2 + 0 = 6$$

Similarly, the binary number 1010 would be

$$1\times 2^3 + 0\times 2^2 + 1\times 2^1 + 0\times 2^0 = 8 + 0 + 2 + 0 = 10$$

The digital computer is based on the binary system of on/off, yes/no, or 1/0. For this reason it may at times be necessary to convert a decimal number (like 12) to a binary number (12 would be 1100).

Example 4.1

The binary number 1110 corresponds to what decimal (base 10) number?

Solution

$$1\times 2^3 + 1\times 2^2 + 1\times 2^1 + 0\times 2^0 = 8 + 4 + 2 = 14$$

The conversion of a decimal number to a binary number can be achieved by the method of remainders as follows. A decimal integer is divided by 2, giving an integer quotient and a remainder. This process is repeated until the quotient becomes 0. The remainders (in the reverse order) form the binary number. The following example illustrates this process.

Example 4.2

Convert decimal number 43 to a binary number.

Solution

	Quotient		Remainder
$43 \div 2 =$	21	+	1
$21 \div 2 =$	10	+	1
$10 \div 2 =$	5	+	0
$5 \div 2 =$	2	+	1
$2 \div 2 =$	1	+	0
$1 \div 2 =$	0	+	1

Answer: $(43)_{10} = (101011)_2$

Conversion of a decimal fraction to a binary fraction is accomplished by successive multiplication by 2. The integer portion of the number after multiplication is the binary digit. The fractional part is repeatedly multiplied until it becomes 0. The integers in the correct order form the binary fraction.

A large number of binary digits is difficult for humans to use. Consequently, for human-machine interaction, other powers of base-2 number systems are used. Octal (base-8) and hexadecimal (base-16) number systems are generally used: the latter is the most common. The hexadecimal system is a shorthand method of representing the value of four binary digits at a time. Since the hexadecimal system requires 16 different characters to represent decimal digits 0 through 15, the letters A, B, C, D, E, and F are used to represent decimal numbers 10 through 15, respectively.

The conversion from binary to hexadecimal is accomplished by grouping binary digits (bits) into groups of four bits starting from the binary point and proceeding to the left and to the right. Each group of four bits is then converted to the corresponding hexadecimal digit. Conversion from decimal to hexadecimal may be carried out in a manner similar to that for decimal to binary conversion, with the divisor 2 (or multiplier in the case of fractions) replaced by 16.

Example 4.3

Convert the base 10 integer 458 to base 16 equivalent value.

	Quotient		Remainder	Hexadecimal Digit
$458 \div 16$	28	+	10	A
$28 \div 16$	1	+	12	C
$1 \div 16$	0	+	1	1

Data Storage

Memory hardware in a modern computer is semiconductor based. Types of memory include random access memory (RAM), read-only memory (ROM), programmable read-only memory (PROM), and erasable programmable read-only memory (EPROM). The amount of RAM (main memory) in a microcomputer is typically 256 MB to 1 GB and is volatile.

For permanent mass data storage, magnetic and optical disk drive units are available and generally included. Magnetic disk drives (hard drives) are made up of several platters each with one or more read/write heads, depending on the density and size of storage and on data access speed. Platters turn at high speed (3000 to 7200 rpm).

Data on a platter are organized into tracks and sectors. Tracks are the concentric storage areas, and sectors are pie-shaped subdivisions of each track. The data are also organized in cylinders, which are the same numbered tracks on all drive platters.

The hard drives are generally fixed, although a number of portable hard drives are now available that connect to a microcomputer (PC) through its universal serial bus (USB) or parallel port. The storage capacity of these fixed drives ranges from 10 GB at the low end to over 240 GB. Portable drives typically have 40-GB storage capacities. Optical disk drives can be read-only (R/O). WORM drives (write once, read many) can be written to by the user. Other drives, such as CD and DVD drives, can be read-only (CD-ROM and DVD-ROM) or read and write (CD-RW and DVD-RW). These types of storage provide a convenient way of transferring large amounts of data between computers. Several other devices, such as memory cards, memory sticks, thumb drives, flash drives, Zip disks, and floppy disks, serve the same purpose of porting data between PCs and in some cases between other digital devices (such as still and video cameras and printers), although with smaller storage capacity.

The storage capacity of a diskette depends on its size, the recording density, number of tracks, and number of sides. Although at one time larger size (8-inch and $5^1/_4$-inch) floppy diskettes were used, the $3^1/_2$-inch rigid diskette is the current standard.

In addition to the storage capacity of these fixed disk drives, several other parameters may be used to describe their performance:

- *Average seek time:* Average time it takes to move a head from one location to a new location.
- *Track-to-track seek time:* The time required to move a head from one track to an adjacent one.
- *Latency or rotational delay:* The time it takes for a head to access a particular sector to read or write. On the average, this is the time required for one-half revolution of the disk.
- *Average access time:* The time required to move to a new sector and read the data. This is the sum of latency and average seek time.

Data Transmission

A computer generates digital (pulse) signals (represented as on/off or 0/1), but often these signals cannot be transmitted on telephone lines over long distances. These signals are converted to analog form (usually tones) and then modulated and transmitted over analog transmission lines. The transmitted signals at the receiving end are demodulated (converted to tone signals) and then converted to digital pulses. The device designed to do this conversion and communication is called a **modem** (derived from modulation-demodulation). Typical modems available for a PC are capable of operating at up to 56 kilobits per second for transmission over voice grade telephone lines. Digital Subscriber Line (DSL) and broadband cable and satellite service use modems that are capable of operating at much higher speeds, usually in megabits per second (Mbps). For transmission and reception of data, modems are required at both ends that are capable of communicating with each other using appropriate protocol.

There are two common methods of communication: asynchronous and synchronous. In an asynchronous system each character is preceded and followed by start and stop bits. Thus, every 8-bit character requires 10 bits for transmission. There is a 20% penalty in transmission rate in overheads due to the required start and stop bits. Asynchronous communication is, however, the most common method in data communication at low data rates.

In a synchronous system the data is transmitted continuously. Start and stop bits are not needed, but synchronous communication requires clock synchronization. This is accomplished by sending special characters for synchronization. Separation of a bit stream into individual characters is accomplished by counting bits from the start of the previous character. Synchronous transmission is approximately 20% faster than asynchronous and is frequently used in large-volume, high-speed data transmission. Characters are sent in blocks and special (synchronization) characters are placed at the beginning of the block and within it to ensure that the receiving clock remains accurate (synchronized). Thus, there is an overhead penalty in this case as well, but it is far less than 20%.

In any transmission system, errors are bound to occur. The methods of ensuring accuracy of transmission and reception are called communication protocols and transmission standards. Frequently, error checking and correcting technique is employed for accuracy. One or more parity bits may be added as a means of checking and correcting. Another method is for both receiver and sender to calculate a block check digit derived from each block of characters sent.

Programming

Computer programming may be thought of as a four-step process:

1. Defining the problem
2. Planning the solution
3. Preparing the program
4. Testing and documenting the program

Once the problem has been carefully defined, the basic programming work of planning the computer solution and preparing the detailed program can proceed. In this section the discussion will be limited to two ways to plan a computer program to solve a problem: algorithmic flowcharts and pseudocode.

Algorithmic Flowcharts

An algorithmic flowchart is a pictorial representation of the step-by-step solution of a problem using standard symbols. Some of the commonly used shapes are shown in Fig. 4.1. Consider the following simple problem.

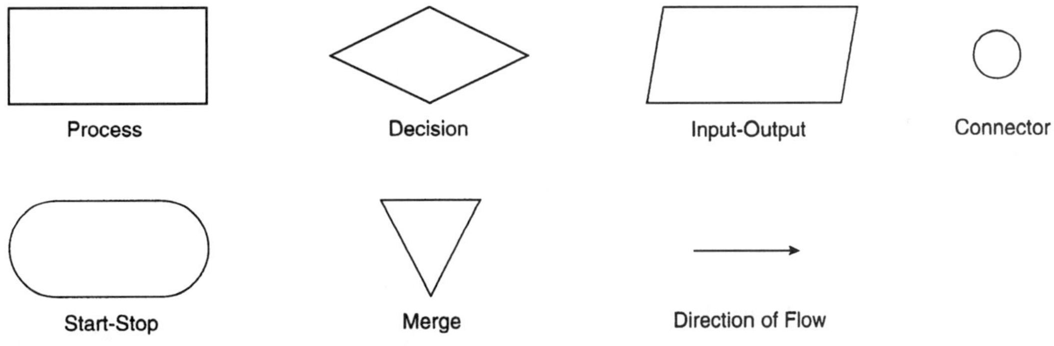

Figure 4.1 Flowchart Symbols

Example 4.4

A present sum of money (P) at an annual interest rate (I), if kept in a bank for N years, would amount to a future sum (F) at the end of that time according to the equation $F = P(1 + I)^N$. Prepare a flowchart for $P = \$100$, $I = 0.07$, and $N =$ 5 years. Then compute and output the values of F for all values of N from 1 to 5.

Solution

Exhibit 1 shows a flowchart for this situation.

Example 4.5

Consider the flowchart in Exhibit 2.

The computation does which of the following?

(a) Inputs hours worked and hourly pay and outputs the weekly paycheck for 40 hours or less.

(b) Inputs hours worked and hourly pay and outputs the weekly paycheck for hours worked including over 40 hours at premium pay.

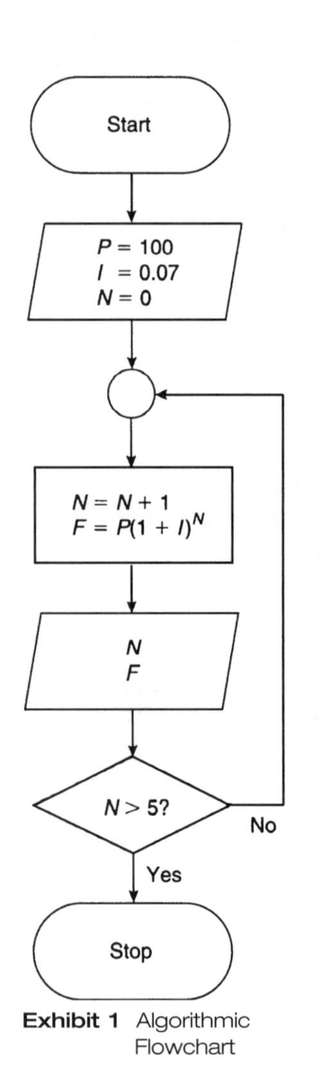

Exhibit 1 Algorithmic Flowchart

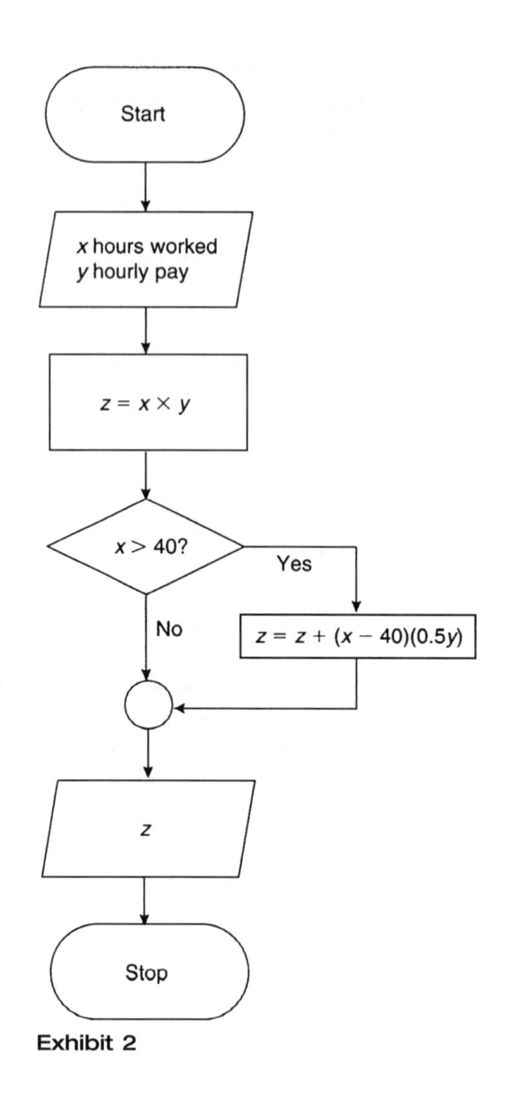

Exhibit 2

Solution

The answer is (b).

Pseudocode

Pseudocode is an English-like-language representation of computer programming. It is carefully organized to be more precise than a simple statement, but may lack the detailed precision of a flowchart or of the computer program itself.

Example 4.6

Prepare pseudocode for the computer problem described in Example 4.4.

Solution

```
INPUT P, I, and N = 0
DOWHILE N < 5
COMPUTE N = N + 1
   F = P (1 + I)^N
OUTPUT N, F
IF N > 5 THEN ENDDO
```

SPREADSHEETS

For today's engineers the ability to create and use spreadsheets is essential. The most popular spreadsheet programs are Microsoft's **Excel**, Novell's **Quattro Pro**, and Lotus **1-2-3**. Each of these programs uses similar construction, methods, operators, and relative references.

Three types of information may be entered into a spreadsheet: text, values, and formulas. Text includes labels, headings, and explanatory text. Values are numbers, times, or dates. Formulas combine operators and values in an algebraic expression.

A **cell** is the intercept of a column and a row. Its location is based upon its column-row location; for example, B3 is the intercept of column B and row 3. Column labels are across the top of the spreadsheet and row labels are on the side. To change a cell entry, the cell must be highlighted using either an address or pointer.

A group of cells may be called out by using a **range**. Cells A1, A2, A3, A4 could be called out using the range reference A1:A4 (or A1..A4). Similarly, the range A2, B2, C2, D2 could use the range reference A2:D2 (or A2..D2).

In order to call out a block of cells, a range callout might be A2:C4 (or A2..C4) and would reference the following cells:

A2 B2 C2

A3 B3 C3

A4 B4 C4

Formulas may include cell references, operators (such as + ,–,*,/), and functions (such as SUM, AVG). The formula SUM(A2:A6) or SUM(A2..A6) would be evaluated as equal to A2 + A3 + A4 + A5 + A6.

Relational References

Most spreadsheet references are relative to the cell's position. For example, if the content of cell A5 contains B4, then the value of A5 is the value of the cell up one and over one. The relational reference is most frequently used in tabulations, as in the following example for an inventory where cost times quantity equals value and the sum of the values yields the total inventory cost.

Inventory		Valuation	
	A	B	C
1 Item	Cost	Quantity	Value
2 box	5.2	2	10.4
3 tie	3.4	3	10.2
4 shoe	2.4	2	4.8
5 hat	1.0	1	1.0
6 Sum			26.4

In C2, the formula is A2*B2, in C3 A3*B3, and so forth. For the summation the function SUM is used; for example, SUM(C2:C5) in C6.

Instead of typing in each cell's formula, the formula can be copied from the first cell to all of the subsequent cells by first highlighting cell C2 then dragging the mouse to include cell C5. The first active cell, C2, would be displayed in the edit window. Typing the formula for C2 as = A2*B2 and holding the control key down and pressing the enter key will copy the relational formula to each of the highlighted cells. Since the call is relational, the formula in cell C3 is evaluated as A3*B3. In C4 the cell is evaluated A4*B4. Similarly, edit operations to include copy or fill operations simplify the duplication of relational formula from previously filled-in cells.

Arithmetic Order of Operation

Operations in equations use the following sequence for precedence: exponentiation, multiplication or division, followed by addition or subtraction. Parentheses in formulas override normal operator order.

Absolute References

Sometimes one must use a reference to a cell that should not be changed, such as a data variable. An absolute reference can be specified by inserting a dollar sign ($) before the column-row reference. If B2 is the data entry cell, then by using B2 as its reference in another cell, the call will always be evaluated to cell B2. Mixed reference can be made by using the dollar sign for only one of the elements of the reference. For example, the reference B$2 is a mixed reference in that the row does not change but the column remains a relational reference.

The power of spreadsheets makes repetitive calculations very easy. In many operations, periods of time are normally new columns. Each element can become a row item with changes in time becoming the columns.

All spreadsheet programs allow for changes in the appearance of the spreadsheet. Headings, borders, or type fonts are usual customizing tools.

NUMERICAL METHODS

This portion of numerical methods includes techniques of finding roots of polynomials by the Routh-Hurwitz criterion and Newton methods, Euler's techniques of numerical integration and the trapezoidal methods, and techniques of numerical solutions of differential equations.

Root Extraction

Routh-Hurwitz Method (without Actual Numerical Results)

Root extraction, even for simple roots (i.e., without imaginary parts), can become quite tedious. Before attempting to find roots, one should first ascertain whether they are really needed or whether just knowing the area of location of these roots will suffice. If all that is needed is knowing whether the roots are all in the left half-plane of the variable (such as is in the s-plane when using Laplace transforms—as is frequently the case in determining system stability in control systems), then one may use the Routh-Hurwitz criterion. This method is fast and easy even for higher-ordered equations. As an example, consider the following polynomial:

$$p_n(x) = \prod_{m=1}^{n}(x - x_m) = x^n + a_1x^{n-1} + a_2x^{n-2} + \cdots + a_{n-1} \tag{4.1}$$

Here, finding the roots, x_m, for $n > 3$ can become quite tedious without a computer; however, if one only needs to know if any of the roots have positive real parts, one can use the Routh-Hurwitz method. Here, an array is formed listing the coefficients of every other term starting with the highest power, n, on a line, followed by a line listing the coefficients of the terms left out of the first row. Following rows are constructed using Routh-Hurwitz techniques, and after completion of the array, one merely checks to see if all the signs are the same (unless there is a zero coefficient—then something else needs to be done) in the first column; if none, no roots will exist in the right half-plane. In case of zero coefficient, a simple technique is used; for details, see almost any text dealing with stability of control systems. A short example follows.

$F(s) = s^3 + 3s^2 + 2s + 10$	Array:	s^3	1	2	Where the s^1 term is formed as
		s^2	3	10	$(3 \times 2 - 10 \times 1)/3 = -\frac{4}{3}$. For details,
$= (s + ?)(s + ?)(s + ?)$		s^1	$-\frac{4}{3}$	0	refer to any text on control
		s^0	10	0	systems or numerical methods.

Here, there are two sign changes: one from 3 to $-\frac{4}{3}$, and one from $-\frac{4}{3}$ to 10. This means there will be two roots in the right half-plane of the s-plane, which yield an unstable system. This technique represents a great savings in time without having to factor the polynomial.

Newton's Method

The use of Newton's method of solving a polynomial and the use of iterative methods can greatly simplify a problem. This method utilizes synthetic division and is based upon the remainder theorem. This synthetic division requires estimating a root at the start, and, of course, the best estimate is the actual root. The root is the correct one when the remainder is zero. (There are several ways of estimating this root, including a slight modification of the Routh-Hurwitz criterion.)

If a $P_n(x)$ polynomial (see Eq. (4.1)) is divided by an estimated factor $(x - x_1)$, the result is a reduced polynomial of degree $n-1$, $Q_{n-1}(x)$, plus a constant remainder of b_{n-1}. Thus, another way of describing Eq. (4.1) is

$$P_n(x)/(x - x_1) = Q_{n-1}(x) + b_{n-1}/(x - x_1) \quad \text{or} \quad P_n(x) = (x - x_1)Q_{n-1}(x) + b_{n-1} \tag{4.2}$$

If one lets $x = x_1$, Eq. (4.2) becomes

$$P_n(x = x_1) = (0)Q_{n-1}(x) + b_{n-1} = b_{n-1} \tag{4.3}$$

Equation (4.3) leads directly to the remainder theorem: "The remainder on division by $(x - x_1)$ is the value of the polynomial at $x = x_1$, $P_n(x_1)$."*

* Gerald & Wheatley, *Applied Numerical Analysis*, 3rd ed., Addison-Wesley, 1985.

Newton's method (actually, the Newton-Raphson method) for finding the roots for an nth-order polynomial is an iterative process involving obtaining an estimated value of a root (leading to a simple computer program). The key to the process is getting the first estimate of a possible root. Without getting too involved, recall that the coefficient of x^{n-1} represents the sum of all of the roots and the last term represents the product of all n roots; then the first estimate can be "guessed" within a reasonable magnitude. After a first root is chosen, find the rate of change of the polynomial at the chosen value of the root to get the next, closer value of the root x_{n+1}. Thus the new root estimate is based on the last value chosen:

$$x_{n+1} = x_n - P_n(x_n)/P_n'(x_n),\quad \text{where } P_n'(x_n) = dP_n(x)/dx \text{ evaluated at } x = x_n \tag{4.4}$$

NUMERICAL INTEGRATION

Numerical integration routines are extremely useful in almost all simulation-type programs, design of digital filters, theory of z-transforms, and almost any problem solution involving differential equations. And since digital computers have essentially replaced analog computers (which were almost true integration devices), the techniques of approximating integration are well developed. Several of the techniques are briefly reviewed below.

Euler's Method

For a simple first-order differential equation, say $dx/dt + ax = af$, one could write the solution as a continuous integral or as an interval type one:

$$x(t) = \int^t [-ax(\tau) + af(\tau)]d\tau \tag{4.5a}$$

$$x(kT) = \int^{kT-T} [-ax + af]d\tau + \int_{kT-T}^{kT} [-ax + af]d\tau = x(kT - T) + A_{\text{rect}} \tag{4.5b}$$

Here, A_{rec} is the area of $(-ax + af)$ over the interval $(kT - T) < \tau < kT$. One now has a choice looking back over the rectangular area or looking forward. The rectangular width is, of course, T. For the forward-looking case, a first approximation for x_1 is**

$$\begin{aligned} x_1(kT) &= x_1(kT - T) + T[ax_1(kT - T) + af(kT - T)] \\ &= (1 - aT)x_1(kT - T) + aTf(kT - T) \end{aligned} \tag{4.5c}$$

Or, in general, for Euler's forward rectangle method, the integral may be approximated in its simplest form (using the notation $t_{k+1} - t_k$ for the width, instead of T, which is $kT–T$) as

$$\int_{t_k}^{t_{k+1}} x(\tau)d\tau \approx (t_{k+1} - t_k)x(t_k) \tag{4.6}$$

** This method is as presented in Franklin & Powell, *Digital Control of Dynamic Systems*, Addison-Wesley, 1980, page 55.

Trapezoidal Rule

This trapezoidal rule is based upon a straight-line approximation between the values of a function, $f(t)$, at t_0 and t_1. To find the area under the function, say a curve, is to evaluate the integral of the function between point a and b. The interval between these points is subdivided into subintervals; the area of each subinterval is approximated by a trapezoid between the end points. It will be necessary only to sum these individual trapezoids to get the whole area; by making the intervals all the same size, the solution will be simpler. For each interval of delta t (i.e., $t_{k+1} - t_k$), the area is then given by

$$\int_{t_k}^{t_{k+1}} x(\tau)\, d\tau \approx (1/2)(t_{k+1} - t_k)[x(t_{k+1}) + x(t_k)] \tag{4.7}$$

This equation gives good results if the delta t's are small, but it is for only one interval and is called the "local error." This error may be shown to be – (1/12)(delta $t)^3 f''$ $(t = \xi_1)$, where ξ_1 is between t_0 and t_1. For a larger "global error" it may be shown that

$$\text{Global error} = -(1/12)(\text{delta } t)^3\ [f''(\xi_1) + f''(\xi_2) + \cdots + f''(\xi_n)] \tag{4.8}$$

Following through on Eq. (4.8) allows one to predict the error for the trapezoidal integration. This technique is beyond the scope of this review or probably the examination; however, for those interested, please refer to pages 249–250 of the previously mentioned reference to Gerald & Wheatley.

NUMERICAL SOLUTIONS OF DIFFERENTIAL EQUATIONS

This solution will be based upon first-order ordinary differential equations. However, the method may be extended to higher-ordered equations by converting them to a matrix of first-ordered ones.

Integration routines produce values of system variables at specific points in time and update this information at each interval of delta time as T (delta $t = T = t_{k+1} - t_k$). Instead of a continuous function of time, $x(t)$, the variable x will be represented with discrete values such that $x(t)$ is represented by $x_0, x_1, x_2, \ldots, x_n$. Consider a simple differential equation as before as, based upon Euler's method,

$$dx/dt + ax = f(t).$$

Now assume the delta time periods, T, are fixed (not all routines use fixed step sizes); then one writes the continous equations as a difference equation where $dx/dt \approx (x_{k+1} - x_k)/T = -ax_k + f_k$ or, solving for the updated value, x_{k+1},

$$x_{k+1} = x_k - Tax_k + Tf_k \tag{4.9a}$$

For fixed increments by knowing the first value of $x_{k=0}$ (or the initial condition), one may calculate the solution for as many "next values" of x_{k+1} as desired for some value of T. The difference equation may be programmed in almost any high-level language on a digital computer; however, T must be small as compared to the shortest time constant of the equation (here, $1/a$).

The following equation—with the "f" term meaning "a function of" rather than as a "forcing function" term as used in Eq. (4.5a)—is a more general form of Eq. (4.9a). This equation is obtained by letting the notation (x_{k+1}) become $y\,[k+1\,\Delta t]$ and is written (perhaps somewhat more confusingly) as

$$y[(k+1)\Delta t] = y(k\Delta t) + \Delta t f[y(k\Delta t), k\Delta t)] \quad \textbf{(4.9b)}$$

Reduction of Differential Equation Order

To reduce the order of a linear time-dependent differential equation, the following technique is used. For example, assume a second-order equation: $x'' + ax' + bx = f(t)$. If we define $x = x_1$ and $x' = x_1' = x_2$, then

$$x_2' + ax_2 + bx_1 = f(t)$$
$$x_1' = x_2 \qquad \text{(by definition)}$$
$$x_2' = -b_{x1} - ax_2 + f(t)$$

This technique can be extended to higher-order systems and, of course, be put into a matrix form (called the state variable form). And it can easily be set up as a matrix of first-order difference equations for solving digitally.

PACKAGED PROGRAMS

Most currently available packaged simulation programs use algorithms not necessarily based upon Euler's methods but more advanced methods such as the Runge-Kutta method. Automatic variable step size methods like Milne's may also be used. However, as mentioned before, these routines are all built into the packaged programs and may be transparent to the user. The user of a specialized program may be without knowledge of the high-level language being employed (except for certain modifications).

GLOSSARY OF COMPUTER TERMS

Accumulators	Registers that hold data, addresses, or instructions for further manipulations in the ALU
Address bus	Two-way parallel path that connects processors and memory containing addresses
AI	Artificial intelligence
Algorithm	A sequence of steps applied to a given data set that solves the intended problem
Alphanumeric data	Data containing the characters a, b, c, …, z, 0, 1, 2, …, 9
ALU	Arithmetic and logic unit
ASCII	American Standard Code for Information Interchange, 7 bit/character (Pronounced AS-key)
Asynchronous	Form of communications in which message data transfer is not synchronous, with the basic transfer rate requiring start/stop protocol
Baud rate	Bits per second
BIOS	Basic input/output system
Bit	0 or 1
Buffer	Temporary storage device
Byte	8 bits
Cache memory	Fast look-ahead memory connecting processors with memory, offering faster access to often-used data
Channel	Logic path for signals or data
CISC	Complex instruction-set control

Clock rate	Cycles per second
Control bus	Separate physical path for control and status information
Control unit	Fetches, decodes instructions to control the operations of registers and ALUs
CPU	Central Processing Unit, the primary processor
Data buffer	Temporary storage of data
Data bus	Separate physical path dedicated for data
Digital	Discrete level or valued quantification, as opposed to analog or continuous valued
Duplex communication	Communications mode where data is transmitted in both directions at the same time
Dynamic memory	Storage that must be continually hardware-refreshed to retain valid information
EBCDIC	Extended Binary Coded Decimal Interchange Code—8 bits/character (pronounced EB-see-dick)
EPROM	Erasable programmable read-only memory
Expert systems	Programs with AI which learn rules from external stimuli
Floppy disk	Removable disk media in various sizes, $5\frac{1}{4}''$, $3\frac{1}{2}''$
Flowchart	Graphical depiction of logic using shapes and lines
Gbyte (GB)	Gigabytes: 1,073,741,824 or 2^{30} bytes
Half-duplex communication	Two-way communications path in which only one direction operates at a time (transmit or receive)
Handshaking	Communications protocol to start/stop data transfer
Hard disk	Disk that has nonremovable media
Hardware	Physical elements of a system
Hexadecimal	Numbering system (base 16) that uses 0–9, A, B, …, F
Hierarchical database	Database organization containing hierarchy of indexes/keys to records
I/O	Input/output devices such as terminal, keyboard, mouse, printer
IR	Instruction register
Kbytes (KB)	Kilobytes: 1024 or 2^{10} bytes
LAN	Local area network
LIFO	Last in–first out
LSI	Large scale integration
Main memory	That memory seen by the CPU
Mbytes (MB)	Megabytes: 1,048, 576 or 2^{20} bytes
Memory	Generic term for random access storage
Microprocessor	Computer architecture with Central Processing Unit in one LSI chip
MODEM	Modulator-demodulator
MOS	Metal oxide semiconductor
Multiplexer	Device that switches several input sources, one at a time, to an output
Nibble	Four bits
Nonvolatile memory	As opposed to volatile memory, does not need power to retain its present state
Number systems	Method of representing computer data as human-readable information
OCR	Optical character recognition
OS	Operating system
OS memory	Memory dedicated to the OS, not usable for other functions
Parallel interface	A character (8-bit) or word (16-bit) interface with as many wires as bits in interface plus data clock wire.
Parity	Method for detecting errors in data: one extra bit carried with data, to make the sum of one bit in a data stream even or odd
PC	Program counter, or personal computer
Peripheral devices	Input/output devices not contained in main processing hardware
Program	A sequence of computer instructions
PROM	Programmable read-only memory
Protocols	Established set of handshaking rules enabling communications
Pseudocode	An English-like way of representing structured programming control structures

RAM	Random access memory
Real time/Batch	Method of program execution: real-time implies immediate execution; batch mode is postponed until run on a group of related activities
Relational database	Database organization that relates individual elements to each other without fixed hierarchical relationships
RISC	Reduced instruction set computer
ROM	Read-only memory
Scratchpad memory	High-speed memory, in either hardware or software
Sequential storage	Memory (usually tape) accessed only in sequential order (n, $n + 1$, …)
Serial interface	Single data stream that encodes data by individual bit per clock period
Simplex communication	One-way communication
Software	Programmable logic
Stacks	Hardware memory organization implementing LIFO access
Static memory	Memory that does not require intermediate refresh cycles to retain state
Structured programming	Use of programming constructs such as Do-While or If-Then, Else
Synchronous	Communications mode in which data and clock are at same rate
Transmission speed	Rate at which data is moved, in baud (bits per second, bps)
Virtual memory	Addressable memory outside physical address bus limits through use of memory mapped pages
Volatile memory	Memory whose contents are lost when power is removed
VRAM	Video memory
Words	8, 16, or 32 bits
WYSIWYG	What you see is what you get
16-bit	Basic organization of data with 2 bytes per word
32-bit	Basic organization of data with 4 bytes per word
64-bit	Basic organization of data with 8 bytes per word
80386	Intel's microprocessor architecture based on 16–data address bus, extended virtual memory, external math coprocessor
80486	Upgrade to 80386 incorporating math coprocessor within VLSI
80586	Intel's microprocessor architecture based on 16-bit data, 32-bit address bus

PROBLEMS

4.1 In spreadsheets, what is the easier way to write B1 + B2 + B3 + B4 + B5?

a. Sum (B1:B5) c. @B1..B5SUM
b. (B1..B5) Sum d. @SUMB2..B5

4.2 The address of the cell located at row 23 and column C is

a. 23C c. C.23
b. C23 d. 23.C

4.3 Which of the following is **not** correct?

a. A CD-ROM may not be written to by a PC.
b. Data stored in a batch processing mode is always up-to-date.
c. The time needed to access data on a disk drive is the sum of the seek time, the head switch time, the rotational delay, and the data transfer time.
d. The methods for storing files of data in secondary storage are sequential file organization, direct file organization, and indexed file organization.

4.4 Which of the following is false?

a. Flowcharts use symbols to represent input/output, decision branches, process statements, and other operations.
b. Pseudocode is an English-like description of a program.
c. Pseudocode uses symbols to represent steps in a program.
d. Structured programming breaks a program into logical steps or calls to subprograms.

4.5 In pseudocode using DOWHILE, the following is true:

a. DOWHILE is normally used for decision branching.
b. The DOWHILE test condition must be false to continue the loop.
c. The DOWHILE test condition tests at the beginning of the loop.
d. The DOWHILE test condition tests at the end of the loop.

4.6 A spreadsheet contains the following formulas in the cells:

	A	B	C
1		A1 +1	B1 +1
2	A1 ^2	B1^2	C1^2
3	Sum (A1:A2)	Sum (B1:B2)	Sum (C1:C2)

If 2 is placed in cell A1, what is the value in cell C3?

a. 12 c. 8
b. 20 d. 28

4.7 A matrix contains the following:

	A	B	C	D
1		3	4	5
2	2	A$2		
3	4			
4	6			

If you copy the formula from B2 into D4, what is the equivalent formula in D4?

a. A$2
b. C4
c. C4
d. C$2

4.8 A processing system is processor limited when sorting on small tables in memory. Which of the following would speed up computations?

a. Adding more main memory
b. Adding virtual memory
c. Adding cache memory
d. Adding peripheral memory

4.9 A small PC processing system is performing large (1-MB) matrix operations that are currently I/O limited since memory is limited to 1MB. Which of the following would speed up computations?

a. Adding more main memory
b. Adding virtual memory
c. Adding cache memory
d. Adding peripheral memory

4.10 Transmission Protocol: Serial, asynchronous, 8-bit ASCII, 1 Start, 1 Stop, 1 Parity bit, 9600 bps. How long will it take for a 1-Kbyte file to be transmitted through the link?

a. 0.85 s
b. 0.96 s
c. 1.07 s
d. 1.17 s

4.11 Transmission Protocol: Serial, synchronous, 8-bit ASCII, 1 Parity bit, 9600 bps. How long will it take for a 1-Kbyte file to be transmitted through the link?

a. 0.85 s
b. 0.96 s
c. 1.07 s
d. 1.17 s

4.12 The hexadecimal number 2DB.A is most nearly equivalent to which decimal number?

a. 731.625
b. 731.10
c. 453.625
d. 341.10

4.13 The decimal number 1938.25 is most nearly equivalent to which hexadecimal number?

a. $(792.25)_{16}$
b. $(792.4)_{16}$
c. $(279.4)_{16}$
d. $(279.04)_{16}$

SOLUTIONS

4.1 a. Sum (B1:B5) or @Sum (B1..B5).

4.2 b.

4.3 b. Batch mode processing always has delays in updating the database.

4.4 c. Pseudocode does not use symbols but uses English-like statements such as IF-THEN and DOWHILE.

4.5 c. IF-THEN is normally used for branching. The DOWHILE test condition must be true to continue branching, and the test is done at the beginning of the loop. The DOUNTIL test is done at the end of the loop.

4.6 b. Plugging 2 into cell A1 of the spreadsheet produces the following matrix:

	A	B	C
1	2	3	4
2	4	9	16
3	6	12	20

The value of C3 is 20.

4.7 d. The formula contains mixed references. The $ implies absolute row reference, whereas the column is relative. The result of any copy would eliminate any answer except for the absolute row 2 entry. The relative column reference A gets replaced by C. The cell contains C$2.

4.8 c. Processor-limited sorting on small tables suggests either speeding up processor cycles or providing faster memory. Since speeding up clock is not an option, making memory faster is the answer. Adding more memory or virtual memory, however, does nothing for small tables. Only adding cache memory would allow the CPU to fetch recently used data without full memory cycles, thereby speeding up the sorting process.

4.9 a. Processing is I/O limited because not all of the matrix can fit into memory. Since this is a large matrix, cache memory probably would not affect processing. The best solution is adding more main memory to fit this problem entirely in memory.

4.10 d.

$$1 \text{ Kbyte} = 2^{10} \text{ bytes} = 1024$$
$$1024 \text{ bytes} + 3 \text{ overhead bits/byte} = 1024 \text{ bytes } (8 \text{ bits/byte} + 3 \text{ bits/byte overhead}) = 11{,}264 \text{ bits}$$
$$\text{Minimum transmission time} = 11{,}264 \text{ bits}/9600 \text{ bps} = 1.17 \text{ s.}$$

4.11 b.

$$1\text{ Kbyte} = 2^{10}\text{ bytes} = 1024$$
$$1024\text{ bytes} + 1\text{ overhead bit/byte} = 1024\text{ bytes (8 bits/byte} + 1\text{ bit/byte overhead)}$$
$$= 9216\text{ bits}$$
$$\text{Minimum transmission time} = 9216\text{ bits/9600 bps} = 0.96\text{ s}$$

4.12 a.

$$(2DB)_{16} = 2 \times 16^2 + 13 \times 16^1 + 11 \times 16^0 = 512 + 208 + 11 = 731$$
$$(.A)_{16} = 10 \times 16^{-1} = .625$$
$$(2DB.A)_{16} = 731.625$$

4.13 b.

	Quotient		Remainder	Hexadecimal Digit
1938 ÷16 =	121	+	2	2
121÷16 =	7	+	9	9
7÷16 =	0	+	7	7

$$(1938)_{10} = (792)_{16}$$

	Integer		Fraction	Hexadecimal Digit
0.25 × 16 =	4	+	0.00	4

$$(.25)_{10} = (.4)_{16}$$
$$(1938.25)_{10} = (792.4)_{16}$$

CHAPTER 5

Probability and Statistics

David R. Arterburn and Wolfgang Baer

OUTLINE

SETS AND SET OPERATIONS

A set is any well-defined list, collection, or class of objects. The objects in a set can be anything: numbers, letters, cards, people, and so on. They are called the **elements**, or **members**, of a set.

The name of a set is usually denoted by a capital letter, such as A, B, Y, Z. The elements of a set are usually denoted by small letters, such as a, b, y, z.

To specify that an element a is a member of a set B, we say "a is in B," which is written

$$a \in B$$

A set is called the **null**, or **empty**, set, denoted by $\emptyset$, if it has no elements. We say the set A is a subset of the set B, written $A \subset B$, if all the members of

A are also in B. The universal set, denoted by U, is the set that contains all the members of the subsets. The **complement** of a set A, denoted as A', consists of all the elements in U that are not in A.

Sets are defined in either of two ways: (1) by listing the members in a tabular form (for example, if A consists of the numbers 1, 3, 5, and 7, we write $A = \{1, 3, 5, 7\}$), or (2) by stating the properties that the members must satisfy in a set-builder form. For example, a set B containing all the odd numbers is written $B = \{x|x \text{ is odd}\}$, where x is an element of the set and the vertical line "|" is read "such that." The full notation is read "B is the set of numbers x such that x is odd."

Example 5.1

Define the set R of all outcomes of the roll of a six-sided die.

Solution

$$R = \{1, 2, 3, 4, 5, 6\}$$

Example 5.2

Let U be the set of integers. Define the complement of $B = \{x|x \text{ is odd}\}$.

Solution

$$B' = \{x|x \text{ is even}\}$$

Set Operations

The **union** of two sets A and B is defined as the set of all elements in either A or B and is traditionally written as $A \cup B$ or, alternatively, as A or B.

The **intersection** of two sets A and B is defined as the set of all elements in both A and B and is traditionally written $A \cap B$.

Set operations are well-behaved mathematically and follow these laws:

identity $\quad A \cup 0 = A \quad A \cup U = U$

$\quad A \cap 0 = 0 \quad A \cap A = A$

complement $\quad A \cup A' = U \quad (A')' = A \quad A \cap A' = 0 \quad U' = 0$

commutative $\quad A \cup B = B \cup A \quad A \cap B = B \cap A$

associative $\quad (A \cup B) \cup C = A \cup (B \cup C)$

$\quad (A \cap B) \cap C = A \cap (B \cap C)$

distributive $\quad A \cup (B \cap C) = (A \cup B) \cap (A \cup C)$

$\quad A \cap (B \cup C) = (A \cap B) \cup (A \cap C)$

de Morgan's Law $\quad (A \cup B)' = A' \cap B'$

$\quad (A \cap B)' = A' \cup B'$

For example, define the following sets:

$U = \{x|x \text{ is a person}\}$

$A = \{a|a \text{ is American}\}$

$F = \{f | f \text{ is French}\}$

$B = \{b | b \text{ is a person with dual French and American citizenship}\}$

Then the union of the sets A and F is the set of all people who are either American or French and is written $A \cup F$. The set of people who are not American is A'. The set of people who are not French is F'.

The set of people who are not French and not American is $A' \cap F'$. This is the same as $(A \cup F)'$, according to de Morgan's Law. In this example, de Morgan's Law means that all the people who are neither American nor French is the same as the set of all the people who are not French and not American.

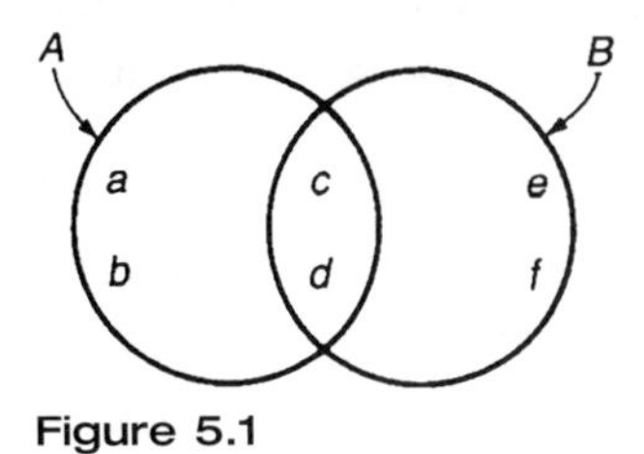

Figure 5.1

Venn Diagrams

A simple and intuitive way to represent these set relationships is known as the **Venn diagram**. Here we represent a set by a closed area and an element in the set by a point in the area. For example, let $A = \{a, b, c, d\}$ and $B = \{c, d, e, f\}$; the sets are represented in a **Venn diagram** as shown in Fig. 5.1.

Relational properties such as the intersection of A and $B = \{c, d\}$ are intuitively understood as the overlap of two regions.

Example 5.3

Draw a Venn diagram of the relationship between the sets U, A, F, and B as defined in the previous section.

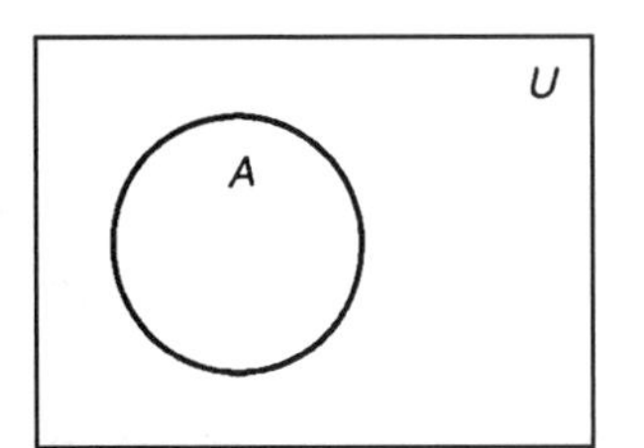

Exhibit 1

Solution

1. $A \subset U$. Americans are a subset of people (Exhibit 1).
2. $A \cup F$ is the set of Americans and French (Exhibit 2).
3. $B = A \cap F$ is the intersection of two sets, it represents Americans who are also French (Exhibit 3).

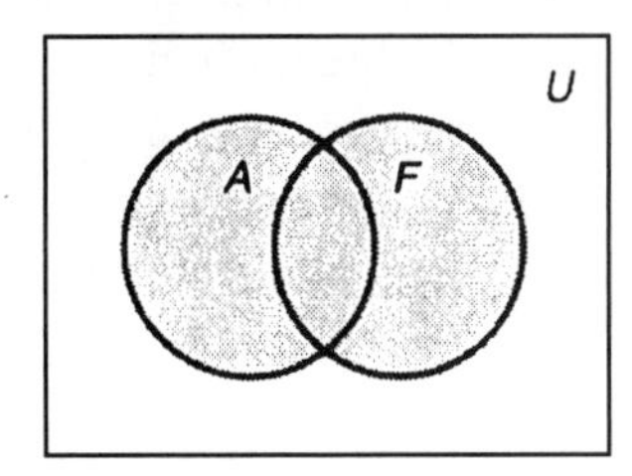

Exhibit 2

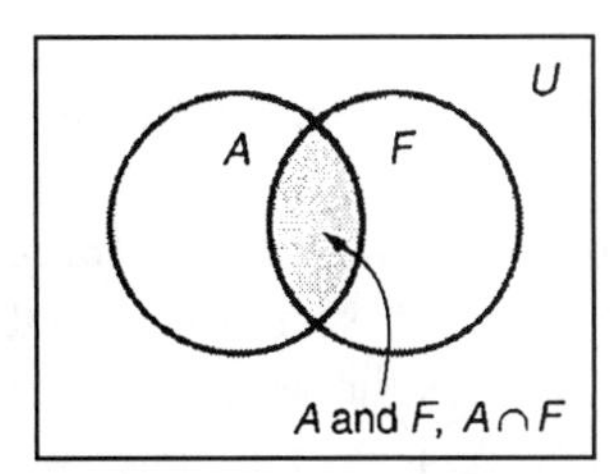

Exhibit 3

Product Sets

An ordered pair of elements a and b is denoted by (a, b). Two ordered pairs (a, b) and (c, d) are equal only if $a = c$ and $b = d$. For example, the ordered pairs (3, 4) and (4, 3) are different.

The product x of two sets A and B is denoted by $A \times B$ and consists of all ordered pairs of elements in A and B.

The two subsets of a product set can be treated like the intersection of two Cartesian coordinate axes. The subsets each act like a dimension, and the product set can be represented as a matrix where the elements of one subset are written along the row of the matrix and those of the other along the column.

For example, the set of results of a single coin toss is either heads (H) or tails (T) and is defined $R = \{H, T\}$. The result of two coin tosses is the product $R \times R$ and is defined

$$R \times R = \{(H, H), (H, T), (T, H), (T, T)\}$$

COUNTING SETS

For discrete probability calculations it is important to count the number of elements in sets of possible outcomes. The primary method is simply to write down all the elements in a set and count them. For example, count the number of elements in the set S of all possible outcomes of a six-sided die throw. The set definition is $S = \{1, 2, 3, 4, 5, 6\}$. Simple counting gives six elements.

Most useful sets are large, and simple counting is too time-consuming. There are several methods for simplifying this task. One can use the product set concept introduced in the last section. If A is a set with n elements and B is a set with m elements, then the product set $A \times B$ has the arithmetic $n \times m$ number of elements. To simplify counting then, first count the sets making up the product set (usually containing a much smaller number of elements) and simply multiply these counts.

Example 5.4

Count the number of possible outcomes for tossing five coins.

Solution

The number of outcomes for a single toss defined by the set $R = \{H, T\}$ is 2. The result of five coin tosses is the product set $R \times R \times R \times R \times R$. The total number of possible outcomes is then the arithmetic product of the number of outcomes in each of the individual five tosses. This is $2 \times 2 \times 2 \times 2 \times 2 = 2^5 = 32$.

Permutations

If A is a set with n elements, a **permutation** of A is an ordered arrangement of A. Given the set $A = \{a, b, c\}$, the order a, b, c of the elements is one permutation. Any other order—for example, b, c, a—is another permutation.

The set B of all permutations of the set A is defined as the set of all arrangements of the three elements. These are

$$B = \{\{a,b,c\}, \{a,c,b\}, \{c,b,a\}, \{b,a,c\}, \{b,c,a\}, \{c,a,b\}\}$$

There are six permutations. This number also can be derived as follows. The number of ways an element can be chosen for the first space is three. Then there are two elements left. One of these can go in the second space. Then there is one element left. This must go in the third space. This gives the formula $3 \times 2 \times 1 = 6$. In general, the number of ways n distinct elements can be arranged is given by

$$n! = n \times (n-1) \times (n-2) \times \cdots 1$$

and is called the **factorial** of the number n. The factorial of 0 is 1 ($0! = 1$).

For example, count the number of ways a standard playing deck can be arranged. Since there 52 distinct cards in a deck, there are 52! different arrangements, or permutations.

Now suppose we have the set of letters L in the word *obtuse* so that $L = \{o, b, t, u, s, e\}$. How many two-letter symbols could be made from this set? Notice that the letters are all distinct. We again count the number of ways the letters can be selected. For the first choice it is six; for the second choice it is five. The two selections are now complete. There are therefore $6 \times 5 = 30$ possibilities.

The general formula for the number of permutations, taking r items from a set of n, is given by

$$P(n, r) = n!/(n - r)!$$

Using this equation, one can express the previous example as $P(6, 2) = 6!/(6 - 2)! = 30$.

Example 5.5

A jeweler has nine different beads and a bracelet design that requires four beads. To find out which looks the best, he decides to try all the permutations. How many different bracelets will he have to try?

Solution

There are $n = 9$ beads. He selects $r = 4$ at a time. The order is important, because each arrangement of r beads on the bracelet makes a different bracelet. So the number of different bracelets is

$$P(9, 4) = 9!/(9 - 4)! = 9 \times 8 \times 7 \times 6 = 3024$$

If the bracelet is a closed circle, there is no discernible difference when it is rotated. Then one observes four identical states for each unique bracelet. This is called ring permutation and is given by the formula

$$P_{\text{ring}}(n, r) = P(n, r)/r$$

There are only 3024/4 = 756 distinct ring bracelets the jeweler can make.

Combinations

When the order of the set of r things that are selected from the set of n things does not matter, we talk about combinations.

Again consider the standard playing deck of 52 cards. How many hands of 5 cards can we get from a deck of 52 cards? Count the number of ways the hands can be drawn. The first draw can be any of the 52 cards. The second draw can only be one of the remaining 51 cards. The third draw can only be one of the remaining 50, the next is one of 49, the last one of 48. So the result is

$$52 \times 51 \times 50 \times 49 \times 48 = 52!/(52 - 5)!$$

This is the formula for permutations discussed in the last section. But the order in which we receive the cards is not important, so many of the hands are the same. In fact there are 5! similar arrangements of cards that make the same hand. The number of distinct hands is

$$(52 \times 51 \times 50 \times 49 \times 48)/(5 \times 4 \times 3 \times 2 \times 1) = 52!/[5! \times (52 - 5)!]$$

The general form r items taken from a set of n items when order is not important is written as the binomial coefficient $C(n, r)$, also written $\binom{n}{r}$, and is given by the formula

$$C(n, r) = \frac{n!}{r!(n - r)!}$$

Example 5.6

There are six skiers staying in a cabin with four bunks. How many combinations of people will be able to sleep in beds?

Solution

$C(6, 4) = 6!/[4! \times (6-4)!] = (6 \times 5 \times 4 \times 3 \times 2 \times 1)/[(4 \times 3 \times 2 \times 1) \times (2 \times 1)] = 15.$

PROBABILITY

Definitions

An **experiment**, or **trial**, is an action that can lead to a measurement.

Sampling is the act of taking a measurement. The **sample space** S is the set of all possible outcomes of an experiment (trial). An event e is one of the possible outcomes of the trial.

If an experiment can occur in n mutually exclusive and equally likely ways, and if m of these ways correspond to an event e, then the probability off the event is given by

$$P\{e\} = m/n$$

Example 5.7

A die is a cube of six faces designated as 1 through 6. The set of outcomes R of one die roll is defined as $R = \{1, 2, 3, 4, 5, 6\}$. If two dice are rolled, define trial, sample space, n, m, and the probability of rolling a seven when adding both dice together.

Solution

The trial is the rolling of two dice. The sample space is all possible outcomes of a two-dice roll, and the event is the outcome that the sum is 7.

The number of all possible outcomes, n, is the number of elements in the product set of the outcome of two dice when each is rolled independently. The product set is $R \times R$ and contains 36 elements.

The number of all possible ways, m, that the (7) event can occur is (1,6), (2,5), (3,4), (4,3), (5,2), and (6,1) for a total of six ways. The probability of rolling a 7 is $P\{7\} = \frac{6}{36} = \frac{1}{6}$.

General Character of Probability

The probability $P\{E\}$ of an event E is a real number in the range 0 through 1. Two theorems identify the range between which all probabilities are defined:

1. If ø is the null set, $P\{ø\} = 0$.
2. If S is the sample space, $P\{S\} = 1$.

The first states that the probability of an impossible event is zero, and the second states that, if an event is certain to occur, the probability is 1.

Complementary Probabilities

If E and E' are complementary events, $P\{E\} = 1 - P\{E'\}$. Complementary events are defined with respect to the sample space. The probability that an event E will happen is complementary to the probability that any of the other possible outcomes will happen.

Example 5.8

If the probability of throwing a 3 on a die is 1/6, what is the probability of not throwing a three?

Solution

E is the probability of not throwing a 3, so $P\{E\} = 1 - P\{E'\} = 1 - \frac{1}{6} = \frac{5}{6}$.

Sometimes the complementary property of probabilities can be used to simplify calculations. This will happen when seeking the probability of an event that represents a larger fraction of the sample space than its complement.

Example 5.9

What is the probability $P\{E\}$ of getting at least one head in four coin tosses?

Solution

The complementary event $P\{E'\}$ to getting at least one head is getting no heads (or all tails) in four tosses. So the probability of getting at least one head is

$$P\{E\} = 1 - (0.5)^4 = 1 - 0.0625 = 0.9375$$

Joint Probability

The probability that a combination of events will occur is covered by joint probability rules. If E and F are two events, the joint probability is given by the rule

$$P\{E \cup F\} = P\{E\} + P\{F\} - P\{E \cap F\} \qquad \text{(Rule 1)}$$

A special case of the joint probability rule can be derived by considering two events, E and F, to be mutually exclusive. In this case the last term in Rule 1 is zero since $P\{E \cap F\} = P\{0\} = 0$. Thus, if E and F are mutually exclusive events,

$$P\{E \cup F\} = P\{E\} + P\{F\} \qquad \text{(Rule 2)}$$

Example 5.10

What is the probability of throwing a 7 or a 10 with two dice?

Solution

We will call the event of throwing a 7 A, and of throwing a 10 B. We know from previous examples that $P\{A\} = \frac{1}{6}$, and we can count outcomes to get $P\{B\} = \frac{1}{12}$. Applying the formula,

$$P\{A \cup B\} = P\{A\} + P\{B\} = \frac{1}{6} + \frac{1}{12} = \frac{1}{4}$$

If two events E and F are independent—that is, if they come from different sample spaces—then the probability that both will happen is given by the rule

$$P\{E \cap F\} = P\{E\} \times P\{F\} \qquad \text{(Rule 3)}$$

Example 5.11

What is the probability of throwing two heads in two coin tosses?

Solution

Call the throwing of one head E, the other F. The probability of throwing a single head is $P\{E\} = \frac{1}{2}$, and $P\{F\} = \frac{1}{2}$. The probability of throwing both heads is

$$P\{E \cap F\} = P\{E\} \times P\{F\} = \frac{1}{2} \times \frac{1}{2} = \frac{1}{4}$$

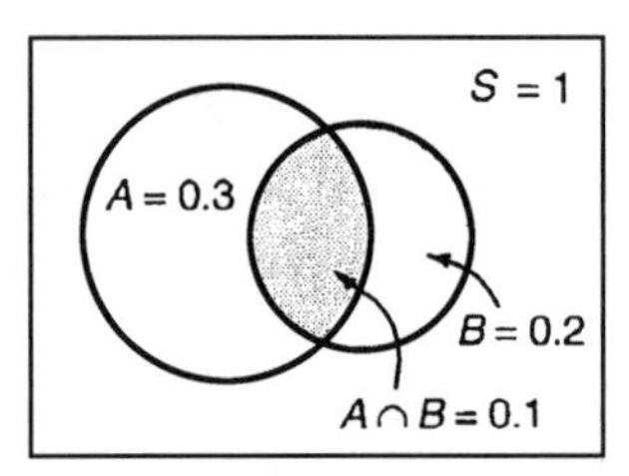

Figure 5.2 Venn diagram of joint probabilities

To visualize joint probabilities, we can use a Venn diagram showing two intersecting events, A and B, as shown in Fig. 5.2. Let the normalized areas of each event represent the probability that the event will occur. For example, think of a random dart thrown at the Venn diagram: What are the chances of hitting one of the areas? Assume the areas correspond to probabilities and are given by $P\{S\} = 1$, $P\{A\} = 0.3$, $P\{B\} = 0.2$, and $P\{A \cap B\}$ is 0.1. The probability of hitting either area A or area B is calculated as the sum of the areas A and B minus the overlap area so it is not counted twice:

$$P\{A \cup B\} = 0.3 + 0.2 - 0.1 = 0.4$$

The result is also equal to the normalized area covered by A and B. The probability of hitting both A and B on one throw is simply the overlap area $P\{A \cap B\} = 0.1$.

If we throw two darts, the area S is used twice and represents two independent sample spaces. Hence Rule 3 applies.

Conditional Probability

The conditional probability of an event E given an event F is denoted by $P\{E|F\}$ and is defined as

$$P\{E|F\} = P\{E \cap F\}/P\{F\} \qquad \text{for } P\{F\} \text{ not zero}$$

Example 5.12

Two six-sided dice, one red and one green, are tossed. What is the probability that the green die shows a 1, given that the sum of numbers on both dice is less than 4?

Solution

Let E be the event "green die shows 1" and let F be the event "sum of numbers shows less than four." Then

$$\begin{aligned} E &= \{(1,1), (1,2), (1,3), (1,4), (1,5), (1,6)\} \\ F &= \{(1,1), (1,2), (2,1)\} \\ E \cap F &= \{(1,1), (1,2)\} \\ P\{E|F\} &= P\{E \cap F\}/P\{F\} = (2/36)/(3/36) = 2/3 \end{aligned}$$

The generalized form of conditional probability is known as Bayes' theorem and is stated as follows: If $E_1, E_2, \ldots, E_n$ are n mutually exclusive events whose union is the sample space S, and E is any arbitrary event such that $P\{E\}$ is not zero, then

$$P\{E_k|E\} = \frac{P\{E_k\} \times P\{E|E_k\}}{\sum_{j=1}^{n} [P\{E_j\} \times P\{E|E_j\}]}$$

RANDOM VARIABLES

The method of random variables is a powerful concept. It casts the set-theory-based probability calculations of previous sections into a functional form and allows the application of standard mathematical tools to probability theory. It is often easy to solve fairly complex probability problems using random variables, although an approach different from the usual one is required.

A random variable, usually denoted by X, is a mapping of the sample space to some set of real numbers. The mapping transforms points of a sample space into points, or more accurately intervals, on the x-axis. The mapping, or random, variable X is called a discrete random variable if it assumes only a denumerable number of values on the x-axis. A random variable is called a continuous random variable if it assumes a continuum of values on the x-axis. The mapping is usually quite easy and intuitive for numerical events but provides no major advantage for nonnumerical discrete sample spaces, where counting remains the major tool.

Example 5.13

Cast the sample space of the outcomes of a roll of a die into random variable form.

Solution

The sample space is the set R defined by $R = \{1, 2, 3, 4, 5, 6\}$. These can easily by written along the x-axis as

$$R = \{1, 2, 3, 4, 5, 6\} \rightarrow 1 \mid 2 \mid 3 \mid 4 \mid 5 \mid 6 \mid x\text{-axis}$$

Probability Density Functions

A probability density function $f(x)$ is a mathematical rule that assigns a probability to the occurrence of the random variable x. Since the random variable is a mapping from trial outcomes, or events, to the numerical intervals on the x-axis, the probability that an event will occur is the area under the probability density function curve over the x interval defining the event.

For a continuous random variable the probability that an event E, mapped into an interval between x_1 and x_2, will occur is defined as

$$\int_{x_1}^{x_2} f(x)\,dx = P\{E\} \qquad \text{for } E \text{ mapped into } (x_1, x_2)$$

For a discrete case the formula is

$$\sum_{i=1}^{n} f(x_i) = P\{E\} \qquad \text{for } E \text{ containing } x_1, x_2, \ldots, x_n$$

It is assumed here that a step interval is associated with each value of x_i; therefore, the equivalent dx in the integral is 1 and is not required in the sum.

Example 5.14

The probability density function of a single six-sided die throw is shown graphically in Exhibit 4. The probability of throwing a 3 is given by the area under the curve over the interval assigned to the numeral 3, which is the step interval from 2.5 to 3.5.

Hence $P\{3\} = f(x) \times 1 = \frac{1}{6} \times 1 = \frac{1}{6}$.

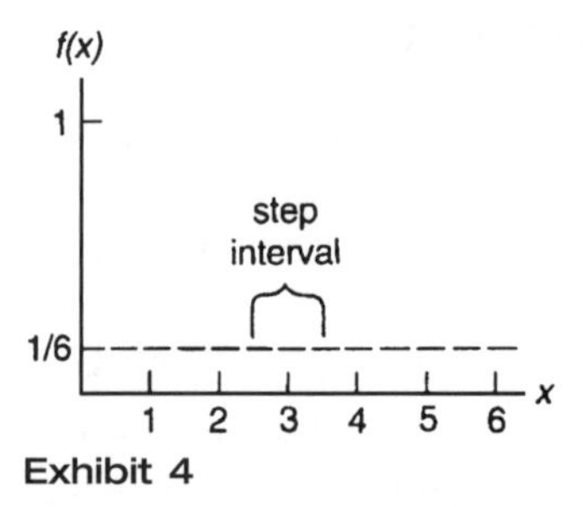

Exhibit 4

Properties of Probability Density Functions

The expected value $E\{X\}$ of a probability density function is also called the mean, and for a discrete case it is given by

$$E\{X\} = \sum x_i \times f(x_i) = u$$

The expected value of a continuous random variable is

$$E\{X\} = \int_{-\inf}^{+\inf} x \times f(x) \times dx = u$$

The expected value for a discrete random variable of a function $g(X)$ is given by

$$E\{g(X)\} = \sum g(x_i) \times f(x_i)$$

The expected value of a continuous random variable is

$$E\{g(X)\} = \int_{-\inf}^{+\inf} g(x) \times f(x) \times dx$$

Of special interest are the functions of the form

$$g(x) = (x - u)^r$$

These are the powers of the random variables around the mean. The expected values of these power functions are called the rth moments about the mean of the distribution, where r is the power. The second moment about the mean is also known as the variance and is calculated as follows:

$$V\{X\} = E\{(x-u)^2\} = E\{(x^2 - 2xu + u^2)\} = E\{x^2\} - E\{2xu\} + E\{u^2\}$$

Since u is a constant, the second term is $2u^2$ and the third term evaluates to u^2; therefore, the second moment about the mean becomes

$$V\{X\} = E\{x^2\} - u^2 = \sigma^2$$

The square root of the variance is signified by the Greek letter sigma and is called the **standard deviation**.

Example 5.15

Calculate the mean and standard deviation of a single die throw.

Solution

This is a discrete function and can be calculated numerically by the discrete formulas given above. The mean, where $f(x_i) = \frac{1}{6}$ (all outcomes are equally likely), is given by

$$u = E\{X\} = \sum_{i=1}^{i=6} x_i \times f(x_i) = (1+2+3+4+5+6)/6 = \frac{21}{6} = 3.5$$

The standard deviation is given by

$$\sigma = \sqrt{V\{x\}} = \sqrt{E\{x^2\} - u^2} = \sqrt{[(1^2+2^2+3^2+4^2+5^2+6^2)/6] - 3.5^2} = 1.7$$

STATISTICAL TREATMENT OF DATA

Whether from the outcome of an experiment or trial, or simply the output of a number generator, we are constantly presented with numerical data. A statistical treatment of such data involves ordering, presentation, and analysis. The tools available for such treatment are generally applicable to a set of numbers and can be applied without much knowledge about the source of the data, although such knowledge is often necessary to make sensible use of the statistical results.

In its raw form, numerical data is simply a list of n numbers denoted by x_i, where $i = 1, 2, 3, \ldots, n$. There is no specific significance associated with the order implicit in the i numbers. They are names for the individuals in the list, although they are often associated with the order in which the raw data was recorded. For example, consider a box of 50 resistors. They are to be used in a sensitive circuit, and their resistances must be measured. The results of the 50 measurements are presented in the following table.

Table of Raw Measurements (Ω)

101	105	110	115	82
86	91	96	117	112
109	103	89	97	98
101	104	99	95	97
85	90	94	112	107
103	94	98	106	98
114	112	108	101	99
93	96	99	104	90
109	106	101	93	92
104	99	109	100	107

Each number is named by the variable x_i, and there are $n = 50$ of them. The numbers range from 82 to 117.

Frequency Distribution

A systematic tool used in ordering data is the frequency distribution. The method requires counting the number of occurrences of raw numbers whose values fall within step intervals. The step intervals (or bins) are usually chosen to (1) be of constant size, (2) cover the range of numbers in the raw data, (3) be small enough in quantity to limit the amount of writing yet not have many empty steps, and (4) be sufficient in quantity so that significant information is not lost.

For example, the aforementioned raw data of measured resistances may be ordered in a frequency distribution table such as Table 5.1. Here the step interval is the event E of a random variable that can be mapped onto the x-axis. The set of eight events is the sample space. If we take a number randomly from the raw measurement set, the probability that it will be in bin 5 is

$$f(E_5) = P\{E_5\} = 10/50 = 0.2$$

Table 5.1 Frequency and cumulative frequency table

Event, E_i	Range, Ω	Frequency	Cumulative Frequency	Probability Density Function, $f(E_i)$
1	80–84	1	1	0.02
2	85–89	3	4	0.06
3	90–94	8	12	0.16
4	95–99	12	24	0.24
5	100–104	10	34	0.20
6	105–109	9	43	0.18
7	110–114	5	48	0.10
8	115–119	2	50	0.04

The last column in Table 5.1 is the probability density function of the distribution. The probability table can be plotted along the x-axis in several ways, as shown in Figs. 5.3 through 5.5.

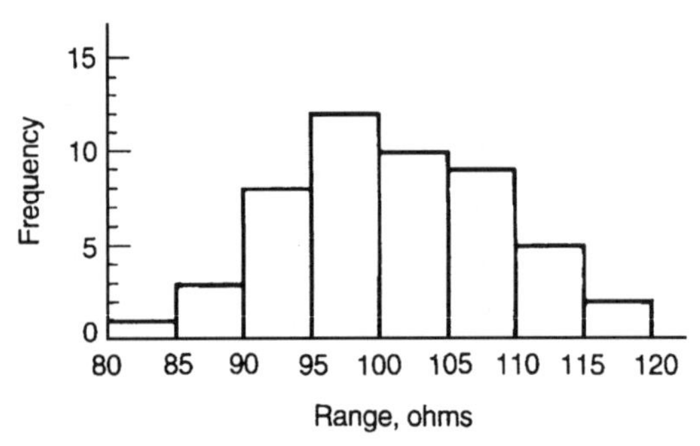

Figure 5.3 Histogram of resistance measurements

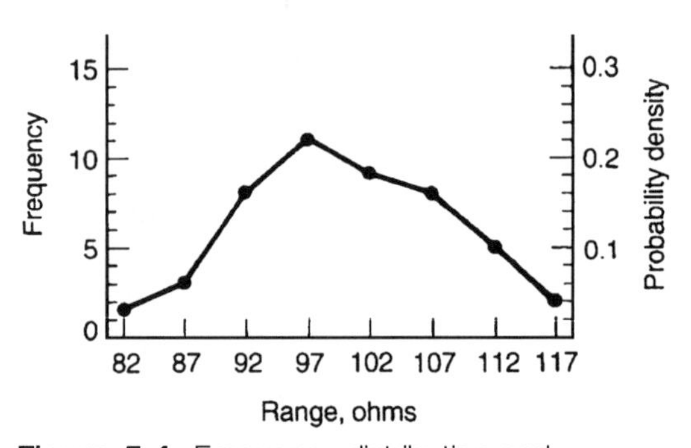

Figure 5.4 Frequency distribution and probability density plot

Standard Statistical Measures

There are several statistical quantities that can be calculated from a set of raw data and its distribution function. Some of the more important ones are listed here, together with the method of their calculation.

Mode The observed value that occurs most frequently; here the mode is bin 4 with a range of 95–99 Ω.

Median The point in the distribution that divides the number of observations such that half of the observations are above and half are below.

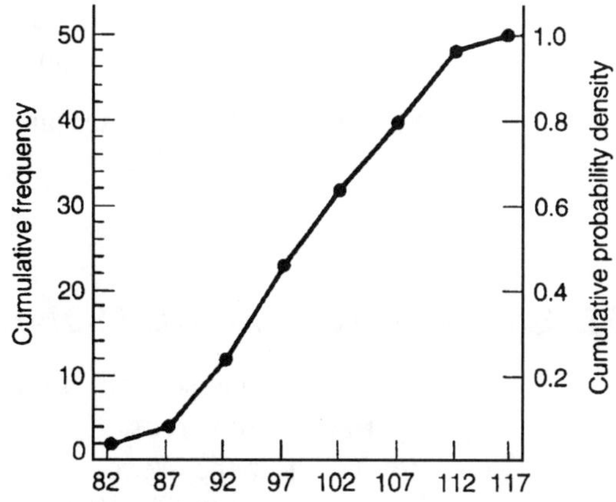

Figure 5.5 Cumulative frequency distribution and cumulative probability density

The median is often the mean of the two middle values; here the median is 4.5 bins, 100 Ω.

Mean The arithmetic mean, or average, is calculated from raw data as

$$\mu = \frac{1}{n}\sum_{i=1}^{n} x_i = 100.6$$

It is calculated from the distribution function as

$$\mu = \sum_{i=1}^{m} b_i \times f(E_i) = 100.4$$

where b_i is the ith event value (for $i = 1$, $b_i = 82$) and m is the number of bins; $f(E_i)$ is the probability density function. (The two averages are not quite the same because of the information lost in assigning the step intervals.)

Standard deviation (a) Computational form for the raw data:

$$\sigma = \sqrt{\frac{1}{n}\left[\left(\sum_{i=1}^{n} x_i^2\right) - n \times \mu^2\right]} = 8.08$$

(b) Computational form for the distribution function:

$$\sigma = \sqrt{\left\{\left[\sum_{i=1}^{m} b_i^2 \times f(E_i)\right] - \mu^2\right\}} = 8.02$$

Sample standard deviation If the data set is a sample of a larger population, then the sample standard deviation is the best estimate of the standard deviation of the larger population.

The computational form for the raw data set is

$$\sigma = \sqrt{\frac{1}{n-1}\left[\left(\sum_{i=1}^{n} x_i^2\right) - n \times \mu^2\right]} = 8.166$$

Sample standard deviations and the use of $(n - 1)$ in the denominator are discussed in the section on sampling.

Skewness This is a measure of the frequency distribution asymmetry and is approximately

$$\text{skewness} \cong 3(\text{mean} - \text{median})/(\text{standard deviation})$$

STANDARD DISTRIBUTION FUNCTIONS

In the previous section, we calculated several general properties of probability distribution functions.

To know the appropriate probability density function for an actual situation, two general methods are available:

1. The probability density function is actually calculated, as was done in the last section, by analyzing the physical mechanism by which experimental events and outcomes are generated and counting the number of ways an individual event occurs.
2. Recognition of an overall similarity between the present experiment and another for which the probability density function is already known permits the known behavior of the function to be applied to the new experiment. This work-saving method is by far the more popular one. Of course, to apply this method, it is necessary to have a repertoire of known probability functions and to understand the problem characteristics to which they apply.

This section lists several popular probability density functions and their characteristics.

Binomial Distribution

The binomial distribution applies when there is a set of discrete binary alternative outcomes. Deriving this distribution function helps one understand the class of problems to which it applies. For example, given a set of n events, each with a probability p of occurring, what is the probability that r of the events will occur and $(n - r)$ not occur?

The probability of one event occurring is p.

The probability of r events occurring is p^r.

The probability of $(n - r)$ events not occurring is $(1 - p)^{n-r}$.

The probability of exactly r events occurring and $(n - r)$ not occurring in a trial is given by the joint probability Rule 3,

$$P[r \cap (n-r)] = p^r \times (1-p)^{n-r}$$

However, there are many ways of choosing r occurrences out of n events. In fact, the number of different ways of choosing r items from a set of n items when order is not important is given by the binomial coefficient $C(n, r)$. The total probability of r occurrences from n trials, given an individual probability of occurrence as p, is thus given by

$$C(n, r) \times p^r \times (1-p)^{n-r} = f(r)$$

This is the **binomial probability density function**.

The mean of this density function is the first moment of the density function, or expected value, and is calculated as

$$E\{x\} = \sum_{r=0}^{n} r \times f(r) = \sum_{r=0}^{n} r \times \frac{n!}{(r)!(n-r)!} \times p^r \times (1-p)^{n-r}$$

This can be rewritten as

$$\sum_{r=1}^{n} \frac{n!}{(r-1)!(n-r)!} \times p^r \times (1-p)^{n-r}$$

We can now factor out the quantity $n \times p$ and let $r - 1 = y$. This can be rewritten as

$$n \times p \times \sum_{y=0}^{n-1} \frac{(n-1)!}{(y)!(n-1-y)!} \times p^y \times (1-p)^{n-1-y} = n \times p \times [p + (1-p)]^{n-1}$$

Since the sum is merely the expansion of a binomial raised to a power, and the number 1 raised to any power is 1, the mean is

$$\mu = n \times p$$

A similar calculation shows the variance is

$$\text{var} = n \times p \times (1-p)$$

The standard deviation is

$$\sigma = \sqrt{\text{var}} = \sqrt{n \times p \times (1-p)}$$

Example 5.16

A truck carrying dairy products and eggs damages its suspension and 5% of the eggs break.

(a) What is the probability that a carton of 12 eggs will have exactly one broken egg?

(b) What is the probability that one or more eggs in a carton will be broken?

Solution

(a) Since an egg is either broken or not broken, the binomial distribution applies. The probability p that an egg is broken is 0.05 and that one is not broken is $(1 - p) =$ 0.95. From the equation for the binomial distribution, with $n = 12$ and $r = 1$,

$$p\{1\} = f(1) = C(12,1) \times 0.05^1 \times 0.95^{11} = 12 \times 0.05 \times 0.57 = 0.34$$

(b) The probability that one or more eggs will be broken can be calculated as the sum of each individual probability:

$$p\{x > 0\} = p\{1\} + p\{2\} + \cdots + p\{12\}$$

However, this requires twelve calculations. The problem can also be solved using the complementary rule:

$$p\{x > 0\} = 1 - p\{0\} = C(12,0) \times 0.05^0 \times 0.95^{12} = 0.95^{12} = 0.54$$

Normal Distribution Function

The normal distribution, or Gaussian distribution, is widely used to represent the distribution of outcomes of experiments and measurements. It is popular because it can be derived from a few empirical assumptions about the errors presumed to cause the distribution of results about the mean. One assumption is that the error is the result of a combination of N elementary errors, each of magnitude e and equally likely to be positive or negative. The derivation then assumes $N \to \infty$ and $e \to 0$ in such a way as to leave the standard deviation constant. This error model is universal, since most experiments are analyzed to eliminate systematic errors. What remains is attributable to errors that are too small to explain systematically, so the normal probability distribution is evoked.

The form of the probability density and distribution functions for the **normal distribution** with a mean μ and variance σ^2 is given by

$$f(x) = \frac{e^{-(x-\mu)^2/2\sigma^2}}{\sigma\sqrt{2\pi}} \qquad -\infty < x < \infty$$

$$F(x) = \int_{-\infty}^{x} \frac{e^{-(t-\mu)^2/2\sigma^2}}{\sigma\sqrt{2\pi}}\, dt$$

The normal distribution is the typical bell-shaped curve shown in Fig. 5.6. Here we see that the curve is symmetric about the mean μ. Its width and height are determined by the standard deviation σ. As σ increases, the curve becomes wider and lower.

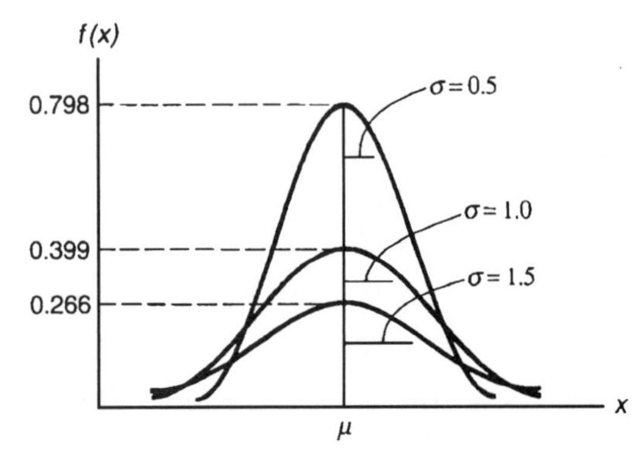

Figure 5.6 Normal distribution curve

Since this function is difficult to integrate, reference tables are used to calculate probabilities in a standard format; then the standard probabilities are converted to the actual variable required by the problem. The relation between the standard variable, z, and a typical problem variable, x, is

$$z = (x - \mu)/\sigma$$

Since μ and σ are constants, the standard probability at a value z is the same as the problem probability for the value at x.

The standard probability density function is

$$f(z) = \frac{1}{\sqrt{2\pi}} \times e^{-z^2/2}$$

Table 5.2 Standard probability table

z	$F(z)$	$f(z)$	z	$F(z)$	$f(z)$
0.0	0.5000	0.3989	2.0	0.9773	0.0540
0.1	0.5398	0.3970	2.1	0.9821	0.0440
0.2	0.5793	0.3910	2.2	0.9861	0.0355
0.3	0.6179	0.3814	2.3	0.9893	0.0283
0.4	0.6554	0.3683	2.4	0.9918	0.0224
0.5	0.6915	0.3521	2.5	0.9938	0.0175
0.6	0.7257	0.3332	2.6	0.9953	0.0136
0.7	0.7580	0.3123	2.7	0.9965	0.0104
0.8	0.7881	0.2897	2.8	0.9974	0.0079
0.9	0.8159	0.2661	2.9	0.9981	0.0060
1.0	0.8413	0.2420	3.0	0.9987	0.0044
1.1	0.8643	0.2179	3.1	0.9990	0.0033
1.2	0.8849	0.1942	3.2	0.9993	0.0024
1.3	0.9032	0.1714	3.3	0.9995	0.0017
1.4	0.9192	0.1497	3.4	0.9997	0.0012
1.5	0.9332	0.1295	3.5	0.9998	0.0009
1.6	0.9452	0.1109	3.6	0.9998	0.0006
1.7	0.9554	0.0940	3.7	0.9999	0.0004
1.8	0.9641	0.0790	3.8	0.9999	0.0003
1.9	0.9713	0.0656	3.9	1.0000	0.0002
			4.0	1.0000	0.0001

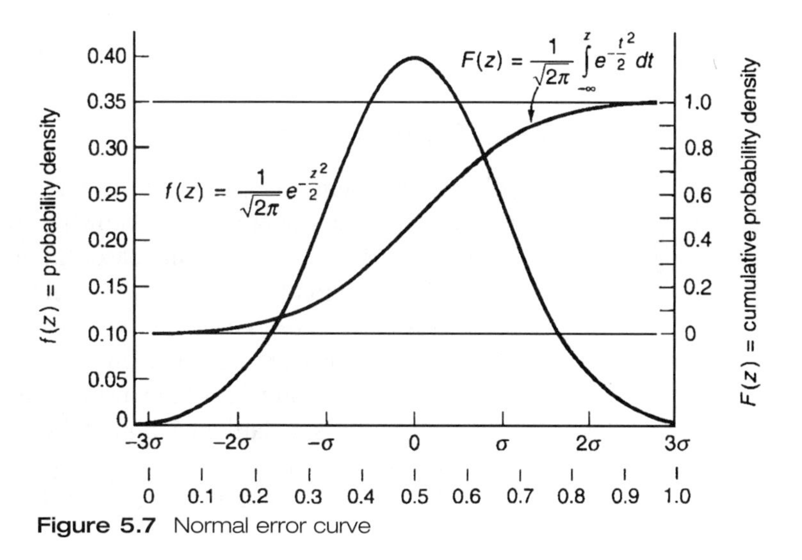

Figure 5.7 Normal error curve

The standard cumulative distribution function is

$$F(z) = \int_{-\infty}^{z} \frac{1}{\sqrt{2\pi}} \times e^{-t^2/2} \times dt$$

The standard probability function is shown graphically in Fig. 5.7, and Table 5.2 shows the corresponding numerical values. The standard probability curve is symmetric about the origin and is given in terms of unit *sigma*. To use the table, remember that the function $F(z)$ is the area under the probability curve from minus infinity to the value z. The area under the curve up to $x = 0$ is therefore 0.5 Also, from symmetry,

$$F(-z) = 1 - F(z)$$

Example 5.17

Find the probability that the standard variable z lies within (a) 1σ, (b) 2σ, and (c) 3σ of the mean.

Solution

(a) The probability is $P_1 = F(1.0) - F(-1.0)$. From the symmetry of F, $F(-1.0) = 1 - F(1.0)$, so

$$\begin{aligned} P_1 &= 2F(1.0) - 1 = 2(0.8413) - 1 \\ &= 0.6826 \end{aligned}$$

(b) In this case, the probability is

$$\begin{aligned} P_2 &= F(2.0) - F(-2.0) \\ &= F(2.0) - [1 - F(2.0)] \\ &= 2F(2.0) - 1 = 2(0.9773) - 1 \\ &= 0.9546 \end{aligned}$$

(c) In the same way,

$$\begin{aligned} P_3 &= 2F(3.0) - 1 \\ &= 2(0.9987) - 1 \\ &= 0.9974 \end{aligned}$$

t-Distribution

The t-distribution is often used to test an assumption about a population mean when the parent population is known to be normally distributed but its standard deviation is unknown. In this case, the inferences made about the parent mean will depend upon the size of the samples being taken.

It is customary to describe the t-distribution in terms of the standard variable t and the number of degrees of freedom ν. The number of degrees of freedom is a measure of the number of independent observations in a sample that can be used to estimate the standard deviation of the parent population; the number of degrees of freedom ν is one less than the sample size ($\nu = n - 1$).

The density function of the t-distribution is given by

$$f(t) = \frac{\Gamma\left(\frac{\nu+1}{2}\right)}{\sqrt{\nu\pi}\,\Gamma\left(\frac{\nu}{2}\right)\left(1 + t^2/\nu\right)^{(\nu+1)/2}}$$

and is provided in Table 5.3. The mean is $m = 0$, and the standard deviation is

$$\sigma = \sqrt{\frac{\nu}{\nu - 2}}$$

Table 5.3 *t*-Distribution; values of $t_{\alpha,v}$

Degrees of Freedom, v	Area of the Tail				
	$\alpha = 0.10$	$\alpha = 0.05$	$\alpha = 0.025$	$\alpha = 0.01$	$\alpha = 0.005$
1	3.078	6.314	12.706	31.821	63.657
2	1.886	2.920	4.303	6.965	9.925
3	1.638	2.353	3.182	4.541	5.841
4	1.533	2.132	2.776	3.747	4.604
5	1.476	2.015	2.571	3.365	4.032
6	1.440	1.943	2.447	3.143	3.707
7	1.415	1.895	2.365	2.998	3.499
8	1.397	1.860	2.306	2.896	3.355
9	1.383	1.833	2.262	2.821	3.250
10	1.372	1.812	2.228	2.764	3.169
11	1.363	1.796	2.201	2.718	3.106
12	1.356	1.782	2.179	2.681	3.055
13	1.350	1.771	2.160	2.650	3.012
14	1.345	1.761	2.145	2.624	2.977
15	1.341	1.753	2.131	2.602	2.947
16	1.337	1.746	2.120	2.583	2.921
17	1.333	1.740	2.110	2.567	2.898
18	1.330	1.734	2.101	2.552	2.878
19	1.328	1.729	2.093	2.539	2.861
20	1.325	1.725	2.086	2.528	2.845
21	1.323	1.721	2.080	2.518	2.831
22	1.321	1.717	2.074	2.508	2.819
23	1.319	1.714	2.069	2.500	2.807
24	1.318	1.711	2.064	2.492	2.797
25	1.316	1.708	2.060	2.485	2.787
26	1.315	1.706	2.056	2.479	2.779
27	1.314	1.703	2.052	2.473	2.771
28	1.313	1.701	2.048	2.467	2.763
29	1.311	1.699	2.045	2.462	2.756
inf.	1.282	1.645	1.960	2.326	2.576

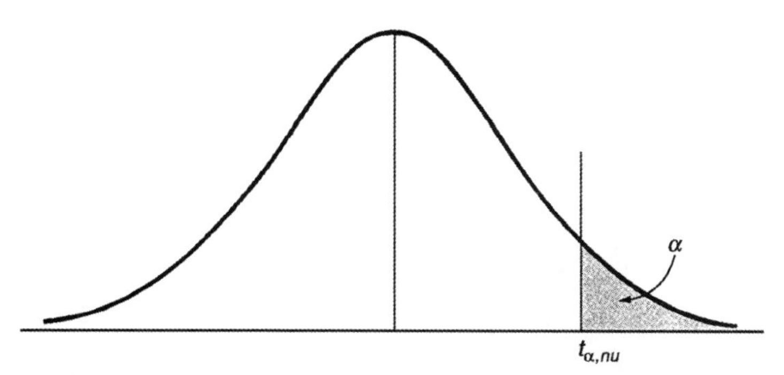

Figure 5.8

Probability questions involving the t-distribution can be answered by using the distribution function $t_{\alpha,v}$ shown in Fig. 5.8. Table 5.3 gives the value of t as a function of the degrees of freedom v down the column and the area (α) of the tail across the top. The t-distribution is symmetric. As an example, the probability of t falling within ± 3.0 when a sample size of 8 ($v = 7$) is selected is one minus twice the tail ($\alpha = 0.01$):

$$P\{-3.0 < t < 3.0\} = 1 - (2 \times 0.01) = 0.98$$

The t-distribution is a family of distributions that approaches the Gaussian distribution for large n.

PROBLEMS

5.1 Define the set of all outcomes for the roll of two dice.

5.2 Draw a Venn diagram showing

The universal set of all people in the United States as U

All the males as M

All the females as F

All the students of both sexes as S

All the students with grades above "B" as A

5.3 What is the probability of drawing a pair of aces in two cards when an ace has been drawn on the first card?

a. 1/13 c. 3/51
b. 1/26 d. 4/51

5.4 An auto manufacturer has three plants (A, B, C). Four out of 500 cars from Plant A must be recalled, 10 out of 800 from Plant B, and 10 out of 1000 from Plant C. Now a customer purchases a car from a dealer who gets 30% of his stock from Plant A, 40% from Plant B, and 30% from Plant C, and the car is recalled. What is the probability it was manufactured in Plant A?

a. 0.0008 c. 0.0125
b. 0.01 d. 0.2308

5.5 There are ten defectives per 1000 times of a product. What is the probability that there is one and only one defective in a random lot of 100?

a. 99×0.01^{99} c. 0.5
b. 0.01 d. 0.99^{99}

5.6 The probability that both stages of a two-stage missile will function correctly is 0.95. The probability that the first stage will function correctly is 0.98. What is the probability that the second stage will function correctly given that the first one does?

a. 0.99 c. 0.97
b. 0.98 d. 0.95

5.7 A standard deck of 52 playing cards is thoroughly shuffled. The probability that the first four cards dealt from the deck will be the four aces is closest to

a. 2.0×10^{-1} c. 4.0×10^{-4}
b. 8.0×10^{-2} d. 4.0×10^{-6}

5.8 In statistics, the standard deviation measures

a. a standard distance c. central tendency
b. a normal distance d. dispersion

5.9 There are three bins containing integrated circuits (ICs). One bin has two premium ICs, one has two regular ICs, and one has one premium IC and one regular IC. An IC is picked at random. It is found to be a premium IC. What is the probability that the remaining IC in that bin is also a premium IC?

a. $\frac{1}{5}$ c. $\frac{1}{3}$
b. $\frac{1}{4}$ d. $\frac{2}{3}$

5.10 How many teams of four can be formed from 35 people?
a. about 25,000 c. about 50,000
b. about 2,000,000 d. about 200,000

5.11 A bin contains 50 bolts, 10 of which are defective. If a worker grabs 5 bolts from the bin in one grab, what is the probability that no more than 2 of the 5 are bad?
a. about 0.5 c. about 0.90
b. about 0.75 d. about 0.95

5.12 How many three-letter codes may be formed from the English alphabet if no repetitions are allowed?
a. 26^3 c. $26 \bullet 25 \bullet 24$
b. 26/3 d. $26^3/3$

5.13 A widget has three parts, A, B, and C, with probabilities of 0.1, 0.2, and 0.25, respectively, of being defective. What is the probability that exactly one of these parts is defective?
a. 0.375 c. 0.95
b. 0.55 d. 0.005

5.14 If three students work on a certain math problem, student A has a probability of success of 0.5; student B, 0.4; and student C, 0.3. If they work independently, what is the probability that no one works the problem successfully?
a. 0.12 c. 0.32
b. 0.25 d. 0.21

5.15 A sample of 50 light bulbs is drawn from a large collection, in which each bulb is good with a probability of 0.9. What is the approximate probability of having less than 3 bad bulbs in the 50?
a. 0.1 c. 0.3
b. 0.2 d. 0.4

SOLUTIONS

5.1 We will write this in ordered pairs:

$$\begin{aligned}R = \{&(1,1),\\ &(1,2), (2,1),\\ &(1,3), (3,1), (2,2),\\ &(1,4), (4,1), (2,3), (3,2)\\ &(1,5), (5,1), (2,3), (3,2), (3,3),\\ &(1,6), (6,1), (5,2), (2, 5), (3,4), (4,3),\\ &(2,6), (6,2), (5,3), (3,5), (4,4),\\ &(3,6), (6,3), (5,4), (4,5),\\ &(4,6), (6,4), (5,5),\\ &(5,6), (6,5),\\ &(6,6)\}\end{aligned}$$

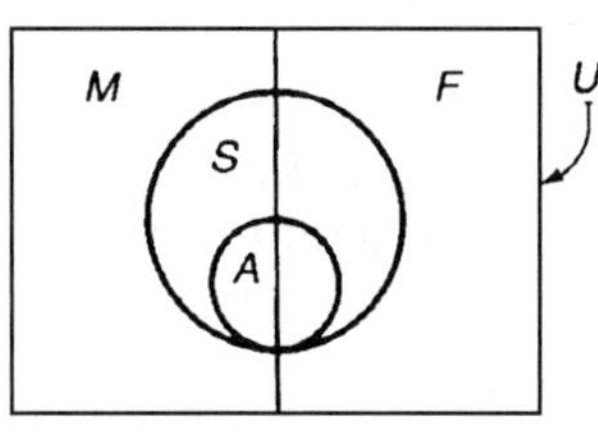

Exhibit 5.2

5.2 See Exhibit 5.2.

5.3 **c.** This is a conditional probability problem. Let B be "draw an ace," and let A be "draw a second ace": $P\{B\} = 1/13$ and $P\{A\} = 3/51$. Then $P\{A|B\} = P\{A\} \times P\{B\}/P\{B\} = 3/51$.

5.4 **d.** This is a Bayes' theorem problem application because partitions are involved. The event E is a recall, with E_1 = Plant A, E_2 = Plant B, and E_3 = Plant C. The conditional probabilities of a recall from Plants E_1, E_2, and E_3 are

$$P(E \mid E_1) = 4/500 = 0.008$$
$$P(E \mid E_2) = 10/800 = 0.0125$$
$$P(E \mid E_3) = 10/1000 = 0.01$$

The probabilities that the dealer had a car from E_1, E_2, or E_3 are $P(E_1) = 0.3$, $P(E_2) = 0.4$, and $P(E_3) = 0.3$. Now applying Bayes' formula gives the probability that the recall was built in Plant A (E_1) as

$$P\{E_1|\text{recall}\} = \frac{P\{E_1\}\times P\{E|E_1\}}{P\{E_1\}\times P\{E \mid E_1\} + P\{E_2\}\times P\{E|E_2\} + P\{E_3\}\times P\{E \mid E_3\}}$$
$$= \frac{0.3\times 0.008}{0.3\times 0.008 + 0.4\times 0.0125 + 0.3\times 0.01} = 0.2308$$

5.5 **d.** The problem involves binomial probability. The probability that one item, selected at random, is defective is

$$p_{\text{defective}} = \frac{10}{1000} = 0.01$$

and the probability that one item is good (not defective) is

$$p_{\text{good}} = 1 - p_{\text{defective}} = 0.99$$

The probability that exactly one defective will be found in a random sample of 100 items is given by the binomial $b(1, 100, 0.01)$, in which

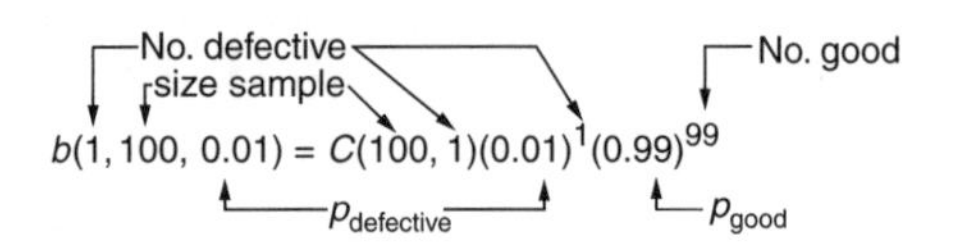

$C(n, r) = \binom{n}{r} = \frac{n!}{(n-r)!r!}$ is the number of combinations of n objects taken r at a time without concern for the order of arrangement. $C(100, 1) = \frac{100!}{99!1!} = 100$, so $b(1, 100, 0.01) = 100(0.01)(0.99)^{99} = 0.99^{99} = 0.3697$.

5.6 c. Here, $P(S_1) = 0.98$ and $P(S_2 \cap S_1) = 0.95$ are given. Hence the conditional probability $P(S_2|S_1)$ is

$$P(S_2 \mid S_1) = \frac{P(S_2 \cap S_1)}{P(S_1)} = \frac{0.95}{0.98} = 0.97$$

5.7 d. The probability of drawing an ace on the first card is 4/52. The probability that the second card is an ace is 3/51. The probability that the third card is an ace is 2/50, and probability for the fourth ace is 1/49. The probability that the first four cards will all be aces is

$$P = \frac{4}{52} \bullet \frac{3}{51} \bullet \frac{2}{50} \bullet \frac{1}{49} = 0.00\,0037 = 3.7 \times 10^{-6}$$

5.8 d.

5.9 d. Since the first IC that is picked is a premium IC, it was drawn from either bin 1 or bin 3. From the distribution of premium ICs, the probability that the premium IC came from bin 1 is $\frac{2}{3}$, and from bin 3 is $\frac{1}{3}$.

In bin 1, the probability that the remaining IC is a premium IC is 1; in bin 3, the probability is 0. Thus, the probability that the remaining IC is a premium IC is

$$\frac{2}{3}(1) + \frac{1}{3}(0) = \frac{2}{3}$$

An alternative solution using Bayes' theorem for conditional probability is

$$P(\text{bin } 1 \mid \text{drew premium}) = \frac{P(\text{bin 1 and premium})}{P(\text{premium})}$$

$$= \frac{P(\text{premium } \mid \text{bin } 1) \bullet P(\text{bin } 1)}{\sum_{i=1}^{3} P(\text{premium} \mid \text{bin } 1) \bullet P(\text{bin } 1)}$$

$$= \frac{1\left(\frac{1}{3}\right)}{1\left(\frac{1}{3}\right) + 0\left(\frac{1}{3}\right) + \frac{1}{2}\left(\frac{1}{3}\right)} = \frac{2}{3}$$

5.10 c. The answer is the binomial coefficient

$$\binom{35}{4} = \frac{35 \bullet 34 \bullet 33 \bullet 32}{4 \bullet 3 \bullet 2 \bullet 1} = 35 \bullet 34 \bullet 11 \bullet 4 = 52{,}360$$

5.11 d. The total number of choices of 5 is $\binom{50}{5}$. Of these, $\binom{40}{5}$ have no bad bolts, $\binom{40}{4} \times \binom{10}{1}$ have one bad bolt, and $\binom{40}{3}\binom{10}{2}$ have two bad bolts. Thus,

$$\frac{\binom{40}{5} + \binom{40}{4}\binom{10}{1} + \binom{40}{3}\binom{10}{2}}{\binom{50}{5}}$$

$$= \frac{\frac{40 \bullet 39 \bullet 38 \bullet 37 \bullet 36}{5 \bullet 4 \bullet 3 \bullet 2} + \frac{40 \bullet 39 \bullet 38 \bullet 37}{4 \bullet 3 \bullet 2} \bullet 10 + \frac{40 \bullet 39 \bullet 38}{3 \bullet 2} \bullet \frac{10 \bullet 9}{2}}{\frac{50 \bullet 49 \bullet 48 \bullet 47 \bullet 46}{5 \bullet 4 \bullet 3 \bullet 2}}$$

$$= \frac{658{,}008 + 913{,}900 + 444{,}600}{2{,}118{,}760} = 0.9517$$

5.12 c. There are 26 choices for the first letter; 25 remain for the second, and 24 for the third. The answer is (c).

5.13 a. The probability that only A is defective is

$$0.1 \times (1 - 0.2) \times (1 - 0.25) = 0.06$$

The probability that only B is defective is

$$(1 - 0.1) \times (0.2) \times (1 - 0.25) = 0.135$$

The probability that only C is defective is

$$(1 - 0.1) \times (1 - 0.2) \times (0.25) = 0.18$$

Now add to find the final probability, which is

$$0.06 + 0.135 + 0.18 = 0.375$$

5.14 d. Simply multiply the complementary probabilities $(1 - 0.5) \times (1 - 0.4) \times (1 - 0.3) = 0.21$.

5.15 a. Apply the binomial distribution. The probability of 0 bad is $(0.9)^{50}$; of 1 bad, $\binom{50}{1}(0.1)(0.9)^{49}$; and of 2 bad, $\binom{50}{1}(0.1)^2(0.9)^{48}$. Adding these, $(0.9)^{48}[(0.9)^2 + 5.0(0.9) + 1225(0.1)^2] = 0.112$.

CHAPTER 6

Ethics

Donald G. Newnan

OUTLINE

Engineers, as members of a profession, are expected to conduct themselves in an ethical manner. This means that engineers must be aware of the standards of professional conduct (the ethical standards) and abide by them.

Codes of ethics have been prepared by the various national engineering societies, the engineering accreditation board (ABET), and the National Council of Examiners for Engineering and Surveying (NCEES). The *NCEES Model Rules of Professional Conduct* is reproduced in the *Fundamentals of Engineering supplied-Reference Handbook*. Thus it is reasonable to assume that this code of ethics will be covered on the exam.

NCEES MODEL RULES OF PROFESSIONAL CONDUCT*

I. Licensee's Obligation to Society
  a. Licensees, in the performance of their services for clients, employers, and customers, shall be cognizant that their first and foremost responsibility is to the public welfare.
  b. Licensees shall approve and seal only those design documents and surveys that conform to accepted engineering and surveying standards and safeguard the life, health, property, and welfare of the public.
  c. Licensees shall notify their employer or client and such other authority as may be appropriate when their professional judgment is overruled under circumstances where the life, health, property, or welfare of the public is endangered.
  d. Licensees shall be objective and truthful in professional reports, statements, or testimony. They shall include all relevant and pertinent information in such reports, statements, or testimony.
  e. Licensees shall express a professional opinion publicly only when it is founded upon an adequate knowledge of the facts and a competent evaluation of the subject matter.

*Reproduced by permission of NCEES, the copyright owner.

f. Licensees shall issue no statements, criticisms, or arguments on technical matters which are inspired or paid for by interested parties, unless they explicitly identify the interested parties on whose behalf they are speaking and reveal any interest they have in the matters.
g. Licensees shall not permit the use of their name or firm name by, nor associate in the business ventures with, any person or firm which is engaging in fraudulent or dishonest business or professional practices.
h. Licensees having knowledge of possible violations of any of these Rules of Professional Conduct shall provide the board with the information and assistance necessary to make the final determination of such violation.
(Section 150, Disciplinary Action, NCEES Model Law)

II. Licensee's Obligation to Employer and Clients
a. Licensees shall undertake assignments only when qualified by education or experience in the specific technical fields of engineering or surveying involved.
b. Licensees shall not affix their signatures or seals to any plans or documents dealing with subject matter in which they lack competence, nor to any such plan or document not prepared under their direct control and personal supervision.
c. Licensees may accept assignments for coordination of an entire project, provided that each design segment is signed and sealed by the licensee responsible for preparation of that design segment.
d. Licensees shall not reveal facts, data, or information obtained in a professional capacity without the prior consent of the client or employer except as authorized or required by law. Licensees shall not solicit or accept gratuities, directly or indirectly, from contractors, their agents, or other parties in connection with work for employers or clients.
e. Licensees shall make full prior disclosures to their employers or clients of potential conflicts of interest or other circumstances which could influence or appear to influence their judgment or the quality of their service.
f. Licensees shall not accept compensation, financial or otherwise, from more than one party for services pertaining to the same project, unless the circumstances are fully disclosed and agreed to by all interested parties.
g. Licensees shall not solicit or accept a professional contract from a governmental body on which a principal or officer of their organization serves as a member. Conversely, licensees serving as members, advisors, or employees of a government body or department, who are the principals or employees of a private concern, shall not participate in decisions with respect to professional services offered or provided by said concern to the governmental body which they serve.
(Section 150, Diciplinary Action, NCEES Model Law)

III. Licensee's Obligation to Other Licensees
a. Licensees shall not falsify or permit misrepresentation of their, or their associates', academic or professional qualifications. They shall not misrepresent or exaggerate their degree of responsibility in prior assignments nor the complexity of said assignments. Presentations incident to the solicitation of employment or business shall not misrepresent pertinent facts concerning employers, employees, associates, joint ventures, or past accomplishments.
b. Licensees shall not offer, give, solicit, or receive, either directly or indirectly, any commission, or gift, or other valuable consideration in order to

secure work, and shall not make any political contribution with the intent to influence the award of a contract by public authority.

c. Licensees shall not attempt to injure, maliciously or falsely, directly or indirectly, the professional reputation, prospects, practice, or employment of other licensees, nor indiscriminately criticize other licensees' work. *(Section 150, Disciplinary Action, NCEES Model Law)*

Example **6.1**

The *NCEES Model Rules of Professional Conduct* allows an engineer to do which one of the following?

(a) Accept money from contractors in connection with work for an employer or client.

(b) Compete with other engineers in seeking to provide professional services.

(c) Accept a professional contract from a governmental body even though a principal or officer of the engineer's firm serves as a member of the governmental body.

(d) As the coordinator of an entire project, an engineer may sign or seal all design segments of the project.

Solution

Although the other items are not allowed by the *Model Rules*, nowhere does it say that an engineer cannot compete with other engineers in seeking to provide professional services. But, of course, he/she should do it in an ethical manner. The answer is (b).

No code of ethics can cover all the kinds of situations an engineer will encounter. It does, however, describe in board terms the ethical principles by which an engineer should be guided.

The examination asks ethics questions concerning relations with clients, peers, and the public. The *NCEES Model Rules of Professional Conduct* are broken down in these same three categories. Thus the exam ethics questions likely are written to see if the examinee has read and generally understands the standards of professional conduct.

PROBLEMS

6.1 Jack, a consulting environmental engineer, was hired by the local garbage company to represent them in a hearing before the local town council. In preparing for an appearance before the town council, Jack worked at the garbage company truck storage yard. As he was starting to leave late one evening he discovered that a garbage company employee was moving garbage trucks to the edge of a steam embankment and dumping the liquid that came from the garbage into the stream. The next morning Jack spoke to the truck storage yard manager about what he had seen. The manager told Jack that they had always done it that way and for him to forget it.

Jack feels very uncomfortable with the situation; he knows the dumping is a clear violation of an environmental regulation. Consider Jack's four alternatives and select the best one.

a. Terminate his consulting with the garbage company.
b. Assume that since it has gone on for a long time there is no major hazard being created and forget what he saw.
c. At the next opportunity advise the local town council of the situation.
d. Write a letter to the garbage company president outlining what he discovered and advising the president that the dumping practice is an environmental regulation violation.

6.2 Dave's employer is a small engineering firm. For one of his first jobs as a young engineer Dave was assigned to write the specifications for and purchase a ten-inch pump. He wrote the specifications and sent a sealed bid request to six firms. Only five firms submitted the bids before the bid request deadline. On the last day Dave called the sixth firm to remind them of the sealed bid request. The estimator indicated he was very busy and did not want to take the time to prepare a sealed bid unless he had a good chance of supplying the pump. Dave wants the sixth quote, and wonders what to do. Which of the following is his best course of action?

a. Since he doesn't plan to open the sealed bids right away, Dave offers to let this last firm submit its quote a week late.
b. Dave opens the five sealed bids and reads them over the phone to the estimator. He says he thinks he can beat them and submit a bid by the sealed bid deadline later that day.
c. Dave already has five bids, so he calls the sixth firm back and tells them there is nothing he can do for them.
d. Dave opens the five bids and finds the lowest one is $8750. He calls the estimator for the sixth firm and says, "I can't tell you the bids, but unless you can get under about $8800 forget about it."

6.3 John, a consulting soils engineer, submitted a draft of his report on soil conditions at a plant site to his client's chief engineer. The chief engineer read the draft of the report and indicated to John that he wanted several portions of the report changed so the planned building could be designed with a smaller, less expensive foundation. The chief engineer told John the allowable soil pressure value he wanted him to recommend in the final report.

Following this meeting John went back to his own office to decide what to do. He believes he has four alternatives. Which one should he select?

a. Submit the draft soil report as the final report without any changes.
b. Reexamine his field test data to see if the chief engineer's desired allowable soil pressure value could be recommended. If so, he would do it. If not, he would leave the report as it is.
c. Make the changes requested by the client's chief engineer and submit the final report.
d. Write to the president of his client's firm and describe the request of the chief engineer. Advise the president that the draft report is being submitted as the final report.

6.4 Jim, a newly hired engineer, was asked by his supervisor to see what he could learn about their competitors' future product plans. This is an important assignment and Jim is anxious to do a good job. He has devised four alternatives. Which one should he select?

a. Jim could call the competitors and tell them he is a college student working on a report for one of his classes. Jim thinks the competitors would be very forthright in this situation.
b. Jim could call up some college buddies who are now working for the competitors. A lunch together could easily result in trading some confidential product plans if Jim worked at it.
c. Jim could call the competitors and simply ask about their future product plans. If the competitors asked who Jim is, he will honestly tell them who his employer is. But if they do not ask, then Jim will not volunteer the information.
d. Jim will contact the competitors and at the beginning of the conversation, explain who he is, his employer, and that the has been assigned to study competitors future product plans.

6.5 When Glenn, a product development engineer, discovered that his manufacturing company was about to lose a major customer to an overseas competitor, he resolved to warn the customer that he might be making a serious mistake. He is considering which one of the following statements to make to the customer:

a. "I read recently that a lot of this kind of product is being assembled in the foreign country by prison labor. I hope you agree it is unethical to use prisoners in that way."
b. "We have some sales literature that describes features of our product and compares it with several competing products. I think you might find this information helpful, so I'm sending you a copy."
c. "I know a number of their engineers because we went to college together. Believe me, they probably have copied our product, as I doubt they would be skillful enough to design one from scratch."
d. "My boss says the overseas competitor will probably give you a low bid to get the initial work, but you will pay a lot more on any subsequent orders."

SOLUTIONS

6.1 d. As a consultant to the garbage company Jack needs to see that responsible management is aware of his liquid dumping discovery. The appropriate next step is to advise the company president. Neither (a) nor (b) is satisfactory as they fail to take appropriate action when some action is clearly called for. Answer (c), on the other hand, is a premature action that may be a public disclosure prior to a thorough evaluation of all the facts. The answer is (d).

6.2 c. In purchasing equipment using a sealed bid method, Dave has the responsibility to treat all bidders fairly. If the sixth bidder indicated he could submit the bid by the morning following the stated bid deadline, this might be considered a reasonable request on this relatively small purchase. Thus Dave might consider it ethical and fair to allow this bidder a few additional hours to submit the quote. A substantial concession (like a week's delay) would not be appropriate unless it were made available to all bidders in a timely way. Opening and giving exact or approximate data to one bidder about competing sealed bids is unethical and unfair. The answer is (c).

6.3 b. Alternative (c) is clearly inappropriate as it would not represent John's professional opinion. Both (a) and (d) might be suitable if John considered the chief engineer's desired allowable soil pressure value obviously wrong. The (b) alternative seems most practical. Soil analysis is not an exact science. A reevaluation of the filed test data might indicate a somewhat larger allowable soil pressure value—possibly even the one the chief engineer wants. Thus the reevaluation and final report would be accurate and professional. The answer is (b).

6.4 d. Jim is immediately faced with a situation where he must recognize what ethical conduct is. Alternatives (a) and (b) are unethical practices. Neither misrepresenting oneself to a competitor nor disclosing an employer's confidential data is appropriate. Between alternatives (c) and (d) the situation is less clear. An ethical engineer must proceed forthrightly, but must he take the path of alternative (d)?

Jim's problem may be his inexperience. Other alternatives are available, such as requesting all available sales literature of the competitor's product. By careful study, and by knowing the basic manufacturing and development capabilities, one can usually correctly predict the direction of products. Further, since this evaluation would naturally help decide the direction of his own product line, then a more careful market analysis of his own situation is certainly warranted. Manufacturers often purchase competitive products and benchmark their manufacturing technology. Since this is an ethics test question, you better choose to be on the safe side. The answer is (d).

6.5 b. Answers (a), (c), and (d) represent malicious comments attempting to injure a competitor's reputation. Answer (b) represents material that has been examined within Glenn's company and has been approved for distribution. Since it was intended for customers or prospective customers, Glenn would appear to be acting in an appropriate and ethical manner if he selects this alternative. The answer is (b).

CHAPTER 7

Statics

Charles E. Smith

OUTLINE

Statics is concerned with the forces of interaction between bodies or within bodies of mechanical systems that have no significant accelerations. Typical engineering problems require the analyst to predict forces induced at certain points by known forces applied at other points.

INTRODUCTORY CONCEPTS IN MECHANICS

Newton's Laws of Motion

Every element of a mechanical system must satisfy **Newton's Second Law of Motion**, which states that the resultant force f acting on the element is related to the acceleration a of the element by

$$f = ma$$

where m represents the mass of the element. This entire chapter deals with the special case in which $a = 0$. **Newton's Third Law** requires that the force exerted on a body A by a body B is of equal magnitude and opposite direction to the force exerted on body B by body A. A careful, unambiguous account of this law is essential for successful analysis of forces; rules for ensuring that such an analysis is done properly will be reviewed in the Equilibrium section.

Newton's Law of Gravitation

Every pair of material elements is attracted toward one another by a pair of *gravitational* forces, the magnitude of which is given by

$$f_g = \frac{\gamma m_1 m_2}{r^2} \tag{7.1}$$

where γ is the **universal gravitational constant** (about 6.7×10^{-11} N•m^2/kg^2), m_1 and m_2 are the masses of the elements, and r is the distance between them. Because very large masses are necessary to make these forces significant, f_g can often be neglected. A notable exception is the force exerted by the earth (which has a mass of about 6×10^{24} kg) on objects near its surface. In this case, the gravitational force has a magnitude given by

$$f_g = mg \tag{7.2}$$

where m is the mass of the attracted object and g (which is related to the earth's mass and radius) has a value that varies between 9.78 and 9.83 N/kg with geographic location. In technically correct terminology, the word *weight* refers to the force of gravity by the earth. However, *weight* has other, closely related meanings that can be a source of serious confusion to the analyst of mechanical systems.

Dimensions and Units of Measurement

Every quantity in mechanics can be expressed in terms of three fundamental quantities. In the SI system of units, these are *mass, length*, and *time*, and units are the kilogram (kg), the meter (m), and the second (s), respectively. In the SI system, Newton's Second Law provides a definition of a fourth unit in terms of the three fundamental ones. The **Newton** (N) is defined as the force required to accelerate a 1-kg body at the rate of one meter per second per second (m/s^2). This can be expressed symbolically as N = kg•m/s^2. See Table 7.1.

Considerable confusion results from the introduction and widespread use (especially outside the United States) of the **kilogram-force (kgf)**, defined to be 9.80665 N.

Errors stemming from force values given in this noncoherent unit can be avoided by converting to newtons before doing further calculation. For example,

Table 7.1 Units common to mechanics

Quantity	SI Unit	Coherent U.S. Engineering Unit	Common Non-coherent Unit
Mass	kilogram (kg)	slug = lbf•s^2/ft ≈ 14.594 kg	pound-mass (lbm) $\approx \frac{1}{32.174}$ slug ≈ 0.4536 kg
Length	meter (m)	foot (ft) = 0.3048 m	inch (in.) = 25.4 mm
Time	second (s)	second (s)	minute (min) = 60 s
Force	newton (N) N = kg•m/s^2	pound-force (lbf) ≈ 4.448 N	kilogram-force (kgf) or kilopound (kp) = 9.80665 N

an estimate of the acceleration imparted by a force of 217 kgf to a 100-kg body would proceed as follows:

$$f = (217 \text{ kgf})\ (9.80665 \text{ N/kgf}) = 2{,}128 \text{ N}$$

$$a = \frac{f}{m} = \frac{2128 \text{ N}}{100 \text{ kg}} = 21.28 \left(\text{N/kg} = \frac{\text{kg} \bullet \text{m/s}^2}{\text{kg}} = \frac{\text{m}}{\text{s}^2} \right)$$

Errors in unit conversion, and many errors of analysis, will be revealed by the practice of appending unit symbols to *every* number and algebraically reducing combinations of symbols resulting from multiplication and division.

VECTOR GEOMETRY AND ALGEBRA

Handling many of the problems that arise in mechanics can be greatly simplified by means of the operations of vector analysis. Their use will be successful if attention is given to the geometric meaning of each operation.

A **vector** represents a physical quantity that can be characterized by a magnitude and a direction in space, which will be taken to be three-dimensional here. We use an arrow to depict each vector, with the length of the arrow proportional to the magnitude represented, and the orientation representing the direction of the vector. Boldface letters are used in text and equations to represent vectors. The **magnitude** of vector **a** is written as |**a**| and sometimes as a.

Addition and Subtraction

The **sum of two vectors** is a vector determined according to the so-called parallelogram law, as illustrated in Fig. 7.1. The sum is the vector represented by the diagonal of the parallelogram formed by the two vectors placed with their tails coincident. The commutative law,

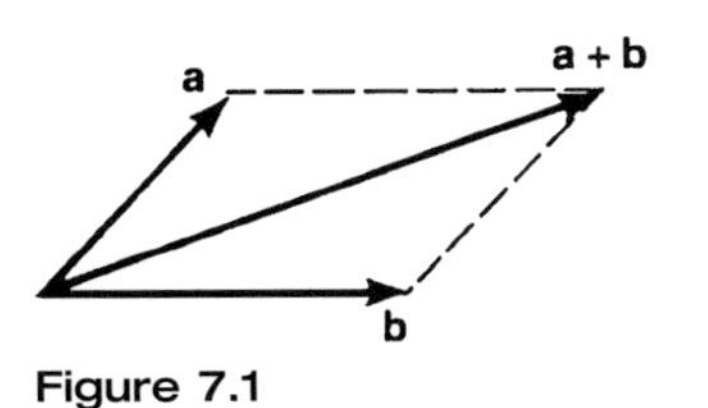

Figure 7.1

$$\mathbf{a} + \mathbf{b} = \mathbf{b} + \mathbf{a}$$

and the associative law,

$$\mathbf{a} + (\mathbf{b} + \mathbf{c}) = (\mathbf{a} + \mathbf{b}) + \mathbf{c}$$

both follow from this definition.

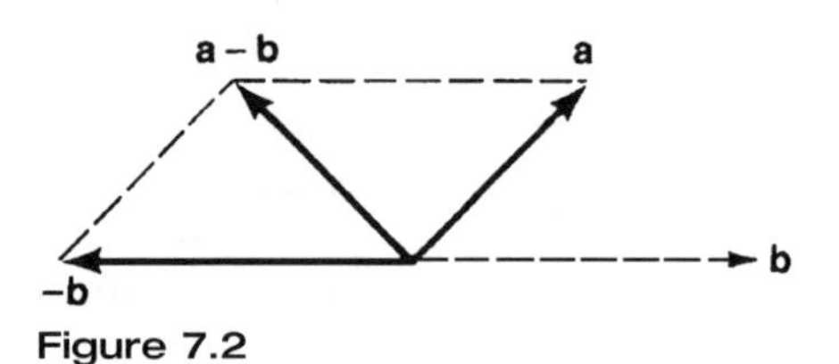

Figure 7.2

The negative of a vector **b** is defined to be of the same magnitude as but of the opposite direction to **b**, and it is written as –**b**. Subtraction of two vectors is then defined by

$$\mathbf{a} - \mathbf{b} = \mathbf{a} + (-\mathbf{b})$$

as indicated in Fig. 7.2.

Multiplication by a Scalar

The product of a scalar, p, and a vector, **a**, is the vector written as $p\mathbf{a}$, and it is defined to have a magnitude of $p\mathbf{a}$, and direction the same as, or the opposite of, **a,** depending on whether p is positive or negative. The following laws,

$$p(q\mathbf{a}) = (pq)\mathbf{a}$$
$$(p+q)\mathbf{a} = p\mathbf{a} + q\mathbf{a}$$
$$p(\mathbf{a}+\mathbf{b}) = p\mathbf{a} + p\mathbf{b}$$

can be readily verified from these definitions.

Addition of two or more vectors is sometimes called the **composition** of the vectors. A reversal of this process is called the **resolution** of a vector, that is, determining a set of vectors (usually in prescribed directions), the sum of which will be the given vector.

Example 7.1

Resolve the vector **f** of magnitude 75 kN into two vectors (components) in the directions of lines L_1 and L_2 shown in Exhibit 1.

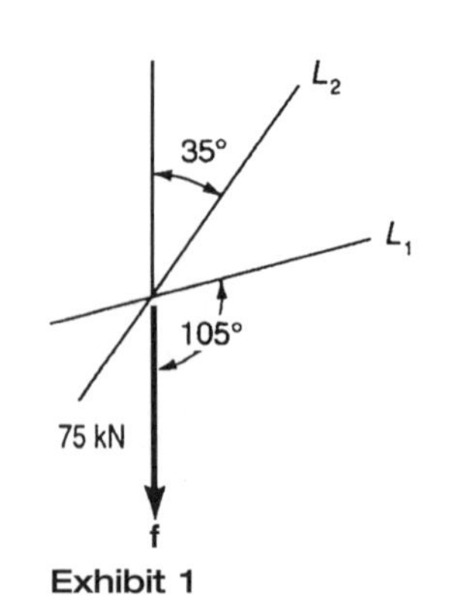

Exhibit 1

Solution

The given vector is to be the diagonal of the parallelogram having the desired components as sides. The parallelogram can be completed by drawing lines through the head of the vector, parallel to the given lines. The magnitudes of the two components can then be determined by applying the trigonometric law of sines:

$$|\mathbf{f}_1| = \frac{\sin 35°}{\sin 40°}(75 \text{ kN}) = 66.9 \text{ kN}$$

$$|\mathbf{f}_2| = \frac{\sin 105°}{\sin 40°}(75 \text{ kN}) = 112.7 \text{ kN}$$

Dot Product

The **dot product** of two vectors **a** and **b** is a scalar (sometimes called the scalar product or inner product) that is equal to the product of the magnitudes of the vectors and the cosine of the angle θ between the vectors. It is written as

$$\mathbf{a} \bullet \mathbf{b} = |\mathbf{a}||\mathbf{b}| \cos\theta \tag{7.3}$$

A special case is the dot product of a vector with itself,

$$\mathbf{a} \bullet \mathbf{a} = |\mathbf{a}||\mathbf{a}| \cos\theta$$

which provides a way of expressing the magnitude of a vector:

$$|\mathbf{a}| = \sqrt{\mathbf{a} \bullet \mathbf{a}} \tag{7.4}$$

The commutative law,

$$\mathbf{a} \bullet \mathbf{b} = \mathbf{b} \bullet \mathbf{a}$$

and the distributive law,

$$\mathbf{a} \bullet (\mathbf{b} + \mathbf{c}) = \mathbf{a} \bullet \mathbf{b} + \mathbf{a} \bullet \mathbf{c}$$

can both be verified from the above definitions.

Unit Vectors and Projections

An extremely useful tool is the **unit vector**. It has a unit magnitude and is designated here by the symbol **e**. Unit vectors can be introduced (defined) by giving the direction in terms of the geometry of the application, or defined in terms of a specified vector by multiplying the vector by the reciprocal of its magnitude. For example, the unit vector in the direction of **a** is given by

$$\mathbf{e}_a = \frac{1}{|\mathbf{a}|}\mathbf{a}$$

Rearranged, this relationship expresses the vector **a** in terms of its magnitude and a unit vector that gives its direction:

$$\mathbf{a} = |\mathbf{a}|\mathbf{e}_a \tag{7.5}$$

The **projection** of vector **a** onto a line L is the vector from the projection of the tail of **a** onto L to the projection of the head of **a** onto L, as indicated in Fig. 7.3. The magnitude of this projection will be the product of $|\mathbf{a}|$ and the cosine of the

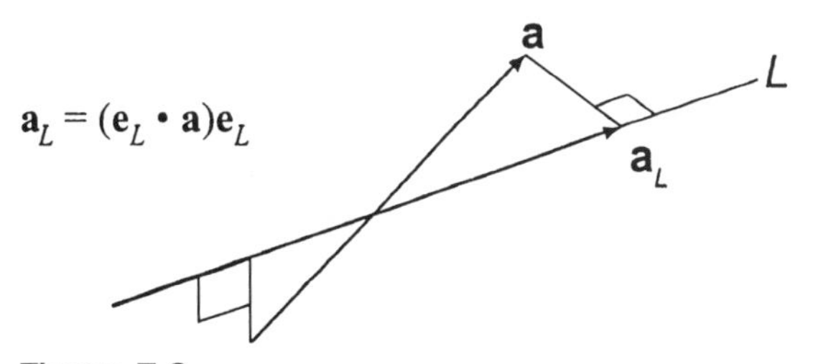

Figure 7.3

angle between **a** and L. With unit vector $\mathbf{e}_L$, defined to be parallel to L, the projection of **a** onto L can be expressed as

$$\mathbf{a}_L = (\mathbf{e}_L \bullet \mathbf{a})\mathbf{e}_L$$

That is, vector projection onto a line in a selected direction can be evaluated by dot-multiplying the vector by a unit vector in that direction. As indicated in the next section, this approach provides the most direct means of obtaining equivalent scalar relationships from a vector relationship.

Vector and Scalar Equations

Many physical laws, such as Newton's Laws of Motion, are best expressed by vector equations. In general, a vector equation can provide up to three independent scalar equations. In Example 7.1, the directions of the lines L_1 and L_2 might be determined by the orientations of two members of a structural truss, and the direction of **f** by a gravitational force. Then the relationship $\mathbf{f}_1 + \mathbf{f}_2 = \mathbf{f}$ might be a requirement of equilibrium and geometry, from which the magnitudes of $\mathbf{f}_1$ and $\mathbf{f}_2$ are to be determined in terms of the magnitude of **f**. With the unit vectors $\mathbf{e}_1$ and $\mathbf{e}_2$ introduced as shown, the equation

$$f_1\mathbf{e}_1 + f_2\mathbf{e}_2 = \mathbf{f}$$

brings the unknowns f_1 and f_2 into evidence and makes available two scalar equations for their determination. A corresponding scalar equation can be obtained by dot-multiplying each member of the vector equation by a selected vector; for example, with the unit vectors already introduced, we can write the two equations

$$f_1\mathbf{e}_1 \bullet \mathbf{e}_1 + f_2\mathbf{e}_1 \bullet \mathbf{e}_2 = \mathbf{e}_1 \bullet \mathbf{f}$$
$$f_1\mathbf{e}_2 \bullet \mathbf{e}_1 + f_2\mathbf{e}_2 \bullet \mathbf{e}_2 = \mathbf{e}_2 \bullet \mathbf{f}$$

reduce them to

$$f_1 + (\cos 40°)f_2 = (75\text{ kN})\ \cos 105°$$
$$(\cos 40°)f_1 + f_2 = (75\text{ kN})\ \cos 145°$$

and solve these for

$$f_1 = 66.9\text{ kN} \qquad f_2 = -112.7\text{ kN}$$

The vector equation could also be dot-multiplied by the unit vectors $\mathbf{e}_a$ and $\mathbf{e}_b$, defined to be perpendicular to f_1 and f_2, as shown in Fig. 7.4. This gives

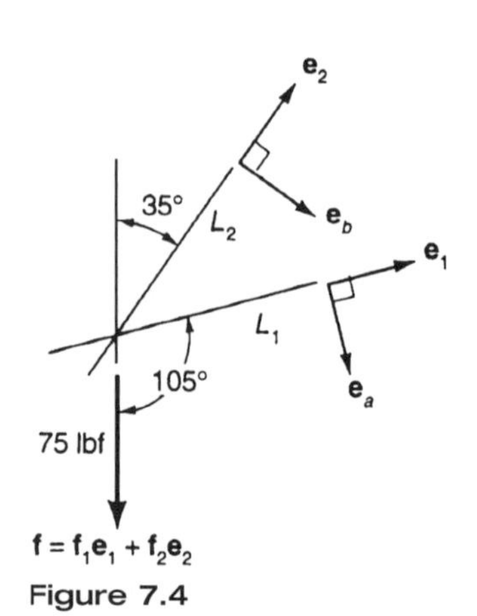

Figure 7.4

$$f_1\mathbf{e}_a \bullet \mathbf{e}_1 + f_2\mathbf{e}_a \bullet \mathbf{e}_2 = \mathbf{e}_a \bullet \mathbf{f}$$
$$f_1\mathbf{e}_b \bullet \mathbf{e}_1 + f_2\mathbf{e}_b \bullet \mathbf{e}_2 = \mathbf{e}_b \bullet \mathbf{f}$$

which reduce to

$$(\cos 130°)f_2 = (75\text{ kN})\ \cos 15°$$
$$(\cos 50°)f_1 = (75\text{ kN})\ \cos 55°$$

The unit vectors $\mathbf{e}_a$ and $\mathbf{e}_b$ are better choices for the purpose of evaluating f_1 and f_2 because each is perpendicular to one of the vectors with unknown magnitudes, so that we can eliminate an unknown from each equation.

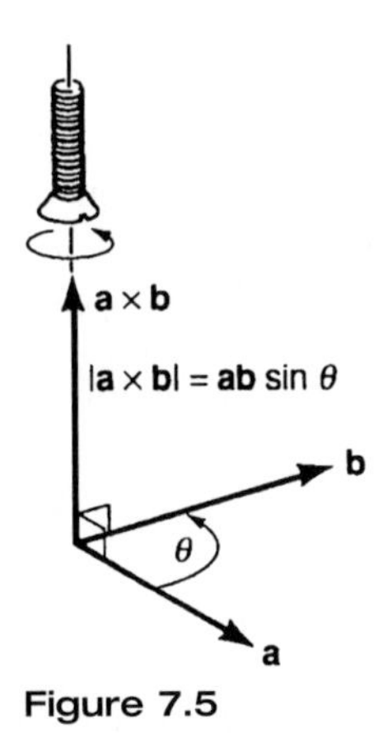

Figure 7.5

The Cross Product

The **cross product** of two vectors **a** and **b** (sometimes called the vector product) is defined to be a vector that is perpendicular to the plane of **a** and **b**, with magnitude equal to the product of the magnitudes of **a** and **b** and the sine of the angle θ between **a** and **b** and with direction determined by the **right-hand rule**. The right-hand rule states that the vector's direction coincides with that of the advancement of a right-hand screw, with the axis of the screw being oriented perpendicular to **a** and **b**, and turned in the direction **a**-toward-**b**, as shown in Fig. 7.5. Note that the cross product is *not* commutative; instead

$$\mathbf{b} \times \mathbf{a} = -\mathbf{a} \times \mathbf{b}$$

However, the definitions of cross-multiplication and addition can be used to show that the distributive law

$$\mathbf{a} \times (\mathbf{b} + \mathbf{c}) = \mathbf{a} \times \mathbf{b} + \mathbf{a} \times \mathbf{c}$$

is valid, and that the cross product is associative with respect to multiplication by scalars:

$$(p\mathbf{a}) \times (q\mathbf{b}) = (pq)(\mathbf{a} \times \mathbf{b})$$

Example 7.2

It is desired to resolve a given vector **a** into two components, one parallel to a second given vector **b** and one perpendicular to **b**. The results are to be expressed in terms of **a** and **b**, using vector operations defined in the preceding pages.

Solution

The component parallel to **b** will be the projection of **a** onto the line parallel to **b**, which is expressible in terms of a unit vector in the direction of **b**:

$$\mathbf{a}_{\parallel} = (\mathbf{e}_b \bullet \mathbf{a})\mathbf{e}_b = \frac{(\mathbf{b} \bullet \mathbf{a})\mathbf{b}}{\mathbf{b} \bullet \mathbf{b}}$$

The component perpendicular to **b** can be determined from the fact that **a** is to be the sum of this and the component just calculated:

$$\mathbf{a}_{\perp} = \mathbf{a} - \mathbf{a}_{\parallel}$$

Alternatively, the perpendicular component can be calculated by

$$\mathbf{a}_{\perp} = \frac{(\mathbf{b} \times \mathbf{a}) \times \mathbf{b}}{\mathbf{b} \bullet \mathbf{b}}$$

Verifying this last relationship from the definition of the cross product provides useful practice in relating the operation to geometry. The vector **a** can now be expressed in terms of the components parallel and perpendicular to **b**:

$$\mathbf{a} = \frac{(\mathbf{b} \bullet \mathbf{a})\mathbf{b}}{\mathbf{b} \bullet \mathbf{b}} + \frac{(\mathbf{b} \times \mathbf{a}) \times \mathbf{b}}{\mathbf{b} \bullet \mathbf{b}}$$

Rectangular Cartesian Components

A special way of resolving vectors consists of forming three mutually perpendicular components. The directions are chosen (usually with consideration of the geometry of the problem at hand) and three mutually perpendicular unit vectors $\mathbf{e}_x$, $\mathbf{e}_y$, and $\mathbf{e}_z$ are defined to be parallel to these directions. The rectangular Cartesian components of a vector **a** are then the projections of **a** onto lines in the selected directions, expressed as

$$\mathbf{a} = a_x\mathbf{e}_x + a_y\mathbf{e}_y + a_z\mathbf{e}_z \tag{7.6}$$

in which $a_x = \mathbf{e}_x \bullet \mathbf{a}$, $a_y = \mathbf{e}_y \bullet \mathbf{a}$, and $a_z = \mathbf{e}_z \bullet \mathbf{a}$. These relationships, together with the associative and distributive laws mentioned previously, lead to the following formulas:

$$\mathbf{a} \bullet \mathbf{b} = a_x b_x + a_y b_y + a_z b_z \tag{7.7}$$

$$|\mathbf{a}| = \sqrt{a_x^2 + a_y^2 + a_z^2} \tag{7.8}$$

$$\begin{aligned} \mathbf{a} \times \mathbf{b} = {} & (a_y b_z - a_z b_y)\mathbf{e}_x \\ & + (a_z b_x - a_x b_z)\mathbf{e}_y \\ & + (a_x b_y - a_y b_x)\mathbf{e}_z \end{aligned} \tag{7.9}$$

Each of these equations depends on the fact that the unit vectors are mutually perpendicular; in addition, the expression for the cross product is valid only if the unit vectors form a *right-handed* set; that is,

$$\mathbf{e}_x = \mathbf{e}_y \times \mathbf{e}_z, \quad \mathbf{e}_y = \mathbf{e}_z \times \mathbf{e}_x, \quad \text{and} \quad \mathbf{e}_z = \mathbf{e}_x \times \mathbf{e}_y$$

Example 7.3

Determine the lengths of the guy lines $O'P$ and $O'Q$, shown in Exhibit 2, and the angle between them.

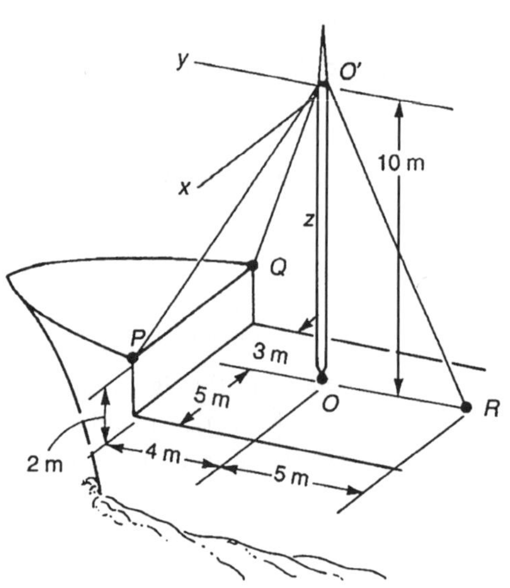

Exhibit 2

Solution

Let position vectors **a** and **b** extend from point O' to points P and Q, respectively, and resolve these into components along the x, y, and z axes shown:

$$\mathbf{a} = (5\text{ m})\mathbf{e}_x + (4\text{ m})\mathbf{e}_y + (8\text{ m})\mathbf{e}_z$$
$$\mathbf{b} = (-3\text{ m})\mathbf{e}_x + (4\text{ m})\mathbf{e}_y + (8\text{ m})\mathbf{e}_z$$

The required lengths can then be determined from Eq. (7.8):

$$|\mathbf{a}| = \sqrt{(5\text{ m})^2 + (4\text{ m})^2 + (8\text{ m})^2} = 10.25\text{ m}$$

$$|\mathbf{b}| = \sqrt{(-3\text{ m})^2 + (4\text{ m})^2 + (8\text{ m})^2} = 9.43\text{ m}$$

The required angle is a factor in the definition of $\mathbf{a} \bullet \mathbf{b}$, and in fact it will be the only unknown in this equation after $\mathbf{a} \bullet \mathbf{b}$ is evaluated. From Eq. (7.7), we have

$$\mathbf{a} \bullet \mathbf{b} = (5\text{ m})(-3\text{ m}) + (4\text{ m})(4\text{ m}) + (8\text{ m})(8\text{ m}) = 65\text{ m}^2$$

Then Eq. (7.3) gives

$$\cos\theta = \frac{\mathbf{a} \bullet \mathbf{b}}{|\mathbf{a}||\mathbf{b}|} = \frac{65\text{ m}^2}{(10.25\text{ m})(9.43\text{ m})} = 0.672$$

from which

$$\theta = 47.7°$$

FORCE SYSTEMS

A body may have several forces acting on it simultaneously. To account for these forces in an organized way, some general properties of a set of forces, or a force system, will prove useful.

Types of Forces

Normally, forces are distributed over some region of the body they act upon; however, some simplification can often be gained without significant loss of accuracy by considering that a force is concentrated at a single point called the **point of application** of the force. However, in some circumstances, it may be necessary to account for the way the forces are distributed over a region of the body. Thus, we make a distinction between **concentrated forces** and **distributed forces**.

A second distinction that is often important is between **surface forces**, or actions that take place where surfaces contact, and **body forces**, which are distributed throughout a body, as in the case of gravity.

Point of Application and Line of Action

In addition to the vector value of a force (that is, magnitude and direction), the point at which a force acts on the body is important to the way the body responds. For this reason, analysis usually requires not only a **force** vector to specify the magnitude and direction of the force, but also a **position** vector to specify the location of the *point of application* of the force. The **line of action** of a force is the line parallel to the force vector and through the point of application. This line is important to the understanding of *moments* of forces.

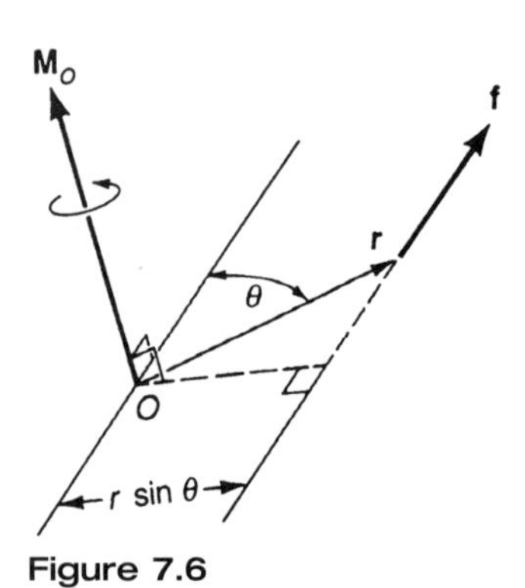

Figure 7.6

Moments of Forces

The **moment** about a point O of a force $\mathbf{f}$ is the vector defined as

$$\mathbf{M}_O = \mathbf{r} \times \mathbf{f} \tag{7.10}$$

where $\mathbf{r}$ is a position vector from O to any point on the line of action of $\mathbf{f}$. Reference to Fig. 7.6 and the definition of the cross product reveal that the magnitude of the moment is

$$|\mathbf{M}_O| = df \tag{7.11}$$

where f is the magnitude of $\mathbf{f}$ and $d = r \sin\theta$ is the perpendicular distance from O to the line of action of $\mathbf{f}$.

Example 7.4

Suppose the guy line $O'P$ in Exhibit 3 has a tension of 800 N. What is the moment about O of the force from this cable acting on the mast?

Solution

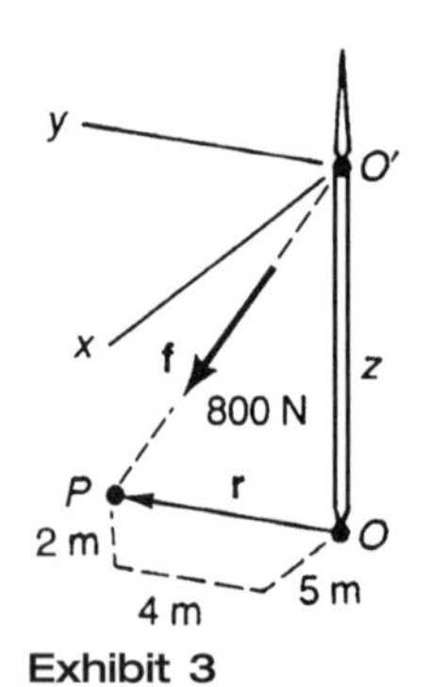

Exhibit 3

Because point P is on the line of action of this force, a position vector, $\mathbf{r}$, from O to P can be used for evaluation of the moment about O. Referring to the axes shown in Exhibit 3, the x-y-z resolution of this vector may be written as

$$\mathbf{r} = (5\text{ m})\mathbf{e}_x + (4\text{ m})\mathbf{e}_y + (-2\text{ m})\mathbf{e}_z$$

The resolution of the force can be determined by multiplying its magnitude by the unit vector in the direction of $O'P$, components of which can be obtained by dividing the vector $\mathbf{a}$ from the preceding example by its magnitude:

$$\begin{aligned}\mathbf{f} &= 800\text{ N}\left(\frac{5}{10.25}\mathbf{e}_x + \frac{4}{10.25}\mathbf{e}_y + \frac{8}{10.25}\mathbf{e}_z\right)\\ &= (390.4\mathbf{e}_x + 312.3\mathbf{e}_y + 624.6\mathbf{e}_z)\text{ N}\end{aligned}$$

The moment can now be evaluated with reference to Eq. (7.9):

$$\begin{aligned}\mathbf{M}_O = {} & [(4\text{ m})(624.6\text{ N}) - (-2\text{ m})(312.3\text{ N})\mathbf{e}_x\\ & + [(-2\text{ m})(390.4\text{ N}) - (5\text{ m})(624.6\text{ N})]\mathbf{e}_y\\ & + [(5\text{ m})(312.3\text{ N}) - (4\text{ m})(390.4\text{ N})]\mathbf{e}_z\\ = {} & (3123\mathbf{e}_x - 3904\mathbf{e}_y)\text{ N}\bullet\text{m}\end{aligned}$$

Observe that point O' is also on the line of action of $\mathbf{f}$, so that a position vector $\mathbf{r} = (-10\text{ m})\mathbf{e}_z$ could have been used instead of the one above, and with less arithmetic.

The moment about an axis Oi is defined as the projection onto the axis of the moment about some point on the axis. To express this, we define the positive sense along the axis with the unit vector $\mathbf{e}_i$ and write

$$\mathbf{M}_{Oi} = (\mathbf{e}_i \bullet \mathbf{M}_O)\mathbf{e}_i \tag{7.12}$$

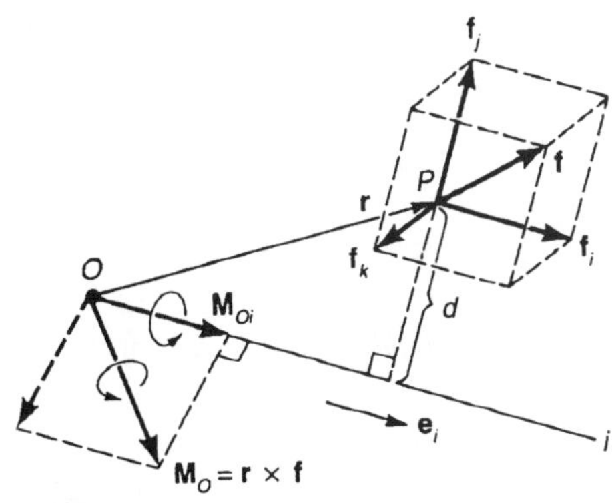

Figure 7.7

Although the moment about the axis can be computed according to this definition, an alternative form is often easier to use and provides a different interpretation. Substitution of Eq. (7.10) into Eq. (7.12) and use of the vector identity $\mathbf{a} \bullet (\mathbf{b} \times \mathbf{c}) = (\mathbf{a} \times \mathbf{b}) \bullet \mathbf{c}$ leads to $\mathbf{M}_{Oi} = [\mathbf{e}_i \times \mathbf{r}) \bullet \mathbf{f}\,]\mathbf{e}_i$

Now if $\mathbf{f}$ is resolved into a component $\mathbf{f}_i$ parallel to the axis Oi, a component $\mathbf{f}_j$ perpendicular to O_i and in the plane of $\mathbf{r}$ and O_i, and a component $\mathbf{f}_k$ perpendicular to this plane, several important facts become apparent. Referring to Fig. 7.7, note that because both $\mathbf{f}_i$ and $\mathbf{f}_j$ are perpendicular to $\mathbf{e}_i \times \mathbf{r}$, neither of these components contributes to $\mathbf{M}_{Oi}{}'$. Also, because $\mathbf{e}_i \times \mathbf{r}$ has the magnitude $d = |\mathbf{r}| \sin \angle \frac{\mathbf{r}}{\mathbf{i}}$, we can express the magnitude of the moment component as

$$|\mathbf{M}_{Oi}| = df_k$$

where d is the perpendicular distance from point P to the axis Oi.

The *sense* of $\mathbf{M}_{Oi}$ (that is, whether it is directed in the positive or negative i-direction) is readily determined from the direction of $\mathbf{f}_k$ and the right-hand rule. Alternatively, the sense may be determined by the sign of the factor $\mathbf{e}_i \bullet \mathbf{M}_O$ in Eq. (7.12). Note that the same value of d would be obtained regardless of where the point O is on the i-axis, and recall that the position vector $\mathbf{r}$ in the definition $\mathbf{M}_O = \mathbf{r} \times \mathbf{f}$ can be from O to any point on the line of action of $\mathbf{f}$. This means that Eq. (7.12) will yield the value $\mathbf{M}_{Oi}$ with $\mathbf{r}$ as a position vector from *any* point on the axis O_i to *any* point on the line of action of $\mathbf{f}$.

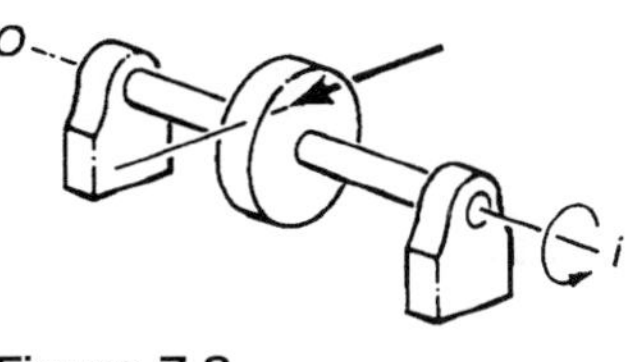

Figure 7.8

The moment about an axis is a measure of the tendency of the force(s) to cause rotation about the axis. For example, if a rotor is mounted in bearings and subjected to a set of forces, the moment of these forces about the axis of the bearings is found to be directly related to the rate of change of rotational speed. Neither forces parallel to the axis nor forces with lines of action passing through the axis will affect the rotation.

Resultant Forces and Moments

If there are several forces $\mathbf{f}_1, \mathbf{f}_2, \ldots, \mathbf{f}_n$, each with its own line of action, the **resultant force** is defined as

$$\mathbf{f} = \sum_{i=1}^{n} \mathbf{f}_i$$

and the **resultant moment** about a point O is defined as

$$\mathbf{M}_O = \sum_{i=1}^{n} \mathbf{r}_i \times \mathbf{f}_i 0$$

where $\mathbf{r}_i$ is a position vector from O to any point on the line of action of $\mathbf{f}_i$.

Couples

A special set of forces, called a **couple**, has zero resultant force but a nonzero resultant moment. An example is a pair of forces of equal magnitude, opposite directions, and separate lines of action.

Moments about Different Points

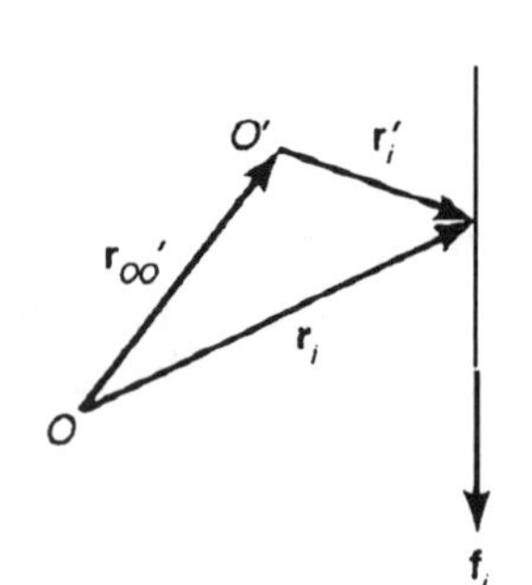

Figure 7.9

The resultant moment of a set of forces about two different points, O and O', are related as follows. With $\mathbf{r}_{oo'}$ designating the position vector from O to O', the position vectors from these two points to a point on the line of action of the force $\mathbf{f}_i$ are shown in Fig. 7.9 and are related by

$$\mathbf{r}_i = \mathbf{r}_{i'} + \mathbf{r}_{oo'}$$

The moment about O can then be expressed as

$$\begin{aligned} \mathbf{M}_O &= \mathbf{r}_1 \times \mathbf{f}_1 + \mathbf{r}_2 \times \mathbf{f}_2 + \cdots + \mathbf{r}_n \times \mathbf{f}_n \\ &= (\mathbf{r}'_1 + \mathbf{r}_{OO'}) \times \mathbf{f}_1 + (\mathbf{r}'_2 + \mathbf{r}_{OO'}) \times \mathbf{f}_2 + \cdots + (\mathbf{r}_n + \mathbf{r}_{OO'}) \times \mathbf{f}_n \\ &= \mathbf{r}'_1 \times \mathbf{f}_1 + \mathbf{r}'_2 \times \mathbf{f}_2 + \cdots + \mathbf{r}'_n \times \mathbf{f}_n + \mathbf{r}_{OO'} \times (\mathbf{f}_1 + \mathbf{f}_1 + \cdots + \mathbf{f}_n) \\ &= \mathbf{M}_{O'} + \mathbf{r}_{OO'} \times \mathbf{f} \end{aligned} \qquad \textbf{(7.13)}$$

That is, the moment about O is equal to that about O' plus the moment that a force equal to the resultant of the given forces would have about O if the line of action of this force passed through O'. For the special case in which $\mathbf{f} = 0$, Eq. (7.13) shows that the moment of a couple about every point is the same.

Equivalent Force Systems

Two sets of forces are said to be **equivalent** if each has the same resultant force and the same resultant moment about some point. If these conditions are met, then Eq. (7.13) can be used to show that the sets will have the same resultant moment about *any* point.

Example 7.5

A 7-kN force acts on the end of the beam, its line of action passing along the center of the web of the channel section. If a bracket were attached to the end of the beam, allowing this force to be applied 80 mm to the left of the center of the web (see Exhibit 4), what horizontal forces applied along the flanges would need to be added so that, together with the displaced 7-kN force, they would form a set equivalent to the 7-kN force acting along the web?

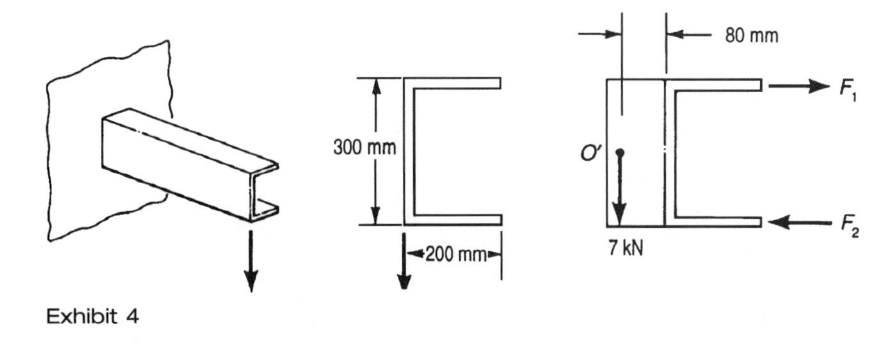

Exhibit 4

Solution

Since the resultants of the two sets must be the same (7-kN downward), the resultant of F_1 and F_2 must be zero, which in turn means that F_1 and F_2 must have equal magnitudes and opposite directions. To determine the magnitude, the resultant moment of each set can be equated; a convenient point about which to evaluate these moments is located where the lines of action of F_2 and the force through O' intersect. The moment about this point of the original force is (0.08 m)(7 kN) = 0.56 kN• m acting clockwise. The resultant moment of the equivalent set about this point is (0.3 m)F_1, also in the clockwise direction. Equivalence requires that (0.3 m)F_1 = 0.56 kN•m from which F_1 = 1.867 kN. Thus, the equivalent set consists of the vertical, 7-kN force through O', together with a couple that has a clockwise-acting moment of 0.56 kN•m.

EQUILIBRIUM

Newton's Laws require for every body or system of bodies that

$$\sum_i \mathbf{f}_i = \sum_j m_j \mathbf{a}_j$$

and

$$\sum_i \mathbf{r}_i \times \mathbf{f}_i = \sum_j \mathbf{r}_j \times m_j \mathbf{a}_j$$

in which $\mathbf{f}_i$ is one of the forces acting on the system (from an *external* source); m_j and $\mathbf{a}_j$ are the mass and acceleration, respectively, of the *j*th material element; $\mathbf{r}_i$ and $\mathbf{r}_j$ are position vectors from *any* selected point to, respectively, a point on the line of action of $\mathbf{f}_i$ and the *j*th material element; and the sums are to include *all* the external forces and material elements. A system will be in static equilibrium whenever the accelerations are all zero. In these cases the laws require that the resultant of all forces from external sources be zero, and that the resultant moment of these forces about any point be zero.

Free-Body Diagrams

In spite of the simplicity of equilibrium relationship, they can easily be misapplied. Experience has repeatedly shown that nearly all such errors stem from lack of attention to the appropriate **free-body diagram**. The free-body diagram must show clearly what body, bodies, or parts thereof are being considered as the system,

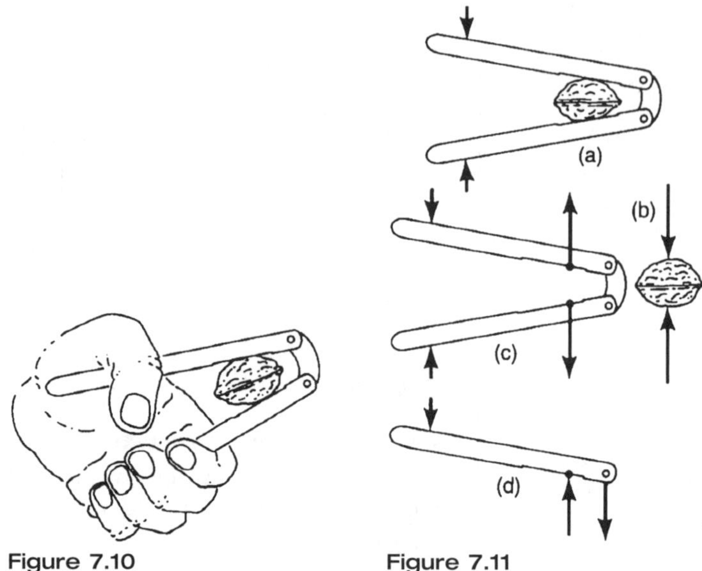

Figure 7.10 Figure 7.11

and all of the forces acting *on* the system from sources *outside* the system. For illustration, consider the device shown in Fig. 7.10. Several different free-bodies are possibly useful and will be constructed.

First, consider the system consisting of the nutcracker together with the walnut. This system is shown isolated from all other objects, with arrows depicting the forces that come from objects *external* to the system. Assuming forces of gravity to be negligible, the only external body that exerts forces on this system is the hand. These forces are shown in Fig. 7.11(a). The interaction between the nut and the cracker is *not* shown on this free-body, since it is internal to this system.

To expose the force tending to break the nut, we might consider free-bodies of the nut and of the nutcracker, shown in Figs. 7.11(b) and (c). In Fig. 7.11(c) the hand and the walnut are external to the nutcracker. Therefore, arrows depicting the forces from the nut as well as from the hand are included in this free-body.

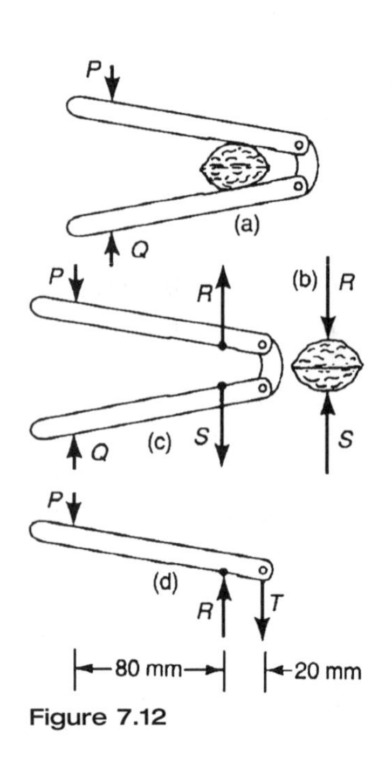

Figure 7.12

Another possibly useful free-body is that of the upper handle. Objects external to this are the hand, the walnut, and the connecting pin. This free-body appears in Fig. 7.11(d).

Here is a summary of the procedure: First, a sketch must be made clearly showing the system to be considered for equilibrium. The system boundary is normally chosen so that it passes through a point where a force interaction of particular interest occurs. Next, *all* forces acting on the system, from bodies *external* to the system, must be properly represented. Force interactions between bodies *within* the system are *not* considered.

To show clearly the physical significance of quantities in equilibrium equations, symbols must accompany the arrows representing the forces that appear in these equations. In Fig. 7.12, the letters P, Q, R, S, and T have been chosen to indicate the magnitudes of several forces. Each letter represents a **scalar** multiplier of a unit vector in the direction of the arrow; since this scalar can take on a positive or negative value, a force in the same or opposite direction indicated by the arrow can be represented. If an analysis leads to the values, say, $P = 80$ N and $R = 400$ N, the forces acting on the upper handle would be 80 N *downward* on the right-hand end and 400 N *upward* where the walnut makes contact at R. If different circumstances led to, say, $P = -15$ N and $R = -75$, these values would imply that the forces acting on the upper handle are 15 N *upward* on the right-hand end and 75 N *downward* where the walnut makes contact. (A little adhesive between the walnut and the nutcracker would make this possible.)

Observe that the forces on the walnut have the same labels as their counterparts on the nutcracker, and the arrows have opposite directions. We have in this way implied satisfaction of Newton's Third Law without further fuss. Other relatively simple aspects of force analysis can be treated as the free-bodies are constructed; for example, unless the walnut is to accelerate, it is evident from a glance at its free-body that $R = S$. Writing this equation could be circumvented by simply labeling both arrows with the same letter.

Equations of Equilibrium

With free-body diagrams properly drawn and labeled, equations of equilibrium can be written for any of the chosen bodies. For example, the vertical force equilibrium of the free-body of Fig. 7.12(a) requires that $Q - P = 0$. Each of the other two equations of force equilibrium, involving components in the horizontal direction and components perpendicular to the plane of the sketch, is the trivial equation $0 = 0$. Moment equilibrium shows that the lines of action of the two forces must coincide, a fact that has already been incorporated into the diagram. Similar analyses of the free-bodies of Fig. 7.12(b) and (c) lead to

$$S - R = 0$$

and

$$Q - P + R - S = 0$$

Now suppose that the reason for this analysis is to obtain an estimate of how hard one must squeeze in order to crack a nut, given the cracking requires 245 N applied to the nut. Then $R = 245$ N, and the three equations above contain the three unknowns, P, Q, and S. Unfortunately, attempts to solve these for P or Q will fail, because the three equations are not independent, because the last equation can be deduced from the first two by addition. Equilibrium of still another body must be considered in order to obtain an independent equation. The free-body of the handle in Fig. 7.12(d) can provide two more equations: the vertical component for force equilibrium,

$$T + P = 245 \text{ N} \qquad \textbf{(a)}$$

and an equation of moment equilibrium. Summing moments about the point of contact with the walnut leads to

$$-(20 \text{ mm})T + (80 \text{ mm})P = 0 \qquad \textbf{(b)}$$

To solve for P, we can multiply Eq. (a) by 20 mm and add Eq. (b) to the result, yielding

$$(100 \text{ mm})P = (20 \text{ mm})\,(245 \text{ N}) \qquad \textbf{(c)}$$

This gives $P = 49$ N.

A more direct analysis stems from considering moments about a different point on the handle, resulting in (c) as the first equation written.

The following examples provide further illustration of the use of the basic laws of static equilibrium.

Example 7.6

Neglecting gravity forces in Exhibit 5(a), except those on the 300-kg load, determine the forces in the cable and in the boom.

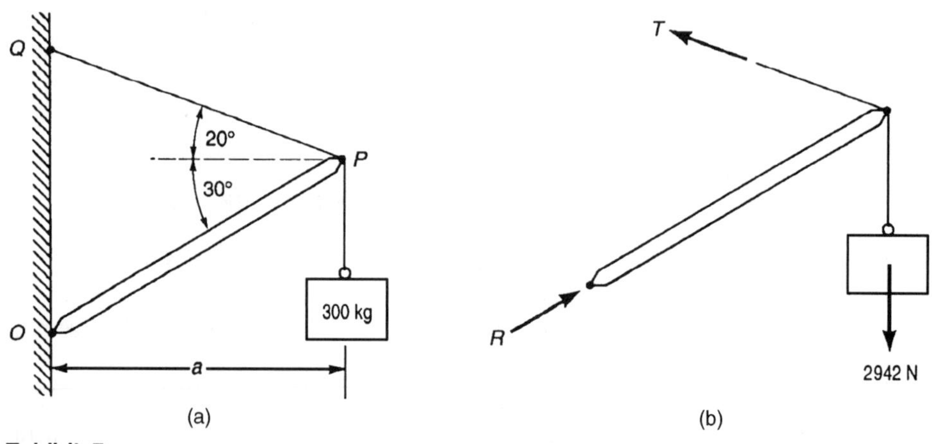

Exhibit 5

Solution

First, a free-body diagram is drawn (Exhibit 5(b)), with the system boundary passing through the support point O and through the cable segment PQ. Vanishing of moment about point P requires that the reaction at O must be directed along the boom. The equation of moments about O can be expressed as

$$(T\sin 20° - 2942\ \text{N})a + (T\cos 20°)(a\tan 30°) = 0$$

from which

$$T = \frac{2942\ \text{kN}}{\sin 20° + \cos 20° \tan 30°} = 3.33\ \text{kN}$$

Equilibrium of horizontal forces,

$$R\cos 30° - T\cos 20° = 0$$

leads to the magnitude of the reaction at O:

$$R = \frac{(3.33\ \text{N})\cos 20°}{\cos 30°} = 3.61\ \text{kN}$$

Example 7.7

Neglecting gravity forces except those on the 2-Mg load shown in Exhibit 6(a), determine the tension in cable AB, which is holding up the crane boom.

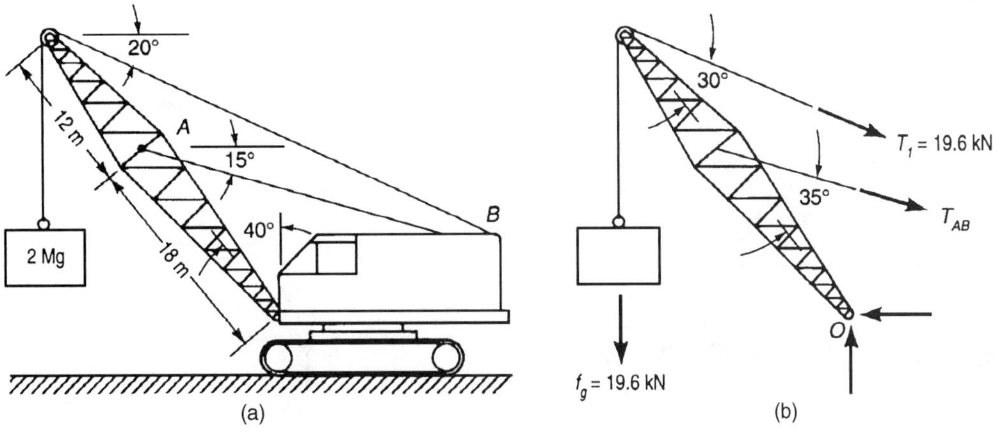

Exhibit 6

Solution

The boundary of the free-body passes through the two upper cable segments and the support point, labeled O in Exhibit 6(b). By considering moment equilibrium of a system consisting of the pulley and a portion of the cable, including the section that is in contact with the pulley (not shown), we find that $T_1 = f_g = 19.6$ kN.

To avoid introducing the unknown reaction at O into the analysis, consider moments of forces about this point. With the radius of the pulley denoted as r, the equation of moments about O is

$$f_g[(30 \text{ m})\sin 40° + r] - T_1[(30 \text{ m})\sin 30° + r] - T_{AB} \sin 35°(18 \text{ m}) = 0$$

With the value of $T_1 = f_g$ substituted, this is readily solved for the tension in the supporting cable:

$$T_{AB} = \frac{(19.6 \text{ kN})(30 \text{ m})(\sin 40° - \sin 30°)}{(18 \text{ m})\sin 35°} = 8.13 \text{ kN}$$

Example **7.8**

Gravity forces on the structural members are negligible compared with P and Q in Exhibit 7(a). Evaluate all the forces acting on each of the three members in the A-frame.

Solution

Free-bodies of the entire frame and of each individual member are shown in Exhibit 7(b)–(e). The roller support at D means that no horizontal force can be transmitted from the ground at that point. From the free-body in view (b) we can consider moments about point E,

$$(3a)Q + (3a \tan 30°)P - (6a \tan 30°)R_D = 0$$

horizontal forces,

$$R_{Ex} - Q = 0$$

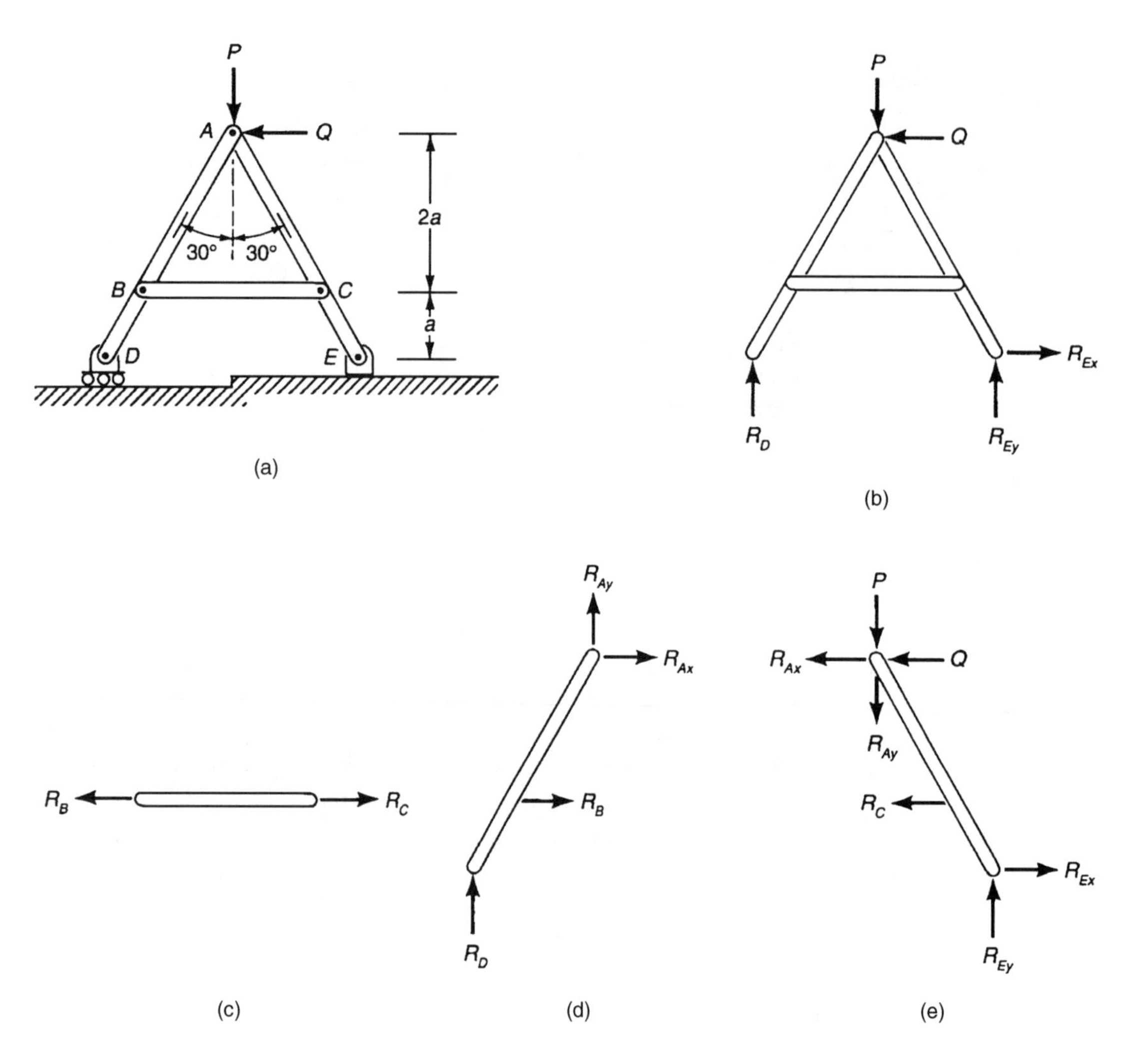

Exhibit 7

and moments about point D:

$$(6a \tan 30°)R_{Ey} + (3a)Q - (3a \tan 30°)P = 0$$

These equations can then be solved for the support reactions:

$$R_D = \frac{1}{2}(P + \cot 30° Q)$$

$$R_{Ex} = Q$$

$$R_{Ey} = \frac{1}{2}(P - \cot 30° Q)$$

As a check, it might be a good idea to consider vertical forces. Moment equilibrium of the free-body in view (c) implies that the lines of action of R_B and R_C are along the bar. Its horizontal equilibrium gives us

$$R_C - R_B = 0$$

Now, turning to the free-body in view (d), we can write equations of moment equilibrium about points A as

$$(2a)R_B - (3a \tan 30°)R_D = 0$$

and horizontal and vertical force equilibrium

$$R_{Ax} + R_B = 0$$
$$R_{Ay} + R_D = 0$$

With values of R_D, R_{Ax}, and R_{Ay} above, these equations give the following values of the remaining unknown reactions:

$$R_B = R_C = \frac{3}{4}(\tan 30° P + Q)$$

$$R_{Ax} = -\frac{3}{4}(\tan 30° P + Q)$$

$$R_{Ay} = -\frac{1}{2}(P + \cot 30°\ Q)$$

Example **7.9**

The cables OA, OB, and OC support a suspended block shown in Exhibit 8. Determine the tension in each cable in terms of the gravitational force f_g.

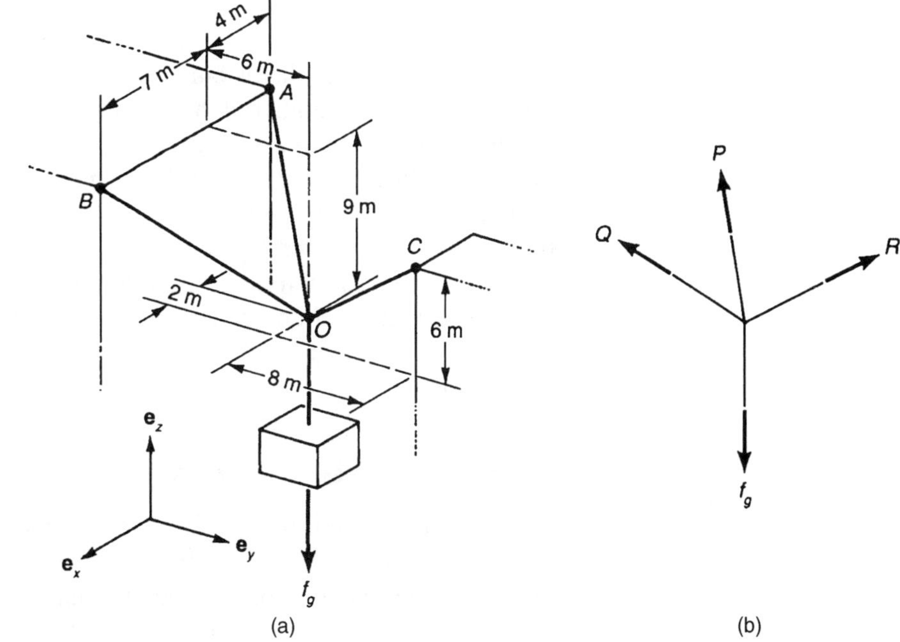

Exhibit 8

Solution

In Exhibit 8(b), the desired tensions P, Q, and R are shown on the free-body of the portion of the structure in the neighborhood of point O. To resolve the forces in the directions of $\mathbf{e}_x$, $\mathbf{e}_y$, and $\mathbf{e}_z$, we need to find the direction cosines between OA, OB, and OC and these directions. This is done, as in Example 7.3, by dividing the projection of the cable onto the axis by the length of the cable; for instance, the direction cosine between OA and $\mathbf{e}_x$ is

$$\cos \angle_{w}^{e_x} = \frac{-4\text{ m}}{\sqrt{(4\text{ m})^2 + (6\text{ m})^2 + (9\text{ m})^2}} = -0.347$$

The results are summarized in the following equations:

$$\mathbf{e}_{OA} = -0.347\mathbf{e}_x - 0.520\mathbf{e}_y + 0.780\mathbf{e}_z$$
$$\mathbf{e}_{OB} = 0.543\mathbf{e}_x - 0.466\mathbf{e}_y + 0.699\mathbf{e}_z$$
$$\mathbf{e}_{OC} = 0.196\mathbf{e}_x + 0.784\mathbf{e}_y + 0.588\mathbf{e}_z$$

Force equilibrium requires that

$$P\mathbf{e}_{OA} + Q\mathbf{e}_{OB} + R\mathbf{e}_{OC} - f_g\mathbf{e}_z = 0$$

Dot-multiplying this equation by $\mathbf{e}_x$, $\mathbf{e}_y$, and $\mathbf{e}_z$ yields the following relations:

$$-0.347P + 0.543Q + 0.196R = 0$$
$$-0.520P - 0.466Q + 0.784R = 0$$
$$0.780P + 0.699Q + 0.588R = f_g$$

which can be solved for the desired forces:

$$P = 0.660f_g$$
$$Q = 0.217f_g$$
$$R = 0.567f_g$$

TRUSSES

A **truss** is a structure that is built with interconnected axial force members. Each such member is a straight rod that can transmit force along its axis. This limitation is the result of interconnections that are all of the ball-and-socket type; that is, they constrain the end points of the connected members against relative position change but allow the members complete freedom to rotate about the connection point. Also, external forces are applied only at these joints.

The symbol T_{IJ} will be used here to denote the tensile force in the member that connects joints I and J. This means that if T_{IJ} takes on a positive value, the member is in tension, and if T_{IJ} takes on a negative value, the member is in compression.

Equations from Joints

One approach to determining the forces in individual members within a truss is to isolate, as a free-body, the portion of the truss in the neighborhood of each joint and write equations of force equilibrium for each. This procedure is illustrated for the truss shown in Fig. 7.13.

First, a free-body is drawn for the joint G (Fig. 7.14), and the corresponding force equilibrium equation is written:

$$T_{GE}\mathbf{e}_{GE} + T_{GF}\mathbf{e}_{GF} - (50\text{ kN})\mathbf{e}_y = 0$$

To evaluate T_{GE}, the equation may be dot-multiplied by $\mathbf{e}_a$, which is defined to be perpendicular to the other unknown force, as shown in Fig. 7.14:

$$\mathbf{e}_a \bullet \mathbf{e}_{GE}T_{GE} + \mathbf{e}_a \bullet \mathbf{e}_{GF}T_{GF} - \mathbf{e}_a \bullet \mathbf{e}_y(50\text{ kN}) = 0$$

$$\left[\left(\frac{1}{\sqrt{17}}\right)\left(-\frac{4}{\sqrt{5}}\right) + \left(\frac{4}{\sqrt{17}}\right)\left(\frac{3}{5}\right)\right]T_{GE} - \frac{4}{\sqrt{17}}(50\text{ kN}) = 0$$

$$T_{GE} = 125\text{ kN}$$

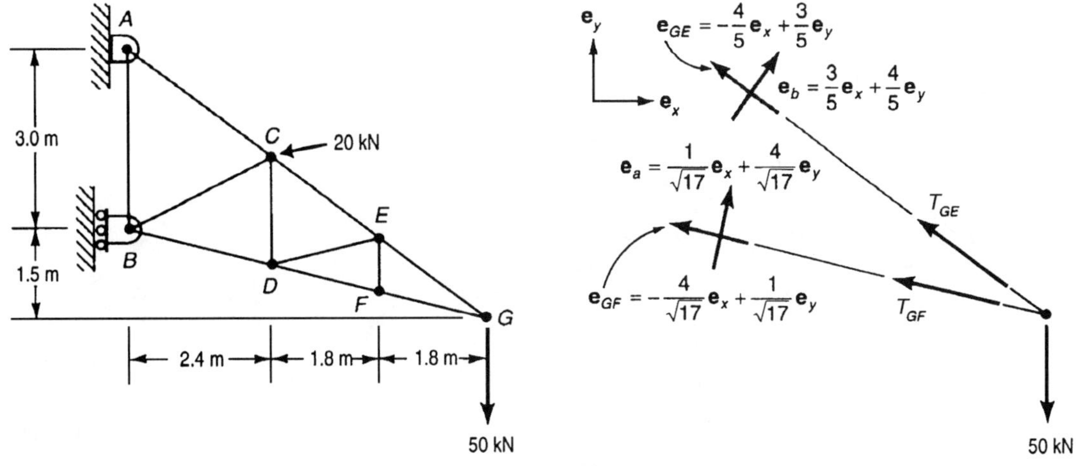

Figure 7.13

Figure 7.14

Dot-multiplication by $\mathbf{e}_b$ will similarly yield the value of T_{GF}:

$$\mathbf{e}_b \bullet \mathbf{e}_{GE} T_{GE} + \mathbf{e}_b \bullet \mathbf{e}_{GF} T_{GF} - \mathbf{e}_b \bullet \mathbf{e}_y (50 \text{ kN}) = 0$$

$$\left[\left(\frac{3}{5}\right)\left(-\frac{4}{\sqrt{17}}\right) + \left(\frac{4}{5}\right)\left(\frac{1}{\sqrt{17}}\right)\right] T_{GF} - \frac{4}{5}(50 \text{ kN}) = 0$$

$$T_{GF} = -103.1 \text{ kN}$$

Next, consider a free-body diagram of the neighborhood of joint F (Fig. 7.15). Considering projections of forces perpendicular to the line DFG, it becomes evident without writing equations that $T_{FE} = 0$. In view of this, and considering forces parallel to DFG, it becomes evident that $T_{FD} = T_{FG} = -103.1$ kN.

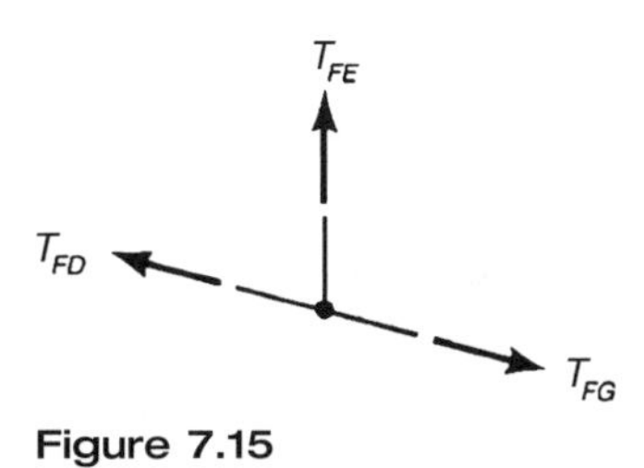

Figure 7.15

We can next proceed to joint E, where T_{ED} and T_{EC} are now the only unknowns. Once these forces are evaluated, the equilibrium equations for joint D contain only two unknowns. Proceeding in this manner, we can evaluate the remainder of the internal forces and the reactions at the supports.

By proper selection of the order in which joints of the truss are considered, it is usually possible to work through a planar truss in the manner indicated. If the values of all the forces are not required, however, use of the equations from joints may be much less efficient than the approach explained next.

Equations from Sections

The forces external to *any* portion of a system in equilibrium have zero resultants of force and moment. This is the basis for the procedure illustrated now. Fig. 7.16

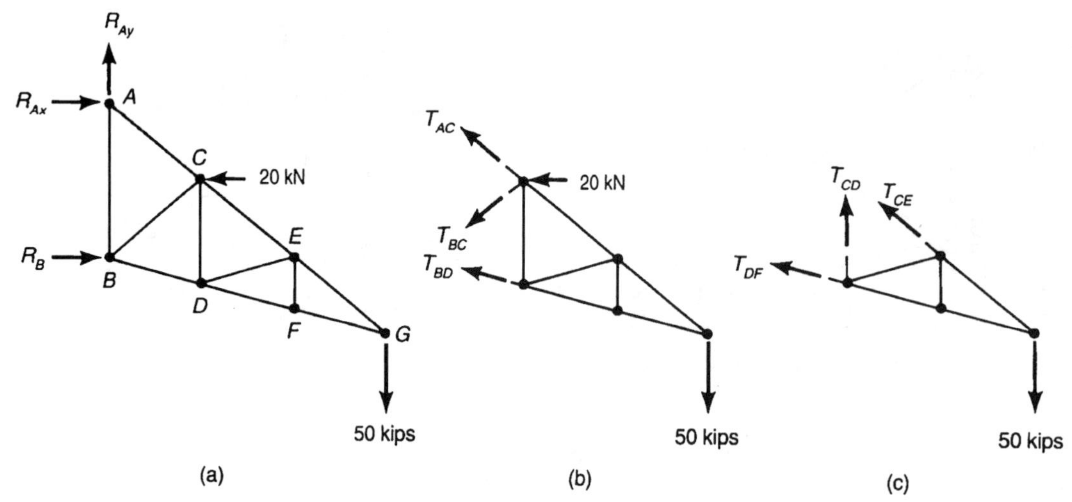

Figure 7.16

shows free-body diagrams of three different portions of the truss in Fig. 7.13. Referring to the free-body of the entire structure, Fig. 7.16(a), we can write the resultant moment about A as

$$(3.0\text{ m})R_B - (1.8\text{ m})(20\text{ kN}) - (6.0\text{ m})(50\text{ kN}) = 0$$

from which

$$R_B = 112\text{ kN}$$

Then summation of horizontal force components gives

$$R_{Ax} = -92\text{ kN}$$

and summation of vertical force components gives

$$R_{Ay} = 50\text{ kN}$$

Next, for the free-body shown in Fig. 7.16(b), we can sum moments about point B,

$$(3.0\text{ m})\frac{4}{5}T_{AC} + (1.2\text{ m})(20\text{ kN}) - (6.0\text{ m})(50\text{ kN}) = 0$$

to obtain

$$T_{AC} = 115\text{ kN}$$

Then we can sum moments about point G,

$$(3.6\text{ m})\left(\frac{1}{\sqrt{5}}T_{BC}\right) + (2.7\text{ m})\left(\frac{2}{\sqrt{5}}T_{BC} + 20\text{ kN}\right) = 0$$

to obtain

$$T_{BC} = 13.42\text{ kN}$$

and sum moments about point C,

$$-(1.8\text{ m})\frac{4}{\sqrt{17}}T_{BD} - (3.6\text{ m})(50\text{ kN}) = 0$$

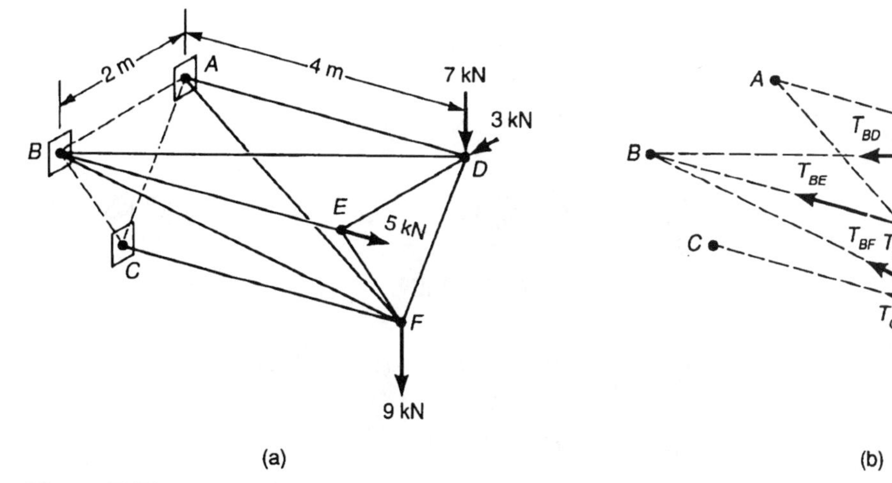

Figure 7.17

to obtain

$$T_{BD} = -103.1 \text{ kN}$$

Now, considering the free-body in Fig. 7.16(c), we can obtain T_{CD} by summing moments about point G:

$$T_{CD} = 0$$

By continuing in this fashion, we can evaluate the remaining forces in a fairly efficient manner.

The strategy in this method is to isolate a portion of the structure, with the boundary passing through the member in which a force is to be evaluated, and, with the free-body completed, to find a direction for force reckoning or an axis for moment reckoning that will yield an equation with as few unknown forces as possible.

For further illustration, consider the truss shown in Fig. 7.17(a). The vertical triangles ABC and DEF are both equilateral, and the rectangle $ABED$ is in a horizontal plane. The force in each of the nine members is to be evaluated.

Consider first the free-body of the portion of the truss shown in Fig. 7.17(b). To analyze this section, we select various axes, about which moments involve only one unknown force. The moment about each axis is evaluated most readily by resolving each force as indicated in Fig. 7.17, keeping in mind that the moment of a force about an axis is zero if the line of action intersects the axis or is parallel to the axis:

$$M_{DF} = \left(\sqrt{3} \text{ m}\right)(T_{BE} - 5 \text{ kN}) = 0$$

$$T_{BE} = 5 \text{ kN}$$

$$M_{CF} = \left(\sqrt{3} \text{ m}\right)\left(\frac{2}{\sqrt{20}} T_{BD} + 3 \text{ kN}\right) - (1 \text{ m})(7 \text{ kN}) = 0$$

$$T_{BD} = \sqrt{5}\left(\frac{7}{\sqrt{3}} - 3\right) \text{kN} = 2.33 \text{ kN}$$

$$M_{FE} = \left(\sqrt{3}\text{ m}\right)\left(T_{AD} + \frac{4}{\sqrt{20}}T_{BD}\right) = 0$$

$$T_{AD} = -2\left(\frac{7}{\sqrt{3}} - 3\right)\text{kN} = 2.08\text{ kN}$$

$$M_{BE} = \left(\sqrt{3}\text{ m}\right)\left(\frac{2}{\sqrt{20}}T_{AF}\right) - (1\text{ m})(9\text{ kN}) - (2\text{ m})(7\text{ kN}) = 0$$

$$T_{AF} = 23\sqrt{\frac{5}{3}}\text{ kN} = 29.7\text{ kN}$$

$$M_{DA} = \left(\sqrt{3}\text{ m}\right)\left(\frac{2}{\sqrt{20}}T_{BF}\right) - (1\text{ m})(9\text{ kN}) = 0$$

$$T_{BF} = 9\sqrt{\frac{5}{3}}\text{ kN} = 11.62\text{ kN}$$

$$M_{BA} = \left(\sqrt{3}\text{ m}\right)T_{CF} + (4\text{ m})(7\text{ kN}) + (4\text{ m})(9\text{ kN}) = 0$$

$$T_{CF} = -\frac{64}{\sqrt{3}}\text{ kN} = -37\text{ kN}$$

To determine the forces in the remaining three members, it is a straightforward matter to isolate joints D and E and sum forces:

$$T_{DF} = -\frac{14}{\sqrt{3}}\text{ kN} = -8.08\text{ kN}$$

$$T_{DE} = T_{EF} = 0$$

As with planar trusses, three-dimensional trusses can be analyzed by writing equations of force equilibrium for each joint, or by considering a larger portion of the truss. The procedure for writing equilibrium equations for a joint is illustrated in the previous section and typically leads to a set of simultaneous equations for the unknown forces. Often, as in the preceding example, the task of solving simultaneous equations can be avoided by considering an entire section of the truss and finding axes about which only one unknown force has a nonzero moment.

COUPLE-SUPPORTING MEMBERS

The loads that a rigid bar can carry are not limited to axial forces. If lateral forces are applied, equilibrium requires that forces across a section have a nonzero moment about the center of the section, as demonstrated in Fig. 7.18. This figure shows a sketch and free-body diagrams: one of the entire and others of two separated portions of the beam. Equilibrium of the portion on the left indicates at once that the forces across the plane of separation must form a lateral force and a couple. The detailed distribution of interaction forces may be fairly complicated and cannot be deduced from equilibrium alone; however, it is often possible to evaluate the resultant force and the moment of the equivalent couple from statics.

Twisting and Bending Moments

In general, the moment of the couple at a section can have any direction, depending on how the external loads are applied. For example, the bracket in Fig. 7.19 has moment components at section A in each of three directions aligned with the axis

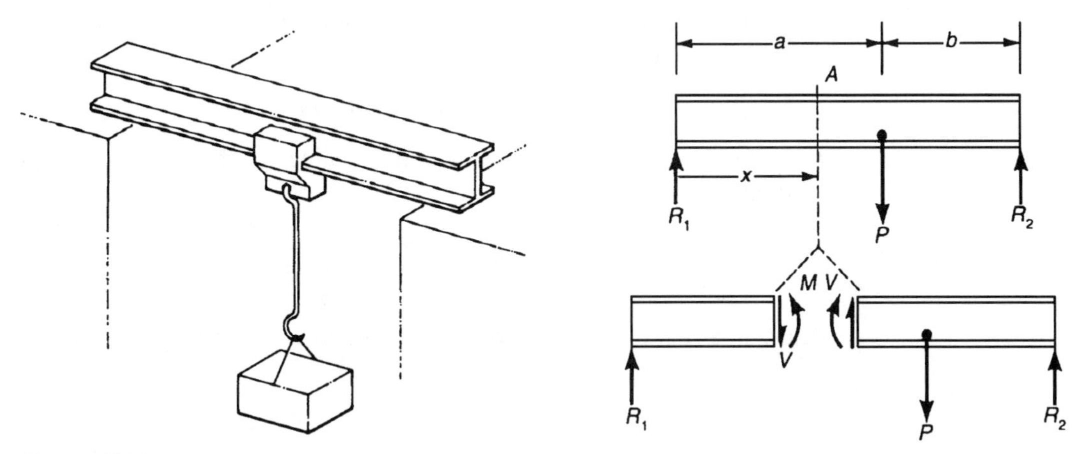

Figure 7.18

of the bracket. The moment vector is usually resolved into a component parallel to the axis of the bar as well as one or two components perpendicular to the axis. (This facilitates the analysis of strength and deformation of the bar.) The component of moment parallel to the axis of the bar is called the **twisting moment**, and the components perpendicular to the axis are called **bending moments**, after the types of deformation they produce, as shown in Fig. 7.20.

The resultant force acting at a section of a bar is similarly resolved. The component parallel to the axis of the rod is called the **axial force**, and the components perpendicular to the axis are called **shearing forces**.

Evaluation of these force and moment components is accomplished by using the same basic ideas already examined: A free-body of a portion of the member on either side of the section of interest is isolated and properly labeled, and equations of equilibrium are written and solved.

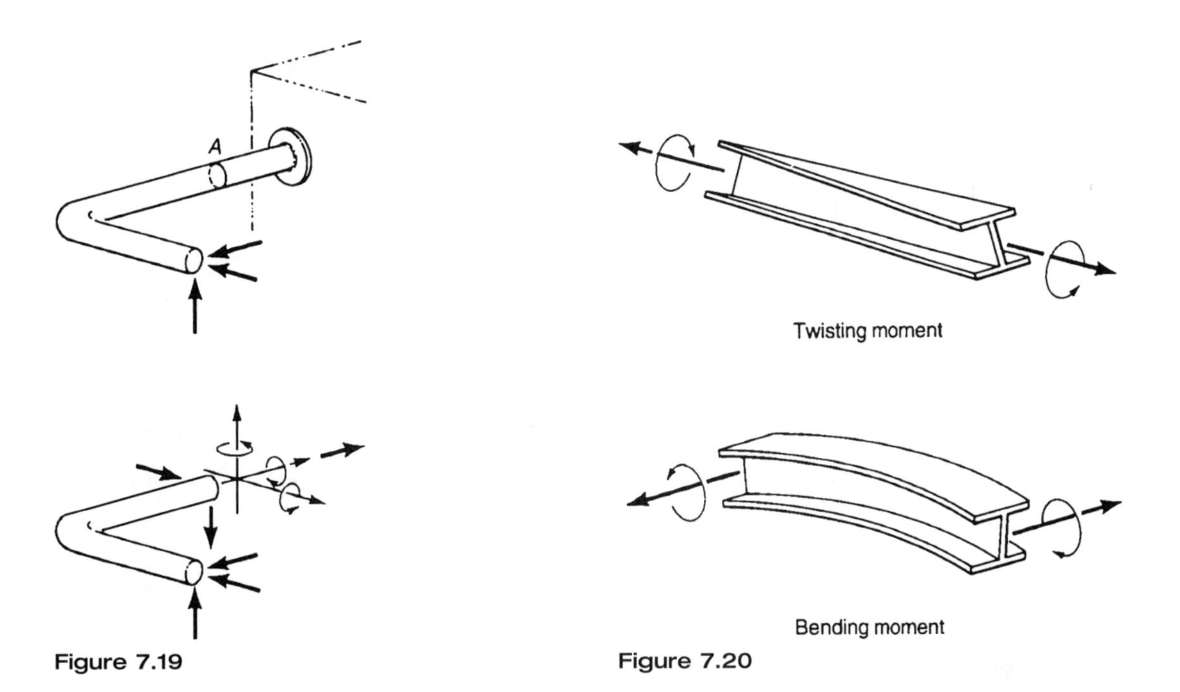

Figure 7.19

Figure 7.20

For example, equilibrium of the portion of the beam to the left of section A in Fig. 7.18 yields

$$V = R_1 \quad \text{and} \quad M = R_1 x$$

and equilibrium of the entire beam gives values of the support reactions in terms of their applied load P as

$$R_1 = \frac{bP}{a+b} \qquad R_2 = \frac{aP}{a+b}$$

Thus, in terms of the applied load, the shear and bending moment are

$$V = \frac{bP}{a+b} \qquad x < a$$

$$M = \frac{bPx}{a+b} \qquad x < a$$

The qualification $x < a$ is necessary because the analysis was done for sections to the left of the applied load. A similar analysis for sections on the other side of the load results in

$$V = \frac{-aP}{a+b} \qquad x > a$$

$$M = \frac{a(a+b-x)P}{a+b} \qquad x > a$$

As another example, consider the force and moment components at the support O for the automobile torsion bar in Fig. 7.21. Force equilibrium of the free-body requires that

$$R + (6.8 \text{ kN})\left(-\frac{8}{17}\mathbf{e}_x + \frac{15}{17}\mathbf{e}_z\right) + (2.5 \text{ kN})\mathbf{e}_y = 0$$

or

$$R = (3.2\mathbf{e}_x - 2.5\mathbf{e}_y - 6.0\mathbf{e}_z) \text{ kN}$$

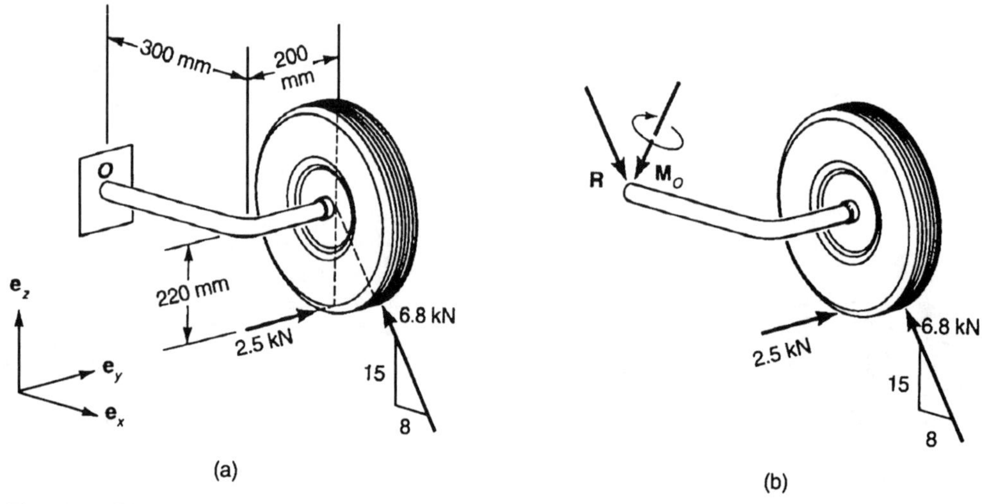

Figure 7.21

Therefore, there is a compressive axial force of 3.2 kN and a resultant shearing force of $\sqrt{(2.5)^2 + (6.0)^2}$ kN = 6.5 kN at the section O. Next, moment equilibrium requires that

$$\mathbf{M}_O + [(0.3 \text{ m})\mathbf{e}_x + (0.2 \text{ m})\mathbf{e}_y] \times [(6.0 \text{ kN})\mathbf{e}_z - (3.2 \text{ kN})\mathbf{e}_x] \\ + [(0.3 \text{ m})\mathbf{e}_x + (0.2 \text{ m})\mathbf{e}_y - (0.22 \text{ m})\mathbf{e}_z] \times (2.5 \text{ kN})\mathbf{e}_y = 0$$

or

$$\mathbf{M}_O = (-1.75\mathbf{e}_x + 1.80\mathbf{e}_y - 1.39\mathbf{e}_z) \text{ kN}\bullet\text{m}$$

This indicates the presence of a twisting moment

$$\mathbf{M}_t = 1.75 \text{ kN}\bullet\text{m}$$

and a bending moment resultant

$$M_b = \sqrt{(1.80)^2 + (1.39)^2} \text{ kN}\bullet\text{m} \\ = 2.27 \text{ kN}\bullet\text{m}$$

SYSTEMS WITH FRICTION

Friction forces act tangentially to the surfaces on which two objects make contact. The ratio of the tangential force to the normal force at the contact surfaces is called the **coefficient of friction**. In general, this ratio depends on several variables, such as the surface materials, the surface finishes, the presence of any surface films, the velocity of sliding, and the temperature. The ratio of tangential force necessary to initiate sliding from a state of rest to the normal force is called the **coefficient of static friction**, whereas the force ratio as sliding continues is called the **coefficient of sliding friction**. Typically, the coefficient of static friction is somewhat greater than the coefficient of sliding friction. Furthermore, a decrease in sliding friction force with an increase in the speed of sliding has been observed for some materials, although this variation is usually small enough that it can be neglected.

In spite of the complexity of the mechanism of friction (Fig. 7.22), the approximation known as *Coulomb friction* has been found to lead to predictions

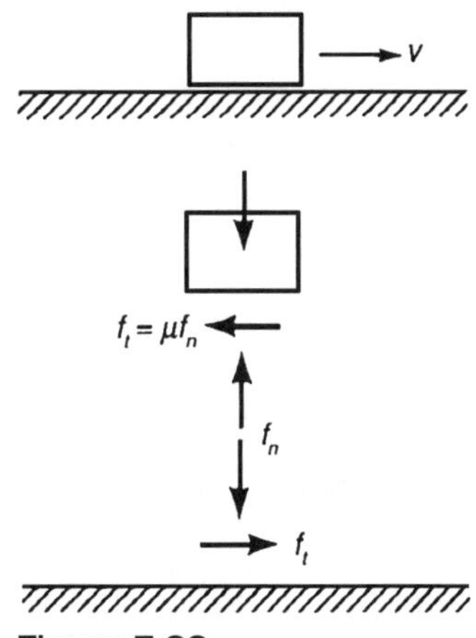

Figure 7.22

of acceptable accuracy for many dry surfaces. The so-called **Coulomb's Law of Friction** states that, whenever sliding takes place, the tangential component of force between the surfaces is proportional to the normal component and acts in a direction to oppose the motion; that is,

$$\mathbf{f}_t = \mu_1 f_n \mathbf{e}_v \qquad v \neq 0 \tag{7.14a}$$

in which $\mathbf{e}_v$ is a unit vector in the direction on the relative velocity, v, of the objects on which $\mathbf{f}_t$ acts, and the coefficient of sliding friction μ_1 depends on the surfaces in contact but not on the magnitude of the normal force or on the velocity. When the surfaces are not sliding, the friction force can have any direction required for equilibrium, but its magnitude is limited by

$$f_t \leq \mu_0 f_n \qquad v = 0 \tag{7.14b}$$

in which μ_0 is the coefficient of static friction. When applied forces induce a friction force that reaches this limit, motion is incipient, meaning that any change tending to increase the friction force will cause acceleration.

Example 7.10

What is the magnitude P of the force required to move the 40-kg block up the 15° incline in Exhibit 9? Also, in the absence of P, will the block remain stationary on the incline? The coefficients of static and sliding friction are both 0.3.

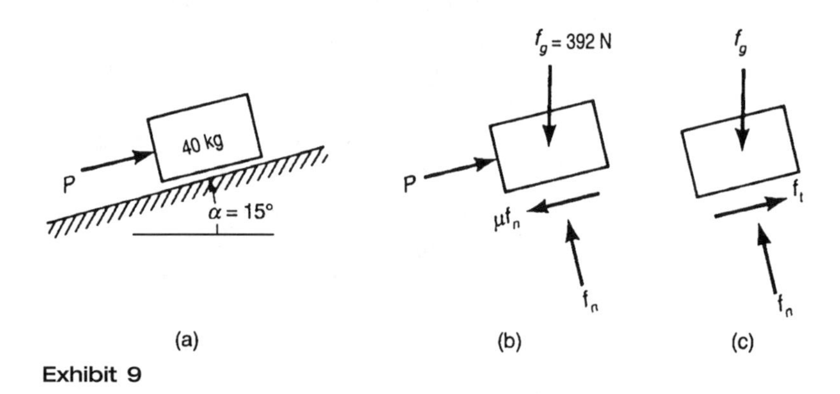

Exhibit 9

Solution

Referring to the free-body diagram (Exhibit 9(b)), we can write equations of equilibrium in the direction normal to the incline,

$$f_n - f_g \cos\ \alpha = 0$$

and in the direction along the incline,

$$P - f_g \sin\ \alpha - \mu f_n = 0$$

Elimination of f_n from these two equations results in

$$\begin{aligned} P &= f_g(\sin\alpha + \mu\ \cos\alpha) \\ &= (392\text{ N})(\sin 15° + 0.30\ \cos 15°) \\ &= 215\text{ N} \end{aligned}$$

The free-body diagram in Exhibit 9(c) depicts the situation in the absence of the force *P*. Observe the reversal of the direction of the friction force. Now, if the block is to remain at rest,

$$f_n = f_g \cos\alpha$$
$$f_t = f_g \sin\alpha$$

But the friction force is limited by

$$f_t \le \mu f_n$$

Substitution of the equilibrium equations into this inequality gives

$$f_g \sin\alpha \le \mu f_g \cos\alpha$$

or

$$\tan\alpha \le \mu$$

Because tan $\alpha = 0.27$, which is less than $\mu = 0.3$, the block will remain at rest.

Example 7.11

In the absence of *P*, the angle of the incline in the previous example is slowly increased until the block begins to slide downward. If the coefficient of static friction is $\mu_0 = 0.47$ and the coefficient of sliding friction is $\mu_1 = 0.44$, what will be the acceleration of the block after it breaks loose?

Solution

Prior to breakaway of the block,

$$f_t \le \mu_0 f_n$$

Or, with the equilibrium relationships from the free-body diagram (Exhibit 9(c)),

$$f_g \sin\alpha \le \mu_0 f_g \cos\alpha$$

The equality occurs when the critical angle, α_c, is reached:

$$\tan\alpha_c = \mu_0 = 0.47$$

from which

$$\alpha_c = 25.2°$$

After the block breaks loose,

$$f_t = \mu_1 f_n$$

Now, the component of acceleration perpendicular to the inclined plane remains zero, so that

$$f_n - f_g \cos\alpha_c = 0$$

But, in the direction parallel to the plane, the static equilibrium relationship must be replaced with Newton's second law,

$$f_g \sin\alpha_c - f_t = ma$$

in which a is the downward tangential acceleration. Combining these relationships leads to

$$a = \frac{f_g}{m}(\sin\alpha_c - \mu_1 \cos\alpha_c)$$

But, since $f_g = mg$,

$$a = g\sin\alpha_c\left(1 - \frac{\mu}{\tan\alpha_c}\right) = g\sin\alpha_c\left(1 - \frac{\mu_1}{\mu_2}\right) = (9.81\ \text{m/s}^2)\sin 25.2°\left(1 - \frac{0.44}{0.47}\right)$$
$$= 0.27\ \text{m/s}^2$$

DISTRIBUTED FORCES

When forces are distributed throughout some region (as in the case of gravity) or over a surface (as in the case of pressure on the wall of a water tank), the fundamental ideas illustrated previously apply. A feature not yet illustrated, however, is the computational detail of summation, which takes the form of integration.

Example 7.12

The beam in Exhibit 10 supports a load that varies in intensity along the length as indicated. The intensity (force per unit length of beam) has the values w_A and w_B at the two ends and varies linearly between these points.

In terms of w_A, w_B, and L, what is the resultant of the downward forces, and what are the magnitudes R_A and R_B of the reactions at the supports?

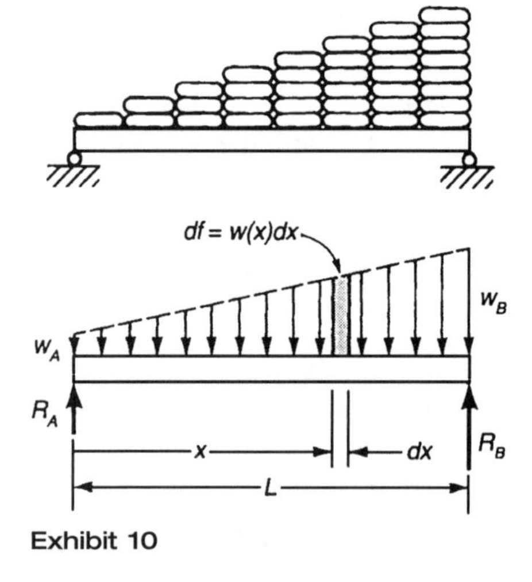

Exhibit 10

Solution

First, we write an expression for the load intensity as a function x, the distance along the span measured from the left-hand support:

$$w(x) = w_A + (w_B - w_A)\frac{x}{L}$$

Next, consider the force in the shaded portion of the load diagram, acting between the points given by x and $x + dx$. The force in this region will be the product of the intensity (force per unit length) and the length dx:

$$df = w(x)\,dx$$

Summing all such forces gives their resultant:

$$
\begin{aligned}
f &= \int w(x)\, dx \\
&= \int_0^L \left[w_A + (w_B - w_A)\frac{x}{L} \right] dx \\
&= \frac{1}{2}(w_A + w_B)L
\end{aligned}
$$

To evaluate R_B, we equate to zero the sum of moments about A of all the forces:

$$
\begin{aligned}
M_A &= R_B L - \int x\, dx \\
&= R_B L - \int_0^L x \left[w_A + (w_B - w_A)\frac{x}{L} \right] dx \\
&= R_B L - \frac{1}{6}(w_A + 2w_B)L^2 = 0
\end{aligned}
$$

This yields

$$
R_B = \frac{1}{6}(w_A + 2w_B)L
$$

To evaluate R_A, we can use the fact that the sum of all vertical forces must be zero:

$$
\begin{aligned}
R_A &= f - R_B \\
&= \frac{1}{2}(w_A + w_B)L - \frac{1}{6}(w_A + 2w_B)L \\
&= \frac{1}{6}(2w_A + w_B)
\end{aligned}
$$

Example 7.13

The uniform, slender, semicircular arch in Exhibit 11 is acted on by gravity and the reactions from the supports.

The free-body diagram shows the desired bending moment, M_b, as the reaction from the other half of the arch. Because of horizontal equilibrium, no axial force exists at this section. The vertical shearing force is also zero at this section because Newton's Third Law would require a shearing force in the opposite direction on the other half of the arch, and this pair of forces would be inconsistent with the symmetry of the system.

Because the arch is slender, the forces of gravity may be treated as distributed along a circular *line*. Let the cross-sectional area be denoted by A and the density (mass per unit volume) by ρ. Then the volume of the shaded element of the arch will be equal to $Aa\, d\theta$, and the magnitude of the force of gravity acting on it will be

$$
df_g = \rho(Aa\, d\theta)g
$$

The resultant of the gravitational forces then has the magnitude

$$
\begin{aligned}
f_g &= \int_0^{\pi/2} \rho A g a\, d\theta \\
&= \frac{1}{2}\pi a \rho A g
\end{aligned}
$$

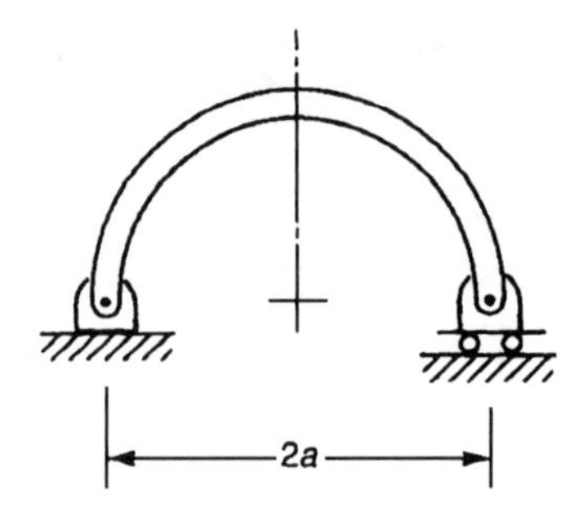

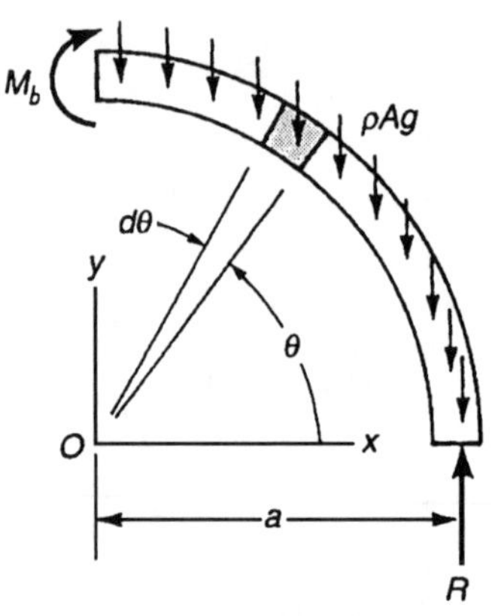

Exhibit 11

and vertical force equilibrium gives the support reaction as

$$R = \frac{1}{2}\pi a \rho A g$$

Now, the resultant moment of the forces of gravity about the point O will be in the clockwise direction and of magnitude

$$\begin{aligned} M_{Og} &= \int a\cos\theta \, df_g \\ &= \int_0^{\pi/2} a\cos\theta \, \rho A g a \, d\theta \\ &= a^2 \rho A g \end{aligned}$$

Finally, moment equilibrium about point O requires that

$$aR - M_{Og} - M_b = 0$$

which, together with the above results, gives

$$M_b = \left(\frac{\pi}{2} - 1\right) a^2 \rho A g$$

In an arch with a large radius, this bending moment could well cause failure of the structure.

Single Force Equivalents

The construction and evaluation of integrals such as those in Examples 7.12 and 7.13 can be circumvented if knowledge of an equivalent discrete force is available. This information, for a variety of special cases, is available in tabulated formulas, these formulas having been determined by integration. Proper use of such formulas requires an understanding of the following concepts.

Center of Mass and Center of Gravity

Consider the forces of gravity distributed throughout an arbitrary body, as shown in Fig. 7.23. The force acting on the element with mass dm will be

$$d\mathbf{f}_g = dm\ g\mathbf{e}_g$$

and the resultant force will be given by

$$\begin{aligned}\mathbf{f}_g &= \int d\mathbf{f}_g \\ &= \int dm\ g\mathbf{e}_g \\ &= \left(\int dm\right) g\mathbf{e}_g = mg\mathbf{e}_g\end{aligned}$$

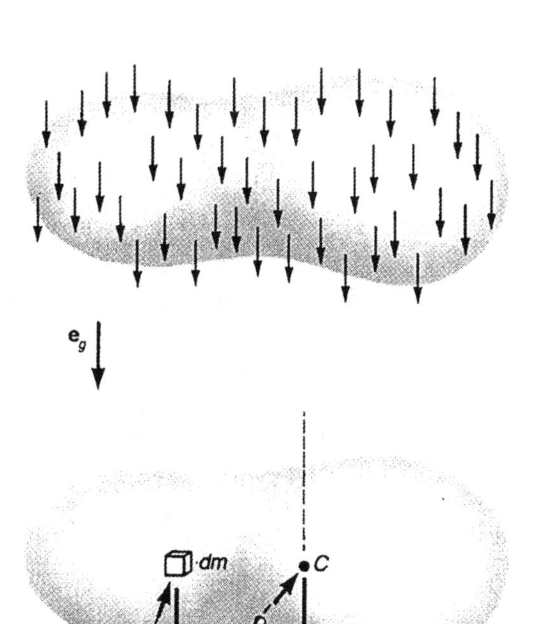

Figure 7.23

where m is the total mass of the body. The resultant moment about a point O is given by

$$\begin{aligned} M_{Og} &= \int \mathbf{r} \times d\mathbf{f}_g \\ &= \int \mathbf{r} \times (g\mathbf{e}_g \; dm) \\ &= \left(\int \mathbf{r} \; dm \right) \times g\mathbf{e}_g \end{aligned}$$

where $\mathbf{r}$ is the position vector locating the mass element. Now, to compose the forces into a single, equivalent resultant, the line of action of the equivalent force $\mathbf{f}_g$ must pass through a point located by the position vector $\mathbf{r}_f$, satisfying the moment equivalence:

$$\mathbf{r}_f \times \mathbf{f}_g = \mathbf{M}_{Og}$$

Thus

$$\begin{aligned} \mathbf{r}_f \times (mg\mathbf{e}_g) &= \left(\int \mathbf{r} \; dm \right) \times g\mathbf{e}_g \\ \mathbf{r}_f \times \mathbf{e}_g &= \bar{\mathbf{r}} \times \mathbf{e}_g \end{aligned}$$

where

$$\bar{\mathbf{r}} = \frac{1}{m} \int \mathbf{r} \; dm \qquad \textbf{(7.15)}$$

The vector $\bar{\mathbf{r}}$ locates an important point C, called the **center of mass** of the body. By the choice of $\mathbf{r}_f = \bar{\mathbf{r}}$, moment equivalence will be satisfied for any $\mathbf{e}_g$, which implies that the line of action of the equivalent force passes through C regardless of how the body is oriented.

The center of mass C is of fundamental importance in the study of dynamics. In statics its significance stems from the fact that the resultant moment about C, of the forces of uniform gravity, is zero. That is, the body could be statically balanced by supporting it at this point only. For this reason it is also called the center of gravity.

Example 7.14

Find the location of the center for gravity for the portion of the arch isolated as a free-body in Example 7.13.

Solution

Using the center of the circle as a reference point, we can locate the shaded mass element with the position vector

$$\mathbf{r} = a \cos\theta \, \mathbf{e}_x + a \sin\theta \, \mathbf{e}_y$$

Then, from the definition for the center of mass, Eq. (7.15),

$$\bar{\mathbf{r}} = \frac{1}{m}\int (a\cos\theta\ \mathbf{e}_x + a\sin\theta\ \mathbf{e}_y)(\rho A a\ d\theta)$$

$$= \frac{\rho A a^2}{m}\int_0^{\pi/2} (\cos\theta\ \mathbf{e}_x + \sin\theta\ \mathbf{e}_y) d\theta$$

$$= \frac{2a}{\pi}(\mathbf{e}_x + \mathbf{e}_y)$$

Centroids

The mass, dm, of the element used in the preceding integrals can be expressed in terms of the density, ρ, and the corresponding element of volume, dV, as $dm = \rho\ dV$. Then the position vector locating the center of mass can be written as

$$\bar{\mathbf{r}} = \frac{\int \mathbf{r}\rho\ dV}{\int \rho\ dV}$$

Now, if the density is uniform throughout the body, ρ can be brought outside the integrals with result

$$\bar{\mathbf{r}} = \frac{1}{V}\int \mathbf{r}\ dV$$

The location of the point C here depends entirely on geometry, because all contributions having to do with material have been canceled. The point C, located according to this equation, is called the **centroid of the volume** V. Similarly, the **centroid of a surface area** A is defined as the point located by the position vector

$$\bar{\mathbf{r}} = \frac{1}{A}\int \mathbf{r}\ dA$$

and the **centroid of a line segment** of length L is defined as the point located by the position vector

$$\bar{\mathbf{r}} = \frac{1}{L}\int \mathbf{r}\ dL$$

The calculation in the Example 14 was for a uniform mass per unit length of the arch. With this density canceled out, the center of mass of the arch segment is also the centroid of a quarter-segment of a circular line.

The integrals $\int \mathbf{r}\ dV, \int \mathbf{r}\ dA$, and $\int \mathbf{r}\ dL$ are called the first moments of the volume, area, and line, respectively, about the reference point O. In carrying out the calculations, it is often convenient to work with one rectangular Cartesian component of the position vector at a time, that is, to evaluate separately the component equivalents obtained by dot-multiplying the vector definition by a unit vector in each direction:

$$\bar{x} = \mathbf{e}_x \bullet \bar{\mathbf{r}} = \frac{1}{V}\int x\, dV$$

$$\bar{y} = \mathbf{e}_y \bullet \bar{\mathbf{r}} = \frac{1}{V}\int y\, dV$$

$$\bar{z} = \mathbf{e}_z \bullet \bar{\mathbf{r}} = \frac{1}{V}\int z\, dV$$

Example 7.15

Determine the location of the centroid of the shaded triangle of Exhibit 12 in terms of the dimensions a, b, and c.

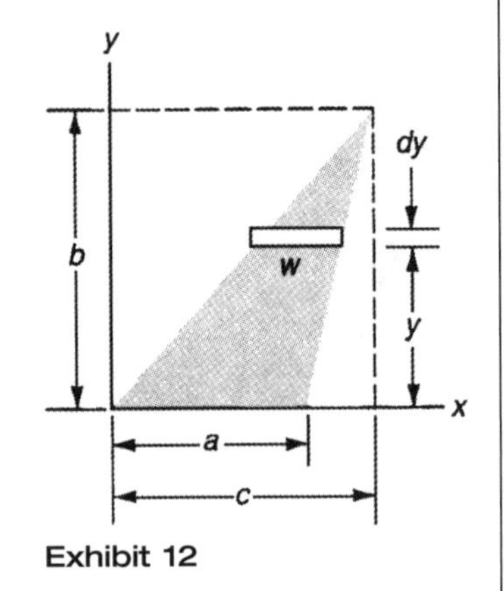

Exhibit 12

Solution

The area and first moment of area can be computed by evaluating the contribution to these quantities from the unshaded element in Exhibit 12 and summing these contributions by integration with respect to y. The width of the element can be expressed in terms of y after observing that the entire triangle is similar to the triangle that lies above the unshaded element:

$$\frac{w}{b-y} = \frac{a}{b}$$

or

$$w = a\left(1 - \frac{y}{b}\right)$$

Thus the area of the element is

$$dA = w\, dy = a\left(1 - \frac{y}{b}\right)dy$$

and the area of the triangle is

$$A = \int_0^b a\left(1 - \frac{y}{b}\right)dy = \frac{1}{2}ab$$

This result is very well known. The center of the element is located on the straight line that connects the apex of the triangle with the midpoint of the base, that is, on the line that has the equation

$$x = \frac{a}{2} + \left(c - \frac{a}{2}\right)\frac{y}{b}$$

Therefore, the x-component of the first moment of area of the triangle is

$$\int x\, dA = \int_0^b \left[\frac{a}{2} + \left(c - \frac{a}{2}\right)\frac{y}{b}\right] a\left(1 - \frac{y}{b}\right)dy = \frac{1}{6}ab(a+c)$$

With this and the above value of area, the x-coordinate of the centroid is determined as

$$\bar{x} = \frac{1}{A}\int x\,dA = \frac{ab(a+c)/6}{ab/2} = \frac{1}{3}(a+c)$$

A similar calculation yields the y-component of the first moment of area as

$$\int y\,dA = \int_0^b ya\left(1-\frac{y}{b}\right)dy = \frac{ab^2}{6}$$

from which we find the y-coordinate of the centroid to be

$$\bar{y} = \frac{1}{A}\int y\,dA = \frac{ab^2/6}{ab/2} = \frac{1}{3}b$$

A composite volume, area, or line may be built up from several parts, for which the location of each centroid is known. In this case, a summation having the same form as the integral definition can readily be shown as valid. In the case of an area, A, that is a composite formed from n areas $A_1, A_2, \ldots, A_n$ and having centroids located by $\bar{\mathbf{r}}_1, \bar{\mathbf{r}}_2, \ldots, \mathbf{r}_n$, the centroid of the composite is located at

$$\bar{\mathbf{r}} = \frac{1}{A}\sum_{i=1}^{n} A_i\bar{\mathbf{r}}_i$$

where

$$A = \sum_{i=1}^{n} A_i$$

Example 7.16

Locate the centroid of the plane area shown shaded in Exhibit 13.

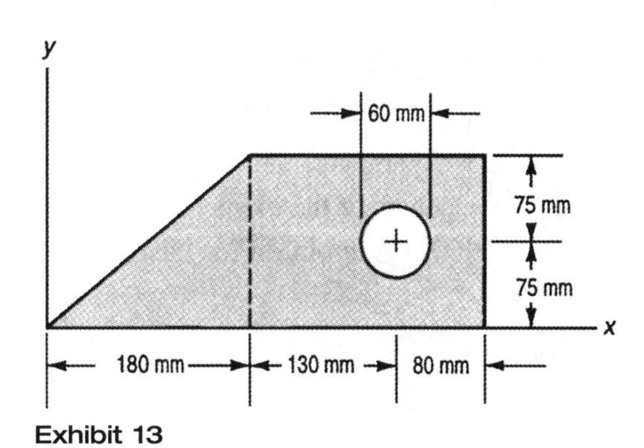

Exhibit 13

Solution

The areas and coordinates of centroids of individual parts are as follows:

	A, m^2	$\bar{x}$, mm	$\bar{y}$, mm
Triangle	0.0135	120	50
Rectangle	0.0315	285	75
Circle	−0.00283	310	75

The coordinates locating the centroid are then

$$\bar{x} = \frac{(120)(0.0135) + (285)(0.0315) + (310)(-0.00283)}{0.0135 + 0.0315 - 0.00283} = 231 \text{ mm}$$

$$\bar{x} = \frac{(50)(0.0135) + (75)(0.0315) + (75)(-0.00283)}{0.0135 + 0.0315 - 0.00283} = 67 \text{ mm}$$

Second Moments of Area

The geometric properties defined in the previous two sections are associated with uniformly distributed forces. Forces that vary *linearly* over a plane area lead to additional geometric properties that are useful in evaluating moment resultants. One example is fluid pressure acting on submerged, plane surfaces; another is the bending stress induced in beams. The latter is covered in the chapter on mechanics of materials. Consider the flat surfaces shown in Fig. 7.24 with a force intensity that varies linearly with the distance from the line Op. With the force per unit area denoted as σ_z and the distance from Op as q, the linear variation is expressed as

$$\sigma_z = kq$$

where k is a proportionality constant. (In the analysis of beam bending, this constant is the product of Young's modulus and the curvature of the deformed

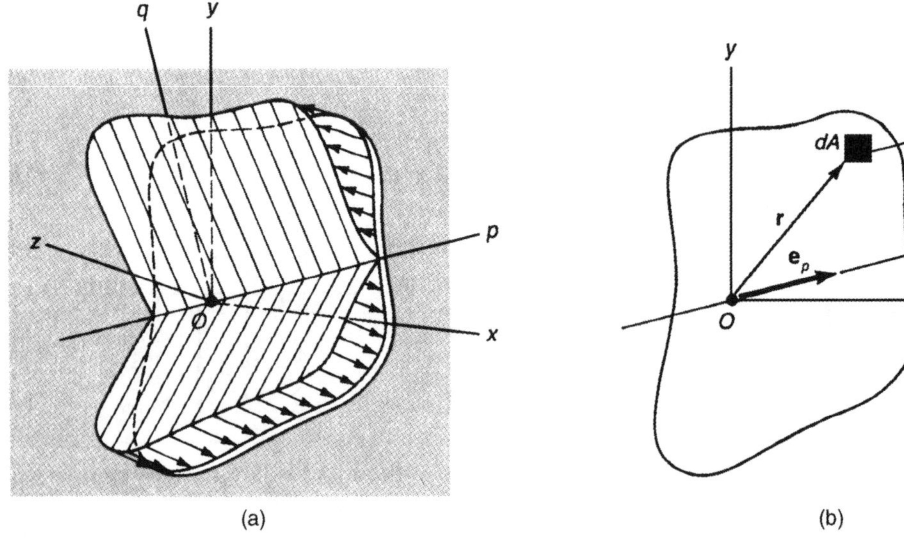

Figure 7.24

axis of the beam.) Observe from Fig. 7.24(b) that the distance from axis Op to the area element dA is given by

$$q = r \sin \angle^{\mathbf{r}} \mathbf{e}_p = (\mathbf{e}_p \times \mathbf{r}) \bullet \mathbf{e}_z$$

so that the force acting on the area element can be expressed as

$$d\mathbf{f} = \sigma_z \, dA \, \mathbf{e}_z = k(\mathbf{e}_p \times \mathbf{r}) dA$$

The resultant moment about O is then

$$\begin{aligned} \mathbf{M}_O &= \int_A \mathbf{r} \times d\mathbf{f} \\ &= k \int_A \mathbf{r} \times (\mathbf{e}_p \times \mathbf{r}) dA \\ &= k \int_A [(\mathbf{r} \bullet \mathbf{r})\mathbf{e}_p - (\mathbf{e}_p \bullet \mathbf{r})\mathbf{r}] dA \end{aligned}$$

In terms of the rectangular Cartesian components in the directions of x and y,

$$\mathbf{r} = x\mathbf{e}_x + y\mathbf{e}_y$$
$$\mathbf{e}_p = p_x\mathbf{e}_x + p_y\mathbf{e}_y$$

The moment can be expressed as

$$\mathbf{M}_O = k \int_A [(x^2 + y^2)\mathbf{e}_p - (xp_x + yp_y)\mathbf{r}] dA$$

Components of the moment are

$$\mathbf{M}_{Ox} = \mathbf{e}_x \bullet \mathbf{M}_O = k \int [(x^2 + y^2)p_x - (xp_x + yp_y)x] dA$$

and

$$\begin{aligned} \mathbf{M}_{Oy} &= k\left[\left(-\int_A xy \, dA\right)p_x + \left(\int_A x^2 dA\right)p_y\right] \\ &= k\left[\left(\int_A y^2 dA\right)p_x + \left(-\int_A xy \, dA\right)p_y\right] \end{aligned}$$

The integrals that appear in these expressions are called **second moments of area** and are fundamental to the relation between the moment and the orientation of the zero-force line Op. The integrals

$$I_{xx} = \int_A y^2 dA \quad \text{and} \quad I_{yy} = \int_A x^2 dA \qquad \textbf{(7.16)}$$

are often called **moments of inertia** of the area about the x and y axes, respectively. (This terminology stems from analogous integrals related to mass distribution in the study of the kinetics of rigid bodies.) The integral

$$I_{xy} = -\int_A xy \, dA \qquad \textbf{(7.17)}$$

is called the product of inertia of the area with respect to the x and y axes. (Many define it as the negative of this definition; then the corresponding terms in the next equation require opposite signs.) The quantities I_{xx}, I_{yy}, and I_{xy} are collectively called *second moments of area*.

In terms of the notation just introduced above, the relationship that gives the moment resultant in terms of the placement of the zero-force line Op is

$$\begin{Bmatrix} M_{Ox} \\ M_{Oy} \end{Bmatrix} = k \begin{bmatrix} I_{xx} & I_{xy} \\ I_{yx} & I_{yy} \end{bmatrix} \begin{Bmatrix} p_x \\ p_y \end{Bmatrix} \tag{7.18}$$

This is fundamental to the analyses of bending moment and deformation of elastic beams that are usually studied in mechanics of materials. Here, we examine only the evaluation of the moments and products of inertia.

The values of I_{xx}, I_{yy}, and I_{xy} depend on the placement of the origin O and the orientation of the x and y axes with respect to the area concerned. For special orientations of the x and y axes, called *principal directions*, the value of I_{xy} is zero. This situation will occur if the orientation is such that the area is symmetric with respect to either the x or y axis. (When no such symmetry exists, the determination of the principal directions requires computations that will not be addressed here.)

Parallel Axis Formulas

Often a value of one of the second moments of area is known for a given placement of the origin coordinates, but another value is needed for a different origin. Substituting the coordinate change indicated in Fig. 7.25 into Eqs. (7.16) and (7.17) leads to the following *parallel axis formulas*, which make this evaluation much easier than carrying out an integration.

$$\begin{aligned} I_{xx}^O &= I_{xx}^C + A\bar{y}^2 \\ I_{yy}^O &= I_{yy}^C + A\bar{x}^2 \\ I_{xy}^O &= I_{xy}^C + A\bar{x}\bar{y} \end{aligned} \tag{7.19}$$

Here, the superscript C indicates the value with the origin at the centroid, and the superscript O indicates an arbitrary origin. A is the area, and (x, y) are the coordinates of C with respect to O.

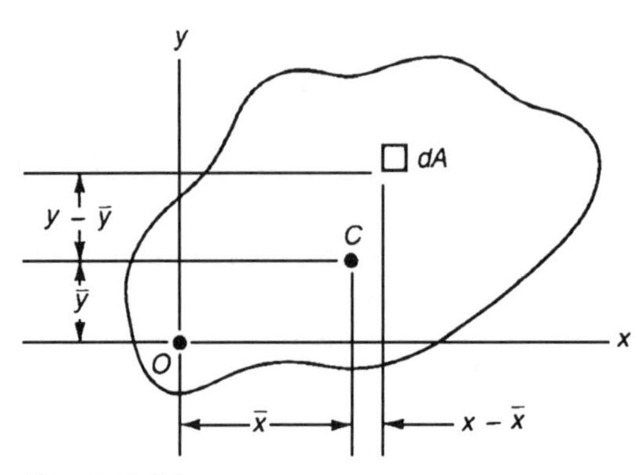

Figure 7.25

Example 7.17

In terms of the dimensions shown in Exhibit 14, evaluate the second moments of area of the triangle with respect to its centroid.

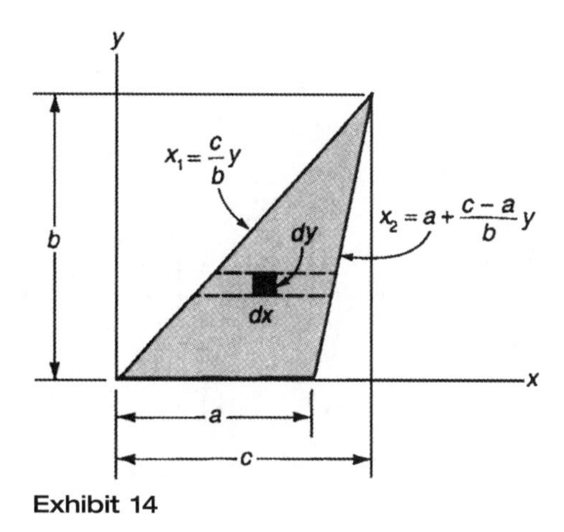

Exhibit 14

Solution

The area of the shaded element is $dx\,dy$. The line that bounds the triangle on the left is

$$x_1 = \frac{c}{b}y$$

and the line that bounds it on the right is

$$x_2 = a + \frac{c-a}{b}y$$

These two equations will form the limits of the first integration (with respect to x) in each case. The moments and product of inertia, with respect to the axes with origin at 0, have the values

$$I_{xx} = \int_0^b \int_{x_1}^{x_2} y^2\,dx\,dy$$

$$= \int_0^b [y^2 x]_{cy/b}^{a+(c-a)y/b}\,dy$$

$$= \int_0^b y^2\left(a - \frac{ay}{b}\right)dy = \frac{ab^3}{12}$$

$$I_{xy} = -\int_0^b \int_{x_1}^{x_2} xy\,dx\,dy$$

$$= -\int_0^b \left[\frac{x^2}{2}\right]_{cy/b}^{a+(c-a)y/b} y\,dy$$

$$= -\frac{a}{2}\int_0^b \left[a + 2(c-a)\frac{y}{b} - (2c-a)\frac{y^2}{b^2} \right] y\, dy$$

$$= -\frac{a}{2}\left[\frac{ab^2}{2} + \frac{2}{3}(c-a)b^2 - \frac{1}{4}(2c-a)b^2 \right]$$

$$= -\frac{ab^2}{24}(a+2c)$$

$$I_{yy} = \int_0^b \int_{x_1}^{x_2} x^2\, dx\, dy$$

$$= \frac{1}{3}\int_0^b [x^3]_{cy/b}^{a+(c-a)y/b}\, dy$$

$$= \frac{ab}{3}\int_0^b \left[a^2\left(1-\frac{y}{b}\right)^3 + 3ac\left(1-\frac{y}{b}\right)^2 \frac{y}{b} + 3c^2\left(1-\frac{y}{b}\right)\frac{y^2}{b^2} \right] \frac{dy}{b}$$

$$= \frac{ab}{12}(a^2 + ac + c^2)$$

As in Example 15, the coordinates of the centroid are $x = (a + c)/3$ and $y = b/3$. Use of the parallel axis formulas (7.19) then gives

$$I_{xx}^C = I_{xx}^O - A\bar{y}^2$$

$$= \frac{ab^3}{12} - \frac{ab}{2}\left(\frac{b}{3}\right)^2 = \frac{ab^3}{36}$$

$$I_{yy}^C = I_{yy}^O - A\bar{x}^2$$

$$= \frac{ab}{12}(a^2 + ac + c^2) - \frac{ab}{c}\left(\frac{a+c}{3}\right)^2 = \frac{ab}{36}(a^2 - ac + c^2)$$

$$I_{xy}^C = I_{xy}^O + A\bar{x}\bar{y}$$

$$= \frac{ab^2}{24}(a+2c) + \frac{ab}{c}\left(\frac{a+c}{3}\right)\left(\frac{b}{3}\right) = \frac{ab^2}{72}(a-2c)$$

PROBLEMS

7.1 A 70-kg astronaut is "floating" inside a spaceship that is in a circular orbit at an altitude of 207 km above the earth, where the gravitational field intensity is 9.2 N/kg. What is the magnitude of the force of gravity on the astronaut?

a. Zero c. 70 kgf
b. 70 N d. 644 N

7.2 In the SI system of units, a pressure of 14.7 $1bf/in.^2$ is

a. 101 kPa c. 6.67 $kg/in.^2$
b. 101 Pa d. 4.77×10^4 kg/m^2

7.3 In the SI system of units, a fuel economy of 29 mi/gal is

a. 12.33 km/L
b. 68.2 km/L
c. 46.7 km/L
d. 46.7 km/gal

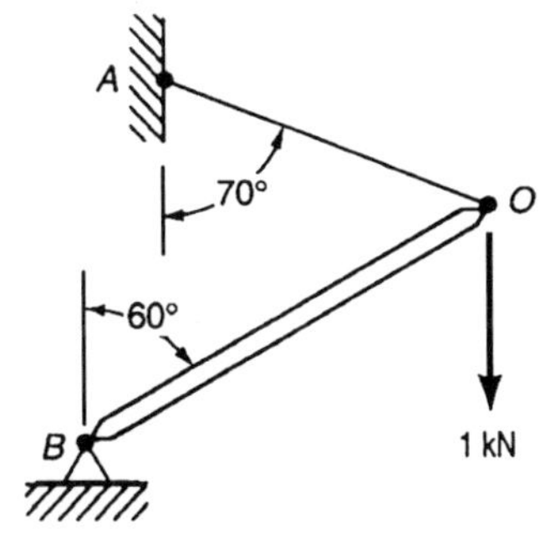

Exhibit 7.4

7.4 The components parallel to *OA* and *OB*, of the vertical 1-kN force in Exhibit 7.4 have magnitudes of

a. 0.34 and 0.50 kN c. 1.13 and 1.23 kN
b. 0.34 and 0.66 kN d. 0.5 and 0.5 kN

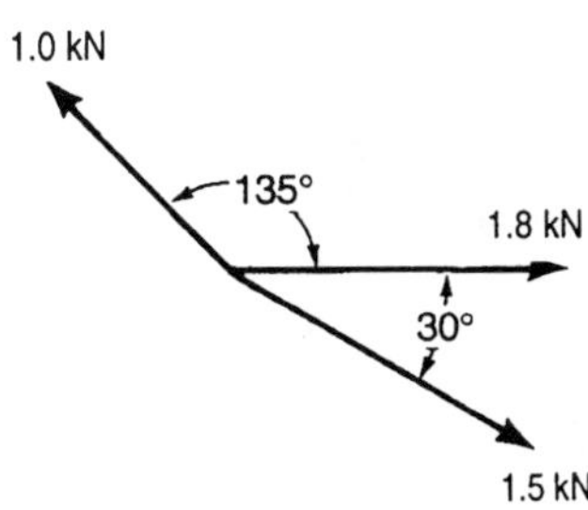

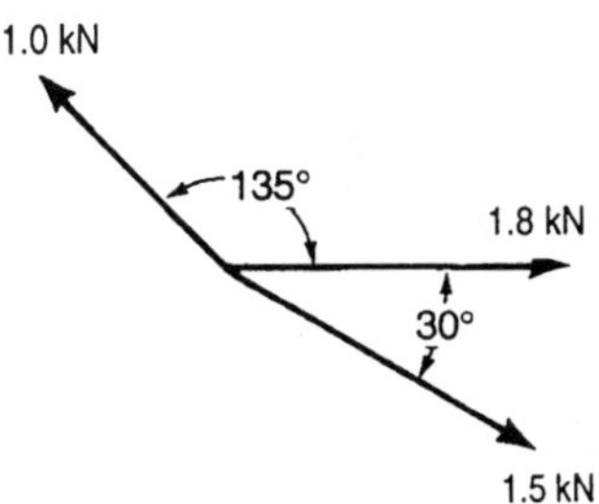

Exhibit 7.5

7.5 The resultant of the three forces in Exhibit 7.5 has a magnitude of

a. 4.3 kN c. 2.3 kN
b. 3.0 kN d. 2.39 kN

Exhibit 7.6

7.6 At a certain instant, the tension in the cable on which the destruction ball is suspended is 9.3 kN (see Exhibit 7.6). At this instant, the acceleration of the ball is

a. 10.36 m/s^2 c. 1.21 m/s^2
b. 5.30 m/s^2 d. 9.81 m/s^2

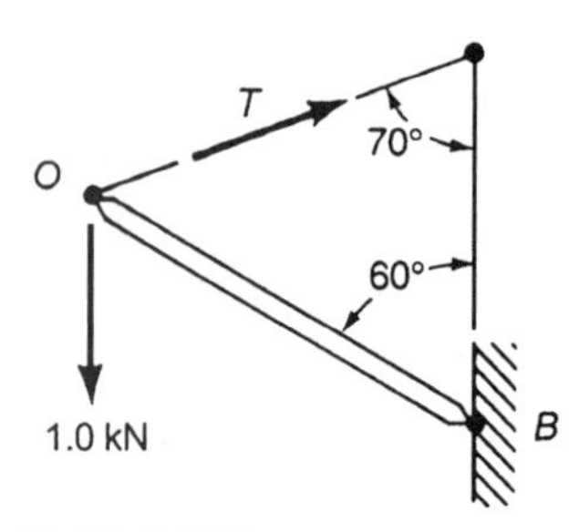

Exhibit 7.7

7.7 If the resultant of the 1-kN vertical force and the tensile force T from the cable is to be in the direction of the boom OB (see Exhibit 7.7), what must be the magnitude of T?

a. 2.9 kN
b. 1.0 kN
c. 1.1 kN
d. 0.94 kN

7.8 Evaluate the magnitude of the resultant force on the doorknob in Exhibit 7.8. The three components are mutually perpendicular.

a. 13 N
b. 19 N
c. 17 N
d. 5 N

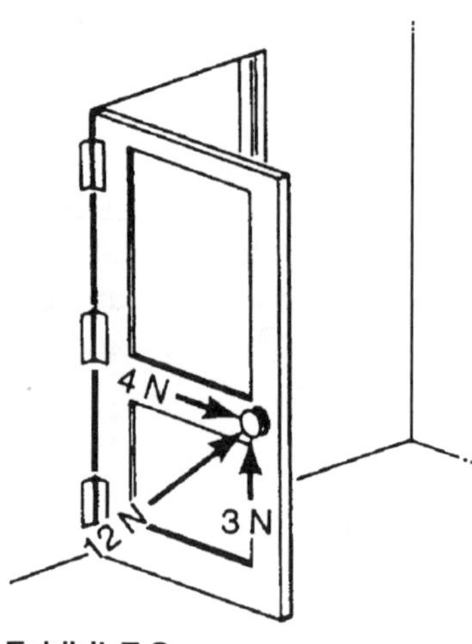

Exhibit 7.8

7.9 To raise the load, the hydraulic cylinder exerts a force of 50 kN in the direction of its axis, AB. For the position shown in Exhibit 7.9, the components of this force parallel and perpendicular to OB are

a. 43.8 and 24.2 kN
b. 41 and 28.7 kN
c. 23.1 and 26.9 kN
d. 25 and 25 kN

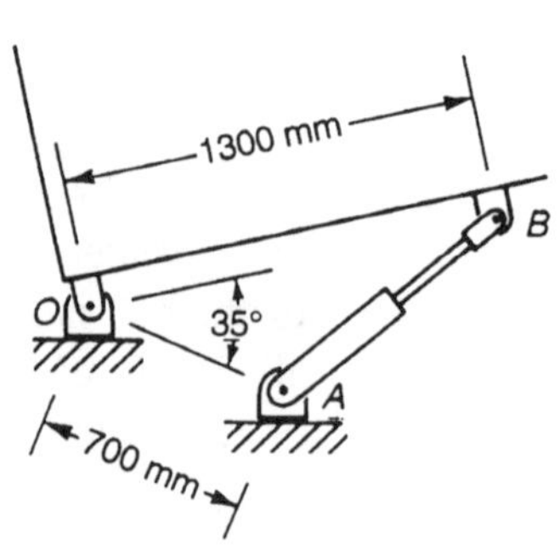

Exhibit 7.9

7.10 Refer to Exhibit 7.10. The moment about *O* of the 250-N force has a magnitude of

a. 63.0 N•m
b. 77.0 N•m
c. 79.5 N•m
d. 23 N•m

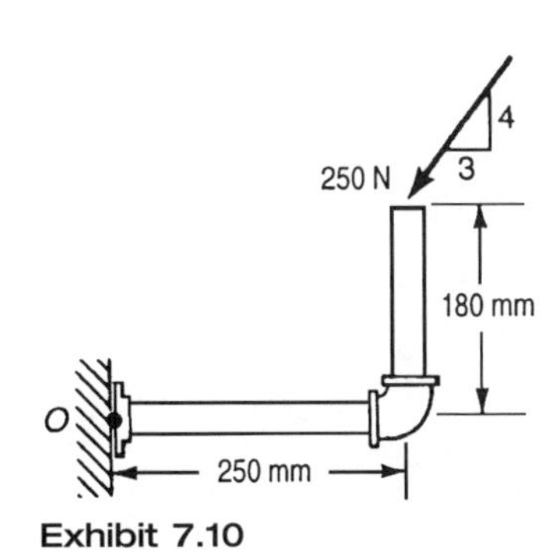

Exhibit 7.10

7.11 The connecting rod in Exhibit 7.11 exerts a force of 4.5 kN on the crank. The moment about *O* of this force has a magnitude of

a. 408 N•m
b. 450 N•m
c. 318 N•m
d. 190 N•m

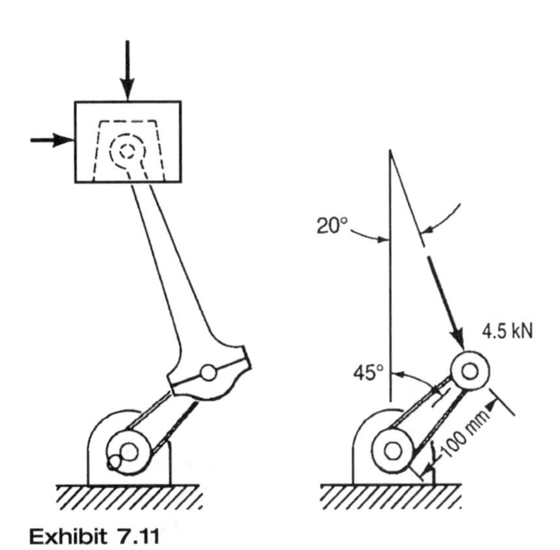

Exhibit 7.11

7.12 The tension in the line *AB* of Exhibit 7.12 is 3.50 kN. What must be the tension in the line *BC* if the moment about *O* of the force that the cable *BC* exerts on the spreader bar *OB* is equal and opposite to that of the force that the cable *AB* exerts on the spreader bar?

a. 3.50 kN
b. 2.86 kN
c. 6.58 kN
d. 4.29 kN

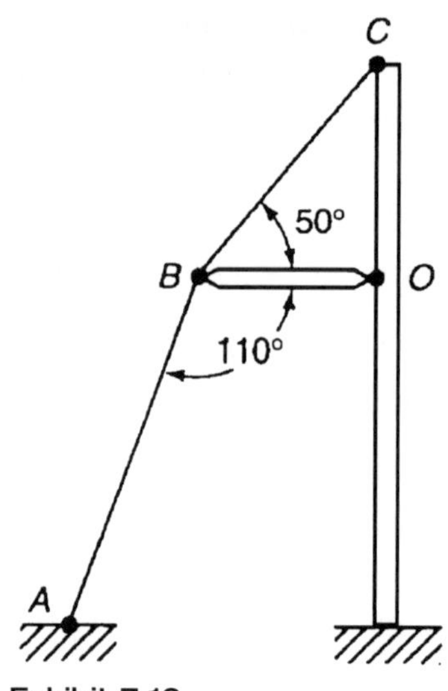

Exhibit 7.12

Exhibit 7.13

7.13 The plumber in Exhibit 7.13 exerts a vertical downward force of 1 kN on the wrench handle. The moment about C of this force has a magnitude of

a. 500 N•m c. 900 N•m
b. 750 N•m d. 1250 N•m

7.14 The moment about the axis CB of the previous problem has a magnitude of

a. 500 N•m c. 900 N•m
b. 750 N•m d. 1250 N•m

7.15 The brake is set on the wheel in Exhibit 7.15, and it will not slip until the moment about the center of the wheel of forces acting on the lug wrench reaches 150 N• m. Will the brake slip?

a. 147.2 N• m; No c. 1335.6 N• m; No
b. 156.6 N• m; Yes d. 313.2 N• m; Yes

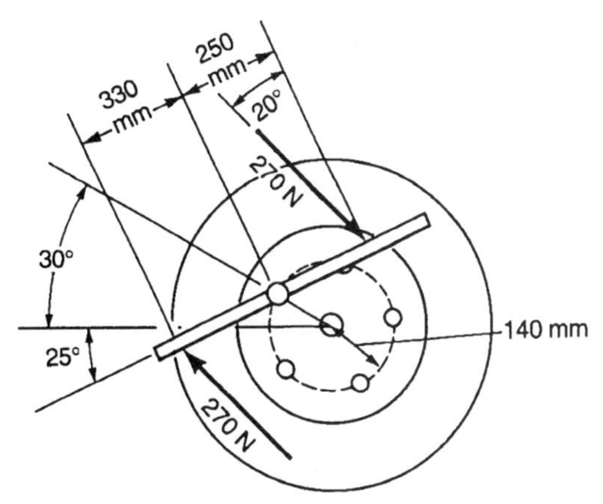

Exhibit 7.15

7.16 The tension in the vertical line AC is 2 kN and that in the line BC is 6 kN (see Exhibit 7.16). The magnitude of the resultant force exerted by the two lines at C is

a. 8.0 kN
b. 6.8 kN
c. 4.0 kN
d. 6.3 kN

Exhibit 7.16

7.17 The moment of **f** about the axis AB in Exhibit 7.17 has the magnitude

a. (144/65 m)f
b. (12/5 m)f
c. (29/13 m)f
d. (12 m)f

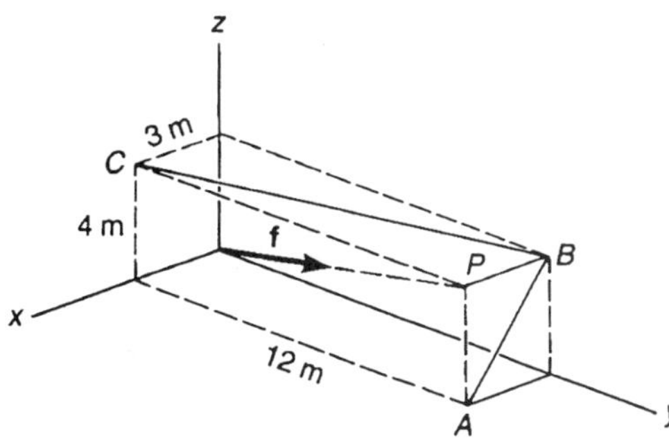

Exhibit 7.17

7.18 The surfaces are smooth where the drum makes contact (see Exhibit 7.18). The reaction at the contact point on the right is

a. 1.77 kN
b. 3.06 kN
c. 2.29 kN
d. 2.70 kN

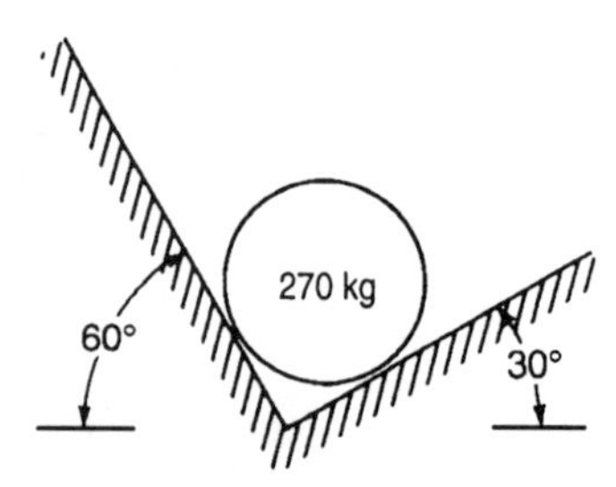

Exhibit 7.18

7.19 Two wheels, each of radius a but of different mass, are connected by the rod of length R in Exhibit 7.19. The assembly is free to roll in the circular trough. The angle θ for equilibrium is given by

a. $\tan\theta = \dfrac{f_{g_1} - f_{g_2}}{f_{g_1} + f_{g_2}} \tan 30°$

b. $\sin\theta = \dfrac{f_{g_1} - f_{g_2}}{f_{g_1} + f_{g_2}} \sin 30°$

c. $\tan\theta = \dfrac{f_{g_1} - f_{g_2}}{f_{g_1} + f_{g_2}} \sin 30°$

d. $\sin\theta = \dfrac{f_{g_1} - f_{g_2}}{f_{g_1} + f_{g_2}} \tan 30°$

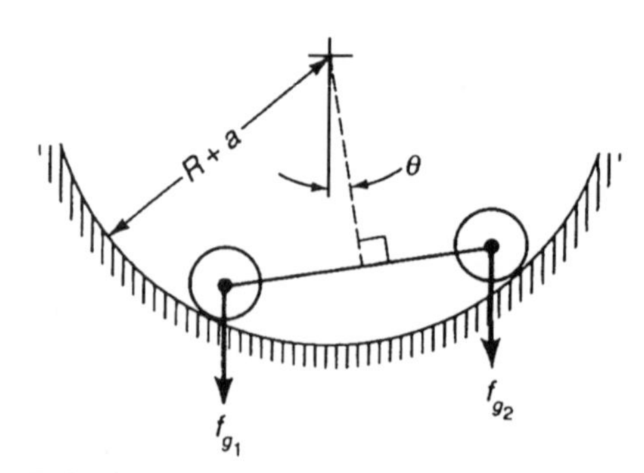

Exhibit 7.19

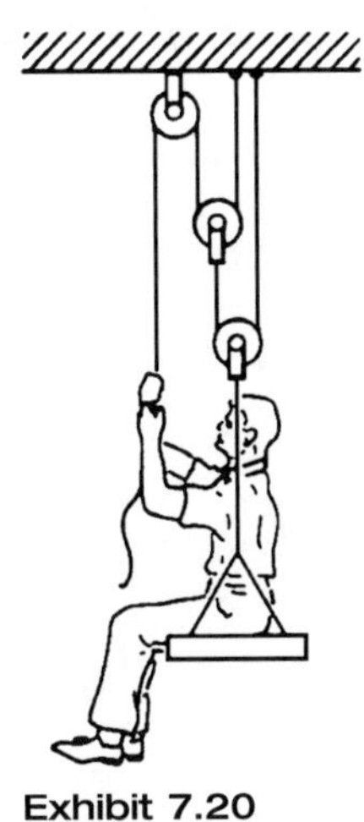

Exhibit 7.20

7.20 Determine the force with which the 80-kg man in Exhibit 7.20 must pull on the rope to support himself. The force is closest to

a. 785 N
b. 628 N
c. 471 N
d. 157 N

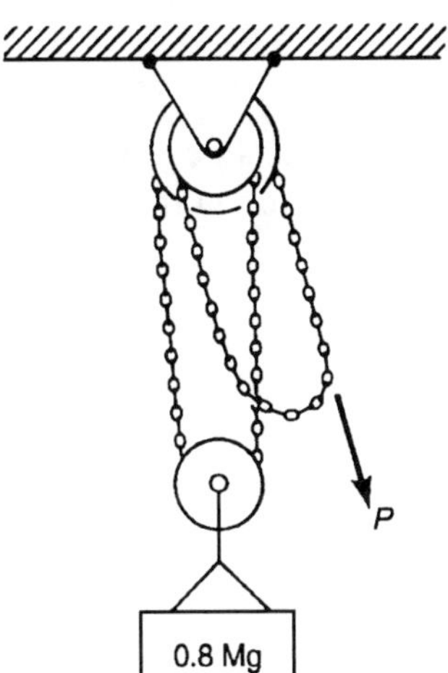

Exhibit 7.21

7.21 Each of the tracks in the upper pulley unit is recessed to fit the chain, so as to prevent slipping (see Exhibit 7.21). The smaller track has a radius equal to 0.9 times that of the larger track. Evaluate the force P necessary to lift the block by means of the differential chain hoist. the force is nearest to

a. 0.392 kN
b. 0.784 kN
c. 3.92 kN
d. 1.96 kN

7.22 The weight of the linkage in Exhibit 7.22 is negligible compared with f_g. What is the value of P necessary to maintain equilibrium?

a. $0.333f_g$
b. $0.500f_g$
c. $0.577f_g$
d. $0.144f_g$

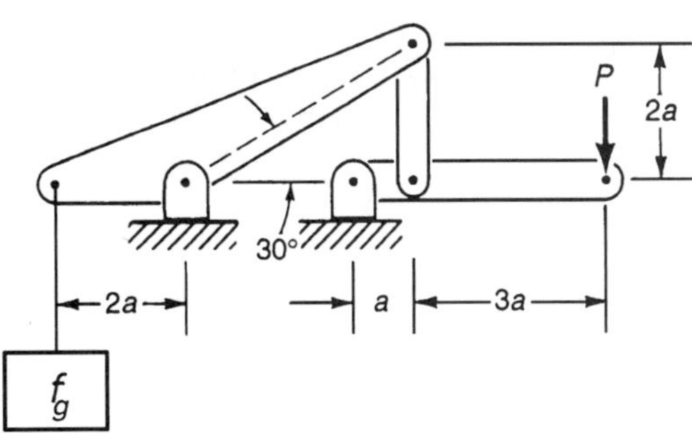

Exhibit 7.22

7.23 Neglecting the mass of the structure of Exhibit 7.23, the tension in the bar *AB*, induced by the 320-kg lifeboat, is nearest to

a. 3.14 kN
b. 0.32 kN
c. 1.57 kN
d. 12.55 kN

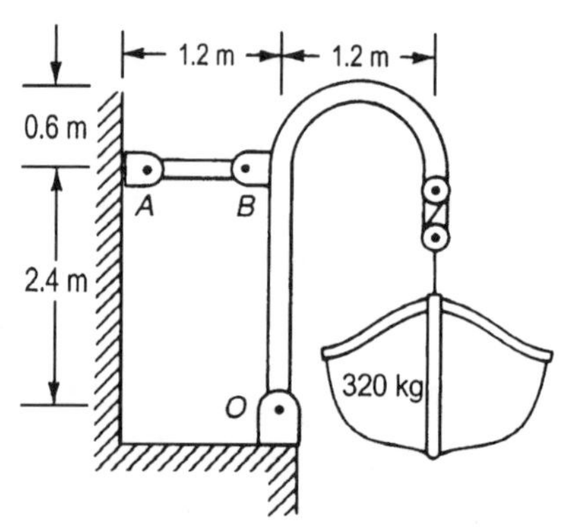

Exhibit 7.23

7.24 Until a clamp is tightened, the drill press table is free to slide along the column (see Exhibit 7.24). Estimate the coefficient of friction required so that the collar will be self-locking against the column under the action of the thrust from the drill. Neglect gravity.

a. 0.21
b. 0.42
c. 0.63
d. 0.84

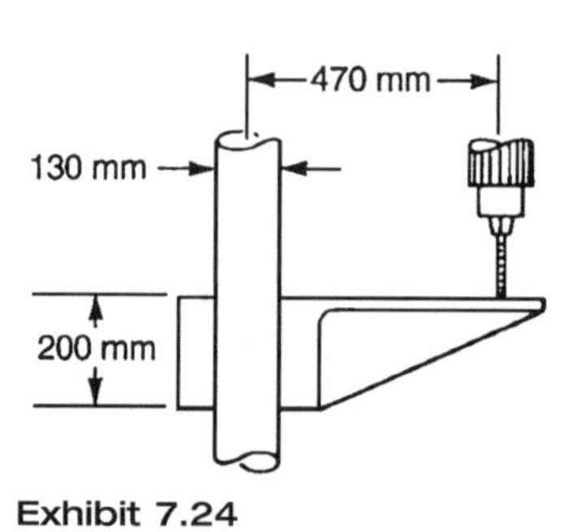

Exhibit 7.24

7.25 The combined reaction at the two rear wheels of the car in Exhibit 7.25 has the magnitude

a. 5.0 kN
b. 4.0 kN
c. 7.0 kN
d. 6.3 kN

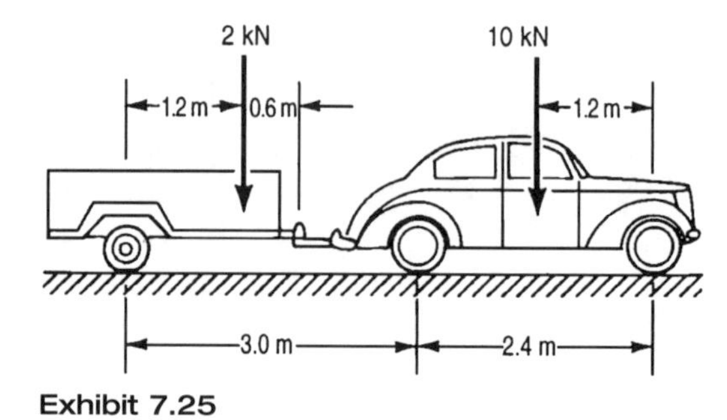

Exhibit 7.25

7.26 In a trailer "load-leveler" hitch, the angle bar slips into the cylindrical socket at A, forming a thrust bearing where the bar bottoms (see Exhibit 7.26(a)). The end, B, is then attached by a short chain to the towing vehicle, as in Exhibit 7.26(b). The pretension in the chain is 1.7 kN. The reaction at C (both wheels) has magnitude

a. 8.34 kN c. 6.20 kN
b. 8.05 kN d. 5.25 kN

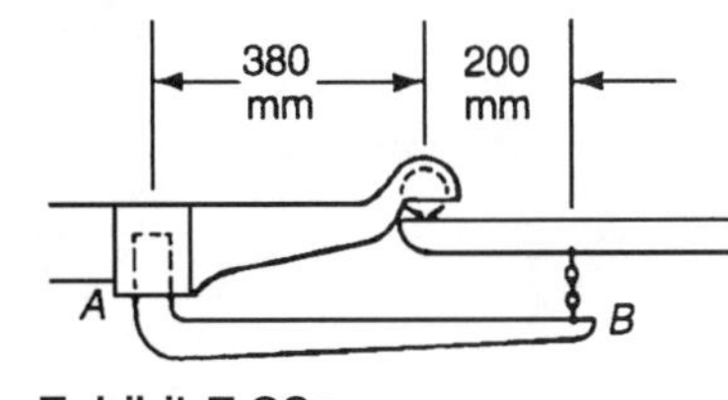

Exhibit 7.26a

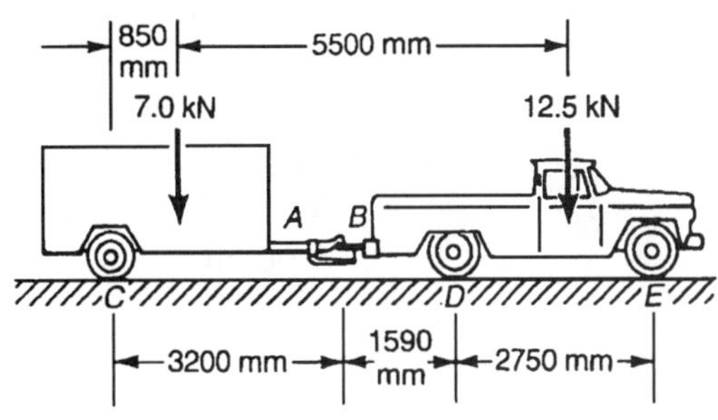

Exhibit 7.26b

7.27 The reaction at the near wheels, D, of the vehicle of problem 7.26 has magnitude

a. 8.34 kN c. 6.20 kN
b. 8.05 kN d. 5.25 kN

7.28 Evaluate the cutting force at C in terms of the force P on the handles of the compound snips (Exhibit 7.28).

a. $2P$ c. $6P$
b. $4P$ d. $8P$

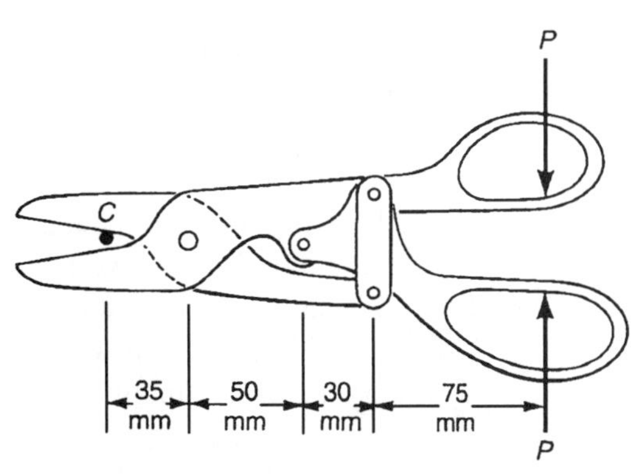

Exhibit 7.28

7.29 The magnitude of the force on the nut at *C* exerted by the jaws of the self-locking pliers (Exhibit 7.29) is

a. 240 N c. 1600 N
b. 480 N d. 2020 N

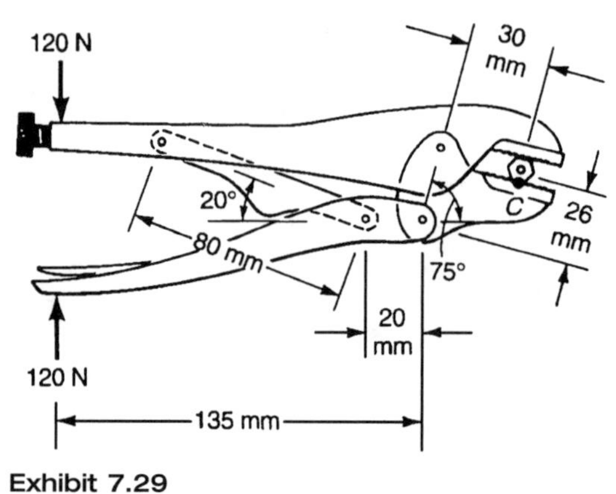

Exhibit 7.29

7.30 The moment of the couple transmitted through the shaft at section *A* (Exhibit 7.30) has magnitude

a. 9.60 N•m c. zero
b. 15.0 N•m d. 30 N•m

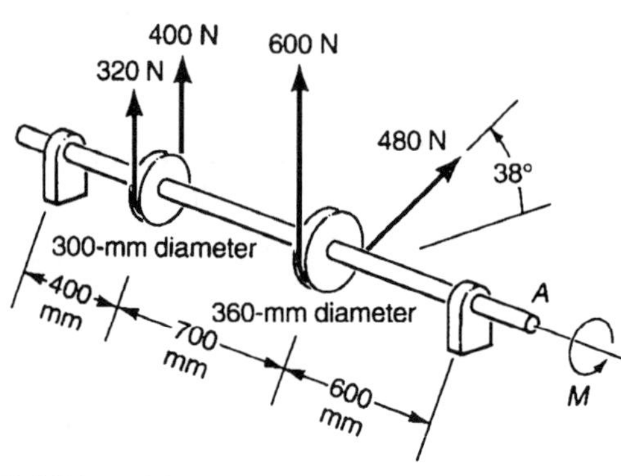

Exhibit 7.30

7.31 The reaction at the bearing near *A* in Problem 7.30 has magnitude

a. 245 N c. 788 N
b. 745 N d. 1296 N

7.32 The slider *A* has a mass of 4 kg and is constrained to slide without friction along the fixed vertical rod (Exhibit 7.32). The mass of the wire *AB* is negligible. The slider *B* is constrained to slide along the horizontal rod without friction. What must be the magnitude *F* of the force applied to the slider *B* to maintain equilibrium?

a. 39.2 N c. 98 N
b. 19.6 N d. 49 N

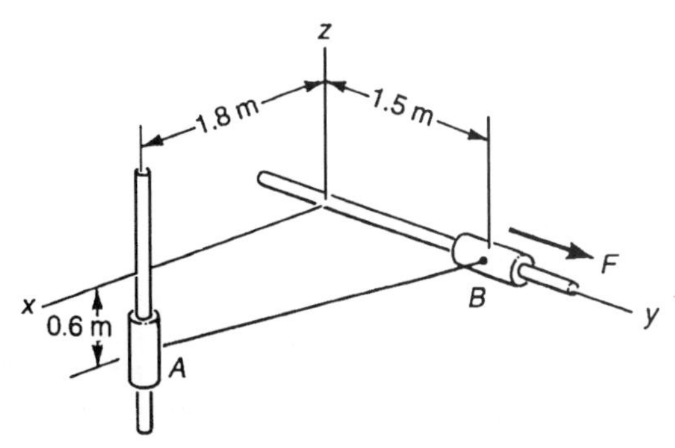

Exhibit 7.32

7.33 The turnbuckle in the guy line *AE* of Exhibit 7.33 is to be tightened such that the vertical component of the reaction at *O* is 800 N. What must be the tension in *AE*?

a. 424 N
b. 212 N
c. 636 N
d. 267 N

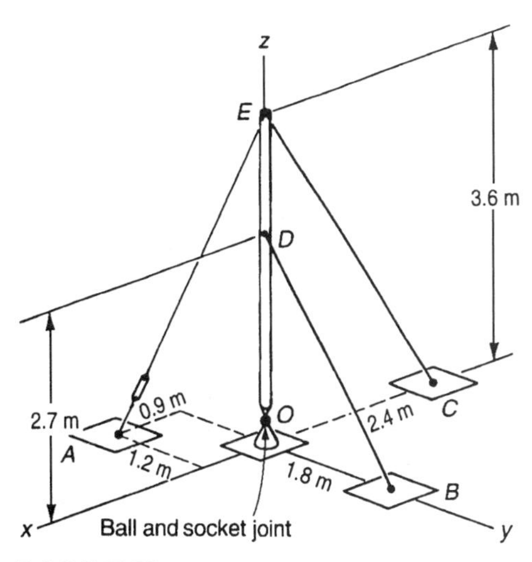

Exhibit 7.33

7.34 Determine the tensions in the cables *AB* and *CD* of Exhibit 7.34. They are closest to

a. 2.27 kN, 1.45 kN
b. 0.75 kN, 0.75 kN
c. 0.45 kN, 1.05 kN
d. 1.87 kN, 1.20 kN

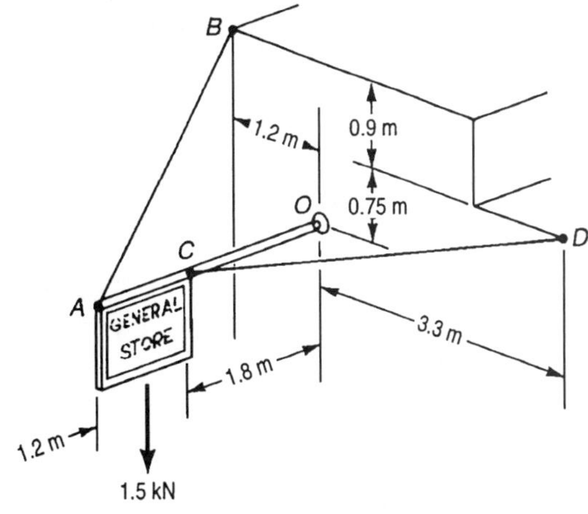

Exhibit 7.34

7.35 The 40-kg rectangular plate in Exhibit 7.35 is held by hinges along its edge *OA* and by the wire *BD*. The gravity force f_g acts through the geometric center of the plate. What is the tension of the wire?

a. 392 N
b. 196 N
c. 36.5 N
d. 358 N

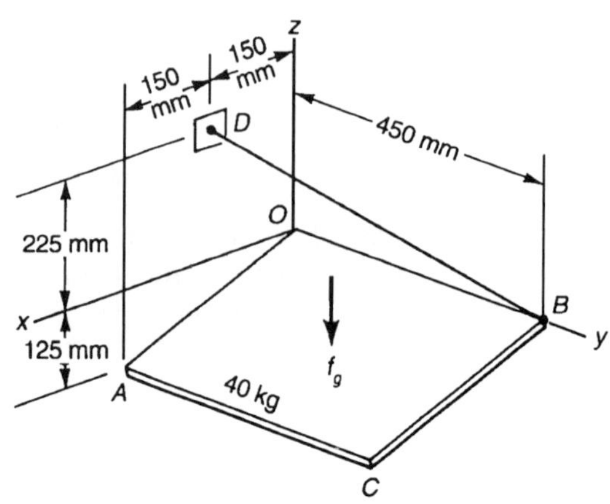

Exhibit 7.35

7.36 The force in the member *BC* of Exhibit 7.36 is

a. 12 kN, tension
b. 17 kN, tension
c. 17 kN, compression
d. 12 kN, compression

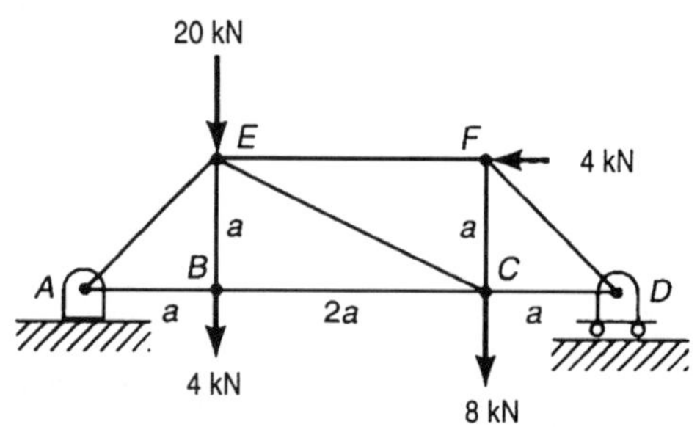

Exhibit 7.36

7.37 The force in the member *ML* of Exhibit 7.37 is

a. 131 kN, tension
b. 131 kN, compression
c. 100 kN, tension
d. 106 kN, tension

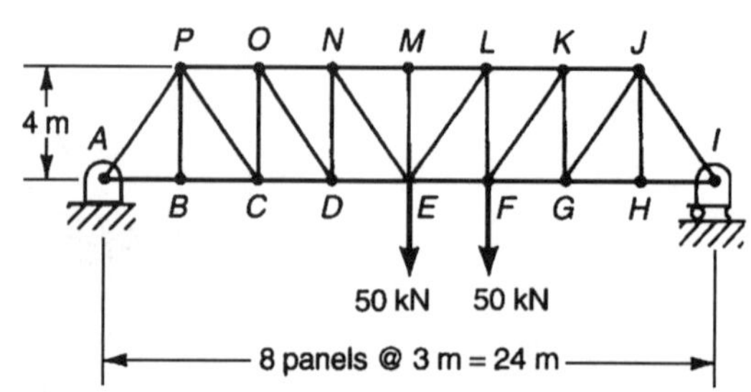

Exhibit 7.37

7.38 The force in the member *CF* of Exhibit 7.38 is
a. 60.6 kN, tension
b. 60.6 kN, compression
c. 43.3 kN, tension
d. 43.3 kN, compression

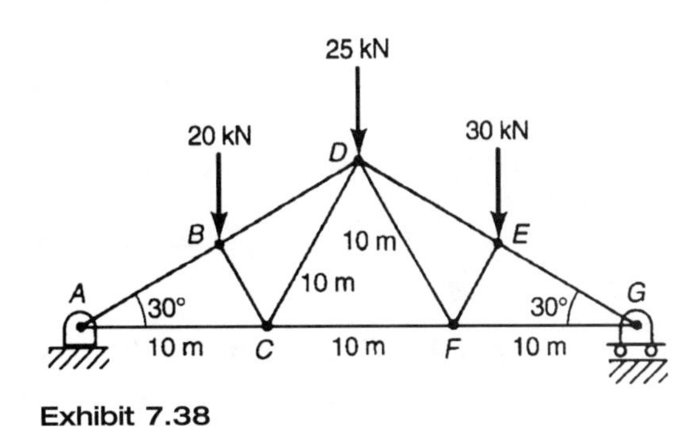

Exhibit 7.38

7.39 Members *AD*, *BE*, and *CF* are perpendicular to the plane *ABC* (see Exhibit 7.39). The force in the member *AD* is
a. 6 kN, compression c. 8 kN, tension
b. 6 kN, tension d. 8 kN, compression

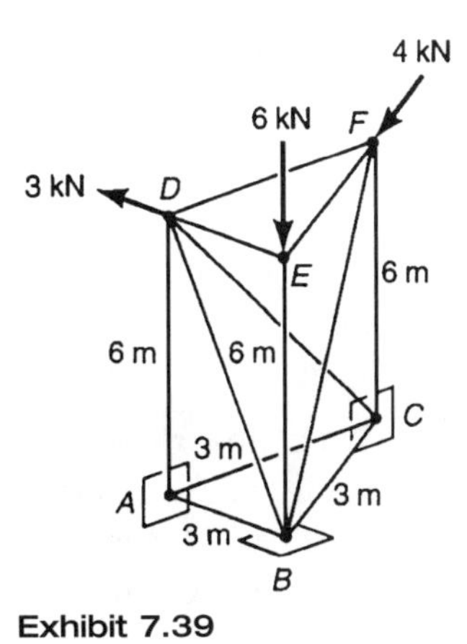

Exhibit 7.39

7.40 The force in member *BE*, shown in Exhibit 7.40 is
a. 0.87*P*, compression
b. *P*, compression
c. 0.87*P*, tension
d. *P*, tension

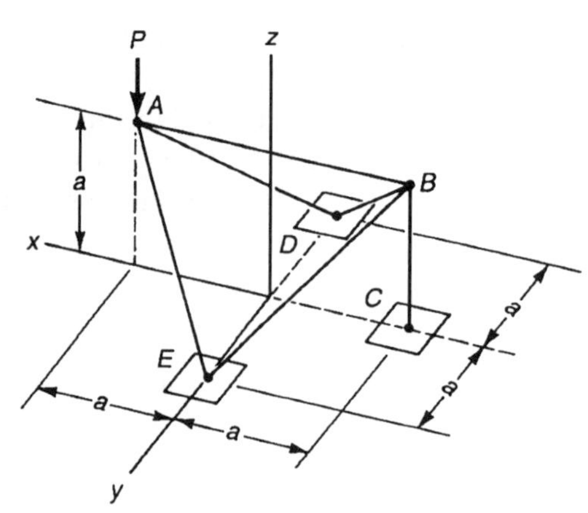

Exhibit 7.40

7.41 The contents of the crate in Exhibit 7.41 are such that the center of gravity is at the geometric center. In the absence of *P*, the crate will

a. remain stationary c. tip over
b. slide downsward d. slide upward

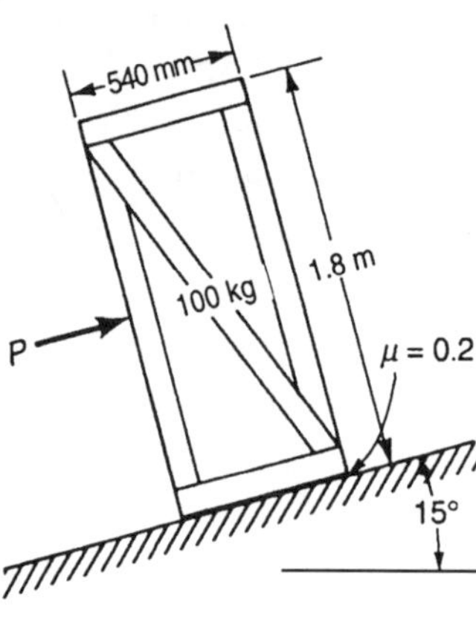

Exhibit 7.41

7.42 In Exhibit 7.41, the greatest distance from the incline to the line of action of *P*, such that the force can slide the crate up the incline without it, is

a. 1.8 m c. 1.35 m
b. 0.9 m d. 1.09 m

7.43 The contents of the crate in Exhibit 7.43 are such that the center of gravity coincides with the geometric center. The greatest value of the angle that will allow the force to slide the crate without tipping it is given by

a. $\tan^{-1}\left(\frac{h}{b}\right)$ c. $\tan^{-1}\left(\frac{1}{\mu}-\frac{2h}{b}\right)$

b. $\tan^{-1}\left(\frac{b}{h}\right)$ d. $\tan^{-1}\left(\frac{1}{\mu}+\frac{2h}{b}\right)$

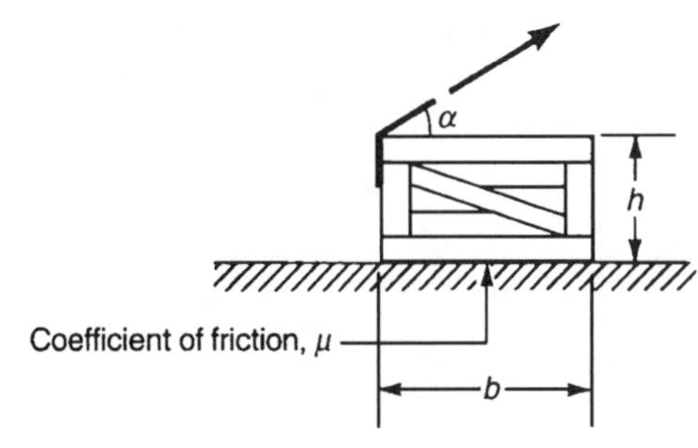

Exhibit 7.43

7.44 If the coefficient of friction between all surfaces in Exhibit 7.44 is 0.27, what will be the minimum value of P necessary to initiate motion?

a. 103 N
b. 81.0 N
c. 40.5 N
d. 51.5 N

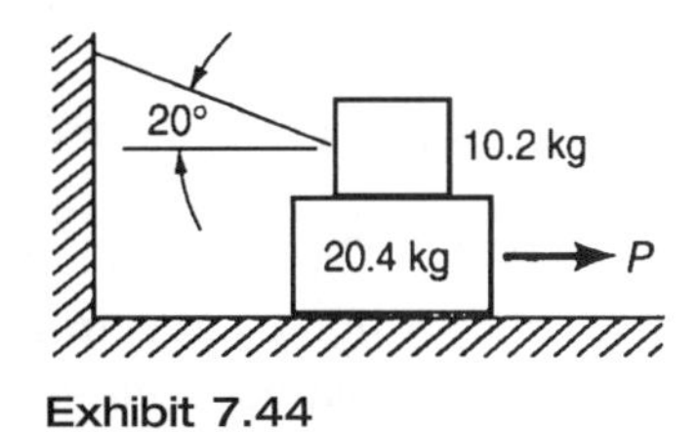

Exhibit 7.44

7.45 If the coefficient of friction between all surfaces in Exhibit 7.45 is 0.27, what will be the minimum value of P necessary to initiate motion?

a. 205 N
b. 103 N
c. 94.5 N
d. 9.63 N

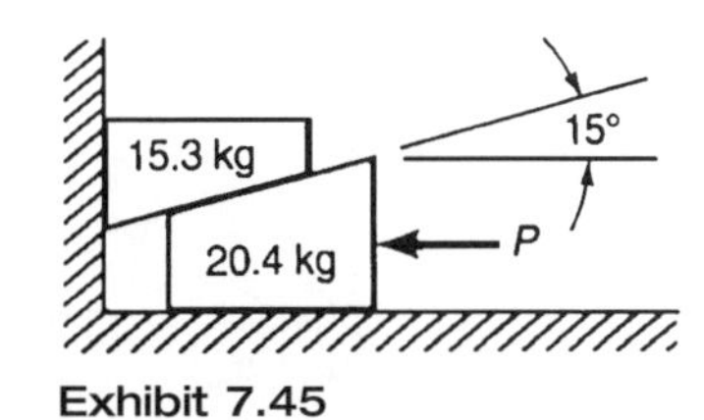

Exhibit 7.45

7.46 The forklift in Exhibit 7.46 is being used to roll the 2-Mg drum up the 40-degree incline while the height of the forks remains constant. The coefficient of friction is 0.45 between the vertical rails and the drum, and 0.30 between the incline and the drum. What horizontal thrust must the vehicle apply to the drum to move it?

a. 85 kN
b. 62.7 kN
c. 78.2 kN
d. 103 kN

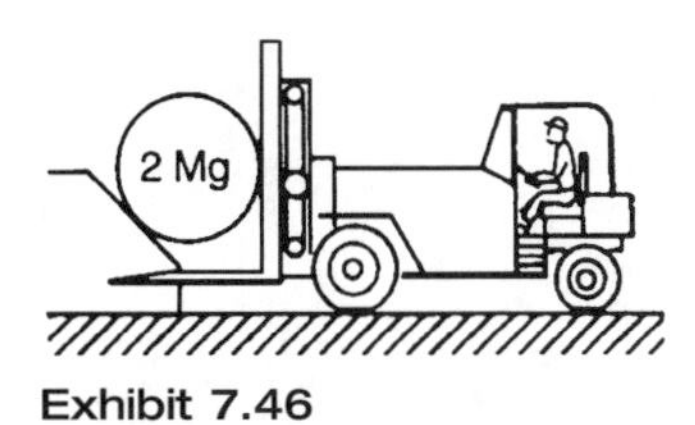

Exhibit 7.46

7.47 Determine the minimum coefficient of friction between the block and the weightless bar in Exhibit 7.47 necessary to prevent collapse.

a. 0.15 c. 0.42
b. 0.27 d. 0.72

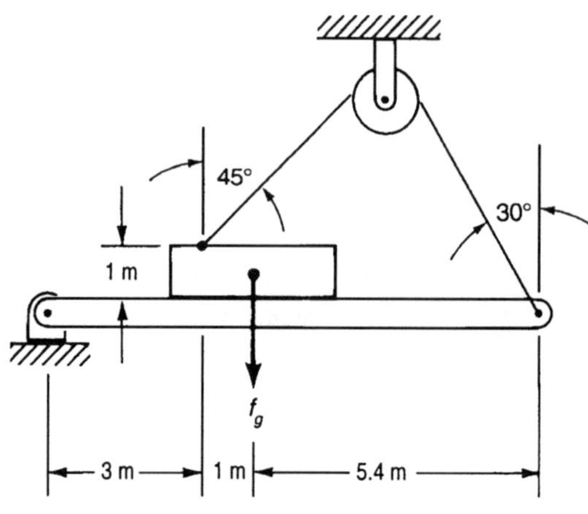

Exhibit 7.47

7.48 The small rollers in Exhibit 7.48 are intended to prevent clockwise rotation of the large drum. The coefficient of friction between the rollers and drum, and between the rollers and walls, is μ. Determine the minimum distance d such that the friction will effect a self-locking mechanism against clockwise rotation. Gravity is negligible.

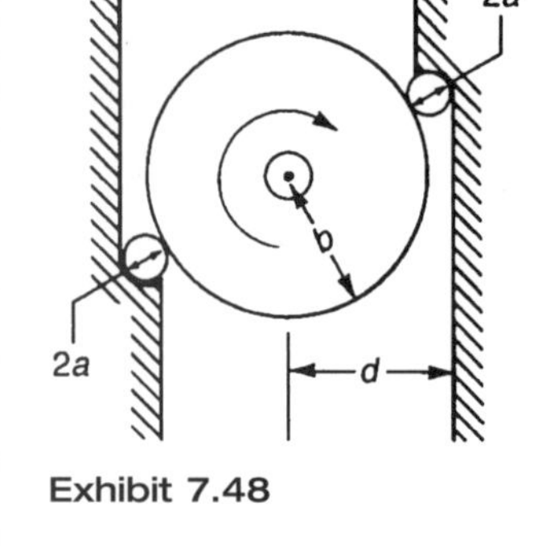

Exhibit 7.48

a. $d > (a+b)\mu$ c. $d > \dfrac{2a+(1-\mu^2)b}{1+\mu^2}$

b. $d > \dfrac{a+b}{\mu}$ d. $d > \dfrac{(a+b)\mu}{1+\mu^2}$

SOLUTIONS

7.1 d. $f_g = (9.2 \text{ N/kg})(70 \text{ kg}) = 644 \text{ Ns}$

7.2 a. $P = (14.7 \text{ lbf/in.}^2)(39.37 \text{ in./m})^2(4.448 \text{ N/lbf}) = 101{,}000 \text{ N/m}^2 = 101 \text{ kPa}$

7.3 a. $\text{Fuel economy} = (29 \text{ mi/gal}) \dfrac{(5280 \text{ ft/mi})(0.3048 \text{ m/ft})}{(1000 \text{ m/km})(3.7854 \text{ L/gal})} = 12.33 \text{ km/L}$

7.4 c.

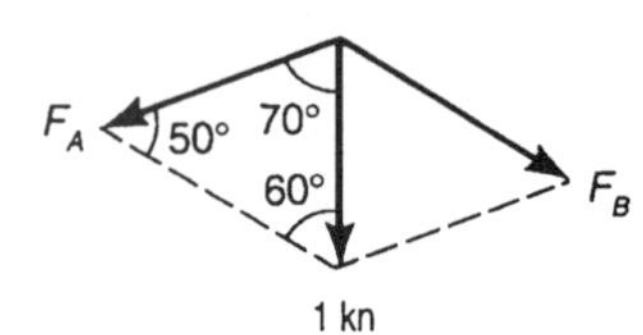

Exhibit 7.4a

$$\frac{F_A}{\sin 60°} = \frac{F_B}{\sin 70°} = \frac{1 \text{kN}}{\sin 50°}$$

$$F_A = \frac{\sin 60°}{\sin 50°}(1 \text{ kN}) = 1.13 \text{ kN}$$

$$F_B = \frac{\sin 70°}{\sin 50°}(1 \text{ kN}) = 1.23 \text{ kN}$$

7.5 d.

$$F_x = 1.8 \text{ kN} + (1.5 \text{ kN})\cos 30° + (1.0 \text{ kN})\cos 135° = 2.39 \text{ kN}$$
$$F_y = (1.0 \text{ kN})\sin 135° - (1.5 \text{ kN})\sin 30° = -0.043 \text{ kN}$$
$$F = \sqrt{F_x^2 + F_y^2} = 2.39 \text{ kN}$$

7.6 b.

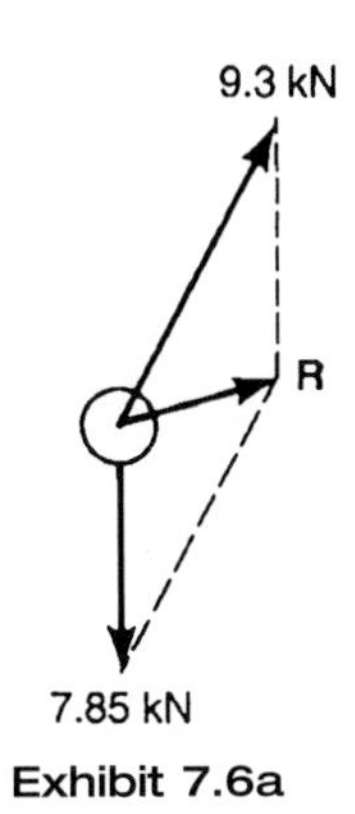

Exhibit 7.6a

The force of gravity on the ball is $f_g = (800 \text{ kg})(9.81 \text{ N/kg}) = 7.85 \text{ kN}$.

$$R^2 = [(9.3)^2 + (7.85)^2 - 2(9.3)(7.85)\cos 27°](\text{kN})^2$$
$$R = 4.24 \text{ kN}$$

$$a = \frac{4.24 \text{ kN}}{0.8 \text{ Mg}} = 5.30 \text{ m/s}^2$$

7.7 c.

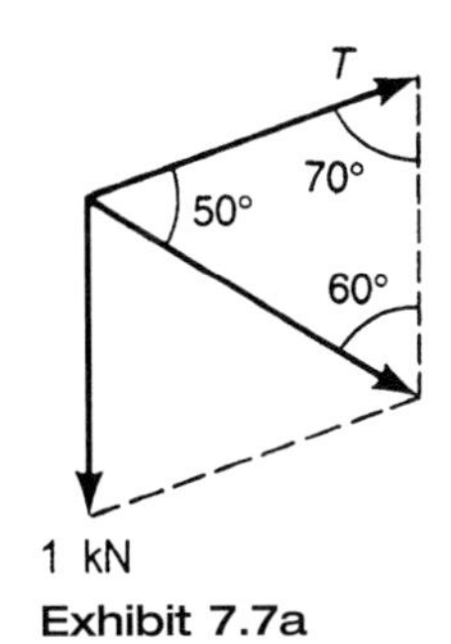

Exhibit 7.7a

$$\frac{T}{\sin 60°} = \frac{1 \text{ kN}}{\sin 50°}$$

$$T = 1.13 \text{ kN}$$

7.8 a.

$$F = \sqrt{(4)^2 + (12)^2 + (3)^2} = 13 \text{ N}$$

7.9 a.

$$\overline{AB} = \sqrt{(0.7)^2 + (1.3)^2 - 2(0.7)(1.3)\cos 35°} \text{ m} = 0.83 \text{ m}$$

$$\frac{\sin {}_B\angle_A^O}{0.7 \text{ m}} = \frac{\sin 35°}{0.83 \text{ m}}; \quad {}_B\angle_A^O = 28.92°$$

$$(50 \text{ kN})\cos 28.9° = 43.8 \text{ kN}$$

$$(50 \text{ kN})\sin 28.9° = 24.2 \text{ kN}$$

7.10 d.

$$M_0 = (0.18 \text{ m})(150 \text{ N}) - (0.25 \text{ m})(200 \text{ N}) = -23.0 \text{ N}\bullet\text{m}$$

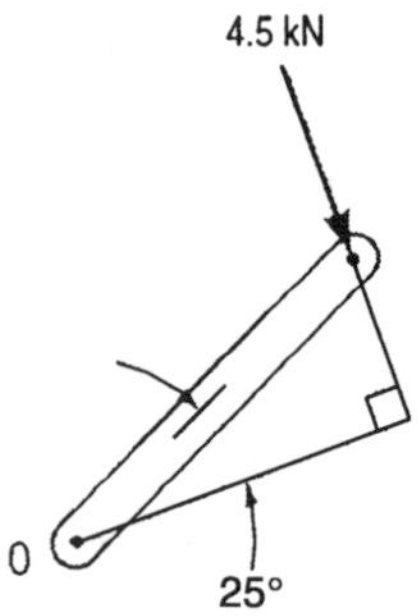

4.5 kN

0

25°

Exhibit 7.11a

7.11 a.

$$M_0 = (0.1\text{ m})\cos 25° \,(4.5\text{ kN}) = 408\text{ N}\bullet\text{m}$$

7.12 d.

$$lT_{BC}\sin 50° = l(3.5\text{ kN})\sin 110°$$

$$T_{BC} = 4.29\text{ kN}$$

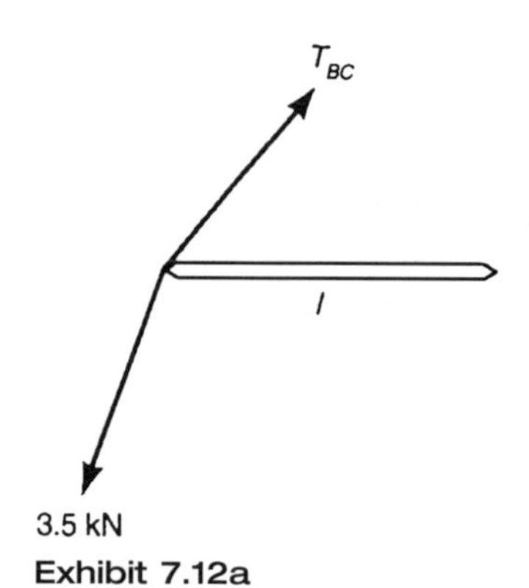

Exhibit 7.12a

7.13 c.

$$r = \sqrt{(0.75\text{ m})^2 + (0.5\text{ m})^2} = 0.9014\text{ m}$$
$$M_C = (0.9014\text{ m})(1\text{ kN}) = 901.4\text{ N}\bullet\text{m}$$

7.14 a.

$$M_{CB} = (0.5\text{ m})(1\text{ kN}) = 500\text{ N}\bullet\text{m}$$

7.15 a.

$$M = (250\text{ mm} + 330\text{ mm})\cos 20° \,(270\text{ N}) = 147.2\text{ N}\bullet\text{m}$$

7.16 b.

$$\overline{C}\,\overline{B} = \sqrt{(2.4\text{ m})^2 + (3.6\text{ m})^2 + (1.2\text{ m})^2} = 4.49\text{ m}$$
$$R_x = \frac{2.4}{4.49}(6\text{ kN}) = 3.207\text{ kN}$$
$$R_y = \frac{3.6}{4.49}(6\text{ kN}) = 4.811\text{ kN}$$
$$R_z = \frac{1.2}{4.49}(6\text{ kN}) + 2\text{ kN} = 3.604\text{ kN}$$
$$R = \sqrt{R_x^2 + R_y^2 + R_z^2} = 6.81\text{ kN}$$

7.17 a.

$$M_{AB} = \mathbf{e}_{AB}\bullet(\mathbf{r}_{AP} \times \mathbf{f})$$

$$= \left(-\frac{3}{5}\mathbf{e}_x + \frac{4}{5}\mathbf{e}_z\right)\bullet\left[(4\text{ m})\mathbf{e}_z \times \left(\frac{3}{13}\mathbf{e}_x + \frac{12}{13}\mathbf{e}_y + \frac{4}{13}\mathbf{e}_z\right)f\right]$$

$$= \begin{vmatrix} -\frac{3}{5} & 0 & \frac{4}{5} \\ 0 & 0 & 4\text{ m} \\ \frac{3}{13} & \frac{12}{13} & \frac{4}{13} \end{vmatrix} f = \left(\frac{144}{65}\text{ m}\right)f$$

7.18 c.

$$R_R - (2648\text{ N})\cos 30° = 0$$
$$R_R = 2.29\text{ kN}$$

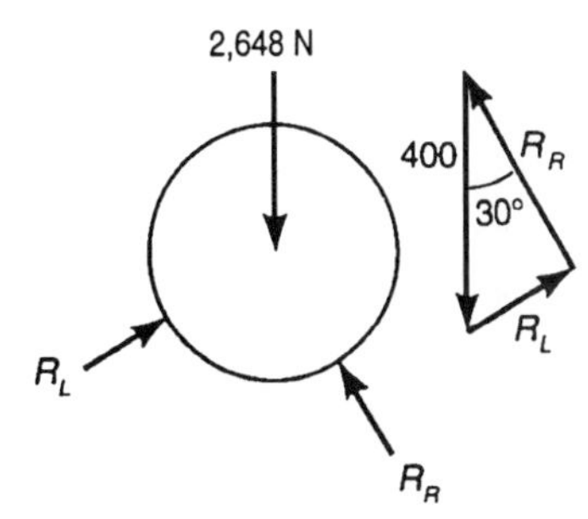

Exhibit 7.18a

7.19 a.

$$M_0 = R\sin(30° - \theta)f_{g_1} - R\sin(30° + \theta)f_{g_2} = 0$$

$$f_{g_1}(\sin 30° \cos\theta - \cos 30° \sin\theta) - f_{g_2}(\sin 30° \cos\theta + \cos\theta + \cos 30° \sin\theta) = 0$$

Dividing by $\cos 30° \cos\theta$,

$$f_{g_1}(\tan 30° - \tan\theta) = f_{g_2}(\tan 30° + \tan\theta)$$

$$\tan\theta = \frac{f_{g_1} - f_{g_2}}{f_{g_1} + f_{g_2}}\tan 30°$$

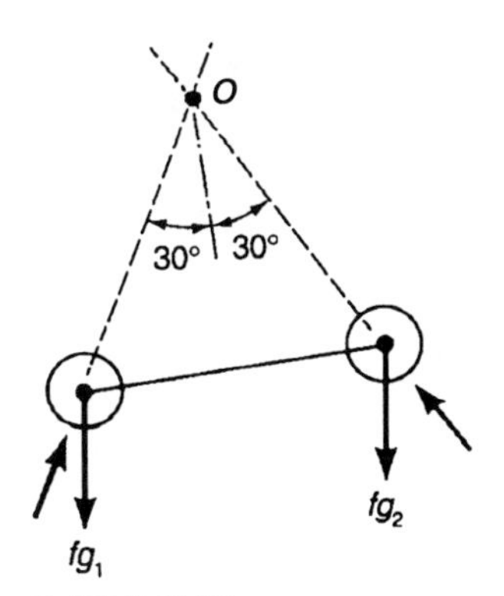

Exhibit 7.19a

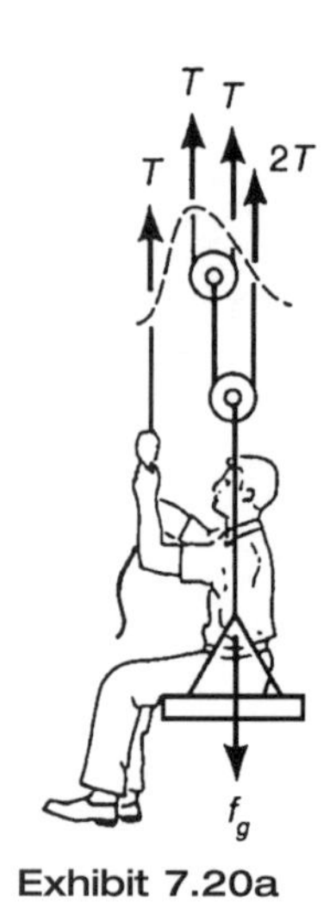

Exhibit 7.20a

7.20 d.

$$T + T + T + 2T - f_g = 0$$

$$T = \frac{1}{5} f_g$$

$$= \frac{1}{5}(80\text{ kg})(9.8\text{ N/kg}) = 157\text{ N}$$

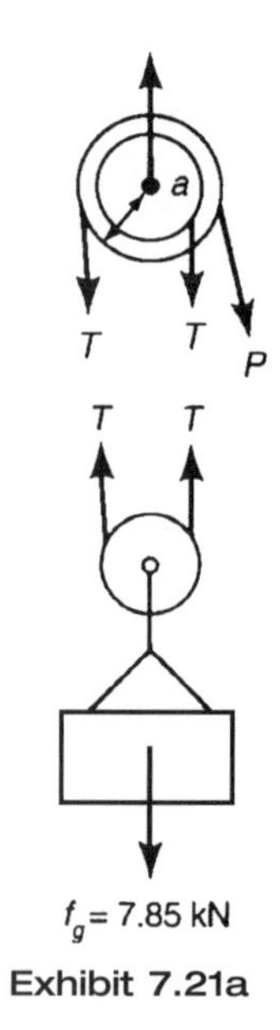

Exhibit 7.21a

7.21 a. The force of gravity on the block is

$$f_g = 0.8\text{ Mg} = 0.8(9.81)\text{ kN} = 7.85\text{ kN}$$
$$aT - 0.9aT - aP = 0$$
$$P = 0.1T$$
$$2T = f_g$$
$$P = 0.1\left(\frac{1}{2} f_g\right) = 0.05 f_g = 0.392\text{ kN}$$

7.22 d.

$$\sum M_A = aQ - 4aP = 0 \quad Q = 4P$$
$$\sum M_B = 2af_g - 2a\cot 30° \quad Q = 0$$
$$f_g = \sqrt{3}\,Q$$
$$P = \frac{Q}{4} = \frac{f_g}{4\sqrt{3}} = 0.144 f_g$$

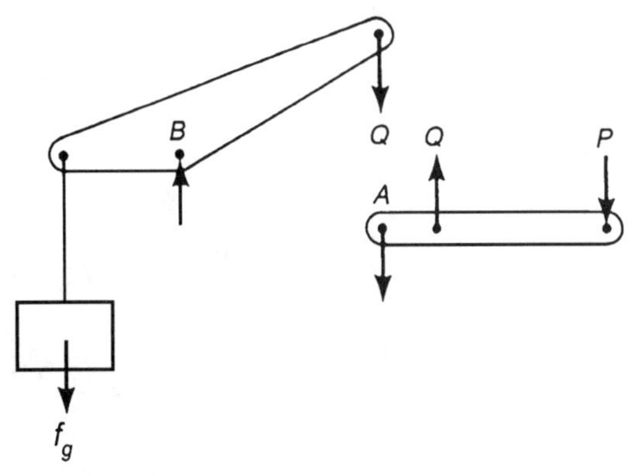

Exhibit 7.22a

7.23 c.

$$\sum M_O = (2.4\text{ m})T_{AB} - (1.2\text{ m})f_g = 0$$

$$T_{AB} = \frac{1.2}{2.4}(320\text{ kg})(9.8\text{ N/kg}) = 1.57\text{ kN}$$

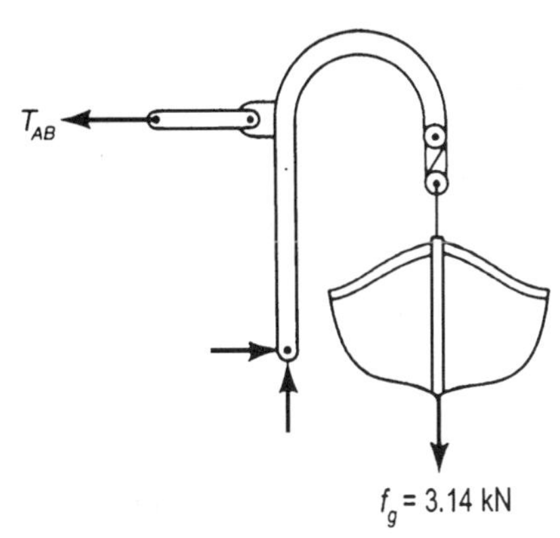

Exhibit 7.23a

7.24 a. Neglecting the force of gravity on the table, $N(200\text{ mm}) + \mu N(130\text{ mm}) = P(535\text{ mm})$ by summing moments about point A. Thus,

$$2\mu N \geq P = \left(\tfrac{200}{535} + \tfrac{130\mu}{535}\right)N$$

$$\mu \geq \frac{\frac{200}{535}}{2 - \frac{130}{535}} = \frac{10}{47} = 0.213$$

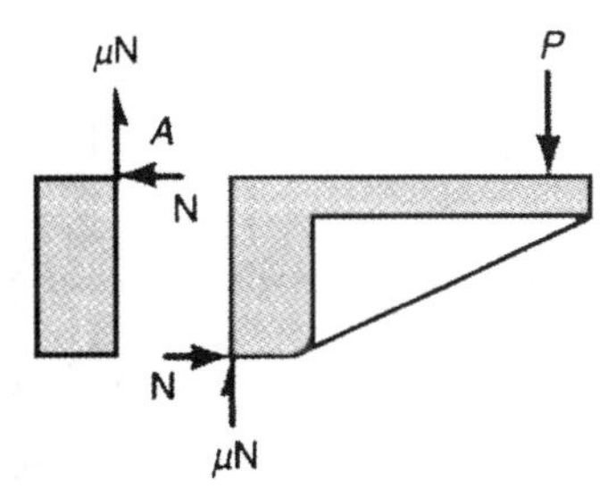

Exhibit 7.24a

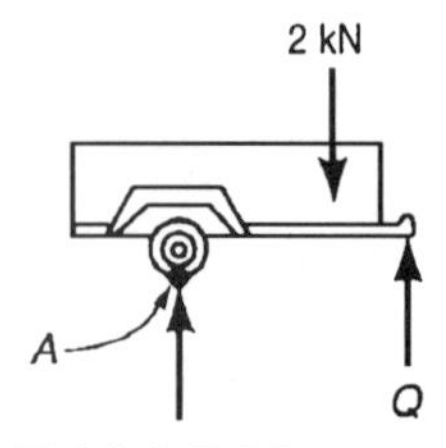

Exhibit 7.25a

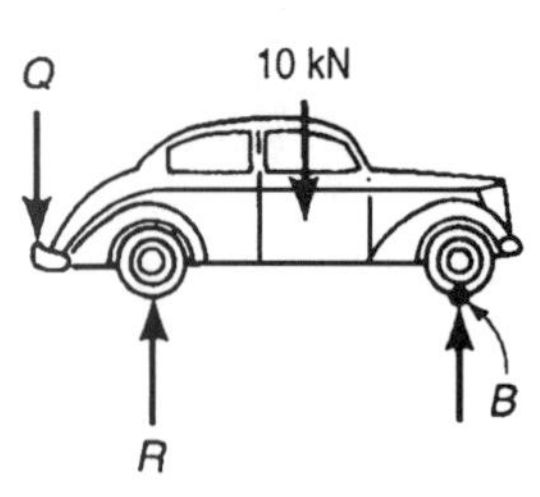

Exhibit 7.25b

7.25 c.

$$\sum M_A = (1.8\text{ m})Q - (1.2\text{ m})(2\text{ kN}) = 0$$

$$Q = \frac{1.2}{1.8}(2\text{ kN}) = 1.33\text{ kN}$$

$$\sum M_B = (3.6\text{ m})Q - (2.4\text{ m})R + (1.2\text{ m})(10\text{ kN}) = 0$$

$$R = \frac{(3.6\text{ m})(1.33\text{ kN}) + (1.2\text{ m})(10\text{ kN})}{2.4\text{ m}} = 7.0\text{ kN}$$

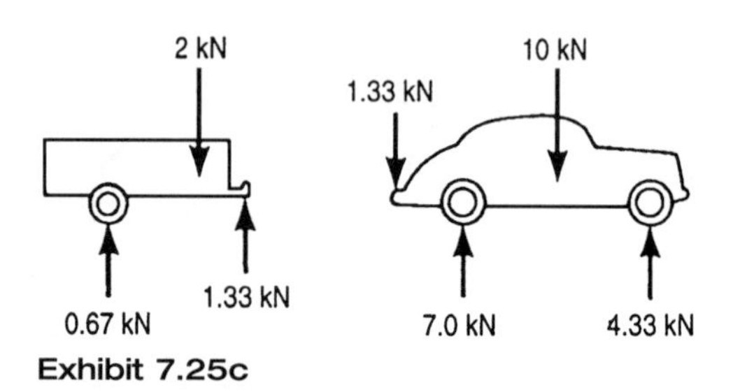

Exhibit 7.25c

7.26 d.

$$\sum M_A = (0.58\text{ m})(1.7\text{ kN}) - M = 0$$

$$M = 0.986\text{ kN}\bullet\text{m}$$

$$\sum F_U = 1.7\text{ kN} - P = 0$$

$$P = 1.7\text{ kN}$$

$$\sum M_C = R(3200\text{ mm}) + 986\text{ N}\bullet\text{m} + (1.7\text{ kN})(2820\text{ mm}) - (7.0\text{ kN})(850\text{ mm}) = 0$$

$$R = 0.053\text{ kN}$$

$$R_C = 7.0\text{ kN} - 1.7\text{ kN} - \text{R} = 5.25\text{ kN}$$

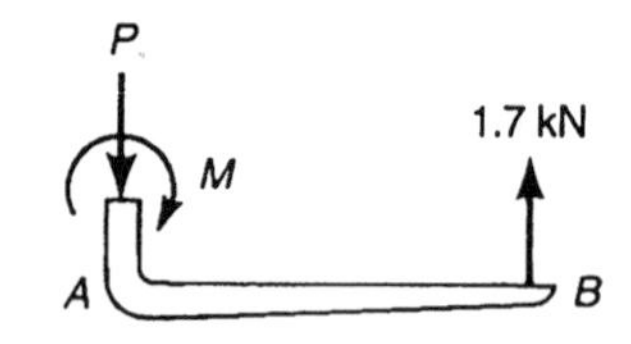

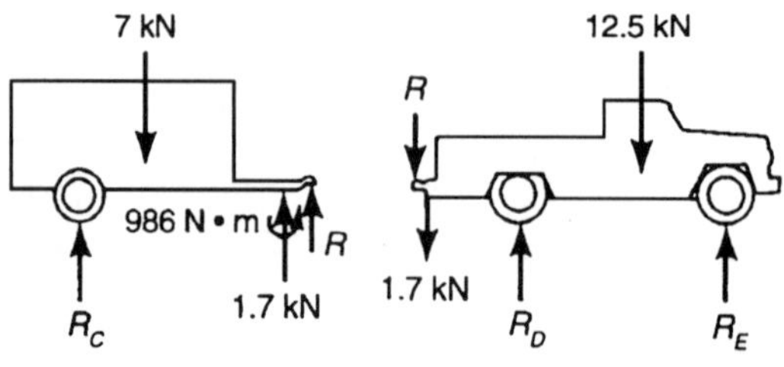

Exhibit 7.26c

7.27 c.

$$\sum M_E = (12.5 \text{ kN})(1190 \text{ mm}) + (1.7 \text{ kN})(4140 \text{ mm}) + R(4340 \text{ mm}) - R_D(2750 \text{ mm}) = 0$$

$$R_D = 8.05 \text{ kN}$$

7.28 d.

$$\sum M_A = (105 \text{ mm})P - (30 \text{ mm})Q = 0$$

$$Q = 3.5P$$

$$\sum M_B = (80 \text{ mm})Q - (35 \text{ mm})R = 0$$

$$R = \frac{80}{35}(3.5P) = 8.0P$$

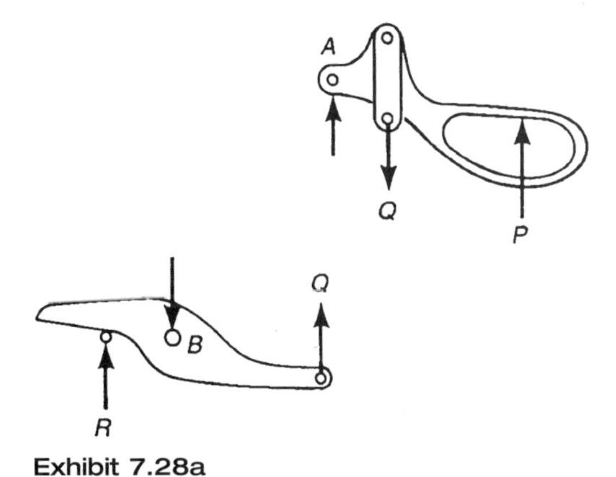

Exhibit 7.28a

7.29 d.

$$h = 115 \tan 20° = 135 \tan(20° - \phi)$$

$$\phi = 20° - \tan^{-1}\left(\frac{115}{135}\tan 20°\right) = 2.77°$$

$$\overline{O}\,\overline{A} = \frac{115 \text{ mm}}{\cos 20°} = 122.4 \text{ mm}$$

$$\sum M_A = \overline{O}\,\overline{A} \sin\phi\, R - (115 \text{ mm})(120 \text{ N}) = 0$$

$$R = \frac{115(120)}{122.4 \sin 2.77°} = 2330 \text{ N}$$

$$\sum M_B = (26 \text{ mm})R\sin(85° + \phi) - (30 \text{ mm})Q = 0$$

$$Q = \frac{26}{30}(2330 \text{ N})\sin 87.77° = 2020 \text{ N}$$

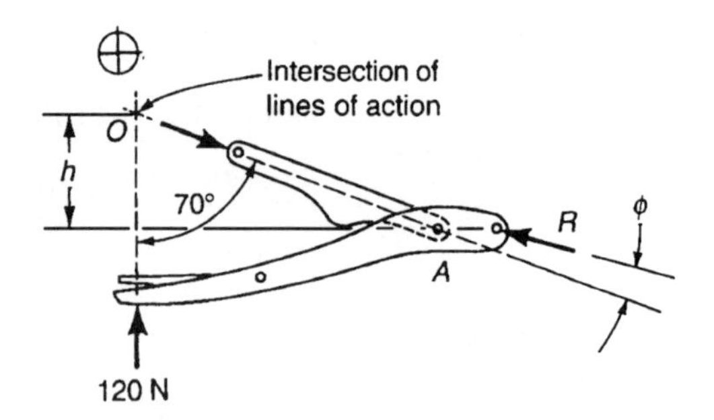

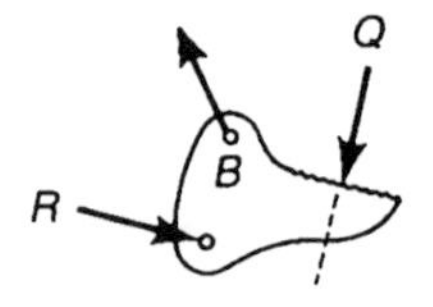

Exhibit 7.29a

7.30 a.

$$M = (600 \text{ N})(0.18 \text{ m}) - (480 \text{ N})(0.18 \text{ m}) + (320 \text{ N})(0.15 \text{ m})$$
$$- (400 \text{ N})(0.15 \text{ m}) = 9.60 \text{ N}\bullet\text{m}$$

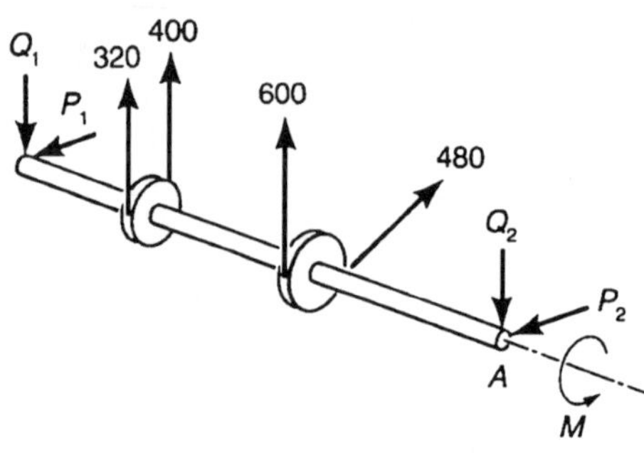

Exhibit 7.30a

7.31 c.

$$P_2(1.7 \text{ m}) = (480 \text{ N})(\cos 38°)(1.1 \text{ m}), \qquad P_2 = 245 \text{ N}$$

$$Q_2(1.7 \text{ m}) = (720 \text{ N})(0.4 \text{ m}) + (600 \text{ N} + 480 \text{ N} \sin 38°)(1.1 \text{ m}), \qquad Q_2 = 749 \text{ N}$$

$$R_2 = \sqrt{P_2^2 + Q_2^2} = 788 \text{ N}$$

7.32 c.

$$\mathbf{T} = T\mathbf{e}_{AB} = T\,\frac{-1.8\mathbf{e}_x + 1.5\mathbf{e}_y + 0.6\mathbf{e}_z}{\sqrt{5.85}}$$

Vertical forces at *A*: $\dfrac{0.6}{\sqrt{5.85}}T = mg$

y-forces at *B*: $F\dfrac{1.5}{\sqrt{5.85}}T$

Combine: $F = \dfrac{1.5}{0.6}mg = 98 \text{ N}$

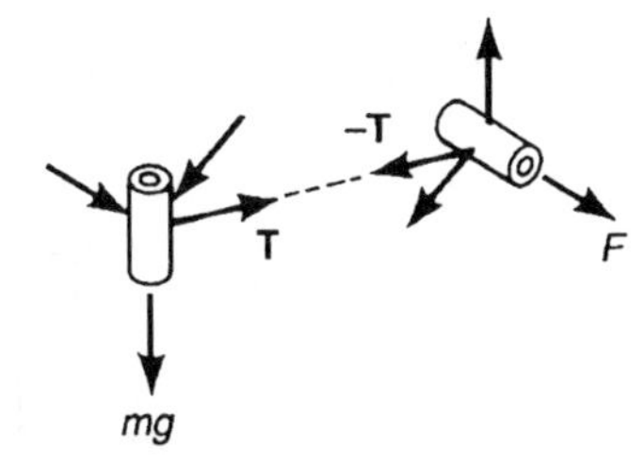

Exhibit 7.32a

7.33 a.

$$d = \frac{4}{5}(1.8\text{ m}) = 1.44\text{ m}$$

$$\mathbf{T}_{AE} = T_{AE}\left(\frac{3}{13}\mathbf{e}_x - \frac{4}{13}\mathbf{e}_y - \frac{12}{13}\mathbf{e}_z\right)$$

$$\sum M_{BC} = -(2.94\text{ m})\left(\frac{12}{13}T_{AE}\right) + (1.44\text{ m})R_z = 0$$

$$T_{AE} = \frac{26}{49}R_z = 424\text{ N}$$

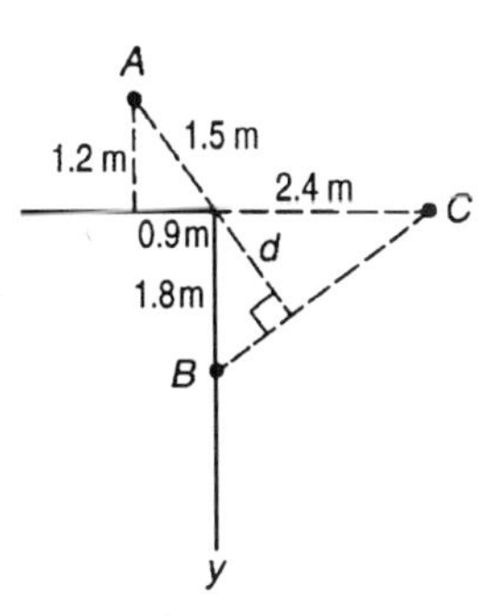

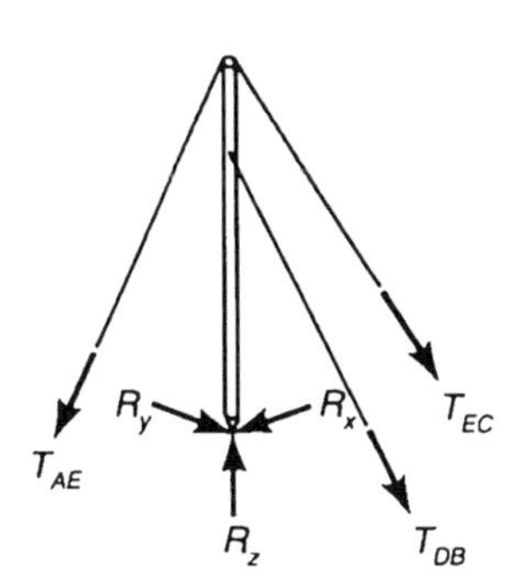

Exhibit 7.33a

7.34 a.

$$\mathbf{e}_{OD} = \frac{3.3\mathbf{e}_y + 0.75\mathbf{e}_z}{\sqrt{11.4525}} \qquad \mathbf{T}_{CD} = T_{CD}\frac{-1.8\mathbf{e}_x + 3.3\mathbf{e}_y + 0.75\mathbf{e}_z}{\sqrt{14.6925}}$$

$$\mathbf{e}_{OB} = \frac{-1.2\mathbf{e}_y + 1.65\mathbf{e}_z}{\sqrt{4.1625}} \qquad \mathbf{T}_{AB} = T_{AB}\frac{-3\mathbf{e}_x - 1.2\mathbf{e}_y + 1.65\mathbf{e}_z}{\sqrt{13.1625}}$$

$$M_{OD} = \mathbf{e}_{OD} \bullet [(3\text{ m})\mathbf{e}_x \times \mathbf{T}_{AB}] + \mathbf{e}_{OD} \bullet [(2.4\text{ m})\mathbf{e}_x \times (-1.5\text{ kN }\mathbf{e}_z)]$$
$$= (-1.5484\text{ m})T_{AB} + 3.5128\text{ kN} \bullet \text{m} = 0$$

$$T_{AB} = \frac{3.5128\text{ kN} \bullet \text{m}}{1.5484\text{ m}} = 2.27\text{ kN}$$

$$M_{OB} = \mathbf{e}_{OB} \bullet [(1.8\text{ m})\mathbf{e}_x \times \mathbf{T}_{CD}] + \mathbf{e}_{OB} \bullet [(2.4\text{ m})\mathbf{e}_x \times (-1.5\text{ kN }\mathbf{e}_z)]$$
$$= (1.4604\text{ m})T_{CD} - 2.1174\text{ kN} \bullet \text{m} = 0$$

$$T_{CD} = \frac{2.1174\text{ kN} \bullet \text{m}}{1.4604\text{ m}} 1.45\text{ kN}$$

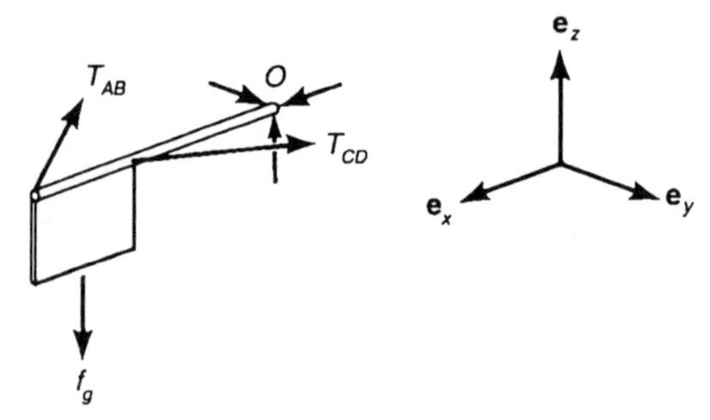

Exhibit 7.34a

7.35 d.

$$M_{OA} = \mathbf{e}_{OA} \bullet (0.45 \text{ m } \mathbf{e}_y \times \mathbf{T}) - (0.225 \text{ m})\frac{12}{13} f_g$$

$$= \frac{0.45 \text{ m}}{13}\left(\frac{46T}{7} - 6f_g\right) = 0$$

$$T = \frac{21}{23} f_g = \frac{21}{23} (40 \text{ kg})(9.8 \text{ N/kg}) = 358 \text{ N}$$

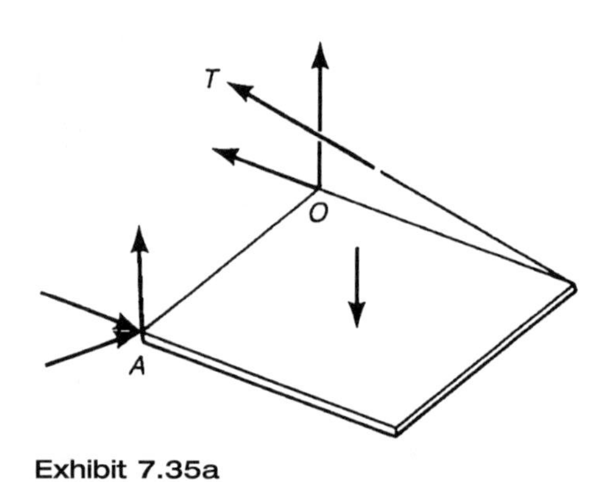

Exhibit 7.35a

7.36 b.

$$\sum M_A = a(4 \text{ kN}) + 4aR_D - a(24 \text{ kN}) - 3a(8 \text{ kN}) = 0$$

$$R_D = 11.0 \text{ kN}$$

$$\sum M_E = 3a(11 \text{ kN}) - 2a(8 \text{ kN}) - aT_{BC} = 0$$

$$T_{BC} = 17 \text{ kN (tension)}$$

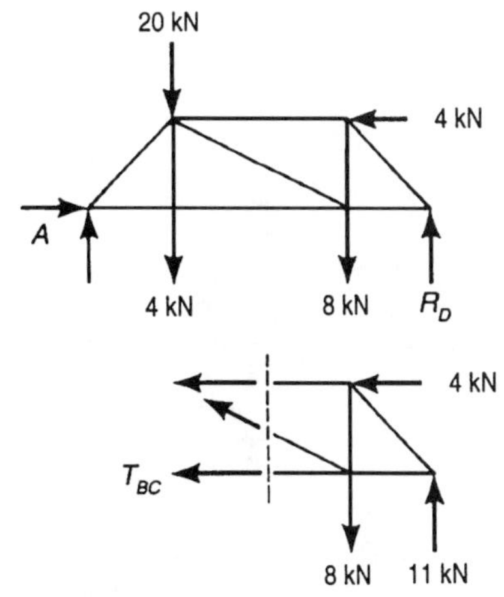

Exhibit 7.36a

7.37 b.

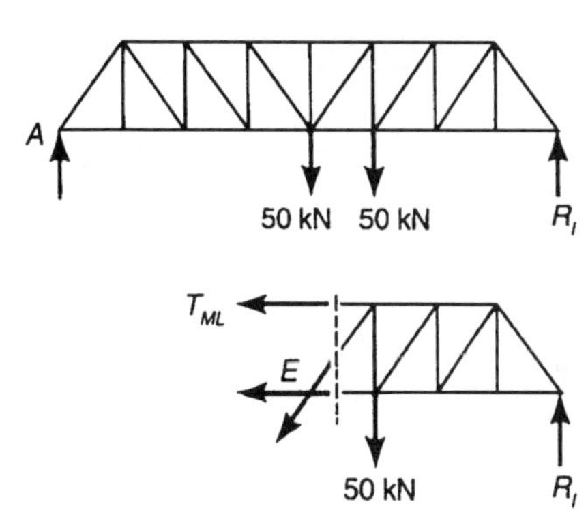

Exhibit 7.37a

$$\sum M_A = (24\text{ m})R_I - (12\text{ m})(50\text{ kN}) - (15\text{ m})(50\text{ kN}) = 0$$
$$R_I = 56.25\text{ kN}$$
$$\sum M_E = (4\text{ m})T_{ML} + (12\text{ m})(56.25\text{ kN}) - (3\text{ m})(50\text{ kN}) = 0$$
$$T_{ML} = -131.25\text{ kN}$$

7.38 c.

$$\sum M_A = (30\text{ m})R_G - (22.5\text{ m})(30\text{ kN}) - (15\text{ m})(25\text{ kN}) - (7.5\text{ m})(20\text{ kN}) = 0$$
$$R_G = 40\text{ kN}$$
$$\sum M_D = (15\text{ m})(40\text{ kN}) - (7.5\text{ m})(30\text{ kN}) - (5\sqrt{3}\text{ m})T_{CF} = 0$$
$$T_{CF} = 43.3\text{ kN}$$

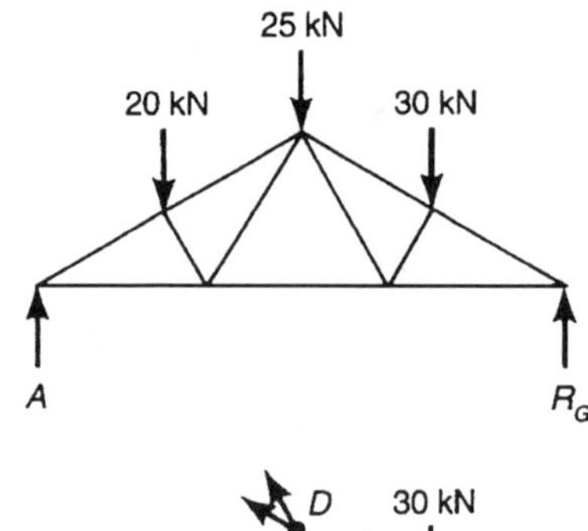

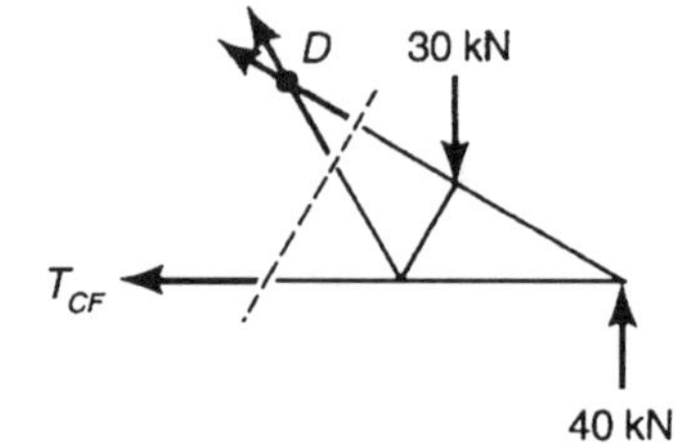

Exhibit 7.38a

7.39 a.

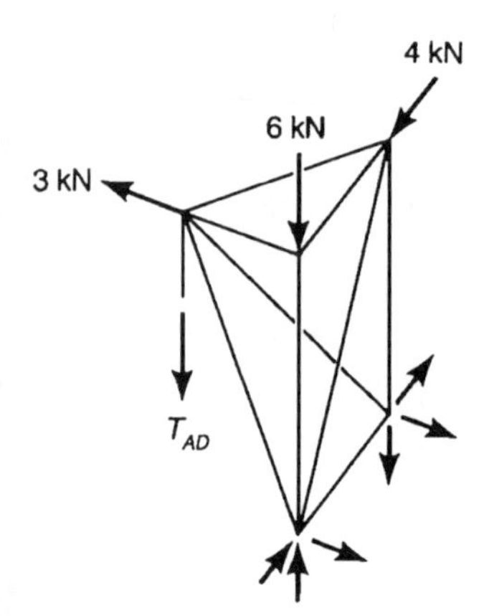

Exhibit 7.39a

$$\sum M_{CB} = (6\text{ m})\frac{\sqrt{3}}{2}(3\text{ kN}) + \left(\frac{3\sqrt{3}}{2}\text{ m}\right)T_{AD} = 0$$

$$T_{AD} = -6\text{ kN (compression)}$$

7.40 a.

$$\sum M_{DE} = aP - aT_{AB} = 0; \qquad T_{AB} = P$$

$$\sum M_{DC} = \mathbf{e}_{CD} \bullet [a\mathbf{e}_z \times (P\mathbf{e}_x + \mathbf{T}_{BE})] = 0$$

$$\mathbf{T}_{BE} = T_{BE}\frac{\mathbf{e}_x + \mathbf{e}_y - \mathbf{e}_z}{\sqrt{3}}$$

$$(\mathbf{e}_{CD} \bullet \mathbf{e}_y)P = -\mathbf{e}_{CD} \bullet \left(\mathbf{e}_z \times T_{BE}\frac{\mathbf{e}_x + \mathbf{e}_y - \mathbf{e}_z}{\sqrt{3}}\right)$$

$$\frac{\mathbf{e}_x - \mathbf{e}_y}{\sqrt{2}} \bullet \mathbf{e}_y P = \frac{\mathbf{e}_x - \mathbf{e}_y}{\sqrt{2}} \bullet \frac{\mathbf{e}_x - \mathbf{e}_y}{\sqrt{3}} T_{BE}$$

$$T_{BE} = -\frac{\sqrt{3}}{2}P, \text{ or } -0.87P \text{ (compression)}$$

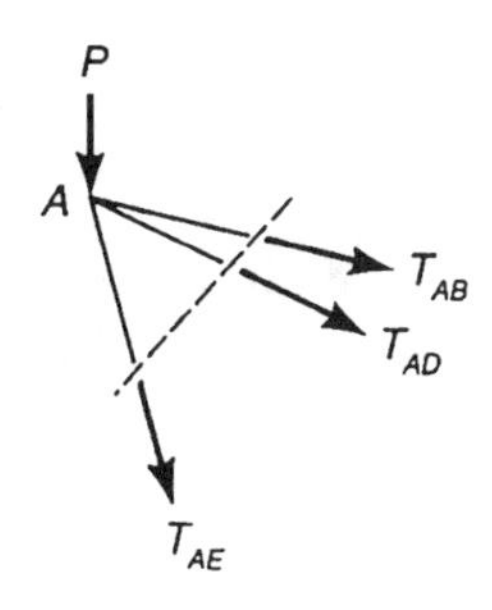

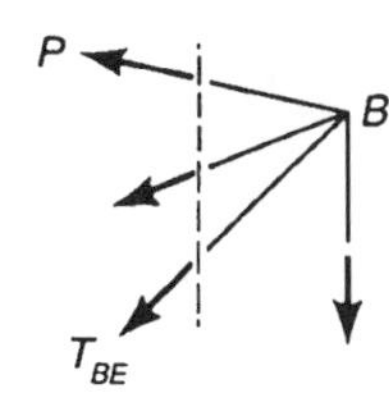

Exhibit 7.40a

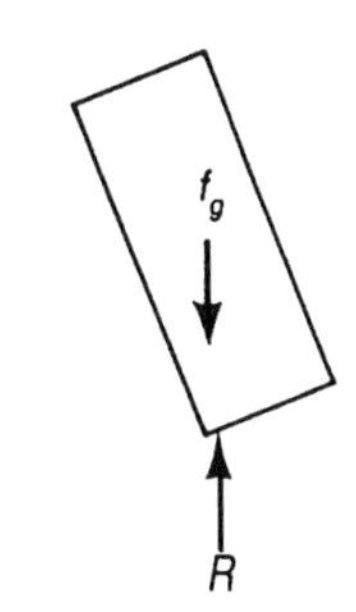

Exhibit 7.41a

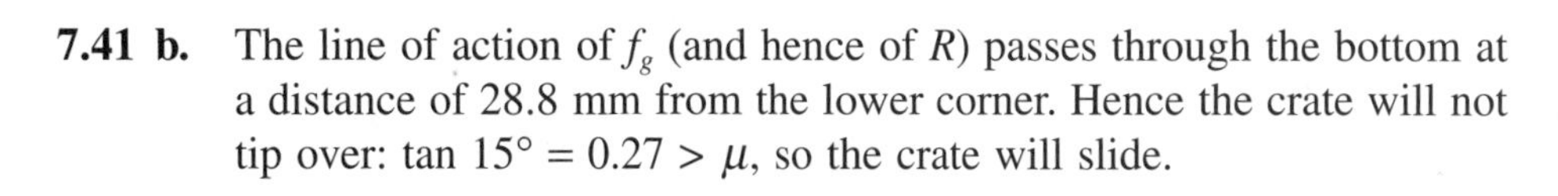

7.41 b. The line of action of f_g (and hence of R) passes through the bottom at a distance of 28.8 mm from the lower corner. Hence the crate will not tip over: $\tan 15° = 0.27 > \mu$, so the crate will slide.

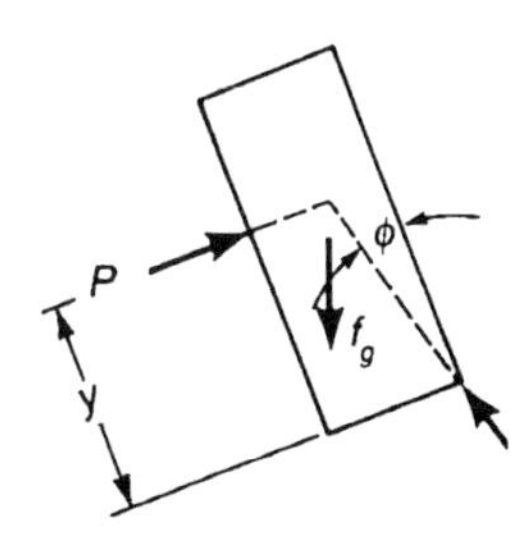

Exhibit 7.42

7.42 d. Vanishing resultant moment requires that the lines of action of the three forces intersect, so that

$$\tan \phi = \mu = \frac{0.27 \text{ m} - (y - 0.9 \text{ m}) \tan 15°}{y}$$

$$y = \frac{0.27 \text{ m} + (0.9 \text{ m}) \tan 15°}{\mu + \tan 15°} = 1.09 \text{ m}$$

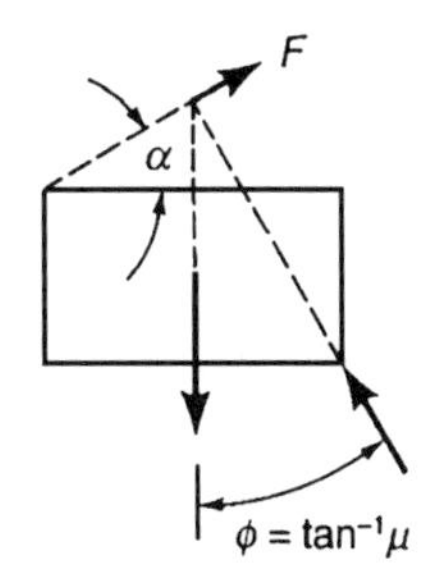

Exhibit 7.43a

7.43 c. Vanishing resultant moment requires that the lines of action of the three forces intersect, so that

$$\frac{b}{2} \tan \alpha = \frac{b}{2} \cot \phi - h = \frac{b}{2\mu} - h$$

$$\alpha = \tan^{-1}\left(\frac{1}{\mu} - \frac{2h}{b}\right)$$

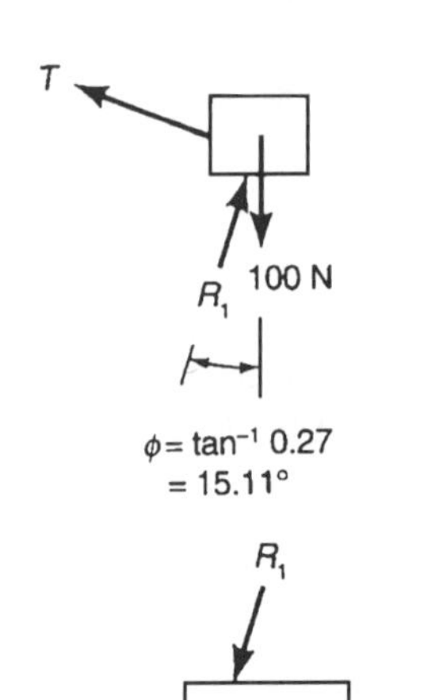

Exhibit 7.44a

7.44 a. Sum forces that are parallel to T on the upper block:

$$R_1 \cos(20° - 15.11°) = (100 \text{ N})\cos 20°; \qquad R_1 = 94.3 \text{ N}$$

Sum forces that are perpendicular to R_2 on the lower block:

$$P \cos 15.11° - (200 \text{ N}) \sin 15.11° - (94.3 \text{ N})\sin 30.22° = 0$$

$$P = 0.27(200 \text{ N}) + \frac{\sin 45.22°}{\cos 15.11°}(206 \text{ N}) = 103 \text{ N}$$

7.45 a.

$$\phi = \tan^{-1}(0.27) = 15.11°$$

Sum forces that are perpendicular to R_0:

$$R_1 \cos(15° + 2\phi) = (150 \text{ N}) \cos \phi; \quad R_1 = 206 \text{ N}$$

Sum forces that are perpendicular to R_2:

$$P \cos \phi - (200 \text{ N}) \sin \phi - R_1 \sin(15° + 2\phi) = 0$$

$$P = 0.27(200 \text{ N}) + \frac{\sin 45.22°}{\cos 15.11°}(206 \text{ N}) = 103 \text{ N s}$$

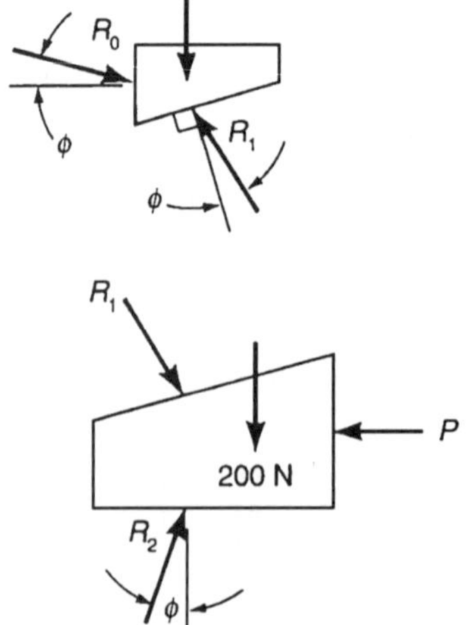

Exhibit 7.45a

7.46 b. Vanishing resultant moment requires that the lines of action of the three forces intersect, so that

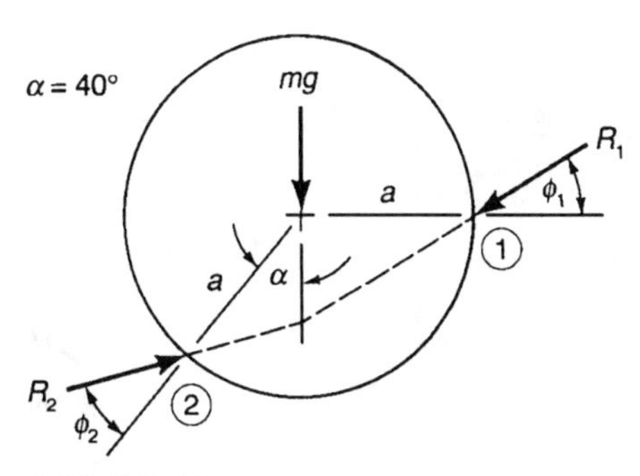

Exhibit 7.46a

$$a \tan \phi_1 = a \cos \alpha - \frac{a \sin \alpha}{\tan(\alpha + \phi_2)}$$

If slip occurs at ②, $\phi_2 = \tan^{-1} 0.3 = 16.70°$.

$$\tan \phi_1 = \cos 40° - \frac{\sin 40°}{\tan 56.70°} = 0.344$$

Since this is less than $\mu_1 = 0.45$, slip *does* occur at ②.

$$\sum M_2 = a \cos \alpha \; R_1 \cos \phi_1 - (a + a \sin \alpha) R_1 \sin \phi_1 - a \sin \alpha \, mg = 0$$

$$R_1 \cos \phi_1 = \frac{mg \sin \alpha}{\cos \alpha - (1 + \sin \alpha) \tan \phi_1} = 3.20 mg = 62.7 \text{ kN}$$

7.47 c. $\sum F_x = 0$ on the block leads to

$$T \sin 45° - \mu N = 0$$
$$-T\sqrt{2} + Ne = 0$$

Combine:

$$e = 2\mu$$

$$(T \cos 30°)(9.4) - N(4 + e) = 0$$

$$\frac{T}{N} = \frac{(4 + e)}{(9.4 \cos 30°)} = \frac{\mu}{\sin 45°}$$

$$\mu = \frac{2}{\frac{9.4 \cos 30°}{2 \sin 45°} - 1} = 0.42$$

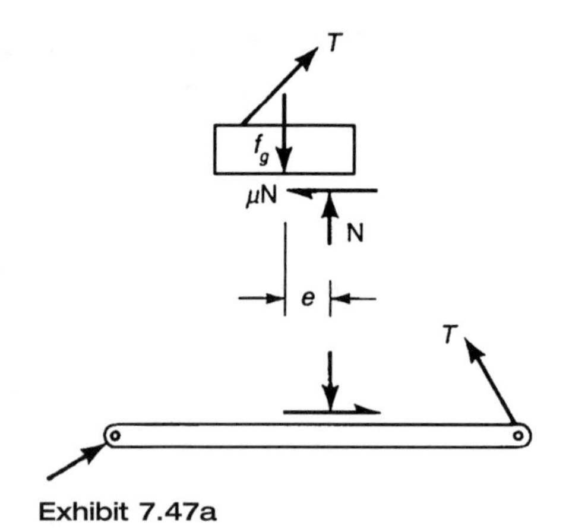

Exhibit 7.47a

7.48 c.

$$d = a + (b + a)\cos 2\phi$$

$$\cos 2\phi = \frac{1 - \tan^2 \phi}{1 + \tan^2 \phi}$$

To prevent slip, $\tan \phi < \mu$, so that

$$d > a + (b + a)\frac{1 - \mu^2}{1 + \mu^2} = \frac{2a + (1 - \mu^2)b}{1 + \mu^2}$$

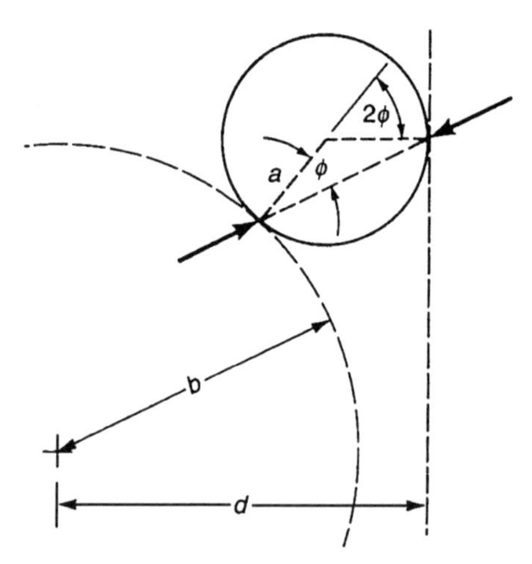

Exhibit 7.48a

CHAPTER 8

Dynamics

Charles Smith and Fidelis O. Eke

OUTLINE

The analysis of a mechanical system having elements under acceleration must consider these accelerations along with the related forces. In such analysis, the *force* side of Newton's second law, $\mathbf{f} = m\mathbf{a}$, and the third law of action and reaction are dealt with in exactly the same manner as in statics. But it is the relationships among positions, velocities, and accelerations that complete the discipline of dynamic analysis. The following two sections review these relationships, and the remainder of the chapter deals with their incorporation into Newton's laws of motion.

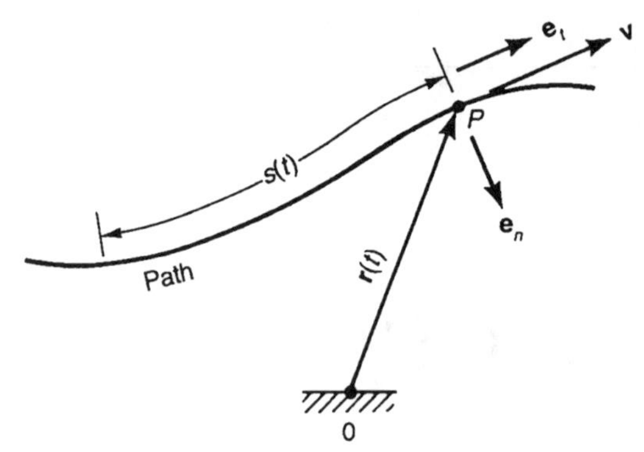

Figure 8.1

KINEMATICS OF A PARTICLE

Consider a point P that moves along a smooth path as indicated in Fig. 8.1. The position of the point may be specified by the vector $\mathbf{r}(t)$, defined to extend from an arbitrarily selected, fixed point 0 to the moving point P. The **velocity** $\mathbf{v}$ of the point is defined to be the derivative with respect to t of $\mathbf{r}(t)$, written as

$$\mathbf{v} = \frac{d\mathbf{r}}{dt} \tag{8.1}$$

Although this definition is sometimes used for evaluation [that is, by differentiating a specific expression for $\mathbf{r}(t)$], it will often be more direct to use other relationships. It follows from the above definition that the velocity vector is tangent to the path of the particle; thus, upon introduction of a unit vector $\mathbf{e}_t$, defined to be tangent to the path, the velocity can also be expressed as

$$\mathbf{v} = v\mathbf{e}_t \tag{8.2}$$

The position of P can also be specified in terms of the distance $s(t)$ traveled along the path from an arbitrarily selected reference point. Then an incremental change in position may be approximated as $\Delta\mathbf{r} \approx \Delta s\mathbf{e}_t$, in which the accuracy increases as the increments Δt and $\Delta\mathbf{r}$ approach zero. This leads to still another way of expressing the velocity as

$$\mathbf{v} = \frac{ds}{dt}\mathbf{e}_t \tag{8.3}$$

The scalar

$$v = \frac{ds}{dt} = \dot{s} \tag{8.4}$$

can be either positive or negative, depending on whether the motion is in the same or the opposite direction as that selected in the definition of $\mathbf{e}_t$.

The **acceleration** of the point is defined as the derivative of the velocity with respect to time:

$$\mathbf{a} = \frac{d\mathbf{v}}{dt} \tag{8.5}$$

A useful relationship follows from application to Eq. (8.2) of the rules for differentiating products and functions of functions:

$$\frac{d\mathbf{v}}{dt} = \dot{v}\mathbf{e}_t + \mathbf{v}\frac{ds}{dt}\frac{d\mathbf{e}_t}{ds}$$

As the direction of $\mathbf{e}_t$ varies, the square of its magnitude, $|\mathbf{e}_t|^2 = \mathbf{e}_t \bullet \mathbf{e}_t$, remains fixed and equal to 1, so that

$$\frac{d}{ds}|\mathbf{e}_t|^2 = \frac{d}{ds}(\mathbf{e}_t \bullet \mathbf{e}_t) = 2\mathbf{e}_t \bullet \frac{d\mathbf{e}_t}{ds} = 0$$

This shows that $d\mathbf{e}_t/ds$ is either zero or perpendicular to $\mathbf{e}_t$. With another unit vector $\mathbf{e}_n$ defined to be in the direction of $d\mathbf{e}_t/ds$, this vector may be expressed as

$$\frac{d\mathbf{e}_t}{ds} = \kappa \mathbf{e}_n$$

The scalar κ is called the local **curvature** of the path; its reciprocal, $\rho = 1/\kappa$, is called the local **radius of curvature** of the path. In the special case in which the path is straight, the curvature, and hence $d\mathbf{e}_t/ds$, are zero. These lead to the following expression for the **acceleration** of the point:

$$\mathbf{a} = \dot{v}\mathbf{e}_t + \frac{v^2}{\rho}\mathbf{e}_n \tag{8.6}$$

The two terms express the **tangential** and **normal** (or **centripetal**) components of acceleration.

If a driver of a car with sufficient capability "steps on the gas," a positive value of $\dot{v}$ is induced, whereas if he "steps on the brake," a negative value is induced. If the path of the car is straight (zero curvature or "infinite" radius of curvature), the entire acceleration is $\dot{v}\mathbf{e}_t$. If the car is rounding a curve, there is an additional component of acceleration directed laterally, toward the center of curvature of the path. These components are indicated in Fig. 8.2, a view of the plane of $\mathbf{e}_t$ and $d\mathbf{e}_t/ds$.

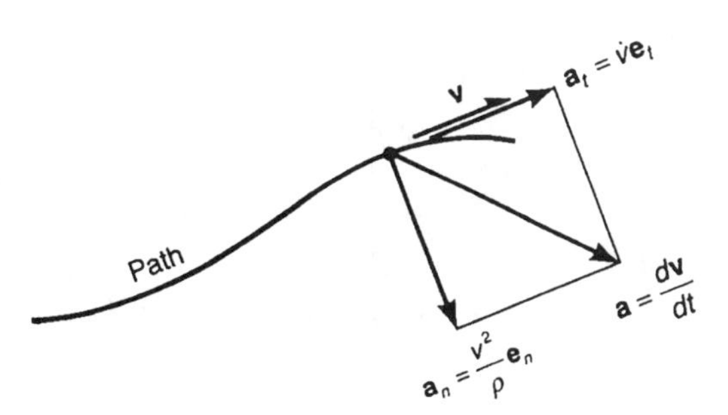

Figure 8.2

Example 8.1

At a certain instant, the velocity and acceleration of a point have the rectangular Cartesian components given by

$$\mathbf{v} = (3.5\mathbf{e}_x - 7.2\mathbf{e}_y + 9.6\mathbf{e}_z)\ \text{m/s}$$
$$\mathbf{a} = (-20\mathbf{e}_x + 20\mathbf{e}_y + 10\mathbf{e}_z)\ \text{m/s}^2$$

At this instant, what are the rate of change of speed dv/dt and the local radius of curvature of the path?

Solution

The rectangular Cartesian components of the unit tangent vector can be determined by dividing the velocity vector by its magnitude:

$$\mathbf{e}_t \frac{\mathbf{v}}{|\mathbf{v}|} = \frac{3.5\mathbf{e}_x - 7.2\mathbf{e}_y + 9.6\mathbf{e}_z}{\sqrt{(3.5)^2 + (-7.2)^2 + (9.6)^2}} = 0.280\mathbf{e}_x - 0.576\mathbf{e}_y + 0.768\mathbf{e}_z$$

The rate of change of speed can then be determined as the projection of the acceleration vector onto the tangent to the path:

$$\begin{aligned}\dot{v} &= \mathbf{e}_t \bullet \mathbf{a} = [(0.280)(-20) + (-0.576)(20) + (0.768)(10)]\ \text{m/s}^2 \\ &= -9.44\ \text{m/s}^2\end{aligned}$$

The negative sign indicates the projection is opposite to $\mathbf{e}_t$ (which was defined by the above equation to be in the same direction as the velocity). This means that the speed is *decreasing* at 9.44 m/s. One sees from Fig. 8.2 that the normal component of acceleration has magnitude

$$a_n = \sqrt{|\mathbf{a}|^2 - \dot{v}^2} = \sqrt{(-20)^2 + (20)^2 + (10)^2 - (-9.44)^2\ \text{m/s}^2} = 28.5\ \text{m/s}^2$$

which, from Eq. (8.6), is related to the speed and radius of curvature by $a_n = v^2/\rho$. Rearrangement of this equation gives the radius of curvature as

$$\rho = \frac{v^2}{a_n} = \frac{[(3.5)^2 + (-7.2)^2 + (9.6)^2]\ \text{m}^2/\text{s}^2}{28.5\ \text{m/s}^2} = 5.48\ \text{m}$$

Relating Distance, Velocity, and the Tangential Component of Acceleration

The basic relationships among tangential acceleration a_t, velocity $v\mathbf{e}_t$, and distance s are

$$\frac{dv}{dt} = a_t \quad \text{or} \quad v = v_0 + \int a_t\, dt \tag{8.7}$$

$$\frac{ds}{dt} = v \quad \text{or} \quad s = s_0 + \int v\, dt \tag{8.8}$$

in which v_0 and s_0 are constants of integration. An alternative relationship comes from writing $dv/dt = (ds/dt)(dv/ds) = v\, dv/ds$:

$$v\frac{dv}{ds} = a_t \quad \text{or} \quad v^2 = v_0^2 + 2\int a_t\, ds \tag{8.9}$$

Equations (8.7) and (8.8) are useful in dealing with the *time* histories of acceleration, velocity, and distance, whereas Eq. (8.9) is helpful in dealing with the manner in which velocity and acceleration vary with distance.

Example 8.2

The variation of tangential acceleration with time is given in Exhibit 1. If a point with an initial velocity of 20 m/s is subjected to this acceleration, what will be its velocity at $t = 6$ s, 10 s, and 15 s, and what will be the values of s at $t = 4$ s, 7.6 s, and 15 s?

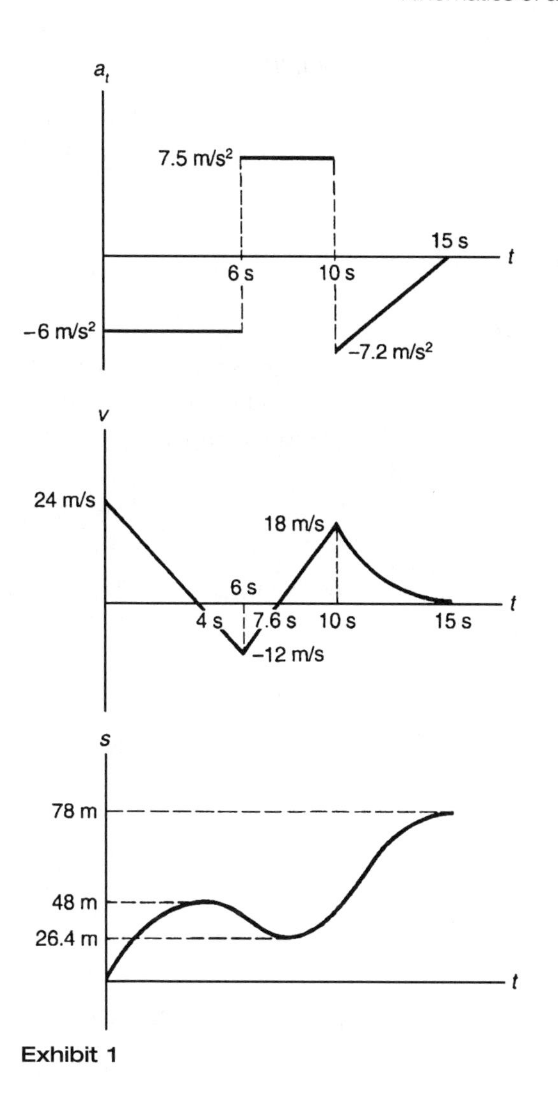

Exhibit 1

Solution

Equation (8.7) has the following graphical interpretations: At each point, the slope of the v-t curve is equal to the ordinate on the a_t-t curve. During any interval, the change in the value of v is equal to the area under the a_t-t curve for the same interval. With these rules and the given initial value of v, the variation of v with t can be plotted, and values of v can be calculated for each point.

The reader should use these rules to verify all details of the v-t curve shown. Equation (8.8) indicates that identical rules for slopes, ordinates, and areas relate the curve of distance s to that of velocity v, so the same procedure can be used to construct the s-t curve from the v-t curve. Again, the reader should verify all details of this curve.

Example **8.3**

The tangential acceleration of the pendulum bob shown in Exhibit 2 varies with position according to $a = -g \sin(s/l)$, in which g is the local acceleration of gravity. If a speed v_0 is imparted at the vertical position (where $s = 0$), what will be the maximum value of s reached?

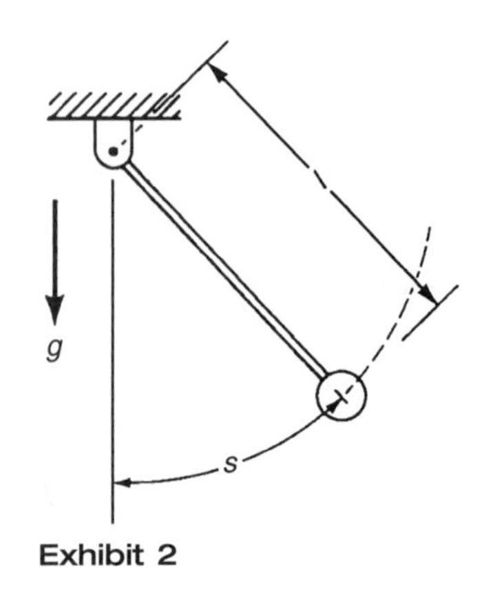

Exhibit 2

Solution

Because the relationship between tangential acceleration and *position* is given, Eq. (8.9) will prove useful. The integrated form leads to

$$v^2 = v_0^2 + 2\int_0^s g\sin\left(\frac{s}{l}\right)ds$$
$$= v_0^2 + 2gl\left(1-\cos\frac{s}{l}\right)$$

which gives the velocity v in the terms of any position s. Since $v = \dot{s}$, the maximum s will occur when $v = 0$, and the corresponding s is easily isolated from the above equation after setting $v = 0$:

$$s_{\max} = l\cos^{-1}\left(1-\frac{v_0^2}{2gl}\right)$$

Observe that if $v_0^2 > 2gl$, no real value of $s_{\max}$ exists, because v never reaches zero in that case.

Constant Tangential Acceleration

When the tangential acceleration is constant, Eqs. (8.7) through (8.9) reduce to

$$v = v_0 + a_t t \tag{8.10}$$

$$s = s_0 + v_0 t + \frac{1}{2}a_t t^2 \tag{8.11}$$

$$v^2 = v_0^2 + 2a_t s \tag{8.12}$$

Rectilinear Motion

In the special case in which the path is a straight line, the unit tangent vector $\mathbf{e}_t$ is constant, and the curvature $1/\rho$ is zero throughout. The acceleration is then given by $(dv/dt)\mathbf{e}_t$, and the subscript on the symbol a_t may be dropped without ambiguity.

Example 8.4

A particle is launched vertically upward with an initial speed of 10 m/s and subsequently moves with constant downward acceleration of magnitude 9.8 m/s^2. What is the maximum height reached by the particle? How long does it take to return to the original launch position? And how fast is it traveling at its return to the launch position?

Solution

In this case the path will be straight and the acceleration is constant. With $\mathbf{e}_t$ defined as upward, the constant scalars appearing in Eqs. (8.10) through (8.12) have the values v_0 = 10 m/s and $a_t = a = -9.8$ m/s^2, so that these equations become

$$v = 10\text{ m/s} - (9.8\text{ m/s}^2)t \tag{i}$$

$$s = (10\text{ m/s})t - \frac{1}{2}(9.8\text{ m/s}^2)t^2 \tag{ii}$$

$$v^2 = (10\text{ m/s})^2 - 2(9.8\text{ m/s}^2)s \tag{iii}$$

The maximum height reached can be obtained by setting $v = 0$ in (iii), which gives

$$s_{\max} = \frac{(10\text{ m/s})^2}{2(9.8\text{ m/s}^2)} = 5.1\text{ m}$$

The time required to reach this height can be obtained by setting $v = 0$ in (i), which gives

$$t_1 = \frac{10 \text{ m/s}}{9.8 \text{ m/s}^2} = 1.02 \text{ s}$$

Finally, setting $s = 0$ in (iii) yields the two values of v that specify the velocity at the launch position:

$$v = \pm 10 \text{ m/s}$$

The positive value gives the upward initial velocity, and the negative value gives the equal-magnitude, downward velocity of the particle when it returns to the launch position.

Rectangular Cartesian Coordinates

Multidimensional motion can be analyzed in terms of components associated with a set of fixed unit vectors $\mathbf{e}_x$, $\mathbf{e}_y$, and $\mathbf{e}_z$, which are defined to be mutually perpendicular. For some aspects of analysis, it is also important that they form a "right-handed" set, or $\mathbf{e}_z = \mathbf{e}_x \times \mathbf{e}_y$, $\mathbf{e}_x = \mathbf{e}_y \times \mathbf{e}_z$, and $\mathbf{e}_y = \mathbf{e}_z \times \mathbf{e}_x$. In terms of these unit vectors, the position, velocity, and acceleration can be expressed as

$$\mathbf{r} = x\mathbf{e}_x + y\mathbf{e}_y + z\mathbf{e}_z$$

$$\mathbf{V} = v_x\mathbf{e}_x + v_y\mathbf{e}_y + v_z\mathbf{e}_z$$

$$\mathbf{A} = a_x\mathbf{e}_x + a_y\mathbf{e}_y + a_z\mathbf{e}_z$$

with

$$v_x = \dot{x}$$
$$a_x = \dot{v}_x = \ddot{x}, \quad \text{etc.}$$

Example 8.5

A wheel rolls without slipping along a straight surface with the orientation of the wheel given in terms of the angle $\theta(t)$. See Exhibit 3. Express the velocity and acceleration of the point P on the rim of the wheel in terms of this angle, its derivatives, and the radius b of the wheel.

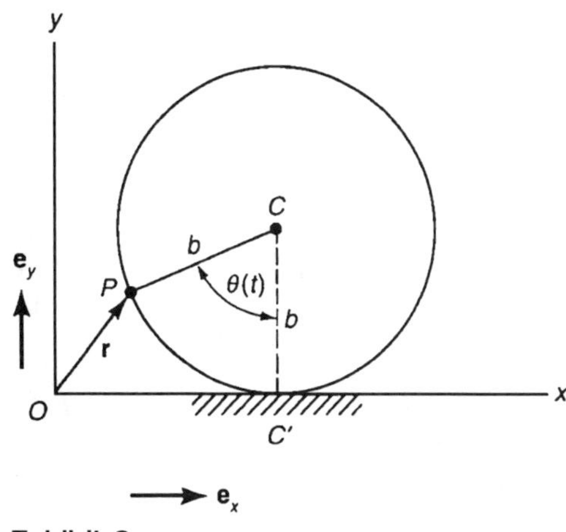

Exhibit 3

Solution

The origin for the x-y coordinates of P is the location of P when $\theta = 0$. Because the wheel rolls without slipping, the distance OC' is equal to the length of the circular arc PC'. The x-coordinate of P is then $OC' - b \sin \theta = b\theta - b \sin\theta$. The y-coordinate of P is that of C (i.e., b) minus $b \cos\theta$. In terms of these coordinates, the position vector from O to P may be expressed as

$$\mathbf{r} = b(\theta - \sin\theta)\mathbf{e}_x + b(1 - \cos\theta)\mathbf{e}_y$$

The velocity is then determined by differentiation of this expression:

$$\mathbf{v} = b\dot{\theta}[(1 - \cos\theta)\mathbf{e}_x + \sin\theta\mathbf{e}_y]$$

The acceleration is determined by another differentiation:

$$\mathbf{a} = b\ddot{\theta}[(1 - \cos\theta)\mathbf{e}_x + \sin\theta\mathbf{e}_y] + b\dot{\theta}^2(\sin\theta\mathbf{e}_x + \cos\theta\mathbf{e}_y)$$

These expressions may be simplified somewhat by rewriting them in terms of the unit vectors $\mathbf{e}_r$ and $\mathbf{e}_\theta$ as defined in Exhibit 4. These unit vectors are given in terms of the original horizontal and vertical unit vectors by

$$\mathbf{e}_r = -\sin\theta\mathbf{e}_x - \cos\theta\mathbf{e}_y$$
$$\mathbf{e}_\theta = -\cos\theta\mathbf{e}_x + \sin\theta\mathbf{e}_y$$

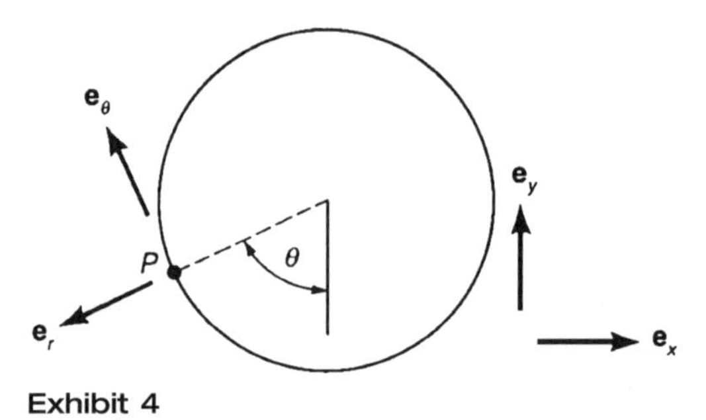

Exhibit 4

The above expressions for velocity and acceleration can now be written as

$$\mathbf{v} = b\dot{\theta}(\mathbf{e}_x + \mathbf{e}_\theta)$$
$$\mathbf{a} = b\ddot{\theta}(\mathbf{e}_x + \mathbf{e}_\theta)b\dot{\theta}^2\mathbf{e}_r$$

Further simplification is possible upon examination of the sum $\mathbf{e}_x + \mathbf{e}_\theta$, shown in Exhibit 5. The magnitude of this sum is $2 \sin(\theta/2)$, and its direction is perpen-

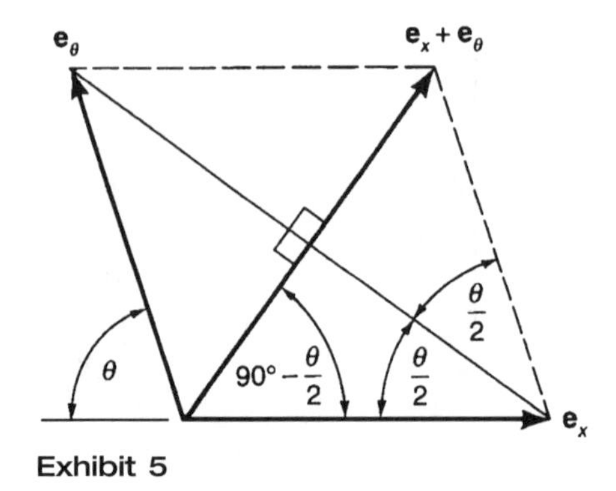

Exhibit 5

dicular to the line connecting points P and C'. The velocity can thus be expressed as $\mathbf{v} = (2b \sin \theta/2\, \dot{\theta})\mathbf{e}_t$ in which the unit tangent vector is perpendicular to the line PC'. The acceleration can be simplified correspondingly to

$$\mathbf{a} = 2b \sin\frac{\theta}{2}\ddot{\theta}\mathbf{e}_t - b\dot{\theta}^2\mathbf{e}_r$$

Several steps were taken to reach the results in Example 5. The position vector was expressed in terms of the geometric constraints on the rolling of the wheel, differentiation led to expressions for the velocity and acceleration, and the introduction of auxiliary unit vectors and several trigonometric relationships simplifyed several expressions.

As mentioned earlier, direct use of the definitions expressed by Eqs. (8.1) and (8.5) may not be the easiest means of evaluating velocities and accelerations. Indeed, we will now review some kinematic relationships for rigid bodies that will make much shorter work of this example.

Circular Cylindrical Coordinates

Figure 8.3 shows a coordinate system that is useful for a number of problems in particle kinematics. The x and y coordinates of the rectangular Cartesian system are replaced with the distance r and the angle ϕ, while the definition of the z-coordinate remains unchanged. Two of the unit vectors associated with the rectangular Cartesian system are also replaced with $\mathbf{e}_r = \cos\phi\mathbf{e}_x + \sin\phi\mathbf{e}_y$ and $\mathbf{e}_\phi = -\sin\phi\mathbf{e}_x + \cos\phi\mathbf{e}_y$.

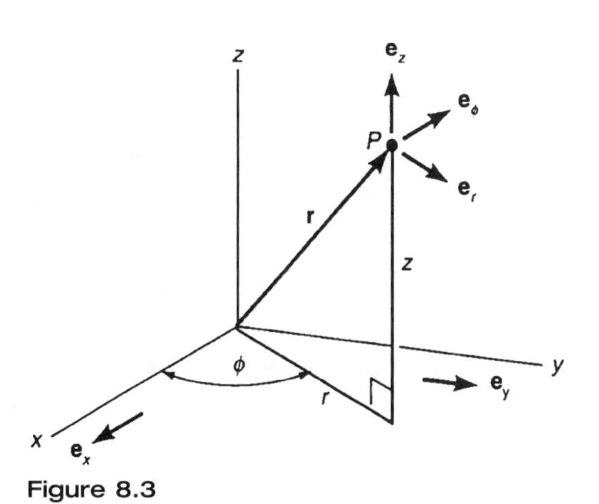

Figure 8.3

Since the angle ϕ varies, these two unit vectors also vary; their derivatives may be obtained by differentiating the above expressions:

$$\frac{d\mathbf{e}_r}{dt} = (-\mathbf{e}_x \sin\phi + \mathbf{e}_y \cos\phi)\frac{d\phi}{dt} = \dot{\phi}\mathbf{e}_\phi$$

$$\frac{d\mathbf{e}_\phi}{dt} = (-\mathbf{e}_x \cos\phi - \mathbf{e}_y \sin\phi)\frac{d\phi}{dt} = -\dot{\phi}\mathbf{e}_r$$

These are used along with the expression $\mathbf{r} = r\,\mathbf{e}_r + z\,\mathbf{e}_z$ for position to obtain expressions for velocity and acceleration:

$$\begin{aligned} \mathbf{v} = \dot{\mathbf{r}} &= \dot{r}\mathbf{e}_r + r\dot{\mathbf{e}}_r + \dot{z}\mathbf{e}_z \\ &= \dot{r}\mathbf{e}_r + r\dot{\phi}\mathbf{e}_\phi + \dot{z}\mathbf{e}_z \end{aligned} \tag{8.13}$$

$$\begin{aligned} \mathbf{a} = \dot{\mathbf{v}} &= \ddot{r}\mathbf{e}_r + \dot{r}\dot{\mathbf{e}}_r + (\dot{r}\dot{\phi} + r\ddot{\phi})\mathbf{e}_\phi + r\dot{\phi}\dot{\mathbf{e}}_\phi + \ddot{z}\mathbf{e}_z \\ &= (\ddot{r} - r\dot{\phi}^2)\mathbf{e}_r + (r\ddot{\phi} + 2\dot{r}\dot{\phi})\mathbf{e}_\phi + \ddot{z}\mathbf{e}_z \end{aligned} \tag{8.14}$$

Example 8.6

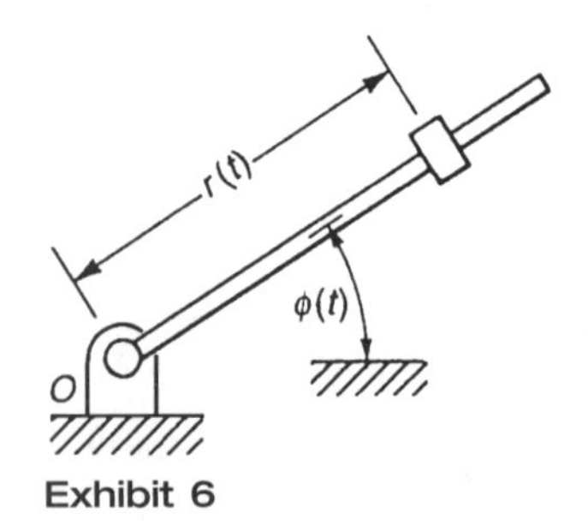

Exhibit 6

In Exhibit 6, the slider moves along the rod as it rotates about the fixed point O. At a particular instant, the slider is 200 mm from O, moving outward at 3 m/s relative to the rod; this relative speed is increasing at 130 m/s^2. At the same instant, the rod is rotating at a constant rate of 191 rpm. Evaluate the velocity and acceleration of the slider, and determine the rate of change of speed of the slider.

Solution

The angular speed of the rod is

$$\dot{\phi} = (191)\frac{2\pi \text{ rad/rev}}{60 \text{ s/min}} = 20.00 \text{ rad/s}$$

and its angular acceleration $\ddot{\phi}$ is zero. Other values to be substituted into Eqs. (8.13) and (8.14) are $r = 0.2$ m, $\dot{r}=3$ m/s, and $\ddot{r}=130$ m/s^2. Substitution into Eqs. (8.13) and (8.14) leads directly to the following radial and transverse components of velocity and acceleration:

$$\begin{aligned} \mathbf{v} &= (3\,\mathbf{e}_r + 4\,\mathbf{e}_\phi) \text{ m/s} \\ \mathbf{a} &= (50\,\mathbf{e}_r + 120\,\mathbf{e}_\phi) \text{ m/s}^2 \end{aligned}$$

Now the radial and transverse components of the unit vector tangent to the path can be obtained by dividing the velocity vector by its magnitude:

$$\mathbf{e}_t = \frac{3\mathbf{e}_r + 4\mathbf{e}_\phi}{\sqrt{(3)^2 + (4)^2}} = 0.6\mathbf{e}_r + 0.8\mathbf{e}_\phi$$

The rate of change of speed is the projection of the acceleration vector onto the tangent to the path, which can be obtained by dot-multiplying the acceleration with the unit tangent vector:

$$\begin{aligned} \dot{v} = a_t &= \mathbf{e}_t \bullet \mathbf{a} \\ &= (0.6)(50 \text{ m/s}^2) + (0.8)(120 \text{ m/s}^2) \\ &= 126 \text{ m/s}^2 \end{aligned}$$

Circular Path

When the path is circular, r is constant, and Eqs. (8.13) and (8.14) reduce to

$$\begin{aligned} \mathbf{v} &= r\dot{\phi}\mathbf{e}_\phi \\ \mathbf{a} &= -r\dot{\phi}^2\mathbf{e}_r + r\ddot{\phi}\mathbf{e}_\phi \end{aligned}$$

Comparing these with Eqs. (8.2) and (8.6) (with $\rho = r$),

$$\mathbf{v} = v\mathbf{e}_t$$

$$\mathbf{a} = \dot{v}\mathbf{e}_t + \frac{v^2}{r}\mathbf{e}_n$$

we see that, for circular path motion, $\mathbf{e}_t = \mathbf{e}_\phi, \mathbf{e}_n = -\mathbf{e}_r$, and

$$v = r\dot{\phi} \tag{8.15}$$

$$a_n = r\dot{\phi}^2 \tag{8.16}$$

Example **8.7**

A satellite is to be placed in a circular orbit over the equator at such an altitude that it makes one revolution around the earth per sidereal day (23.9345 hours). The gravitational acceleration is $(3.99 \times 10^{14}\ \mathrm{m^3/s^2})/r^2$, where r is the distance from the center of the earth. What is the altitude at which the satellite must be placed to achieve this period of orbit?

Solution

The angular speed of the line from the center of the earth to the satellite is

$$\dot{\phi}\frac{2\pi\ \mathrm{rad}}{(23.9345\ \mathrm{h})(3600\ \mathrm{s/h})} = 7.292\times10^{-5}\ \mathrm{rad/s}$$

The acceleration has no tangential component, but the radial component in terms of the orbit radius and the angular speed will be $a_n = r(7.292 \times 10^{-5}\mathrm{s}^{-1})^2$. This acceleration is imparted by the earth's gravitational attraction, so that $r(7.292 \times 10^{-5}\mathrm{s}^{-1})^2 = (3.99 \times 10^{14}\ \mathrm{m^3/s^2})/r^2$. This equation is readily solved for r, resulting in

$$r = \sqrt[3]{\frac{3.99\times10^{14}\mathrm{m^3/s^2}}{(7.292\times10^{-5}\mathrm{s}^{-1})^2}} = 42.2\times10^6\ \mathrm{m}$$

The altitude will then be the difference between this value and the size of earth's radius, which is about 6.4×10^6 m: Altitude = 35.8×10^6 m.

RIGID BODY KINEMATICS

The analysis of numerous mechanical systems rests on the assumption that the bodies making up the system are *rigid*. If the forces involved and the materials and geometry of the bodies are such that there is little deformation, the resulting predictions can be expected to be quite accurate.

The Constraint of Rigidity

If a body is **rigid**, the distance between each pair of points remains constant as the body moves. This constraint may be expressed in terms of a position vector $\mathbf{r}_{PQ}$ from a point P of the body to a point Q of the body, as indicated in Fig. 8.4. If the magnitude of $\mathbf{r}_{PQ}$ is constant, then

$$\frac{d}{dt}|\mathrm{r}_{PQ}|^2 = \frac{d}{dt}(\mathbf{r}_{PQ}\bullet\mathbf{r}_{PQ}) = 2\mathbf{r}_{PQ}\bullet\frac{d\mathbf{r}_{PQ}}{dt} = 0 \tag{i}$$

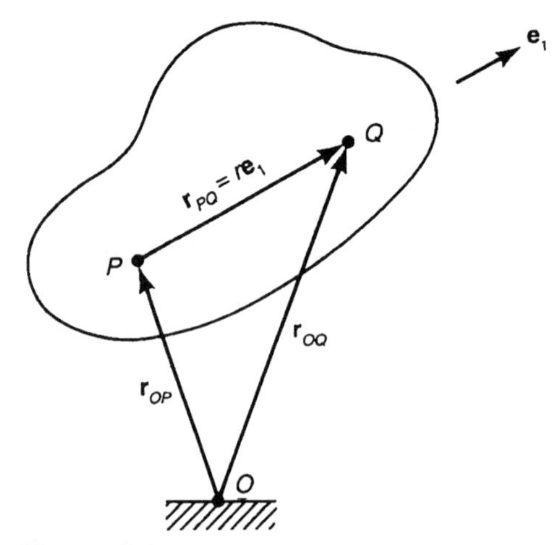

Figure 8.4

which indicates that $\dot{\mathbf{r}}_{PQ}$ is perpendicular to $\mathbf{r}_{PQ}$. Now, with a selected, fixed point designated as O, and vectors $\mathbf{r}_{OP}$ and $\mathbf{r}_{OQ}$ defined as indicated in Fig. 8.4, differentiation of the vector relationship $\mathbf{r}_{PQ} = \mathbf{r}_{OQ} - \mathbf{r}_{OP}$ leads to the relationship

$$\frac{d\mathbf{r}_{PQ}}{dt} = \mathbf{v}_Q - \mathbf{v}_P \quad \textbf{(ii)}$$

in which $\mathbf{v}_P$ and $\mathbf{v}_Q$ designate the velocities of P and Q, respectively. Finally, if we define $\mathbf{e}_1$ to be the unit vector in the direction of $\mathbf{r}_{PQ}$, so that

$$\mathbf{r}_{PQ} = r\mathbf{e}_1 \quad \textbf{(iii)}$$

then substitution of (ii) and (iii) into (i) leads to

$$2r\mathbf{e}_1 \bullet (\mathbf{v}_Q - \mathbf{v}_P) = 0$$

or

$$\mathbf{e}_1 \bullet \mathbf{v}_Q = \mathbf{e}_1 \bullet \mathbf{v}_P \quad \textbf{(8.17)}$$

This shows that *the projections of the velocities of any two points of a rigid body onto the line connecting the two points must be equal*. This is intuitively plausible; otherwise the distance between the points would be changing. This frequently provides the most direct way of evaluating the velocities of various points within a mechanism.

Example **8.8**

As the crank OQ in Exhibit 7 rotates clockwise at 200 rad/s, the piston P moves vertically. What will be the velocity of the piston at the instant when the angle θ is 50 degrees?

Solution

Since point Q must follow a circular path, its speed may be determined from Eq. (8.15): $v_Q = (0.075 \text{ m})(200 \text{ s}^{-1}) = 15$ m/s, with the direction of $\mathbf{v}_Q$ as indicated in the figure. Because the cylinder wall constrains the piston, its velocity is vertical. The connecting rod PQ is rigid, so the velocities of the points P and Q must

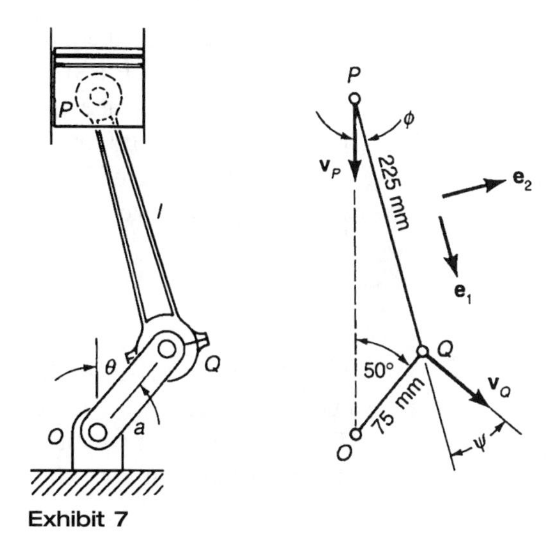

Exhibit 7

satisfy $v_P \cos\phi = v_Q \cos\psi$. The trigonometric rule of sines, applied to the triangle OPQ, gives

$$\sin\phi = \frac{a}{l}\sin\theta = \frac{75}{225}\sin 50°$$

which yields $\phi = 14.8°$. The other required angle is then $\psi = 90° - \theta - \phi = 25.2°$. Once these angles are determined, the constraint equation yields the speed of the piston:

$$v_P = \frac{\cos\psi}{\cos\phi} v_Q = 14.04 \text{ m/s}$$

The Angular Velocity Vector

If a rigid body is in *plane motion*, that is, if the velocities of all points of the body lie in a fixed plane, then its orientation may be specified by the angle θ between two fixed lines, one of which passes through the body, as indicated in Fig. 8.5. The rate of change of this angle is central to the analysis of the velocities of various points of the body.

To determine this relationship, consider Fig. 8.6, which shows a position vector from the point P to point Q, both fixed in the moving body. Two configurations are shown, one at time t and another after an arbitrary change during a time increment

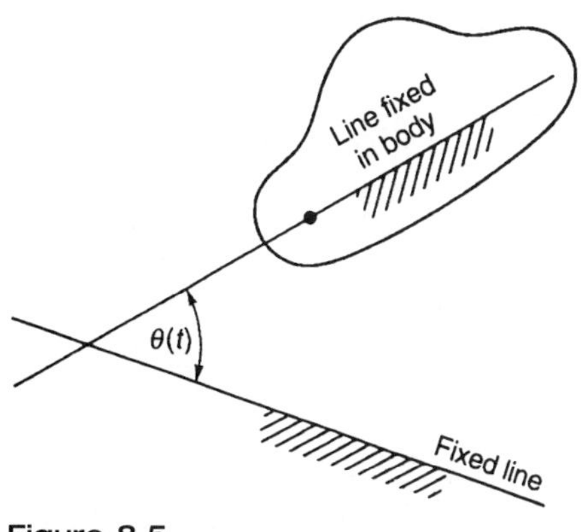

Figure 8.5

Δt. $\mathbf{e}_1$ is defined to be the unit vector in the direction of $\mathbf{r}_{PQ}(t)$, and $\mathbf{e}_2$ is defined to be the unit vector of P and Q, 90° counterclockwise from $\mathbf{e}_1$. $\mathbf{e}_1$ and $\mathbf{e}_2$ are further assumed to lie in the plane of motion; this assumption is convenient but not limiting. The vector diagram in Fig. 8.6 shows the change in $\mathbf{r}_{PQ}$ to be given by the approximation $\Delta\mathbf{r} \approx r\Delta\mathbf{e}_2$. Dividing both sides by the time increment Δt and letting this increment approach zero leads to the relation

$$\frac{d\mathbf{r}_{PQ}}{dt} = r\dot{\theta}\mathbf{e}_2$$

The scalar $\dot{\theta}$ will be denoted also by ω. Note from the definition of θ that a positive value of ω indicates a counterclockwise rotation, whereas a negative value of ω indicates a clockwise rotation. Another useful form of this relation may be written in terms of the **angular velocity vector,**

$$\boldsymbol{\omega} = \omega\,\mathbf{e}_3$$

where $\mathbf{e}_3$ is defined to be $\mathbf{e}_1 \times \mathbf{e}_2$, oriented perpendicular to the plane of Fig. 8.6. With this definition,

$$\frac{d\mathbf{r}_{PQ}}{dt} = \boldsymbol{\omega}\times\mathbf{r}_{PQ}$$

Note that this relation is valid for *any* two points in the body. (A pair of points different from those shown in Fig. 8.6 might give rise to a different angle, but as the body moves *changes* in this angle would equal *changes* in θ.)

It can be shown that for the most general motion of a rigid body (not restricted to planar motion) there also exists a unique angular velocity vector for which the same relation holds. However, in nonplanar motion, the angular velocity ω is not straight forwardly related to the rate of change of an angle, and its calculation requires a more extensive analysis than in the case of planar motion.

When $\dot{\mathbf{r}}_{PQ}$ is replaced with $\mathbf{v}_Q - \mathbf{v}_P$ according to Eq. (8.ii) of the previous section, the important velocity relationship

$$\mathbf{v}_Q = \mathbf{v}_P + \omega\times\mathbf{r}_{PQ} \quad \textbf{(8.20)}$$

is obtained, which, for planar motion, becomes

$$\mathbf{v}_Q = \mathbf{v}_P + r\omega\mathbf{e}_2 \quad \textbf{(8.21)}$$

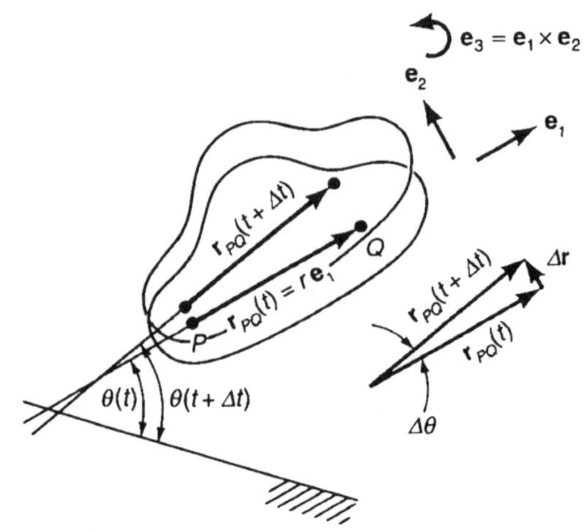

Figure 8.6

In the special case in which $\omega = 0$, this indicates that all points have the same velocity, a motion called **translation**. In the special case in which $\mathbf{v}_P = \mathbf{0}$, the motion is simply rotation about a fixed axis through P. Thus, in the general case, the two terms on the right of Eq. (8.20) can be seen to express a superposition of a translation and a rotation about P. But since P can be selected *arbitrarily*, there are as many combinations of a translation and a corresponding "center of rotation" as the analyst wishes to consider!

In all of the these cases, the angular velocity is a property of the *body's* motion, and Eq. (8.21) relates the velocities of *any* two points of a body experiencing planar motion. Dot-multiplication of each member of Eq. (8.21) with $\mathbf{e}_2$ leads to the following means of evaluating the angular velocity of a plane motion in terms of the velocities of two points:

$$\omega = \frac{\mathbf{e}_2 \bullet \mathbf{v}_Q - \mathbf{e}_2 \bullet \mathbf{v}_P}{r} \tag{8.22}$$

That is, ω will be the difference between the magnitudes of the projections of the velocities of P and Q onto the perpendicular to the line connecting P and Q, divided by the distance between P and Q.

Example **8.9**

What will be the angular velocity of the connecting rod in Example 8.8, at the instant when the angle θ is 50 degrees?

Solution

Referring to Exhibit 7 for the definition of $\mathbf{e}_2$, we see that

$$\omega = \frac{v_Q \sin\psi + v_Q \sin\phi}{l}$$

$$= \frac{(15 \text{ m/s})\sin 25.2° + (14.04 \text{ m/s})\sin 14.8°}{0.225 \text{ m}} = 44.3 \text{ rad/s}$$

The positive value indicates that the rotation is counterclockwise at this instant.

Instantaneous Center of Zero Velocity

For planar motion with $\omega \neq 0$, there always exists a point C' of the body (or an imagined extension of the body) that has zero velocity. If point P of Eq. (8.21) is selected to be this special point, the equation reduces to $\mathbf{v}_Q = \mathbf{v}_{C'} + \omega \times \mathbf{r}_{C'Q} = r\omega\mathbf{e}_2$ where r is now the distance from C' to Q and $\mathbf{e}_2$ is perpendicular to the line connecting C' and Q. This latter property can be used to locate C' if the directions of the velocities of two points of the body are known.

Example **8.10**

What is the location of the instantaneous center C' of the connecting rod in Examples 8 and 9? Use this to verify the previously-determined values of the angular velocity of the connecting rod and the velocity of point P.

Solution

The velocity of any point of the connecting rod must be perpendicular to the line from C' to that point. Hence C' must lie at the point of intersection of the horizontal

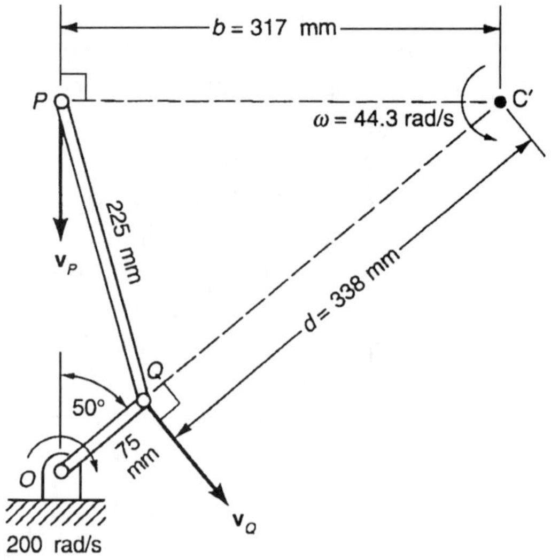

Exhibit 8

line through P and the line through Q perpendicular to $\mathbf{v}_Q$ (i.e., on the line through O and Q), as shown in Exhibit 8. The pertinent distances can be found as follows:

$$OP = (75 \text{ mm}) \cos 50° + (225 \text{ mm}) \cos 14.8° = 266 \text{ mm}$$
$$PC' = OP \tan 50° = 317 \text{ mm}$$
$$QC' = OP \sec 50° - 75 \text{ mm} = 338 \text{ mm}$$

The angular velocity of the connecting rod is then

$$\omega = \frac{v_Q}{QC'} = \frac{15 \text{ m/s}}{0.339 \text{ m}} = 44.3 \text{ rad/s}$$

and the velocity of P is then

$$v_P = PC'\omega = (0.317 \text{ m})(44.\ 3 \text{ s}^{-1}) = 14.04 \text{ m/s}$$

in agreement with values the previously obtained.

Example 8.11

Using the properties of the instantaneous center, determine the velocity of the point P on the rim of the rolling wheel in Example 8.5.

Solution

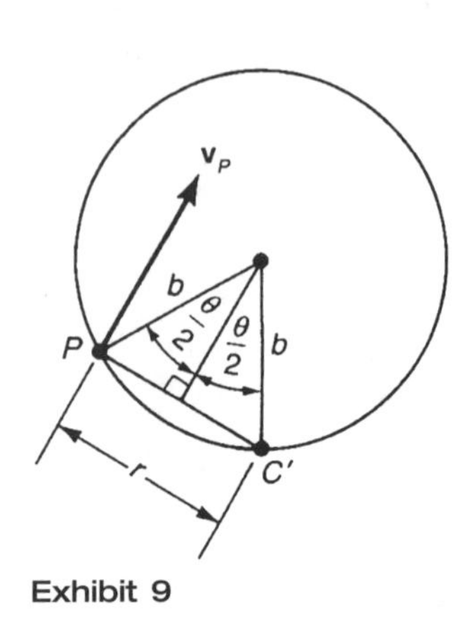

Exhibit 9

Since the wheel rolls without slipping, the point of the wheel in contact with the flat surface has zero velocity and is therefore its instantaneous center. The angular speed of the wheel is $\dot{\theta}$, and the distance from C' to P is readily determined from Exhibit 9:

$$r = 2b \sin\frac{\theta}{2}$$

The velocity of point P then has the magnitude

$$v_P = r\omega = 2b \sin\frac{\theta}{2}\dot{\theta}$$

and the direction shown in Exhibit 9. This direction should be evident by inspection once it is realized that a positive $\dot{\theta}$ corresponds to clockwise rotation. The reader may find it instructive to recall the conventions for the choice of $\mathbf{e}_2$ and positive ω used in the derivation leading to Eq. (8.21) and verify the agreement. Note the simplicity of this analysis as compared with the one expressing the position of P in a rectangular cartesian coordinate system.

Accelerations in Rigid Bodies

Formally differentiating Eq. (8.20) and substituting for $\dot{\mathbf{r}}_{PQ}$ using Eq. (8.19) leads to

$$\mathbf{a}_Q = \mathbf{a}_P + (\boldsymbol{\alpha} \times \mathbf{r}_{PQ}) + \boldsymbol{\omega} \times (\boldsymbol{\omega} \times \mathbf{r}_{PQ}) \tag{8.23}$$

in which the vector $\boldsymbol{\alpha} = d\omega/dt$ is called the **angular acceleration** of the body. For planar motion, $\boldsymbol{\alpha} = \alpha\mathbf{e}_3 = \dot{\omega}\mathbf{e}_3$ and $\boldsymbol{\omega} \times (\boldsymbol{\omega} \times \mathbf{r}) = -\omega^2\mathbf{r}$ so that

$$\mathbf{a}_Q = \mathbf{a}_P + r\alpha\mathbf{e}_2 - r\omega^2\mathbf{e}_1 \tag{8.24}$$

where $\mathbf{e}_1$ and $\mathbf{e}_2$ are defined as indicated in Fig. 8.6.

Equivalent relationships, analogous to Eq. (8.17) and Eq. (8.22) for velocity, can be obtained by dot-multiplying this equation by $\mathbf{e}_1$ and by $\mathbf{e}_2$:

$$\mathbf{e}_1 \bullet \mathbf{a}_Q = \mathbf{e}_1 \bullet \mathbf{a}_P - r\omega^2 \tag{8.25}$$

$$\alpha = \frac{\mathbf{e}_2 \bullet \mathbf{a}_Q - \mathbf{e}_2 \bullet \mathbf{a}_P}{r} \tag{8.26}$$

Example 8.12

If the speed of the crank in Examples 8–10 is constant, what are the acceleration $\mathbf{a}_P$ of the piston and the angular acceleration $\boldsymbol{\alpha}$ of the connecting rod at the instant when the angle θ is 50 degrees (Exhibit 10)?

Solution

When the crank speed is constant, the acceleration of Q is entirely centripetal, of magnitude

$$a_Q = (0.075 \text{ m})(200 \text{ s}^{-1})^2 = 3000 \text{ m/s}^2$$

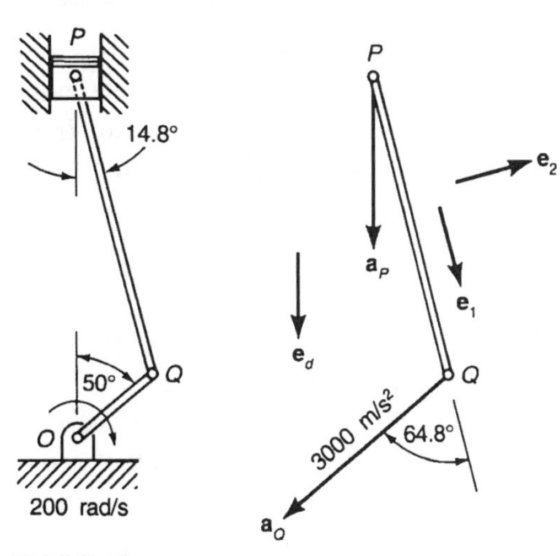

Exhibit 10

and directed toward the center of curvature O of the path of Q. The acceleration of P is vertically upward or downward. To determine the direction, we define a downward unit vector $\mathbf{e}_d$ and let $\mathbf{a}_p = a_p\mathbf{e}_d$ (see Exhibit 10). A positive value of a_P then indicates a downward acceleration and a negative value an upward acceleration. These expressions for a_Q and $\mathbf{a}_P$ are substituted into Eq. (8.25), along with the previously determined angular velocity of the rod, giving

$$(3000 \text{ m/s}^2)\cos 64.8° = a_P \cos 14.8° - (0.225 \text{ m})(44.3 \text{ s}^{-1})^2$$

which yields

$$a_P = 1779 \text{ m/s}^2$$

The angular acceleration $\boldsymbol{\alpha}$ of the rod can then be determined from Eq. (8.26):

$$\alpha = \frac{(3000 \text{ m/s}^2)\cos 154.8° - (1779 \text{ m/s}^2)\cos 104.8°}{0.225 \text{ m}} = -10\ 050 \text{ rad/s}^2$$

The negative value indicates that the angular acceleration is clockwise; that is, the 44.3-rad/s counterclockwise angular velocity is rapidly decreasing at this instant.

NEWTON'S LAWS OF MOTION

Every element of a mechanical system must satisfy Newton's second law of motion, that is, the resultant force **f** acting on the element is related to the acceleration **a** of the element by

$$\mathbf{f} = m\mathbf{a} \tag{8.27}$$

in which m represents the mass of the element. Newton's third law requires that the force exerted on a body A by a body B is of equal magnitude and opposite direction to the force exerted on body B by body A. These laws and their logical consequences provide the basis for relating motions to the forces that cause them.

Applications to a Particle

A **particle** is an idealization of a material element in which its spatial extent is disregarded, so that the motion of all of its parts is completely characterized by the path of a geometric *point*. When the accelerations of various parts of a system differ significantly, the system is considered to be composed of a number of particles and analyzed as described in the next section.

Example 8.13

An 1800-kg aircraft in a loop maneuver follows a circular path of radius 3 km in a vertical plane. At a particular instant, its velocity is 210 m/s directed 25 degrees above the horizontal as shown in Exhibit 11. If the engine thrust is 16 kN greater than the aerodynamic drag force, what is the rate of change of the aircraft's speed, the magnitude of the aircraft's acceleration, and the aerodynamic lift force?

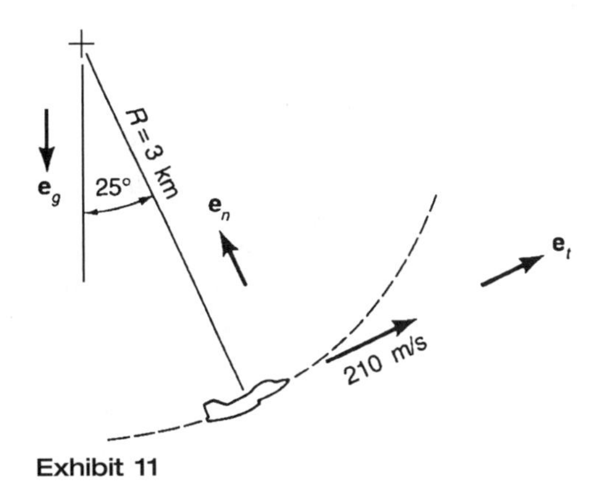

Exhibit 11

Solution

Since the dimensions of the aircraft are small compared with the radius of the path, all of its material elements can be considered to have essentially the same motion so treating the aircraft as a particle as described above is reasonable.

The forces acting on the aircraft are shown on the free-body diagram, Exhibit 12. The thrust **T**, the drag **D**, and the lift **L** all result from aerodynamic pressure from the surrounding air and engine gas. The lift is defined to be the component of the total force that is perpendicular to the flight path, and arises primarily from the wings. The force of gravity, $m\mathbf{g}$, is the only other force arising from a source external to the free body. The left-hand side of Eq. (8.27) is the resultant of these forces whereas the right-hand side is obtained from Eq. (8.6). Thus, Newton's second law is written in this case as

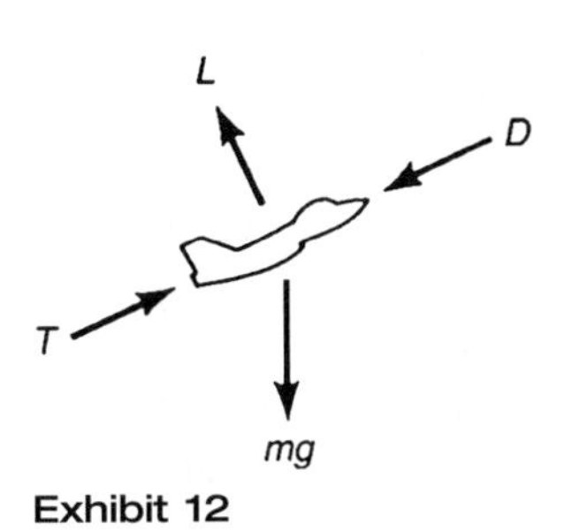

Exhibit 12

$$(T-D)\mathbf{e}_t + L\mathbf{e}_n + mg\mathbf{e}_g = m\left(\dot{v}\mathbf{e}_t + \frac{v^2}{R}\mathbf{e}_n\right)$$

Two independent equations arise from this two-dimensional vector equation. Dot multiplication with $\mathbf{e}_t$ yields $(T-D) + mg\mathbf{e}_t \bullet \mathbf{e}_g = m\dot{v}$ which leads to the rate of change of speed:

$$\dot{v} = \frac{T-D}{m} - g\sin 25° = \frac{16{,}000\ \text{N}}{1800\ \text{kg}} - (9.81\ \text{m/s}^2)\sin 25° = 4.74\ \text{m/s}^2$$

Dot-multiplication with $\mathbf{e}_n$ yields

$$L + mg\mathbf{e}_n \bullet \mathbf{e}_g = \frac{mv^2}{R}$$

which then allows us to determine the magnitude of lift force,

$$L = m\left(g\cos 25° + \frac{v^2}{R}\right) = (1800\ \text{kg})\left[(9.81\ \text{m/s}^2)\cos 25° + \frac{(210\ \text{m/s})^2}{3000\ \text{m}}\right] = 42.5\ \text{kN}$$

The magnitude of the acceleration is then determined by combining the tangential and normal components found above:

$$|\mathbf{a}| = \sqrt{(4.74\ \text{m/s}^2)^2 + \left[\frac{(210\ \text{m/s})^2}{3000\ \text{m}}\right]^2} = 15.45\ \text{m/s}^2$$

Example 8.14

Two blocks are interconnected by an inextensible, massless line through the pulley arrangement shown in Exhibit 13. The inertia and friction of the pulleys are negligible. The coefficient of friction between the block of mass m_1 and the horizontal surface is μ. What is the acceleration of the block of mass m_2 as it moves downward?

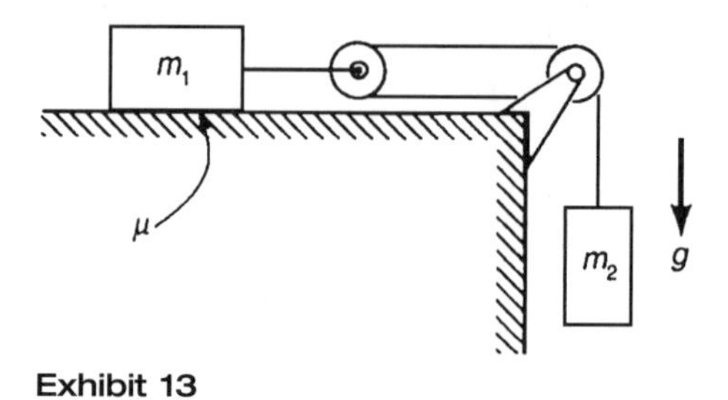

Exhibit 13

Solution

Since the motion of each block is a translation, the acceleration of each element in a block is the same; hence their spatial extensions may be ignored and each block may be treated as a particle.

Moment equilibrium of each pulley requires that the tension be the same in each part of the longer line around the pulleys. Denoting this tension by T, the two free-body diagrams in Exhibit 14 are used to write expressions for Newton's second law. Since the acceleration of the block on the left has no vertical component, $R_1 - m_1g = 0$. Denoting its rightward acceleration by a_1, consideration of

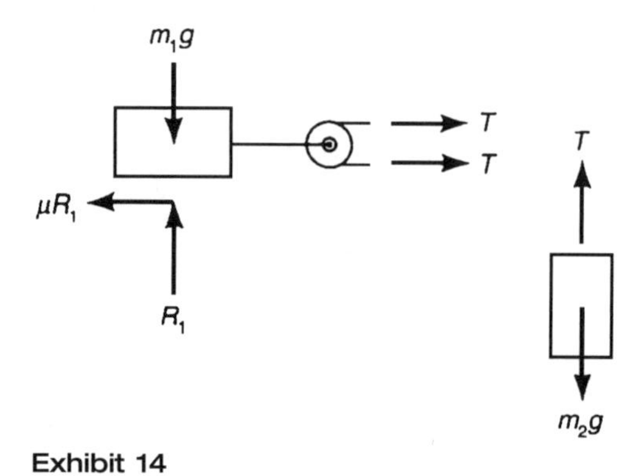

Exhibit 14

the horizontal forces leads to $2T - \mu R_1 = m_1a_1$. Denoting the downward acceleration of the other block by a_2, application of Newton's second law to the other free body yields $m_2g - T = m_2a_2$. To relate the two accelerations, consider the pulley connected to the horizontal block. The instantaneous center of this pulley is at its rim and directly below its center. Thus, the speed of the pulley's upper rim is twice that of its center, so that the speed of the vertically moving block is twice that of the horizontally moving block. Since this is true at all times, we have $a_2 = 2a_1$.

Eliminating R, T, and a_1 from these four equations now leads to

$$a_2 = \frac{4m_2 - 2\mu m_1}{4m_2 + m_1} g$$

Observe that if $m_2 > \mu m_1/2$, the acceleration is downward, whereas if $m_2 < \mu m_1/2$, the acceleration is upward, implying that the downward velocity will reach zero, after which time the friction force will no longer equal μR_1.

Systems of Particles

A mechanical **system** is any collection of material elements of fixed identity whose motion we may wish to consider. Such a system is treated as a collection of particles in which the individual particles must obey Newton's laws of motion.

It proves to be very useful to separate the forces acting on a system into those arising from sources outside the system and arising from the interaction between members of the system. That is, the resultant force on the ith particle is written as

$$\mathbf{f}_i = \mathbf{f}_{ie} + \sum_j \mathbf{f}_{i/j}$$

in which $\mathbf{f}_{ie}$ represents the resultant of all forces on the ith particle arising from sources external to the system, and $\mathbf{f}_{i/j}$ represents the force exerted on ith particle by the jth particle. With this notation, Newton's third law may be expressed as $\mathbf{f}_{j/i} = -\mathbf{f}_{i/j}$.

Now, each particle moves according to Newton's second law:

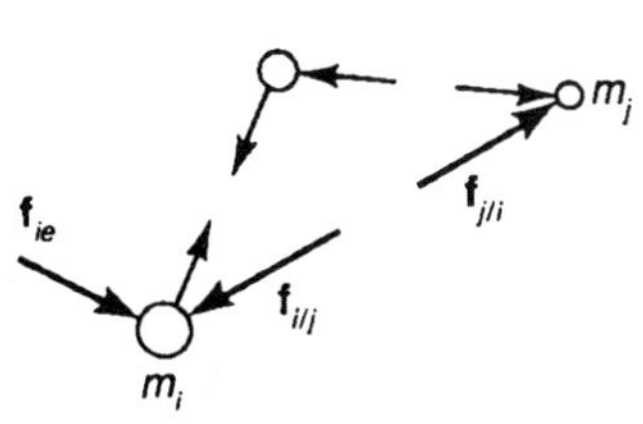

Figure 8.7

$$\mathbf{f}_{ie} + \sum_j \mathbf{f}_{i/j}$$

in which m_i denotes the mass of the ith particle and $\mathbf{a}_i$ its acceleration. There are as many such equations as there are particles in the system; if all such equations are added, the result is

$$\sum_i \mathbf{f}_{ie} + \sum_i \sum_j \mathbf{f}_{i/j} = \sum_i m_i \mathbf{a}_i$$

In view of Newton's third law, the internal forces can be grouped as pairs of oppositely-directed forces of equal magnitude, and so their sum vanishes, leaving

$$\sum_i \mathbf{f}_{ie} = \sum_i m_i \mathbf{a}_i \qquad \textbf{(8.28a)}$$

That is, the resultant of all *external* forces is equal to the sum of the products of the individual masses and their corresponding accelerations.

Linear Momentum and Center of Mass

The right-hand member of Eq. (8.28a) can be expressed alternatively in terms of the **linear momentum** of the system, which is defined as

$$\mathbf{p} = \sum_i m_i \mathbf{v}_i \qquad \textbf{(8.29)}$$

in which $\mathbf{v}_i$ denotes the velocity of the ith particle. Differentiation of this equation results in

$$\frac{d\mathbf{p}}{dt} = \sum_i m_i \mathbf{a}_i$$

which is the same expression appearing in Eq. (8.28a). Hence an alternative to Eq. (8.28a) is

$$\sum_i \mathbf{f}_{ie} = \frac{d\mathbf{p}}{dt} \tag{8.28b}$$

which states that the sum of the external forces is equal to the time rate of change of the linear momentum of the system.

The **center of mass** of the system is a point C located, relative to an arbitrarily-selected reference point O, by the position vector $\mathbf{r}_C$ which satisfies the defining equation

$$m\mathbf{r}_C = \sum_i m_i \mathbf{r}_i \tag{8.30}$$

in which m denotes the total mass of the system and $\mathbf{r}_i$ is a position vector from O to the ith particle. Differentiation of this equation leads to

$$m\mathbf{v}_C = \sum_i m_i \mathbf{v}_i \tag{8.31}$$

which shows that the linear momentum is the product of the total mass and the velocity of the center of mass. Another differentiation yields

$$m\mathbf{a}_C = \sum_i m_i \mathbf{a}_i$$

which provides still another way of expressing Eq. (8.28a):

$$\sum_i \mathbf{f}_{ie} = m\mathbf{a}_C \tag{8.28c}$$

This is sometimes called the **principle of motion of the mass center**. It indicates that the mass center responds to the resultant of external forces exactly as would a single particle having a mass equal to the total mass of the system.

Example 8.15

A motor inside the case shown in Exhibit 15 drives the eccentric rotor at a constant angular speed ω. The distance from the rotor bearing to its center of mass is e, the mass of the rotor is m_r, and the mass of the non-rotating housing is $m - m_r$. (That is, the total mass of the rotor and housing together is m.) The housing is

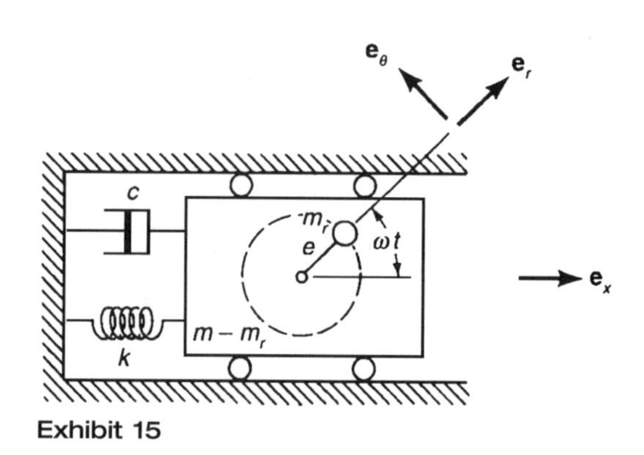

Exhibit 15

free to translate horizontally, constrained by the rollers, and under the influence of a spring of stiffness k and a dashpot which transmits a force to the housing of magnitude c times the speed of the housing in the direction opposite to that of the velocity of the housing. Write the differential equation that governs the extension $x(t)$ of the spring from its relaxed position.

Solution

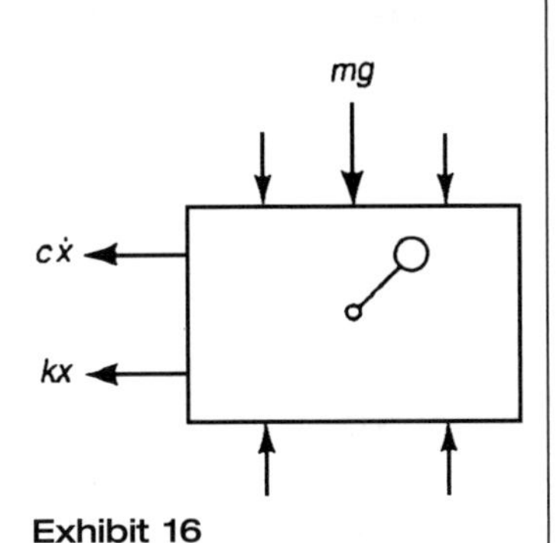

Exhibit 16

Consider the system consisting of the housing and rotor together. The free-body diagram (Exhibit 16) shows forces acting on this system from sources *external* to it. It does *not* include the torque necessary to maintain constant rotor speed nor the reaction at the bearing, these being internal, action-reaction pairs.

Note that when x is positive (the spring extended), the force exerted by the spring on the housing acts to the left, and when x is negative (the spring compressed), this force acts to the right. Both situations are depicted properly by the label kx on the arrow; that is, this indicates that the force equals $-kx\mathbf{e}_x$ in all cases. The same consideration applies to the arrow and label representing the force from the dashpot. The sum of all external forces then is

$$\sum_i \mathbf{f}_{ie} = -(kx + c\dot{x})\mathbf{e}_x + f_y\mathbf{e}_y$$

The acceleration of the housing is simply $\ddot{x}\mathbf{e}_x$, while that of the mass center of the rotor can be most readily determined by using Eq. (8.24), letting P be the center of the bearing and noting that $\alpha = 0$. This gives the acceleration of the mass center of the rotor as

$$\mathbf{a} = \ddot{x}\mathbf{e}_x - e\omega^2\mathbf{e}_r$$

Substitution into Eq. (8.28a) results in

$$-(kx + c\dot{x})\mathbf{e}_x + f_y\mathbf{e}_y = (m - m_r)\ddot{x}\mathbf{e}_x + m_r(\ddot{x}\mathbf{e}_x - e\omega^2\mathbf{e}_r)$$

The forces f_y are neither known nor of interest for our purpose; they may be eliminated from the equation by dot-multiplying each member with $\mathbf{e}_x$, which leads to

$$-(kx + c\dot{x}) = m\ddot{x} - m_r e\omega^2 \cos\,\omega t$$

A "standard" form of this equation is obtained by placing the dependent variable and its derivatives on one side and the known function of time on the other:

$$m\ddot{x} + c\dot{x} + kx = m_r e\omega^2 \cos\,\omega t$$

The same result can be obtained using either Eq. (8.28b) or (8.28c).

Impulse and Momentum

The integral with respect to time of the resultant of external forces is called the **impulse** of this resultant force:

$$\mathbf{g} = \int_{t_1}^{t_2} \sum_i \mathbf{f}_{ie}\,dt$$

Integration of both members of Eq. (8.28b) results in the following integrated form:

$$\mathbf{g} = \mathbf{p}(t_2) - \mathbf{p}(t_1) \tag{8.32}$$

This states that impulse of the resultant external force is equal to the change in momentum of the system. Since it is a vector equation, we may obtain up to three independent relationships from it.

A special case occasionally arises, in which one or more components of the impulse are absent. The corresponding components of momentum then remain constant, and are said to be **conserved**.

Example **8.16**

A rocket is simulated by a vehicle that is accelerated by the action of the passenger throwing rocks in the rearward direction as the vehicle moves along the roadway as shown in Exhibit 17. At a certain time, the mass of the vehicle, passenger, and supply of rocks is m, and all are moving at speed v. The passenger then launches a rock of mass m_1 with a rearward velocity of magnitude v_e *relative to the vehicle*. What is the increase Δv in the speed of the vehicle resulting from this action?

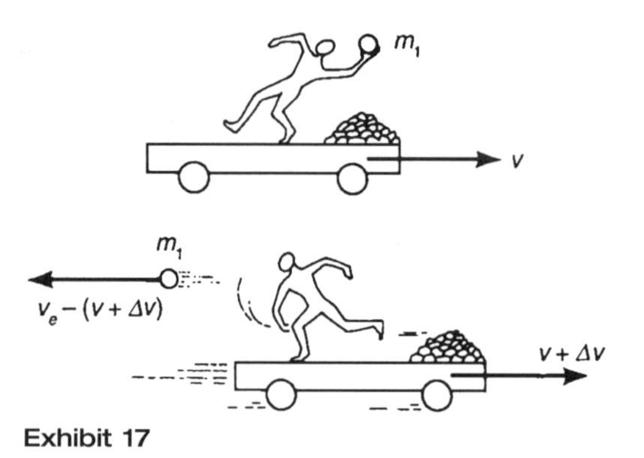

Exhibit 17

Solution

Assuming there is negligible friction at the wheels, the system consisting of passenger, vehicle, and rocks has no external forces acting on it in the direction of travel. The horizontal component of momentum is therefore conserved; that is, it is the same after the rock is launched as it was prior to the launching. After this is written in detail as

$$(m - m_1)(v + \Delta v) - m_1[v_e - (v + \Delta v)] = mv$$

the equation can then be solved for the increase in vehicle speed:

$$\Delta v = \frac{m_1}{m} v_e$$

Moments of Force and Momentum

Equations (8.28) and (8.29) are valid regardless of the lines of action of the forces $\mathbf{f}_{ie}$. For example, the acceleration of the center of mass and the change in momentum of the system shown in Fig. 8.8 will be the same for each of the

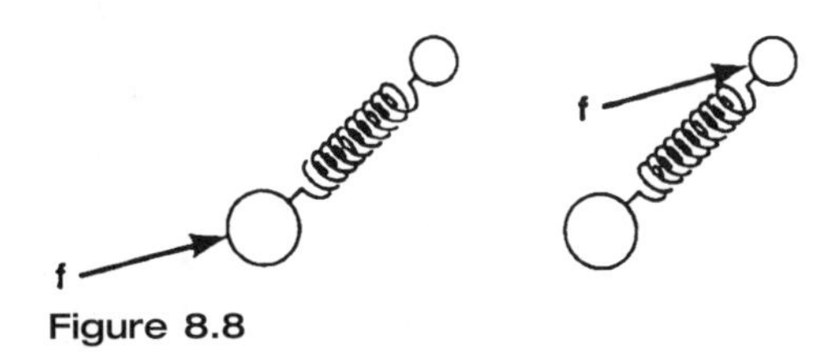

Figure 8.8

different points of application of the force. However, there are characteristics of the motions induced by these forces that *do* depend on the lines of action, some of which are revealed by considering *moments* of forces.

Consider again the external and internal forces acting on two typical particles of a system (Fig. 8.9). Let the position vectors $\mathbf{r}_i$ and $\mathbf{r}_j$ locate the ith and jth particles, respectively, with respect to a selected point O. If the equation expressing Newton's second law for the ith particle is cross-multiplied by $\mathbf{r}_i$ and all such equations are added, the result is

$$\sum_i \mathbf{r}_i \times \mathbf{f}_{ie} + \sum_i \sum_j \mathbf{r}_i \times \mathbf{f}_{i/j} = \sum_i \mathbf{r}_i \times m_i \mathbf{a}_i$$

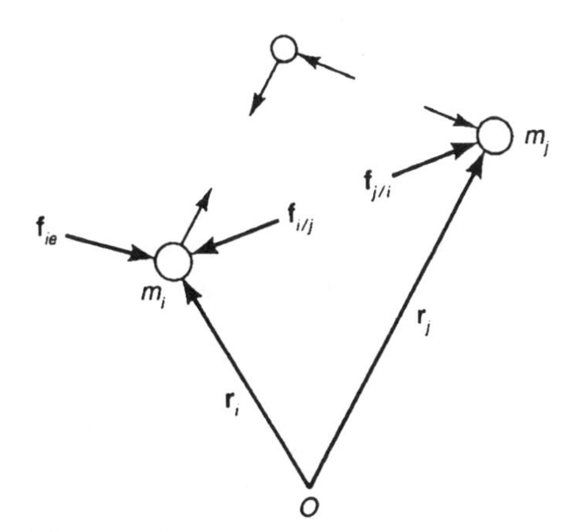

Figure 8.9

Now, if the forces $\mathbf{f}_{i/j}$ and $\mathbf{f}_{j/i} = -\mathbf{f}_{i/j}$ have a common line of action, then

$$\mathbf{r}_i \times \mathbf{f}_{i/j} + \mathbf{r}_j \times \mathbf{f}_{j/i} = 0$$

That is, the **moments** of the members of each action-reaction pair cancel one another, leaving the **moment equation** for the system:

$$\mathbf{M}_O = \sum_i \mathbf{r}_i \times m_i \mathbf{a}_i \qquad \textbf{(8.33a)}$$

in which the moment of the external forces is evaluated as in the previous chapter:

$$\mathbf{M}_O = \sum_i \mathbf{r}_i \times \mathbf{f}_{ie}$$

Example 8.17

The pendulum in Exhibit 18 consists of a stiff rod of negligible mass with two masses attached, and it swings in the vertical plane about the frictionless hinge at O under

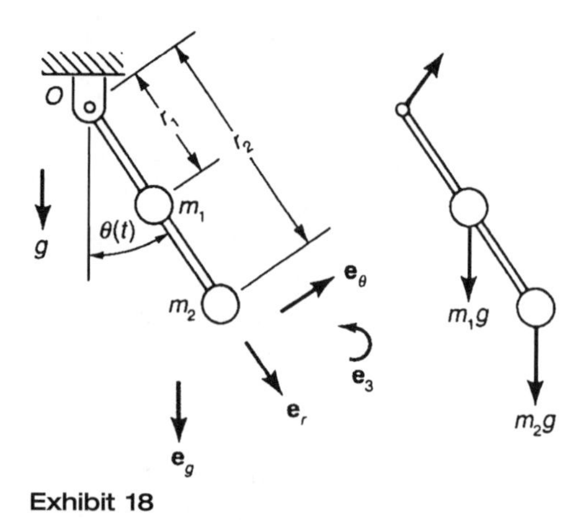

Exhibit 18

the influence of gravity. What will be the angular acceleration of the pendulum in terms of angular displacement θ and the other parameters indicated in the sketch?

Solution

The free-body diagram shows the forces external to the system consisting of the rod together with the two particles. Since the reaction at the support is unknown and of no interest for our purpose, a good strategy would be to consider moments about this point. Referring to the free-body diagram, we evaluate the resultant moment as usual:

$$\mathbf{M}_O = (r_i\mathbf{e}_r)\times(m_1g\mathbf{e}_g) + (r_2\mathbf{e}_r)\times(m_2g\mathbf{e}_g) = -(m_1r_1 + m_2r_2)g\sin\theta\,\mathbf{e}_3$$

Since each particle follows a circular path with center at O, their accelerations may be expressed as $\mathbf{a}_i = r_i\dot{\theta}^2\mathbf{e}_r + r_i\ddot{\theta}\mathbf{e}_\theta$. The right-hand member of the moment law Eq. (8.33a), is then evaluated in this case as

$$\begin{aligned}\sum \mathbf{r}_i \times m_i\mathbf{a}_i &= (r_1\mathbf{e}_r)\times m_1r_1(-\dot{\theta}^2\mathbf{e}_r + \ddot{\theta}\mathbf{e}_\theta) + (r_2\mathbf{e}_r)\times m_2r_2(-\dot{\theta}^2\mathbf{e}_r + \ddot{\theta}\mathbf{e}_\theta)\\ &= (m_1r_1^2 + m_2r_2^2)\ddot{\theta}\mathbf{e}_3\end{aligned}$$

Substitution into the moment law, Eq. (8.33a), results in

$$-(m_1r_1 + m_2r_2)g\sin\theta\,\mathbf{e}_3 = (m_1r_1^2 + m_2r_2^2)\ddot{\theta}\,\mathbf{e}_3$$

or

$$\ddot{\theta} = -\frac{m_1r_1 + m_2r_2}{m_1r_1^2 + m_2r_2^2}\,g\sin\theta$$

The **moment of momentum** or **angular momentum about point** O is defined as

$$\mathbf{H}_O = \sum_i \mathbf{r}_i \times m_i\mathbf{v}_i \tag{8.34}$$

Now, if O is fixed in the inertial frame, then $\mathbf{v}_i = \mathbf{r}_i$, and it follows that

$$\frac{d\mathbf{H}_O}{dt} = \sum_i (\dot{\mathbf{r}}_i \times m_i\mathbf{v}_i + \mathbf{r}_i \times m_i\dot{\mathbf{v}}_i) = \sum_i \mathbf{r}_i \times m_i\mathbf{a}_i$$

This provides an alternative way of writing the moment law, as

$$\mathbf{M}_O = \frac{d\mathbf{H}_O}{dt} \tag{8.33b}$$

In the preceding example, the angular momentum about O is

$$\mathbf{H}_O = (r_1\mathbf{e}_r)\times(m_1 r_1\dot{\theta}\mathbf{e}_\theta) + (r_2\mathbf{e}_r)\times(m_2 r_2\dot{\theta}\mathbf{e}_\theta) = \left(m_1 r_1^2 + m_2 r_2^2\right)\dot{\theta}\mathbf{e}_3$$

Differentiating this expression and substituting for the right-hand member of Eq. (8.33b) leads to the result achieved using Eq. (8.33a).

As with forces and linear momentum, there are situations in which the moment of external forces vanishes. Then, Eq. (8.33b) implies that the angular momentum about O remains constant, or is *conserved.*

Example **8.18**

Suppose the pendulum of Example 8.17 is suspended at rest when it is struck by a small projectile, which becomes imbedded in the lower ball (Exhibit 19). What angular velocity ω is imparted to the pendulum?

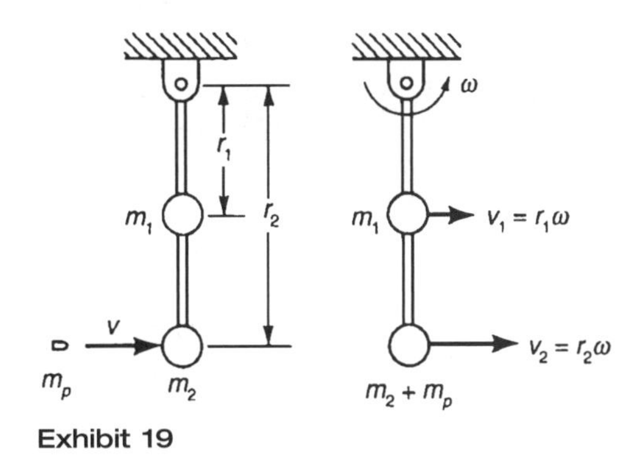

Exhibit 19

Solution

During the collision, which can induce a large reaction at the support as well as between the projectile and ball, the moment of forces external to the system—the pendulum and the projectile—will be zero. Hence, the angular momentum of this system prior to impact will equal that immediately after impact:

$$r_2 m_p v\mathbf{e}_3 = r_1 m_1 v_1\mathbf{e}_3 + r_2(m_2 + m_p)v_2\mathbf{e}_3 = [m_1 r_1^2 + (m_2 + m_p)r_2^2]\omega\mathbf{e}_3$$

Hence,

$$\omega = \frac{r_2 m_p v}{m_1 r_1^2 + (m_2 + m_p)r_2^2}$$

WORK AND KINETIC ENERGY

The integration in Eq. (8.9) has extensive implications which will be examined in this section.

A Single Particle

If the form of the tangential acceleration indicated in Eq. (8.9) is merged with Eq. (8.6), the result can be used to express Newton's second law as

$$\mathbf{f} = m\left(v\frac{dv}{ds}\mathbf{e}_t + \frac{v^2}{\rho}\mathbf{e}_n\right) \tag{i}$$

Now, if each member is dot-multiplied by an increment of change of position, $d\mathbf{r} = ds\mathbf{e}_t$, and the resulting scalars are integrated, they become

$$\int_{\mathbf{r}_1}^{\mathbf{r}_2} \mathbf{f} \bullet d\mathbf{r} = \int_{s_1}^{s_2} f_t \, ds = W_{1-2}$$

and

$$m\int_{v_1}^{v_2} v\,dv = \tfrac{1}{2} m v_2^{\,2} - \tfrac{1}{2} m v_1^{\,2}$$

The integral W_{1-2} is called the **work** done on the particle by the force **f** as the particle moves from position 1 to position 2. The scalar $T = \frac{1}{2} mv^2$ is called the **kinetic energy** of the particle. Since (i) holds throughout any interval, a consequence of Newton's second law is the **work-kinetic energy relationship:**

$$W_{1-2} = T_2 - T_1 \tag{8.35}$$

When enough information is available to permit evaluation of the work integral, this provides a useful way of predicting the change in the speed of the particle.

Example 8.19

A 3.5-Mg airplane is to be launched from the deck of an aircraft carrier with the aid of a steam-powered catapult. The force that the catapult exerts on the aircraft varies with the distance s along the deck as shown in Exhibit 20. If other forces are negligible, what value of the constant f_0 is necessary for the catapult to accelerate the aircraft from rest to a speed of 160 km/h at the end of the 30-m travel?

$$f(s) = \frac{f_0}{1 + \frac{s}{30 \text{ m}}}$$

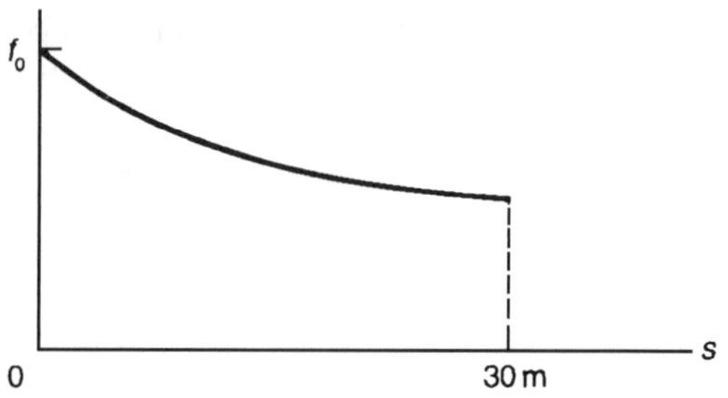

f_0

0 30 m s

Exhibit 20

Solution

Letting d stand for the 30-m travel, the work done on the aircraft will be

$$W = \int_0^d \frac{f_0 ds}{1+\frac{s}{d}} = (f_0 d)\ln\left(1+\frac{s}{d}\right)\Bigg|_0^d = (f_0 d)\ln 2 = (20.8\,\text{m})f_0$$

This will equal the change in kinetic energy, which is initially zero and increase to

$$\frac{1}{2}(3500 \text{ kg})\left[\left(160\frac{\text{km}}{\text{h}}\right)\frac{1000 \text{ m/km}}{3600 \text{ s/h}}\right]^2 = 3.46\times 10^6 \text{J}$$

The work-kinetic energy relationship

$$(20.8\text{ m})f_0 = 3.46\times10^6\text{ N}\bullet\text{m}$$

implies that the constant f_0 must have the value

$$f_0 = 166\text{ kN}$$

Work of a Constant Force

A commonly encountered force of constant magnitude and direction is that of gravity near the earth's surface. When a constant force acts on a particle as it moves, the work done by the force can be evaluated as indicated in Fig. 8.10. The increment of work as the particle undergoes an increment $d\mathbf{r}$ of displacement can be expressed as

$$dW = \mathbf{f}_0 \bullet d\mathbf{r} = |\mathbf{f}_0||d\mathbf{r}|\cos \measuredangle_{\mathbf{f}_0}^{d\mathbf{r}} = f_0 dq$$

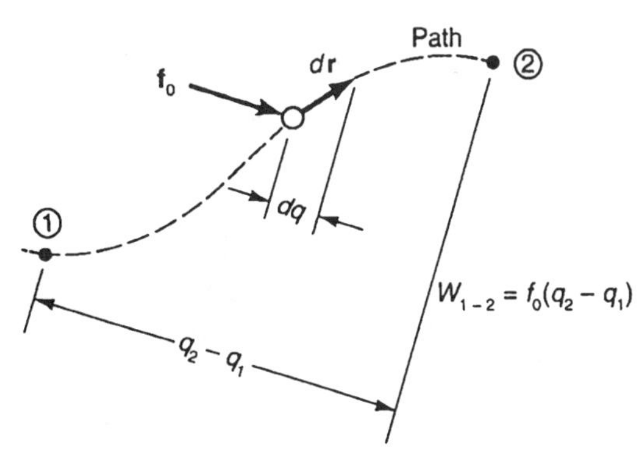

Figure 8.10

in which dq is the component of the displacement increment that is parallel to the force. Since the force is constant,

$$W_{1-2} = f_0\int_1^2 dq = f_0(q_2 - q_1) \qquad \textbf{(8.36)}$$

Thus, any movement of the particle that is perpendicular to the direction of the force has no effect on the work. In other words, the work is the same as would have been done if the particle had moved rectilinearly through a distance of $(q_2 - q_1)$ in a direction parallel to the force.

Example 8.20

How fast must the toy race-car be traveling at the bottom of the hill to be able to coast to the top of the hill (see Exhibit 21)?

Solution

As the car moves up the hill, the work done by the force of gravity will be $W_g = -mgh$. The work done by friction forces may be negligible if the wheels are well made. If this is the case, the work done by all forces is approximately that due to

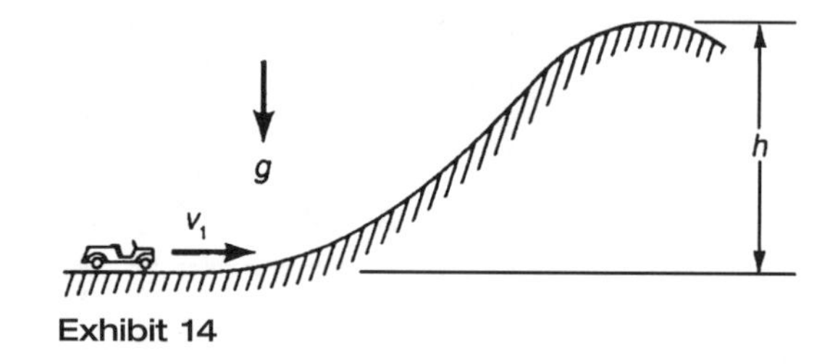

Exhibit 14

gravity. The speed of the car as it nears the hilltop can approach zero, so the work-kinetic energy equation may be written as

$$-mgh = T_2 - T_1 = 0 - \frac{1}{2}mv_1^2$$

which implies a minimum required speed of

$$v_1 = \sqrt{2gh}$$

With friction, the required speed will be somewhat greater.

Distance-Dependent Central Force

A force that remains directed toward or away from a fixed point is called a **central force**. Examples of forces for which the magnitude depends only on the distance from the particle to a fixed point are the force of gravitational attraction and the force from an elastic, tension-compression member with one end pinned to a fixed support. Figure 8.11 shows a particle P moving with such a central force acting on it; the dependence on distance is expressed by the function $f(r)$, with the convention that a positive value of f indicates an attractive force and a negative value of f a repulsive force.

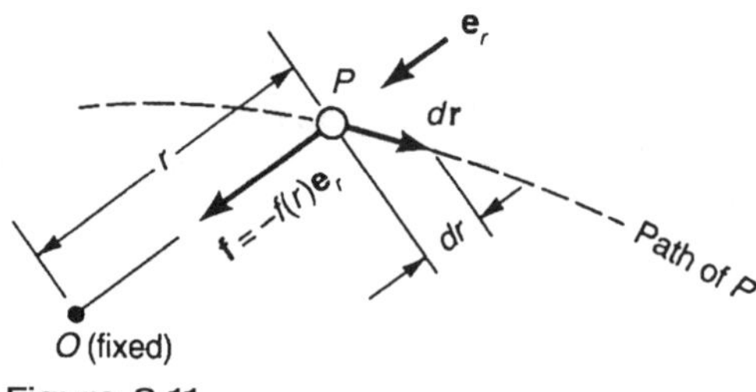

Figure 8.11

Now, the increment of work may be written as

$$dW = \mathbf{f} \bullet d\mathbf{r} = -f(r)\mathbf{e}_r \bullet d\mathbf{r}$$

Referring to the figure, we see that $\mathbf{e}_r \bullet d\mathbf{r} = |d\mathbf{r}| \cos \measuredangle_{er}^{dr}$, is equal to the change dr in radial distance r. Thus,

$$W_{1-2} = -\int_{r_1}^{r_2} f(r)dr \qquad \textbf{(8.37)}$$

Similar to the case of the constant force, the work done by a central force through an arbitrary motion is the same as would be done for a rectilinear motion, but in the radial direction.

Example 8.21

The elastic spring in Exhibit 22 has a linear force-displacement characteristic; that is, it exerts a force equal to the stiffness k times the amount it is stretched from its relaxed length l_0. As the particle moves from position 1 to position 2, what is the work done by the spring force on the particle?

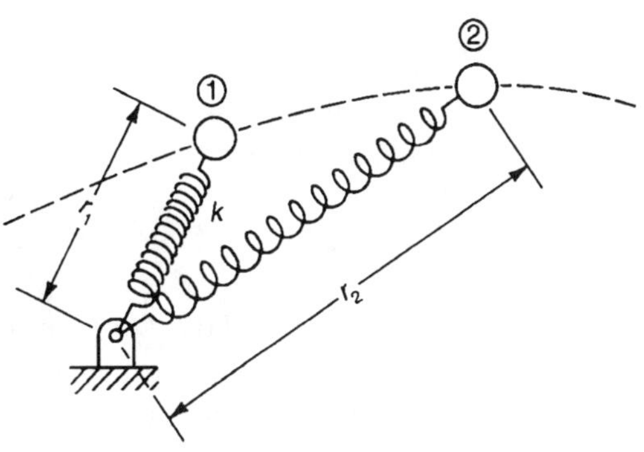

Exhibit 22

Solution

This is a case of a central distance-dependent force with $f(r) = -k(r - l_0)$. Equation (8.37) then becomes

$$W_{1-2} = -k\int_{r_1}^{r_2}(r - l_0)\,dr$$

A more convenient form results if we introduce the amount of spring extension as $\delta = (r - l_0)$. The integral then becomes

$$W_{1-2} = -k\int_{\delta_1}^{\delta_2}\delta\,d\delta = -\frac{k}{2}\left(\delta_2^2 - \delta_1^2\right)$$

Example 8.22

A 0.6-kg puck slides on a horizontal surface without friction under the influence of the tension in a light cord that passes through a small hole at O (see Exhibit 23). A spring under the surface imparts a tension in the cord that is proportional to

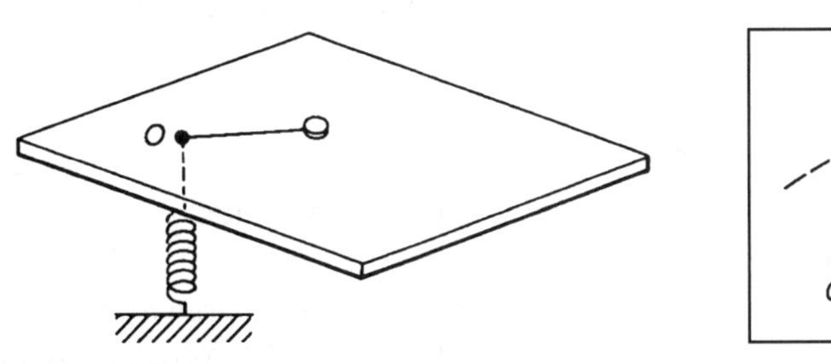

Exhibit 23

the distance from the hole to the puck; its stiffness is k = 30 N/m. At a certain instant, the puck is 200 mm from the hole and moving at 2 m/s in the direction indicated in the top view. If the spring is in its relaxed position when the puck is at the hole, what is the maximum distance from the hole reached by the puck?

Solution

With the initial and maximum distances denoted by r_1 and r_2, respectively, the work done on the puck by the force from the cord from the initial position to that of maximum distance will be

$$W_{1-2} = -\frac{k}{2}\left(r_2^2 - r_1^2\right)$$

Since this is the only force that does work, this value must equal the change in kinetic energy:

$$-\frac{k}{2}\left(r_2^2 - r_1^2\right) = \frac{m}{2}\left(v_2^2 - v_1^2\right)$$

The moment about O of the force from the cord is zero, so the angular momentum about the hole is conserved. Because there is no radial component of velocity at the maximum distance, the angular momentum there is simply $r_2 m v_2$. Hence,

$$r_1 m v_1 \sin 45° = r_2 m v_2$$

These two equations contain the unknowns r_2 and v_2. Isolating v_2 from the latter, substituting this expression into the energy equation, and rearranging leads to the equation

$$\left(\frac{r_2}{r_1}\right)^4 - \left(1 + \frac{m v_1^2}{k r_1^2}\right)\left(\frac{r_2}{r_1}\right)^2 + \frac{m v_1^2}{k r_1^2}\sin^2 45° = 0$$

When the given values are substituted, the quadratic formula yields

$$\left(\frac{r_2}{r_1}\right) = 1.618$$

as the largest root so that the maximum distance reached is $r_2 = 1.618$ (200 mm) = 324 mm.

Example 8.23

A torpedo expulsion device operates by means of gas expanding within a tube that holds the torpedo. When test-fired with the tube firmly anchored, a 550-kg torpedo leaves the tube at 20 m/s. In operation, a 30-Mg submarine is traveling at 5 m/s when it expels a 550-kg torpedo in the forward direction. What are the speeds of the submarine and torpedo immediately after expulsion?

Solution

Considering the two-body system consisting of the submarine and torpedo, let us assume that the external forces remain in balance during expulsion. Then $W_e = 0$. Assuming also that the gas pressure depends only on the position of the torpedo relative to the submarine, the work W_{12} done by the internal forces will be the same during actual operation as during the test-firing. The work-kinetic energy relationship for the test-firing yields $W_{12} = \frac{1}{2}(550 \text{ kg})(20 \text{ m/s})^2 = 110$ kJ and for the operating condition the relationship is

$$110 \text{ kJ} = \frac{1}{2}(30{,}000 \text{ kg})\left[v_1^2 - (5 \text{ m/s})^2\right] + \tfrac{1}{2}(550 \text{ kg})\left[v_2^2 - (5 \text{ m/s})^2\right]$$

Also, if the external forces are in balance, momentum will be conserved:

$$(30{,}550 \text{ kg})(5 \text{ m/s}) = (30{,}000 \text{ kg})v_1 + (550 \text{ kg})v_2$$

These two relationships give the desired speeds as

$$v_1 = 4.6 \text{ m/s}, \qquad v_2 = 24.8 \text{ m/s}$$

The 19.8-m/s boost in speed given the torpedo is slightly less than when it is fired from the firmly-anchored tube. However, the speed of the torpedo relative to the submarine is

$$v_2 - v_1 = 20.2 \text{ m/s}$$

or slightly higher than in the fixed-tube test.

Two special cases of the work done by internal forces are of interest. The simpler is that in which the particles are constrained so that the distances between all pairs remain fixed; that is, the case of a rigid body. In this case, all the $dr_{i/j}$ are zero and the work-kinetic energy equation, Eq. (8.38), reduces to $W_e = \Delta T$.

Another special case occurs when the force T_{ij} depends only on the distance $r_{i/j}$. That is, the force is not a function of relative velocity or previous history of deformation. Then the work integral $-\int T_{ij} dr_{i/j}$ is a function only of the distance between the particles and does not depend on the manner in which the particles move to reach a particular configuration. This would be the case, for example, with elastic spring interconnections or gravitational interactions. In this case, we can define the potential functions

$$V_{ij}(r_{i/j}) = \int_{(r_{i/j})_0}^{r_{i/j}} T_{ij}(\rho_{i/j}) d\rho_{i/j}$$

and if their sum is denoted by

$$V = \sum_{i-j} V_{ij}$$

the work-energy integral becomes

$$W_e = \Delta T + \Delta V$$

That is, when the internal forces are all conservative, the work done by the external forces is equal to the change in total mechanical energy within the system.

KINETICS OF RIGID BODIES

If a system of particles is structurally constrained so that the distance between every pair of particles remains constant as the system moves, it forms a rigid body. Thus, the laws of kinetics in the previous section are applicable, along with the kinematics relationships reviewed earlier. Of the kinematics relationships, Eq. (8.28c), will be useful in the form given, whereas the moment equation, Eq. (8.33a), must be specialized to relate accelerations to angular velocity and angular acceleration.

In the general (three-dimensional) case, both the moment equation and kinematics relationships become considerably more complicated than they are for planar motion, and will be outside the scope of this review.

Moment Relationships for Planar Motion

Figure 8.12 shows the plane of the motion of a rigid body, with a point P (to be selected by the analyst for moment reference) along with an element of mass dm, which in the following is analogous to the mass m_i in the earlier analysis of a set of particles. The summation appearing in Eq. (8.33a) will be written as an integral in this case, because the body is viewed as having continuously-distributed mass. The element of mass is located relative to P by the position vector $\mathbf{r} = r\mathbf{e}_1 + z\mathbf{e}_3$. Its acceleration is related to that of point P through Eq. (8.24), and the moment equation, Eq. (8.33a), may be written as

$$\begin{aligned}\mathbf{M}_P &= \int_m (r\mathbf{e}_1 + z\mathbf{e}_3)\times(\mathbf{a}_P + r\alpha\mathbf{e}_2 - r\omega^2\mathbf{e}_1)dm \\ &= \left(\int_m \mathbf{r}\, dm\right)\times\mathbf{a}_P + \left(\int_m r^2 dm\right)\alpha\mathbf{e}_3 - \alpha\int_m zr\mathbf{e}_1\, dm - \omega^2\int_m zr\mathbf{e}_2\, dm\end{aligned}$$

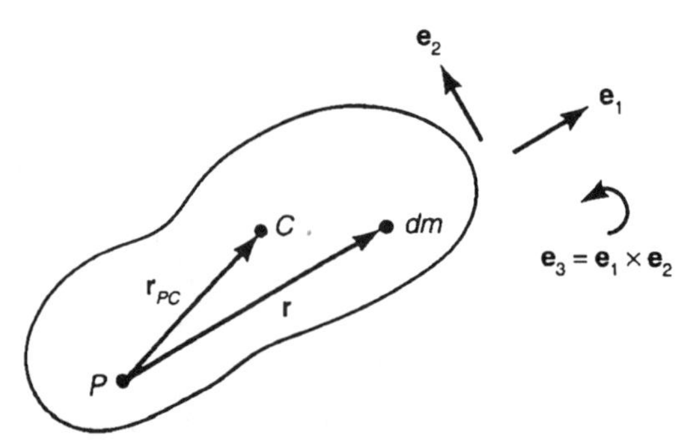

Figure 8.12

If the body's mass is distributed symmetrically with respect to the plane of motion through P, the last two integrals in the last line will vanish; without such symmetry, these terms indicate the possibility of components of moment in the plane of motion. Thus, even for *plane* motion, the distribution of mass may imply forces *perpendicular* to the plane of motion. These will not be pursued in detail here, but the reader must be aware of this possibility. The other two integrals will be of concern. The first is exactly the expression one would write to determine the location of the center of mass from point P:

$$\int_m \mathbf{r}dm = m\mathbf{r}_{PC}$$

The second integral is called the **moment of inertia** of the body about the axis through P and perpendicular to the plane of motion:

$$\int_m r^2 dm = I_P$$

The moment about this axis is thus related to accelerations through

$$M_{P3} = \mathbf{e}_3 \bullet \mathbf{M}_P = I_P\alpha + \mathbf{e}_3 \bullet (m\mathbf{r}_{PC}\times\mathbf{a}_P) \tag{8.40a}$$

By using Eq. (8.24) to relate the acceleration of P to that of the mass center C, it is possible to express this moment law in the alternative form

$$M_{P3} = I_C\alpha + \mathbf{e}_3 \bullet (m\mathbf{r}_{PC}\times\mathbf{a}_C) \tag{8.40b}$$

where I_C is the moment of inertia of the body about an axis through C perpendicular to the plane of motion. The two moments of inertia are related through the **parallel axis formula**

$$I_P = I_C + md^2$$

in which d is the distance between P and C.

Two special cases warrant attention. If P is chosen to be the mass center C, then $\mathbf{r}_{PC} = 0$, and the relationship is

$$M_{C3} = I_C\alpha$$

If the body is hinged about a fixed support and P is selected to be on the axis of the hinge, then

$\mathbf{a}_P = 0$ and the relationship is

$$M_{P3} = I_P\alpha$$

These last two relationships indicate the moment of inertia of the body is the property that provides resistance to changes in the angular velocity, much as mass provides resistance to changes in the velocity of a particle. For bodies of simple geometry, the integrals have been evaluated in terms of mass and the geometry, and results can be found in tabulated summaries. More complicated bodies can require tedious work to estimate the moment of inertia, or there are experiments based on the implications of Eq. (8.40) that can be used to determine it. It is common to specify the moment of inertia by giving the mass of the body and its **radius of gyration**, k_p, defined by $I_P = mk_P^2$.

Example 8.24

A 23-kg rotor has a 127 mm radius of gyration about its axis of rotation. What average torque about its fixed axis of rotation is required to bring the rotor from rest to a speed of 200 rpm in 6 seconds?

Solution

The moment of inertia of the rotor is

$$I = (23 \text{ kg})(0.127 \text{ m})^2 = 0.371 \text{ kg} \bullet \text{m}^2$$

and the average angular acceleration is

$$\alpha = \frac{(200 \text{ rpm})}{6 \text{ s}} \frac{(2\pi \text{ rad/r})}{(60 \text{ s/min})} = 3.49 \text{ rad/s}^2$$

For fixed-axis rotation,

$$M = I\alpha = (0.371 \text{ kg} \bullet \text{m}^2)(3.49 \text{ s}^{-2}) = 1.29 \text{ N} \bullet \text{m}$$

Example 8.25

The car with rear-wheel drive in Exhibit 24 has sufficient power to cause the drive wheels to slip as it accelerates. The coefficient of friction between the drive wheels and roadway is μ. What is the acceleration in terms of g, μ, and the dimensions shown?

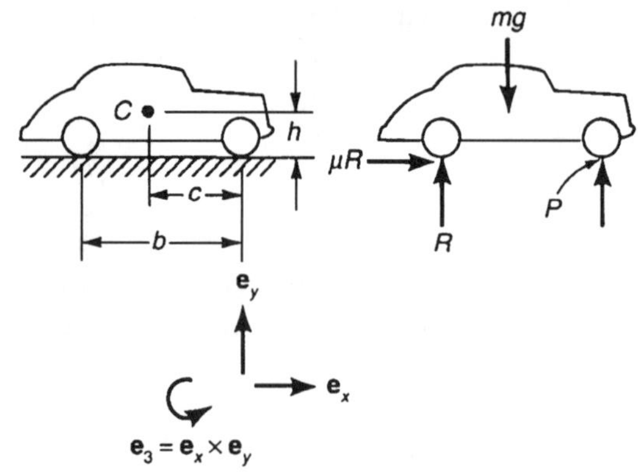

Exhibit 24

Solution

Assuming the car does not rotate, every point will have the same acceleration, $\mathbf{a} = a_x\mathbf{e}_x$. The free-body diagram shows the forces external to the car, with the label on the horizontal force at the drive wheels accounting for the fact that the friction limit has been reached there. Equation (8.28c) implies that $\mu R = ma_x$. Since the reaction R in this equation is unknown, another relationship must be introduced. Of several that could be written (for example, forces in another direction, moments about a selected point), it would be best if no additional unknowns are introduced. Thus, to avoid bringing the unknown reaction at the front wheels into the analysis, consider moments about point P, which are related by the moment law Eq. (8.40b):

$$-bR + c\,mg = I(0) + \mathbf{e}_3 \bullet [m(-c\mathbf{e}_x + h\mathbf{e}_y) \times a_x\mathbf{e}_x] = -mha_x$$

Eliminating R between this equation and the friction equation leads to

$$a_x = \frac{\mu c}{b - \mu h} g$$

Example **8.26**

The uniform slender rod in Exhibit 25 slides along the wall and floor under the effects of gravity. If friction is negligible, what is the angular acceleration in terms of g, l, and the angle θ?

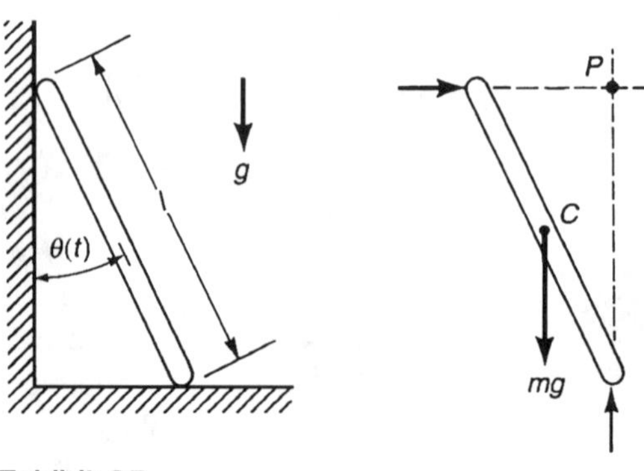

Exhibit 25

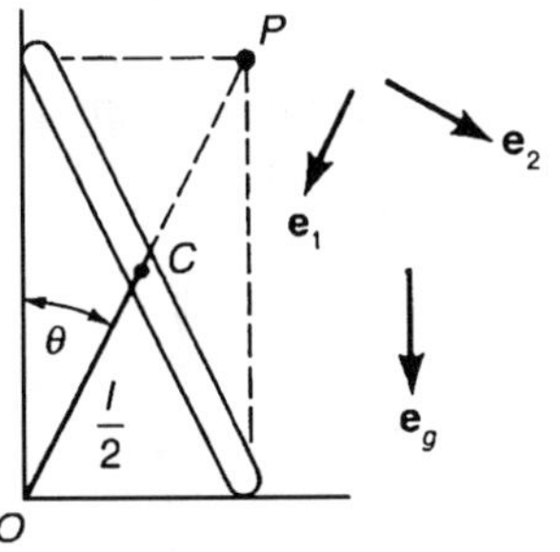

Exhibit 26

Solution

As indicated on the free-body diagram, the reactions from the wall and floor are horizontal and vertical since there is no friction. Because neither of these is known, a good strategy would be to avoid dealing with them; to this end, consider moments about point *P*, which will be related by Eq. (8.40b). To evaluate the acceleration of the mass center *C*, observe that it follows a circular path with radius *l*/2 and center at *O* (see Exhibit 26). Thus, the acceleration of *C* is

$$\mathbf{a}_C = \frac{l}{2}(\dot{\theta}^2\mathbf{e}_1 + \ddot{\theta}\mathbf{e}_2)$$

Referring to the free-body diagram, we can write the moment of forces about *P* as

$$\mathbf{M}_P = \frac{l}{2}\mathbf{e}_1 \times mg\mathbf{e}_g = \frac{1}{2}mgl\sin\theta\mathbf{e}_3$$

Any of a number of references gives the moment of inertia of a slender, uniform rod about an axis through its center as

$$I_C = \frac{1}{12}ml^2$$

Substituting the above into the moment law, Eq. (8.40b), yields

$$\frac{1}{2}mgl\sin\theta = \frac{1}{12}ml^2\ddot{\theta} + \mathbf{e}_3 \bullet \left[m\frac{l}{2}\mathbf{e}_1 \times \frac{l}{2}(\dot{\theta}^2\mathbf{e}_1 + \ddot{\theta}\mathbf{e}_2)\right] = \frac{ml^2}{3}\ddot{\theta}$$

which leads to the desired angular acceleration:

$$\ddot{\theta} = \frac{3g}{2l}\sin\theta$$

Example **8.27**

The uniform, slender beam of length l is suspended by the two wires in the configuration shown in Exhibit 27 when the wire on the left is cut. Immediately after the wire is severed (that is, while the velocities of all points are still zero), what is the tension in the remaining wire?

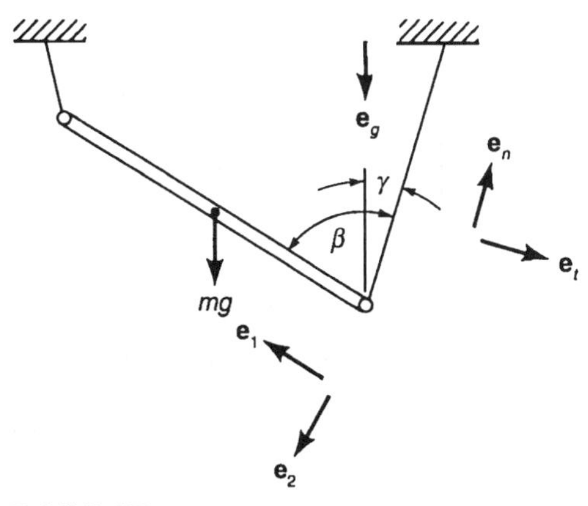

Exhibit 27

Solution

The free-body diagram (Exhibit 28) shows the desired force and the only other force acting on the beam, that of gravity. Summing moments about the center of mass gives a relationship between the desired tension and the angular acceleration of the bar:

$$\frac{l}{2}T\sin\beta = \frac{ml^2}{12}\alpha \qquad \textbf{(i)}$$

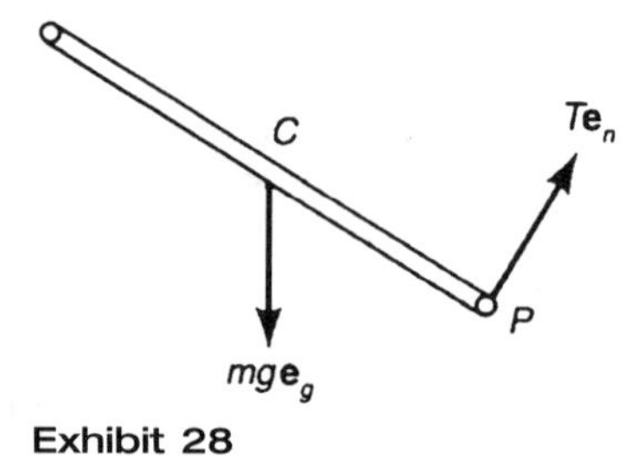

Exhibit 28

We also know that the sum of all forces is related to the acceleration of the mass center by

$$T\mathbf{e}_n + mg\mathbf{e}_g = m\mathbf{a}_C \qquad \textbf{(ii)}$$

Since the end P is constrained by the wire to follow a circular path, its acceleration may be expressed by

$$\mathbf{a}_P = a_t\mathbf{e}_t + \frac{v_P^2}{R}\mathbf{e}_n \qquad \textbf{(iii)}$$

in which a_t is another unknown quantity. Finally, this acceleration is related to that of the center of mass by

$$\mathbf{a}_C = \mathbf{a}_P + \frac{l}{2}\alpha\mathbf{e}_2 - \frac{l}{2}\omega^2\mathbf{e}_1 \qquad \textbf{(iv)}$$

Since velocities are still zero, the centripetal terms v_P^2/R and $\frac{l}{2}\omega^2$ are both zero. With this simplification, Eqs. (ii), (iii), and (iv) readily combine to give

$$T\mathbf{e}_n + mg\mathbf{e}_g = m\left(a_t\mathbf{e}_t + \frac{l}{2}\alpha\mathbf{e}_2\right)$$

To avoid dealing with the unknown a_t, we may dot-multiply each term in this equation by $\mathbf{e}_n$ with the result

$$T + mg\mathbf{e}_n \bullet \mathbf{e}_g = \frac{1}{2}ml\alpha\mathbf{e}_n \bullet \mathbf{e}_2$$

Referring to the specified geometry, the dot-products are evaluated as

$$\mathbf{e}_n \bullet \mathbf{e}_g = \cos\,(180° - \gamma) = -\cos\,\gamma$$

$$\mathbf{e}_n \bullet \mathbf{e}_2 = \cos\,(90° + \beta) = -\sin\,\beta$$

and the equation can be written as

$$T + \frac{1}{2} ml\alpha \sin\beta = mg\cos\gamma \tag{v}$$

Now α is readily eliminated by substituting the expression for α obtained from Eq. (i) into Eq. (v), leading to the desired value of the tension:

$$T = \frac{mg\cos\gamma}{1 + 3\sin^2\beta}$$

Work and Kinetic Energy

If a rigid body has a number of forces $f_1, f_2, \ldots, f_n$ applied at points $P_1, P_2, \ldots, P_n$, the time rate at which these forces do work on the body (that is, the power transmitted to the body) can be evaluated as

$$\frac{dW}{dt} = \mathbf{f}_1 \bullet \frac{d\mathbf{r}_1}{dt} + \mathbf{f}_2 \bullet \frac{d\mathbf{r}_2}{dt} + \cdots + \mathbf{f}_n \bullet \frac{d\mathbf{r}_n}{dt} = \sum_i \mathbf{f}_i \bullet \mathbf{v}_i$$

But $\mathbf{v}_i$, the velocity of point P_i, can be related to the velocity of a selected point P of the body:

$$\mathbf{v}_i = \mathbf{v}_P + \omega \times \mathbf{r}_{Pi}$$

so that the power can also be expressed as

$$\frac{dW}{dt} = \sum_i \mathbf{f}_i \bullet (\mathbf{v}_P + \omega \times \mathbf{r}_{Pi}) = \left(\sum_i \mathbf{f}_i\right) \bullet \mathbf{v}_P + \left(\sum_i \mathbf{r}_{Pi} \times \mathbf{f}_i\right) \bullet \omega = \mathbf{f} \bullet \mathbf{v}_P + \mathbf{M}_P \bullet \omega \tag{8.41}$$

in which **f** is the resultant of all of the forces. P may be selected as any point of the body, and M_P is the resultant moment about P. For example, as a rotor turns about a fixed axis, there may be forces from the support bearings in addition to an accelerating torque about the axis of rotation. If the point P is selected to be on the axis of rotation, $\mathbf{v}_P$ will be zero, and the power transmitted to the rotor (which will, of course, induce a change in its kinetic energy) is simply the dot product of the torque and the angular velocity. Negative or positive values are possible, depending on the angle between $\mathbf{M}_P$ and ω (that is, whether the moment component is in the same or the opposite direction as the rotation).

The kinetic energy of a rigid body is the sum of the kinetic energies of its individual elements, whose velocities can be related to the velocity of a selected point P and the angular velocity ω. Referring to Fig. 8.12, we write this for plane motion as

$$\begin{aligned} T &= \frac{1}{2}\int_m |\mathbf{v}|^2\, dm = \frac{1}{2}\int_m (\mathbf{v}_P + \omega \times \mathbf{r}) \bullet (\mathbf{v}_P + \omega \times \mathbf{r})\, dm \\ &= \frac{1}{2}\left(\int_m dm\right) v_P^2 + \mathbf{v}_P \bullet \left[\omega \times \left(\int_m \mathbf{r}\, dm\right)\right] + \frac{1}{2}\int_m |\omega \times \mathbf{r}|^2 dm \end{aligned}$$

The integral in the first term is simply the mass m of the body. The integral in the second term is related to the mass and position of the center of mass by

$$\int_m \mathbf{r} dm = m\mathbf{r}_{PC}$$

For plane motion, $\omega \times \mathbf{r} = \omega r \mathbf{e}_2$, so the last term becomes

$$\int_m |\omega \times \mathbf{r}|^2 dm = \omega^2 \int_m r^2 dm = I_P \omega^2$$

The expression for kinetic energy for plane motion of a rigid body is then

$$T = \frac{1}{2} m v_P^2 + m\mathbf{v}_P \bullet (\omega \times \mathbf{r}_{PC}) + \frac{1}{2} I_P \omega^2 \tag{8.42}$$

Example 8.28

The wheel in Exhibit 29 is released from rest and rolls down the hill with sufficient friction to prevent slipping. Its mass is m, its radius is r, and its central radius of gyration is k. After the center of the wheel has dropped a vertical distance h, what is the speed of the center of mass of the wheel?

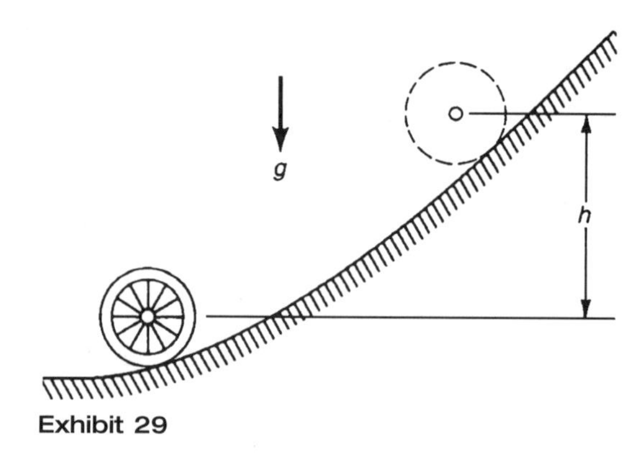

Exhibit 29

Solution

Since the velocity of the contact point is zero, the work of the force there is zero. The work of the force of gravity is then the total work done on the wheel and is simply $W = mgh$. Since the contact point is the instantaneous center, the speed of the center of the wheel is readily related to the angular velocity, $\omega = v_C/r$. The kinetic energy can be written from Eq. (8.2), with P selected as any point on the wheel. If we choose P to be the center of mass,

$$T = \frac{1}{2} m v_C^2 + 0 + \frac{1}{2} m k^2 \omega^2 = \frac{1}{2}\left(1 + \frac{k^2}{r^2}\right) m v_C^2$$

If, instead, we choose P to be the instantaneous center, the kinetic energy is

$$T = \frac{1}{2} m(0)^2 + 0 + \frac{1}{2}(mk^2 + mr^2)\omega^2 = \frac{1}{2}\left(1 + \frac{k^2}{r^2}\right) m v_C^2$$

Since the work must equal the change in kinetic energy,

$$mgh = \frac{1}{2}\left(1 + \frac{k^2}{r^2}\right) m v_C^2$$

and the speed of the center will be

$$v_C = \sqrt{\frac{2gh}{1 + \frac{k^2}{r^2}}}$$

SELECTED SYMBOLS AND ABBREVIATIONS

Symbol or Abbreviation	Description
$\mathbf{a}$	acceleration
$\mathbf{a}_t$	tangential component of acceleration
$\mathbf{e}_t$	unit vector tangent to path
$\mathbf{e}_n$	unit vector in principal normal direction
$\mathbf{e}_i$	unit vector in direction indicated by the specific value of *i*
$\mathbf{f}$	resultant force
g	gravitational field intensity
$\mathbf{g}$	impulse resultant force
$\mathbf{H}_o$	angular momentum about *O*
I_P	moment of inertia about *P*
κ	local curvature of path
k_P	radius of gyration about *P*
M_i	mass of *i*th particle
M	total mass
$\mathbf{M}_P$	moment of forces about *P*
N	coefficient of kinetic fraction
ρ	radius of curvature of path
$\mathbf{r}$	position vector
s	distance along path
t	time
$\boldsymbol{T}$	kinetic energy
$\mathbf{v}$	velocity
$\boldsymbol{W}$	work
α	angular acceleration
μ	coefficient of sliding friction
ω	angular velocity

PROBLEMS

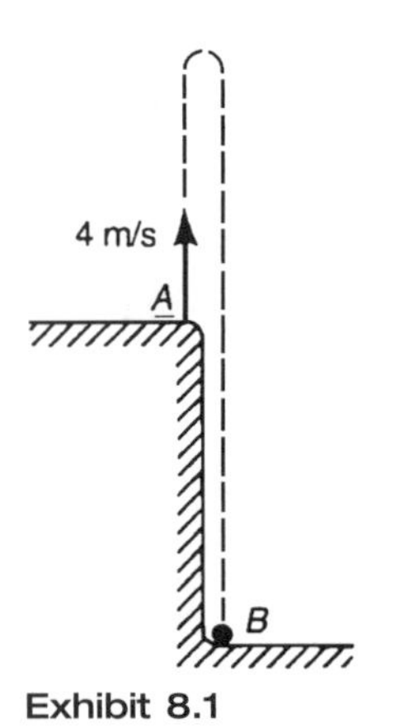

Exhibit 8.1

8.1 A particle is thrown vertically upward from the edge A of the ditch shown in Exhibit 8.1. If the initial velocity is 4 m/s, and the particle is known to hit the bottom, B, of the ditch exactly 6 seconds after it was released at A, determine the depth of this ditch. Neglect air resistance.

a. 24.0 m
b. 152.6 m
c. 200 m
d. 176.6 m

8.2 The slider P in Exhibit 8.2 is driven by a complex mechanism in such a way that (i) it remains on a straight path throughout; (ii) at the instant $t = 0$, the slider is located at A, (iii) at any general instant of time, the velocity of P is given by $v = (3t^2 - t + 2)$ m/s. Determine the distance of P from point O when $t = 2$ s.

a. 26 m
b. 10 m
c. 6 m
d. 12 m

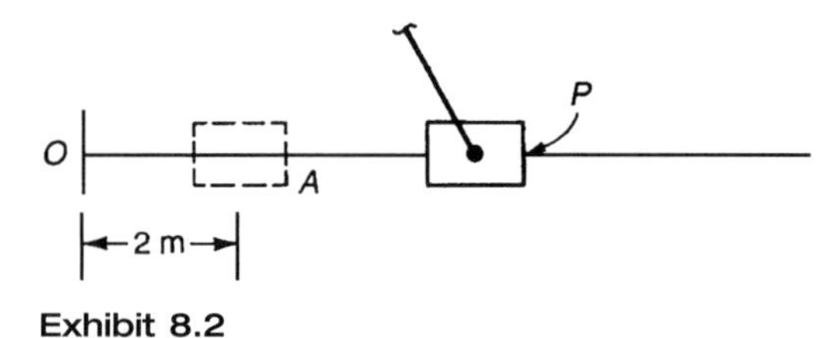

Exhibit 8.2

8.3 A particle in rectilinear motion starts from rest and maintain the acceleration profile shown in Exhibit 8.3. The displacement of the particle in the first 8 seconds is

a. 4 m
b. 28 m
c. 24 m
d. 20 m

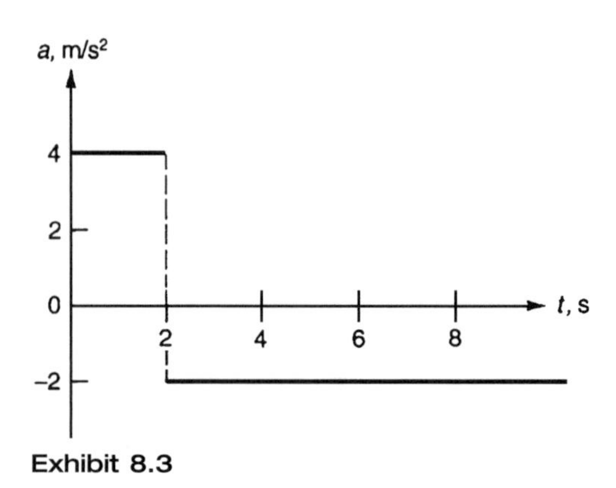

Exhibit 8.3

8.4 A ball is thrown by a player (Exhibit 8.4) from a position 2 m above the ground surface with a velocity of 40 m/s inclined at 60° to the horizontal. Determine the maximum height, H, the ball will attain.

a. 63.2 m
b. 61.2 m
c. 30.6 m
d. 31 m

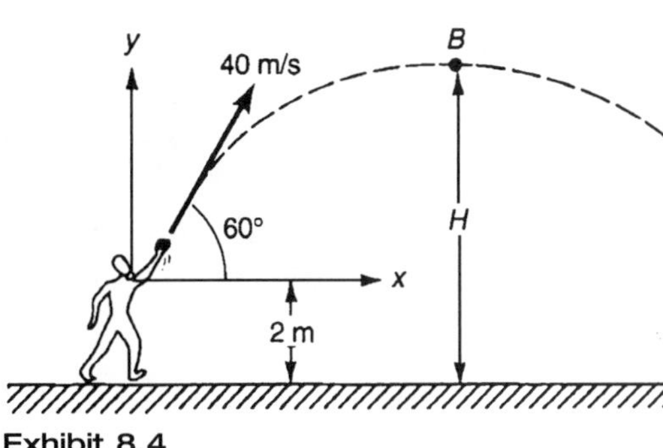

Exhibit 8.4

8.5 A golf ball (Exhibit 8.5) is struck horizontally from point A of an elevated fairway. Determine the initial speed that must be imparted to the ball if the ball is to strike the base of the flag stick on the green 140 meters away. Neglect air friction.

a. 34.3 m/s
b. 103 m/s
c. 90 m/s
d. 19.2 m/s

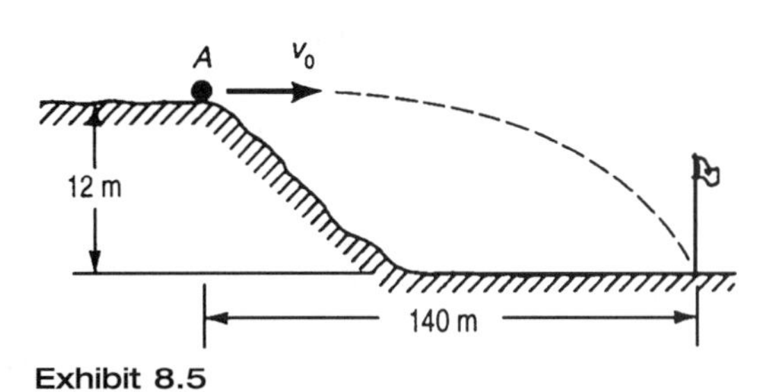

Exhibit 8.5

8.6 In Exhibit 8.6, the rod R rotates about a fixed axis at O. A small collar B is forced down the rod (towards O) at a constant speed of 3 m/s relative to the rod. If the value of θ at any given instant is $\theta = (t^2 + t - 2)$ rad, find the magnitude of the acceleration of B at time $t = 1$ second, when B is known the be 1 meter away from O.

a. 8.0 m/s^2
b. 20.2 m/s^2
c. 18.4 m/s^2
d. 3.0 m/s^2

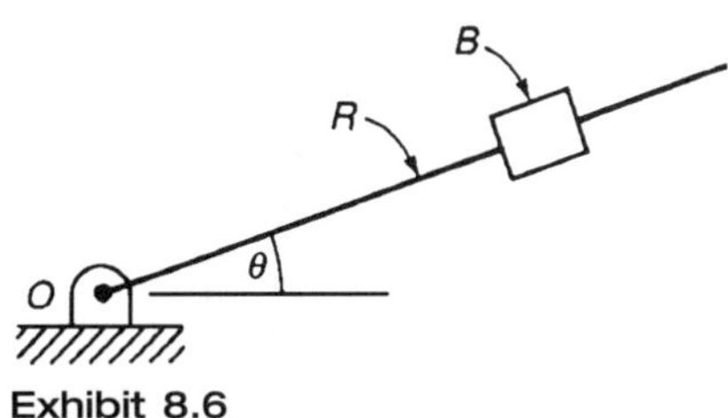

Exhibit 8.6

8.7 A rocket (Exhibit 8.7) is fired vertically upward from a launching pad at B, and its flight is tracked by radar from point A. Find the magnitude of the velocity of the rocket when $\theta = 45°$ if $\dot{\theta} = 0.1$ rad/s.

a. 36 m/s
b. 180 m/s
c. 90 m/s
d. 360 m/s

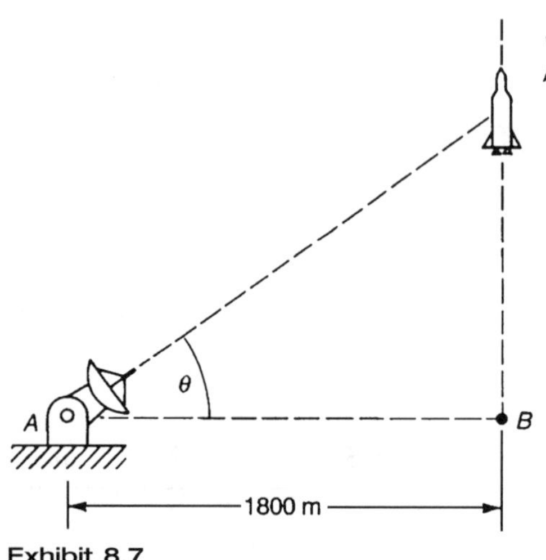

Exhibit 8.7

8.8 A particle is given an initial velocity of 50 m/s at an angle of 30° with the horizontal as shown in Exhibit 8.8. What is the radius of curvature of its path at the highest point, C?

a. 19.5 m
b. 255 m
c. 221 m
d. 191 m

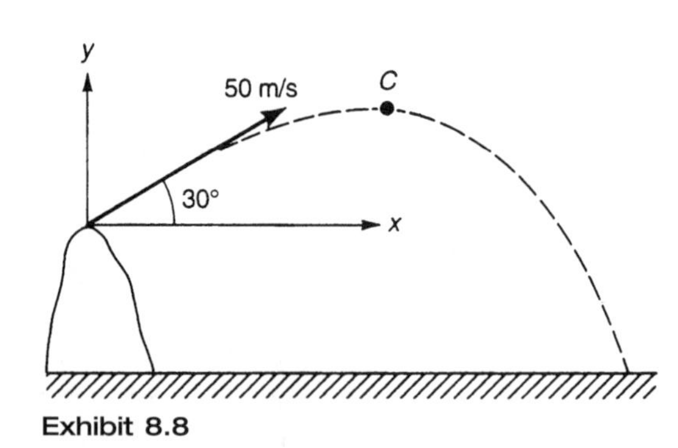

Exhibit 8.8

8.9 An automobile moves along a curved path that can be approximated by a circular arc of radius 110 meters. The driver keeps his foot on the accelerator pedal in such a way that the speed increases at the constant rate of 3 m/s^2. What is the total acceleration of the vehicle at the instant when its speed is 20 m/s?

a. 22.0 m/s^2
b. 3.6 m/s^2
c. 3.0 m/s^2
d. 4.7 m/s^2

8.10 A pilot testing an airplane at 800 kph wishes to subject the aircraft to a normal acceleration of 5 g's in order to fulfill the requirements of an on-board experiment. Find the radius of the circular path that would allow the pilot to do this.

a. 502 m
c. 1007 m
b. 3308 m
d. 1453 m

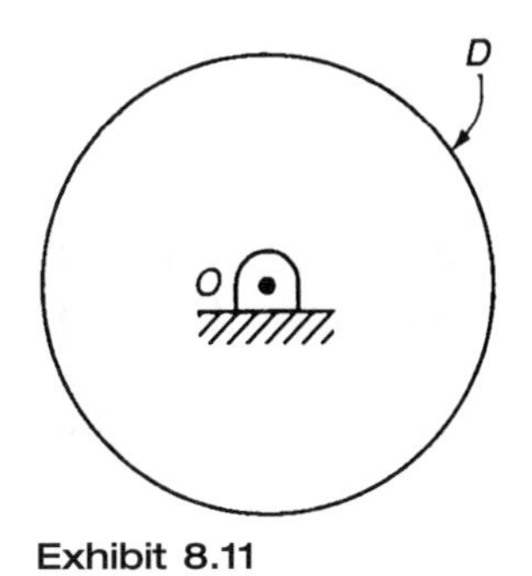

Exhibit 8.11

8.11 At the instant $t = 0$, the disk D in Exhibit 8.11 is spinning about a fixed axis through O at an angular speed of 300 rev/min. Bearing friction and other effects are known to slow the disk at a rate that is k times its instantaneous angular speed, where k is a constant with the value $k = 1.2\ s^{-1}$. Determine when (from $t = 0$) the disk's spin rate is cut in half.

a. 6.5 s
c. 0.8 s
b. 13.1 s
d. 0.6 s

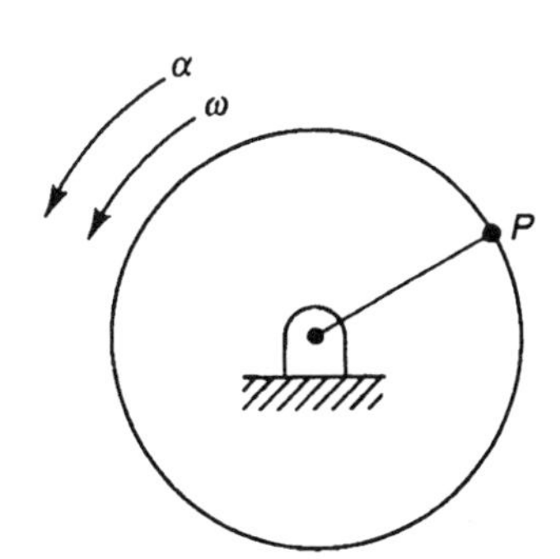

Exhibit 8.12

8.12 In Exhibit 8.12, a flywheel 2 m in radius is brought uniformly from rest up to an angular speed of 300 rpm in 30 s. Find the speed of a point P on the periphery 5 seconds after the wheel started from rest.

a. 10.5 m/s
c. 62.8 m/s
b. 5.2 m/s
d. 100.0 m/s

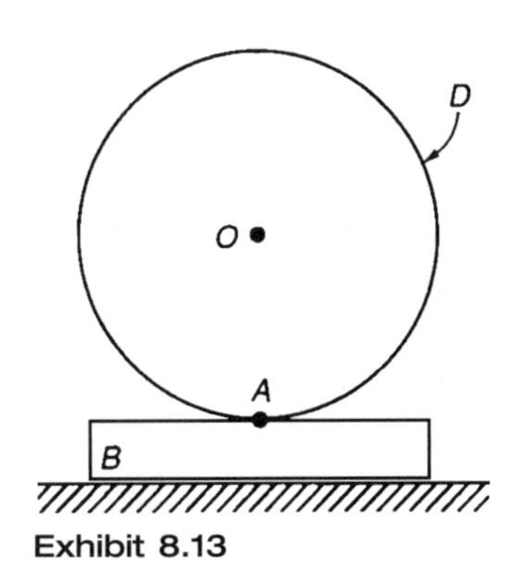

Exhibit 8.13

8.13 The block B (Exhibit 8.13) slides along a straight path on a horizontal floor with a constant velocity of 2 m/s to the right. At the same time, the disk, D, of 3-m diameter rolls without slip on the block. If the velocity of the center, O, of the disk is directed to the left and remains constant at 1 m/s, determine the angular velocity of the disk.

a. 0.3 rad/s counterclockwise
b. 2.0 rad/s counterclockwise
c. 0.7 rad/s counterclockwise
d. 0.7 rad/s clockwise

8.14 In Exhibit 8.14, the disk, D, rolls without slipping on a horizontal floor with a constant clockwise angular velocity of 3 rad/s. The rod, R, is hinged to D at A, and the end, B, of the rod touches the floor at all times. Determine the angular velocity of R when the line OA joining the center of the disk to the hinge at A is horizontal as shown.

a. 0.6 rad/s counterclockwise
b. 0.6 rad/s clockwise
c. 3.0 rad/s counterclockwise
d. 3.0 rad/s clockwise

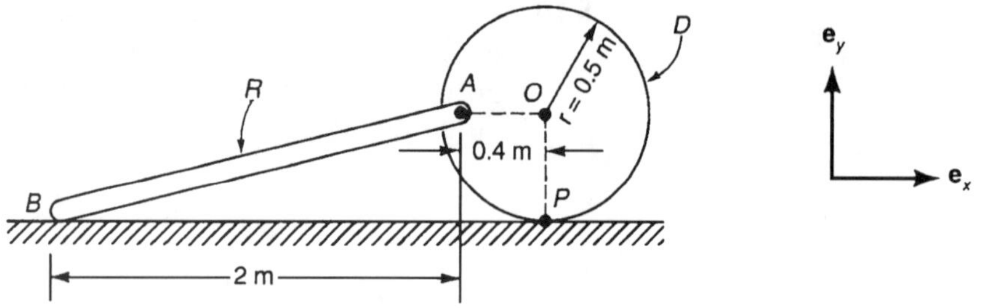

Exhibit 8.14

8.15 The fire truck in Exhibit 8.15 is moving forward along a straight path at the constant speed of 50 km/hr. At the same time, its 2-meter ladder *OA* is being raised so that the angle θ is given as a function of time by $\theta = (0.5t^2 - t)$ rad, where t is in seconds. The magnitude of the acceleration of the tip of the ladder when $t = 2$ seconds is

a. 0
b. 4.0 m/s^2
c. 2.0 m/s^2
d. 2.8 m/s^2

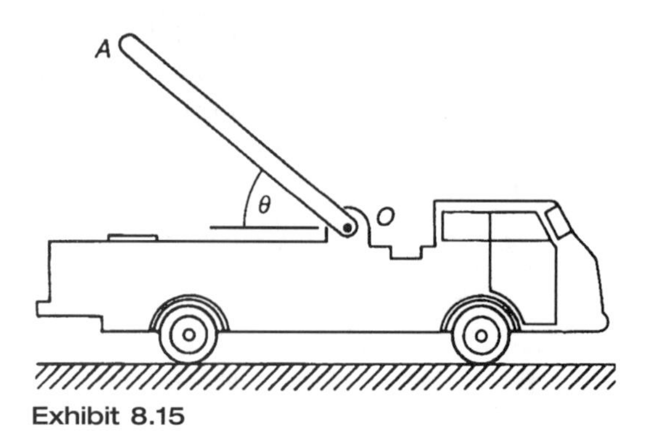

Exhibit 8.15

8.16 In Exhibit 8.16, the block *B* is constrained to move along a horizontal rectilinear path with a constant acceleration of 2 m/s^2 to the right. The slender rod, *R*, of length 2 m is pinned to *B* at *O* and can swing freely in the vertical plane. At the instant when $\theta = 0°$ (rod is vertical), the angular velocity of the rod is zero but its angular acceleration is 2.5 m/s^2 clockwise. Find the acceleration of the midpoint *G* of the rod at this instant ($\theta = 0°$).

a. 3.0 m/s^2 ←
b. 0.5 m/s^2 →
c. 2.5 m/s^2 ←
d. 2.5 m/s^2 →

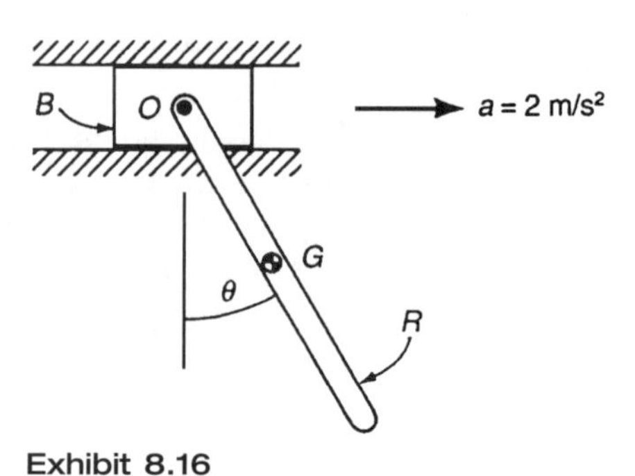

Exhibit 8.16

8.17 The block, B, in Exhibit 8.17, contains a square-cut circular groove, A particle, P, moves in this groove in the clockwise direction and maintains a constant speed of 6 m/s relative to the block. At the same time, the block slides to the right on a straight path at the constant speed of 10 m/s. Find the magnitude of the absolute velocity of P at the instant when $\theta = 30°$.

a. 8.7 m/s c. 4 m/s
b. 16 m/s d. 14 m/s

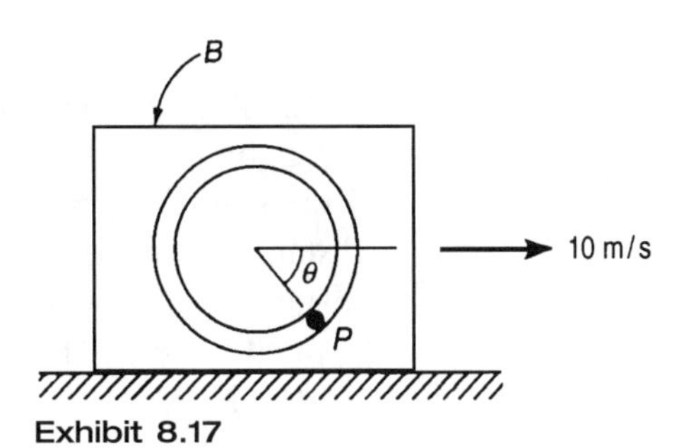

Exhibit 8.17

8.18 In Exhibit 8.18, a pin moves with a constant speed of 2 m/s along a slot in a disk that is rotating with a constant clockwise angular velocity of 5 rad/s. Calculate the absolute acceleration of this pin when it reaches the position C (directly above O). The unit vectors $\mathbf{e}_x$ and $\mathbf{e}_y$ are fixed to the disk.

a. $17.5\mathbf{e}_y$ m/s^2 c. $-2.5\mathbf{e}_y$ m/s^2
b. $-17.5\mathbf{e}_y$ m/s^2 d. $-22.5\mathbf{e}_y$ m/s^2

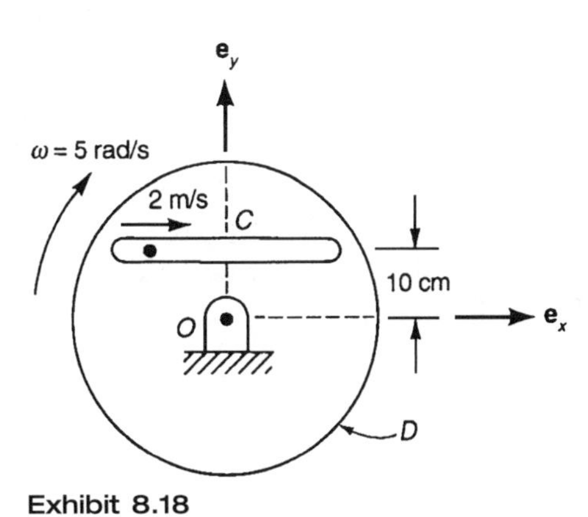

Exhibit 8.18

8.19 In Exhibit 8.19, a particle P of mass 5 kg is launched vertically upward from the ground with an initial velocity of 10 m/s. A constant upward thrust $T = 100$ newtons is applied continuously to P, and a downward resistive force $R = 2z$ newtons also acts on the particle, where z is the height of the particle above the ground. Determine the maximum height attained by P.

a. 6.0 m c. 15.8 m
b. 45.5 m d. 55.5 m

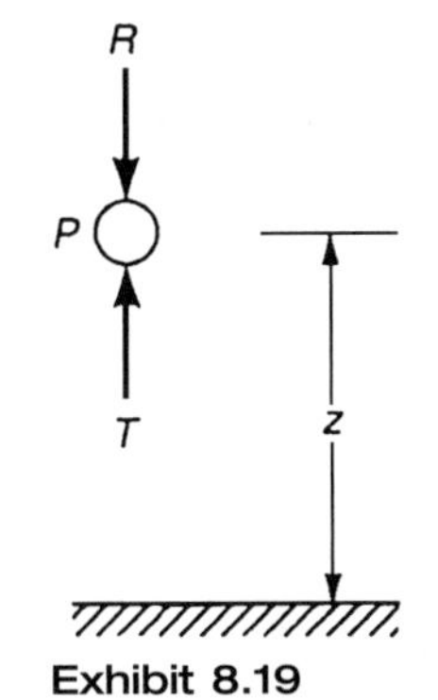

Exhibit 8.19

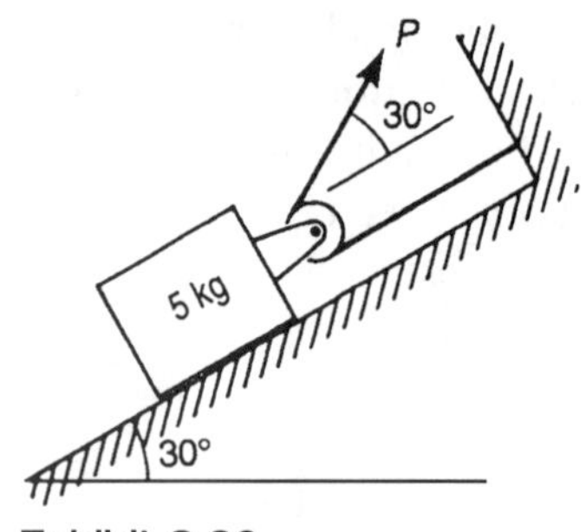

Exhibit 8.20

8.20 Determine the force P required to give the block shown in Exhibit 8.20 an acceleration of 2 m/s^2 up the incline. The coefficient of kinetic friction between the block and the incline is 0.2.

a. 39.2 N c. 44.6 N
b. 21.9 N d. 49.8 N

8.21 In Exhibit 8.21 the rod R rotates in the vertical plane about a fixed axis through the point O with a constant counterclockswise angular velocity of 5 rad/s. A collar B of mass 2 kg slides down the rod (toward O) so that the distance between B and O decreases at the constant rate of 1 m/s. At the instant when $\theta = 30°$ and $r = 400$ mm, determine the magnitude of the applied force P. The coefficient of kinetic friction between B and R is 0.1.

a. 9.9 N c. 10.5 N
b. 11.9 N d. 0.3 N

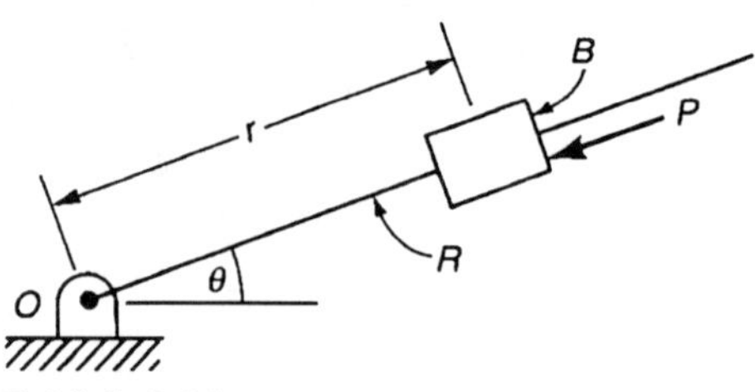

Exhibit 8.21

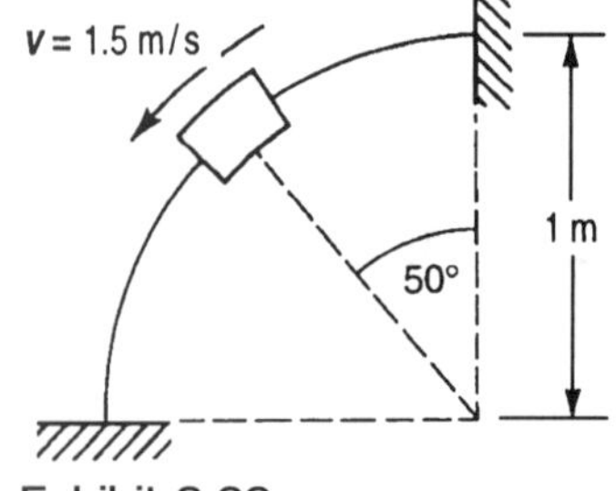

Exhibit 8.22

8.22 The 3-kg collar in Exhibit 8.22 slides down the smooth circular rod. In the position shown, its velocity is 1.5 m/s. Find the normal force (contact force) the rod exerts on the collar.

a. 12.2 N↖ c. 19.2 N↖
b. 24.4 N↘ d. 12.2 N↘

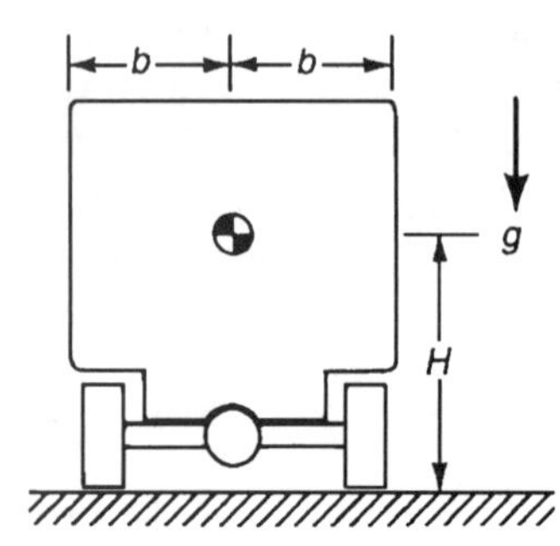

Exhibit 8.23

8.23 Fork lift vehicles, Exhibit 8.23, tend to roll over if they are driven too fast while turning. For a vehicle of mass m with a mass center that describes a circle of radius R, find the relationship between the forward speed u and the vehicle dimensions and path radius at the onset of tipping.

a. $u = (Rgb/H)^{0.5}$ c. $u = (gh)^{0.5}$
b. $u = (RgH/b)^{0.5}$ d. $u = (bg)^{0.5}$

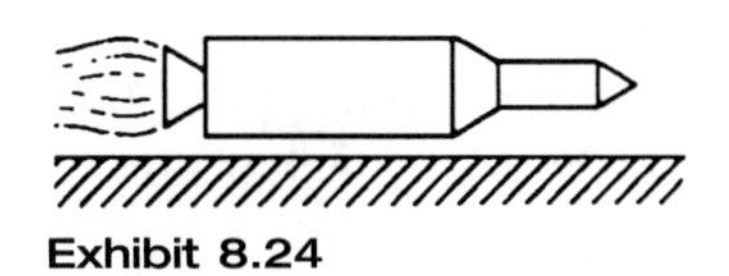
Exhibit 8.24

8.24 A toy rocket of mass 1 kg is placed on a horizontal surface, and the engine is ignited (Exhibit 8.24). The engine delivers a force equal to $(0.25 + 0.5t)$ N, where t is time in seconds, and the coefficient of friction between the rocket and the surface is 0.01. Determine the velocity of the rocket 7 seconds after ignition.

a. 14.0 m/s
b. 3.7 m/s
c. 13.3 m/s
d. 26.3 m/s

8.25 A 2000-kg pickup truck is traveling backward down a 10° incline at 80 km/hr when the driver notices through his rear-view mirror an object on the roadway. He applies the brakes, and this results in a constant braking (retarding) force of 4000 N. How long does it take the truck to stop?

a. 11.1 s
b. 74.9 s
c. 2.3 s
d. 13.0 s

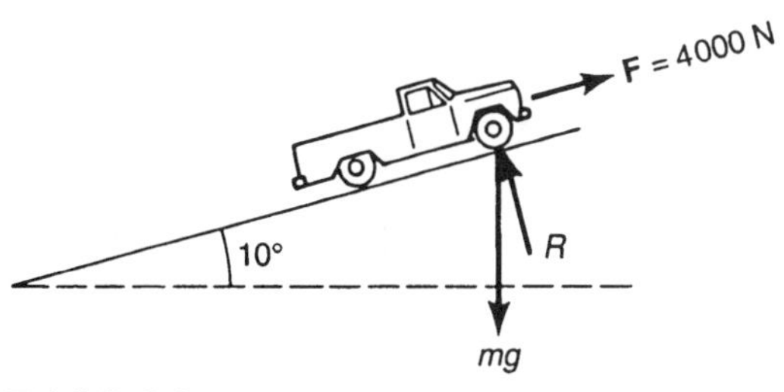

Exhibit 8.25

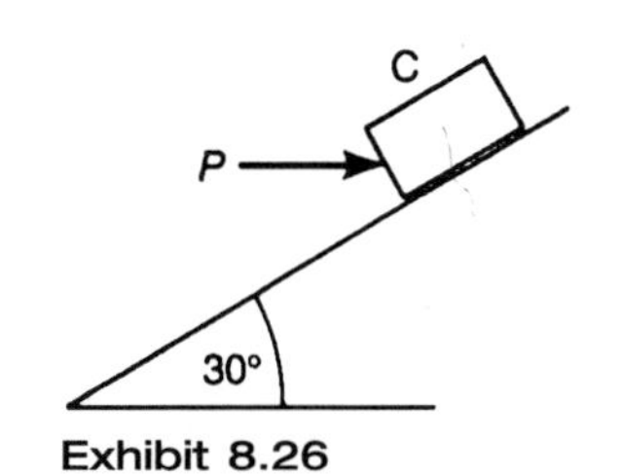

Exhibit 8.26

8.26 In Exhibit 8.26, a particle C of mass 2 kg is sliding down a smooth incline with a velocity of 3 m/s when a horizontal force $P = 15$ N is applied to it. What is the distance traveled by C between the instant when P is first applied and the instant when the velocity of C becomes zero?

a. 1.7 m
b. 5.6 m
c. 2.8 m
d. 0.9 m

8.27 A particle moves in a vertical plane along the path ABC shown in Exhibit 8.27. The portion AB of the path is a quarter-circle of radius r and is smooth. The portion BC is horizontal and has a coefficient of friction μ. If the particle has mass m, and is released from rest at A, determine the horizontal distance H that the particle will travel along BC before coming to rest.

a. μr
b. $2r$
c. r/μ
d. $2r/\mu$

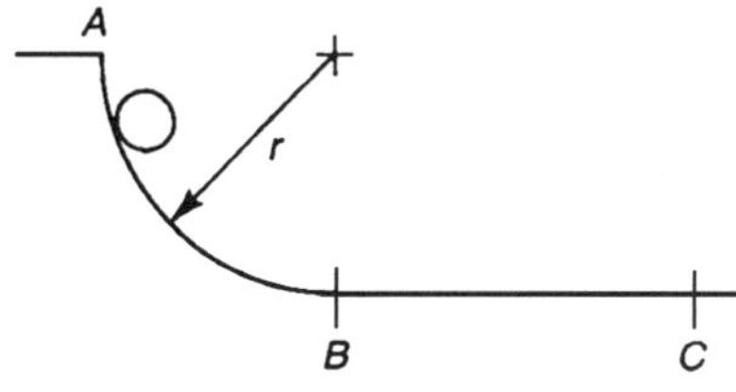

Exhibit 8.27

8.28 In Exhibit 8.28, a block of mass 2 kg is pressed against a linear spring of constant $k = 200$ N/m through a distance Δ on a horizontal surface. When the block is released at A, it travels along the straight horizontal path ADB and traverses point B with a velocity of 1 m/s. If the coefficient of kinetic friction between the block and the floor is 0.2, find Δ.

a. 0.22 m c. 0.26 m
b. 0.12 m d. 0.08 m

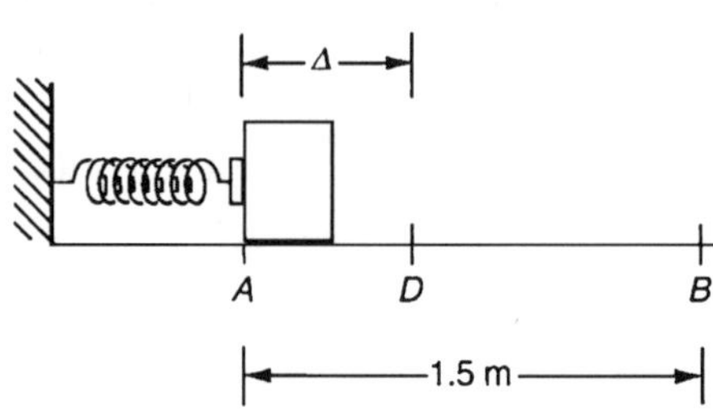

Exhibit 8.28

8.29 In Exhibit 8.29, a 6-kg block is released from rest on a smooth inclined plane as shown. If the spring constant $k = 1000$ N/m, determine how far the spring is compressed. Assume the acceleration of gravity, $g = 10$ m/s^2.

a. 0.40 m c. 0.83 m
b. 0.45 m d. 3.96 m

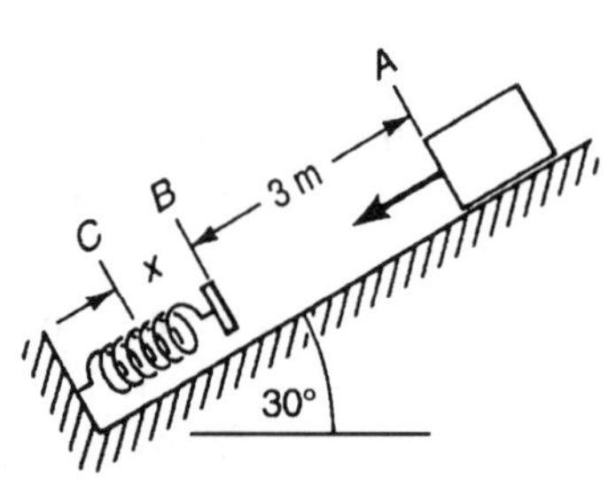

Exhibit 8.29

8.30 A train of joyride cars full of children in an amusement park is pulled by an engine along a straight-level track. It then begins to climb up a 5° slope. At a point B, 50 m up the grade when the velocity is 32 km/h, the last car uncouples without the driver noticing (Exhibit 8.30). If the total mass of the car with its passengers is 500 kg and the track resistance is 2% of the total vehicle weight, calculate the total distance up the grade where the car stops at point C.

a. 260 m c. 48.7 m
b. 37.6 m d. 87.6 m

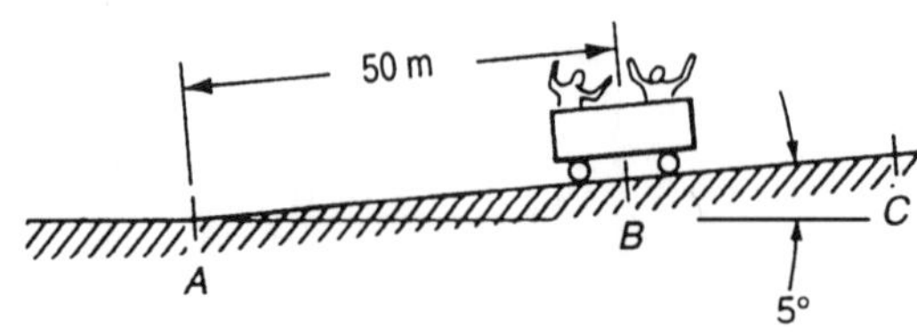

Exhibit 8.30

8.31 Two identical rods, each of mass 4 kg and length 3 m, are rigidly connected as shown in Exhibit 8.31. Determine the moment of inertia of the rigid assembly about an axis through the point A and perpendicular to the plane of the paper.

a. 19 kg-m^2 c. 18 kg-m^2
b. 23 kg-m^2 d. 15 kg-m^2

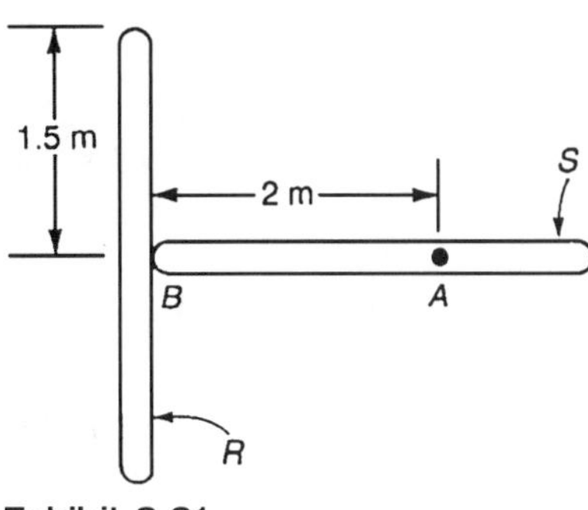

Exhibit 8.31

8.32 A torque motor, represented by the box in Exhibit 8.32, is to drive a thin steel disk of radius 2 m and mass 1.5 kg around its shaft axis. Ignoring the bearing friction about the shaft and the shaft mass, find the angular speed of the disk after applying a constant motor torque of 5 N-m for 5 seconds. The initial angular velocity of the shaft is 1 rad/s.

a. 8.3 rad/s c. 5.2 rad/s
b. 7.3 rad/s d. 9.3 rad/s

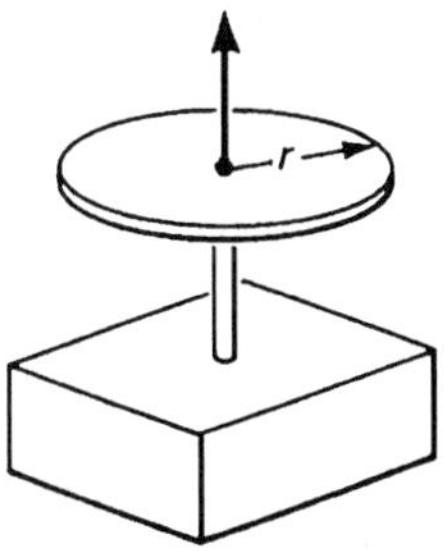

Exhibit 8.32

8.33 In Exhibit 8.33, the uniform slender rod R is hinged to a block B that can slide horizontally. Determine the horizontal acceleration α that must be given to B in order to keep the angle θ constant at 10°, balancing the rod in a tilted position.

a. 1.73 m/s^2 c. 9.81 m/s^2
b. 0 d. 9.66 m/s^2

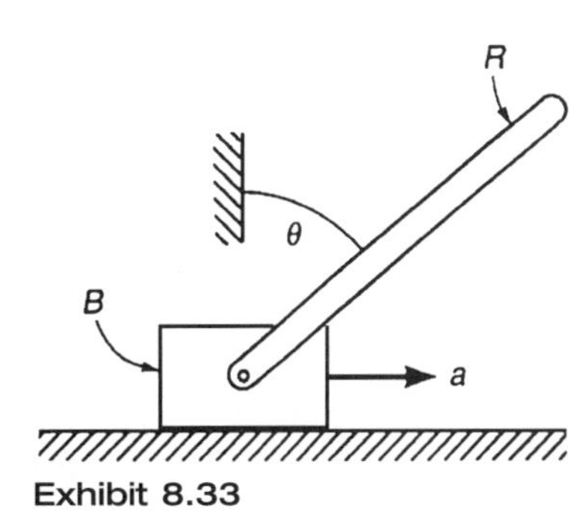

Exhibit 8.33

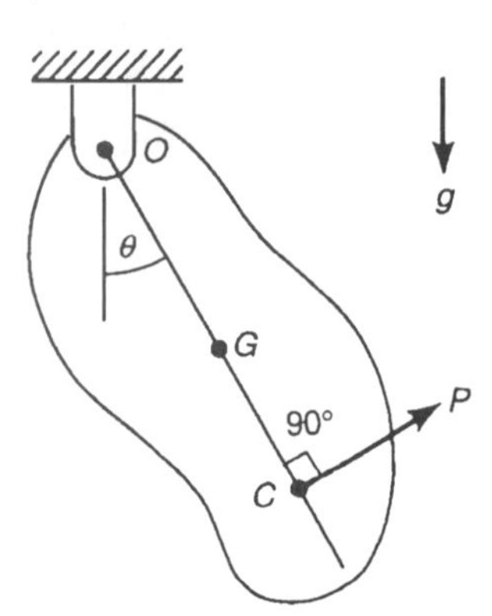

Exhibit 8.34

8.34 In Exhibit 8.34, a force P of constant magnitude is applied to the physical pendulum at point C and remains perpendicular to OC at all times. The pendulum moves in the vertical plane and has mass 3 kg; its mass center is located at G, and the distances are $OG = 1.5$ m, $OC = 2$ m. Also, $P = 10$ N and the radius of gyration of the pendulum about an axis through C and perpendicular to the plane of motion is 0.8 m. Determine the angular acceleration of the pendulum when $\theta = 30°$.

a. 5.31 rad/s^2 counterclockwise
b. 5.31 rad/s^2 clockwise
c. 0.26 rad/s^2 counterclockwise
d. 0.26 rad/s^2 clockwise

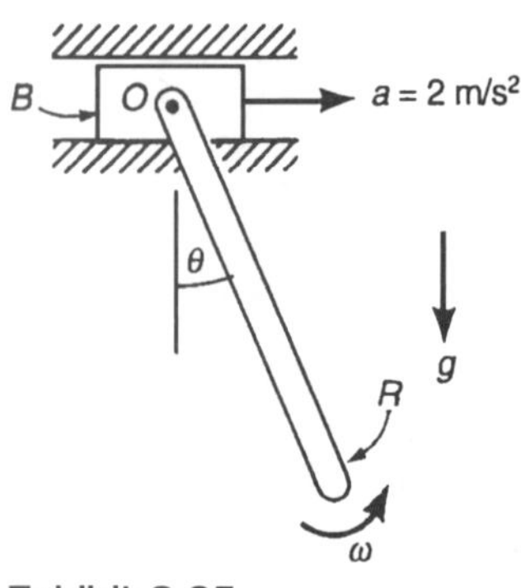

Exhibit 8.35

8.35 In Exhibit 8.35, the block B moves along a straight horizontal path with a constant acceleration of 2 m/s^2 to the right. The uniform slender rod R of mass 1 kg and length 2 m is connected to B through a frictionless hinge and swings freely about O as B moves. Determine the horizontal component of the reaction force at O on the rod when $\theta = 30°$ and $\omega = 2$ rad/s counterclockwise.

a. 4.31 N → c. 6.31 N →
b. 4.31 N ← d. 6.31 N ←

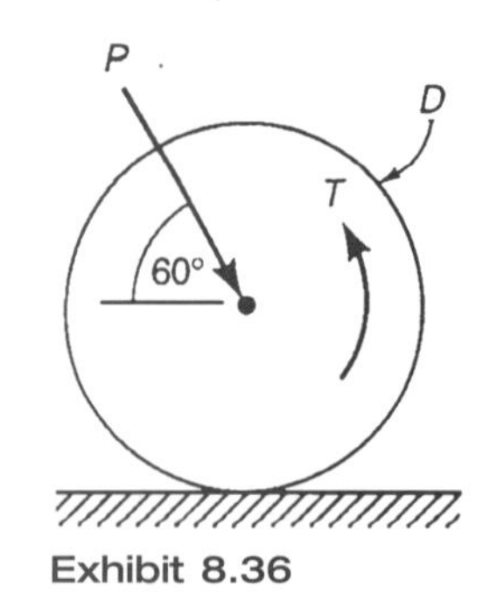

Exhibit 8.36

8.36 In Exhibit 8.36, a homogeneous cylinder rolls without slipping on a horizontal floor under the influence of a force $P = 6$ N and a torque $T =$ 0.5 N-m. The cylinder has radius 1 m and mass 2 kg. If the cylinder started from rest, what is its angular velocity after 10 seconds?

a. 8.3 rad/s c. 1.7 rad/s
b. 6.8 rad/s d. 0.68 rad/s

Exhibit 8.37

8.37 A slender rod of length 2 m and mass 3 kg is released from rest in the horizontal position (Exhibit 8.37) and swings freely (no hinge friction). Find the angular velocity of the rod when it passes a vertical position.

a. 4.43 rad/s c. 7.68 rad/s
b. 3.84 rad/s d. 5.43 rad/s

8.38 The uniform slender bar of mass 2 kg and length 3 m is released from rest in the near-vertical position as shown in Exhibit 8.38, where the torsional spring is undeformed. The rod is to rotate clockwise about O and come gently to rest in the horizontal position. Determine the stiffness k of the torsional spring that would make this possible. The hinge is smooth.

a. 47.8 N-m/rad c. 0.7 N-m/rad
b. 37.5 N-m/rad d. 23.8 N-m/rad

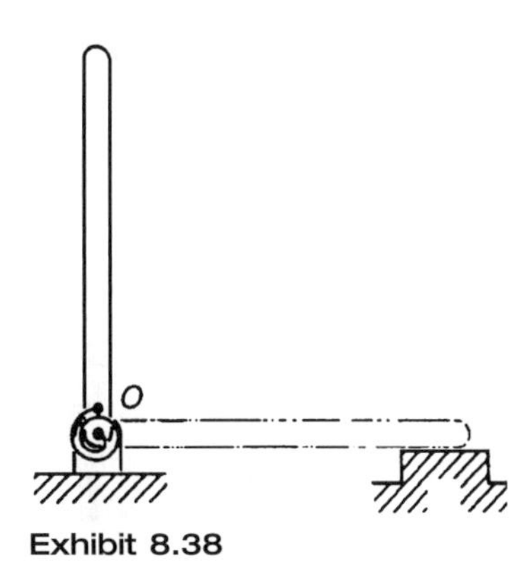

Exhibit 8.38

8.39 A solid homogeneous cylinder is released from rest in the position shown in Exhibit 8.39 and rolls without slip on a horizontal floor. The cylinder has a mass of 12 kg. The spring constant is 2 N/m, and the unstretched length of the spring is 3 m. What is the angular velocity of the cylinder when its center is directly below the point *O*?

a. 1.33 rad/s
b. 1.63 rad/s
c. 1.78 rad/s
d. 2.31 rad/s

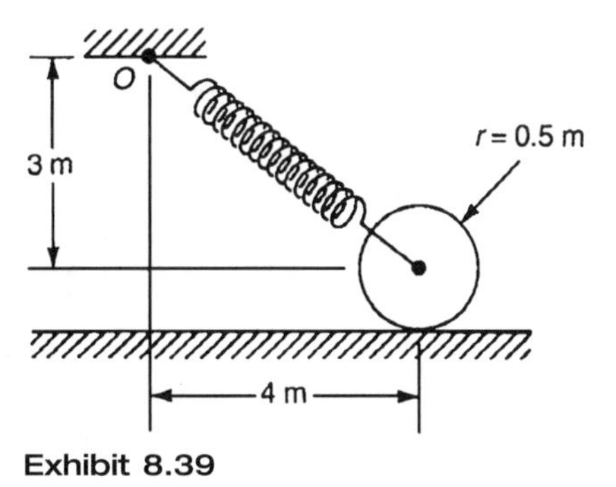

Exhibit 8.39

SOLUTIONS

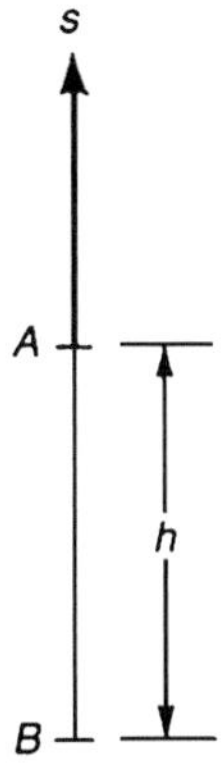

8.1 **b.** Let the depth of the ditch be h, and set up a vertical s-axis, positive upwards with the origin at *A* Exhibit 8.1a. Then, for motion between *A* and *B*, $s_0 = 0$, $s = -h$, $v_0 = 4$ m/s, $a = -9.81$ m/s^2, and $t = 6$ s. Substituting these values in the relationship $s = s_0 + v_0 t + (at^2)/2$ yields $-h = 0 + 4(6) - 9.81(6)^2/2$ or $h = 152.6$ m.

Exhibit 8.1a

8.2 **d.** Set up a horizontal s-axis with origin at O, and positive to the right (Exhibit 8.2a). Then, $s_0 = 2$ m, and $v = ds/dt$, so

$$\int_{s_0}^{s} ds = \int_0^t v\,dt = \int_0^t (3t^2 - t + 2)\,dt$$

or $s = s_0 + t^3 - t^2/2 + 2t$. Substituting values into the equation yields $s(2\text{s}) = 12$ m.

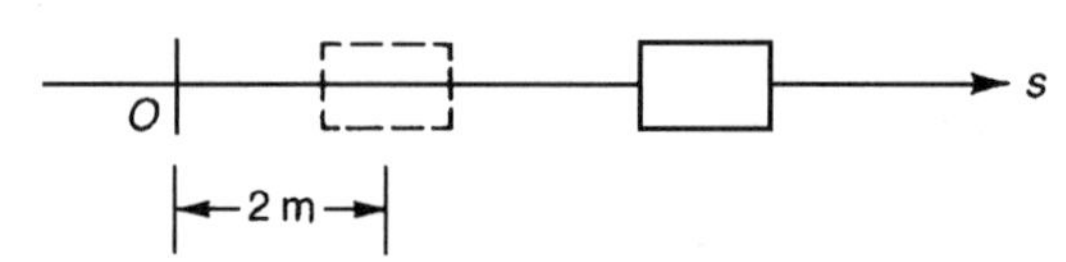

Exhibit 8.2a

8.3 **d.** The velocity-time curve for this particle is shown below the acceleration curve in Exhibit 8.3a. (Velocity at any given instant t_1 equals the area under the acceleration curve from time 0 to the time t_1, plus the initial velocity). Any area above the $a = 0$ line is counted as positive, and any area below is counted as negative.

The displacement D is the total area under the velocity curve between $t = 0$ and $t = 8$ s. Hence, $D = 0.5(6)(8) - 0.5(2)(4) = 20$ m.

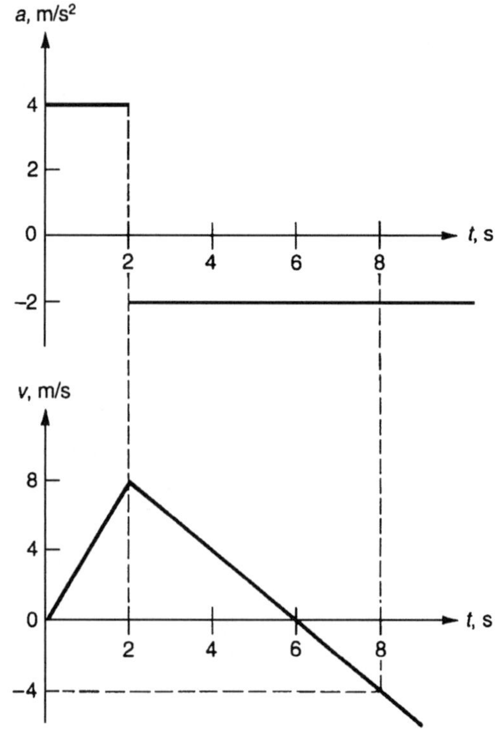

Exhibit 8.3a

8.4 **a.** When the ball reaches its highest position, B, the vertical component of the velocity of the ball is zero. Applying Eq. (8.12) vertically,

$$v_y^2 = v_{yo}^2 + 2a_y(y - y_0)$$
$$0 = (40 \sin 60°)^2 - 2(9.81)(y - y_0)$$

Thus, $y - y_0 = 61.2$ m and $H = 63.2$ m.

8.5 **c.** Refer to Exhibit 8.5a.

Horizontal Motion:

$$x = x_0 + v_{x0}t + (a_x t^2)/2$$
$$140 = 0 + v_0 t + 0$$
$$t = 1.56 \text{ s}$$

Vertical Motion:

$$y = y_0 + y_{y0}t + (a_y t^2)/2$$
$$-12 = 0 + 0 - 0.5(9.81)t^2$$
$$v_o = 140/\text{t} = 140/1.56 = 89.7 \text{ m/s}$$

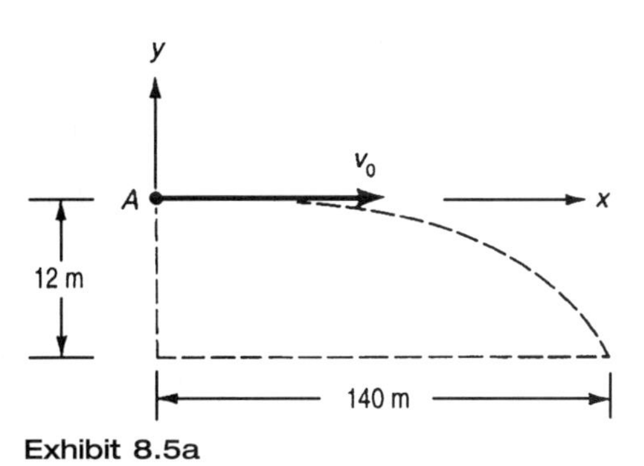

Exhibit 8.5a

8.6 **c.** In Exhibit 8.6a, we have

$$r = 1 \text{ m}; \qquad \dot{r} = -3 \text{ m/s}; \qquad \ddot{r} = 0$$

Since $\theta = (t^2 + t - 2)$ rad, differentiation gives

$$\dot{\theta} = (2t + 1) \text{ rad/s; at } t = 1 \text{ s, } \dot{\theta} = 2(1) + 1 = 3 \text{ rad/s}$$

Also, $\ddot{\theta} = 2 \text{ rad/s}^2$

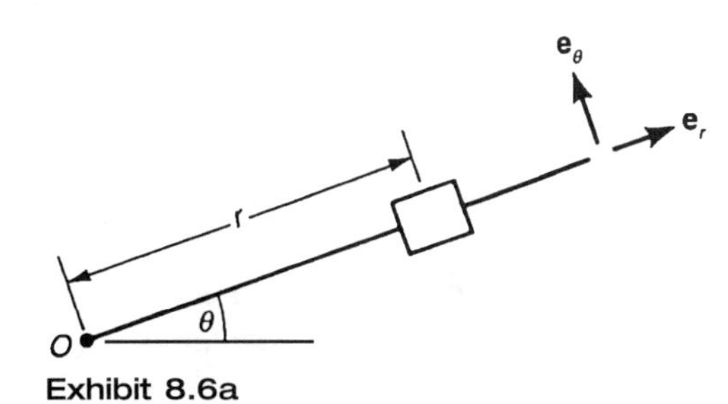

Exhibit 8.6a

Acceleration of the collar is therefore given by

$$\begin{aligned}\mathbf{a} &= (\ddot{r} - r\dot{\theta}^2)\mathbf{e}_r + (r\ddot{\theta} + 2\dot{r}\dot{\theta})\mathbf{e}_\theta \\ &= [0 - 1(3)^2]\mathbf{e}_r + [1(2) + 2(-3)(3)]\mathbf{e}_\theta \ \mathbf{m/s^2} \\ &= [-9\mathbf{e}_r - 16\mathbf{e}_\theta]\ \mathbf{m/s^2}\end{aligned}$$

Hence, |a| = $[(-9)^2 + (-16)^2]^{0.5} = 18.4\ \text{m/s}^2$.

8.7 **d.** Refer to Exhibit 8.7a. The velocity is

$$v = \left(v_r^2 + v_\theta^2\right)^{0.5} \quad \text{where } v_r = \dot{r} \text{ and } v_\theta = r\dot{\theta}$$

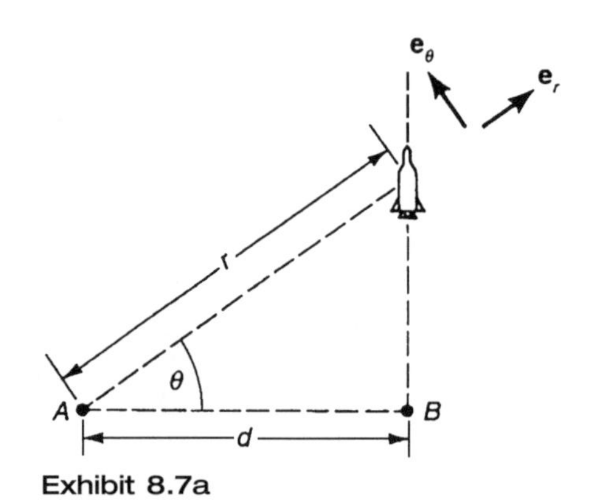

Exhibit 8.7a

Now, $r = d/\cos\theta$ so

$$v_r = \dot{r} = \frac{d\dot{\theta}\sin\theta}{\cos^2\theta} = \frac{d\dot{\theta}\tan\theta}{\cos\theta}$$

and

$$v_\theta r\dot{\theta} = \frac{d\dot{\theta}}{\cos\theta}$$

Thus

$$v^2 = v_r^2 + v_\theta^2 = \frac{d^2\dot{\theta}^2(\tan^2\theta + 1)}{\cos^2\theta} = \frac{d^2\dot{\theta}^2}{\cos^4\theta}; \quad \ddot{r} = 0$$

and

$$v = \frac{d\dot{\theta}}{\cos^2\theta} = \frac{1800(0.1)(2)}{1}\ \text{m/s} = 360\ \text{m/s}$$

8.8 **d.** At the highest point, the vertical component of velocity is zero, and the acceleration is normal to the path. Thus, $v = v_x = v_{x0} + a_x t = 50$ cos 30° + 0. Also, the normal component of the acceleration is $a_n = -a_y = 9.81\ \text{m/s}^2$. But $a_n = v^2/\rho$. Hence $\rho = v^2/a_{n,} = (50\cos 30°)^2/9.81 = 191$ m.

8.9 **d.** The tangential acceleration is given as $a_t = 3\ \text{m/s}^2$; the normal acceleration is $a_n = v^2/\rho = [(20)^2/110]\ \text{m/s}^2 = 3.6\ \text{m/s}^2$. Hence, the total acceleration is $a = [(3)^2 + (3.6)^2]^{0.5}\ \text{m/s}^2 = 4.7\ \text{m/s}^2$.

8.10 **c.** The normal acceleration is given by $a_n = v^2/\rho$. So, $\rho = v^2/a_n$. Substituting values (converted to consistent units), we obtain

$$\rho = \frac{[800(1000)]^2}{[60(60)]^2(5)(9.81)} = 1007 \text{ m}$$

8.11 **d.** At any given instant, the angular acceleration of the disk is $\alpha = -k\omega = d\omega/dt$. So,

$$\int_{\omega_0}^{\omega} \frac{d\omega}{\omega} = -k\int_0^t dt$$

$$\ln \omega|_{\omega_0}^{\omega} = \ln(\omega/\omega_0) = -kt \text{ and } t = -(1/k)\ln(\omega/\omega_0) = -(1/1.2)\ln(0.5) = 0.6\,s.$$

8.12 **a.** Initially, $\omega_0 = 0$. At t = 30 s, $\omega = 300$ rpm = $[300\,(2\pi)/60]$ rad/s = 10π rad/s. For uniformly accelerated rotational motion,

$$\omega = \omega_0 + \alpha t$$
$$\alpha = (\omega - \omega_0)/t = (10\pi - 0)/30 \text{ rad/s}^2 = \pi/3 \text{ rad/s}^2$$
$$\omega(5) = \omega_0 + \alpha t = 0 + (\pi/3)(5) \text{ rad/s} = 5\pi/3 \text{ rad/s}$$
$$v_P(5) = \omega(5)r = (5\pi/3)\ (2) \text{ m/}s = 10.5 \text{ m/s}$$

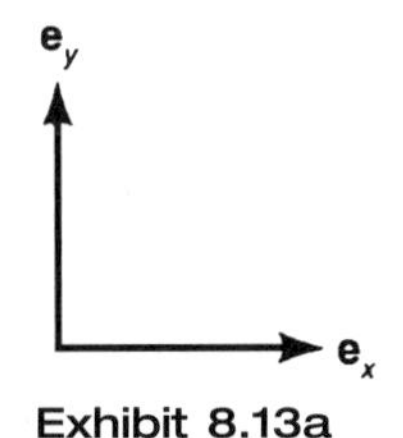

Exhibit 8.13a

8.13 **b.** Adopt the coordinate system of Exhibit 8.13a. Rolling of the disk without slip on B implies that the velocity of the point A, viewed as a point on D, equals the velocity of A viewed as a point on B. Hence, $\mathbf{v}_A = 2\mathbf{e}_x$ m/s. Since A and O are points of the same rigid body D, $\mathbf{v}_O = \mathbf{v}_A + \omega \times \mathbf{r}_{AO}$, or $-1\mathbf{e}_x = 2\mathbf{e}_x + \omega\mathbf{e}_z \times (1.5)\mathbf{e}_y = (2 - 1.5\omega)\mathbf{e}_x$. Hence, $-1 = 2 - 1.5\omega$ or $\omega = 2$ rad/s. The positive sign indicates a counterclockwise rotation.

8.14 In the current configuration, $\mathbf{v}_P = 0$. P and A are points on D, so

$$\mathbf{v}_A = \mathbf{v}_P + \omega_D \times \mathbf{r}_{PA} = 0 - 3\mathbf{e}_z \times (0.5\mathbf{e}_y - 0.4\mathbf{e}_x)$$
$$= (1.5\mathbf{e}_x + 1.2\mathbf{e}_y) \text{ m/s}$$

Similarly, because B and A are points on R, $\mathbf{v}_B = \mathbf{v}_A + \omega_R \times \mathbf{r}_{AB}$. Therefore,

$$v_B\mathbf{e}_x = 1.5\mathbf{e}_x + 1.2\mathbf{e}_y + \omega_R\mathbf{e}_z \times (-2\mathbf{e}_x - 0.5\mathbf{e}_y)$$
$$= (1.5 + 0.5\omega_R)\mathbf{e}_x + (1.2 - 2\omega_R)\mathbf{e}_y$$

Equating the coefficients of $\mathbf{e}_y$, yields $0 = 1.2 - 2\omega_R$, or $\omega_R = 0.6$ rad/s. The positive sign indicates that ω_R is in the positive e_z direction, so the rotation is counterclockwise.

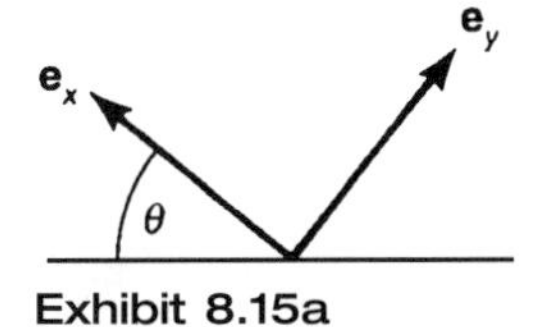

Exhibit 8.15a

8.15 **d.** Referring to the coordinate system in Exhibit 8.15a, the accelerations are

$$\mathbf{a}_A = \mathbf{a}_O + \alpha \times \mathbf{r}_{OA} + \omega \times (\omega \times \mathbf{r}_{OA})$$
$$\mathbf{a}_O = 0 \text{ (constant velocity)}$$

Since $\theta = (0.5t^2 - t)$ rad, $\dot{\theta} = (t-1)$ rad/s and $\ddot{\theta} = 1$ rad/s^2.
So, at t = 2 s, $\omega = \dot{\theta}\mathbf{e}_z = 1\mathbf{e}_z$ rad/s. Since $a = \ddot{\theta}\mathbf{e}_z = 1\mathbf{e}_z$ rad/s^2,

$$\mathbf{a}_A = 0 + \mathbf{e}_z \times (2\mathbf{e}_x) + \mathbf{e}_z \times [\mathbf{e}_z \times (2\mathbf{e}_x)] = (2\mathbf{e}_y - 2\mathbf{e}_x) \text{ m/s}^2$$

Hence, $|\mathbf{a}_A| = (2^2 + 2^2)^{0.5}$ m/s^2 = 2.8 m/s^2.

8.16 **b.** With the coordinate system in Exhibit 8.16a, we can write

$$\mathbf{a}_O = 2\mathbf{e}_x \text{ m/s}^2;\ \omega = 0;\ \alpha = -2.5\mathbf{e}_z \text{ rad/s}^2$$

Now, $\mathbf{a}_G = \mathbf{a}_O + \mathbf{a} \times \mathbf{r}_{OG} + \omega \times (\omega \times \mathbf{r}_{OG})$ where $\mathbf{r}_{OG} = -1\mathbf{e}_y$ m. Hence

$$\mathbf{a}_G = [2\mathbf{e}_x - 2.5\mathbf{e}_z \times (-1)\mathbf{e}_y + 0] \text{ m/s}^2 = -0.5\mathbf{e}_x \text{ m/s}^2$$

$\mathbf{e}_y$ $\mathbf{e}_x$

Exhibit 8.16a

8.17 **a.** We know that $\mathbf{v}_P = \mathbf{v}_{P/B} + \mathbf{v}_{P'}$ where $\mathbf{v}_{P/B}$ is the velocity of P relative to B and $\mathbf{v}_{P'}$ is the velocity of the point P' of the block that coincides with P at the instant under consideration (coincident point velocity). Here, $|\mathbf{v}_{P/B}| = 6$ m/s and $|\mathbf{v}_{P'}| = 10$ m/s (velocity of block). Because v_P is the vector sum of $\mathbf{v}_{P/B}$ and $\mathbf{v}_{P'}$, as shown in Exhibit 8.17a, we can use the law of cosines,

$$(\mathbf{v}_P)^2 = 10^2 + 6^2 - 2(10)(6)\cos 60° = 76$$
$$\mathbf{v}_P = 8.7 \text{ m/s}$$

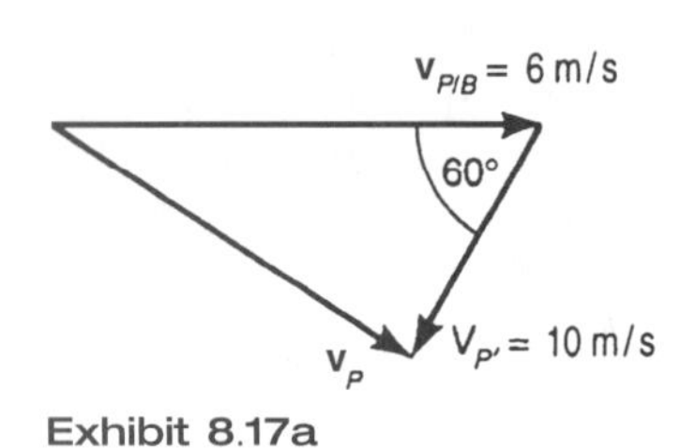

Exhibit 8.17a

8.18 **d.**

$$\mathbf{a}_P = \mathbf{a}_{P/D} + \mathbf{a}_{P¢} + \mathbf{a}_C$$

Here,
$\mathbf{a}_{P/D}$ = relative acceleration = 0
$\mathbf{a}_{P'}$ = acceleration of the point of D that is coincident with P at the instant under consideration:

$$\mathbf{a}_{P'} = \mathbf{a}_O + \boldsymbol{\alpha} \times \mathbf{r}_{OC} + \omega \times (\omega \times \mathbf{r}_{OC})$$
$$= 0 + 0 + (-5\mathbf{e}_z) \times [(-5\mathbf{e}_z) \times (0.1\mathbf{e}_y)]$$
$$= -2.5\mathbf{e}_y \text{ m/s}^2$$

$\mathbf{a}_C$ = Coriolis acceleration = $2\omega \times \mathbf{v}_{P/D}$
$\mathbf{a}_C = 2(-5\mathbf{e}_z) \times 2\mathbf{e}_x = -20\mathbf{e}_y \text{ m/s}^2$
Finally,

$$\mathbf{a}_P = [-2.5\mathbf{e}_y - 20\mathbf{e}_y] \text{ m/s}^2 = -22.5\mathbf{e}_y \text{ m/s}^2$$

8.19 **d.** The free-body diagram is shown in Exhibit 8.19a. Apply Newton's second law:

$$\sum F_z = ma_z$$
$$T - R - mg = ma$$
$$a = [(T - R)/m] - g = [(100 - 2z)N/5 \text{ kg}] - 9.81 \text{ m/s}^2$$

so

$$a = 10.2 - 0.4z = \frac{v\,dv}{dz}$$

and

$$\int_0^H (10.2 - 0.4z)\,dz = \int_{10}^0 v\,dv$$

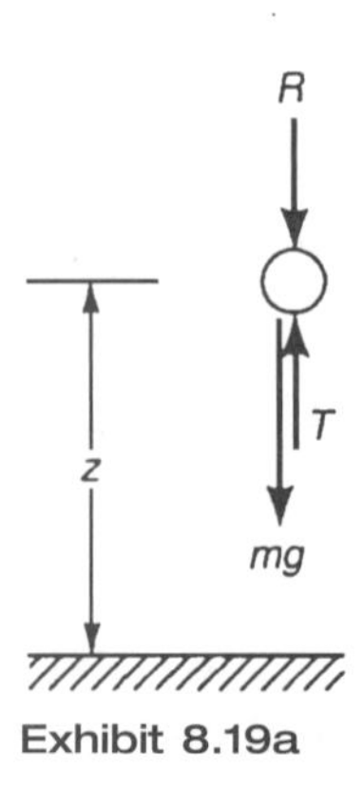

Exhibit 8.19a

where H is the highest height attained. Note also that $v = 0$ at this height. Integration yields,

$$10.2H - 0.2H^2 = [v^2/2]_{10}^{0} = -50$$

or

$$0.2H^2 - 10.2H - 50 = 0$$

Solving this quadratic (and discarding the negative value) gives the maximum height,

$$H = 55.5 \text{ m}$$

8.20 **b.** Refer to Exhibit 8.20a. Apply Newton's second law in the x and y directions.

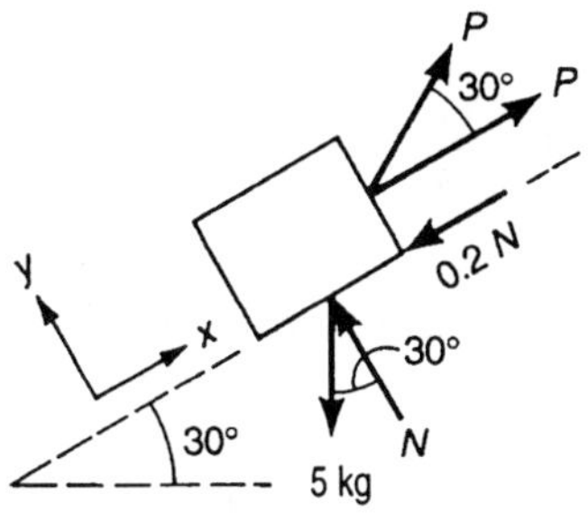

$$\sum F = ma_x \quad \textbf{(i)}$$

$$P + P\cos 30° - 0.2N - 5(9.81)\sin 30° = 5(2)$$

$$\sum F_y = 0$$

$$N + P\sin 30° - 5(9.81)\cos 30° = 0 \quad \textbf{(ii)}$$

Solving Eqs (i) and (ii) simultaneously yields $P = 21.9$ N.

Exhibit 8.20a

8.21 **a.** In Exhibit 8.21a, apply Newton's second law in the radial and transverse directions.

$$\sum F_\theta = ma_\theta$$

$$N - mg\cos\theta = m(r\ddot{\theta} + 2\dot{r}\dot{\theta}) \quad \textbf{(i)}$$

$$\sum F_r = ma_r$$

$$\mu N - P - mg\sin\theta = m(\ddot{r} - r\dot{\theta}^2) \quad \textbf{(ii)}$$

Substitute values into Eqs. (i) and (ii):

$$N - 2(9.81)\cos 30° = 2[0 + 2(-1)(5)] \quad \textbf{(iii)}$$

$$0.1N - P - 2(9.81)\sin 30° = 2[0 - 0.4(5)^2] \quad \textbf{(iv)}$$

From Eq. (iii), N = −3.0 newtons. Subsituting this value into Eq. (iv) gives $P = 9.9$N.

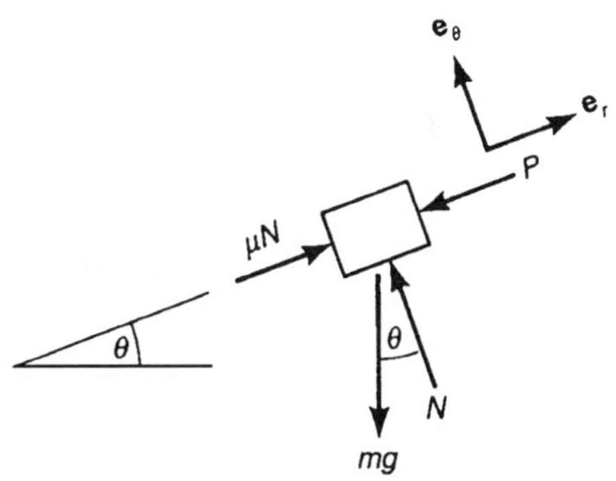

Exhibit 8.21a

8.22 **a.** Apply Newtons' second law to the diagrams in Exhibit 8.22:

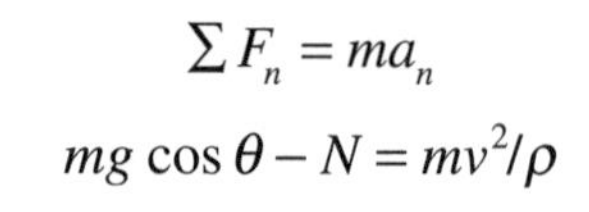

$$\Sigma F_n = ma_n$$

$$mg \cos \theta - N = mv^2/\rho$$

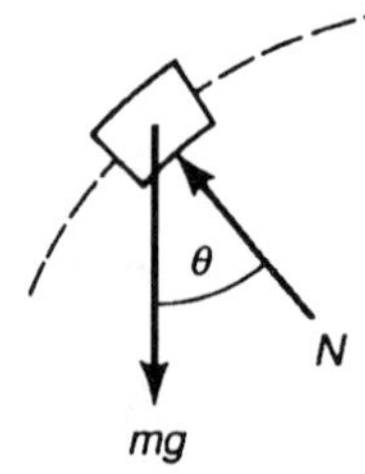

Exhibit 8.22a

or

$$\begin{aligned} N &= m[g\cos\theta - v^2/\rho] \\ &= (3)[9.81\cos 50° - 1.5^2/1] \\ &= 12.2 \text{ N} \end{aligned}$$

The positive sign indicates that N is directed as shown in the free-body diagram, Exhibit 8.22a.

8.23 **a.** At the onset of tipping, the free-body diagram and the intertia force diagram are as shown in Exhibit 8.23a. Take moments about point A:

$$mgb = m(u^2/R)H$$

Hence,

$$u = (Rgb/H)^{0.5}$$

FBD at Onset of Tipping

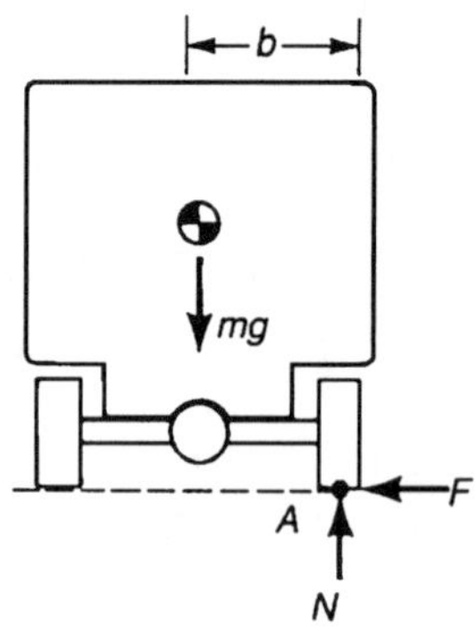

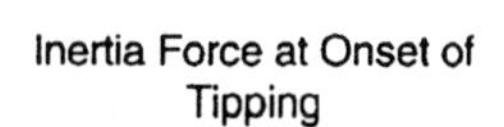

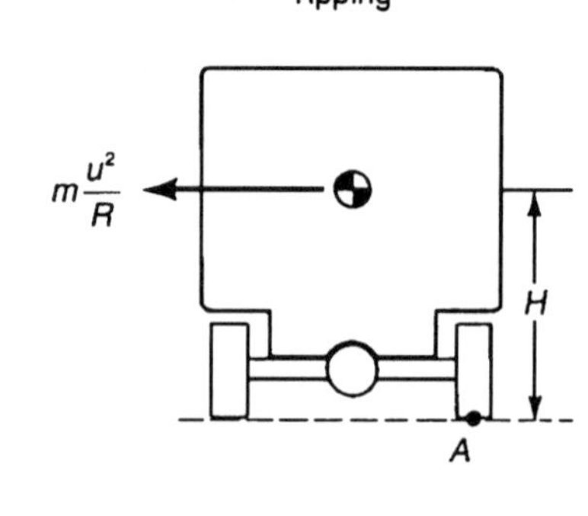

Exhibit 8.23a

8.24 **c.** In Exhibit 8.24a, the sum of the forces in the vertical direction yields

$$N = mg$$

Apply the impulse-momentum principle in the horizontal direction [Eq. (8.32)]:

$$\int_{t_1}^{t_2} F_{\text{horiz}}\, dt = mv_2 - mv_1$$

Thus

$$\int_0^7 [0.25 + 0.5t - (0.01)(9.81)]dt = (1)v_2$$

or

$$0.25t + 0.25t^2 - 0.098t\Big|_0^7 = v_2 = 13.3 \text{ m/s}$$

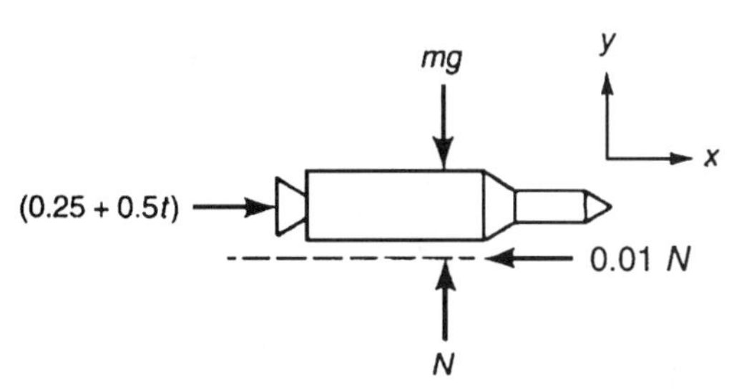

Exhibit 8.24a

8.25 **b.** Apply the impulse-momentum principle between the instant t_1 when the brakes are applied and the instant t_2 when the truck comes to a stop:

$$\int_{t_1}^{t_2} \mathbf{F}dt = m\mathbf{v}_2 - m\mathbf{v}_1$$

In the direction tangent to the road surface,

$$\int_0^t (mg \sin 10° - F)dt = 0 - mv_1$$

or

$$[2000(9.81) \ \sin 10° - 4000]t = -2000\frac{80(1000)}{60(60)}$$

so that $t = 74.9$ s. The answer is (b), regardless of how silly the problem is!

8.26 **c.** Refer to Exhibit 8.26a. The subscript 1 is used for the instant when the force P is used first applied, and the subscript 2 is used for the instant when the block comes to rest. Apply the work-energy principle between 1 and 2:

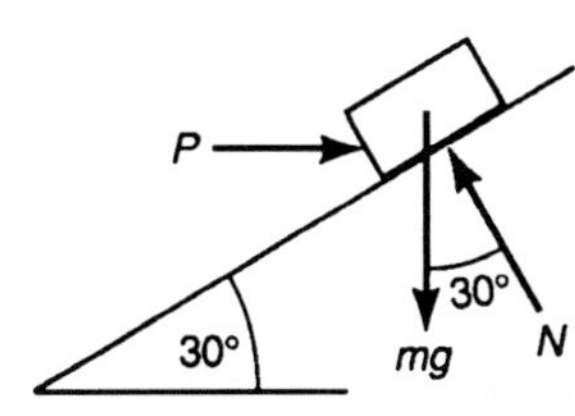

Exhibit 8.26a

$$W_{1-2} = T_2 - T_1$$
$$(mg\sin 30° - P\cos 30°)\Delta x = 0 - \tfrac{1}{2}mv_1^2$$

which yields

$$\Delta x = \frac{\frac{1}{2}mv_1^2}{P\cos 30° - mg\sin 30°} = \frac{0.5(2)3^2}{15\cos 30° - 2(9.81)\sin 30°} = 2.83 \text{ m}$$

8.27 **c.** Consult Exhibit 8.27a. Apply the work-energy principle between A and B, and then between B and C.

$A \rightarrow B$: $W_{A-B} = T_B - T_A$

$$mgr = \frac{1}{2}mv_B^2 - 0 \qquad \text{(i)}$$

$B \rightarrow C$: $W_{B-C} = T_C - T_B$

$$N = mg \qquad \text{(ii)}$$

$$-\mu NH = 0 - \frac{1}{2}mv_B^2 \qquad \text{(iii)}$$

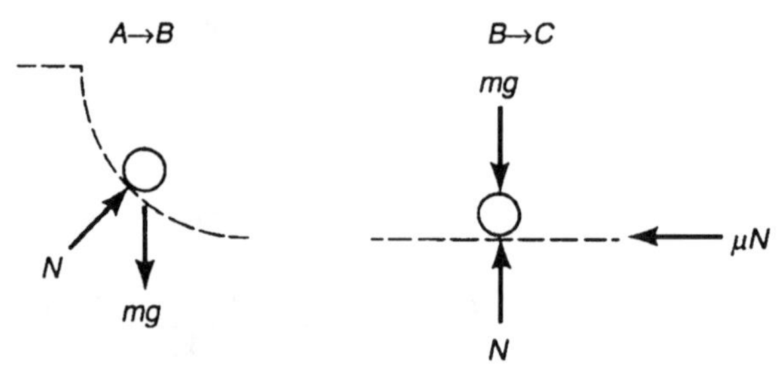

Exhibit 8.27a

Substituting Eqs. (i) and (ii) into Eq. (iii),

$$-\mu mgH = -mgr, \quad \text{or} \quad H = r/\mu$$

8.28 **c.** Let D be the position at which the spring has its natural (unstretched) length. Apply the work-energy principle (Exhibit 8.28a) from A to D:

$$W_{A-D} = T_D - T_A$$

$$\frac{1}{2}k\Delta^2 - \mu N\Delta = \frac{1}{2}mv_D^2 - 0$$

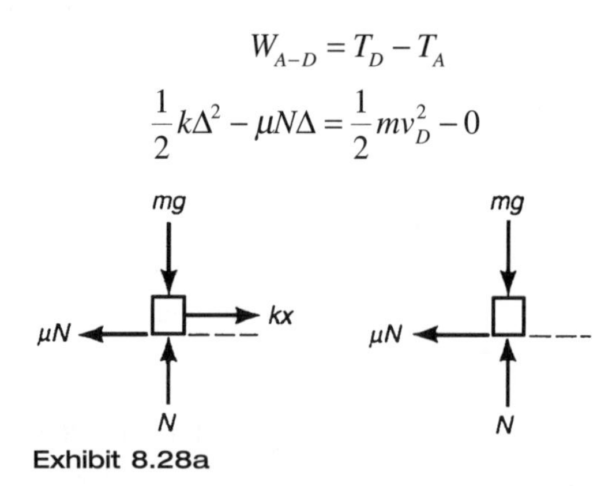

Exhibit 8.28a

Since $N = mg$, we have

$$\frac{1}{2}mv_D^2 = \frac{1}{2}k\Delta^2 - \mu mg\Delta \qquad \text{(i)}$$

Now apply the work-energy principle from D to B:

$$W_{D-B} = T_B - T_D$$

$$-\mu N(1.5 - \Delta) = \frac{1}{2}mv_B^2 - \frac{1}{2}mv_D^2$$

Again, since $N = mg$, and $\frac{1}{2}mv_D^2$ is given by Eq. (i), we have

$$-\mu mg(1.5 - \Delta) = \frac{1}{2}mv_B^2 - \frac{1}{2}k\Delta^2 + \mu mg\Delta$$

or

$$\Delta\sqrt{\frac{2\left(1.5\mu mg+0.5mv_B^2\right)}{k}}=0.26\text{ m}$$

8.29 **b.** Apply the work-energy principle between A and B, and then between B and C.

$$W_{A-B}=T_B-T_A$$

With the forces shown in Exhibit 8.29a,

$$mg\sin\,30°(3)=\frac{1}{2}mv_B^2-0$$

$$W_{B-C}=T_C-T_B$$

$$-\frac{1}{2}kx^2+mg\sin\,30°(x)=0-\frac{1}{2}mv_B^2=-mg\sin\,30°(3)$$

Substituting values, we obtain the quadratic equation $500x^2-30x-90=0$, which can be solved to yield $x=0.45$ m.

A→B
30°
mg
N

B→C
kx
30°
mg
N

Exhibit 8.29a

8.30 **d.** Using the diagram in Exhibit 8.30a, apply the work-energy principle between B and C:

$$W_{B-C}=T_C-T_B$$

$$-mg\ \sin\,5°(x)-0.02\ mgx=\ 0-\frac{1}{2}mv_B^2$$

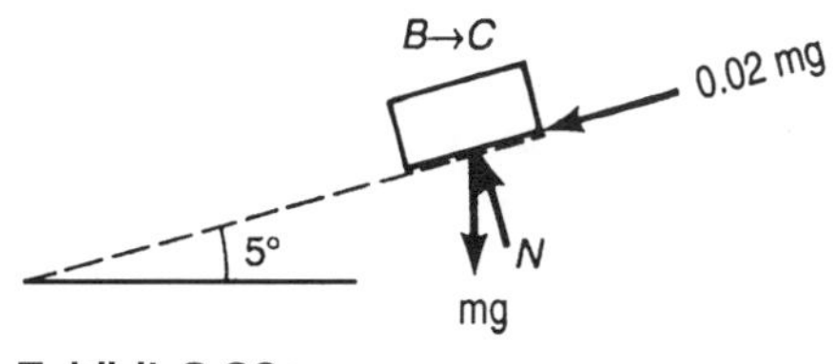

Exhibit 8.30a

or

$$x=\frac{\frac{1}{2}mv_B^2}{mg\sin 5°+0.02\ mg}=\frac{0.5(500)\left(\frac{32\bullet 1000}{60\bullet 60}\right)^2}{500(9.81)\sin 5°+0.02(500)9.81}=37.6\text{ m}$$

Total distance up the grade is (50 + 37.6) m = 87.6 m.

8.31 **b.** Consult Exhibit 8.31a. The moment of inertia of each rod about its mass center is

$$I_{R/B}=I_{S/O}=\frac{1}{12}ml^2=\frac{1}{12}(4)(3)^2=3\text{ kg-m}^2$$

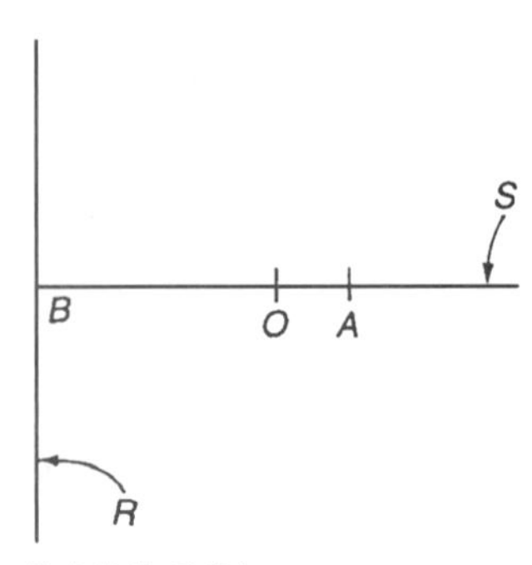

Exhibit 8.31a

Here, O is the mass center of S, and B is the mass center of R. Apply the parallel axes theorem:

$$I_{R/A} = I_{R/B} + m(2)^2 = 3 + 4(2)^2 = 19 \text{ kg-m}^2$$

$$I_{S/A} = I_{S/O} + m(2-1.5)^2 = 3 + 4(0.5)^2 = 4 \text{ kg-m}^2$$

And, for the assemblage,

$$I_A = I_{R/A} + I_{S/A} = (19+4)\text{kg-m}^2 = 23 \text{ kg-m}^2$$

8.32 **d.** Since $M = I\alpha = (1/2)mr^2\alpha$, $\alpha = M/(0.5mr^2)$. Substituting values, we have $\alpha = 1.67$ rad/s^2. Because this angular acceleration is constant, the final angular velocity is given by

$$\omega = \omega_0 + \alpha t = 1 + 1.67(5) = 9.3 \text{ rad/s}$$

8.33 **a.** When the desired configuration is achieved, the rod is in translation. The free-body diagram and the inertia force diagram for the rod are shown in Exhibit 8.33a. Taking moments about point O,

$$mg(l/2) \sin \theta = ma(l/2) \cos \theta$$

where l is the length of the rod. Thus,

$$a = g \tan \theta = 9.81 \tan 10° = 1.73 \text{ m/s}^2$$

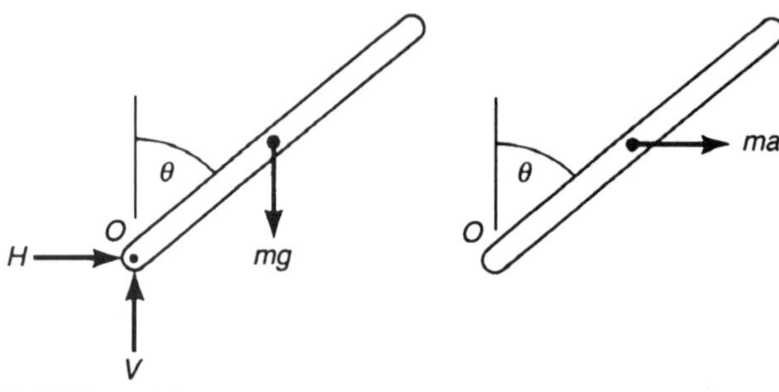

Exhibit 8.33a

8.34 **d.** From the diagrams in Exhibit 8.34a, and taking moments about O,

$$PH - mgl \sin\theta = I_G\alpha + ml^2\alpha$$

so that

$$\alpha = (PH - mgl \sin\theta)/(I_G + ml^2)$$

Now,

$$I_C = I_G + m(H-l)^2$$

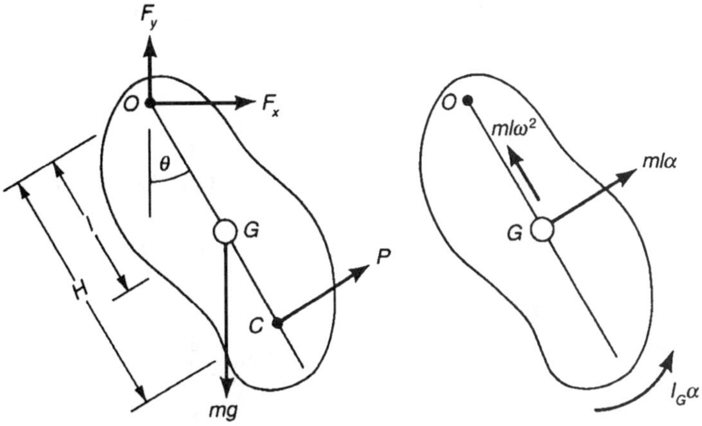

Exhibit 8.34a

from the parallel axis theorem. Thus,

$$I_G = I_C - m(H-l)^2 = mk^2 - m(H-l)^2$$

and

$$\alpha = \frac{PH - mgl\,\sin\theta}{m[k^2-(H-l)^2]+ml^2}$$

Substitute values to get

$$\alpha = \frac{10(2)-3(9.81)(1.5)(0.5)}{3[(0.8)^2-(0.5)^2+(1.5)^2]} = 0.26 \text{ rad/s}^2$$

8.35 **b.** C is the center of mass of the rod in Exhibit 8.35a. Summing moments about O gives

$$-mg(l/2)\sin\theta = (1/12)ml^2\alpha + m(l/2)\,^2\alpha + ma(l/2)\cos\theta \quad \textbf{(i)}$$

Summing forces along the horizontal, gives

$$H = ma + m(l/2)\,\alpha\cos\theta - m\omega^2(l/2)\sin\theta \quad \textbf{(ii)}$$

From Eq. (i),

$$\alpha = -\frac{3}{2}\frac{(g\sin\theta\cos\theta)}{l}$$

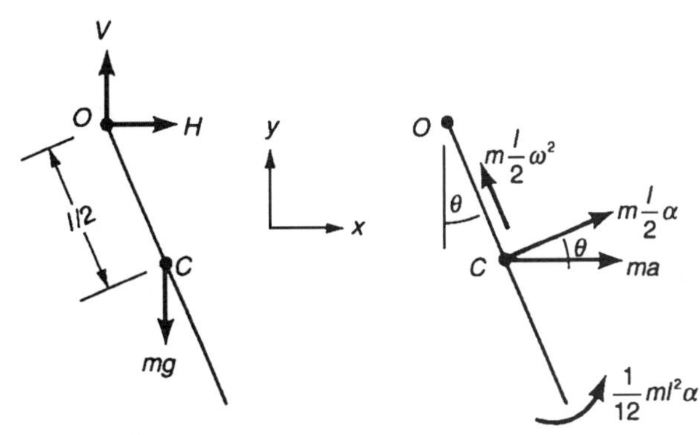

Exhibit 8.35a

Substituting values,

$$\alpha = -4.98 \text{ rad/s}^2 \qquad \textbf{(iii)}$$

Substituting Eq. (iii) and the given values into Eq. (ii) yields

$$H = (1)(2) + (1)(1)(-4.98)\cos 30° - (1)(2)^2(1)\sin 30° = -4.31 \text{ N}$$

8.36 **a.**

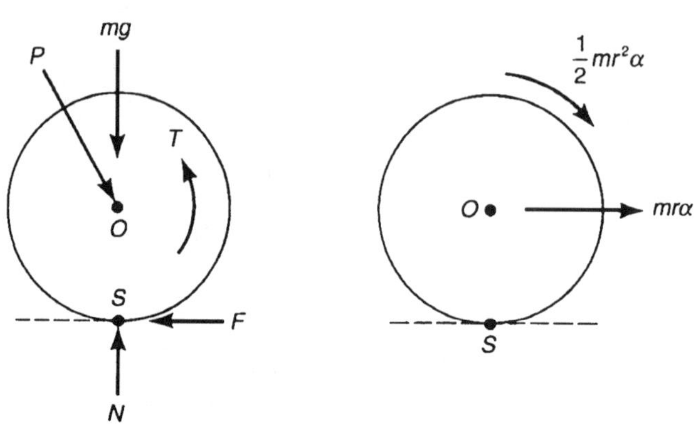

Exhibit 8.36a

Using the free-body diagram shown in Exhibit 8.36a, take moments about the contact point S:

$$rP\cos 60° - T = mr^2\alpha + (1/2)\,mr^2\alpha = (3/2)\,mr^2\alpha$$

or

$$\alpha = (rP\cos 60° - T)/(1.5mr^2) = \text{constant}$$

Substituting values, we find

$$\alpha = 0.83 \text{ rad/s}^2$$

With a constant angular acceleration, the angular velocity is

$$\omega = \omega_0 + \alpha t = 0 + 0.83(10) = 8.33 \text{ rad/s}$$

8.37 **b.** Exhibit 8.37a shows the forces acting on the rod as it swings from position 1 (horizontal) to position 2 (vertical). The work-energy principle gives

$$W_{1\to 2} = T_2 - T_1$$

That is,

$$mg\frac{l}{2} = \frac{1}{2}I_A\omega_2^2 - 0 = \frac{1}{2}\bullet\frac{1}{3}ml^2\omega_2^2$$

and

$$\omega^2 = \sqrt{\frac{3g}{l}} = \sqrt{\frac{(3)(9.81)}{2}} = 3.84 \text{ rad/s}$$

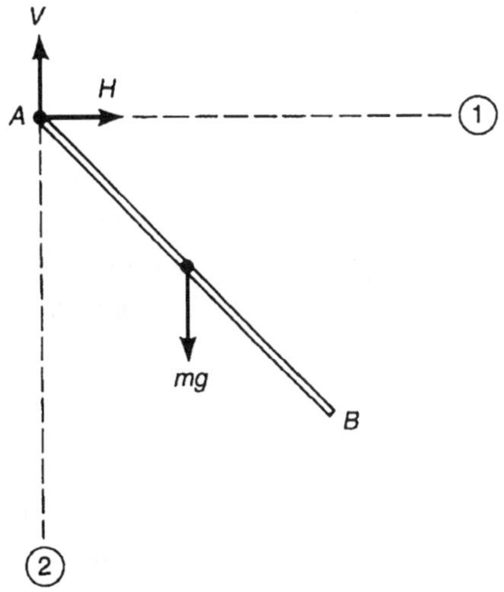

Exhibit 8.37a

8.38 **d.** Exhibit 8.38a shows the forces and torque acting on the rod as it rotates from position 1 (vertical) to position 2 (horizontal). The work-energy relation gives

$$W_{1-2} = T_2 - T_1$$

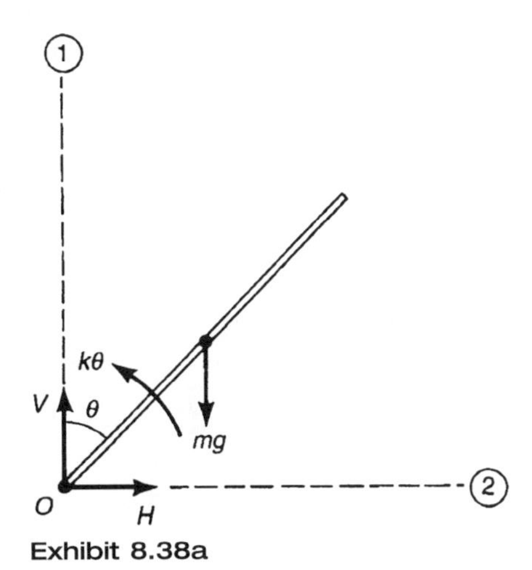

Exhibit 8.38a

That is,

$$mg\frac{l}{2} + \frac{l}{2}k\left(\theta_1^2 - \theta_2^2\right) = \frac{1}{2}I_0\omega_2^2 - \frac{1}{2}I_0\omega_1^2$$

Now, $\theta_1 = 0$, $\theta_2 = \pi/2\text{k}$, and $\omega_2 = \omega_2 = 0$. Thus

$$k = \frac{mgl}{\theta_2^2 - \theta_2^2} = \frac{2(9.8)(3)}{(\pi/2)^2} = 23.8 \text{ N-m/rad}$$

8.39 The forces acting on the cylinder during this motion are shown in Exhibit 8.39a. Applying the work-energy principle,

$$W_{1-2} = T_2 - T_1$$

F and R do no work because their point of application has zero velocity (rolling without slip); mg does no work because its point of application moves perpendicular to the force. Work done by the spring force is

$$W_{sp} \frac{1}{2}k\left(\Delta_1^2 - \Delta_2^2\right) = W_{1-2}$$

where

$$\Delta_1 = [(3^2 + 4^2)^{0.5} - 3]\text{ m} = 2\text{ m}, \quad \text{and} \quad \Delta_2 = 0$$

$$T_1 = 0, \text{ and } T_2 = \frac{1}{2}m(v_P)^2 + \frac{1}{2}I_P\omega^2 = \frac{1}{2}m(\omega r)^2 + \frac{1}{2}\bullet\frac{1}{2}mr^2\omega^2 = \frac{3}{4}mr^2\omega^2$$

Substituting into the work-energy principle yields

$$W_{1-2} = \frac{1}{2}k\Delta_t^2 = \frac{3}{4}mr^2\omega^2$$

and

$$\omega = \left(\frac{2}{3}\frac{k}{m}\right)^{0.5}\frac{\Delta_1}{r} = \left(\frac{2(2)}{3(12)}\right)^{0.5} \bullet \frac{2}{0.5} = 1.33\text{ rad/s}$$

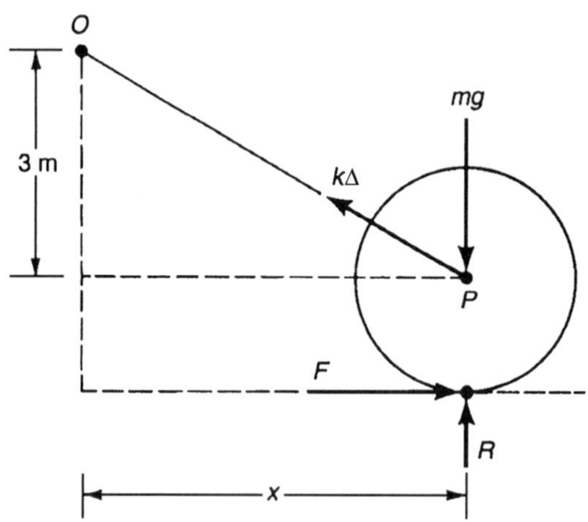

Exhibit 8.39a

CHAPTER 9

Mechanics of Materials

James R. Hutchinson

OUTLINE

Mechanics of materials deals with the determination of the internal forces (stresses) and the deformation of solids such as metals, wood, concrete, plastics and composites. In mechanics of materials there are three main considerations in the solution of problems:

1. Equilibrium
2. Force-deformation relations
3. Compatibility

Equilibrium refers to the equilibrium of forces. The laws of statics must hold for the body and all parts of the body. Force-deformation relations refer to the relation of the applied forces to the deformation of the body. If certain forces are applied, then certain deformations will result. Compatibility refers to the compatibility of deformation. Upon loading, the parts of a body or structure must not come apart. These three principles will be emphasized throughout.

AXIALLY LOADED MEMBERS

If a force P is applied to a member as shown in Fig. 9.1(a), then a short distance away from the point of application the force becomes uniformly distributed over the area as shown in Fig. 9.1(b). The force per unit area is called the axis or normal stress and is given the symbol σ. Thus,

$$\sigma = \frac{P}{A} \tag{9.1}$$

The original length between two points A and B is L as shown in Fig. 9.1(c). Upon application of the load P, the length L grows by an amount ΔL. The final length is $L + \Delta L$ as shown in Fig. 9.1(d). A quantity measuring the intensity of deformation and being independent of the original length L is the strain ε, defined as

$$\varepsilon = \frac{\Delta L}{L} = \frac{\delta}{L} \tag{9.2}$$

where ΔL is denoted as δ.

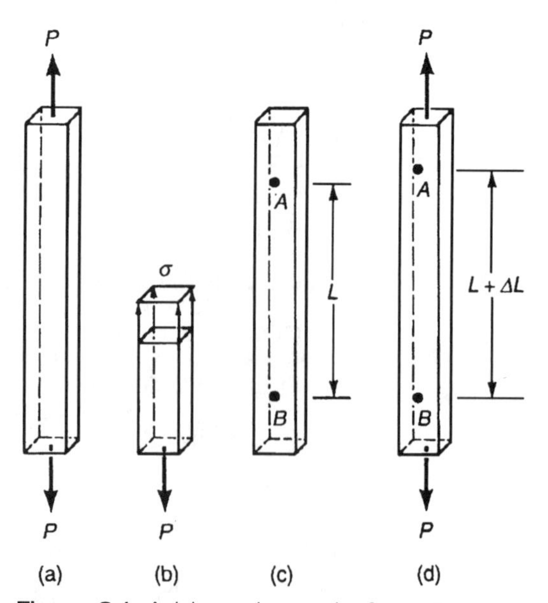

Figure 9.1 Axial member under force P

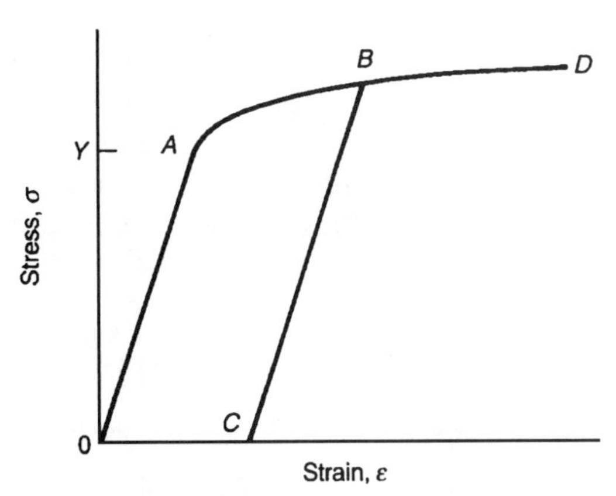

Figure 9.2 Stress-strain curve for a typical material

The relationship between stress and strain is determined experimentally. A typical plot of stress versus strain is shown in Fig. 9.2. On initial loading, the plot is a straight line until the material reaches yield at a stress of *Y*. If the stress remains less than yield then subsequent loading and reloading continues along that same straight line. If the material is allowed to go beyond yield, then during an increase in the load the curve goes from *A* to *D*. If unloading occurs at some point *B*, for example, then the material unloads along the line *BC* which has approximately the same slope as the original straight line from 0 to *A*. Reloading would occur along the line *CB* and then proceed along the line *BD*. It can be seen that if the material is allowed to go into the plastic region (*A* to *D*) it will have a permanent strain offset on unloading.

Modulus of Elasticity

The region of greatest concern is that below the yield point. The slope of the line between 0 and *A* is called the modulus of elasticity and is given the symbol *E*, so

$$\sigma = E\varepsilon \tag{9.3}$$

This is Hooke's Law for axial loading; a more general form will be considered in a later section. The modulus of elasticity is a function of the material alone and not a function of the shape or size of the axial member.

The relation of the applied force in a member to its axial deformation can be found by inserting the definitions of the axial stress [Eq. (9.1)] and the axial strain [Eq. (9.2)] into Hooke's Law [Eq. (9.3)], which gives

$$\frac{P}{A} = E\frac{\delta}{L} \tag{9.4}$$

or

$$\delta = \frac{PL}{AE} \tag{9.5}$$

In the examples that follow, wherever it is appropriate, the three steps of (1) Equilibrium, (2) Force-Deformation, and (3) Compatibility will be explicitly stated.

Example 9.1

The steel rod shown in Exhibit 1 is fixed to a wall at its left end. It has two applied forces. The 3 kN force is applied at the Point B and the 1 kN force is applied at the Point C. The area of the rod between A and B is A_{AB} = 1000 mm^2, and the area of the rod between B and C is A_{BC} = 500 mm^2. Take E = 210 GPa. Find (a) the stress in each section of the rod and (b) the horizontal displacement at the points B and C.

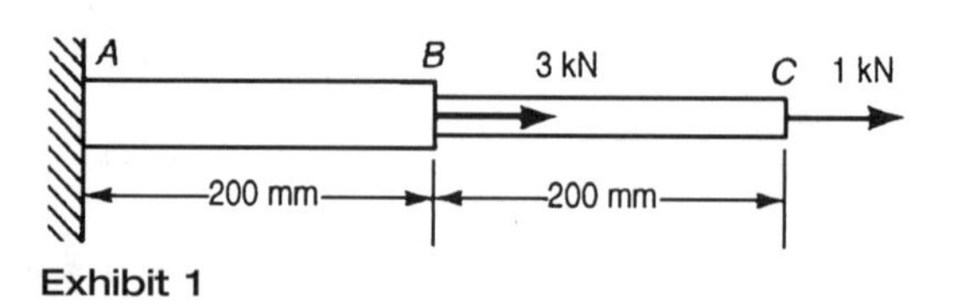

Exhibit 1

Solution—Equilibrium

Draw free-body diagrams for each section of the rod (Exhibit 2). From a summation of forces on the member BC, F_{BC} = 1 kN. Summing forces in the horizontal direction on the center free-body diagram, F_{BA} = 3 + 1 = 4 kN. Summing forces on the left free-body diagram gives $F_{AB} = F_{BA}$ = 4 kN. The stresses then are

$$\sigma_{AB} = 4 \text{ kN/1000 mm}^2 = 4 \text{ MPa}$$
$$\sigma_{BC} = 1 \text{ kN/500 mm}^2 = 2 \text{ MPa}$$

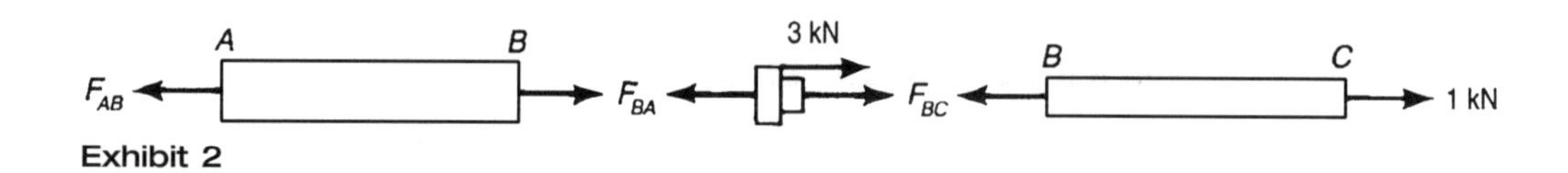

Exhibit 2

Solution—Force-Deformation

$$\delta_{AB} = \left(\frac{PL}{AE}\right)_{AB} = \frac{(4 \text{ kN})(200 \text{ mm})}{(1000 \text{ mm}^2)(210 \text{ GPa})} = 0.00381 \text{ mm}$$

$$\delta_{BC} = \left(\frac{PL}{AE}\right)_{BC} = \frac{(1 \text{ kN})(200 \text{ mm})}{(500 \text{ mm}^2)(210 \text{ GPa})} = 0.001905 \text{ mm}$$

Solution—Compatibility

Draw the body before loading and after loading (Exhibit 3).

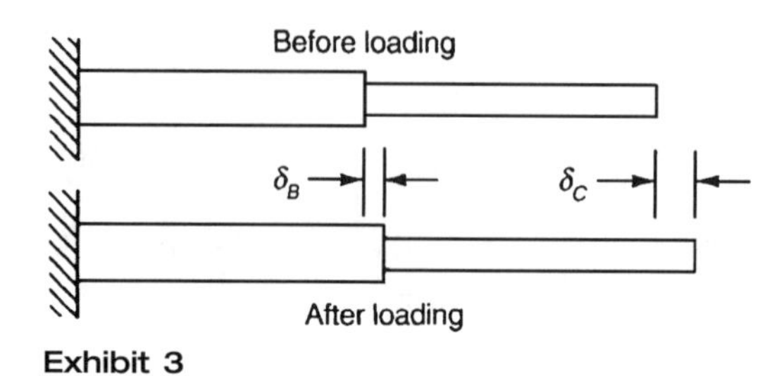

Exhibit 3

It is then obvious that

$$\delta_B = \delta_{AB} = 0.00381 \text{ mm}$$

$$\delta_C = \delta_{AB} + \delta_{BC} = 0.00381 + 0.001905 = 0.00571 \text{ mm}$$

In this first example the problem was statically determinate, and the three steps of Equilibrium, Force-Deformation, and Compatibility were independent steps. The steps are not independent when the problem is statically indeterminate, as the next example will show.

Example 9.2

Consider the same steel rod as in Example 9.1 except that now the right end is fixed to a wall as well as the left (Exhibit 4). It is assumed that the rod is built into the walls before the load is applied. Find (a) the stress in each section of the rod, and (b) the horizontal displacement at the point *B*.

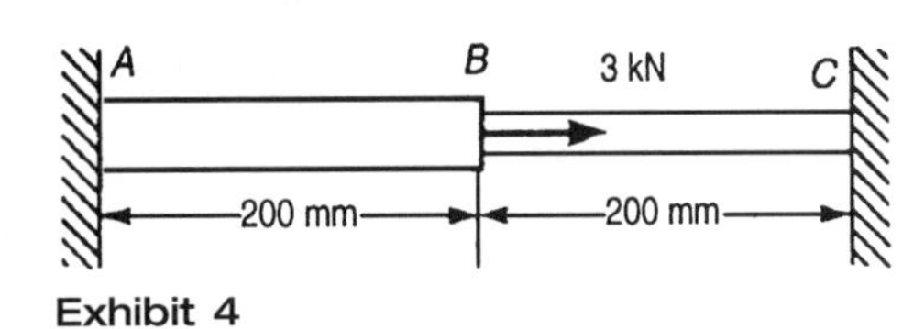

Exhibit 4

Solution—Equilibrium

Draw free-body diagrams for each section of the rod (Exhibit 5). Summing forces in the horizontal direction on the center free-body diagram

$$-F_{AB} + F_{BC} + 3 = 0$$

It can be seen that the forces cannot be determined by statics alone so that the other steps must be completed before the stresses in the rods can be determined.

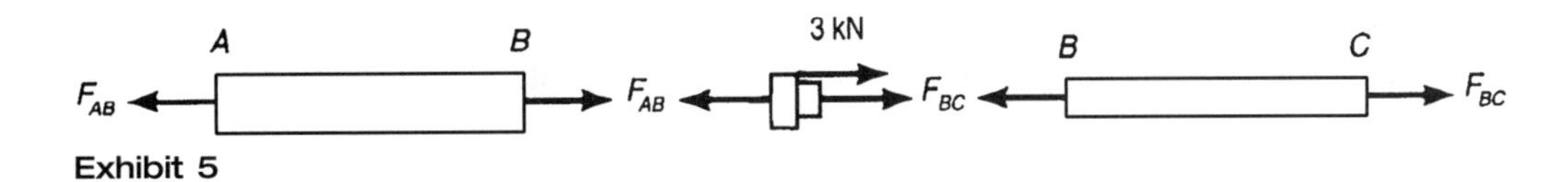

Exhibit 5

Solution—Force-Deformation

$$\delta_{AB} = \left(\frac{PL}{AE}\right)_{AB} = \frac{F_{AB}L}{A_{AB}E}$$

$$\delta_{BC} = \left(\frac{PL}{AE}\right)_{BC} = \frac{F_{BC}L}{A_{BC}}$$

The equilibrium, force-deformation, and compatibility equations can now be solved as follows (see Exhibit 6). The force-deformation relations are put into the compatibility equations:

$$\frac{F_{AB}L}{2A_{BC}E} = -\frac{F_{BC}L}{A_{BC}E}$$

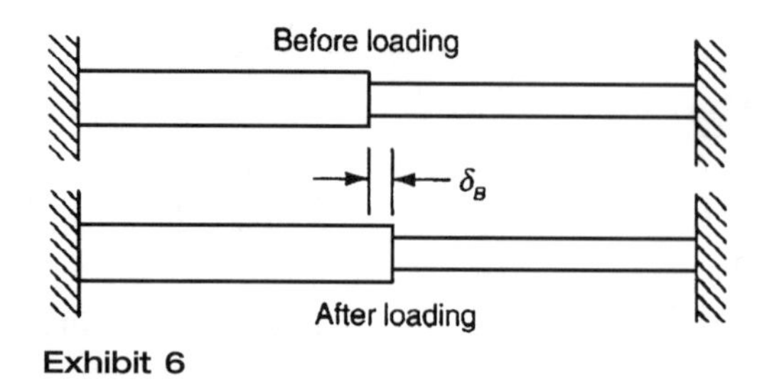

Exhibit 6

Then, $F_{AB} = -2F_{BC}$. Insert this relationship into the equilibrium equation

$$-F_{AB} + F_{BC} + 3 = 0 = 2F_{BC} + F_{BC} + 3; \quad F_{BC} = -1 \text{ kN and } F_{AB} = 2 \text{ kN}$$

The stresses are

$$\sigma_{AB} = 2 \text{ kN/1000 mm}^2 = 2 \text{ MPa (tension)}$$
$$\sigma_{BC} = -1 \text{ kN/500 mm}^2 = -2 \text{ MPa (compression)}$$

The displacement at B is

$$\delta_A = \delta_{AB} = F_{AB}L/(AE) = (2 \text{ kN})(200 \text{ mm})/[(1000 \text{ mm}^2)(210 \text{ GPa})] = 0.001905 \text{ mm}$$

Poisson's Ratio

The axial member shown in Fig. 9.1 also has a strain in the lateral direction. If the rod is in tension, then stretching takes place in the longitudinal direction while contraction takes place in the lateral direction. The ratio of the magnitude of the lateral strain to the magnitude of the longitudinal strain is called Poisson's ratio v.

$$v = -\frac{\text{Lateral strain}}{\text{Longitudinal strain}} \quad \textbf{(9.6)}$$

Poisson's ratio is a dimensionless material property that never exceeds 0.5. Typical values for steel, aluminum, and copper are 0.30, 0.33, and 0.34, respectively.

Example 9.3

A circular aluminum rod 10 mm in diameter is loaded with an axial force of 2 kN. What is the decrease in diameter of the rod? Take E = 70 GN/m^2 and v = 0.33.

Solution

The stress is $\sigma = P/A = 2 \text{ kN}/(\pi 5^2 \text{ mm}^2) = 0.0255 \text{ GN/m}^2 = 25.5 \text{ MN/m}^2$
The longitudinal strain is $\varepsilon_{\text{lon}} = \sigma/\text{E} = (25.5 \text{ MN/m}^2)/(70 \text{ GN/m}) = 0.000364$
The lateral strain is $\varepsilon_{\text{lat}} = -v\,\varepsilon_{\text{lon}} = -0.33(0.000364) = -0.000120$
The decrease in diameter is then $-\text{D}\,\varepsilon_{\text{lat}} = -(10 \text{ mm})(-0.000120) = 0.00120 \text{ mm}$

Thermal Deformations

When a material is heated, expansion forces are created. If it is free to expand, the thermal strain is

$$\varepsilon_t = \alpha(t - t_0) \tag{9.7}$$

where α is the linear coefficient of thermal expansion, t is the final temperature and t_0 is the initial temperature. Since strain is dimensionless, the units of α are °F^{-1} or °C^{-1} (sometimes the units are given as in/in/°F or m/m/°C which amounts to the same thing). The total strain ε_T is equal to the strain from the applied loads plus the thermal strain. For problems where the load is purely axial, this becomes

$$\varepsilon_T = \frac{\sigma}{E} + \alpha(t - t_0) \tag{9.8}$$

The deformation δ is found by multiplying the strain by the length L

$$\delta = \frac{PL}{AE} + \alpha L(t - t_0) \tag{9.9}$$

Example 9.4

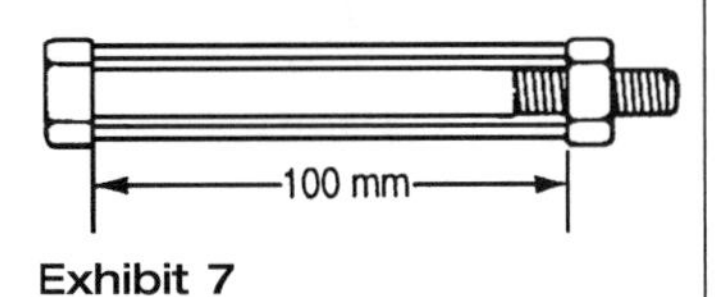

Exhibit 7

A steel bolt is put through an aluminum tube as shown in Exhibit 7. The nut is made just tight. The temperature of the entire assembly is then raised by 60°C. Because aluminum expands more than steel, the bolt will be put in tension and the tube in compression. Find the force in the bolt and the tube. For the steel bolt, take E = 210 GPa, $\alpha = 12 \times 10^{-6}$ °C^{-1} and A = 32 mm^2. For the aluminum tube, take E = 69 GPa, $\alpha = 23 \times 10^{-6}$ °C^{-1} and A = 64 mm^2.

Solution—Equilibrium

Draw free-body diagrams (Exhibit 8). From equilibrium of the bolt head it can be seen that $P_B = P_T$.

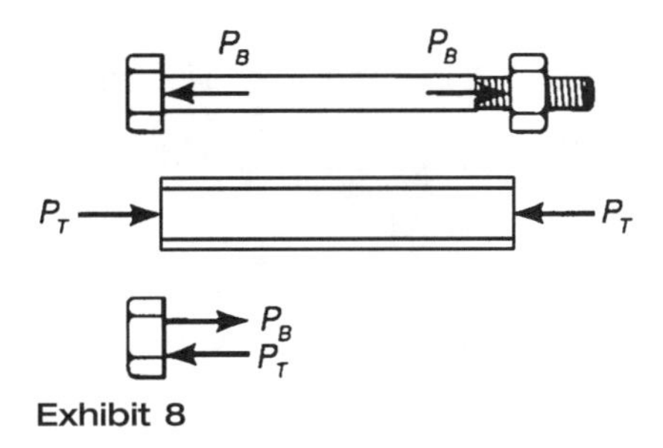

Exhibit 8

Solution—Force-Deformation

Note that both members have the same length and the same force, P.

$$\delta_B = \frac{PL}{A_B E_B} + \alpha_B L(t - t_0)$$

$$\delta_T = -\frac{PL}{A_T E_T} + \alpha_T L(t - t_0)$$

The minus sign in the second expression occurs because the tube is in compression.

Solution—Compatibility

The tube and bolt must both expand the same amount, therefore,

$$\delta_B = \delta_T$$

$$\delta_B = \delta_T = \frac{P \times (100\text{ mm})}{32\text{ mm}^2 \times 210\text{ GPa}} + 12 \times 10^{-6}\frac{1}{°\text{C}} \times 100\text{ mm} \times 60\text{ °C}$$

$$= -\frac{P \times (100\text{ mm})}{64\text{ mm}^2 \times 69\text{ GPa}} + 23 \times 10^{-6}\frac{1}{°\text{C}} \times 100\text{ mm} \times 60\text{ °C}$$

Solving for P gives $P = 1.759$ kN.

Variable Load

In certain cases the load in the member will not be constant but will be a continuous function of the length. These cases occur when there is a distributed load on the member. Such distributed loads most commonly occur when the member is subjected to gravitation, acceleration or magnetic fields. In such cases, Eq. (9.5) holds only over an infinitesimally small length $L = dx$. Eq. (9.5) then becomes

$$d\delta = \frac{P(x)}{AE}dx \quad \textbf{(9.10)}$$

or equivalently

$$\delta = \int_0^L \frac{P(x)}{AE}dx \quad \textbf{(9.11)}$$

Example 9.5

An aluminum rod is hanging from one end. The rod is 1 m long and has a square cross-section 20 mm by 20 mm. Find the total extension of the rod resulting from its own weight. Take $E = 70$ GPa and the unit weight $\gamma = 27$ kN/m^3.

Solution—Equilibrium

Draw a free-body diagram (Exhibit 9). The weight of the section shown in Exhibit 9 is

$$W = \gamma V = \gamma Ax = P$$

which clearly yields P as a function of x, and Eq. (9.11) gives

$$\delta = \int_0^L \frac{\gamma Ax}{AE}dx = \frac{\gamma}{E}\int_0^L xdx = \frac{\gamma L^2}{2E} = \frac{\left(27\frac{\text{kN}}{\text{m}^3}\right)(1\text{m})^2}{2\left(70\frac{\text{GN}}{\text{m}^2}\right)} = 0.1929\ \mu\text{m}$$

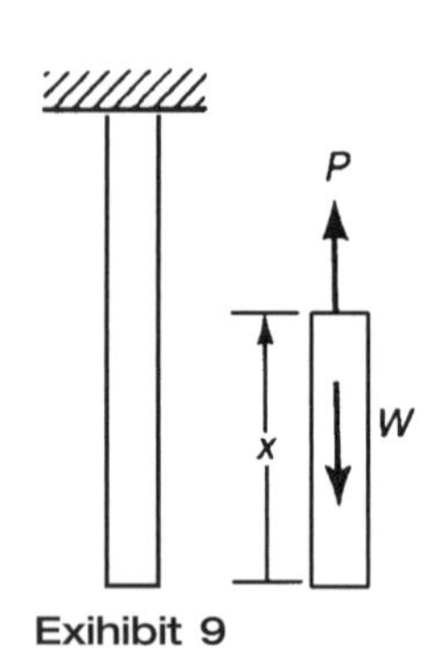

Exihibit 9

THIN-WALLED CYLINDER

Consider the thin-walled circular cylinder subjected to a uniform internal pressure q as shown in Fig. 9.3. A section of length a, is cut out of the vessel in (a). The cut-out portion is shown in (b). The pressure q can be considered as acting across

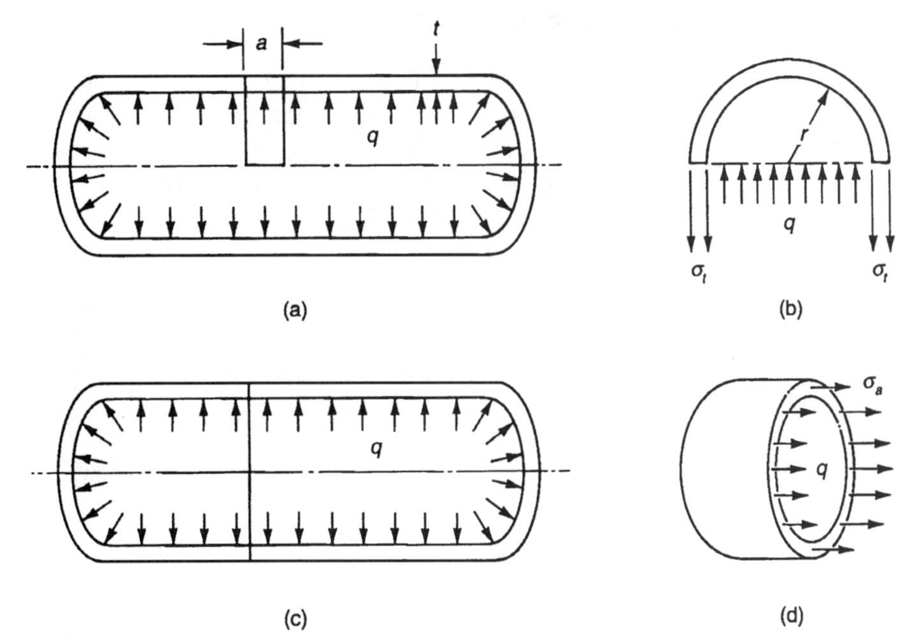

Figure 9.3

the diameter as shown. The tangential stress σ_t is assumed constant through the thickness. Summing forces in the vertical direction gives

$$qDa - 2\sigma_t ta = 0 \tag{9.12}$$

$$\sigma_t = \frac{qD}{2t} \tag{9.13}$$

where D is the inner diameter of the cylinder and t is the wall thickness. The axial stress σ_a is also assumed to be uniform over the wall thickness. The axial stress can be found by making a cut through the cylinder as shown in (c). Consider the horizontal equilibrium for the free-body diagram shown in (d). The pressure q acts over the area πr^2 and the stress σ_a acts over the area πDt which gives

$$\sigma_a \pi Dt = q\pi\left(\frac{D}{2}\right)^2 \tag{9.14}$$

so

$$\sigma_a = \frac{qD}{4t} \tag{9.15}$$

Example 9.6

Consider a cylindrical pressure vessel with a wall thickness of 25 mm, an internal pressure of 1.4 MPa, and an outer diameter of 1.2 m. Find the axial and tangential stresses.

Solution

$$q = 1.4 \text{ MPa};\ D = 1200 - 50 = 1150 \text{ mm};\ t = 25 \text{ mm}$$

$$\sigma_t = \frac{qD}{2t} = \frac{1.4 \text{ MPa} \times 1150 \text{ mm}}{2 \times 25 \text{ mm}} = 32.2 \text{ MPa}$$

$$\sigma_a = \frac{qD}{4t} = \frac{1.4 \text{ MPa} \times 1150 \text{ mm}}{4 \times 25 \text{ mm}} = 16.1 \text{ MPa}$$

GENERAL STATE OF STRESS

Stress is defined as force per unit area acting on a certain area. Consider a body that is cut so that its area has an outward normal in the x direction as shown in Fig. 9.4. The force ΔF that is acting over the area ΔA_x can be split into its components ΔF_x, ΔF_y, and ΔF_z. The stress components acting on this face are then defined as

$$\sigma_x = \lim_{\Delta A_x \to 0} \frac{\Delta F_x}{\Delta A_x} \tag{9.16}$$

$$\tau_{xy} = \lim_{\Delta A_x \to 0} \frac{\Delta F_y}{\Delta A_x} \tag{9.17}$$

$$\tau_{xz} = \lim_{\Delta A_x \to 0} \frac{\Delta F_z}{\Delta A_x} \tag{9.18}$$

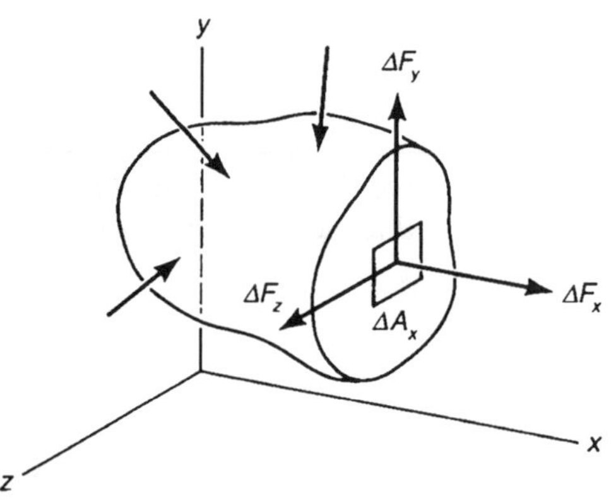

Figure 9.4 Stress on a face

The stress component σ_x is the normal stress. It acts normal to the x face in the x direction. The stress component τ_{xy} is a shear stress. It acts parallel to the x face in the y direction. The stress component τ_{xy} is also a shear stress and acts parallel to the x face in the z direction. For shear stress, the first subscript indicates the *face* on which it acts, and the second subscript indicates the *direction* in which it acts. For normal stress, the single subscript indicates both face and direction. In the general state of stress, there are normal and shear stresses on all faces of an element as shown in Fig. 9.5.

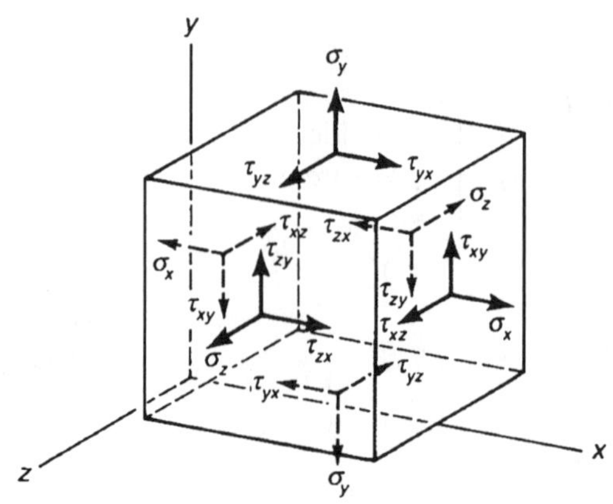

Figure 9.5 Stress at a point (shown in positive directions)

From equilibrium of moments around axes parallel to x, y, and z and passing through the center of the element in Fig. 9.5, it can be shown that the following relations hold

$$\tau_{xy} = \tau_{yx}; \qquad \tau_{yz} = \tau_{zy}; \qquad \tau_{zx} = \tau_{xz} \tag{9.19}$$

Thus, at any point in a body the state of stress is given by six components: $\sigma_x, \sigma_y, \sigma_z, \tau_{xy}, \tau_{yz}$, and τ_{zx}. The usual sign convention is to take the components shown in Fig. 9.5 as positive. One way of saying this is that normal stresses are positive in tension. Shear stresses are positive on a positive face in the positive direction. A **positive face** is defined as a face with a positive outward normal.

PLANE STRESS

In elementary mechanics of materials, we usually deal with a state of plane stress in which only the stresses in the x-y plane are non-zero. The stress components σ_z, τ_{xz}, and τ_{yz} are taken as zero.

Mohr's Circle—Stress

In plane stress, the three components σ_x, τ_y, and τ_{xy} define the state of stress at a point, but the components on any other face have different values. To find the components on an arbitrary face, consider equilibrium of the wedges shown in Fig. 9.6.

Summation of faces in the x′ and y′ directions for the wedge shown in Fig. 9.6(a) gives

$$\sum F_{x'} = 0 = \sigma_{x'}\Delta A - \sigma_x \Delta A(\cos\theta)^2 - \sigma_y \Delta A(\sin\theta)^2 - 2\tau_{xy}\Delta A \sin\theta\cos\theta \tag{9.20}$$

$$\sum F_{y'} = 0 = \tau_{x'y'}\Delta A + (\sigma_x - \sigma_y)\Delta A \sin\theta\cos\theta - \tau_{xy}\Delta A\,[(\cos\theta)^2 - (\sin\theta)^2] \tag{9.21}$$

Canceling ΔA from each of these expressions and using the double angle relations gives

$$\sigma_{x'} = \frac{\sigma_x + \sigma_y}{2} + \frac{\sigma_x - \sigma_y}{2}\cos 2\theta + \tau_{xy}\sin 2\theta \tag{9.22}$$

$$\tau_{x'y'} = -\frac{\sigma_x - \sigma_y}{2}\sin 2\theta + \tau_{xy}\cos 2\theta \tag{9.23}$$

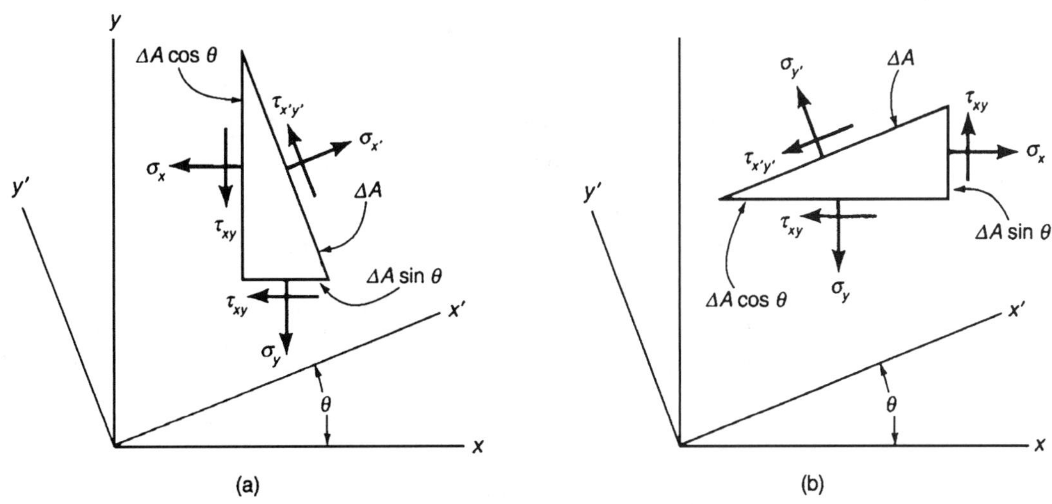

Figure 9.6 Stress on an arbitrary face

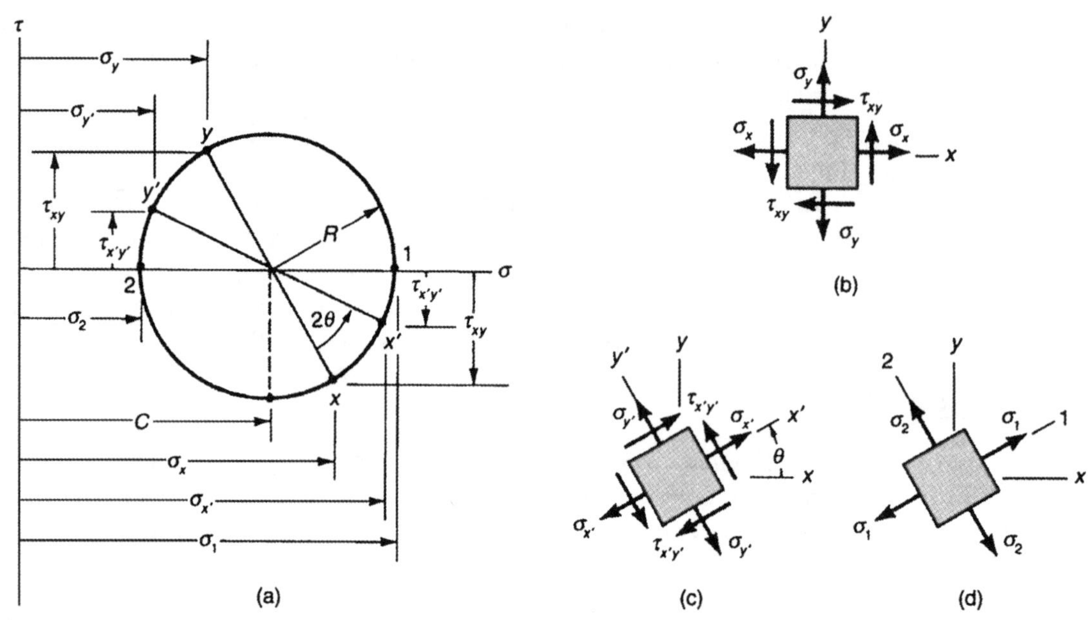

Figure 9.7 Mohr's circle for the stress at a point

Similarly, summation of forces in the y′ direction for the wedge shown in Fig. 9.6(b) gives

$$\sigma_{y'} = \frac{\sigma_x + \sigma_y}{2} - \frac{\sigma_x - \sigma_y}{2}\cos 2\theta - \tau_{xy}\sin 2\theta \qquad \textbf{(9.24)}$$

Equations (9.22), (9.23), and (9.24) are the parametric equations of Mohr's circle; Fig. 9.7(a) shows the general Mohr's circle; Fig. 9.7(b) shows the stress on the element in an *x*-*y* orientation; Fig. 9.7(c) shows the stress in the same element in an *x*′-*y*′ orientation; and Fig. 9.7(d) shows the stress on the element in the 1-2 orientation. Notice that there is always an orientation (for example, a 1-2 orientation) for which the shear stress is zero. The normal stresses σ_1 and σ_2 on these 1-2 faces are the principal stresses, and the 1 and 2 axes are the principal axes of stress. In three-dimensional problems the same is true. There are always three mutually perpendicular faces on which there is no shear stress. Hence, there are always three principal stresses.

To draw Mohr's circle knowing σ_x, σ_y, and τ_{xy},

1. Draw vertical lines corresponding to σ_x and σ_y as shown in Fig. 9.8(a) according to the signs of σ_x and σ_y (to the right of the origin if positive and to the left if negative).

2. Put a point on the σ_x vertical line a distance τ_{xy} below the horizontal axis if τ_{xy} is positive (above if τ_{xy} is negative) as in Fig. 9.8(a). Name this point *x*.

3. Put a point on the σ_y vertical line a distance τ_{xy} in the opposite direction as on the σ_x vertical line also as shown in Fig. 9.8(a). Name this point *y*.

4. Connect the two points *x* and *y*, and draw the circle with diameter *xy* as shown in Fig. 9.8(b).

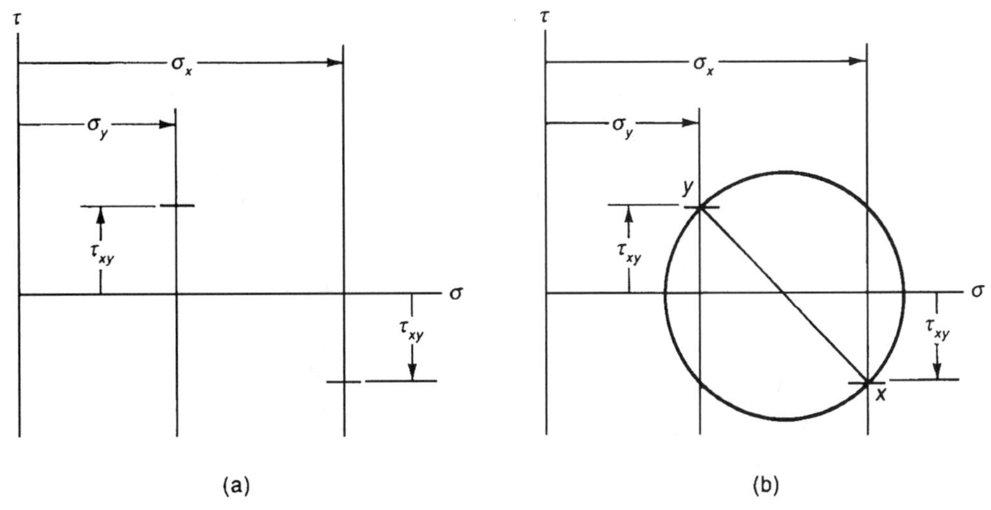

Figure 9.8 Constructing Mohr's circle

Upon constructing Mohr's circle you can now rotate the *xy* diameter through an angle of 2θ to a new position $x'y'$, which can determine the stress on any face at that point in the body as shown in Fig. 9.7. Note that rotations of 2θ on Mohr's circle correspond to θ in the physical plane; also note that the direction of rotation is the same as in the physical plane (that is, if you go clockwise on Mohr's circle, the rotation is also clockwise in the physical plane). The construction can also be used to find the principal stresses and the orientation of the principal axes.

Problems involving stress transformations can be solved with Eqs. (9.22), (9.23), and (9.24), from construction of Mohr's circle, or from some combination. As an example of a combination, it can be seen that the center of Mohr's circle can be represented as

$$C = \frac{\sigma_x + \sigma_y}{2} \tag{9.25}$$

The radius of the circle is

$$R = \sqrt{\left(\frac{\sigma_x - \sigma_y}{2}\right)^2 + \tau_{xy}^{\ 2}} \tag{9.26}$$

The principal stresses then are

$$\sigma_1 = C + R; \qquad \sigma_2 = C - R \tag{9.27}$$

Example **9.7**

Given $\sigma_x = -3$ MPa; $\sigma_y = 5$ MPa; $\tau_{xy} = 3$ MPa. Find the principle stresses and their orientation.

Solution

Mohr's circle is constructed as shown in Exhibit 10. The angle 2θ was chosen as the angle between the *y* axis and the 1 axis clockwise from *y* to 1 as shown in the circle. The angle θ in the physical plane is between the y axis and the 1 axis also clockwise from y to 1. The values of σ_1, σ_2, and 2θ can all be scaled from the circle. The values can also be calculated as follows:

$$R = \sqrt{\left(\frac{\sigma_x - \sigma_y}{2}\right)^2 + \tau_{xy}^{\ 2}} = \sqrt{\left(\frac{-3-5}{2}\right)^2 + 3^2} = 5 \text{ MPa}$$

$$C = \frac{\sigma_x + \sigma_y}{2} = \frac{-3+5}{2} = 1 \text{ MPa}$$

$$\sigma_1 = C + R = 6 \text{ MPa}$$

$$\sigma_2 = C - R = -4 \text{ MPa}$$

$$2\theta = \tan^{-1}(3/5) = 30.96°; \qquad \theta = 15.48°$$

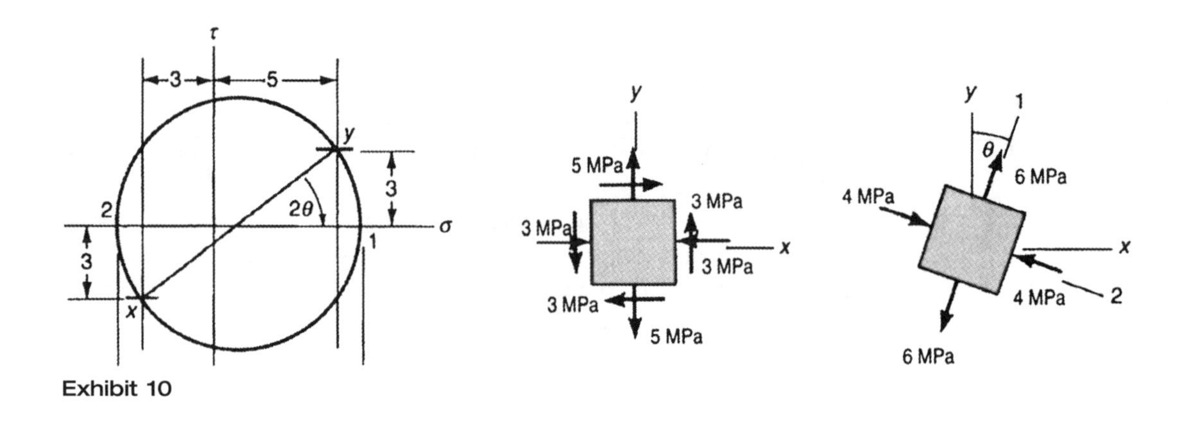

Exhibit 10

STRAIN

Axial strain was previously defined as

$$\varepsilon = \frac{\Delta L}{L} \tag{9.28}$$

In the general case, there are three components of axial strain, ε_x, ε_y, and ε_z. Shear strain is defined as the decrease in angle of two originally perpendicular line segments passing through the point at which strain is defined. In Fig. 9.9, AB is vertical and BC is horizontal. They represent line segments that are drawn before loading. After loading, points A, B, and C move to A', B', and C', respectively. The angle between $A'B'$ and the vertical is α, and the angle between B' and C' and the horizontal is β. The original right angle has been decreased by $\alpha + \beta$, and the shear strain is

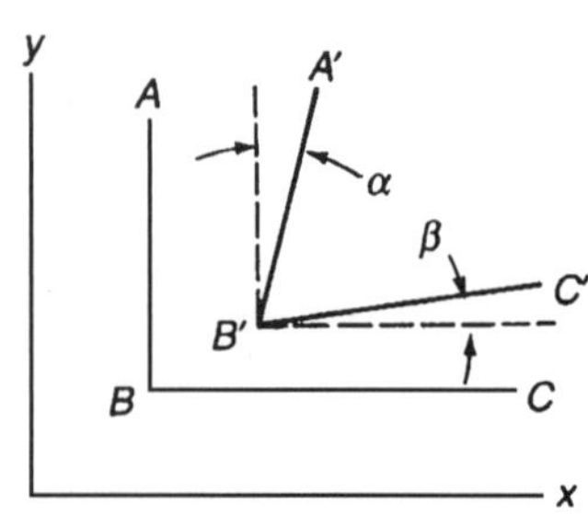

Figure 9.9 Definition of shear strain

$$\gamma_{xy} = \alpha + \beta \tag{9.29}$$

In the general case, there are three components of shear strain, γ_{xy}, γ_{yz}, and γ_{zx}.

Plane Strain

In two dimensions, strain undergoes a similar rotation transformation as stress. The transformation equations are

$$\varepsilon_{x'} = \frac{\varepsilon_x + \varepsilon_y}{2} + \frac{\varepsilon_x - \varepsilon_y}{2} \cos 2\theta + \frac{\gamma_{xy}}{2} \sin 2\theta \tag{9.30}$$

$$\frac{\gamma_{x'y'}}{2} = -\frac{\varepsilon_x - \varepsilon_y}{2} \sin 2\theta + \frac{\gamma_{xy}}{2} \cos 2\theta \tag{9.31}$$

$$\varepsilon_{y'} = \frac{\varepsilon_x + \varepsilon_y}{2} - \frac{\varepsilon_x - \varepsilon_y}{2} \cos 2\theta - \frac{\gamma_{xy}}{2} \sin 2\theta \tag{9.32}$$

These equations are the same as Eq. (9.22), (9.23), and (9.24) for stress, except that the σ_x has been replaced with ε_x, σ_y with ε_y, and τ_{xy} with $\gamma_{xy}/2$. Therefore, Mohr's circle for strain is treated the same way as that for stress, except for the factor of two on the shear strain.

Example 9.8

Given that $\varepsilon_x = 600\ \mu$; $\varepsilon_y = -200\ \mu$; $\gamma_{xy} = -800\ \mu$, find the principal strains and their orientation. The symbol μ signifies 10^{-6}.

Solution

From the Mohr's circle shown in Exhibit 11, it is seen that $2\theta = 45°$; so, $\theta = 22.5°$ clockwise from x to 1. The principal strains are $\varepsilon_1 = 766\ \mu$ and $\varepsilon_2 = -366\ \mu$.

The principal strains can also be found by computation in the same way as principal stresses,

$$R = \sqrt{\left(\frac{\varepsilon_x - \varepsilon_y}{2}\right)^2 + \left(\frac{\gamma_{xy}}{2}\right)^2} = \sqrt{\left(\frac{600+200}{2}\right)^2 + \left(\frac{-800}{2}\right)^2} = 565.7\ \mu$$

$$C = \frac{\varepsilon_x + \varepsilon_y}{2} = \frac{600-200}{2} = 200\ \mu$$

$$\varepsilon_1 = C + R = 766\ \mu$$

$$\varepsilon_2 = C - R = -366\ \mu$$

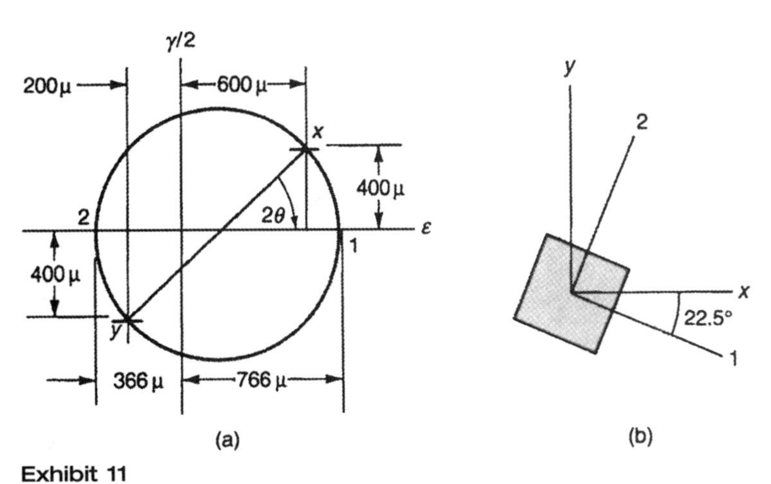

Exhibit 11

HOOKE'S LAW

The relationship between stress and strain is expressed by Hooke's Law. For an isotropic material it is

$$\varepsilon_x = \frac{1}{E}(\sigma_x - v\sigma_y - v\sigma_z) \quad \textbf{(9.33)}$$

$$\varepsilon_y = \frac{1}{E}(\sigma_y - v\sigma_z - v\sigma_x) \quad \textbf{(9.34)}$$

$$\varepsilon_z = \frac{1}{E}(\sigma_z - v\sigma_x - v\sigma_y) \quad \textbf{(9.35)}$$

$$\gamma_{xy} = \frac{1}{G}\tau_{xy} \quad \textbf{(9.36)}$$

$$\gamma_{yz} = \frac{1}{G}\tau_{yz} \quad \textbf{(9.37)}$$

$$\gamma_{zx} = \frac{1}{G}\tau_{zx} \quad \textbf{(9.38)}$$

Further, there is a relationship between E, G, and v which is

$$G = \frac{E}{2(1+v)} \quad \textbf{(9.39)}$$

Thus, for an isotropic material there are only two independent elastic constants. An **isotropic material** is one that has the same material properties in all directions. Notable exceptions to isotropy are wood- and fiber-reinforced composites.

Example 9.9

A steel plate in a state of plane stress is known to have the following strains: $\varepsilon_x = 650\,\mu$, $\varepsilon_y = 250\,\mu$, and $\gamma_{xy} = 400\,\mu$. If $E = 210$ GPa and $v = 0.3$, what are the stress components, and what is the strain ε_z?

Solution

In a state of plane stress, the stresses $\sigma_z = 0$, $\tau_{xz} = 0$ and $\tau_{yz} = 0$. From Hooke's law,

$$\varepsilon_x = \frac{1}{E}(\sigma_x - v\sigma_y - 0)$$

$$\varepsilon_y = \frac{1}{E}(\sigma_y - v\sigma_x - 0)$$

Inverting these relations gives

$$\sigma_x = \frac{E}{1-v^2}(\varepsilon_x + v\varepsilon_y) = \frac{210 \text{ GPa}}{1-0.3^2}[650\,\mu + 0.3(250\,\mu)] = 167.3 \text{ Mpa}$$

$$\sigma_y = \frac{E}{1-v^2}(\varepsilon_y + v\varepsilon_x) = \frac{210 \text{ GPa}}{1-0.3^2}[250\,\mu + 0.3(650\,\mu)] = 102.7 \text{ Mpa}$$

From Hooke's law, the strain γ_{xy} is

$$\gamma_{xy} = \frac{\tau_{xy}}{G}; \qquad G = \frac{E}{2(1+v)}; \qquad \tau_{xy} = \frac{E\gamma_{xy}}{2(1+v)} = \frac{(210 \text{ GPa})\,(400\ \mu)}{2(1+0.3)} = 32.3 \text{ MPa}$$

The strain in the z direction is

$$\varepsilon_z = \frac{1}{E}(0 - \nu\sigma_x - \nu\sigma_y) = \frac{-\nu}{E}(\sigma_x + \sigma_y)$$
$$= \frac{-0.3}{210\ \text{GPa}}(167.3\ \text{MPa} + 102.7\ \text{MPa}) = -386\ \mu$$

TORSION

Torsion refers to the twisting of long members. Torsion can occur with members of any cross-sectional shape, but the most common is the circular shaft. Another fairly common shaft configuration, which has a simple solution, is the hollow, thin-walled shaft.

Circular Shafts

Fig. 9.10(a) shows a circular shaft before loading; the r-θ-z cylindrical coordinate system is also shown. In addition to the outline of the shaft, two longitudinal lines, two circumferential lines, and two diametral lines are shown scribed on the shaft. These lines are drawn to show the deformed shape loading. Fig. 9.10(b) shows the shaft after loading with a torque T. The **double arrow notation** on T indicates a moment about the z axis in a right-handed direction. The effect of the torsion is that each cross-section remains plane and simply rotates with respect to other cross-sections. The angle ϕ is the twist of the shaft at any position z. The rotation $\phi(z)$ is in the θ direction.

The distance b shown in Fig. 9.10(b) can be expressed as $b = \phi r$ or as $b = \gamma z$. The shear strain for this special case can be expressed as

$$\gamma_{\phi z} = r\frac{\phi}{z} \qquad \textbf{(9.40)}$$

For the general case where ϕ is not a linear function of z the shear strain can be expressed as

$$\gamma_{\phi z} = r\frac{d\phi}{dz} \qquad \textbf{(9.41)}$$

$d\phi/dz$ is the twist per unit length or the rate of twist.

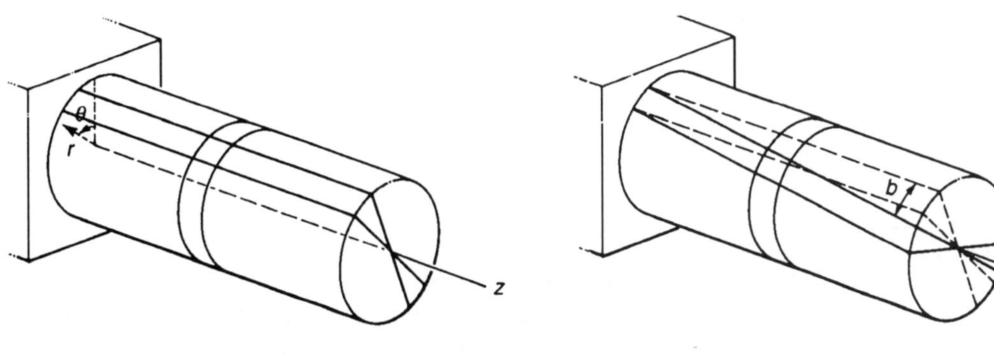

(a) Before loading

(b) After loading

Figure 9.10 Torsion in a circular shaft

The application of Hooke's Law gives

$$\tau_{\phi z} = G\gamma_{\phi z} = Gr\frac{d\phi}{dz} \tag{9.42}$$

The torque at the distance z along the shaft is found by summing the contributions of the shear stress at each point in the cross-section by means of an integration

$$T = \int_A \tau_{\phi z} r\, dA = G\frac{d\phi}{dz}\int_A r^2 dA = GJ\frac{d\phi}{dz} \tag{9.43}$$

where J is the polar moment of inertia of the circular cross-section. For a solid shaft with an outer radius of r_o the polar moment of inertia is

$$J = \frac{\pi r_o^4}{2} \tag{9.44}$$

For a hollow circular shaft with outer radius r_o and inner radius r_i, the polar moment of inertia is

$$J = \frac{\pi}{2}\left(r_o^4 - r_i^4\right) \tag{9.45}$$

Note that the J that appears in Eq. (9.43) is the polar moment of inertia only for the special case of circular shafts (either solid or hollow). For any other cross-section shape, Eq. (9.43) is valid only if J is redefined as a torsional constant *not equal* to the polar moment of inertia. Eq. (9.42) can be combined with Eq. (9.43) to give

$$\tau_{\phi z} = \frac{Tr}{J} \tag{9.46}$$

The maximum shear stress occurs at the outer radius of the shaft and at the location along the shaft where the torque is maximum.

$$\tau_{\phi z\,\max} = \frac{T_{\max} r_o}{J} \tag{9.47}$$

The angle of twist of the shaft can be found by integrating Eq. (9.43)

$$\phi = \int_0^L \frac{T}{GJ}dz \tag{9.48}$$

For a uniform circular shaft with a constant torque along its length, this equation becomes

$$\phi = \frac{TL}{GJ} \tag{9.49}$$

Example 9.10

The hollow circular steel shaft shown in Exhibit 12 has an inner diameter of 25 mm, an outer diameter of 50 mm, and a length of 600 mm. It is fixed at the left end and subjected to a torque of 1400 N • m as shown in Exhibit 12. Find the maximum shear stress in the shaft and the angle of twist at the right end. Take G = 84 GPa.

1000 N•m

Exhibit 12

Solution

$$J = \frac{\pi}{2}\left(r_o^4 - r_i^4\right) = \frac{\pi}{2}[(25\text{ mm})^4 - (12.5\text{ mm})^4] = 575 \times 10^3\,\text{mm}^4$$

$$\tau_{\theta z\,\max} = \frac{T_{\max} r_o}{J} = \frac{(1400\text{N} \bullet \text{m})(25\text{ mm})}{575 \times 10^3\,\text{mm}^4} = 60.8\text{ MPa}$$

$$\phi = \frac{TL}{GJ} = \frac{(1400\text{N} \bullet \text{m})(600\text{ mm})}{(84\text{ GPa})(575 \times 10^3\,\text{mm}^4)} = 0.01738\text{ rad}$$

Hollow, Thin-Walled Shafts

In hollow, thin-walled shafts, the assumption is made that the shear stress τ_{sz} is constant throughout the wall thickness τ. The shear flow q is defined as the product of τ_{sz} and τ. From a summation of forces in the z direction, it can be shown that q is constant—even with varying thickness. The torque is found by summing the contributions of the shear flow. Fig. 9.11 shows the cross-section of the thin-walled tube of nonconstant thickness. The z coordinate is perpendicular to the plane of the paper. The shear flow q is taken in a counter-clockwise sense. The torque produced by q over the element ds is

$$dT = qr\,ds$$

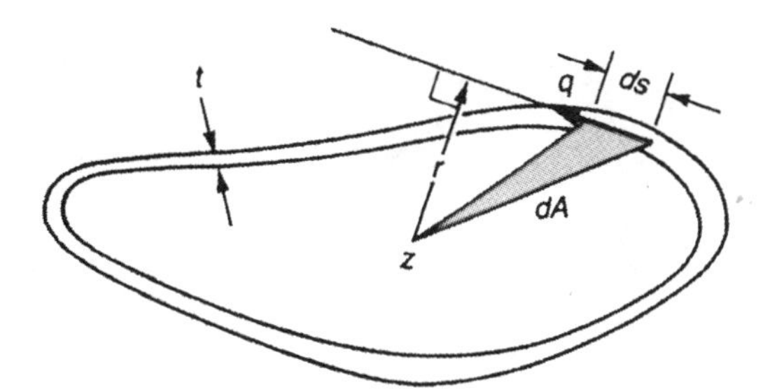

Figure 9.11 Cross-section of thin-walled tube

The total torque is, therefore,

$$T = \oint qr\,ds = q\oint r\,ds \tag{9.50}$$

The area dA is the area of the triangle of base ds and height r,

$$dA = \tfrac{1}{2}(\text{base})(\text{height}) = \frac{r\,ds}{2} \tag{9.51}$$

so that

$$\oint r\,ds = 2A_m \tag{9.52}$$

where A_m is the area enclosed by the wall (including the hole). It is best to use the centerline of the wall to define the boundary of the area, hence A_m is the mean area. The expression for the torque is

$$T = 2A_m q \tag{9.53}$$

and from the definition of q the shear stress can be expressed as

$$\tau_{sz} = \frac{T}{2A_m t} \tag{9.54}$$

Example 9.11

A torque of 10 kN•m is applied to a thin-walled rectangular steel shaft whose cross-section is shown in Exhibit 13. The shaft has wall thicknesses of 5 mm and 10 mm. Find the maximum shear stress in the shaft.

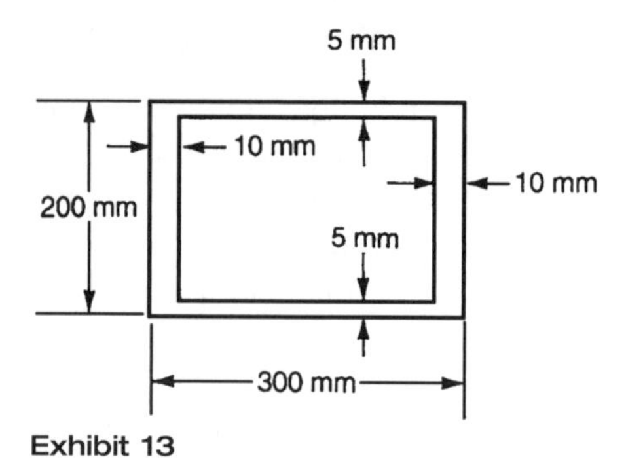

Exhibit 13

Solution

$$A_m = (200 - 5)(300 - 10) = 56{,}550 \text{ mm}^2$$

The maximum shear stress will occur in the thinnest section, so $t = 5$ mm.

$$\tau_{sz} = \frac{T}{2A_m t} = \frac{10 \text{ kN} \bullet \text{m}}{2(56{,}550 \text{ mm}^2)(5 \text{ mm})} = 17.68 \frac{\text{MN}}{\text{m}^2}$$

BEAMS

Shear and Moment Diagrams

Shear and moment diagrams are plots of the shear forces and bending moments, respectively, along the length of a beam. The purpose of these plots is to clearly show maximums of the shear force and bending moment, which are important in the design of beams. The most common sign convention for the shear force and bending moment in beams is shown in Fig. 9.12. One method of determining the shear and moment diagrams is by the following steps:

1. Determine the reactions from equilibrium of the entire beam.
2. Cut the beam at an arbitrary point.

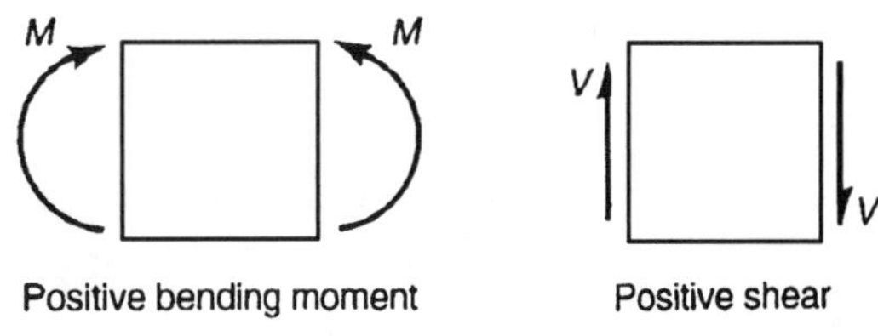

Figure 9.12 Sign convention for bending moment and shear

3. Show the unknown shear and moment on the cut using the positive sign convention shown in Fig. 9.12.
4. Sum forces in the vertical direction to determine the unknown shear.
5. Sum moments about the cut to determine the unknown moment.

Example **9.12**

For the beam shown in Exhibit 14, plot the shear and moment diagram.

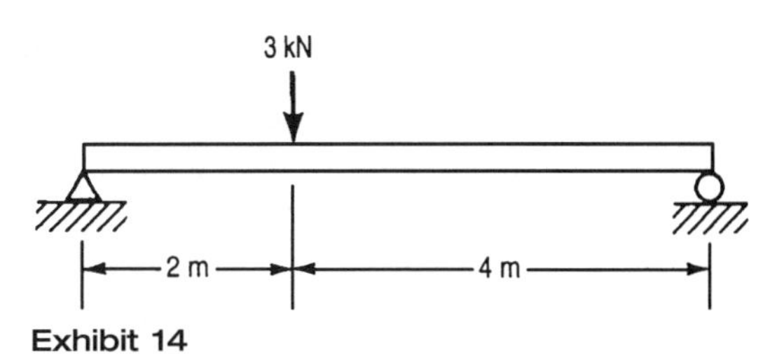

Exhibit 14

Solution

First, solve for the unknown reactions using the free-body diagram of the beam shown in Exhibit 15(a). To find the reactions, sum moments about the left end,

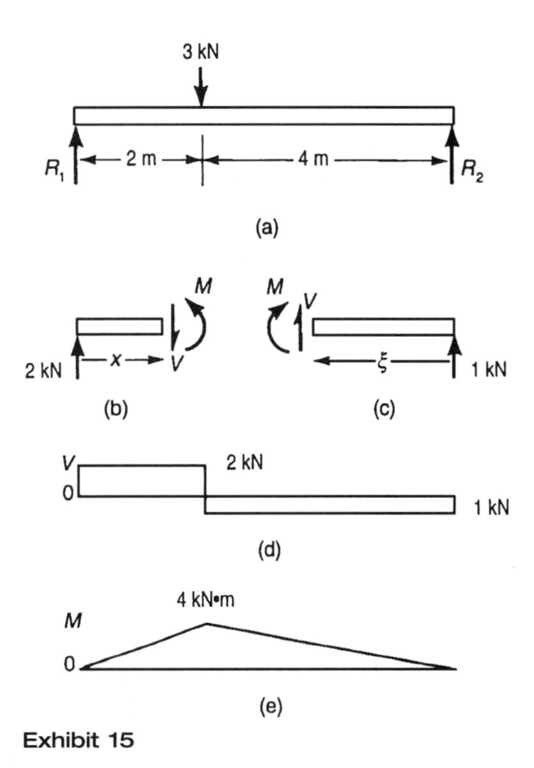

Exhibit 15

which gives

$$6R_2 - (3)(2) = 0 \quad \text{or} \quad R_2 = 6/6 = 1 \text{ kN}$$

Sum forces in the vertical direction to get

$$R_1 + R_2 = 3 = R_1 + 1 \quad \text{or} \quad R_1 = 2 \text{ kN}$$

Cut the beam between the left end and the load as shown in Exhibit 15(b). Show the unknown moment and shear on the cut using the positive sign convention shown in Fig. 9.12. Sum the vertical forces to get

$$V = 2 \text{ kN (independent of } x\text{)}$$

Sum moments about the cut to get

$$M = R_1 x = 2x$$

Repeat the procedure by making a cut between the right end of the beam and the 3-kN load, as shown in Exhibit 15(c). Again, sum vertical forces and sum moments about the cut to get

$$V = 1 \text{ kN (independent of } \xi\text{), and } M = 1\xi$$

The plots of these expressions for shear and moment give the shear and moment diagrams shown in Exhibit 15(d) and 15(e).

It should be noted that the shear diagram in this example has a jump at the point of the load and that the jump is equal to the load. This is always the case. Similarly, a moment diagram will have a jump equal to an applied concentrated moment. In this example, there was no concentrated moment applied, so the moment was everywhere continuous.

Another useful way of determining the shear and moment diagram is by using differential relationships. These relationships are found by considering an element of length Δx of the beam. The forces on that element are shown in Fig. 9.13. Summation of forces in the y direction gives

$$q\Delta x + V - V - \frac{dV}{dx}\Delta x = 0 \tag{9.55}$$

which gives

$$\frac{dV}{dx} = q \tag{9.56}$$

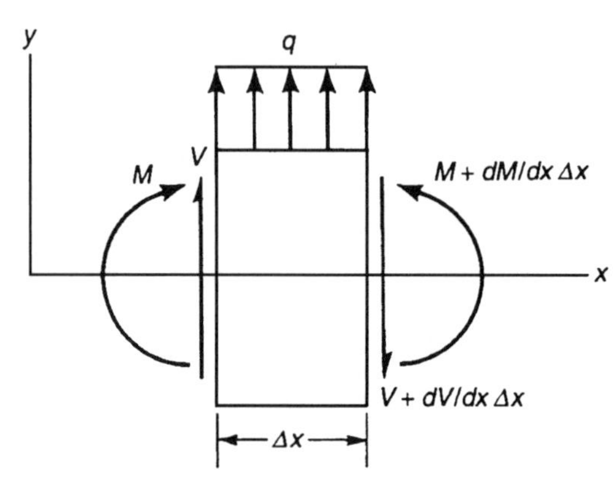

Figure 9.13

Summing moments and neglecting higher order terms gives

$$-M + M + \frac{dM}{dx}\Delta x - V\Delta x = 0 \tag{9.57}$$

which gives

$$\frac{dM}{dx} = V \tag{9.58}$$

Integral forms of these relationships are expressed as

$$V_2 - V_1 = \int_{x_1}^{x_2} q\,dx \tag{9.59}$$

$$M_2 - M_1 = \int_{x_1}^{x_2} V\,dx \tag{9.60}$$

Example 9.13

The simply supported uniform beam shown in Exhibit 16 carries a uniform load of w_0. Plot the shear and moment diagrams for this beam.

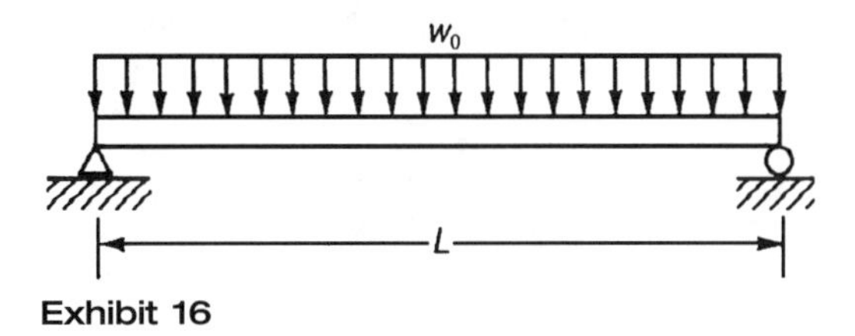

Exhibit 16

Solution

As before, the reactions can be found first from the free-body diagram of the beam shown in Exhibit 17(a). It can be seen that, from symmetry, $R_1 = R_2$. Summing vertical forces then gives

$$R = R_1 = R_2 = \frac{w_0 L}{2}$$

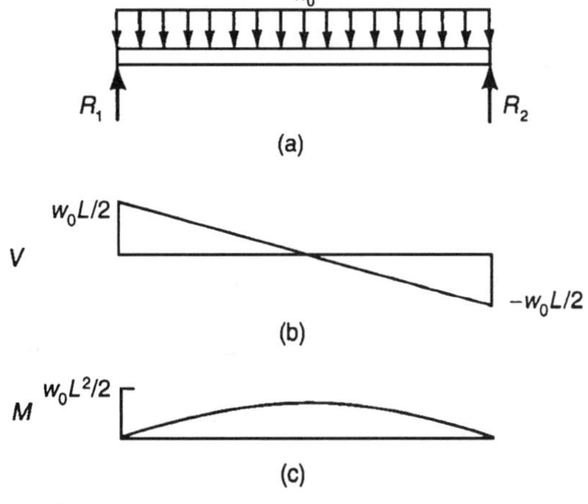

Exhibit 17

The load $q = -w_0$, so Eq. (9.59) reads

$$V = V_0 - \int_0^x w_0\, dx = \frac{w_0 L}{2} - w_0 x$$

Noting that the moment at $x = 0$ is zero, Eq. (9.60) gives

$$M = M_0 - \int_0^x \left(\frac{w_0 L}{2} - w_0 x \right) dx = 0 + \frac{w_0 L x}{2} - \frac{w_0 x^2}{2} = \frac{w_0 x}{2}(L - x)$$

It can be seen that the shear diagram is a straight line, and the moment varies parabolically with x. Shear and moment diagrams are shown in Exhibit 17(b) and Exhibit 17(c). It can be seen that the maximum bending moment occurs at the center of the beam where the shear stress is zero. The maximum bending moment always has a relative maximum at the place where the shear is zero because the shear is the derivative of the moment, and relative maxima occur when the derivative is zero.

Often it is helpful to use a combination of methods to find the shear and moment diagrams. For instance, if there is no load between two points, then the shear diagram is constant, and the moment diagram is a straight line. If there is a uniform load, then the shear diagram is a straight line, and the moment diagram is parabolic. The following example illustrates this method.

Example 9.14

Draw the shear and moment diagrams for the beam shown in Exhibit 18(a).

Solution

Draw the free-body diagram of the beam as shown in Exhibit 18(b). From a summation of the moments about the right end,

$$10\,R_1 = (4)(7) + (3)(2) = 34; \quad \text{so } R_1 = 3.4\,\text{kN}$$

From a summation of forces in the vertical direction,

$$R_2 = 7 - 3.4 = 3.6\,\text{kN}$$

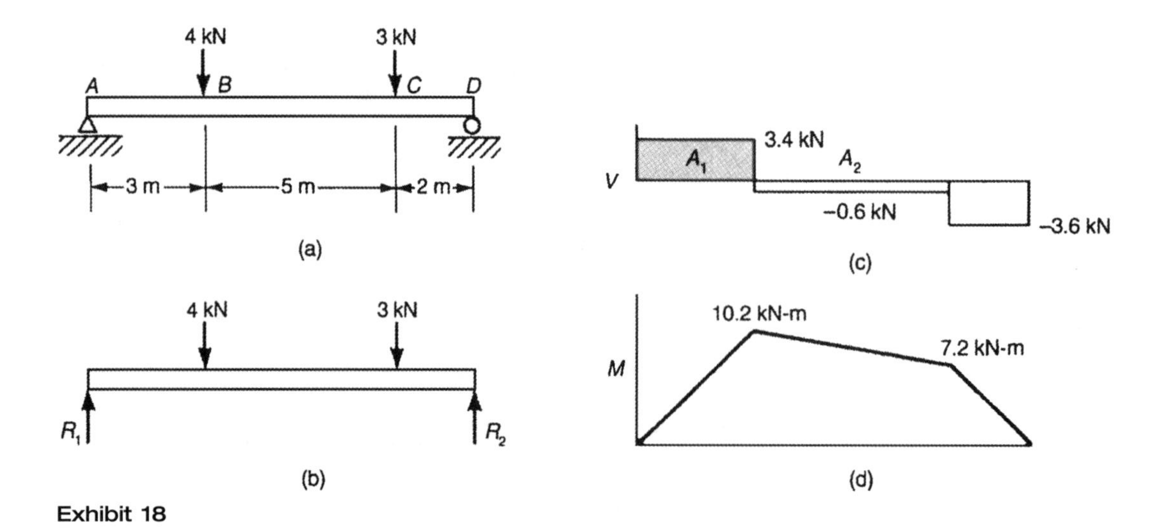

Exhibit 18

The shear in the left portion is 3.4 kN, the shear in the right portion is –3.6 kN and the shear in the center portion is 3.4 – 4 = –0.6 kN. This is sufficient information to draw the shear diagram shown in Exhibit 18(c). The moment at A is zero, so the moment at B is the shaded area A_1 and the moment at C is $A_1 - A_2$.

$$M_B = A_1 = (3.4\,\text{kN})(3\,\text{m}) = 10.2\,\text{kN}\bullet\text{m}$$
$$M_C = A_1 - A_2 = (3.4\,\text{kN})(3\,\text{m}) - (0.6\,\text{kN})(5\,\text{m}) = 7.2\,\text{kN}\bullet\text{m}$$

The moments at A and D are zero, and the moment diagram consists of straight lines between the points A, B, C, and D. There is, therefore, enough information to plot the moment diagram shown in Exhibit 18(d).

Stresses in Beams

The basic assumption in elementary beam theory is that the beam cross-section remains plane and perpendicular to the neutral axis as shown in Fig. 9.14 when the beam is loaded. This assumption is strictly true only for the case of pure bending (constant bending moment and no shear) but gives good results even when shear is taking place. Figure 9.14 shows a beam element before as well as after loading. It can be seen that there is a line of length ds that does not change length upon deformation. This line is called the neutral axis. The distance y is measured from this neutral axis. The strain in the x direction is $\Delta L/L$. The change in length $\Delta L = -yd\phi$ and the length is ds, so

$$\varepsilon_x = -y\frac{d\phi}{ds} = -\frac{y}{\rho} = -\kappa y \qquad \textbf{(9.61)}$$

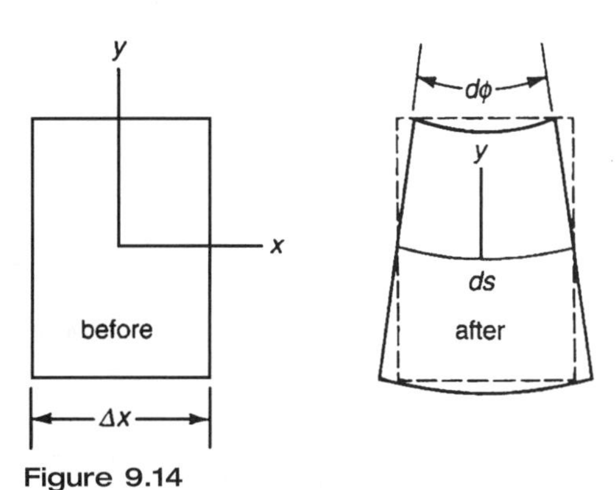

Figure 9.14

where ρ is the radius of curvature of the beam and κ is the curvature of the beam. Assuming that σ_y and σ_z are zero, Hooke's Law yields

$$\sigma_x = -E\kappa y \qquad \textbf{(9.62)}$$

The axial force and bending moment can be found by summing the effects of the normal stress σ_x,

$$P = \int_A \sigma_x \, dA = -E\kappa \int_A y \, dA \qquad \textbf{(9.63)}$$

$$M = -\int_A y\sigma_x \, dA = E\kappa \int_A y^2 dA = EI\kappa \qquad \textbf{(9.64)}$$

where I is the moment of inertia of the beam cross-section. If the axial force is zero (as is the usual case) then the integral of $y\,dA$ is zero. That means that y is measured from the centroidal axis of the cross-section. Since y is also measured from the neutral axis, the neutral axis coincides with the centroidal axis. From Eq. (9.62) and (9.64), the bending stress σ_x can be expressed as

$$\sigma_x = -\frac{My}{I} \tag{9.65}$$

The maximum bending stress occurs where the magnitude of the bending moment is a maximum and at the maximum distance from the neutral axis. For symmetrical beam sections the value of $y_{\max} = \pm C$ where C is the distance to the extreme fiber so the maximum stress is

$$\sigma_x = \pm\frac{MC}{I} = \pm\frac{M}{S} \tag{9.66}$$

where S is the section modulus ($S = I/C$).

Example 9.15

A 100 mm × 150 mm wooden cantilever beam is 2 m long. It is loaded at its tip with a 4-kN load. Find the maximum bending stress in the beam shown in Exhibit 19. The maximum bending moment occurs at the wall and is $M_{\max} = 8$ kN • m.

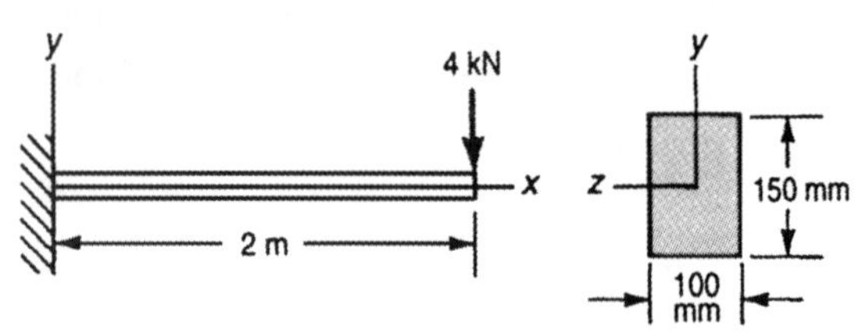

Exhibit 19

Solution

$$I = \frac{bh^3}{12} = \frac{100(150)^3}{12} = 28.1\times10^6 \text{ mm}^4$$

$$\sigma_{x\max} = \frac{|M|_{\max}c}{I} = \frac{(8\,\text{kN}\bullet\text{m})(75\,\text{mm})}{28.1\times10^6\text{ mm}^4} = 21.3 \text{ MPa}$$

Shear Stress

To find the shear stress, consider the element of length Δx shown in Fig. 9.15(a). A cut is made in the beam at $y = y_1$. At that point the beam has a thickness b. The shaded cross-sectional area above that cut is called A_1. The bending stresses acting on that element are shown in Fig. 9.15(b). The stresses are slightly larger at the right side than at the left side so that a force per unit length q is needed for equilibrium. Summation of forces in the x direction for the free-body diagram shown in Fig. 9.15(b) gives

$$-F = q\Delta x = \int_{A_1}\sigma\,dA - \int_{A_1}\left(\sigma + \frac{d\sigma}{dx}\Delta x\right)dA = -\int_{A_1}\frac{d\sigma}{dx}\Delta x\,dA \tag{9.67}$$

From the expression for the bending stress ($\sigma = -My/I$) it follows that

$$\frac{d\sigma}{dx} = -\left(\frac{dM}{dx}\right)\frac{y}{I} = -V\frac{y}{I} \tag{9.68}$$

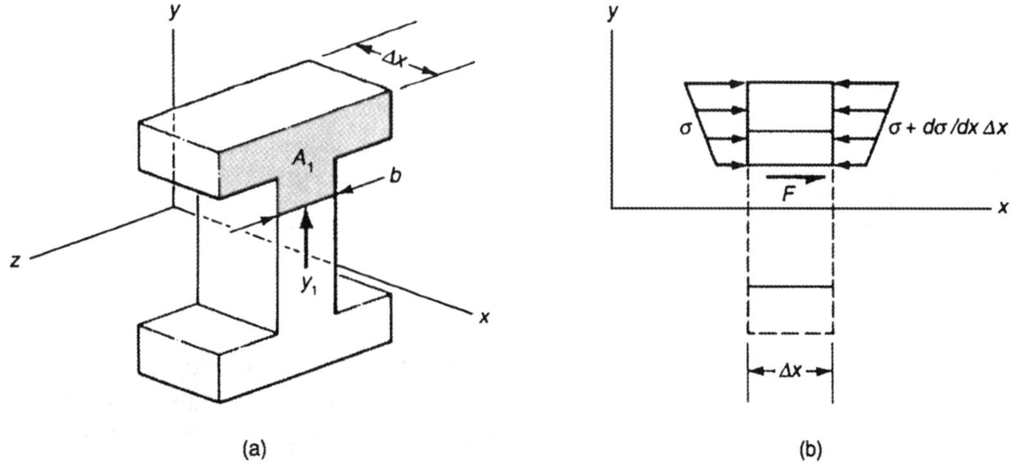

Figure 9.15 Shear stress in beams

Substituting Eq. (9.68) into Eq. (9.67) gives

$$q = \frac{V}{I}\int_{A_1} y\,dA = \frac{VQ}{I} \tag{9.69}$$

If the shear stress τ is assumed to be uniform over the thickness b then $\tau = q/b$ and the expression for shear stress is

$$\tau = \frac{VQ}{Ib} \tag{9.70}$$

where V is the shear in the beam, Q is the moment of area above (or below) the point in the beam at which the shear stress is sought, I is the moment of inertia of the entire beam cross-section, and b is the thickness of the beam cross-section at the point where the shear stress is sought. The definition of Q from Eq. (9.69) is

$$Q = \int_{A_1} y\,dA = A_1 y \tag{9.71}$$

Example 9.16

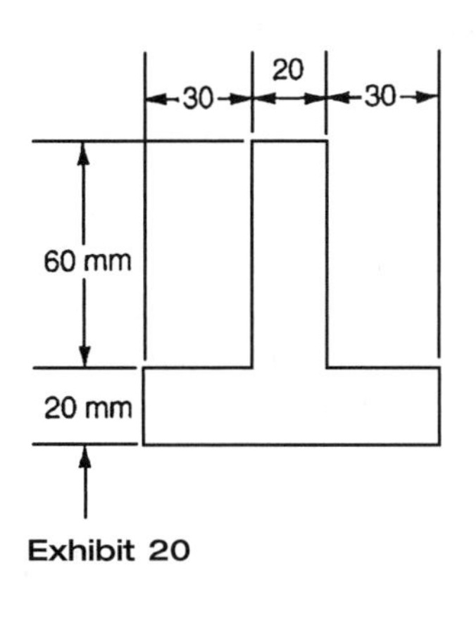

Exhibit 20

The cross-section of the beam shown in Exhibit 20 has an applied shear of 10 kN. Find (a) the shear stress at a point 20 mm below the top of the beam and (b) the maximum shear stress from the shear force.

Solution

The section is divided into two parts by the dashed line shown in Exhibit 21(a). The centroids of each of the two sections are also shown in Exhibit 21(a). The centroid of the entire cross-section is found as follows

$$\bar{y} = \frac{\sum_{n=1}^{N} \bar{y}_n A_n}{\sum_{n=1}^{N} A_n} = \frac{(60)(20)(30+20)+(80)(20)(10)}{(60)(20)+(80)(20)} = 27.14 \text{ mm (from bottom)}$$

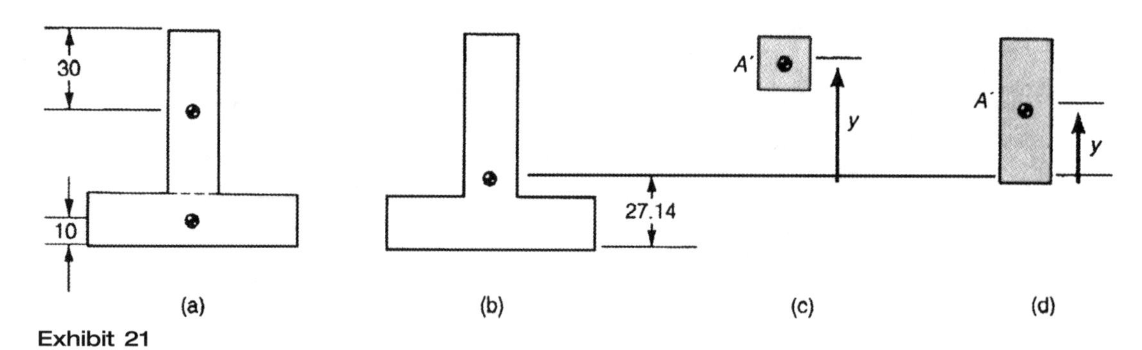

Exhibit 21

Exhibit 21(b) shows the location of the centroid.

The moment of inertia of the cross-section is found by summing the moments of inertia of the two sections taken about the centroid of the entire section. The moment of inertia of each part is found about its own centroid; then the parallel axis theorem is used to transfer it to the centroid of the entire section.

$$
\begin{aligned}
I &= \sum_{n=1}^{N} I_n + A_n \bar{y}_n \\
&= \frac{(20)(60)^3}{12} + (20)(60)(50-27.14)^2 + \frac{(80)(20)^3}{12} - (20)(80)(27.14-10)^2 \\
&= 1.510 \times 10^6 \text{ mm}^4
\end{aligned}
$$

For the point 20 mm below the top of the beam, the area A' and the distance y are shown in Exhibit 21(c). The distance y is from the neutral axis to the centroid of A'. The value of Q is then

$$Q = \int_{A'} y\,dA = A'y = (20)(20)(70-27.14) = 17{,}140\,\text{mm}^3$$

$$\tau = \frac{VQ}{Ib} = \frac{(10\text{kN})(17{,}140\,\text{mm}^3)}{(1.510\times 10^6\,\text{mm}^4)(20\,\text{mm})} = 0.00568\frac{\text{kN}}{\text{mm}^2} = 5.68\,\text{MPa}$$

The maximum Q will be at the centroid of the cross-section. Since the thickness is the same everywhere, the maximum shear stress will appear at the centroid. The maximum moment of area Q_{max} is

$$Q = \int_{A'} y\,dA = A'y_1 = (20)(80-27.14)\frac{(80+27.14)}{2} = 56{,}600\,\text{mm}^3$$

$$\tau = \frac{VQ}{IB} = \frac{(10\text{kN})(56{,}600\,\text{mm}^3)}{(1.510\times 10^6\,\text{mm}^4)(20\,\text{mm})} = 0.01875\frac{\text{kN}}{\text{mm}^2} = 18.75\,\text{MPa}$$

Deflection of Beams

The beam deflection in the y direction will be denoted as y, while most modern texts use v for the deflection in the y direction. The *FE Supplied-Reference Handbook* uses the older notation. The main assumption in the deflection of beams is that the slope of the beam is small. The slope of the beam is dy/dx. Since the slope is small, the slope is equal to the angle of rotation in radians.

$$\frac{dy}{dx} = \text{rotation in radians} \qquad \textbf{(9.72)}$$

Because the slope is small it also follows that

$$\kappa = \frac{1}{\rho} \approx \frac{d^2y}{dx^2} \tag{9.73}$$

From Eq. (9.62) this gives

$$\frac{d^2y}{dx^2} = \frac{M}{EI} \tag{9.74}$$

This equation, together with two boundary conditions, can be used to find the beam deflection. Integrating twice with respect to x gives

$$\frac{dy}{dx} = \int \frac{M}{EI} dx + C_1 \tag{9.75}$$

$$y = \iint \frac{M}{EI} dx + C_1 x + C_2 \tag{9.76}$$

where the constants C_1 and C_2 are determined from the two boundary conditions. Appropriate boundary conditions are on the displacement y or on the slope dy/dx. In the common problems of uniform beams, the beam stiffness EI is a constant and can be removed from beneath the integral sign.

Example 9.17

The uniform cantilever beam shown in Exhibit 22(a) has a constant, uniform, downward load w_0 along its length. Find the deflection and slope of this beam.

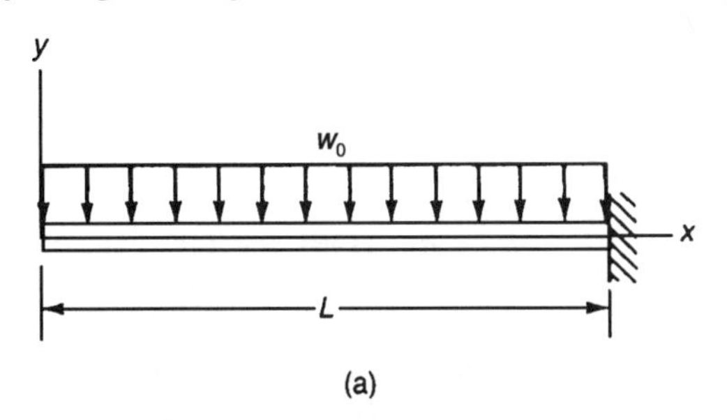

(a)

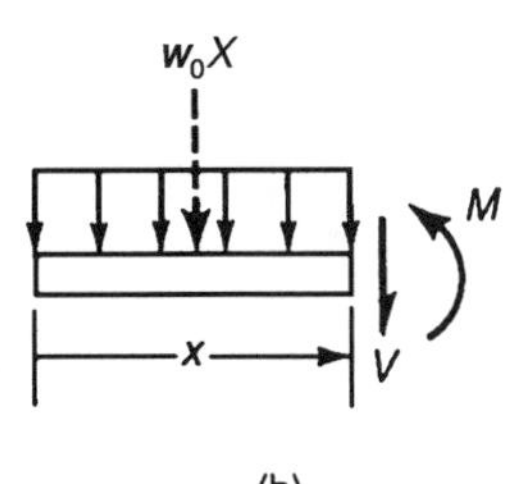

(b)

Exhibit 22

Solution

The moment is found by drawing the free-body diagram shown in Exhibit 23(b). The uniform load is replaced with the statically equivalent load w_0x at the position $x/2$. Moments are then summed about the cut giving

$$M = -w_0 \frac{x^2}{2}$$

Integrating twice with respect to x,

$$\frac{dy}{dx} = \int \frac{M}{EI} dx + C_1 = \frac{1}{EI} \int \left(-w_0 \frac{x^2}{2} \right) dx + C_1 = -\frac{1}{6} \frac{w_0 x^3}{EI} + C_1$$

$$y = \int \left(-\frac{1}{6} \frac{w_0 x^3}{EI} \right) dx + C_1 x + C_2 = -\frac{1}{24} \frac{w_0 x^4}{EI} + C_1 x + C_2$$

At $x = L$ the displacement and slope must be zero so that

$$y(L) = 0 = -\frac{1}{24} \frac{w_0 L^4}{EI} + C_1 L + C_2$$

$$\frac{dy}{dx}(L) = 0 = -\frac{1}{6} \frac{w_0 L^3}{EI} + C_1$$

Therefore,

$$C_1 = \frac{1}{6} \frac{w_0 L^3}{EI}; \qquad C_2 = -\frac{1}{8} \frac{w_0 L^4}{EI}$$

Inserting C_1 and C_2 into the previous expressions gives

$$y = -\frac{w_0}{24EI} (x^4 - 4xL^3 + 3L^4)$$

$$\frac{dy}{dx} = \frac{w_0}{6EI} (L^3 - x^3)$$

Fourth-Order Beam Equation

The second-order beam Eq. (9.74) can be combined with the differential relationships between the shear, moment, and distributed load. Differentiate Eq. (9.74) with respect to x, and use Eq. (9.58).

$$\frac{d}{dx} \left(EI \frac{d^2 y}{dx^2} \right) = \frac{dM}{dx} = V \tag{9.77}$$

Differentiate again with respect to x and use Eq. (9.56).

$$\frac{d^2}{dx^2} \left(EI \frac{d^2 y}{dx^2} \right) = \frac{dV}{dx} = q \tag{9.78}$$

For a uniform beam (that is, constant EI) the fourth-order beam equation becomes

$$EI \frac{d^4 y}{dx^4} = q \tag{9.79}$$

This equation can be integrated four times with respect to x. Four boundary conditions are required to solve for the four constants of integration. The boundary

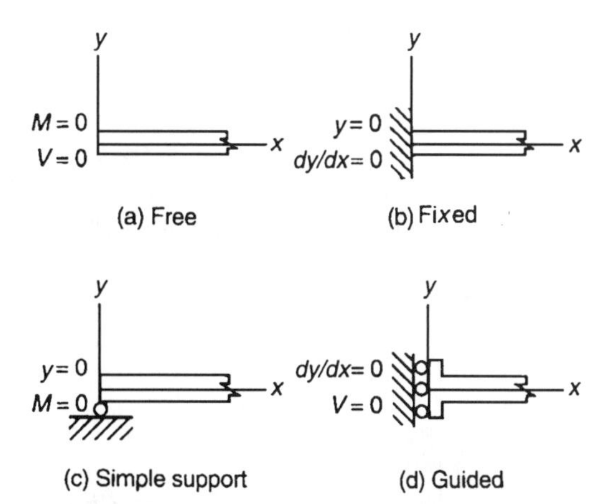

Figure 9.16 Boundary conditions for beams

conditions are on the displacement, slope, moment, and/or shear. Fig. 9.16 shows the appropriate boundary conditions on the end of a beam, even with a distributed loading. If there is a concentrated force or moment applied at the end of a beam, that force or moment enters the boundary condition. For instance, an upward load of P at the left end for the free or guided beam would give $V(0) = P$ instead of $V(0) = 0$.

Example 9.18

Consider the uniformly loaded uniform beam shown in Exhibit 24. The beam is clamped at both ends. The uniform load w_0 is acting downward. Find an expression for the displacement as a function of x.

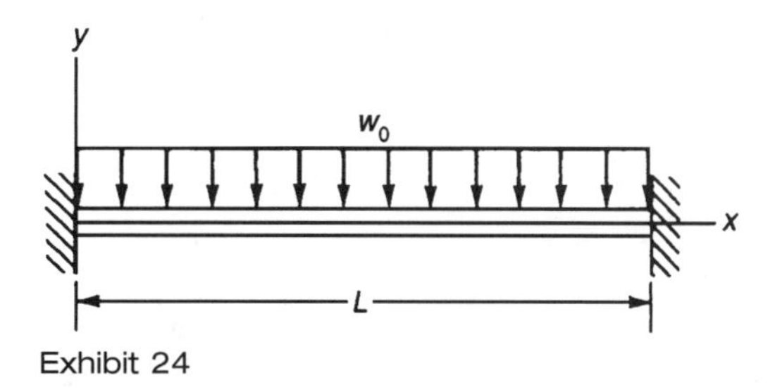

Exhibit 24

Solution

The differential equation is

$$EI\frac{d^4y}{dx^4} = q = -w_0(\text{constant})$$

Integrate four times with respect to x.

$$V = EI\frac{d^3y}{dx^3} = -w_0x + C_1$$

$$M = EI\frac{d^2y}{dx^2} = -w_0\frac{x^2}{2} + C_1x + C_2$$

$$EI\frac{dy}{dx} = -w_0\frac{x^3}{6} + C_1\frac{x^2}{2} + C_2x + C_3$$

$$EIy = -w_0\frac{x^4}{24} + C_1\frac{x^3}{6} + C_2\frac{x^2}{2} + C_3x + C_4$$

The four constants of integration can be found from four boundary conditions. The boundary conditions are

$$y(0)=0; \quad \frac{dy}{dx}(0)=0; \quad y(L)=0; \quad \frac{dy}{dx}(L)=0$$

These lead to the following:

$$EIy(0)=0=C_4$$

$$EI\frac{dy}{dx}(0)=0=C_3$$

$$EIy(L)=0=-w_0\frac{L^4}{24}+C_1\frac{L^3}{6}+C_2\frac{L^2}{2}$$

$$EI\frac{dy}{dx}(L)=0=-w_0\frac{L^3}{6}+C_1\frac{L^2}{2}+C_2L$$

Solving the last two equations for C_1 and C_2 gives

$$C_1=\frac{1}{2}w_0L; \qquad C_2=-\frac{1}{12}w_0L^2$$

Inserting these values into the equation for y gives

$$y=-\frac{w_0x^2}{EI}\left(\frac{1}{24}x^2-\frac{1}{12}xL+\frac{1}{24}L^2\right)$$

Some solutions for uniform beams with various loads and boundary conditions are shown in Table 9.1.

Table 9.1 Deflection and slope formulas for beams

<table>
<tr><th></th><th>Beam</th><th>Deflection, v</th><th>Slope, v'</th></tr>
<tr><td>1.</td><td>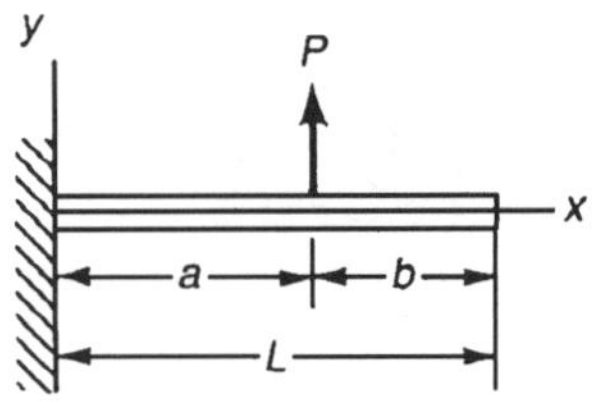</td><td>For $0 \le x \le a$
$y=\frac{Px^2}{6EI}(3a-x)$
For $a \le x \le L$
$y=\frac{Pa^2}{6EI}(3x-a)$</td><td>For $0 \le x \le a$
$\frac{dy}{dx}=\frac{px}{2EI}(2a-x)$
For $a \le x \le L$
$\frac{dy}{dx}=\frac{Pa^2}{2EI}a$</td></tr>
<tr><td>2.</td><td>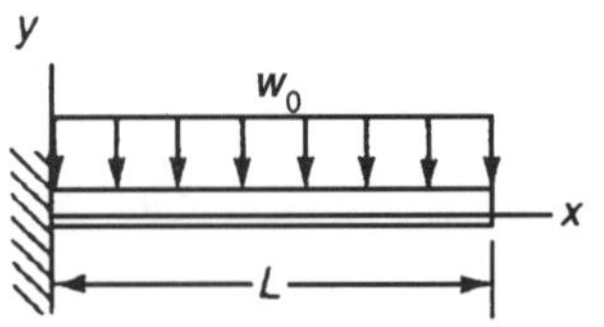</td><td>$y=-\frac{w_0x^2}{24\,EI}(x^2-4Lx+6L^2)$</td><td>$\frac{dy}{dx}=-\frac{w_0x}{6EI}(x^2-12Lx+12L^2)$</td></tr>
<tr><td>3.</td><td>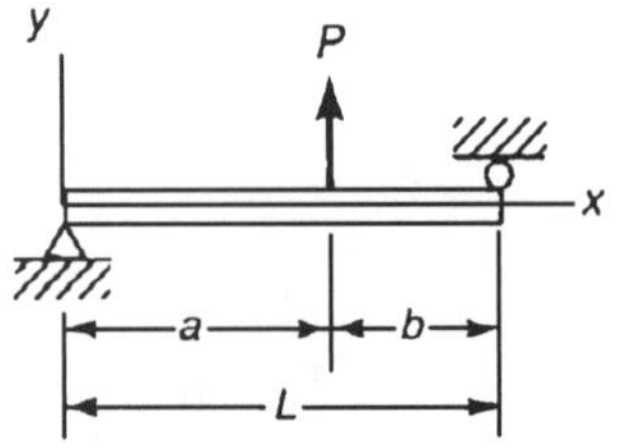</td><td>For $0 \le x \le a$
$y=\frac{Pbx}{6LEI}(L^2-b^2-x^2)$
For $a \le x \le L$
$y=\frac{Pa(L-x)}{6LEI}(2Lx-a^2-x^2)$</td><td>For $0 \le x \le a$
$\frac{dy}{dx}=\frac{Pb}{6LEI}(L^2-b^2-3x^2)$
For $a \le x \le L$
$\frac{dy}{dx}=\frac{Pa}{6LEI}(2L^2+a^2-6Lx+3x^2)$</td></tr>
<tr><td>4.</td><td>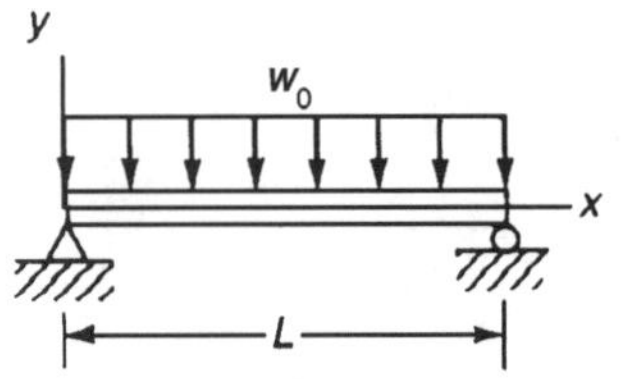</td><td>$y=-\frac{w_0x}{24EI}(L^3-2Lx^2+x^3)$</td><td>$\frac{dy}{dx}=-\frac{w_0}{24EI}(L^3-6Lx^2+4x^3)$</td></tr>
</table>

Table 9.1 Deflection and slope formulas for beams (*Continued*)

Beam	Deflection, v	Slope, v'
5.	$y = -\dfrac{M_0 x}{6EIL}(L^2 - x^2)$	$\dfrac{dy}{dx} = -\dfrac{M_0}{6EIL}(L^2 - 3x^2)$

Superposition

In addition to the use of second-order and fourth-order differential equations, a very powerful technique for determining deflections is the use of superposition. Because all of the governing differential equations are linear, solutions can be directly superposed. Use can be made of tables of known solutions, such as those in Table 9.1, to form solutions to many other problems. Some examples of superposition follow.

Example 9.19

Find the maximum displacement for the simply supported uniform beam loaded by two equal loads placed at equal distances from the ends as shown in Exhibit 25.

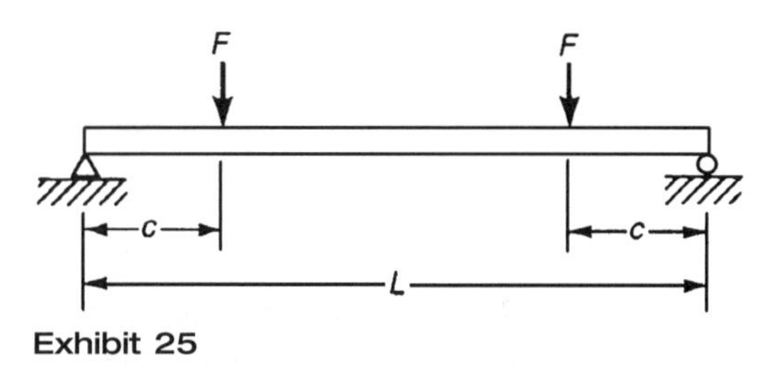

Exhibit 25

Solution

The solution can be found by superposition of the two problems shown in Exhibit 26. From the symmetry of this problem, it can be seen that the maximum deflection will be at the center of the span. The solution for the beam shown in Exhibit 26(a) is found as Case 3 in Table 9.1. In Exhibit 26(a) the center of the span is to the left of the load F so that the formula from the table for $0 \le x \le a$ is chosen. In the formula, $x = L/2$, $c = b$, and $P = -F$ so that

$$y_a\left(\frac{L}{2}\right) = \frac{Pbx}{6LEI}(L^2 - b^2 - x^2) = -\frac{Fc\left(\frac{L}{2}\right)}{6LEI}\left[L^2 - c^2 - \left(\frac{L}{2}\right)^2\right] = \frac{Fc}{48EI}(3L^2 - 4c^2)$$

The central deflection of the beam in Exhibit 26(b) will be the same, so the maximum downward deflection, Δ, will be

$$\delta = -2y_a\left(\frac{L}{2}\right) = \frac{Fc}{24EI}(3L^2 - 4c^2)$$

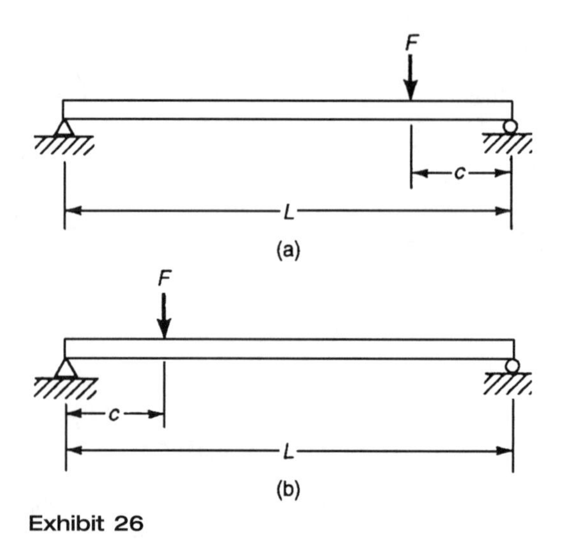

Exhibit 26

Example 9.20

Find an expression for the deflection of the uniformly loaded, supported, cantilever beam shown in Exhibit 27.

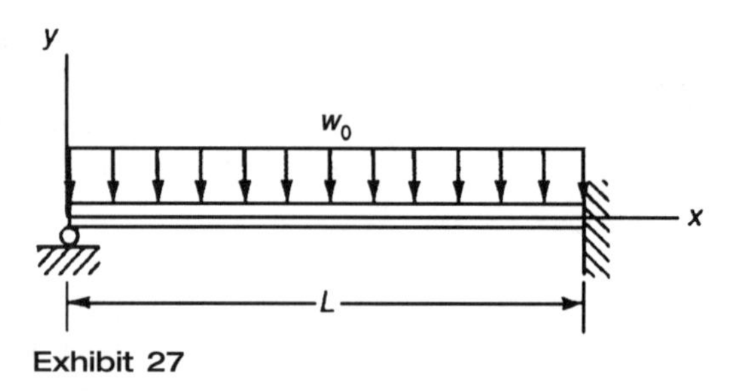

Exhibit 27

Solution

Superpose Case 4 and 5 as shown in Exhibit 28 so that the moment M_0 is of the right magnitude and direction to suppress the rotation at the right end. The rotation

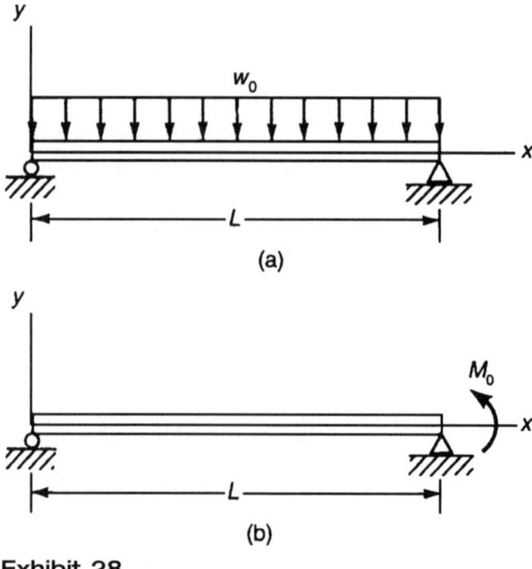

Exhibit 28

for each case from Table 9.1 is

$$\left(\frac{dy}{dx}\right)_4\Bigg|_{x=L} = -\frac{w_0}{24EI}(L^3 - 6L^3 + 4L^3) = \frac{w_0 L^3}{24EI}$$

$$\left(\frac{dy}{dx}\right)_5\Bigg|_{x=L} = -\frac{M_0}{6EIL}(L^2 - 3L^2) = \frac{M_0 L}{3EI}$$

Setting the rotation at the end equal to zero gives

$$\left(\frac{dy}{dx}\right)_4\Bigg|_{x=L} + \left(\frac{dy}{dx}\right)_5\Bigg|_{x=L} = 0 = -\frac{w_0 L^3}{24EI} + \frac{M_0 L}{3EI}$$

$$M_0 = -\frac{w_0 L^2}{8}$$

Substituting this expression into the formulas in the table and adding gives

$$y = -\frac{w_0 x}{24EI}(L^3 - 2Lx^2 + x^3) + \frac{w_0 L^2}{8}\frac{x}{6EI}(L^2 - x^2) = -\frac{w_0 x}{48EI}(L^3 - 3Lx^2 + 2x^3)$$

COMBINED STRESS

In many cases, members can be loaded in a combination of bending, torsion, and axial loading. In these cases, the solution of each portion is exactly as before; the effects of each are simply added. This concept is best illustrated by an example.

Example 9.21

In Exhibit 29, there is a thin-walled, aluminum tube *AB*, which is attached to a wall at *A*. The tube has a rectangular cross-section member *BC* attached to it. A vertical load is placed on the member *BC* as shown. The aluminum tube has an outer diameter of 50 mm and a wall thickness of 3.25 mm. Take P = 900 N, a = 450 mm, and b = 400 mm. Find the state of stress at the top of the tube at the point *D*. Draw Mohr's circle for this point, and find the three principal stresses.

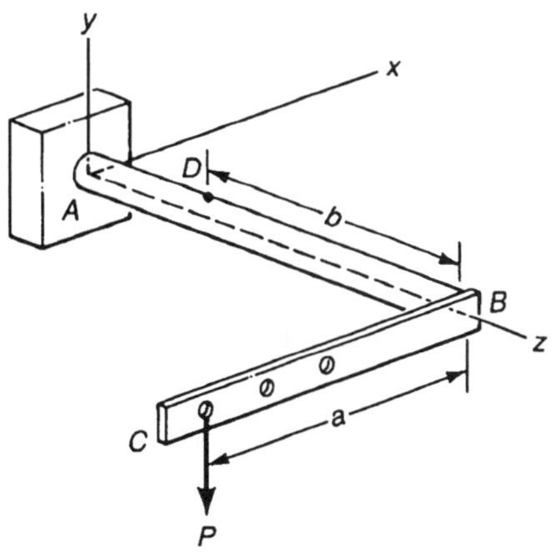

Exhibit 29

Solution

Cut the tube at the Point D. Draw the free-body diagram as in Exhibit 30(a). From that free-body diagram, a summation of moments at the cut about the z axis gives

$$T = Pa = (900\text{ N})(450\text{ mm}) = 405\text{ N}\bullet\text{m}$$

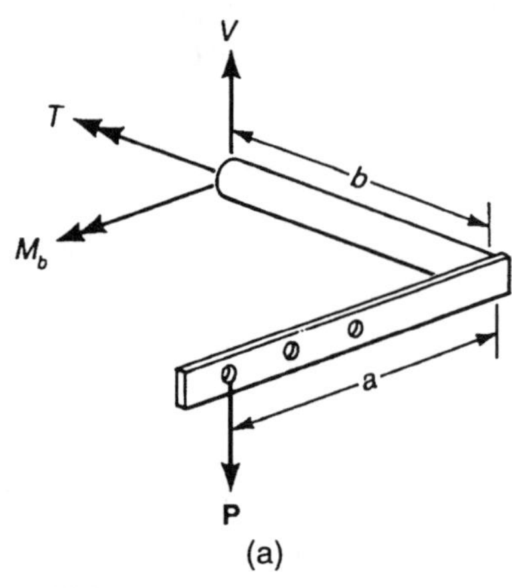

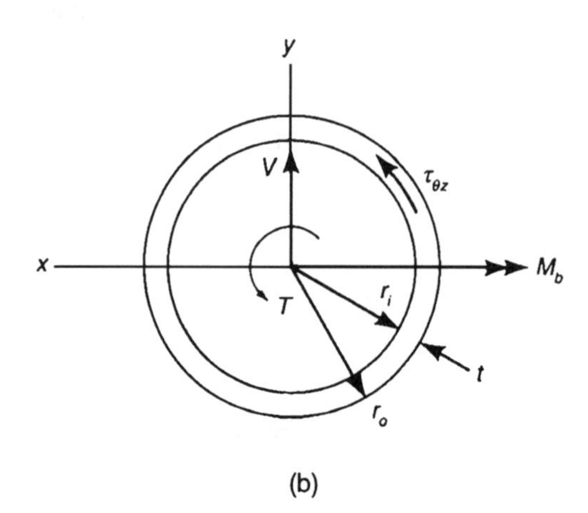

Exhibit 30

A summation of moments at the cut about an axis parallel with the x axis gives

$$M_b = Pb = (900\text{ N})(400\text{ mm}) = 360\text{ N}\bullet\text{m}$$

A summation of vertical forces gives

$$V = P$$

Exhibit 30(b) shows the force and moments acting on the cross-section. The bending and shearing stresses caused by these loads are

$$\sigma_z = \frac{M_b y}{I_{xx}} \qquad (\text{from } M_b)$$

$$\tau_{\theta z} = \frac{Tr}{I_z} \qquad (\text{from } T)$$

$$\tau_{\theta z} = \frac{VQ}{I_{xx} b} \qquad (\text{from } V)$$

The shearing stress attributed to V will be zero at the top of the beam and can be neglected. The moments of inertia are

$$I_{xx} = \frac{\pi\left(r_o^4 - r_i^4\right)}{4} = \frac{\pi(25^4 - 21.75^4)}{4} = 131\times10^3\text{ mm}^4$$

$$I_z = \frac{\pi\left(r_o^4 - r_i^4\right)}{2} = 2I_{xx} = 262\times10^3\text{ mm}^4$$

At the top of the tube $r = 25$ mm and $y = 25$ mm, so the stresses are

$$\sigma_z = \frac{M_b y}{I_{xx}} = \frac{(360\text{ N}\bullet\text{m})(25\text{mm})}{131\times10^3\text{ mm}^4} = 68.7\text{ MPa}$$

$$\tau_{\theta z} = \frac{Tr}{I_z} = \frac{(405\text{ N}\bullet\text{m})(25\text{mm})}{262\times10^3\text{ mm}^4} = 38.6\text{ MPa}$$

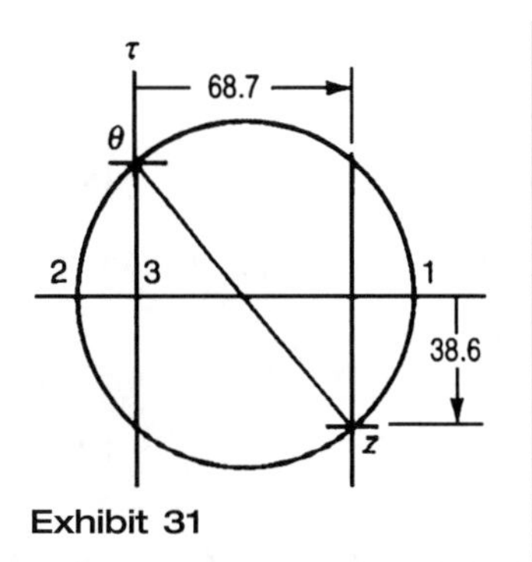

Exhibit 31

The Mohr's circle plot for this is shown in Exhibit 31.

$$R = \sqrt{\left(\frac{\sigma_z - \sigma_\theta}{2}\right)^2 + \tau_{\theta z}^2} = \sqrt{\left(\frac{68.7 - 0}{2}\right)^2 + 38.6^2} = 51.7\,\text{MPa}$$

$$C = \frac{\sigma_z \sigma_\theta}{2} = \frac{68.7}{2} = 34.4\,\text{MPa}$$

$$\sigma_1 = C + R = 86.1\,\text{MPa}$$

$$\sigma_2 = C - R = -17.3\,\text{MPa}$$

Because this is a state of plane stress, the third principal stress is

$$\sigma_3 = 0$$

COLUMNS

Buckling can occur in slender columns when they carry a high axial load. Fig. 9.17(a) shows a simply supported slender member with an axial load. The beam is shown in the horizontal position rather than in the vertical position for convenience. It is assumed that the member will deflect from its normally straight configuration as shown. The free-body diagram of the beam is shown in Fig. 9.17(b). Figure 9.17(c) shows the free-body diagram of a section of the beam. Summation of moments on the beam section in Fig. 9.17(c) yields

$$M + Py = 0 \tag{9.80}$$

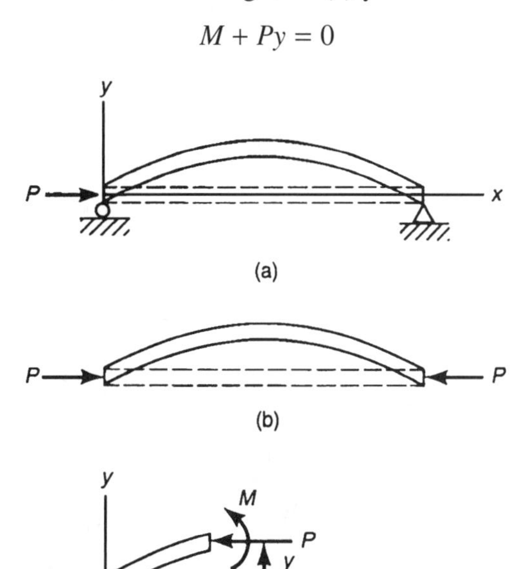

Figure 9.17 Buckling of simply supported column

Since M is equal to EI times the curvature, the equation for this beam can be expressed as

$$\frac{d^2y}{dx^2} + \lambda y = 0 \tag{9.81}$$

where

$$\lambda^2 = \frac{P}{EI} \tag{9.82}$$

The solution satisfying the boundary conditions that the displacement is zero at either end is

$$v = \sin(\lambda x), \text{ where } \lambda = n\pi/L \; n = 1, 2, 3\ldots \quad \textbf{(9.83)}$$

The lowest value for the load P is the buckling load, so $n = 1$ and the critical buckling load, or Euler buckling load, is

$$P_{\text{cr}} = \frac{\pi^2 EI}{L^2} \quad \textbf{(9.84)}$$

For other than simply supported boundary conditions, the shape of the deflected curve will always be some portion of a sine curve. The simplest shape consistent with the boundary conditions will be the deflected shape. Fig. 9.18 shows a sine curve and the beam lengths that can be selected from the sine curve. The critical buckling load can be redefined as

$$P_{\text{cr}} = \frac{\pi^2 EI}{L_e^2} = \frac{\pi^2 EI}{(kL)^2} = \frac{\pi^2 E}{(kL/r)^2} \quad \textbf{(9.85)}$$

where the radius of gyration r is defined as $\sqrt{I/A}$. The ratio L/r is called the slenderness ratio.

From Fig. 9.18, it can be seen that the values for L_e and k are as follows:

For simple supports: $L = L_e$; $L_e = L$; $k = 1$

For a cantilever: $L = 0.5L_e$; $L_e = 2L$; $k = 2$

For both ends clamped: $L = 2L_e$; $L_e = 0.5L$; $k = 0.5$

For supported-clamped: $L = 1.43L_e$; $L_e = 0.7L$; $k = 0.7$

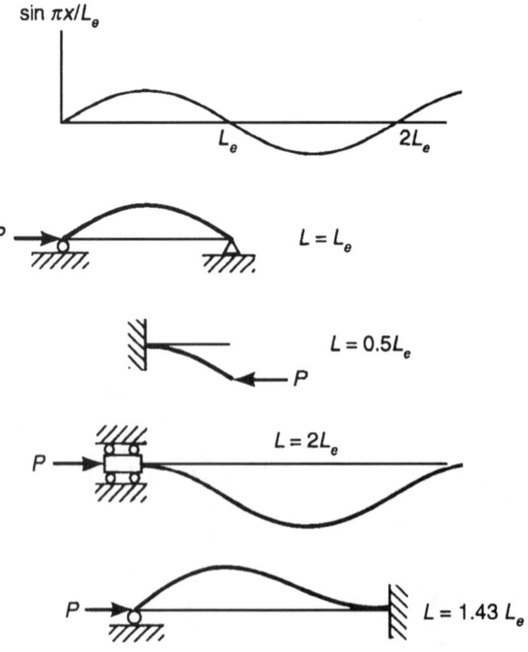

Figure 9.18 Buckling of columns with various boundary conditions

In dealing with buckling problems, keep in mind that the member must be slender before buckling is the mode of failure. If the beam is not slender, it will fail by yielding or crushing before buckling can take place.

Example 9.22

A steel pipe is to be used to support a weight of 130 kN as shown in Exhibit 32. The pipe has the following specifications: $OD = 100$ mm, $ID = 90$ mm, $A = 1500$ mm^2, and $I = 1.7 \times 10^6$ mm^4. Take $E = 210$ GPa and the yield stress $Y = 250$ MPa. Find the maximum length of the pipe.

Solution

First, check to make sure that the pipe won't yield under the applied weight. The stress is

$$\sigma = \frac{P}{A} = \frac{130\,\text{kN}}{1500\,\text{mm}^2} = 86.7\,\text{MPa} < Y$$

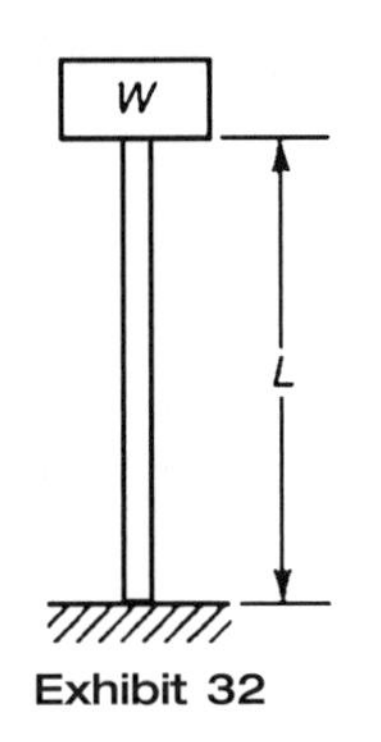

Exhibit 32

This stress is well below the yield, so buckling will be the governing mode of failure. This is a cantilever column, so the constant k is 2. The critical load is

$$P_{\text{cr}} = \frac{\pi^2 EI}{(2L)^2}$$

Solving for L gives

$$L = \pi\sqrt{\frac{EI}{4P}} = \pi\sqrt{\frac{(210\,\text{GPa})(1.7\times10^6\,\text{mm}^4)}{4(130\,\text{kN})}} = 2.60\,\text{m}$$

The maximum length is 2.6 m.

SELECTED SYMBOLS AND ABBREVIATIONS

Symbol or Abbreviation	Description
σ	stress
ε	strain
v	Poisson's ratio
kip	kilopound
E	modulus of elasticity
δ	deformation
W	weight
P	load
P, p	pressure
I	moment of inertia
τ	shear stress
T	torque
A	area
M	moment
V	shear
L	length
F	force

PROBLEMS

9.1 The stepped circular aluminum shaft in Exhibit 9.1 has two different diameters: 20 mm and 30 mm. Loads of 20 kN and 12 kN are applied at the end of the shaft and at the step. The maximum stress is most nearly

a. 23.4 MPa c. 28.3 MPa
b. 26.2 MPa d. 30.1 MPa

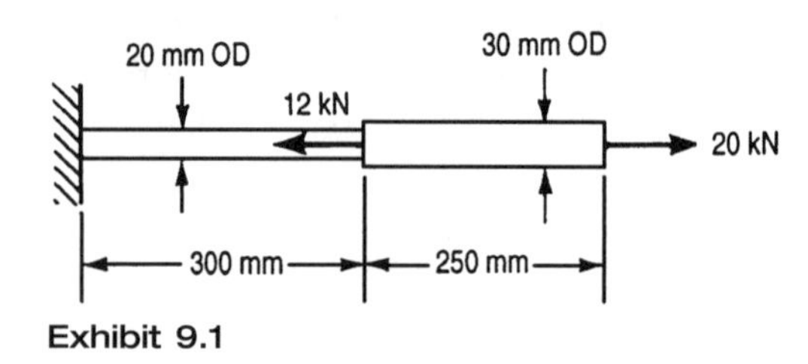

Exhibit 9.1

9.2 For the same shaft as in Problem 9.1 take $E = 69$ GPa. The end deflection is most nearly

a. 0.18 mm c. 0.35 mm
b. 0.21 mm d. 0.72 mm

9.3 The shaft in Exhibit 9.3 is the same aluminum stepped shaft considered in Problems 9.1 and 9.2, except now the right-hand end is also built into a wall. Assume that the member was built in before the load was applied. The maximum stress is most nearly

a. 12.2 MPa c. 13.1 MPa
b. 12.7 MPa d. 15.2 MPa

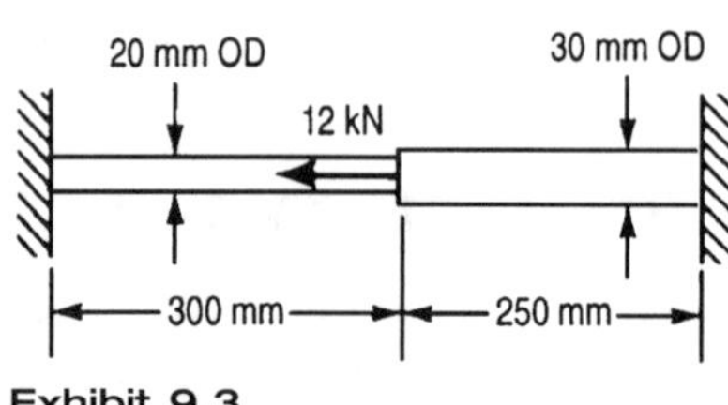
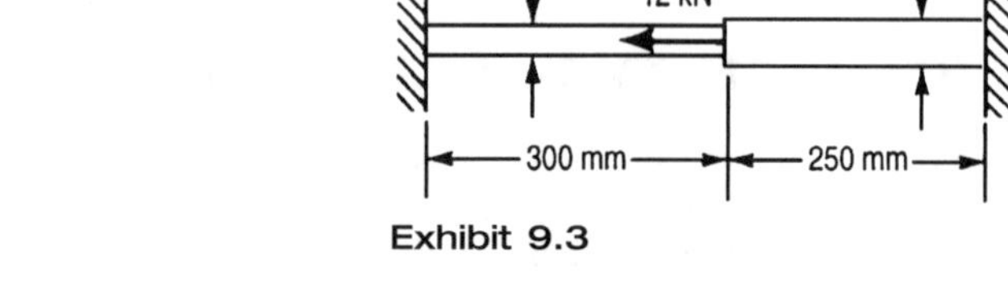

Exhibit 9.3

9.4 For the same shaft as in Problem 9.3 the deflection of the step is most nearly

a. 0.038 mm c. 0.064 mm
b. 0.042 mm d. 0.086 mm

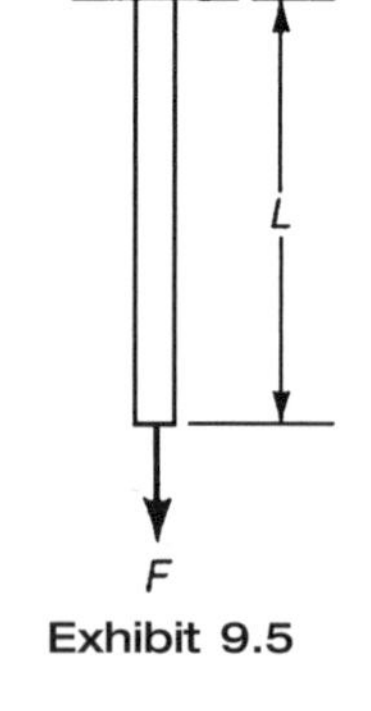

Exhibit 9.5

9.5 The uniform rod shown in Exhibit 9.5 has a force F at its end which is equal to the total weight of the rod. The rod has a unit weight γ. The total deflection of the rod is most nearly

a. $1.00\ \gamma L^2/E$ c. $1.50\ \gamma L^2/E$
b. $1.25\ \gamma L^2/E$ d. $1.75\ \gamma L^2/E$

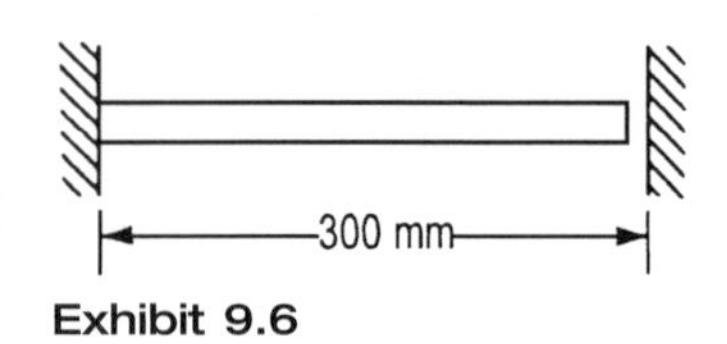

Exhibit 9.6

9.6 At room temperature, 22°C, a 300-mm stainless steel rod (Exhibit 9.6) has a gap of 0.15 mm between its end and a rigid wall. The modulus of elasticity $E = 210$ GPa. The coefficient of thermal expansion $\alpha = 17 \times 10^{-6}$/°C. The area of the rod is 650 mm^2. When the temperature is raised to 100 °C, the stress in the rod is most nearly

a. 175 MPa (tension) c. −17.5 MPa (compression)
b. 0 MPa d. −175 MPa (compression)

9.7 A steel cylindrical pressure vessel is subjected to a pressure of 21 MPa. Its outer diameter is 4.6 m, and its wall thickness is 200 mm. The maximum principal stress in this vessel is most nearly

a. 183 MPa
b. 221 MPa
c. 362 MPa
d. 432 MPa

9.8 A pressure vessel shown in Exhibit 9.8 is known to have an internal pressure of 1.4 MPa. The outer diameter of the vessel is 300 mm. The vessel is made of steel; $v = 0.3$ and $E = 210$ GPa. A strain gage in the circumferential direction on the vessel indicates that, under the given pressure, the strain is 200×10^{-6}. The wall thickness of the pressure vessel is most nearly

a. 3.2 mm
b. 4.3 mm
c. 6.4 mm
d. 7.8 mm

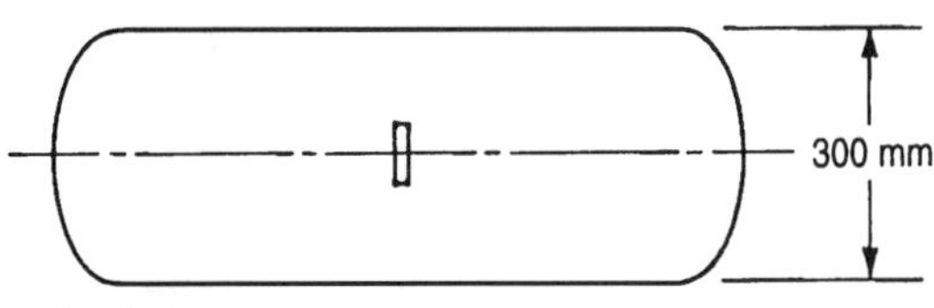

Exhibit 9.8

9.9 An aluminum pressure vessel has an internal pressure of 0.7 MPa. The vessel has an outer diameter of 200 mm and a wall thickness of 3 mm. Poisson's ratio is 0.33 and the modulus of elasticity is 69 GPa for this material. A strain gage is attached to the outside of the vessel at 45° to the longitudinal axis as shown in Exhibit 9.9. The strain on the gage would read most nearly

a. 40×10^{-6}
b. 60×10^{-6}
c. 80×10^{-6}
d. 160×10^{-6}

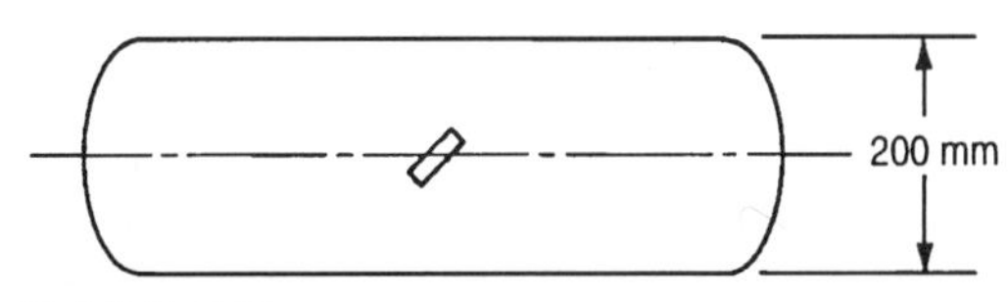

Exhibit 9.9

9.10 If $\sigma_x = -3$ MPa, $\sigma_y = 5$ MPa, and $\tau_{xy} = -3$ MPa, the maximum principal stress is most nearly

a. 4 MPa
b. 5 MPa
c. 6 MPa
d. 7 MPa

9.11 Given that $\sigma_x = 5$ MPa, $\sigma_y = -1$ MPa, and the maximum principal stress is 7 MPa, the shear stress τ_{xy} is most nearly

a. 1 MPa
b. 2 MPa
c. 3 MPa
d. 4 MPa

9.12 Given $\varepsilon_x = 800\ \mu$, $\varepsilon_y = 200\ \mu$, and $\gamma_{xy} = 400\ \mu$, the maximum principal strain is most nearly

a. 840 μ
b. 860 μ
c. 900 μ
d. 960 μ

9.13 A steel plate in a state of plane stress has the same strains as in Problem 9.12: $\varepsilon_x = 800\ \mu$, $\varepsilon_y = 200\ \mu$, and $\gamma_{xy} = 400\ \mu$. Poisson's ratio $v = 0.3$ and the modulus of elasticity $E = 210$ GPa. The maximum principal stress in the plane is most nearly

a. 109 MPa
b. 132 MPa
c. 173 MPa
d. 208 MPa

9.14 A stepped steel shaft shown in Exhibit 9.14 has torques of 10 kN • m applied at the end and at the step. The maximum shear stress in the shaft is most nearly

a. 760 MPa
b. 810 MPa
c. 870 MPa
d. 930 MPa

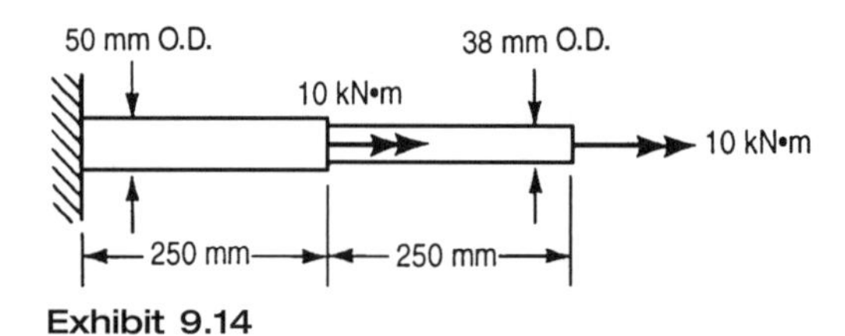

Exhibit 9.14

9.15 The shear modulus for steel is 83 MPa. For the same shaft as in Problem 9.14, the rotation at the end of the shaft is most nearly

a. 0.014°
b. 0.14°
c. 1.4°
d. 14°

9.16 The same stepped shaft as in Problems 9.14 and 9.15 is now built into a wall at its right end before the load is applied (Exhibit 9.16). The maximum stress in the shaft is most nearly

a. 130 MPa
b. 200 MPa
c. 230 MPa
d. 300 MPa

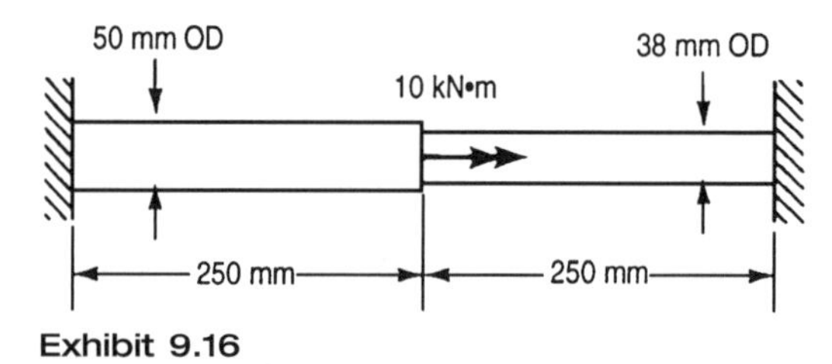

Exhibit 9.16

9.17 For the same shaft as in Problem 9.16, the rotation of the step is most nearly

a. 0.2°
b. 1.1°
c. 1.8°
d. 2.1°

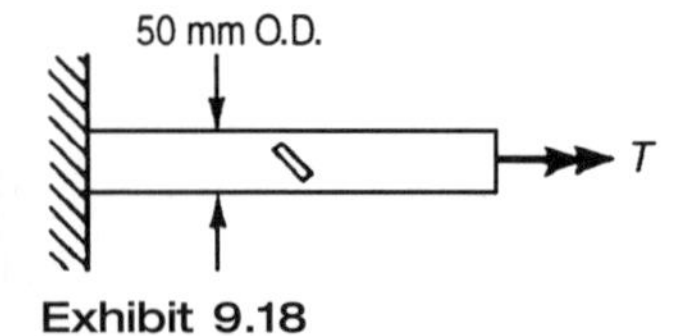

Exhibit 9.18

9.18 A strain gage shown in Exhibit 9.18 is placed on a circular steel shaft which is being twisted with a torque T. The gage is inclined 45° to the axis. If the strain reads $\varepsilon_{45} = 245\ \mu$, the torque is most nearly

a. 1000 N•m
b. 1230 N•m
c. 1570 N•m
d. 2635 N•m

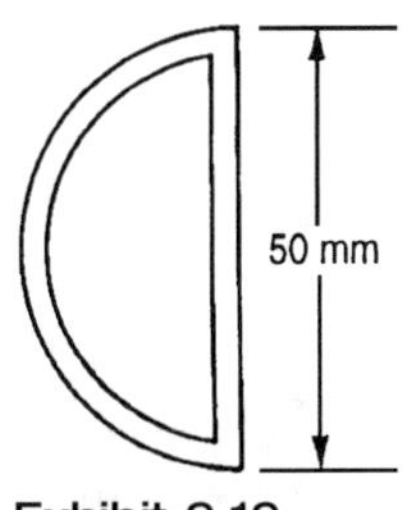

Exhibit 9.19

9.19 A shaft whose cross section is in the shape of a semicircle is shown in Exhibit 9.19 and has a constant wall thickness of 3 mm. The shaft carries a torque of 300 N • m. Neglecting any stress concentrations at the corners, the maximum shear stress in the shaft is most nearly

a. 32 MPa
b. 48 MPa
c. 59 MPa
d. 66 MPa

9.20 The maximum magnitude of shear in the beam shown in Exhibit 9.20 is most nearly

a. 40 kN
b. 50 kN
c. 60 kN
d. 75 kN

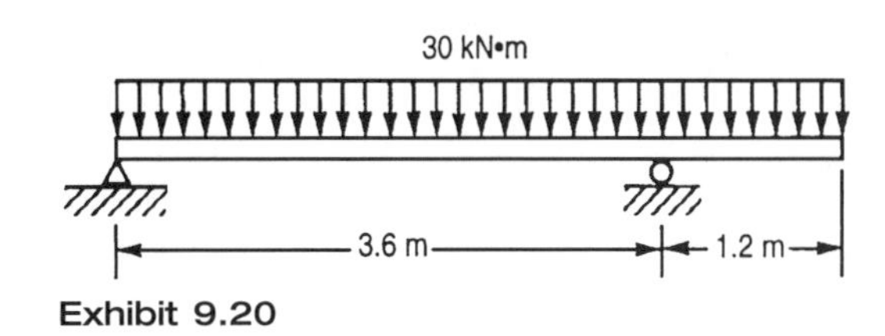

Exhibit 9.20

9.21 For the same beam as in Problem 9.20, the magnitude of the largest bending moment is most nearly

a. 21.0 kN • m
b. 26.3 kN • m
c. 38.4 kN • m
d. 42.1 kN • m

9.22 The shear diagram shown in Exhibit 9.22 is for a beam that has zero moments at either end. The maximum concentrated force on the beam is most nearly

a. 60 kN upward
b. 30 kN upward
c. 0
d. 30 kN downward

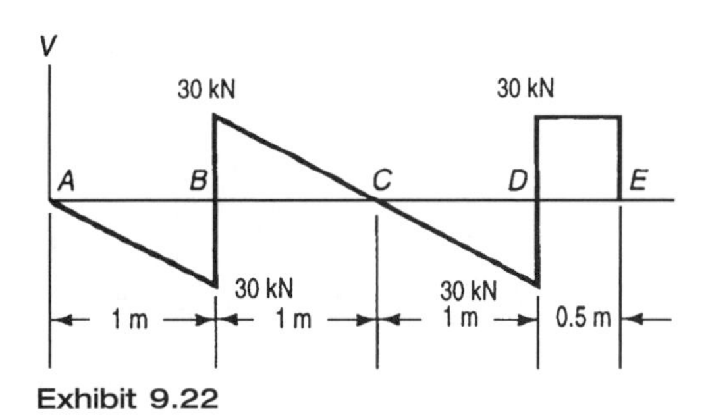

Exhibit 9.22

9.23 For the same beam as in Problem 9.22 the largest magnitude of the bending moment is most nearly

a. 0
b. 8 kN • m
c. 12 kN • m
d. 15 kN • m

9.24 The 4-m long, simply supported beam shown in Exhibit 9.24 has a section modulus $Z = 1408 \times 10^3$ mm^3. The allowable stress in the beam is not to exceed 100 MPa. The maximum load, w (including its own weight), that the beam can carry is most nearly:

a. 50 kN • m
b. 40 kN • m
c. 60 kN • m
d. 70 kN • m

Exhibit 9.24

9.25 The standard wide flange beam shown in Exhibit 9.25 has a moment of inertia about the z axis of $I = 365 \times 10^6\ \text{mm}^4$. The maximum bending stress is most nearly

a. 4.5 MPa c. 6.5 MPa
b. 5.0 MPa d. 8 MPa

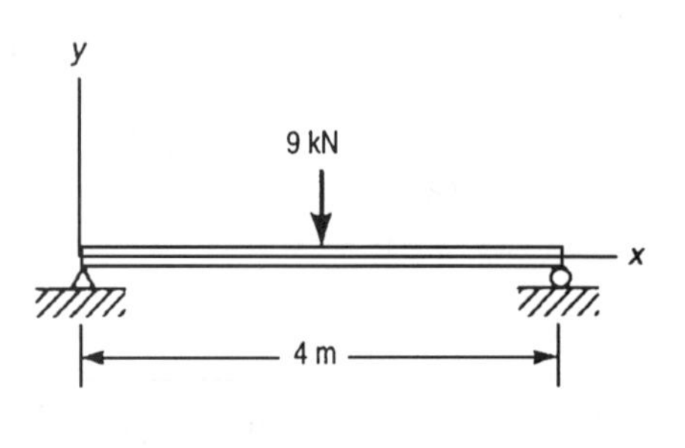

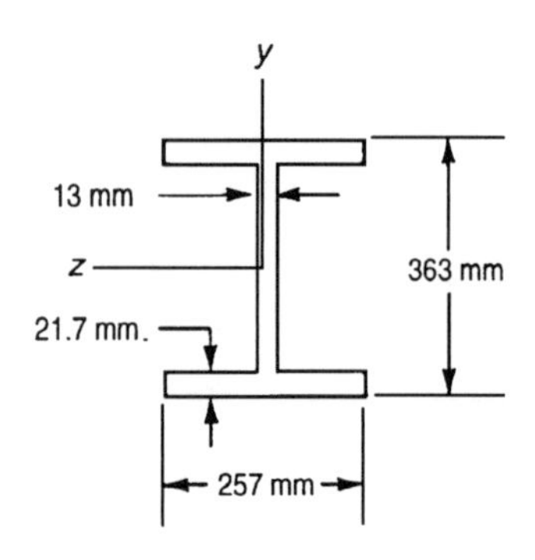

Exhibit 9.25

9.26 For the same beam as in Problem 9.25, the maximum shear stress τ_{xy} in the web is most nearly

a. 1 MPa c. 2.0 MPa
b. 1.5 MPa d. 2.5 MPa

9.27 The deflection at the end of the beam shown in Exhibit 9.27 is most nearly

a. $0.330\ FL^3/EI$ (downward) c. $0.410\ FL^3/EI$ (downward)
b. $0.380\ FL^3/EI$ (downward) d. $0.440\ FL^3/EI$ (downward)

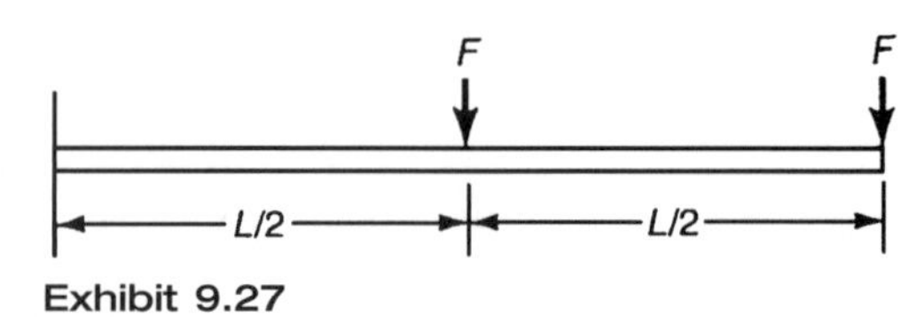

Exhibit 9.27

9.28 A uniformly loaded beam (Exhibit 9.28) has a concentrated load wL at its center that has the same magnitude as the total distributed load w. The maximum deflection of this beam is most nearly

a. $0.029\ wL^4/EI$ (downward) c. $0.043\ wL^4/EI$ (downward)
b. $0.034\ wL^4/EI$ (downward) d. $0.056\ wL^4/EI$ (downward)

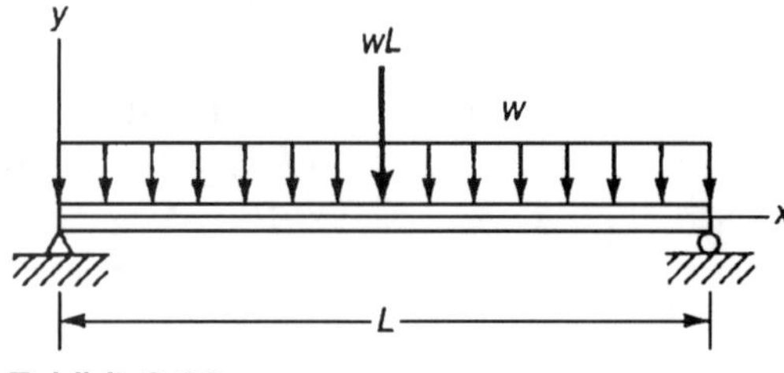

Exhibit 9.28

9.29 The reaction at the center support of the uniformly loaded beam shown in Exhibit 9.29 is most nearly

a. $0.525wL$ c. $0.575\,wL$
b. $0.550wL$ d. $0.625\,wL$

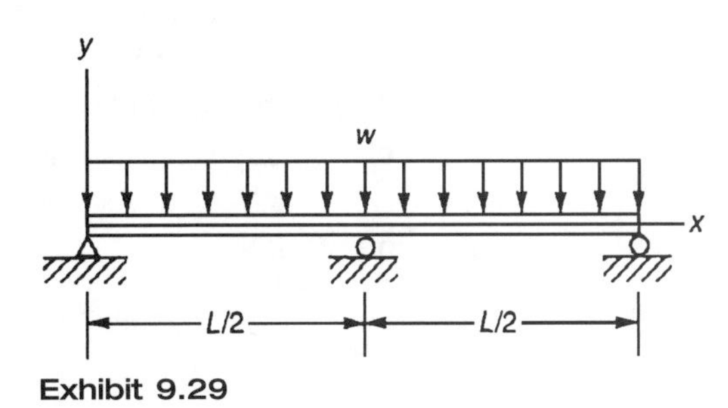

Exhibit 9.29

9.30 A solid circular rod has a diameter of 25 mm (Exhibit 9.30). It is fixed into a wall at A and bent 90° at B. The maximum bending stress in the section BC is most nearly

a. 21.7 MPa c. 32.6 MPa
b. 29.3 MPa d. 45.7 MPa

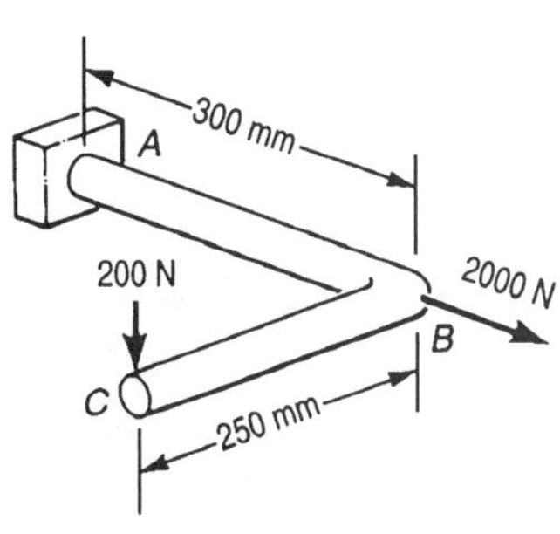

Exhibit 9.30

9.31 For the same member as in Problem 9.30 the maximum bending stress in the section AB is most nearly

a. 21 MPa c. 31 MPa
b. 25 MPa d. 39 MPa

9.32 For the same member as in Problem 9.30 the maximum shear stress due to torsion in the section AB is most nearly

a. 15.2 MPa c. 17.4 MPa
b. 16.3 MPa d. 18.5 MPa

9.33 For the same member as in Problem 9.30, the maximum stress due to the axial force in the section AB is most nearly

a. 4 MPa c. 6 MPa
b. 5 MPa d. 8 MPa

9.34 For the same member as in Problem 9.30, the maximum principal stress in the section *AB* is most nearly

a. 17 MPa
b. 27 MPa
c. 39 MPa
d. 44 MPa

9.35 A truss is supported so that it can't move out of the plane (Exhibit 9.35). All members are steel and have a square cross section 25 mm by 25 mm. The modulus of elasticity for steel is 210 GPa. The maximum load *P* that can be supported without any buckling is most nearly

a. 14 kN
b. 25 kN
c. 34 kN
d. 51 kN

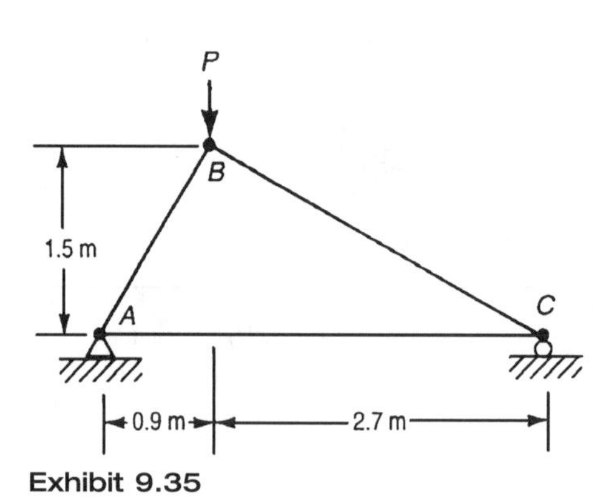

Exhibit 9.35

9.36 A beam is pinned at both ends (Exhibit 9.36). In the *x*-*y* plane it can rotate about the pins, but in the *x*-*z* plane the pins constrain the end rotation. In order to have buckling equally likely in each plane, the ratio *b*/*a* is most nearly

a. 0.5
b. 1.0
c. 1.5
d. 2.0

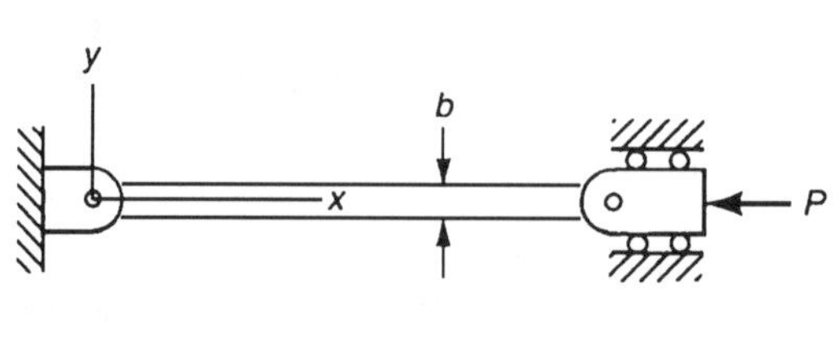

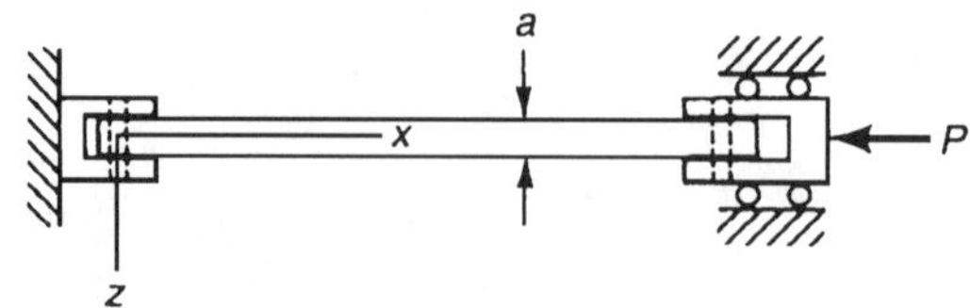

Exhibit 9.36

SOLUTIONS

9.1 c. Draw free-body diagrams. Equilibrium of the center free-body diagram gives

$$F_1 = 20 - 12 = 8 \text{ kN}$$

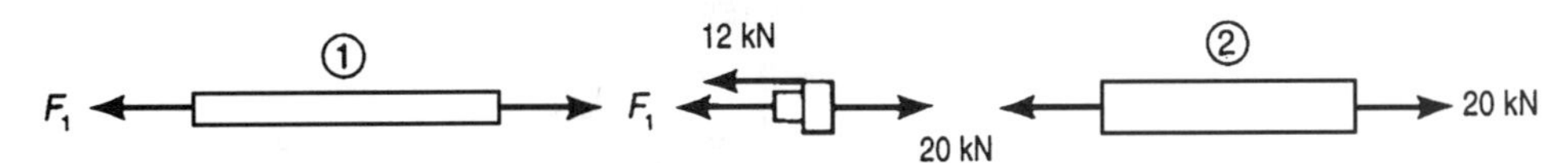

Exhibit 9.1a

The areas are

$$A_1 = \pi r^2 = \pi(10 \text{ mm})^2 = 314 \text{ mm}^2$$
$$A_2 = \pi r^2 = \pi(15 \text{ mm})^2 = 707 \text{ mm}^2$$

The stresses are

$$\sigma_1 = \frac{P}{A} = \frac{8\,\text{kN}}{314\,\text{mm}^2} = 25.5\,\text{MPa}$$
$$\sigma_2 = \frac{P}{A} = \frac{20\,\text{kN}}{707\,\text{mm}^2} = 28.3\,\text{MPa}$$

9.2 b. The force-deformation equations give

$$\delta_1 = \frac{P_1 L_1}{A_1 E_1} = \frac{(8\,\text{kN})(300\,\text{mm})}{(314\text{ mm}^2)(69\,\text{GPa})} = 0.1107\,\text{mm}$$
$$\delta_2 = \frac{P_2 L_2}{A_2 E_2} = \frac{(20\,\text{kN})(250\,\text{mm})}{(707\,\text{mm}^2)(69\,\text{GPa})} = 0.1025\,\text{mm}$$

Compatibility of deformation gives

$$\delta_{\text{end}} = \delta_1 + \delta_2 = 0.1107 + 0.1025 = 0.213\,\text{mm}$$

9.3 c. Draw the free-body diagrams. From the center free-body diagram, summation of forces yields

$$F_2 = 12 \text{ kN} + F_1$$

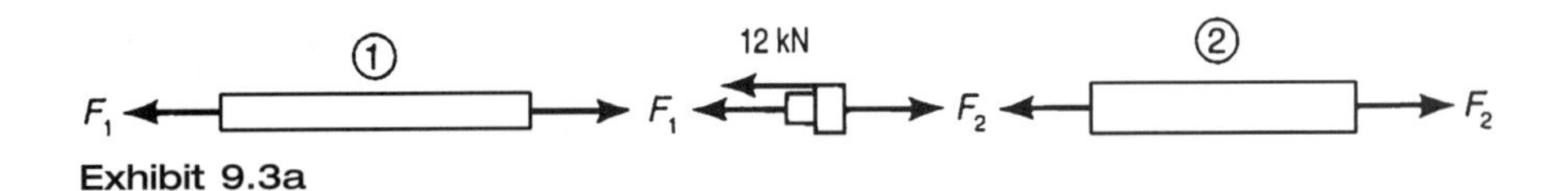

Exhibit 9.3a

Force-deformation relations are

$$\delta_1 = \frac{P_1 L_1}{A_1 E_1} = \frac{(F_1\,\text{kN})(300\,\text{mm})}{(314\,\text{mm}^2)(69\,\text{GPa})} = 0.01384\,F_1$$
$$\delta_2 = \frac{P_2 L_2}{A_2 E_2} = \frac{(F_2\,\text{kN})(200\,\text{mm})}{(707\,\text{mm}^2)(69\,\text{GPa})} = 0.00410\,F_2$$

Compatibility gives

$$\delta_{\text{end}} = 0 = \delta_1 + \delta_2 = 0.01384\,F_1 + 0.00410\,F_2$$

Substitution of the equilibrium relation, $F_2 = 12 + F_1$, into the above equation gives

$$0 = 0.01384\ F_1 + 0.00410\ (12 + F_1)$$
$$F_1 = -2.74\ \text{kN},\ F_2 = 9.26\ \text{kN}$$

The stresses are

$$\sigma_1 = \frac{P}{A} = \frac{(-2.74\,\text{kN})}{(314\,\text{mm}^2)} = -8.73\,\text{MPa}; \qquad \sigma_2 = \frac{P}{A} = \frac{(9.26\,\text{kN})}{(707\,\text{mm}^2)} = 13.10\,\text{MPa}$$

9.4 a. The same three-step process as in Problem 9.3 must be carried out. Since this process has already been completed, the results can be used. The deflection can be expressed as

$$\delta = \delta_1 = -\delta_2 = 0.01384\ F_1 = 0.01384\ (-2.74) = -0.0380\ \text{mm}$$

9.5 c. Draw a free-body diagram. Summation of forces in the vertical direction gives

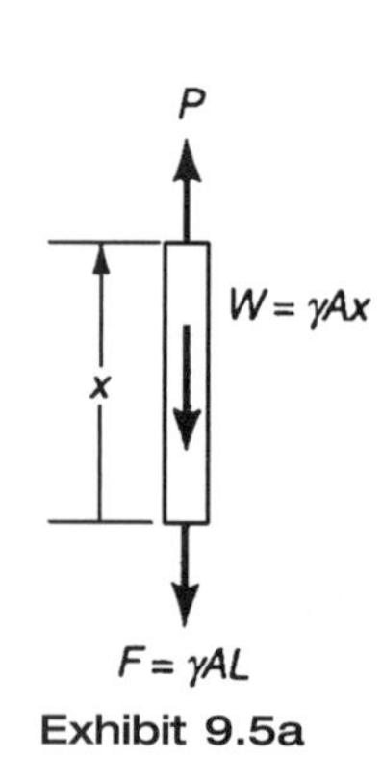

Exhibit 9.5a

$$P = \gamma AL + \gamma Ax$$
$$\delta = \int_0^L \frac{P}{AE}\,dx = \int_0^L \frac{\gamma A(L+x)}{AE}\,dx = \frac{\gamma}{E}\left(L^2 + \frac{L^2}{2}\right) = \frac{3\gamma L^2}{2E}$$

9.6 d. A force will develop in the rod if it attempts to grow more than 0.15 mm. Assuming that it does grow that amount, the displacement is

$$\delta = \frac{PL}{AE} + \alpha L(t - t_o) = 0.15\,\text{mm}$$
$$= \frac{P(300\,\text{mm})}{(650\,\text{mm})(210\,\text{GPa})} + \left(17\times10^{-6}\frac{1}{°\text{C}}\right)(300\,\text{mm})(100\,°\text{C} - 22\,°\text{C})$$
$$0.15\ \text{mm} = 0.00220\,P + 0.3978$$
$$P = -112.7\ \text{kN}$$

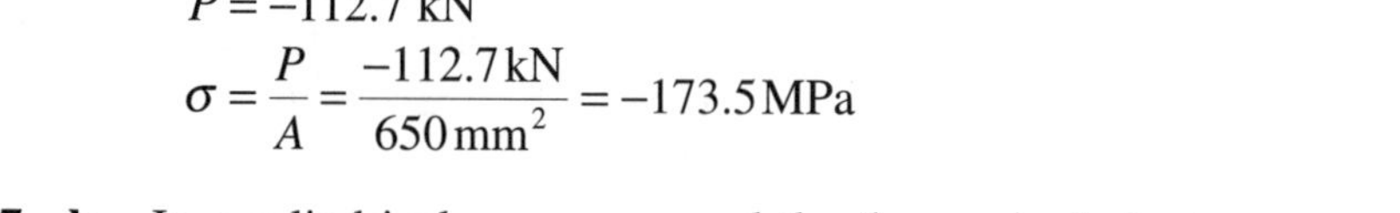

$$\sigma = \frac{P}{A} = \frac{-112.7\,\text{kN}}{650\,\text{mm}^2} = -173.5\,\text{MPa}$$

9.7 b. In a cylindrical pressure vessel the three principal stresses are

$$\sigma_t = qD/2t; \quad \sigma_a = qD/4t; \quad \sigma_r \approx 0$$

The maximum is σ_t, which gives

$$\sigma_t = \frac{qD}{2t} = \frac{(21\,\text{MPa})(4600\,\text{mm} - 400\,\text{mm})}{2(200\,\text{mm})} = 221\,\text{MPa}$$

9.8 b. The stresses in the pressure vessel are, as in the last problem,

$$\sigma_t = \frac{qD}{2t}; \qquad \sigma_a = \frac{qD}{4t}; \qquad \sigma_r \approx 0$$

The wall thickness is usually thin enough so that it can be assumed that

$$D_i \approx D_o$$

The tangential strain can be found from Hooke's law:

$$\varepsilon_t = \frac{1}{E}(\sigma_t - v\sigma_\theta - v\sigma_r) = \frac{1}{E}\left(\frac{qD}{2t} - v\frac{qD}{4t} - v0\right) = 0.425\frac{qD}{Et}$$

The thickness is then

$$t = 0.425\frac{pD}{E\varepsilon_t} = 0.425\frac{(1.4\,\text{MPa})(300\,\text{mm})}{(210\,\text{GPa})(200\times10^{-6})} = 4.25\,\text{mm}$$

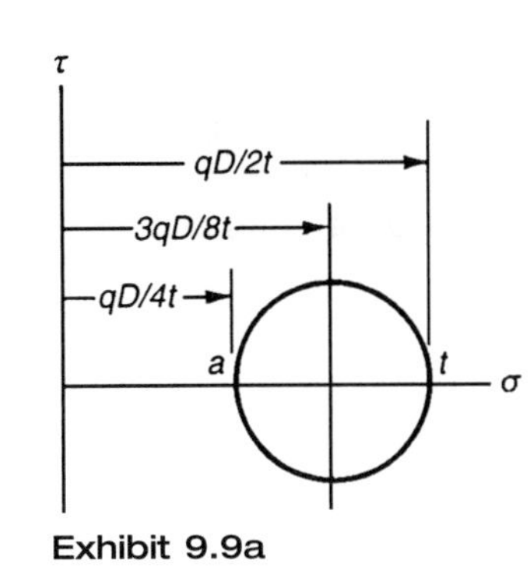

Exhibit 9.9a

9.9 d. Draw Mohr's circle for the stress. At 45° in the physical plane (90° on Mohr's circle) the two normal stresses are $3qD/8t$. Hooke's law gives

$$\varepsilon_{45} = \frac{1}{E}(\sigma_{45} - v\sigma_{-45} - v\sigma_r) = \frac{1}{E}\left(\frac{3qD}{8t} - v\frac{3qd}{8t} - v0\right)$$

$$= \frac{(1-v)}{E}\left(\frac{3qD}{8t}\right)$$

$$\varepsilon_{45} = \frac{(1-0.33)}{(69\,\text{GPa})}\frac{3(0.7\,\text{MPa})(200\,\text{mm} - 6\,\text{mm})}{8(3\,\text{mm})} = 164.8\times10^{-6}$$

9.10 c. Draw Mohr's circle. The maximum principal stress is 6 MPa. As an alternative,

$$R = \sqrt{\left(\frac{\sigma_x - \sigma_y}{2}\right)^2 + \tau_{xy}^2} = \sqrt{\left(\frac{-3-5}{2}\right)^2 + (-3)^2} = 5$$

$$C = \frac{\sigma_x + \sigma_y}{2} = \frac{-3+5}{2} = 1$$

$$\sigma_1 = R + C = 6\ \text{MPa}$$

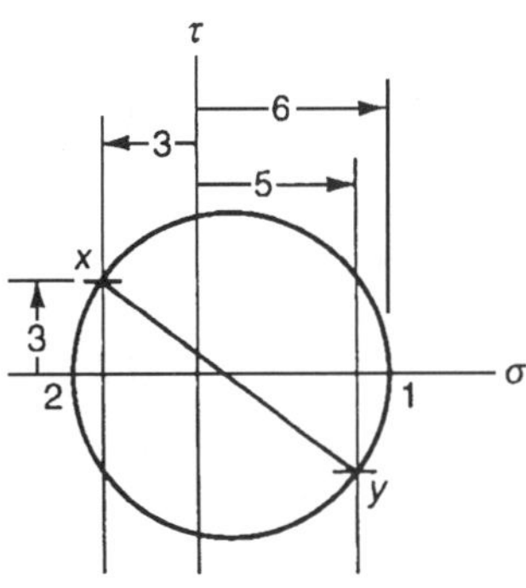

Exhibit 9.10a

9.11 d. Draw Mohr's circle. The center of the circle is

$$C = \frac{\sigma_x + \sigma_y}{2} = \frac{5-1}{2} = 2$$

The radius is then $R = 7 - 2 = 5$. The shear stress can be found from the Mohr's circle or from the expression

$$R = \sqrt{\left(\frac{\sigma_x - \sigma_y}{2}\right)^2 + \tau_{xy}^2}; \qquad \tau_{xy}^2 = R^2 - \left(\frac{\sigma_x - \sigma_y}{2}\right)^2$$

Exhibit 9.11a

In either case the shear stress $\tau_{xy} = 4$ MPa.

Exhibit 9.12a

9.12 b. Draw Mohr's circle. ε_1 can be scaled from the circle or computed as follows,

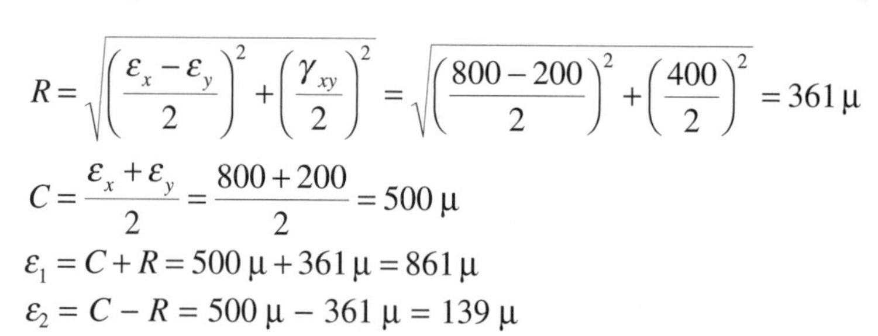

$$R = \sqrt{\left(\frac{\varepsilon_x - \varepsilon_y}{2}\right)^2 + \left(\frac{\gamma_{xy}}{2}\right)^2} = \sqrt{\left(\frac{800 - 200}{2}\right)^2 + \left(\frac{400}{2}\right)^2} = 361\,\mu$$

$$C = \frac{\varepsilon_x + \varepsilon_y}{2} = \frac{800 + 200}{2} = 500\,\mu$$

$$\varepsilon_1 = C + R = 500\,\mu + 361\,\mu = 861\,\mu$$

$$\varepsilon_2 = C - R = 500\,\mu - 361\,\mu = 139\,\mu$$

ε_1 is the maximum.

9.13 d. Problems of this type can be done by using Hooke's law first and then Mohr's circle or by using Mohr's circle first and then applying Hooke's law. Since Mohr's circle was already drawn for this problem in the previous solution, the second approach will be followed. The principal strains were found to be the following: $\varepsilon_1 = 861\ \mu$; $\varepsilon_2 = 139\ \mu$. Hooke's law in plane stress is

$$\varepsilon_1 = \frac{1}{E}(\sigma_1 - v\sigma_2); \qquad \varepsilon_2 = \frac{1}{E}(\sigma_2 - v\sigma_1)$$

Inverting these relationships gives

$$\sigma_1 = \frac{E}{1 - v^2}(\varepsilon_1 + v\varepsilon_2); \qquad \sigma_2 = \frac{E}{1 - v^2}(\varepsilon_2 + v\varepsilon_1)$$

The maximum principal stress is σ_1, which is

$$\sigma_1 = \frac{E}{1 - v^2}(\varepsilon_1 + v\varepsilon_2) = \frac{210\,\text{GPa}}{1 - 0.3^2}[861 \times 10^{-6} + 0.3(139 \times 10^{-6})] = 208\ \text{MPa}$$

9.14 d. Draw free-body diagrams. The torque in shaft 1 is $T_1 = 10 + 10 = 20$ kN • m. The torque in shaft 2 is $T_2 = 10$ kN • m.

$$\tau_1 = \frac{T_1 r_1}{J_1} = \frac{(20\,\text{kN} \bullet \text{m})(25\,\text{mm})}{0.5\pi(25\,\text{mm})^4} = 815\ \text{MPa}$$

$$\tau_2 = \frac{T_2 r_2}{J_2} = \frac{(10\,\text{kN} \bullet \text{m})(19\,\text{mm})}{0.5\pi(19\,\text{mm})^4} = 928\ \text{MPa}$$

The largest stress is 928 MPa.

Exhibit 9.14a

9.15 d. From the force-deformation relations,

$$\phi_1 = \frac{T_1L_1}{GJ_1} = \frac{(20\,\text{kN} \bullet \text{m})(250\,\text{mm})}{(83\,\text{GPa})0.5\pi(25\,\text{mm})^4} = 0.982\,\text{rad} = 5.63°$$

$$\phi_2 = \frac{T_2L_2}{GJ_2} = \frac{(10\,\text{kN} \bullet \text{m})(250\,\text{mm})}{(83\,\text{GPa})0.5\pi(19\,\text{mm})^4} = 0.1471\,\text{rad} = 8.43°$$

From compatibility,

$$\phi = \phi_1 + \phi_2 = 5.63° + 8.43° = 14.06°$$

9.16 d. Draw the free-body diagrams. Equilibrium of the center free body gives

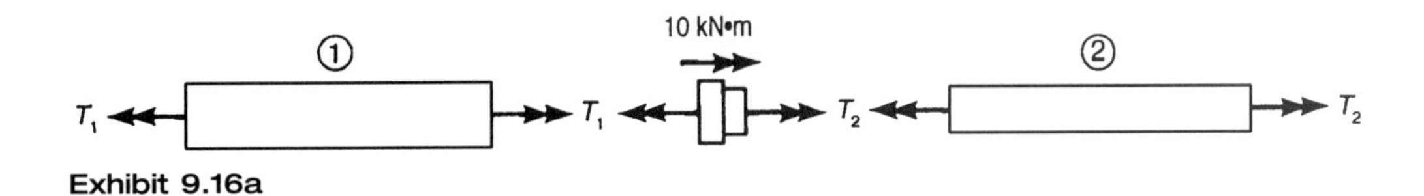

Exhibit 9.16a

$$T_1 = T_2 + 10$$

The force-deformation relations are

$$\phi_1 = \frac{T_1L_1}{GJ_1} = \frac{(10+T_2)(250\,\text{mm})}{(83\,\text{GPa})0.5\pi(25\,\text{mm})^4} = 49.1\times10^{-3} + 4.91\times10^{-3}T_2$$

$$\phi_2 = \frac{T_2r_2}{J_2} = \frac{T_2(250\,\text{mm})}{(83\,\text{GPa})\ 0.5\pi(19\,\text{mm})^4} = 14.7\times10^{-3}\,T_2$$

Compatibility requires that

$$\phi_1 + \phi_2 = 0 = 49.1 \times 10^{-3} + (4.91 \times 10^{-3} + 14.7 \times 10^{-3})T_2$$

Solving for the torques gives

$$T_2 = \frac{5.305}{2.207} = -2.50\,\text{kN} \bullet \text{m}$$

$$T_1 = T_2 + 10 = -2.50 + 10 = 7.50\,\text{kN} \bullet \text{m}$$

The stresses then are

$$\tau_1 = \frac{T_1r_1}{J_1} = \frac{(7.50\,\text{kN} \bullet \text{m})(25\,\text{mm})}{0.5\pi(25\,\text{mm})^4} = 306\,\text{MPa}$$

$$\tau_2 = \frac{T_2r_2}{J_2} = \frac{(-2.5\,\text{kN} \bullet \text{m})(19\,\text{mm})}{0.5\pi(19\,\text{mm})^4} = -232\,\text{MPa}$$

9.17 d. The same three-step process as in Problem 9.16 must be carried out. Since this process has already been completed, the results can be used. The rotation can be expressed as

$$\phi = \phi_1 = -\phi_2 = \frac{T_1 L_1}{G J_1} = \frac{(7.50\text{kN} \bullet \text{m})(250\,\text{mm})}{(83\,\text{GPa})0.5\pi(25\,\text{mm})^4} = 0.0368 \text{ rad} = 2.11°$$

9.18 a. For a torsion problem, the shear strain is

$$\gamma_{\phi z} = \frac{\tau_{\phi z}}{G} = \frac{Tr}{GJ}$$

Other shear strains in the $r - \phi$ orientation are zero. Mohr's circle for this state of strain is shown in Exhibit 9.18(a). From Mohr's circle,

$$\varepsilon_{45} = \frac{\gamma_{\phi z}}{2} = \frac{\tau_{\phi z}}{2G} = \frac{Tr}{2GJ}$$

$$T = \frac{2GJ\varepsilon_{45}}{r} = \frac{2(83\,\text{GPa})[0.5\pi(25\,\text{mm})^4](245 \times 10^{-6})}{25\,\text{mm}} = 998\,\text{N} \bullet \text{m}$$

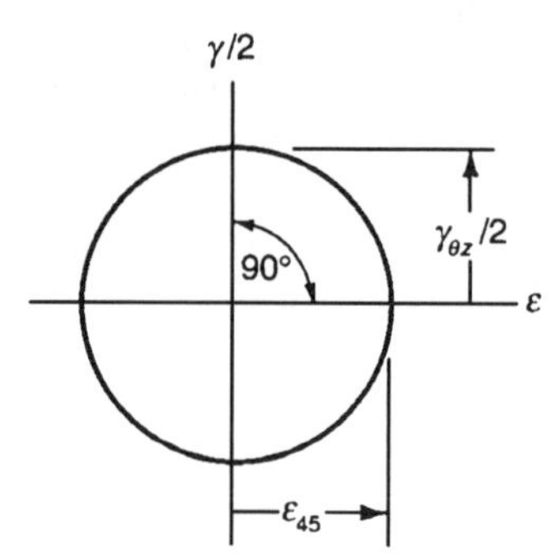

Exhibit 9.18a

9.19 d. In thin-walled shafts the shear stress is

$$\tau_{sz} = \frac{T}{2At}$$

The area A is the cross-sectional area of the shaft including the hole so,

$$A = \frac{\pi r^2}{2} = \frac{\pi(25\,\text{mm} - 3\,\text{mm})^2}{2} = 760 \text{ mm}^2$$

$$\tau_{sz} = \frac{T}{2At} = \frac{300\,\text{N} \bullet \text{m}}{2(760\,\text{mm}^2)(3\,\text{mm})} = 65.7 \text{ MPa}$$

9.20 c. Draw the free-body diagram of the beam, replacing the distributed load with its statically equivalent loads. Summation of moments about the left end gives

$$0 = -3.6\,R_2 + (108)(1.8) + (36)(4.2)$$
$$R_2 = 96 \text{ kN}$$

Summation of forces in the vertical direction gives

$$R_1 = 144 - 96 = 48 \text{ kN}$$

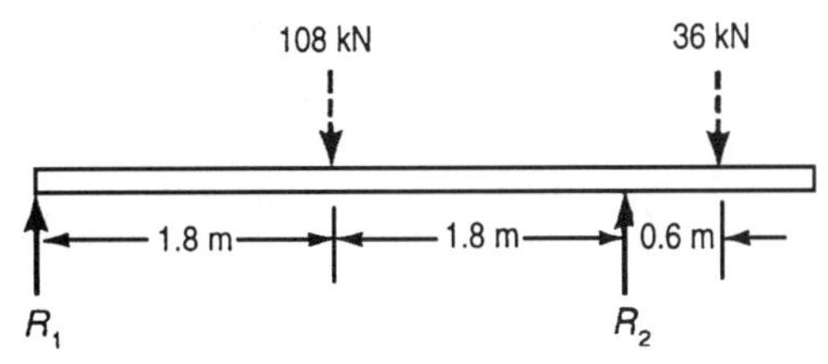

Exhibit 9.20a

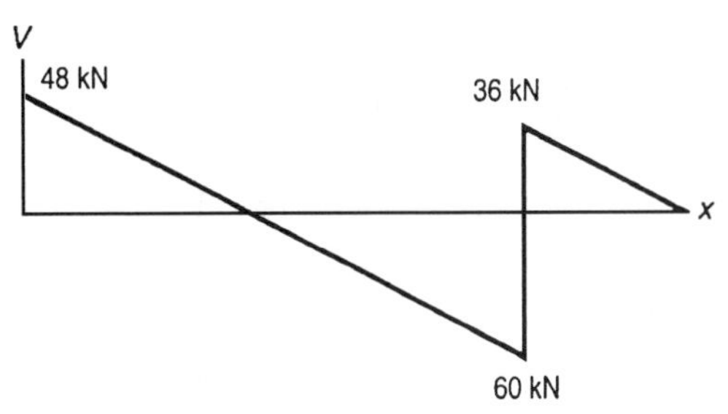

Exhibit 9.20b

This is enough information to plot the shear diagram (Exhibit 9.20b). The largest magnitude of shear is 60 kN.

9.21 c. The maximum bending moment occurs where the shear is zero. From the shear diagram, the distance to the zero from the left end can be found by similar triangles.

$$\frac{48}{x} = \frac{108}{3.6}; \qquad x = \frac{(48)(3.6)}{108} = 1.6\,\text{m}$$

The areas of the shear diagrams are the changes in moment (Exhibit 9.21a).

$$A_1 = \frac{(48\,\text{kN})(1.6\,\text{m})}{2} = 38.4\,\text{kN} \bullet \text{m}$$

$$A_2 = \frac{(36\,\text{kN})(1.2\,\text{m})}{2} = 21.6\,\text{kN} \bullet \text{m}$$

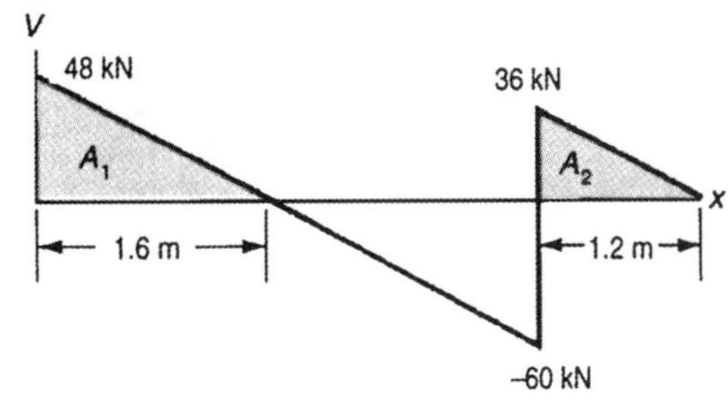

Exhibit 9.21a

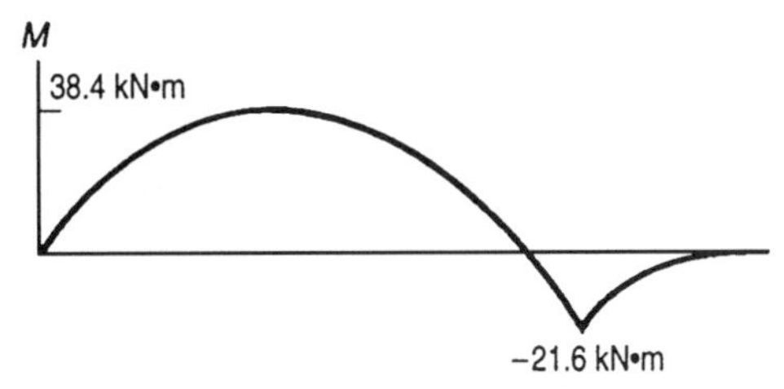

Exhibit 9.21b

The moment diagram is shown in Exhibit 9.21b. The maximum bending moment is 38.4 kN • m.

9.22 a. There is a jump at *B* and *D* of 60 kN upward and a downward jump of 30kN at *E*. These jumps correspond to concentrated forces.

9.23 d. The areas of the shear diagrams (Exhibit 9.23) are the changes in moment. Since the moments are zero on either end,

$$M_B = A_1 = \frac{(30\,\text{kN})(1\,\text{m})}{2} = 15\ \text{kN} \bullet \text{m}$$

$$M_C = A_1 + A_2 = \frac{(30\,\text{kN})(1\,\text{m})}{2} + \frac{(30\,\text{kN})(1\,\text{m})}{2} = 0$$

$$\text{M}_D = A_3 = (30\,\text{kN})(-0.5\,\text{m}) = -15\ \text{kN} \bullet \text{m}$$

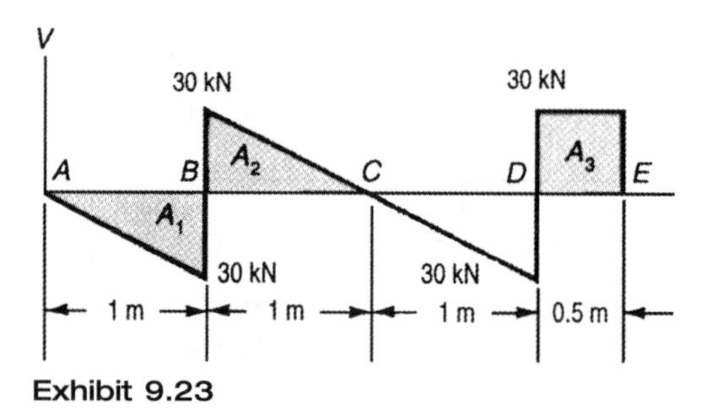

Exhibit 9.23

The largest magnitude of the bending moment is therefore 15 kN • m.

9.24 d. It is obvious that each support will carry half of the load, so the reactions are *wL*/2. The shear diagram is shown in Exhibit 9.24a. The maximum bending moment is

$$M = A_1 = \frac{1}{2}\left(\frac{wL}{2}\right)\left(\frac{L}{2}\right) = \frac{wL^2}{8}$$

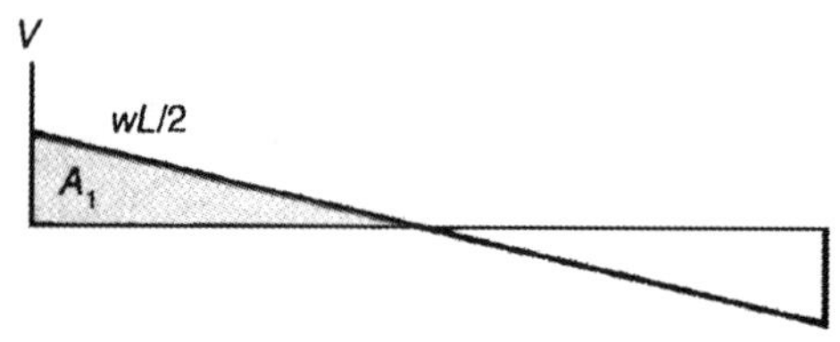

Exhibit 9.24a

The maximum bending stress is

$$\sigma_{\text{max}} = \frac{M}{Z} = \frac{wL^2}{8Z} = 100\,\text{MPa}$$

$$w = \frac{(100\,\text{MPa})\,8Z}{L^2} = \frac{(100\,\text{MPa})(8)(1408 \times 10^3\,\text{mm}^3)}{(4\,\text{m})^2} = 70.4\,\text{kN/m}$$

9.25 a. Draw the free-body diagram and the shear and moment diagrams as shown in Exhibit 9.25a. The maximum bending stress is

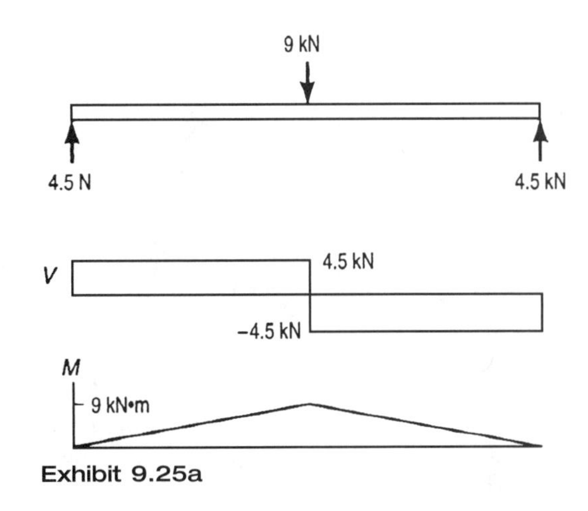

Exhibit 9.25a

$$\sigma_{max} = \frac{M_{max}c}{I} = \frac{(9\,\text{kN}\bullet\text{m})\left(\frac{363\,\text{mm}}{2}\right)}{(365\times10^{6}\ \text{mm}^{4})} = 4.48\ \text{MPa}$$

9.26 a. From the previous problem, the maximum shear in the beam is 4.5 kN. The maximum shearing stress will take place at the centroid (Exhibit 9.26), so a cut must be made there in order to calculate Q. The moment of the area Q is, therefore,

$$Q = A_1\bar{y}_1 + A_2\bar{y}_2$$

$$Q = (257\,\text{mm})(21.7\,\text{mm})\left(\frac{363\,\text{mm}}{2} - \frac{21.7\,\text{mm}}{2}\right) + \cdots + \left(\frac{363\,\text{mm}}{2} - 21.7\,\text{mm}\right)$$

$$\times(13\,\text{mm})\left(\frac{\frac{363\,\text{mm}}{2} - 21.7\,\text{mm}}{2}\right)$$

$$Q = 1.117\times10^{6}\,\text{mm}^{3}$$

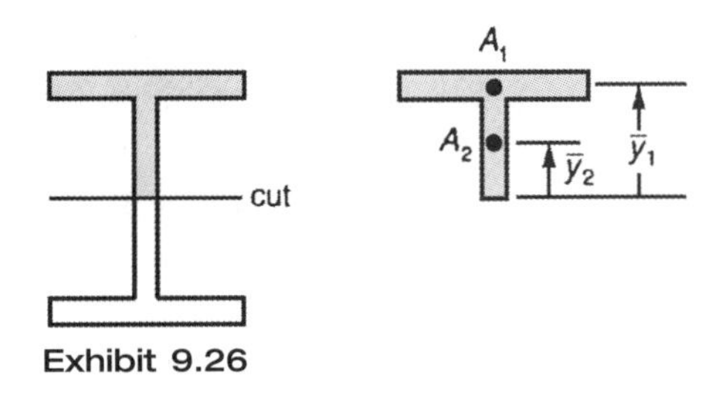

Exhibit 9.26

The maximum shear stress is then

$$\tau = \frac{VQ}{Ib} = \frac{(4.5\,\text{kN})(1.117\times10^{6}\ \text{mm}^{3})}{(365\times10^{6}\,\text{mm}^{4})(13\,\text{mm})} = 1.060\,\text{MPa}$$

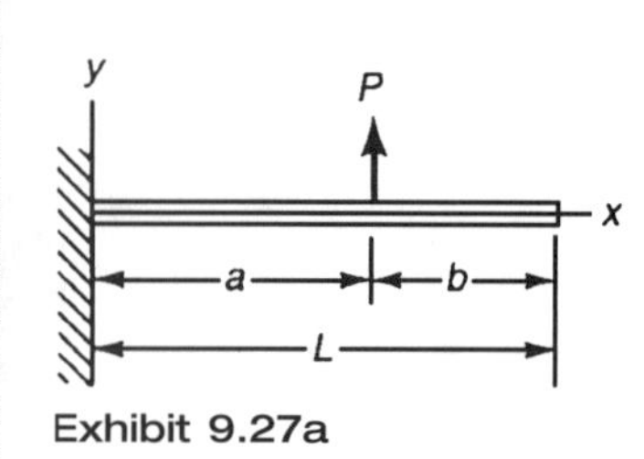

Exhibit 9.27a

9.27 d. From Table 9.1, Beam Type 1 (Exhibit 9.27a), for $a \le x \le L$

$$y = \frac{Pa^2}{6EI}(3x - a)$$

For the load at the half-way point, $a = L/2$, $x = L$, and $P = -F$. For the load at the end, $a = L$, $x = L$, and $P = -F$. Therefore,

$$y = \frac{-F\left(\frac{L}{2}\right)^2}{6EI}\left[3L - \left(\frac{L}{2}\right)\right] + \frac{-FL^2}{6EI}(3L - L) = -0.4375\frac{FL^3}{EI}$$

9.28 b. The maximum deflection for this beam will take place at the center of the beam. This problem can be solved with the superposition of the following cases from Table 9.1 (Exhibits 9.28a and b).

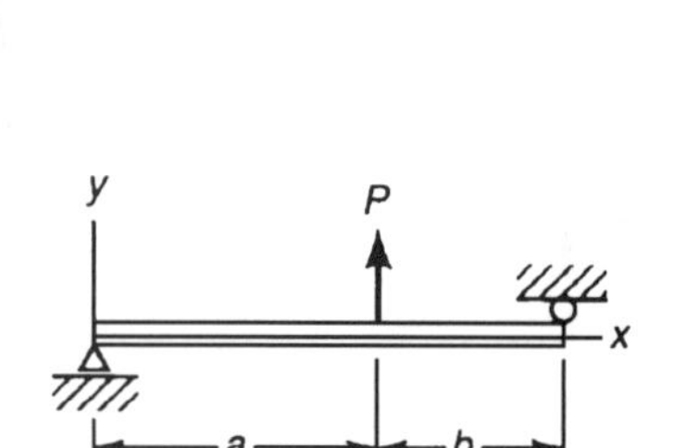

Exhibit 9.28a

For $0 \le x \le a$,

$$y = \frac{Pbx}{6LEI}(L^2 - b^2 - x^2)$$

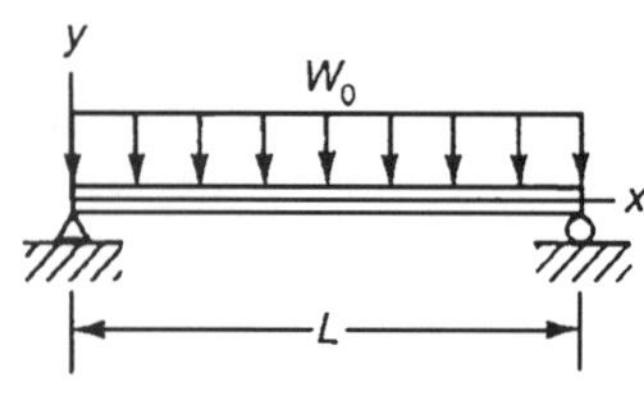

Exhibit 9.28b

For this problem, $P = -wL$, $a = b = L/2$, and $x = L/2$.

$$y = -\frac{wx}{24EI}(L^3 - 2Lx^2 + x^3)$$

For this problem, $x = L/2$. The total deflection is, therefore,

$$y = \frac{Pbx}{6LEI}(L^2 - b^2 - x^2) - \frac{wx}{24EI}(L^3 - 2Lx^2 + x^3)$$

$$y = \frac{(-wL)\left(\frac{L}{2}\right)\left(\frac{L}{2}\right)}{6LEI}\left[L^2 - \left(\frac{L}{2}\right)^2 - \left(\frac{L}{2}\right)^2\right] - \frac{w\left(\frac{L}{2}\right)}{24EI}\left[L^3 - 2L\left(\frac{L}{2}\right)^2 + \left(\frac{L}{2}\right)^3\right]$$

$$v = -0.0339\frac{wL^4}{EI} = -\frac{13wL^4}{384EI}$$

9.29 d. This problem can be solved from superposition of the same two cases as used in Problem 9.28. For the concentrated load solution, $b = L/2$ and P is left as an unknown. In both, $x = L/2$. The center support means the beam does not deflect in the center. Therefore,

$$y = 0 = \frac{P\left(\frac{L}{2}\right)\left(\frac{L}{2}\right)}{6LEI}\left[L^2 - \left(\frac{L}{2}\right)^2 - \left(\frac{L}{2}\right)^2\right] - \frac{w\left(\frac{L}{2}\right)}{24EI}\left[L^3 - 2L\left(\frac{L}{2}\right)^2 + \left(\frac{L}{2}\right)^3\right]$$

$$y = 0 = \frac{PL^3}{48EI} - \frac{5wL^4}{384EI}$$

$$P = \frac{5}{8}wL$$

9.30 c. Draw the free-body diagram (Exhibit 9.30a). From a summation of moments about the cut at *B*, the maximum bending moment in *BC* is the moment $M = 200\text{ N} \times 250\text{ mm}$ or 50 kN • m. The maximum bending stress is

$$\sigma = \frac{Mc}{I} = \frac{(50\,\text{kN} \bullet \text{mm})(12.5\,\text{mm})}{0.25\pi(12.5\,\text{mm})^4} = 32.6\text{ MPa}$$

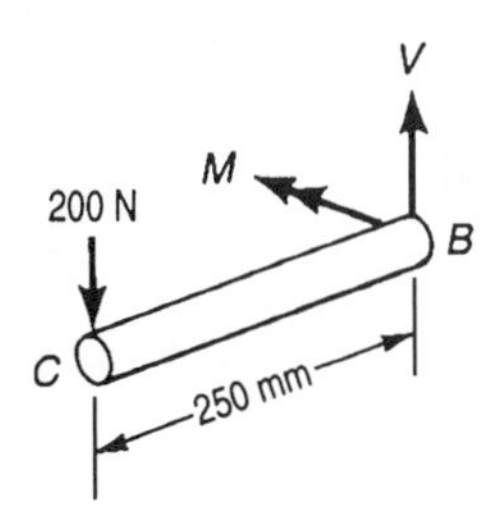

Exhibit 9.30a

9.31 d. Draw the free-body diagram (Exhibit 9.31). The maximum stresses in section *AB* will occur at *A*. Summation of forces in the vertical direction gives $V_A = 200$ N. Summation of forces along the direction of the rod *AB* gives $P = 2000$ N. Summation of moments along the rod *AB* gives $T_A = 200\text{ N} \times 250\text{ mm}$ or 50 kN • mm. Summation of moments at the cut perpendicular to the rod *AB* gives $M_A = 200\text{ N} \times 300\text{ mm} = 600$ kN • mm. The maximum bending stress is

$$\sigma = \frac{Mc}{I} = \frac{(60\,\text{kN} \bullet \text{mm})(12.5\,\text{mm})}{0.25\pi(12.5\,\text{mm})^4} = 39.1\text{ MPa}$$

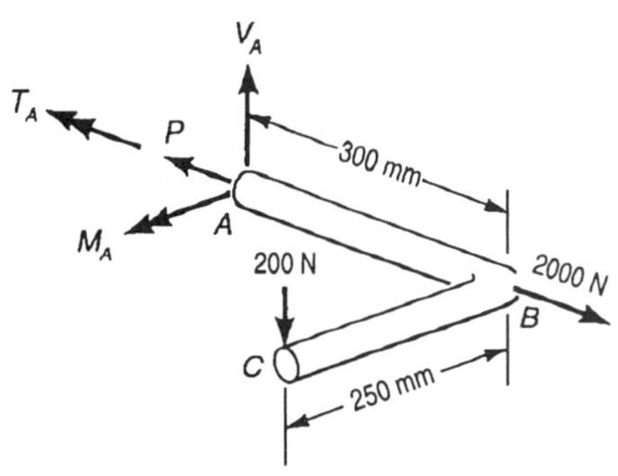

Exhibit 9.31

9.32 b. From the free-body diagram in Problem 9.31, the maximum torque is 50 kN • mm. The maximum shear stress is, therefore,

$$\tau_{\max} = \frac{T_{\max} r_0}{J} = \frac{(50\,\text{kN} \bullet \text{mm})(12.5\,\text{mm})}{0.5\pi(12.5\,\text{mm})^4} = 16.30\text{ MPa}$$

9.33 a. From the free-body diagram in Problem 9.31, the maximum axial force is 2000 N. The maximum stress due to this force is, therefore,

$$\sigma = \frac{P}{A} = \frac{2000\,\text{N}}{\pi(12.5\,\text{mm})^2} = 4.07\,\text{MPa}$$

9.34 d. The stresses were found in the previous three problems. There is an axial stress due to both bending and axial loads. This stress is

$$\sigma = 39.1 \text{ MPa} + 4.07 \text{ MPa} = 43.2 \text{ MPa}$$

The shear stress is 16.3 MPa. These are the only non-zero stresses. The maximum principal stress can be calculated as follows,

$$R = \sqrt{\left(\frac{\sigma_x - \sigma_y}{2}\right)^2 + \tau_{xy}^2} = \sqrt{\left(\frac{43.2 - 0}{2}\right)^2 + 16.3^2} = 22.0 \text{ MPa}$$

$$C = \frac{\sigma_x + \sigma_y}{2} = \frac{43.2 + 0}{2} = 21.6 \text{ MPa}$$

$$\sigma_1 = C + R = 22.0 + 21.6 = 43.6 \text{ MPa}$$

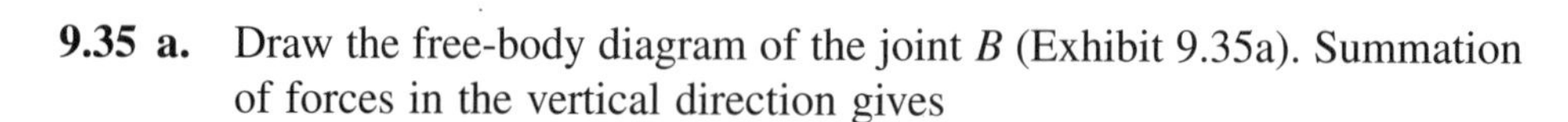

9.35 a. Draw the free-body diagram of the joint B (Exhibit 9.35a). Summation of forces in the vertical direction gives

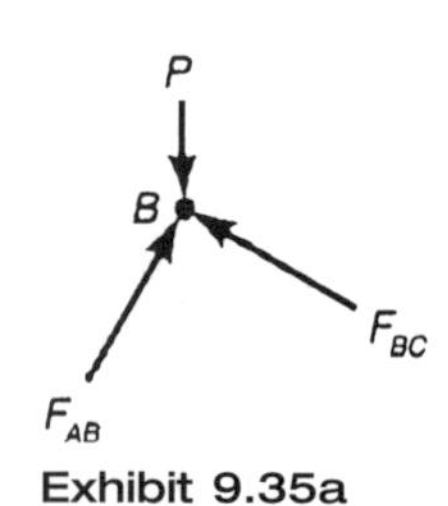

Exhibit 9.35a

$$P = F_{AB}\frac{1.5}{\sqrt{1.5^2 + 0.9^2}} + F_{BC}\frac{1.5}{\sqrt{2.7^2 + 1.5^2}}$$

Summation of forces in the horizontal direction gives

$$F_{AB}\frac{0.9}{\sqrt{1.5^2 + 0.9^2}} = F_{BC}\frac{2.7}{\sqrt{2.7^2 + 1.5^2}}$$

Solving for F_{AB} and F_{BC} gives

$$F_{AB} = 0.875P; \quad F_{BC} = 0.515P$$

The member AC is in tension and does not need to be considered. The moment of inertia for both members is

$$I = \frac{bh^3}{12} = \frac{(25\,\text{mm})(25\,\text{mm})^3}{12} = 32{,}600\,\text{mm}^4$$

The critical buckling load for member AB is

$$P_{\text{cr}} = F_{AB} = \frac{\pi^2 EI}{L^2} = \frac{\pi^2 (210\,\text{GPa})(32{,}600^4)}{(1.5\,\text{m})^2 + (0.9\,\text{m})^2} = 22.0 \text{ kN}$$

The load P for buckling to occur in AB is

$$P = \frac{22\,\text{kN}}{0.875} = 25.2 \text{ kN}$$

The critical buckling load for member BC is

$$P_{\text{cr}} = F_{BC} = \frac{\pi^2 EI}{L^2} = \frac{\pi^2 (210\,\text{GPa})(32{,}600\,\text{mm}^4)}{[(2.7\,\text{m})^2 + (1.5\,\text{m})^2]} = 7.07 \text{ kN}$$

The load P for buckling to occur in BC is

$$P = \frac{7.07\,\text{kN}}{0.515} = 13.7\ \text{kN}$$

9.36 d. To buckle in the x-y plane the critical buckling load is

$$P_{\text{cr}} = \frac{\pi^2 EI}{L^2} = \frac{\pi^2 E\left(\frac{ab^3}{12}\right)}{L^2}$$

To buckle in the x-z plane the critical buckling load is

$$P_{\text{cr}} = \frac{4\pi^2 EI}{L^2} = \frac{4\pi^2 E\left(\frac{ba^3}{12}\right)}{L^2}$$

Equating these two representations of P_{cr} gives

$$ab^3 = 4ba^3; \quad b^2 = 4a^2; \quad \frac{b}{a} = 2$$

CHAPTER 10

Fluid Mechanics

Gary Crossman

OUTLINE

Fluid mechanics is the study of fluids at rest or in motion. The topic is generally divided into two categories: *liquids* and *gases*. Liquids are considered to be incompressible, and gases are compressible. The treatment of incompressible fluids and compressible fluids each have their own groups of equations. However, there are times when a gas may be treated as incompressible (or at least uncompressed). For example, the flow of air through a heating duct is one such case. This chapter will concentrate on incompressible fluids.

FLUID PROPERTIES

Thermodynamic properties are important in incompressible fluid mechanics. Those of particular importance are density, specific gravity, specific weight, viscosity and pressure. Temperature is also important but is primarily used in finding other properties such as density and viscosity in tables or graphs.

Density

The **density**, ρ, is the mass per unit volume and is the reciprocal of the specific volume, a property used in thermodynamics:

$$\rho = \frac{m}{V} = \frac{1\,\text{kg}}{v\,\text{m}^3}$$

Specific Gravity

The **specific gravity,** SG, is defined by the following equation

$$\text{SG} = \frac{\rho\left(\frac{\text{kg}}{\text{m}^3}\right)}{\frac{1000\text{ kg}}{\text{m}^3}}$$

where 1000 kg/m^3 is the density of water at 4°C.

In many cases the specific gravity of a liquid is known or found from tables and must be converted to density using this equation.

Specific Weight

The **specific weight**, γ, of a fluid is its weight per unit volume and is related to the density as follows:

$$\gamma = \frac{W}{V} = \rho\left(\frac{g}{g_c}\right)\frac{\text{N}}{\text{m}^3}$$

where g = local acceleration of gravity, $\frac{\text{m}}{\text{s}^2}$, and g_c = gravitational constant:

$$g_c = \frac{\text{kg}\bullet\text{m}}{\text{N}\bullet\text{s}^2}$$

The density of water at 4°C is 1,000 kg/m^3. Its specific weight at sea level (g = 9.81 m/s^2) is calculated as follows:

$$\gamma = \rho\frac{g}{g_c} = 1000\frac{\text{kg}}{\text{m}^3} = \frac{9.81\frac{\text{m}}{\text{s}^2}}{\frac{\text{kg}\bullet\text{m}}{\text{N}\bullet\text{s}^2}} = 9810\frac{\text{N}}{\text{m}^3}$$

The density, specific gravity, and specific weight of a liquid are generally considered to be constant, with little variation, over a wide temperature range.

Viscosity

The **viscosity** of a fluid is a measure of its resistance to flow; the higher the viscosity the more resistance to flow. Water has a relatively low viscosity, and heavy fuel oils have a high viscosity. The **dynamic (absolute) viscosity**, μ, of a fluid is defined as the ratio of shearing stress to the rate of shearing strain. In equation form:

$$\mu = \frac{\tau}{\frac{dV}{dy}} \frac{\text{N}\bullet\text{s}}{\text{m}^2}\left(\frac{\text{kg}}{\text{m}\bullet\text{s}}\right)$$

where τ = shearing stress (force per unit area), N/m^2, and dV/dy = rate of shearing strain, $1/s$.

Fluids may be classified as Newtonian or non-Newtonian. Newtonian fluids are those in which dV/dy in the above equation can be considered to be constant for a given temperature. Thus the shearing stress, τ (horizontal force divided by the surface area), of a plate on a thin layer, δ, of a Newtonian fluid, as shown in Fig. 10.1, may be found from

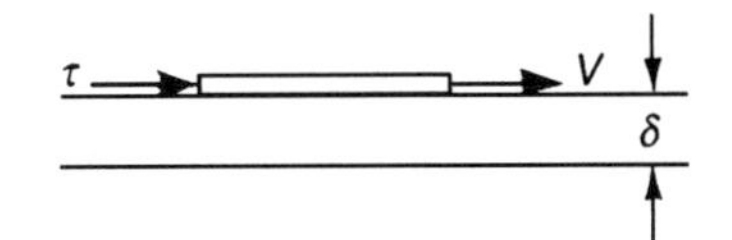

Figure 10.1

$$\tau = \mu\frac{dV}{dy} = \frac{\mu V}{\delta}$$

where V = velocity, m/s, and δ = thickness, m.

Most common fluids such as water, oil, gasoline, and alcohol are classified as Newtonian fluids.

The **kinematic viscosity** is defined by

$$\nu = \frac{\mu}{\rho} \frac{\text{m}^2}{\text{s}}$$

Both the dynamic and kinematic viscosities are highly dependent on temperature. The viscosity of most liquids decreases significantly (orders of magnitude) with increases in temperature, while the viscosity of gases increases mildly with increases in temperature. The viscosity of any gas is less than the viscosity of any liquid. Viscosities are generally found in tables and graphs.

The definition of viscosity assists in the development of the engineering definition of a fluid as follows:

> A fluid is a substance that will deform readily and continuously when subjected to a shear force, no matter how small the force.

Pressure

Pressure, p, is the force per unit area of a fluid on its surroundings or vice versa. Pressure may be specified using two different datums. Absolute pressure, P_{abs}, is measured from absolute zero or a complete vacuum (void). At absolute zero there are no molecules and a negative absolute pressure does not exist. Absolute pressures are needed for ideal gas relations and in compressible fluid mechanics. Gage pressure, p_{gage}, on the other hand, uses local atmospheric pressure as its datum. Gage pressures may be positive (above atmospheric pressure) or negative (below atmospheric pressure). Negative gage pressure is also called vacuum. A complete vacuum occurs at a negative gage pressure that is equivalent to the atmospheric pressure or at absolute zero.

The relationship between absolute pressure and gage pressure is as follows:

$$P_{abs} = p_{gage} + p_{atm} \frac{N}{m^2}(\text{Pa})$$

Actually, the pressure is usually expressed in kN/m^2 or kPa but should be converted to these units for use in most equations.

Example 10.1

A pressure gage measures 50 kPa vacuum in a system. What is the absolute pressure if the atmospheric pressure is 101 kPa?

Solution

Change vacuum to a negative gage pressure:

$$p_{abs} = -50 \text{ kPa} + 101 \text{ kPa} = 51 \text{ kPa} \bullet 1000 = 51{,}000 \text{ Pa}$$

Most pressure measuring devices measure gage pressure. For incompressible fluid dynamics, gage pressure may be used in most equations. This capability simplifies equations significantly when one or more pressures in the system are atmospheric or $p_{gage} = 0$.

Surface Tension

Surface tension is another property of liquids. It is the force that holds a water droplet or mercury globule together, since the cohesive forces of the liquid are more than the adhesive forces of the surrounding air. The surface tension (or surface tension coefficient), σ, of liquids in air is available in tables and can be used to calculate the internal pressure, p, in a droplet from

$$p = \frac{4\sigma}{d}$$

where σ = surface tension of the liquid, kN/m, and d = droplet diameter, m. Values of surface tension for various liquids are found in tables as a function of the surrounding medium (air, etc.) and the temperature.

Surface tension is also the property that causes a liquid to rise (or fall) in a capillary tube. The amount of rise (or fall) depends on the liquid and the capillary tube material. When *adhesive* forces dominate, the liquid will rise—as with water. When cohesive forces dominate, it will fall—as with mercury. The capillary rise, h, can be calculated from the following equation:

$$h = \frac{4\sigma \cos\beta}{\gamma d}$$

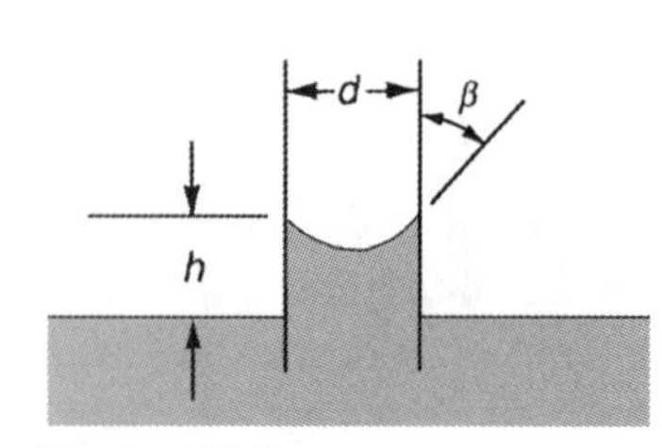

Figure 10.2

where β = angle made by the liquid with the tube wall, and d = diameter of capillary tube, as shown in Fig. 10.2.

The angle, β, varies with different liquid/tube material combinations and is found in tables. β is within the range 0 to 180°. When $\beta > 90°$, h will be negative.

FLUID STATICS

Pressure-Height Relationship

For a static liquid, the pressure increases with depth (decreases with height) according to the following relationship

$$p_2 - p_1 = -\gamma(Z_2 - Z_1) = \gamma h$$

where h = depth from Point 1 to Point 2.

If p_1 is at the surface of a liquid that is open to the atmosphere then, the gage pressure at Point 2 is found from

$$p_2 = p = \gamma h$$

Example 10.2

Calculate the gage pressure at a depth of 100 meters in seawater, for which γ = 10.1 kN/m^3.

Solution

$$p = \gamma h = \left(\frac{10.1\ \text{kN}}{\text{m}^3}\right)(100\ \text{m}) = 1010\frac{\text{kN}}{\text{m}^2} = 1010\ \text{kPa}$$

Manometers

A manometer is a device used to measure moderate pressure differences using the pressure-height relationship. The simplest manometer is the U-tube shown in Fig. 10.3. The pressure difference between System 1 and System 2 is found from

$$p_1 - p_2 = \gamma_m h_m + \gamma_2 h_2 - \gamma_1 h_1$$

where γ_m, γ_1, and γ_2 = specific weight of manometer fluid, fluid in System 1, and fluid in System 2, respectively, and h_m, h_1, h_2 = depths as shown. If the fluids in both systems are gases and the manometer fluid is any liquid, then $\gamma_m >>> \gamma_1$ or γ_2 and the equation simplifies to

$$p_1 - p_2 = \gamma_m h_m = \gamma h$$

If system 2 were the atmosphere ($p_2 = 0_{\text{gage}}$) then

$$p_1 = \gamma h = \text{gage pressure in System 1.}$$

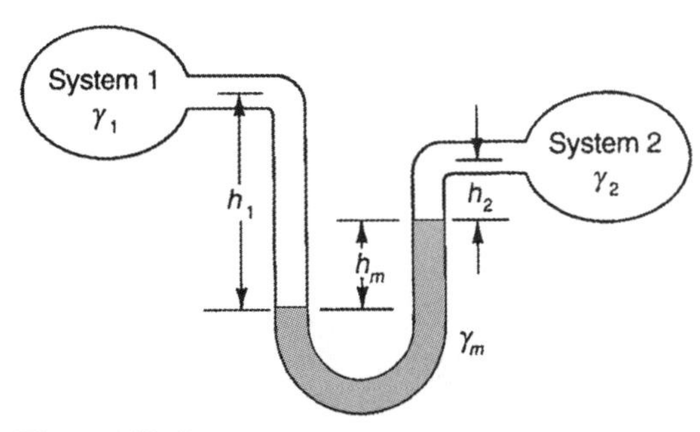

Figure 10.3

If System 1 were the atmosphere ($p_1 = 0$) then

$$p_2 = -\gamma h = \text{gage pressure in System 2.}$$

The gage pressure in System 2 would be negative or a vacuum. Manometers are commonly used to measure system pressures between –101.3 kPa and +101.3 kPa. In many cases, particularly where the gage pressure is negative, the pressure may be given in millimeters of a fluid, and the equation above used to convert it to standard units.

Example 10.3

A system gage pressure is given as 500 millimeters of mercury vacuum (mm Hg vac). What is the gage pressure in kPa? The specific gravity of mercury is 13.6.

Solution

The pressure is $p = p_2 = -\gamma h$ since vacuum is a negative gage pressure:

$$\gamma_m = (13.6)\left(9.81\frac{\text{kN}}{\text{m}^3}\right) = 133.4\frac{\text{kN}}{\text{m}^3}$$

$$p = -\gamma h = -133.4\frac{\text{kN}}{\text{m}^3} \times 0.5\,\text{m} = -66.7\text{ kPa}$$

The conversion factor from millimeters of mercury to N/m^2 (pascals) is 133.4.

A barometer is a special type of mercury manometer. In this case one leg of the U-tube is very wide. If we can adjust the scale on the narrow leg so that zero is at the level of the large leg, then the narrow leg will read the atmospheric pressure impinging on the wide leg corrected by the vapor pressure of the mercury. A barometer is shown in Fig. 10.4.

There are several other types of manometers. A compound manometer consists of more than one U-tube in series between one system and another. The equation for $p_1 - p_2$ may be developed by starting at System 2 and adding γh's going downward and subtracting γh's going upward until System 1 is reached as follows:

$$p_2 + \sum \gamma h \text{ (downward)} - \sum \gamma h \text{ (upward)} = p_1$$

An inclined manometer is used to measure small pressure differentials. The measurement along the manometer must be multiplied by the sine of the angle of incline. An inclined manometer is generally "single leg," similar to the barometer previously described, and is shown in Fig. 10.5. The pressure difference is found

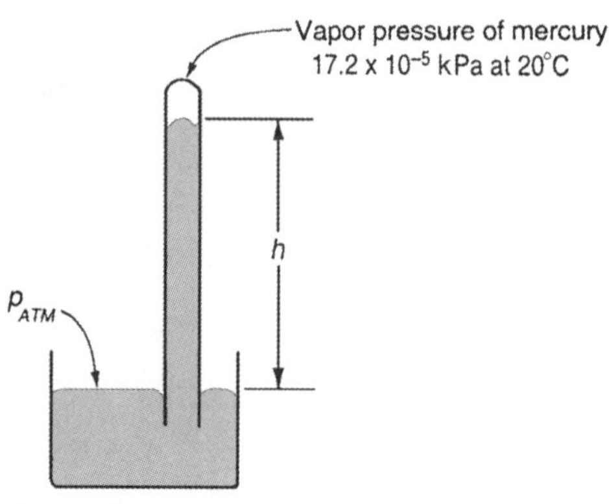

Figure 10.4

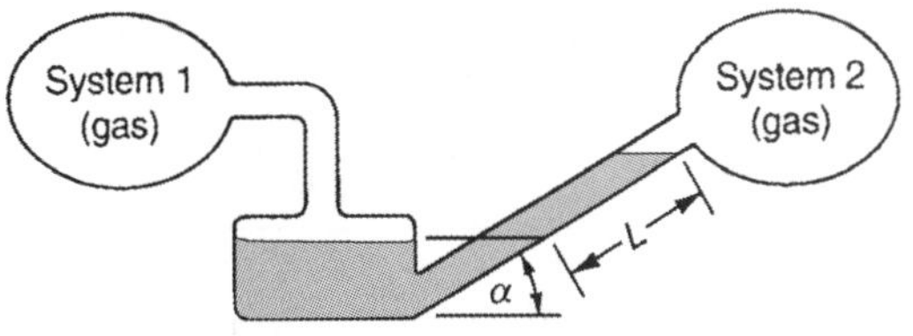

Figure 10.5

from $p_2 - p_1 = \gamma_m L \sin \alpha$, where L = length along manometer leg and α = angle of inclination.

Forces on Flat Submerged Surfaces

A flat surface of arbitrary shape below a liquid surface is shown in Fig. 10.6. The resultant force, F, on one side of the flat surface acts perpendicular to the surface. Its magnitude and location may be calculated from the following equations:

$$F = (p_0 + \gamma h_c)A$$

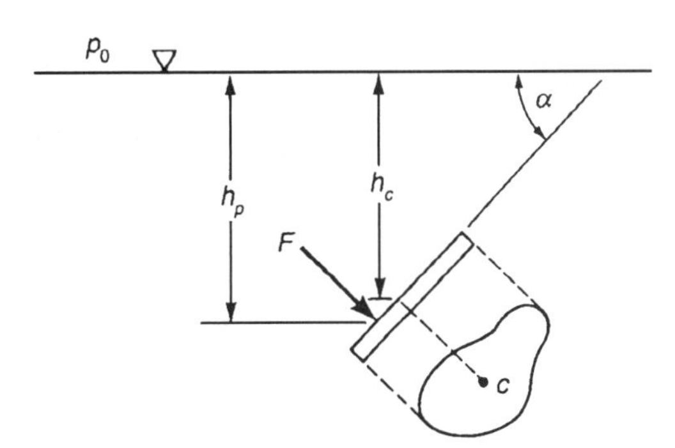

Figure 10.6

and

$$h_p = h_c + \frac{I_c \sin^2 \alpha}{\left(\frac{p_0}{\gamma} + h_c\right)A}$$

where

F = resultant force on the flat surface, N
p_0 = gage pressure on the surface, Pa
γ = specific weight of the fluid, N/m^3
h_c = vertical distance from fluid surface to the centroid of the flat surface area, m
A = area of flat surface, m^2
h_p = vertical distance from fluid surface to the center of pressure of the flat surface (where the equivalent, concentrated force acts), m
I_c = moment of inertia of the flat surface about a horizontal axis through its centroid, m^4
α = angle that the inclined flat surface makes with the horizontal surface

The values of h_c and I_c for common geometric shapes such as rectangles, triangles, and circles may be determined from existing tables. Typical values are presented in Table 10.1.

Table 10.1 Areas, centroids, and moments of inertia for selected areas

Section	Area of Section, A	Distance to Centroidal Axis, $\bar{y}$	Moment of Inertia about Centroidal Axis, I_c
Rectangle 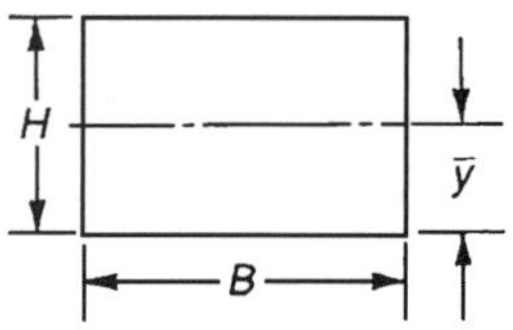	BH	$H/2$	$BH^3/12$
Triangle 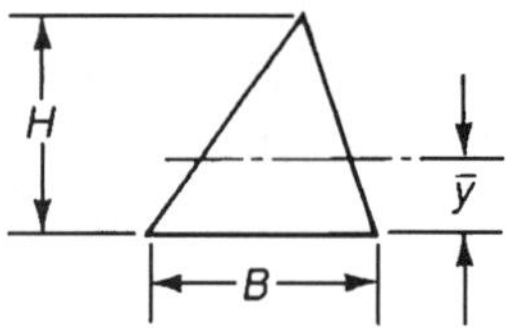	$BH/2$	$H/3$	$BH^3/36$
Circle 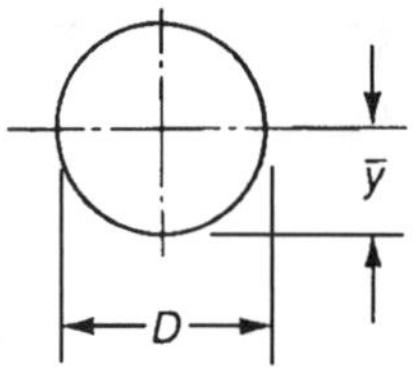	$\pi D^2/4$	$D/2$	$\pi D^4/64$
Ring 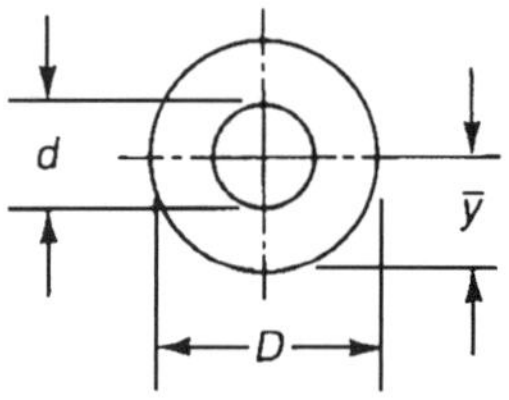	$\frac{\pi(D^2-d^2)}{4}$	$D/2$	$\frac{\pi(D^4-d^4)}{64}$
Semicircle 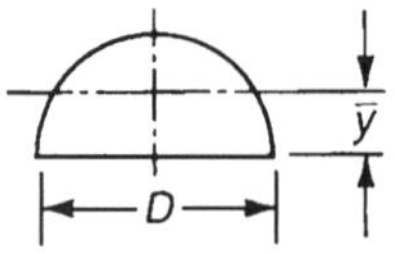	$\pi D^2/8$	$0.212D$	$(6.86\times10^{-3})\,D^4$
Quadrant 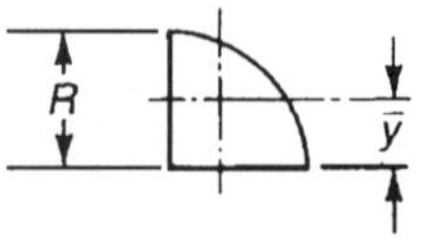	$\pi D^2/16$ $\pi R^2/4$	$0.212D$ $0.424R$	$(3.43\times10^{-3})D^4$ $(5.49\times10^{-2})R^4$
Trapezoid 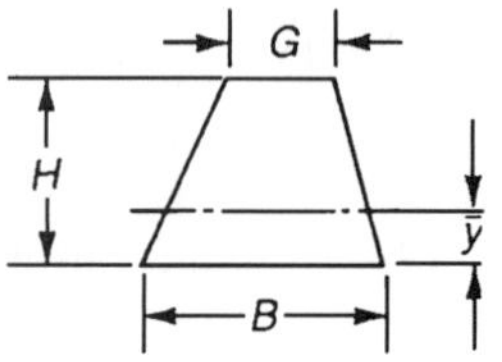	$\frac{H(G+B)}{2}$	$\frac{H(2G+B)}{3(G+B)}$	$\frac{H^3(G^2+4GB+B^2)}{36(G+B)}$

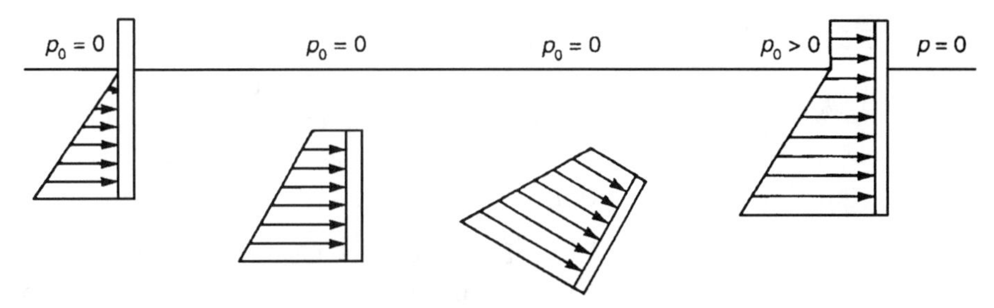

Figure 10.7

For the common case when p_0 is atmospheric pressure ($p_0 = 0$), the equations simplify to

$$F = \gamma h_c A$$

and

$$h_p = h_c + \frac{I_c \sin^2 \alpha}{h_c A}$$

From the above equations it is apparent that the center of pressure is always below the centroid except when the surface is horizontal ($\alpha = 0$). In that case the center of pressure is at the centroid. The deeper the flat surface is located below the fluid surface, the closer the center of pressure is to the centroid.

The pressure profile on the flat surface is generally trapezoidal (triangular, if the flat surface pierces the surface of a fluid exposed to atmospheric pressure). The slope of the pressure profile is equivalent to the specific weight of the fluid. Examples are shown in Fig. 10.7.

Example 10.4

A vertical side of a saltwater tank contains a round viewing window 60 cm in diameter with its center five meters below the liquid surface. If the specific weight of the saltwater is 10 kN/m^3, find the force of the water on the window and where it acts.

Solution

Assume atmospheric pressure on the liquid surface, $p_0 = 0$.

$$d = 60\,\text{cm} = 0.6\text{ m}, \quad A = \frac{\pi(0.6)^2}{4} = 0.283\text{ m}^2$$

$$F = \gamma h_c A = 10\frac{\text{kN}}{\text{m}^3} \times 5\text{ m} \times 0.283\text{ m}^2 = 14.14\text{ kN}$$

$$I_c = \frac{\pi d^4}{64} = \frac{\pi(0.6\text{ m})^4}{64} = 0.00636\text{ m}^4$$

$$\alpha = 90°, \quad \sin\alpha = 1$$

$$h_p = h_c + \frac{I_c \sin^2 \alpha}{h_c A} = 5\,\text{m} + \frac{0.00636\text{ m}^4(1)^2}{5\,\text{m} \bullet 0.283\text{ m}^2} = 5.0045\text{ m}$$

In many cases problems involving fluid forces on flat surfaces are combined with a statics problem. The fluid force is just another force to be added into the statics equation.

Example 10.5

In the previous example, suppose the circular window were hinged at the top with some sort of clamp at the bottom (Exhibit 1). What force, *P*, would be required of the clamp to keep the window closed?

Solution

From Example 10.4 calculations, the force of the water is 14.14 kN located 5.0045 m below the fluid surface. The hinge is located 5 m – 0.6 m/2 = 4.7 m below the water surface. Thus the force location is 5.0045 m – 4.7 m = 0.3045 m below the hinge.

Summing moments about hinge,

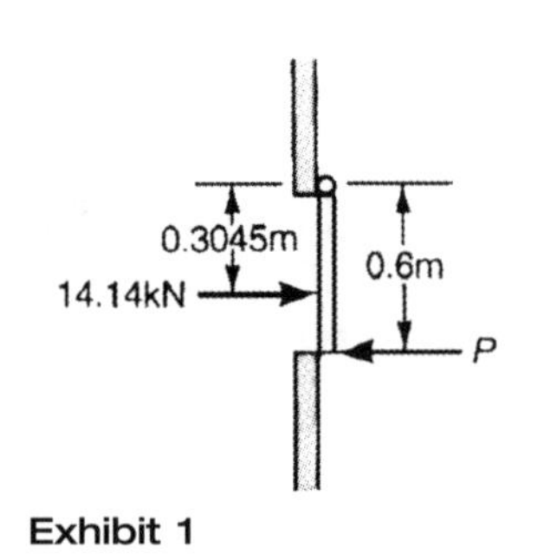

Exhibit 1

$$\sum M_H = 0.6\,\text{m} \bullet P - 0.3045\,\text{m} \bullet 14.14\,\text{kN} = 0$$

$$P = \frac{0.3045\,\text{m} \bullet 14.14\,\text{kN}}{0.6\,\text{m}} = 7.18\,\text{kN}$$

Buoyancy

In addition to the force of gravity, or weight, all objects submerged in a fluid are acted on by a buoyant force, F_B. The buoyant force acts upward and is equal to the weight of the fluid displaced by the object. This is known as Archimedes' Principle. The upward buoyant force also acts through the center of gravity (or centroid) of the displaced volume, known as the center of buoyancy, *B*. Thus

$$F_B = \gamma_f V_D$$

where F_B = buoyant force, N; γ_f = specific weight of the fluid, N/m^3; and V_D = volume displaced by the object, m^3.

For a freely floating object (no external forces) the weight of the object (acting downward) is equal to the buoyant force on the object (acting upward) or

$$W = F_B = \gamma_f V_D$$

This equation is useful in determining what part of an object will float below the surface of a liquid. For objects partially submerged in a liquid and a gas, the buoyant force of the gas is usually neglected. However, the buoyant force on a totally submerged body in a gas is very important in the study of balloons, dirigibles, etc.

Example 10.6

A wooden cube that is 15 centimeters on each side with a specific weight of 6300 N/m^3 is floating in fresh water (γ = 9,810 N/m^3) (Exhibit 2). What is the depth of the cube below the surface?

Solution

$$W = F_B = \gamma_f V_D$$

There are actually two buoyant forces on the cube, that of the water on the volume below the surface and that of the air on the volume above the surface. Neglecting the buoyant force of the air and rearranging the equation

$$V_D = \frac{W}{\gamma_f} = \frac{\gamma_C V_C}{\gamma_f} = \frac{(6300\ \text{N/m}^3)(0.15\ \text{m})^3}{9810\ \text{N/m}^3} = 0.00217\ \text{m}^3$$

$$V_D = (0.15)^2 \bullet d = 0.00217\ \text{m}^3$$

$$d = \frac{0.00217\ \text{m}^3}{.0225\ \text{m}^2} = .0964\ \text{m} = 9.64\ \text{cm}$$

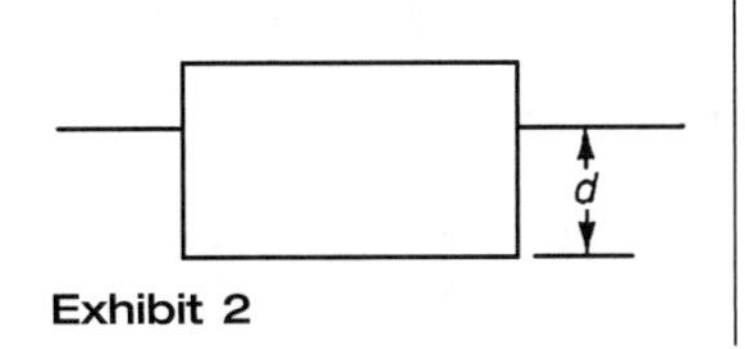

Exhibit 2

Neutral buoyancy exists when the buoyant force equals the weight when an object is completely submerged. The object will remain at whatever location it is placed below the fluid surface.

In the case of an object floating at the interface of two liquids, the total buoyant force is equal to the sum of the buoyant forces on the object created by each fluid on that part that is immersed. When external forces also act on a submerged or partially submerged object, they must be included in the force balance on the object. The force balance equation then becomes

$$W + \sum F_{\text{ext}}(\text{down}) = F_B + \sum F_{\text{ext}}(\text{up})$$

If weight is added to an object internally, or possibly on top of a partially-submerged object, it will only affect the weight of the object. But if the weight is added externally, beneath the surface of the fluid, its buoyant force as well as its weight must be considered.

Example 10.7

If, in Example 10.6, a concrete weight (anchor) is added to the bottom of the cube externally, what anchor volume, V_A, would be required to make the cube float neutrally (below the surface.) The specific weight of the concrete, γ_c, is 24 kN/m^3.

Solution

Let the subscript *C* denote the properties of the cube and subscript *A* denote those of the anchor. Summing forces vertically,

$$W_C + W_A = F_{BC} + F_{BA}$$

$$\gamma_C V_C + \gamma_A V_A = \gamma_f V_D + \gamma_f V_A$$

Solving for V_A,

$$V_A = \frac{\gamma_f V_D - \gamma_C V_C}{\gamma_A - \gamma_f}$$

But for neutral buoyancy, the displaced volume, V_D, is equal to the total volume of the cube, V_C, and

$$V_A = \frac{(\gamma_f - \gamma_C)\bullet V_C}{\gamma_A - \gamma_f} = \frac{(9810 - 6300)\frac{\text{N}}{\text{m}^3}\bullet(0.15\,\text{m})^3}{(24{,}000 - 9810)\frac{\text{N}}{\text{m}^3}} = 8.34 \times 10^{-4}\ \text{m}^3$$

THE FLOW OF INCOMPRESSIBLE FLUIDS

The Continuity Equation

Most problems in fluid mechanics involve steady flow, meaning that the amount of mass in a system does not change with time. This is generally written as

$$\dot{m}_1 = \dot{m}_2 = \dot{m} = \text{mass rate}$$

where the subscript 1 denotes the entrance and the subscript 2 denotes the exit of the system. The mass rate may be written in terms of fluid properties:

$$\dot{m} = \rho A V$$

where ρ = fluid density, A = cross sectional area of flow, and V = average velocity of the fluid. Thus,

$$\rho_1 A_1 V_1 = \rho_2 A_2 V_2$$

and, since $\rho_1 = \rho_2 = \rho$ for an incompressible fluid, then

$$A_1 V_1 = A_2 V_2 = Q = \text{volume flow rate.}$$

This equation is useful in determining one velocity when another velocity in the system is known.

Example 10.8

Water is flowing in a 5 centimeter diameter pipe at a velocity of 5 m/s (Exhibit 3). The pipe expands to a 10-centimeter diameter pipe. Find the velocity in the 10-centimeter diameter pipe and the flow rate in liters (L) per minute.

Solution

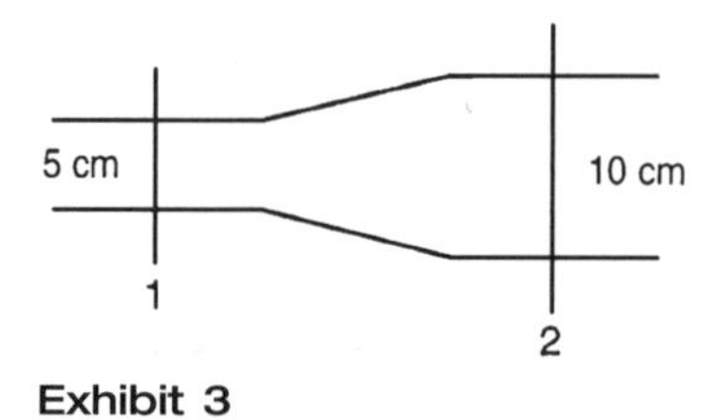

Exhibit 3

$$A_1 V_1 = A_2 V_2$$

$$V_2 = \frac{A_1}{A_2} V_1 = \frac{\pi d_1^2/4}{\pi d_2^2/4} V_1 = \left(\frac{d_1}{d_2}\right)^2 V_1 = \left(\frac{5\text{ cm}}{10\text{ cm}}\right)^2 \left(5\frac{\text{m}}{\text{s}}\right) = 1.25\text{ m/s}$$

$$Q = A_1 V_1 = \frac{\pi d_1^2}{4} V_1 = \frac{\pi (.05\text{ m})^2}{4} \bullet 5\frac{\text{m}}{\text{s}} = 0.00982\frac{\text{m}^3}{\text{s}}$$

$$Q = 0.00982\frac{\text{m}^3}{\text{s}} \frac{1{,}000\text{ L}}{\text{m}^3} \bullet \frac{60\text{ s}}{\text{min}} = 589\text{ L/min}$$

In most cases the flow area may be calculated using the diameter. In some cases the nominal pipe size is known, such as a 4-inch Schedule 40 pipe. The exact inside dimensions of Schedule 40 and other pipes as well as dimensions for steel and copper tubing can be found in existing tables.

Reynolds Number

The **Reynolds number**, Re, is a dimensionless flow parameter which helps describe the nature of flow. It is sometimes defined as the ratio of dynamic forces to viscous forces. In terms of other fluid properties, it is defined as

$$\mathrm{Re} = \frac{\rho V d}{\mu} = \frac{Vd}{\nu}$$

where V = fluid velocity, and d = characteristic length (diameter for pipes)

If the Reynolds number is below 2300, flow is laminar and occurs in layers with no mixing of adjacent fluid. Re = 2300 is known as the critical Reynolds number. Above the critical Reynolds number mixing begins to occur, and the flow becomes turbulent. As the Reynolds number increases, the flow becomes more turbulent.

For pipe flow with a circular cross section the Reynolds number may also be calculated from

$$\mathrm{Re} = \frac{4\rho Q}{\pi d \mu} = \frac{4Q}{\pi d \nu}$$

The Reynolds number is a significant indicator of the influence of friction on the flow that occurs in pipes and other conduits as well as through flow meters. It is also important in the application of dynamic similarity to modeling and many other areas of fluid mechanics.

Example 10.9

For the pipe in Example 10.8, calculate the Reynolds number in the 5-centimeter diameter section of pipe. The kinematic viscosity of the water is 1.12×10^{-6} m^2/s.

Solution

$$\mathrm{Re} = \frac{Vd}{\nu} = \frac{5\frac{\mathrm{m}}{\mathrm{s}} \bullet (0.05\,\mathrm{m})}{1.12 \times 10^{-6}\,\frac{\mathrm{m}^2}{\mathrm{s}}} = 2.2 \times 10^5$$

The flow is well into the turbulent regime.

The Energy Equation

The energy equation in fluid mechanics is similar to that used in thermodynamics. Instead of each energy term having the traditional units such as kJ/kg, energy is expressed in meters of head. For instance, kinetic energy is called velocity head. The general energy equation between two points in a system for incompressible steady flow (mass and energy in the system or at a point do not vary with time) is given by the following expression:

$$\frac{P_1}{\gamma} + \frac{V_1^2}{2g} + Z_1 + h_A - h_R = \frac{p_2}{\gamma} + \frac{V_2^2}{2g} + Z_2 + h_f$$

where

$\frac{p_1}{\gamma}, \frac{p_2}{\gamma}$ = pressure heads at Points 1 and 2

$\frac{V_1^2}{2g}, \frac{V_2^2}{2g}$ = velocity heads at Points 1 and 2

Z_1, Z_2 = potential or elevation heads at Points 1 and 2

h_A, h_R = the head added (pump) or removed (turbine) mechanically

h_f = head loss from friction in the pipe and fittings between Points 1 and 2

The energy equation in fluid mechanics assumes no heat transfer or changes in temperature. This equation, including its reduced forms, will solve most energy-related problems in fluid mechanics when used in conjunction with the continuity equation.

Bernoulli's Equation

Whereas Bernoulli's equation is usually derived from momentum principles using vector calculus, it can also be produced from the energy equation by introducing two additional restrictions to those of incompressible, steady flow and no heat transfer. If we restrict the energy equation to systems with no mechanical energy addition or removal (no pump or turbine) and with no (or negligible) friction losses, Bernoulli's equation is produced,

$$\frac{p_1}{\gamma} + \frac{V_1^2}{2g} + Z_1 = \frac{p_2}{\gamma} + \frac{V_2^2}{2g} + Z_2$$

Bernoulli's equation can be used to solve a variety of problems.

Example 10.10

Referring again to Example 10.8, calculate the pressure just after the expansion to the 10-centimeter diameter pipe if the pressure in the 5-centimeter pipe is 300 kPa. Friction is negligible. The specific weight of water is 9.81 kN/m^3.

Solution

For the horizontal orientation $Z_1 = Z_2$ and Bernoulli's equation reduces to

$$\frac{p_1}{\gamma} + \frac{V_1^2}{2g} = \frac{p_2}{\gamma} + \frac{V_2^2}{2g}$$

$$p_2 = p_1 + \gamma\left(\frac{V_1^2 - V_2^2}{2g}\right)$$

$$p_2 = 300\,\text{kPa} + 9.81\frac{\text{kN}}{\text{m}^3}\left[\frac{(5^2 - 1.25^2)\frac{\text{m}^2}{\text{s}^2}}{2 \bullet 9.81\frac{\text{m}}{\text{s}^2}}\right] = 311.7\,\text{kPa}$$

Other Forms of the Energy Equation

An important rearrangement of the energy equation is to solve for the head added by a pump or removed by a turbine. For the head added by a pump

$$h_A = \frac{p_2 - p_1}{\gamma} + \frac{V_2^2 - V_1^2}{2g} + Z_2 - Z_1 + h_f$$

Example 10.11

A pump is being used to deliver 130 L/min of hot water from a tank through 15 meters of 2.5-cm diameter, smooth pipe, exiting through a 1.0 cm diameter nozzle 3 meters above the level of the tank as shown in Exhibit 4. The head loss from friction of the pipe is 8.33 m. The specific weight of the hot water is 9.53 kN/m^3. Calculate the head delivered to the water by the pump.

Solution

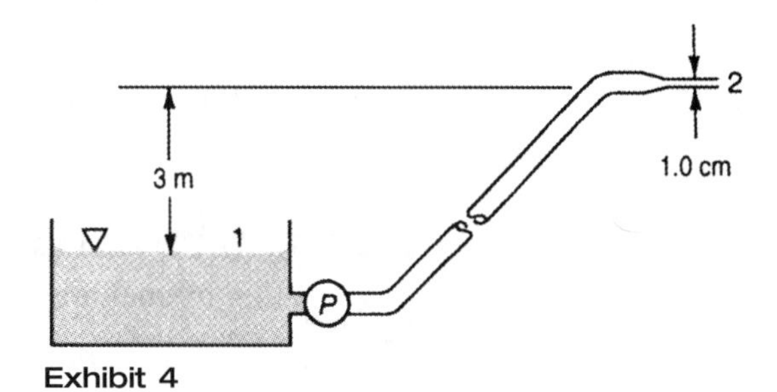

Exhibit 4

Select Points 1 and 2 as shown.

$$h_A = \frac{p_2 - p_1}{\gamma} + \frac{V_2^2 - V_1^2}{2g} + Z_2 - Z_1 + h_f$$

$$p_2 = p_1 = 0, \quad V_1 = 0, \quad Z_2 = 3\text{ m}, \quad Z_1 = 0, \quad h_f = 8.33\text{ m}$$

$$V_2 = \frac{Q}{A_2} = \frac{130 \dfrac{\text{L}}{\text{min}} \bullet \dfrac{1\text{ m}^3}{1000\text{ L}} \bullet \dfrac{\text{min}}{60\text{ s}}}{\dfrac{\pi(.01)^2}{4}\text{m}^2} = 27.6\frac{\text{m}}{\text{s}}$$

$$h_A = \frac{\left(27.6\dfrac{\text{m}}{\text{s}}\right)^2}{2 \bullet 9.81\dfrac{\text{m}}{\text{s}^2}} + 3\text{ m} + 8.33\text{ m} = 50.2\text{ m}$$

In this problem the head of the pump serves three purposes: to increase the velocity of the water, raise its level, and overcome friction.

Pump and Turbine Power and Efficiency

The power delivered by a pump to a fluid or removed by a turbine from the fluid is given by the following

$$P = \gamma Q h_A = \gamma Q h_R$$

The term γQ is the weight rate of flow. In the SI system, the units of power will usually be kN-m/s or kilowatts.

In selecting a pump or turbine, its efficiency is important. The efficiency may be calculated from

$$\eta_P = P/\dot{W} \bullet 100$$
$$\eta_T = \dot{W}/P \bullet 100$$

η_p, η_T = pump and turbine efficiency, respectively, %, P = fluid power, and $\dot{W}$ = mechanical (or shaft) power actually delivered to the pump or by the turbine.

Example 10.12

From Example 10.11, calculate the power delivered to the water by the pump. If the efficiency of the pump is 60%, calculate the mechanical power delivered to the pump ($\dot{W}$).

Solution

$$P = \gamma Q h_A = 9.53\frac{\text{kN}}{\text{m}^3}\bullet 130\frac{\text{L}}{\text{min}}\bullet\frac{1\,\text{m}^3}{1000\,\text{L}}\bullet\frac{\text{min}}{60\,\text{s}}\bullet 50.2\text{ m} = 1.04\text{ kW}$$

$$\eta = \frac{P}{\dot{W}}\bullet 100 \quad \text{or} \quad \dot{W} = \frac{P\bullet 100}{\eta} = \frac{1.04\text{ kW}\bullet 100}{60} = 1.73\text{ kW}$$

Head Loss from Friction in Pipes

Most of the terms in the energy equation will be known or calculated from the energy equation in conjunction with the continuity equation. Even the head loss from friction may be calculated if all other parameters are known. For example, if one wished to know the friction loss in a particular horizontal length of pipe, or in a fitting with equal entrance and exit areas, it could be calculated from the energy equation. Thus, $Z_1 = Z_2$, and since $A_1 = A_2$, then $V_1 = V_2$ and the energy equation becomes

$$\frac{p_1}{\gamma} = \frac{p_2}{\gamma} + h_f$$

If the pressure drop ($p_1 - p_2$) were known or measured, then

$$h_f = \frac{p_1 - p_2}{\gamma}$$

For most applications of the energy equation, the head loss from friction must be known and substituted into the energy equation to solve for an unknown pressure or height, pump or turbine head, or flow rate. For pipe flow, the head loss may be calculated from the Darcy equation:

$$h_f = f\frac{L}{d}\frac{V^2}{2g}$$

where h_f is the head loss from friction in a pipe of length L and diameter d, and f is the friction factor which is a function of the Reynolds number, Re, and the pipe relative roughness, ε/d.

The friction factor, f, can be found from the Moody diagram where f is plotted as a function of the Reynolds number and appears as a family of curves for different values of relative roughness, ε/d. ***Relative roughness*** is the roughness factor of the pipe, ε, divided by the pipe diameter, d. Typical roughness factors are shown in Table 10.2. Glass and plastic have the smallest roughness factors and are shown by the "smooth" curve on the Moody diagram. The Moody diagram is presented in Fig. 10.8.

For Reynolds numbers below 2300 (laminar flow) the friction factor is independent of the relative roughness and may be calculated from the following relationship

$$f = \frac{64}{\text{Re}}$$

Table 10.2 Moody pipe roughness

Material	Roughness, ε (m)
Glass, plastic	Smooth
Copper, brass, lead (tubing)	1.5×10^{-6}
Cast iron—uncoated	2.4×10^{-4}
Cast iron—asphalt coated	1.2×10^{-4}
Commercial steel or welded steel	4.6×10^{-5}
Wrought iron	4.6×10^{-5}
Riveted steel	1.8×10^{-3}
Concrete	1.2×10^{-3}

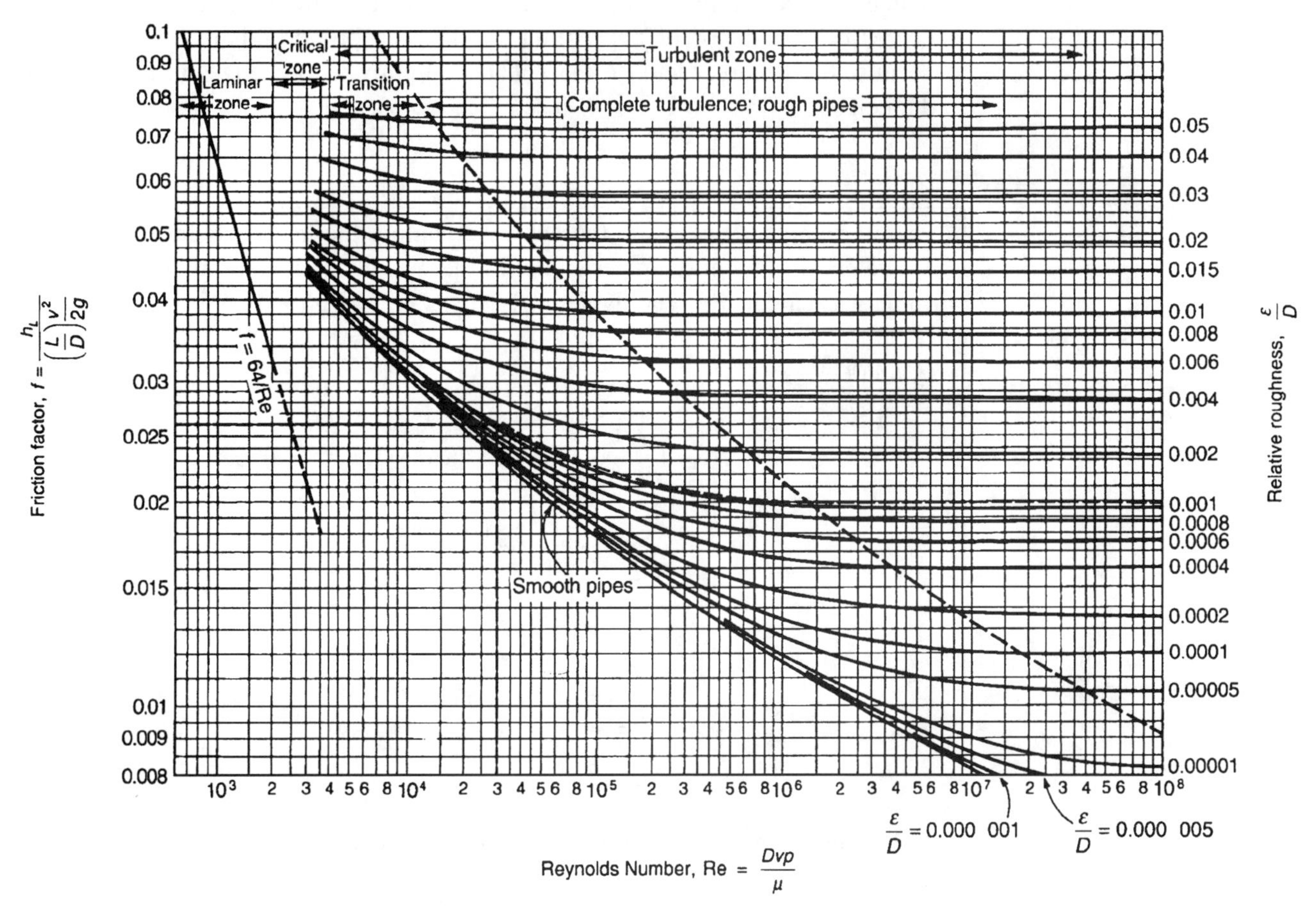

Figure 10.8 Moody diagram used with permission from *Transactions of the ASME*, 1944, vol. 66 by L. F. Moody

Substituting this value into the Darcy equation allows direct calculation of the head loss for laminar flow:

$$h_f = \frac{32\mu LV}{\gamma d^2}$$

This relation is known as the Hagen-Poiseuille equation. It may also be written in terms of flow rate, Q where $V = Q/A = 4Q/\pi d^2$. Then

$$h_f = \frac{128\mu LQ}{\pi\gamma d^4}$$

Thus, the head loss from friction for laminar flow can be seen to be proportional to flow rate and inversely proportional to the diameter raised to the fourth power. If the diameter is doubled, and the flow remains laminar, the head loss will decrease by a factor of 16.

If the head loss from friction for turbulent flow is rewritten in terms of flow rate, Q, substitution of $4Q/\pi d^2$ for V yields

$$h_f = \frac{8fLQ^2}{\pi^2 g d^5}$$

Thus, the head loss from friction for turbulent flow is approximately proportional to the square of the flow rate and approximately inversely proportional to the diameter raised to the fifth power. If the diameter is doubled, the head loss will decrease by a factor of approximately 32. The proportionality is approximate because the friction factor, f, may change slightly with changes in flow rate and diameter.

Example 10.13

From Example 10.11 using the energy equation, the head loss from friction for 15 meters of 2.5-cm diameter smooth pipe was given as 8.33 m. Show how this value was calculated. The dynamic viscosity, ν, is 3.56×10^{-7} m^2/s.

Solution

$$h_f = f\frac{L}{d}\frac{V^2}{2g}, \quad L = 15\text{ m},\ d = 2.5\text{ cm} = .025\text{ m}$$

$$V = \frac{Q}{A} = \frac{130\dfrac{\text{L}}{\text{min}} \bullet \dfrac{1\text{ m}^3}{1000\text{ L}} \bullet \dfrac{\text{min}}{60\text{ s}}}{\dfrac{\pi(.025)^2}{4}\text{m}^2} = 4.41\text{ m/s}$$

$$\text{Re} = \frac{Vd}{\nu} = \frac{4.41\frac{\text{m}}{\text{s}} \bullet .025\text{ m}}{3.56 \times 10^{-7}\frac{\text{m}^2}{\text{s}}} = 3.1 \times 10^5$$

Now enter the Moody diagram for this Reynolds number, and reflect off the ε/d = smooth line to read $f = 0.014$
Hence

$$h_f = 0.014 \bullet \frac{15\text{ m}}{.025\text{ m}} \bullet \frac{\left(4.41\frac{\text{m}}{\text{s}}\right)^2}{2 \bullet 9.81\frac{\text{m}}{\text{s}^2}} = 8.33\text{ m}$$

Minor Losses

Flow losses from friction in pipe fittings, contractions, and enlargements are collectively known as minor losses. In problems where the pipe length is large, the minor losses in the system may be neglected. Minor losses are generally denoted by one of three methods:

(1) An equivalent length of pipe, L_e, is chosen, and the amount of the loss is calculated from

$$h_f = f\frac{L_e}{d}\frac{V^2}{2g}$$

(2) An equivalent length of pipe in diameters, $(L/d)_e$, is chosen, and the amount of loss is calculated from

$$h_f = f\left(\frac{L}{d}\right)_e\frac{V^2}{2g}$$

(3) A loss coefficient, C (or K), is chosen, and the amount of loss is calculated from

$$h_f = C\frac{V^2}{2g}$$

In more recent years the use of loss coefficients (C values) has become predominant and will be used in Fundamentals of Engineering exams. The C values may be a function of flow rate, fitting geometry, and/or diameter ratio (as in the case of contractions or enlargements) but in many cases may be considered constant over a wide range of conditions.

The equivalent lengths and C values for fittings and contractions are tabulated in catalogues, handbooks, and manuals. Representative values are listed in Table 10.3 and are illustrated in Fig. 10.9.

The total friction loss in a piping system is the sum of the pipe losses and all minor losses.

Table 10.3 Resistance in valves and fittings expressed as equivalent length in pipe diameters $(L/d)_e$, and loss coefficient, C

Type	Equivalent Length in Pipe Diameters, $(L/d)_e$	Loss Coefficient, C
Globe valve—fully open	340	6.80
Angle valve—fully open	145	2.90
Gate valve—fully open	13	0.26
Check valve—swing type	135	2.70
Check valve—ball type	150	3.00
Butterfly valve—fully open	40	0.80
90° standard elbow	30	0.60
90° long-radius elbow	20	0.40
90° street elbow	50	1.00
45° standard elbow	16	0.32
45° street elbow	26	0.52
Close return bend	50	1.00
Standard tee with flow-through run	20	0.40
Standard tee with flow-through branch	60	1.20

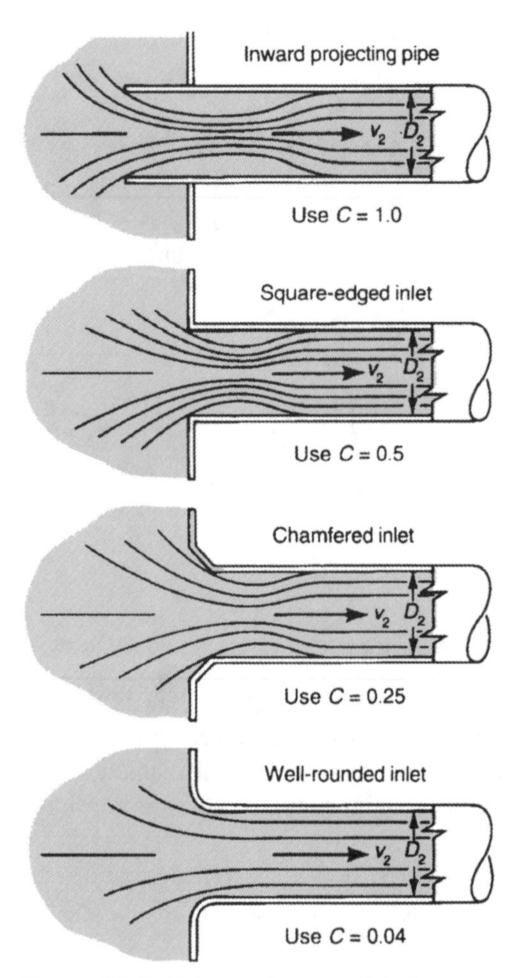

Exit Loss: $h_l = 1.0(v_1^2/2g)$

V_1 2 $V_2 = 0$ 1

Figure 10.9 Entrance loss coefficients

Example **10.14**

For Example 10.13, suppose there were three elbows ($C = 0.6$), two gate valves ($C = 0.26$), one globe valve ($C = 6.8$), and a square-edged entrance from the tank ($C = 0.5$). Calculate the head loss created by the minor losses for the existing flow conditions. Recalculate the head delivered by the pump considering both pipe and local losses.

Solution

Minor losses:

$$h_{fm} = \underbrace{3C\frac{V^2}{2g}}_{\text{elbows}} + \underbrace{2C\frac{V^2}{2g}}_{\text{gate vlvs.}} + \underbrace{C\frac{V^2}{2g}}_{\text{globe vlv.}} + \underbrace{C\frac{V^2}{2g}}_{\text{entrance}}$$

$$h_{fm} = [3(0.6) + 2(0.26) + 6.8 + 0.5]\frac{V^2}{2g}$$

$$h_{fm} = (9.62)\frac{\left(4.41\frac{\text{m}}{\text{s}}\right)^2}{2 \bullet 9.81\frac{\text{m}}{\text{s}^2}} = 9.54 \text{ m}$$

Total losses:

$$h_f = 8.33\,\text{m(pipe)} + 9.54\,\text{m(minor)} = 17.87\,\text{m}$$

$$h_A = \frac{\left(27.6\frac{\text{m}}{\text{s}}\right)^2}{2 \bullet 9.81\frac{\text{m}}{\text{s}^2}} + 3\,\text{m} + 17.87\,\text{m} = 59.7\,\text{m}$$

Problems to determine the pump or turbine head, a pressure, or an elevation are the simplest to solve. Since the flow rate and pipe diameter are known, the Reynolds number and relative roughness can be calculated directly; the friction factor can then be found from the Moody diagram, and the head loss is calculated for substitution into the energy equation. Problems where the flow rate is being sought—and other parameters including the pipe diameter are known—require a single iteration process (for turbulent flow). Since the Reynolds number cannot be initially calculated, an initial friction factor must be assumed—then corrected—to determine the flow rate. Problems where the pipe diameter is being sought, but the flow rate and other parameters are known, require an iteration process. After simplification of the energy equation, different pipe diameters are assumed and friction factors are determined from the Moody diagram, then both are substituted into the energy equation until the equation is satisfied.

Flow in Noncircular Conduits

The same fundamental equations for Reynolds number, relative roughness, and head loss from friction may be used for noncircular conduits. In place of the diameter, an equivalent diameter (or characteristic length) is used. The equivalent diameter is defined by

$$d_e = 4R_H = 4\frac{A}{WP}$$

where d_e = equivalent diameter, R_H = hydraulic radius = $\frac{A}{WP}$, A = cross-sectional area, and WP = wetted perimeter.

Example 10.15

Calculate the equivalent diameter of a rectangular conduit 0.6 meters wide and 0.3 meters high.

Solution

$$A = 0.6\,\text{m} \bullet 0.3\,\text{m} = 0.18\,\text{m}^2$$

$$WP = 2(0.6\,\text{m} + 0.3\,\text{m}) = 1.8\,\text{m}$$

$$d_e = 4\frac{A}{WP} = 4 \bullet \frac{0.18\,\text{m}^2}{1.8\,\text{m}} = 0.4\,\text{m}$$

Parallel Pipe Flow

The text to this point has addressed only flow in series piping systems; that is, all flow was considered to go through each pipe and fitting. If the flow divides into

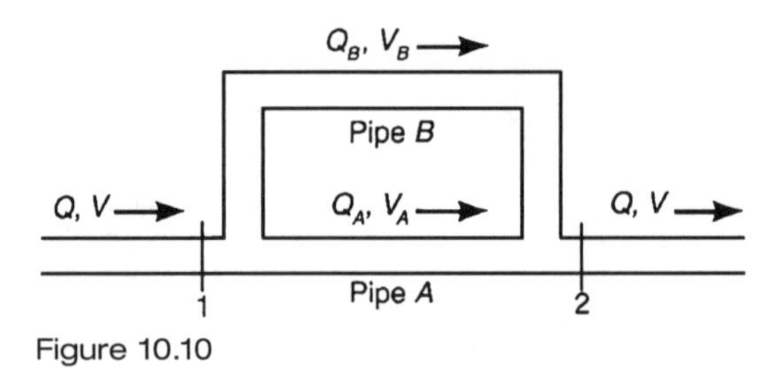

Figure 10.10

two parallel branches and returns again to a single pipe as shown in Fig. 10.10, the flow will divide so that the head loss in each branch is the same, or

$$h_{f1-2} = h_{fA} = h_{fB}$$

In the case when minor losses can be neglected, this relation becomes

$$h_{f1-2} = f_A \frac{L_A}{d_A} \frac{V_A^2}{2g} = f_B \frac{L_B}{d_B} \frac{V_B^2}{2g}$$

In addition, the continuity equation requires that

$$Q = QA + QB$$

Generally the flow rate, Q, or velocity, V, is known. Thus we have two equations and two unknowns, V_A and V_B. Values of f_A and f_B can be estimated and the two equations are then solved simultaneously for V_A and V_B. Using V_A and V_B to calculate the Reynolds numbers, corrected values of f_A and f_B can be found from the Moody diagram, and V_A and V_B are then recalculated. Then the head loss can be found from the above equation. For a flow that divides into three parallel branches, the same analysis can be used, where

$$h_{f1-2} = h_{fA} = h_{fB} = h_{fC}$$
$$Q = Q_A + Q_B + Q_C$$

In this case, we have three equations, and three unknowns.

FORCES ATTRIBUTABLE TO CHANGE IN MOMENTUM

The force created by the change in momentum of a fluid undergoing steady flow is given by the impulse momentum equation

$$F = \Delta(\dot{m}V) = \rho Q V_2 - \rho Q V_1$$

where F = resultant force on the fluid stream, ρQ = mass rate of flow, V_1, V_2 = inlet and exit velocities, respectively, $\rho Q V_1$ = momentum per second at the inlet, and $\rho Q V_2$ = momentum per second at the exit.

Forces on Bends

The magnitude and direction of the resultant force in flow through a bend will depend on the change in velocity magnitude and/or the change in direction of the flow. If the flow occurs in a two-dimensional bend, the equation can be rewritten in scalar form. Using Fig. 10.11,

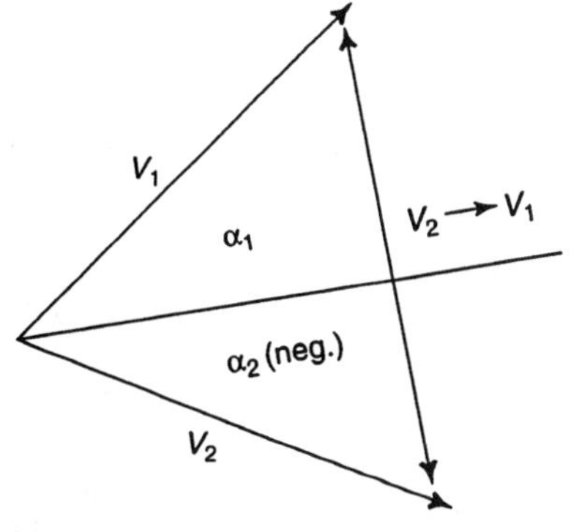

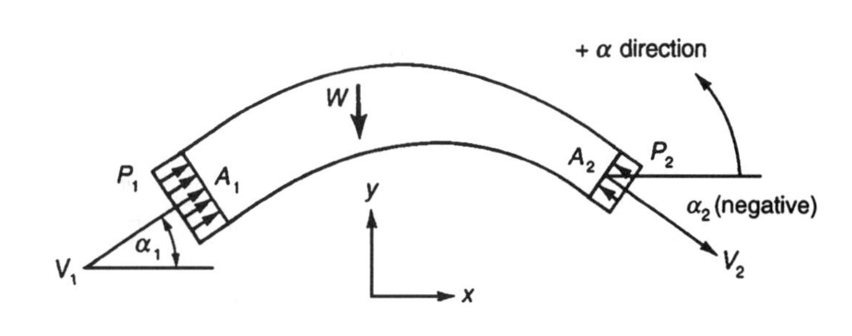

Figure 10.11

$$F_x = p_2A_2 \cos\alpha_2 - p_1A_1 \cos\alpha_1 + \rho Q(V_2 \cos\alpha_2 - V_1 \cos\alpha_1)$$

$$F_y - W = p_2A_2 \sin\alpha_2 - p_1A_1 \sin\alpha_1 + \rho Q(V_2 \sin\alpha_2 - V_1 \sin\alpha_1)$$

$$F = \sqrt{F_x^2 + F_y^2}$$

where

F_x, F_y = resultant force component on the fluid by the bend in the x and y directions, respectively,

p_1, p_2 = inlet and exit pressure on the fluid in the bend,

A_1, A_2 = inlet and exit cross-sectional areas, respectively, and

α_1, α_2 = angles that the direction of flow makes with the positive x-direction at the entrance and exit, respectively.

The equation may be simplified somewhat if the x-direction is chosen as the direction of the entering flow. The $\alpha_1 = 0$ and α_2 would be measured relative to that x-axis and

$$F_x = p_2A_2 \cos\alpha_2 - p_1A_1 + \rho Q(V_2 \cos\alpha_2 - V_1)$$

$$F_y - W = p_2A_2 \sin\alpha_2 + \rho QV_2 \sin\alpha_2$$

In many cases the weight of the fluid in the bend may not be significant. In addition to the impulse-momentum equations, it may be necessary simultaneously to utilize the energy and continuity equations in solving such problems.

Example 10.16

The 45° reducing bend discharges 0.008 m^3/s of water to the atmosphere, as shown in Exhibit 5. The entrance diameter of the bend is 50 mm, and the exit diameter is 30 mm. Neglect the small elevation change, the weight of the fluid in the bend, and friction. Calculate the magnitude and force of the water on the bend. The density of the water is 1000 kg/m^3, and its specific weight is 9810 N/m^3.

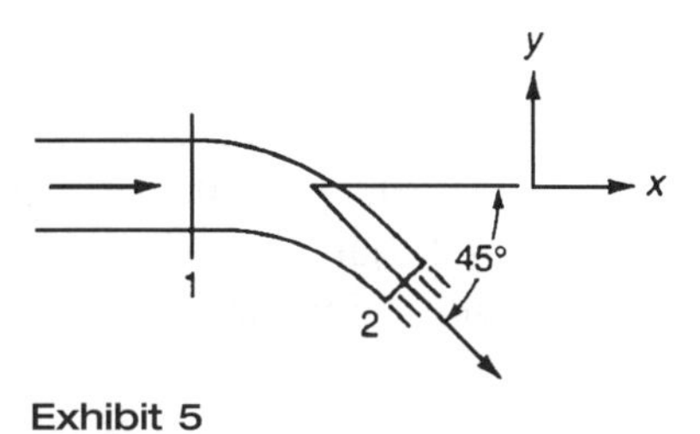

Exhibit 5

Solution

Select Sections 1 and 2 as shown. Then $\alpha_1 = 0$, $\cos\alpha_1 = 1$, $\sin\alpha_1 = 0$, $\alpha_2 = -45°$, $\cos\alpha_2 = 0.707$, $\sin\alpha_2 = -0.707$, $p_2 = 0$, and the force-momentum equation becomes

$$F_x = -p_1A_1 + \rho Q[V_2(0.707) - V_1]$$
$$F_y = \rho Q[V_2(-0.707)]$$
$$\rho Q = 1000\frac{\text{kg}}{\text{m}^3} \bullet 0.008\frac{\text{m}^3}{\text{s}} = 8.0\frac{\text{kg}}{\text{s}}$$
$$A_1 = \frac{\pi(0.05\,\text{m})^2}{4} = 0.00196\,\text{m}^2, \quad A_2 = \frac{\pi(0.03\,\text{m})^2}{4} = 0.00071\ \text{m}^2$$
$$V_1 = \frac{Q}{A_1} = \frac{0.008\ \text{m}^3/\text{s}}{0.00196\ \text{m}^2} = 4.08\,\text{m/s}, \quad V_2 = \frac{0.008\,\text{m}^3/\text{s}}{0.00071\,\text{m}^2} = 11.3\ \text{m/s}$$

The pressure p_1 may be found from the energy equation:

$$\frac{p_1}{\gamma} + \frac{V_1^2}{2g} + Z_1 = \frac{p_2}{\gamma} + \frac{V_2^2}{2g} + Z_2$$

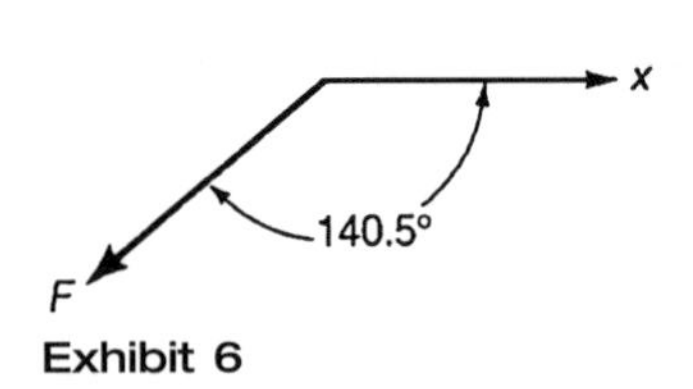

Exhibit 6

$$p_1 = \gamma\left(\frac{V_2^2 - V_1^2}{2g}\right) = 9810\frac{\text{N}}{\text{m}^3} \bullet \frac{\left(11.3^2 - 4.08^2\right)\frac{\text{m}^2}{\text{s}^2}}{2 \bullet 9.81\frac{\text{m}}{\text{s}^2}} = 55{,}500\frac{\text{N}}{\text{m}^2}$$

$$F_x = -55{,}500\frac{\text{N}}{\text{m}^2} \bullet 0.00196\ \text{m}^2 + 8.0\frac{\text{kg}}{\text{s}}\left(11.3\frac{\text{m}}{\text{s}} \bullet 0.707 - 4.08\frac{\text{m}}{\text{s}}\right)$$
$$F_x = -108.8\,\text{N} + 31.3\,\text{N} = -77.5\,\text{N}$$
$$F_y = 8.0\frac{\text{kg}}{\text{s}} \bullet 11.3\frac{\text{m}}{\text{s}}(-0.707) = -63.9\,\text{N}$$
$$F = \sqrt{(-77.5)^2 + (-63.9)^2} = 100.4\,\text{N}$$
$$\text{at } \theta = -90.0° - \arctan\frac{-77.5}{-63.9} = -140.5°$$

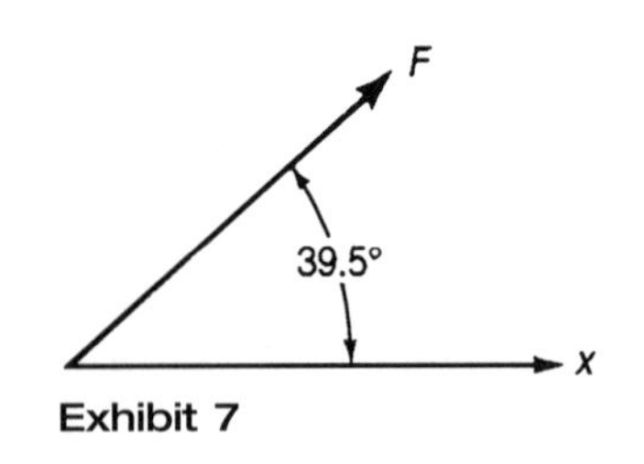

Exhibit 7

This is the force of the bend on the fluid (Exhibit 6). The force of the fluid on the bend is equal and opposite (Exhibit 7).

Jet Engine and Rocket Thrust

The impulse-momentum relationship may be used to calculate the thrust created by a jet engine. Since the inlet and exit pressures are zero, the thrust may be calculated from a combination of equations for both the air and fuel:

$$F = \rho_a Q_a (V_2 - V_1) + \rho_f Q_f V_2$$

where V_1 = entering velocity relative to the engine (for engines in flight, this is the velocity of the aircraft), V_2 = exit velocity relative to the engine, ρ_a, ρ_f = density of the air and fuel, respectively, Q_a, Q_f = flow rate of the air and fuel, respectively, and F = thrust.

This equation assumes that the fuel enters the engine perpendicular to the thrust direction and leaves with the exhaust flow.

In the case of rocket propulsion, since the fuel and oxidizer initially are at rest, the equation for thrust becomes

$$F = \rho_m Q_m V_2$$

where ρ_m = density of the fuel-oxidizer mixture, and Q_m = flow rate of the fuel-oxidizer mixture.

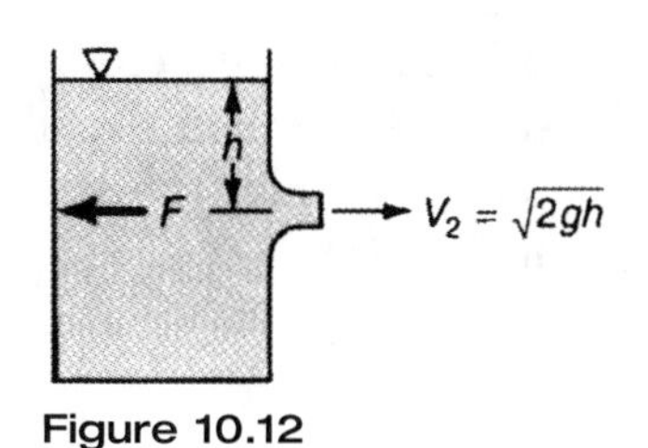

Figure 10.12

A particular example of jet propulsion is shown Fig. 10.12. The thrust is given by

$$F = \rho Q V_2 = \rho(A_2 V_2)V_2 = \rho A_2 V_2^2$$

but by Bernoulli's equation $V_2 = \sqrt{2gh}$. Substituting,

$$F = \rho A_2(2gh) = 2\gamma A_2 h$$

where γ = specific weight of the liquid, A_2 = area of nozzle exit, h = height of fluid surface above the nozzle, and F = thrust or propulsion force.

Forces on Stationary Vanes

The impulse-momentum equation can also be used to determine forces *on* stationary vanes. Consider the stationary vane shown in Fig. 10.13.

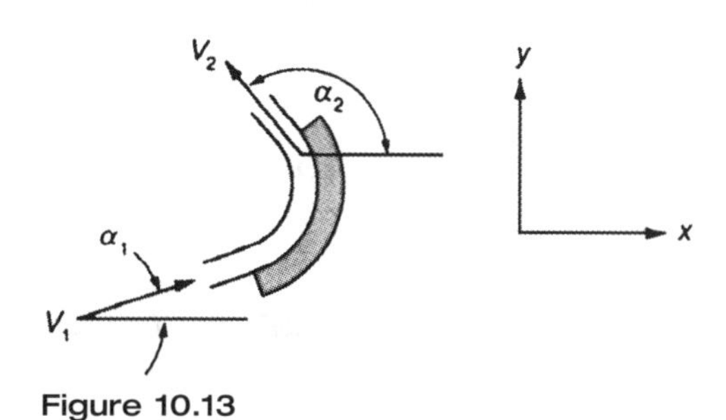

Figure 10.13

Since the fluid is open to the atmosphere, $p_1 = p_2 = 0$. If friction is neglected, it can be shown from Bernoulli's equation that $V_1 = V_2 = V$. The impulse-momentum equation reduces to

$$F_x = \rho Q(V_1 \cos \alpha_1 - V_2 \cos \alpha_2) = \rho QV(\cos \alpha_1 - \cos \alpha_2)$$
$$F_y = \rho QV(\sin \alpha_1 - \sin \alpha_2)$$

where F_x, F_y = *force of the fluid on the vane* in the x and y directions, respectively, α_1, α_2 = angle between the positive x-direction and the entrance and exit velocities, respectively, ρQ = mass flow rate, and V = fluid velocity.

If we reorient the vane so that the entrance velocity is in the x-direction, as shown in Fig. 10.14, the equation can be rewritten as

$$F_x = \rho QV(1 - \cos \alpha_2)$$
$$F_y = -\rho QV \sin \alpha_2$$

Figure 10.14

Example 10.17

Water impinges upon a stationary vane with a velocity of 15 m/s through a cross-sectional area of 6 square centimeters. The vane is oriented so that the fluid enters the vane cavity in the x-direction and leaves at an angle of 60° (Exhibit 8). The density of the water is 1000 kg/m^3. Calculate the force on the vane.

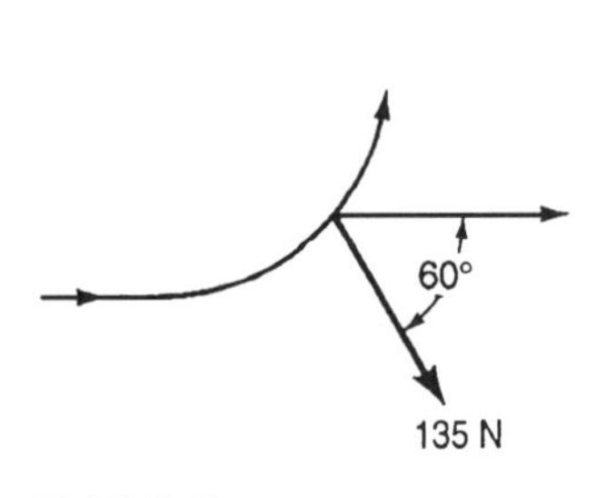

Exhibit 8

Solution

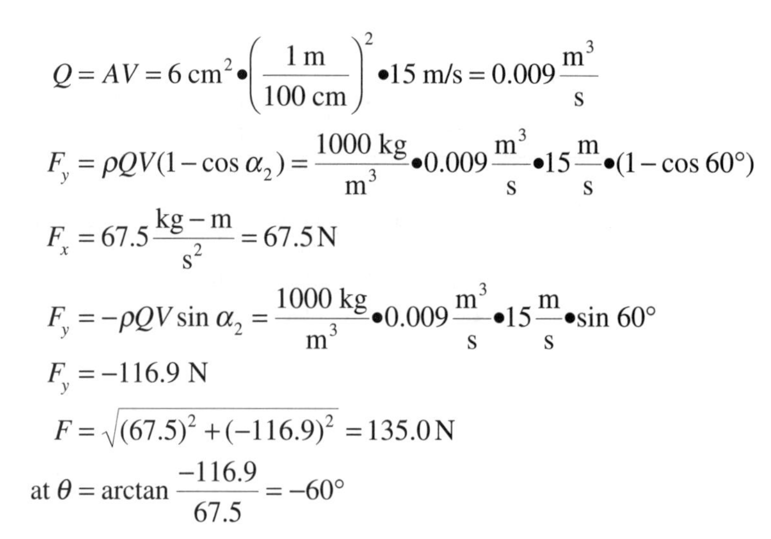

$$Q = AV = 6\ \text{cm}^2 \bullet \left(\frac{1\ \text{m}}{100\ \text{cm}}\right)^2 \bullet 15\ \text{m/s} = 0.009\frac{\text{m}^3}{\text{s}}$$

$$F_y = \rho Q V(1-\cos\alpha_2) = \frac{1000\ \text{kg}}{\text{m}^3}\bullet 0.009\frac{\text{m}^3}{\text{s}}\bullet 15\frac{\text{m}}{\text{s}}\bullet(1-\cos 60°)$$

$$F_x = 67.5\frac{\text{kg}-\text{m}}{\text{s}^2} = 67.5\,\text{N}$$

$$F_y = -\rho Q V\sin\alpha_2 = \frac{1000\ \text{kg}}{\text{m}^3}\bullet 0.009\frac{\text{m}^3}{\text{s}}\bullet 15\frac{\text{m}}{\text{s}}\bullet\sin 60°$$

$$F_y = -116.9\ \text{N}$$

$$F = \sqrt{(67.5)^2 + (-116.9)^2} = 135.0\,\text{N}$$

$$\text{at } \theta = \arctan\frac{-116.9}{67.5} = -60°$$

Forces on Moving Vanes

If the fluid enters the vane cavity in the x-direction and the vane is also moving in the x-direction with a velocity, v, as shown in Fig. 10.15(a), then the force of the fluid on the moving vane in the direction of motion may be calculated from

$$F_x = \rho Q'(V-v)(1-\cos\alpha_2)$$
$$F_y = -\rho Q'(V-v)\sin\alpha_2$$
$$Q' = A(V-v)$$

where v = velocity of the vane, and $V - v$ = fluid velocity relative to the vane.

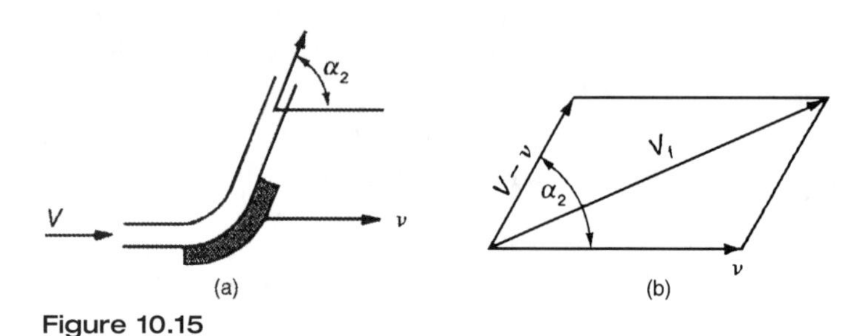

Figure 10.15

The final direction and magnitude of the jet is shown by V_f in the vector diagram [Fig. 10.15(b)].

An impulse turbine contains a series of moving vanes as described above, one immediately replacing another as the turbine rotor rotates. See Fig. 10.16.

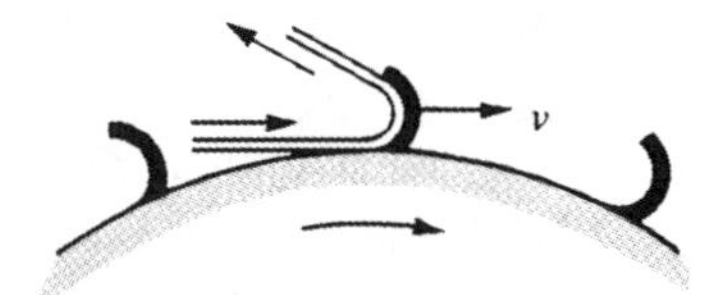

Figure 10.16

The x-direction force of the fluid on the series of vanes is given by

$$F = \rho Q(V - v)(1 - \cos \alpha_2)$$

The power delivered to the turbine, P, is given by

$$P = Fv = \rho Qv(V - v)(1 - \cos \alpha_2)$$

The maximum power for a given discharge angle, α_2, occurs when $V = 2v$ and is given by

$$P_{\max} = \frac{\rho Q V^2}{4}(1 - \cos \alpha_2)$$

The discharge angle which produces the maximum power possible is $\alpha_2 = 180°$. This maximum power is calculated from

$$P_{\max} = \frac{\rho Q V^2}{2} = g_c \gamma Q \frac{V^2}{2g}$$

VELOCITY AND FLOW MEASURING DEVICES

Pitot Tubes

For liquids flowing at relatively low pressures the mean static pressure may be measured using a piezometric tube indicated in Fig. 10.17 by h_1. The stagnation pressure is indicated by h_2.

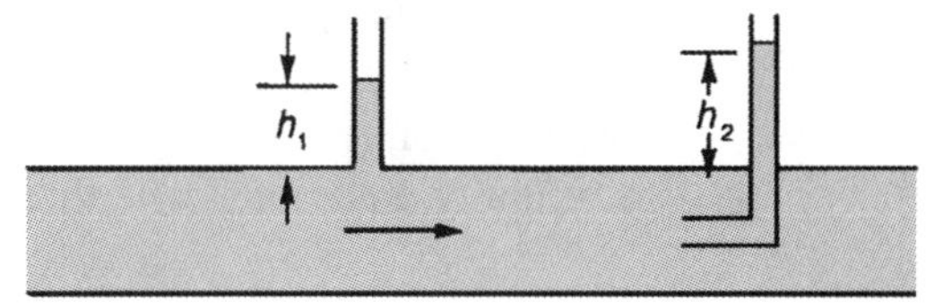

Figure 10.17

The relationship of each measurement to pressure, specific weight, and velocity is shown in the following two equations:

$$h_1 = \frac{p}{\gamma}, \quad h_2 = \frac{p_s}{\gamma} = \frac{p}{\gamma} + \frac{V^2}{2g}$$

Combining these relations, the velocity in the duct is

$$V = \sqrt{2g\left(\frac{p_s - p}{\gamma}\right)} = \sqrt{2g(h_2 - h_1)}$$

where V = velocity, γ = specific weight, p = static pressure, and p_s = stagnation pressure.

The combination of the two tubes as a single device is known as a pitot tube. If a manometer is connected between the static and stagnation pressure taps, velocities at moderate pressures may be calculated using the following equation:

$$V = \sqrt{2gh_m\left(\frac{\gamma_m}{\gamma} - 1\right)}$$

where h_m = height indicated by the manometer, and γ_m = specific weight of manometer fluid.

Example 10.18

A pitot tube is used to measure the mean velocity in a pipe where water is flowing. A manometer containing mercury is connected to the pitot tube and indicates a height of 150 mm. The specific weights of the water and mercury are 9810 N/m^3 and 133,400 N/m^3, respectively. Calculate the velocity of the water.

Solution

$$V = \sqrt{2gh_m\left(\frac{\gamma_m}{\gamma} - 1\right)}$$

$$V = \sqrt{2 \bullet 9.81\,\frac{\text{m}}{\text{s}^2} \bullet 0.15\,\text{m} \bullet \left(\frac{133{,}400\,\frac{\text{N}}{\text{m}^3}}{9810\,\frac{\text{N}}{\text{m}^3}} - 1\right)} = 6.09\,\frac{\text{m}}{\text{s}}$$

The pitot tube equation may also be used for compressible fluids with Mach numbers less than or equal to 0.3.

Flow Meters

There are three commonly used meters that measure flow rate in fluid systems: venturi meters, flow nozzles, and orifice meters. All three operate on the same basic principle, their equations being developed by combining the Bernoulli and continuity equations. The three meters are shown in Fig. 10.18.

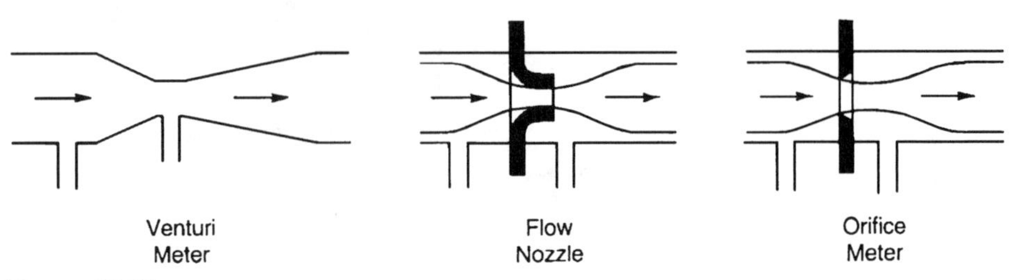

Figure 10.18

The equation for flow rate is given by

$$Q = \frac{c_v c_c}{\sqrt{1 - c_c^2 \left(\frac{A_2}{A_1}\right)^2}} \bullet A_2 \bullet \sqrt{2g\left(\frac{p_1 - p_2}{\gamma} + Z_1 - Z_2\right)}$$

or if a manometer is used between the pressure taps

$$Q = \frac{c_v c_c}{\sqrt{1 - c_c^2 \left(\frac{A_2}{A_1}\right)^2}} \bullet A_2 \bullet \sqrt{2gh_m\left(\frac{\gamma_m}{\gamma} - 1\right)}$$

where

Q = flow rate
$p_1 - p_2$ = pressure difference between a point before the entrance to the meter, and the point of narrowest flow cross-section in the meter
$Z_1 - Z_2$ = height difference between a point before the entrance to the meter, and the point of narrowest flow cross-section in the meter
A_1 = area of entrance
A_2 = area of narrowest flow cross section, except in the orifice, where it is the orifice area
h_m = height indicated by manometer
γ_m = specific weight of manometer fluid
γ = specific weight of fluid
c_v = coefficient of velocity
c_c = coefficient of contraction

For the orifice meter (and sometimes the flow nozzle) the coefficient terms in the equation are combined as follows:

$$c = \frac{c_v c_c}{\sqrt{1 - c_c^2 \left(\frac{A_2}{A_1}\right)^2}}$$

and the flow rate equation is then written

$$Q = cA_2\sqrt{2g\left(\frac{p_1 - p_2}{\gamma} + Z_1 - Z_2\right)} = cA_2\sqrt{2gh_m\left(\frac{\gamma_m}{\gamma} - 1\right)}$$

where c = orifice (or flow nozzle) coefficient.

The following values of c_c, c_v, and c are used for the various flow meters:

Venturi:	$c_c = 1,\ 0.95 < c_v < 0.99$;	c_v (nominal) = 0.984
Flow nozzle:	$c_c = 1,\ 0.95 < c_v < 0.99$;	c_v (nominal) = 0.98
Orifice:	$c_c = 0.62,\ c_v = 0.98$;	c (nominal) = 0.61

Actual values of c for orifice meters (and flow nozzles) vary with the diameter ratio, $d_0{:}d_1$, and the Reynolds number and are found on existing graphs. Curves for the values of c_v for venturi meters and flow nozzles are also available.

Flow From a Tank

The flow from a tank through various types of exit configurations can be calculated by using the energy equation and experimentally determined configuration coefficients. Consider a tank as shown in Fig. 10.19. For frictionless flow, the flowrate can be calculated using the energy and continuity equations which reduce to

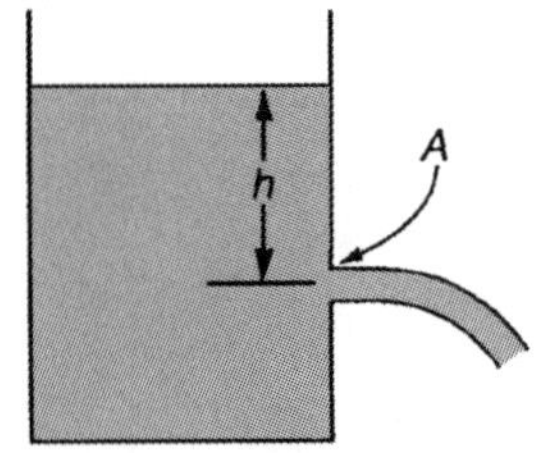

$$Q = AV = A\sqrt{2gh}$$

Figure 10.19

Considering friction, the flow rate may be calculated from

$$Q = cA\sqrt{2gh}, \quad c = c_v c_c$$

where Q = flow rate from the tank, h = height of water level above the exit, and c = coefficient of discharge for the exit.

If the friction in the exit is neglected, then $c = 1$.

For a sharp edged orifice, $c_v = 0.98$, $c_c = 0.62$, and $c = 0.61$.

For a rounded exit, $c_v = 0.98$, $c_c = 1.00$, and $c = 0.98$.

For a short tube exiting from the tank, $c_v = 0.80$, $c_c = 1.00$, and $c = 0.80$.

For a re-entrant pipe, $c_v = 0.98$, $c_c = 0.52$, and $c = 0.51$.

For the special case when the flow from the tank discharges beneath the surface of the same fluid outside the tank, the flow rate from the tank is given by

$$Q = cA\sqrt{2gh(h_1 - h_2)}$$

where h_1 = height of fluid above exit in tank, and h_2 = height of fluid above exit outside tank.

SIMILARITY AND DIMENSIONLESS NUMBERS

The Reynolds number is a dimensionless number defined as the ratio of inertial forces to viscous forces. To test a model of some prototype, such as an air foil or length of pipe, the Reynolds number of the model must be equal to the Reynolds number of the prototype, or

$$(\mathrm{Re})_m = (\mathrm{Re})_p$$

$$\left(\frac{\rho V l}{\mu}\right)_m = \left(\frac{\rho V l}{\mu}\right)_p$$

where l = characteristic length.

A model is generally smaller geometrically than its prototype. Thus, if the characteristic length, l, of a model is to be one-tenth that of the prototype then one or more of the other terms in the Reynolds number must be adjusted to retain the Reynolds number the same for the model. For instance, the velocity, V, could be increased by a factor of ten using the same fluid or the fluid could be changed (liquid to gas) such that μ/ρ decreases by a factor of 10 with the same velocity, or a combination of the two. This condition is known as **dynamic similarity** of the prototype and model.

In fact, there are several independent force ratios that should be maintained in developing a model, depending on what forces are predominant in the situation.

These force ratios involve pressure, inertia, viscosity, gravity, elasticity, and surface tension. The force ratios required for dynamic similarity are defined as follows, where the subscript m denotes model and p denotes prototype:

Inertia Force/Pressure Force Ratio

$$\left(\frac{\rho V^2}{p}\right)_m = \left(\frac{\rho V^2}{p}\right)_p$$

Inertia Force/Viscous Force Ratio

$$\left(\frac{\rho Vl}{\mu}\right)_m = \left(\frac{\rho Vl}{\mu}\right)_p = \text{Reynolds number, Re}$$

Inertia Force/Gravity Force Ratio

$$\left(\frac{V^2}{lg}\right)_m = \left(\frac{V^2}{lg}\right)_p = \text{Froude number, } \mathcal{F}$$

Inertia Force/Elastic Force Ratio

$$\left(\frac{\rho V^2}{E}\right)_m = \left(\frac{\rho V^2}{E}\right)_p = \text{Cauchy number, } C_a$$

where E = modulus of elasticity of fluid.

Inertia Force/Surface Tension Force

$$\left(\frac{\rho l V^2}{\sigma}\right)_m = \left(\frac{\rho l V^2}{\sigma}\right)_p = \text{Weber number, } W_e$$

In many applications, one or more of the force ratios may be neglected because the forces are negligible.

INCOMPRESSIBLE FLOW OF GASES

The relationships thus far developed are primarily for the flow of incompressible fluids (liquids). Many are also applicable to the flow of compressible fluids (gases), for example, the continuity equation, equation for the Reynolds number, and so forth. The energy equation for incompressible flow also may be used under the following conditions:

1. The change in pressure in the pipe length is less than 10 percent of the inlet pressure. The density and specific weight at inlet conditions (pressure and temperature) should be used.
2. The change in pressure in the pipe length is between 10 and 40 percent of the inlet pressure. The density and specific weight at the average of the inlet and outlet conditions should be used. In some problems the outlet (or inlet) pressure is sought. In this case, the inlet (or outlet) conditions are used to find initial values of density and specific weight, and the approximate outlet (or inlet) pressure is then calculated. An iterative process ensues.

It may be necessary to utilize the Perfect Gas Law from thermodynamics to calculate various properties, particularly the density of the gas given the pressure and temperature. The Perfect Gas Law may be written in the following form

$$\rho = \frac{P}{RT}$$

where $R = \overline{R}/\text{MW}$ = gas constant
MW = gas molecular weight
$\overline{R}$ = universal gas constant

It should also be noted that the speed of sound, c, in a perfect gas is given by

$$c = \sqrt{kRT}$$

where k = ratio of specific heats, c_p/c_v
c_p = specific heat at constant pressure
c_v = specific heat at constant volume

It is apparent from the equation above that the speed of sound (acoustic velocity) in a gas depends only on its temperature.

The mach number Ma is the ratio of the actual fluid velocity to the speed of sound:

$$\text{Ma} = V/c$$

The accuracy of utilizing incompressible fluid flow equations for the flow of gases decreases with increasing velocities and their use is not recommended for mach numbers greater than 0.2.

SELECTED SYMBOLS AND ABBREVIATIONS

Symbol or Abbreviation	Description
α	angle with horizontal
β	angle with vertical
C	loss coefficient
C_a	Cauchy number
c	discharge coefficient, speed of sound
c_c	contraction coefficient
c_p	gas specific heat at constant pressure
c_v	velocity coefficient, gas specific heat at constant volume
δ	small thickness
d	diameter
d_e	equivalent diameter
ε	roughness factor
F_B	buoyant force
f	Darcy friction factor
$\mathcal{F}$	Froude number
γ	specific weight
g_c	gravitational constant
h	depth of fluid

(continued)

Symbol or Abbreviation	Description
h_A	pump head
h_f	head loss from friction
h_R	turbine head
I	moment of inertia
k	ratio of specific heats
l	characteristic length
$\dot{m}$	mass flow rate
Ma	mach number
MW	molecular weight
μ	dynamic viscosity
η	efficiency
ν	kinematic viscosity
P	power to or from fluid
p	pressure
Q	volume rate of flow
ρ	density
R	gas constant
$\overline{R}$	universal gas constant
Re	Reynolds number
R_H	hydraulic radius
σ	surface tension
s	seconds
SG	specific gravity
τ	shear stress
T	absolute temperature, °R, °K
V	fluid velocity, volume
V_D	displaced volume
v	vane velocity, specific volume
$\dot{W}$	mechanical (shaft) power
W_c	Weber number
WP	Wetted perimeter
Z	elevation

PROBLEMS

10.1 Kinematic viscosity may be expressed in units of

a. m^2/s c. kg • s/m
b. s^2/m d. kg/s

10.2 The absolute viscosity of a fluid varies with pressure and temperature and is defined as a function of

a. density and angular deformation rate
b. density, shear stress, and angular deformation rate
c. density and shear stress
d. shear stress and angular deformation rate

10.3 An open chamber rests on the ocean floor in 50 m of sea water (SG = 1.03). The air pressure in kilopascals that must be maintained inside to exclude water is nearest to

a. 318 c. 505
b. 431 d. 661

10.4 What is the static gage pressure in pascals in the air chamber of the container in Exhibit 10.4? The specific weight of the water is 9810 N/m^3.

a. −14,700 Pa c. 0
b. −4500 Pa d. +4500 kPa

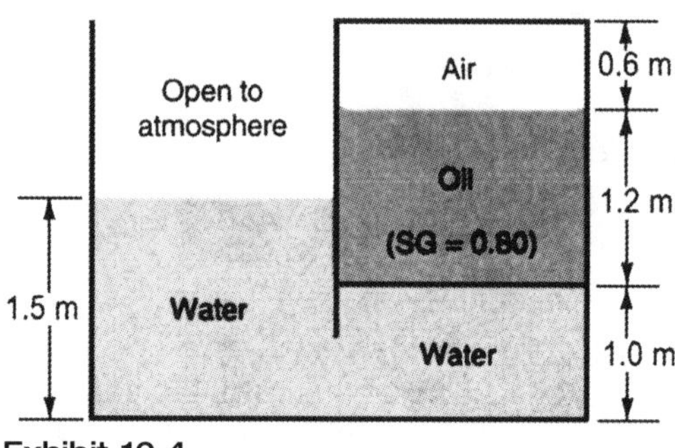

Exhibit 10.4

10.5 The pressure in kilopascals at a depth of 100 meters in fresh water is nearest to

a. 268 kPa c. 981 kPa
b. 650 kPa d. 1,620 kPa

10.6 What head, in meters of air, at ambient conditions of 100 kPa and 20 °C, is equivalent to 15 kPa?

a. 49 c. 257
b. 131 d. 1282

10.7 With a normal barometric pressure at sea level, the atmospheric pressure at an elevation of 1200 meters is nearest to:

a. 87.3 kPa c. 115.3 kPa
b. 83 kPa d. 101.3 kPa

10.8 The funnel in Exhibit 10.8 is full of water. The volume of the upper part is 0.165m^3 and of the lower part is 0.057m^3. The force tending to push the plug out is

a. 1.00 kN
b. 1.47 kN
c. 1.63 kN
d. 2.00 kN

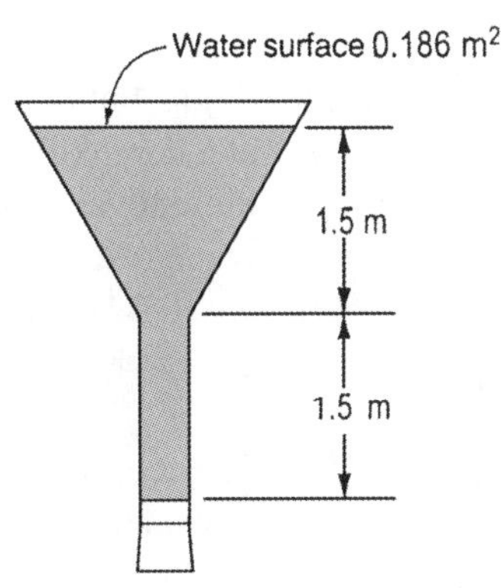

Exhibit 10.8

10.9 An open-topped cylindrical water tank has a horizontal circular base 3 meters in diameter. When it is filled to a height of 2.5 meters, the force in Newtons exerted on its base is nearest to:

a. 17,340
b. 34,680
c. 100,000
d. 170,000

10.10 A cubical tank with 1.5 meter sides is filled with water (see Exhibit 10.10). The force, in kilonewtons, developed on one of the vertical sides is nearest to

a. 4.1
b. 8.3
c. 16.5
d. 33.0

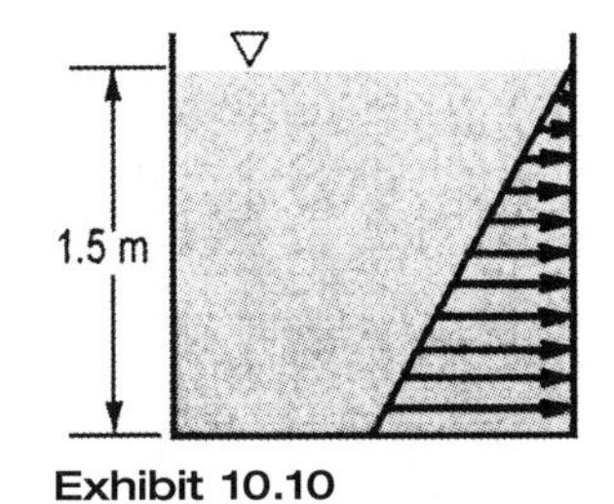

Exhibit 10.10

10.11 A conical reducing section (see Exhibit 10.11) connects an existing 10-centimeter diameter pipeline with a new 5-centimeter diameter line. At 700 kPa under no-flow conditions, what tensile force in kilonewtons is exerted on the reducing section?

a. 5.50
b. 2.07
c. 1.37
d. 4.13

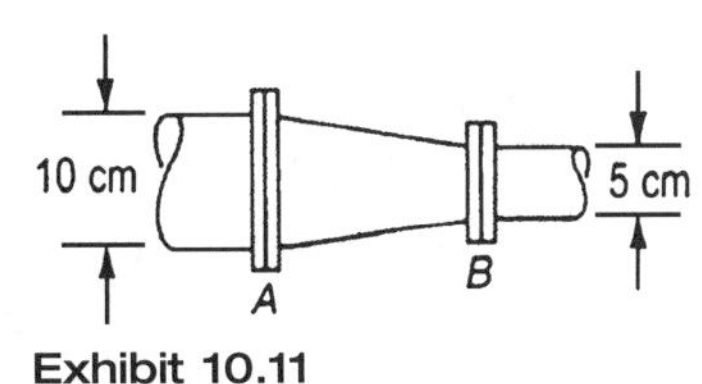

Exhibit 10.11

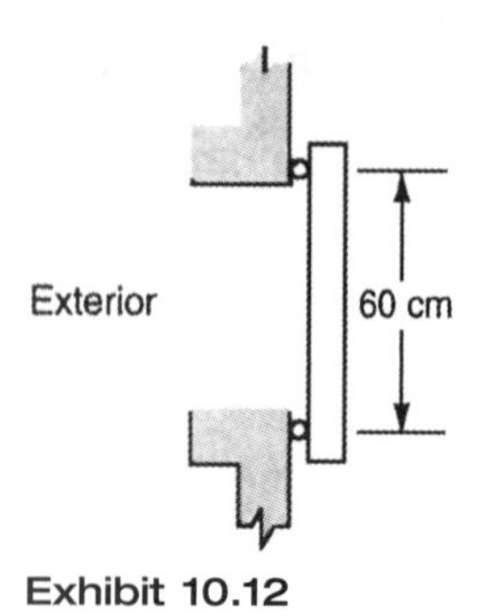

Exhibit 10.12

10.12 A circular access (see Exhibit 10.12) port 60 cm in diameter seals an environmental test chamber that is pressurized to 100 kPa above the external pressure. What force in newtons does the port exert upon its retaining structure?

a. 7100 c. 14,100
b. 9500 d. 28,300

10.13 A gas bubble rising from the ocean floor is 2.5 centimeters in diameter at a depth of 15 meters. Given that the specific gravity of seawater is 1.03, the buoyant force in newtons being exerted on the bubble at this instant is nearest to

a. 0.0413 c. 0.164
b. 0.0826 d. 0.328

10.14 The ice in an iceberg has a specific gravity of 0.922. When floating in seawater (SG = 1.03), the percentage of its exposed volume is nearest to

a. 5.6 c. 8.9
b. 7.4 d. 10.5

10.15 A cylinder of cork is floating upright in a container partially filled with water. A vacuum is applied to the container that partially removes the air within the vessel. The cork will

a. rise somewhat in the water
b. sink somewhat in the water
c. remain stationary
d. turn over on its side

10.16 A floating cylinder 8 cm in diameter and weighing 9.32 newtons is placed in a cylindrical container that is 20 cm in diameter and partially full of water. The increase in the depth of water when the float is placed in it is

a. 10 cm c. 3 cm
b. 5 cm d. 2 cm

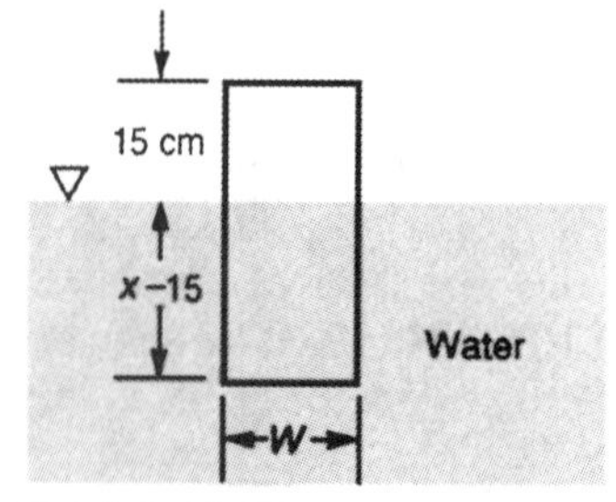

Exhibit 10.17

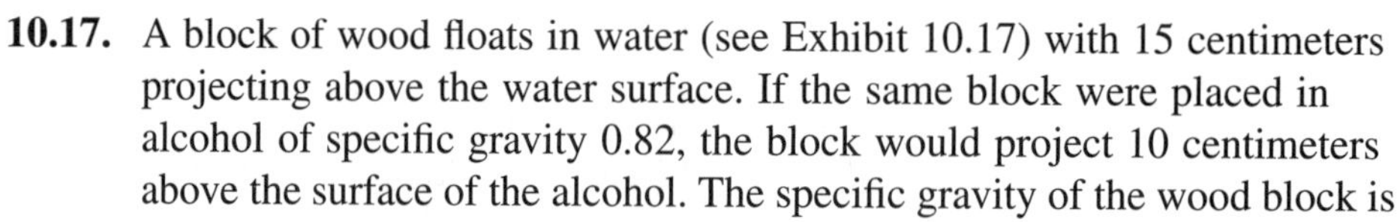

10.17. A block of wood floats in water (see Exhibit 10.17) with 15 centimeters projecting above the water surface. If the same block were placed in alcohol of specific gravity 0.82, the block would project 10 centimeters above the surface of the alcohol. The specific gravity of the wood block is

a. 0.67
b. 3.00
c. 0.55
d. 0.60

10.18 The average velocity in a full pipe of incompressible fluid at Section 1 in Exhibit 10.18 is 3 m/s. After passing through a conical section that reduces the stream's cross-sectional area at Section 2 to one-fourth of its previous value, the velocity at Section 2, in m/s, is

a. 1.0 c. 3
b. 1.5 d. 12

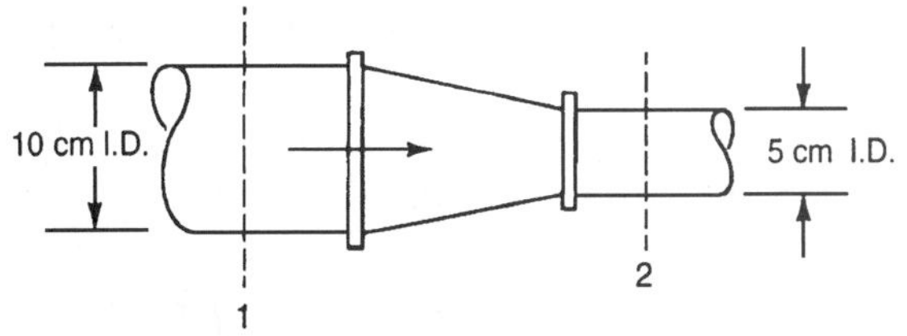

Exhibit 10.18

10.19 Refer to Exhibit 10.18. If the static pressure at Section 1 is 700 kPa and the 10 cm diameter pipe is full of water undergoing steady flow at an average velocity of 10 m/s at A, the mass flow rate in kg/s at Section 2 is nearest to

a. 10.0 c. 78.5
b. 19.5 d. 98.6

10.20 Air flows in a long length of 2.5 cm diameter pipe. At one end the pressure is 200 kPa, the temperature is 150°C and the velocity is 10 m/s. At the other end, the pressure has been reduced by friction and heat loss to 130 kPa. The mass flow rate in kg/s at any section along the pipe is nearest to

a. 0.008 c. 0.126
b. 0.042 d. 0.5

10.21 Water flows through a long 1.0 cm I.D. hose at 10 liters per minute. The water velocity in m/s is nearest to

a. 1 c. 4.24
b. 2.12 d. 21.2

10.22 Gasoline ($\rho = 800$ kg/m^3) enters and leaves a pump system with the energy in N•m/N of fluid that is shown in the following table:

	Entering	Leaving
Potential energy, Z meters above datum	1.5	4.5
Kinetic energy, $V^2/(2g_c)$	1.5	3.0
Flow energy, p/γ	9.0	45
Total energy	12.0	52.5

The pressure increase in kPa between the entering and leaving streams is nearest to

a. 283 c. 722
b. 566 d. 803

10.23 Use the data of Problem 10.22. If the volume flow rate of the gasoline (800 kg/m^3) is 55 liters per minute, the theoretical pumping power, in kW, is nearest to

a. 0.3 c. 3.2
b. 0.5 d. 300

10.24 Water flowing in a pipe enters a horizontal venturi meter whose throat area at B is 1/4 that of the original and final cross-sections at A and C, as shown in Exhibit 10.24. Continuity and energy conservation demand that which one of the following be true?

a. The pressure at B is increased.
b. The velocity at B is decreased.
c. The potential energy at C is decreased.
d. The flow energy at B is decreased.

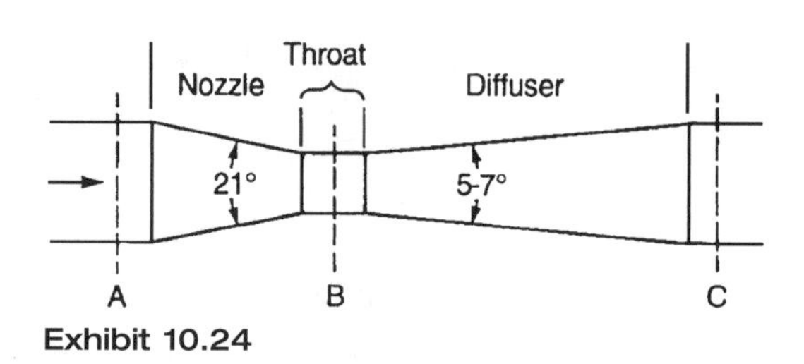

Exhibit 10.24

10.25 Given the energy data in N•m/N shown below, existing at two sections across a pipe transporting water in steady flow.

	Section A	Section B
Potential energy	20	40
Kinetic energy	15	15
Flow energy	100	75
Total	135	130

What frictional head loss in feet has occurred?

a. 0 c. 130
b. 5 d. 265

10.26 The power in kilowatts required in the absence of losses to pump water at 400 liters per minute from a large reservoir to another large reservoir 120 meters higher is nearest to

a. 5.85 c. 15.70
b. 7.85 d. 30.00

10.27 The theoretical velocity generated by a 10-meter hydraulic head is

a. 3 m/s c. 14 m/s
b. 10 m/s d. 16.4 m/s

10.28 What is the static head corresponding to a fluid velocity of 10 m/sec?

a. 5.1 m c. 16.4 m
b. 10.2 m d. 50 m

10.29 The elevation to which water will rise in a piezometer tube is termed the
a. stagnation pressure
b. the energy grade line
c. the hydraulic grade line
d. friction head

10.30 A stream of fluid with a mass flow rate of 30 kg/s and a velocity of 6 m/s to the right has its direction reversed 180° in a "U" fitting. The net dynamic force in N exerted by the fluid on the fitting is nearest to
a. 180
b. 360
c. 2030
d. 4300

10.31 The thrust in newtons generated by an aircraft jet engine on takeoff, for each 1 kg/s of exhaust products whose velocity has been increased from essentially 0 to 150 m/s, is nearest to
a. 150
b. 1300
c. 3600
d. 7100

10.32 For the configuration in Exhibit 10.32, compute the velocity of the water in the 300-meter branch of the 15-cm diameter pipe. Assume the friction factors in the two pipes are the same and that the incidental losses are equal in the two branches. The velocity in m/s is
a. 10.0
b. 4.2
c. 1.8
d. 3.7

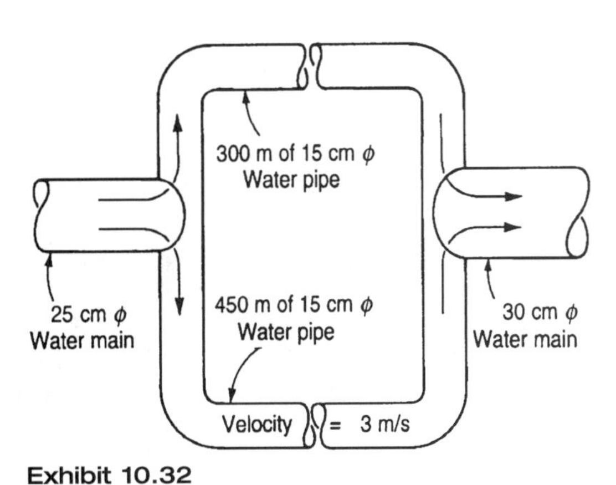

Exhibit 10.32

10.33 Which of the following statements most nearly approximates conditions in turbulent flow?
a. Fluid particles move along smooth, straight paths.
b. Energy loss varies linearly with velocity.
c. Energy loss varies as the square of the velocity.
d. Newton's law of viscosity governs the flow.

10.34 For turbulent flow of a fluid in a pipe, all of the following are true *except:*
a. The average velocity will be nearly the same as at the pipe center.
b. The energy lost to turbulence and friction varies with kinetic energy.
c. Pipe roughness affects the friction factor.
d. The Reynolds number will be less than 2300.

10.35 If the fluid flows in parallel, adjacent layers and the paths of individual particles do not cross, the flow is said to be
a. laminar
b. turbulent
c. critical
d. dynamic

10.36 Which of the following constitutes a group of parameters with the dimensions of power?
a. ρAV
b. pAV
c. $\frac{DV\rho}{\mu}$
d. $\frac{\rho V^2}{P}$

10.37 At or below the critical velocity in small pipes or at very low velocities, the loss of head from friction
a. varies linearly with the velocity
b. can be ignored
c. is infinitely large
d. varies as the velocity squared

10.38 The Moody diagram in Exhibit 10.38 is a log-log plot of friction factor vs. Reynolds number. Which of the lines *A-D* represents the friction factor to use for turbulent flow in a smooth pipe of low roughness ratio (ε/D):
a. *A*
b. *B*
c. *C*
d. *D*

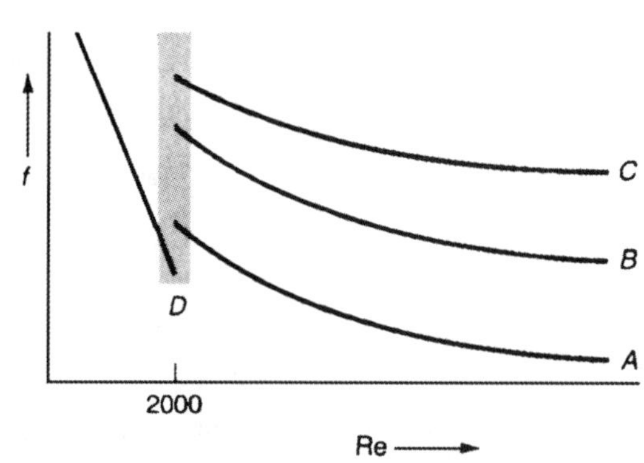

Exhibit 10.38

10.39 A 60-cm water pipe carries a flow of 0.1 m^3/s. At Point *A* the elevation is 50 meters and the pressure is 200 kPa. At Point *B*, 1200 meters downstream from *A*, the elevation is 40 meters and the pressure is 230 kPa. The head loss, in feet, between *A* and *B* is
a. 6.94
b. 15.08
c. 20.88
d. 100.2

10.40 Entrance losses between tank and pipe, or losses through elbows, fittings and valves are generally expressed as functions of

a. kinetic energy
b. pipe diameter
c. friction factor
d. volume flow rate

10.41 A 5-cm diameter orifice discharges fluid from a tank with a head of 5 meters. The discharge rate, Q, is measured at 0.015 m^3/s. The actual velocity at the *vena contracta* (v.c.) is 9 m/s. The coefficient of discharge, C_D, is nearest to

a. 0.62 c. 0.99
b. 0.77 d. 0.86

10.42 At normal atmospheric pressure, the maximum height in meters to which an nonvolatile fluid of specific gravity 0.80 may be siphoned is nearest to

a. 4.0 c. 10.3
b. 6.4 d. 12.9

10.43 The water flow rate in a 15-centimeter diameter pipe is measured with a differential pressure gage connected between a static pressure tap in the pipe wall and a pitot tube located at the pipe centerline. Which volume flow rate, Q in cubic meters per second, results in a differential pressure of 7 kPa?

a. 0.005 c. 0.50
b. 0.066 d. 1.00

10.44 The hydraulic formula $CA\sqrt{2gh}$ is used to find the

a. discharge through an orifice
b. velocity of flow in a closed conduit
c. length of pipe in a closed network
d. friction factor of a pipe

10.45 The hydraulic radius of an open-channel section is defined as

a. the wetted perimeter divided by the cross-sectional area
b. the cross-sectional area divided by the total perimeter
c. the cross-sectional area divided by the wetted perimeter
d. one-fourth the radius of a circle with the same area

10.46 To calculate a Reynolds number for flow in open channels and in cross-sections, one must utilize hydraulic radius, R, and modify the usual expression for circular cross-sections which is

$$\mathrm{Re} = \frac{DV\rho}{\mu} = \frac{VD}{\nu}$$

where D = diameter, V = velocity, ρ = density, μ = absolute viscosity, and ν = kinematic viscosity.

Which of the following modified expressions for Re is applicable to flow in open or non-circular cross-section?

a. $\frac{RD}{v}$ c. $\frac{2RD}{v}$

b. $\frac{RV\rho}{\mu}$ d. $\frac{4RV}{v}$

SOLUTIONS

10.1 **a.** Kinematic viscosity $v = \frac{\text{absolute viscosity}}{\text{density}} = \frac{\mu}{\rho}$

Units of absolute viscosity: N•s/m^2 or kg/m-s

Units of density: kg/m^3

The dimensions of kinematic viscosity would be

$$v = \frac{\mu\ (\text{kg/m•s})}{\rho\ (\text{kg/m}^3)} = \frac{\text{m}^2}{\text{s}}$$

10.2 **d.** The absolute viscosity is proportional to the shear stress (τ) divided by the angular deformation rate. Density is not involved in the definition. The rate of angular deformation $\cong \frac{dV}{dy}$.

$$\text{Thus, } \tau = \mu \frac{dV}{dy}.$$

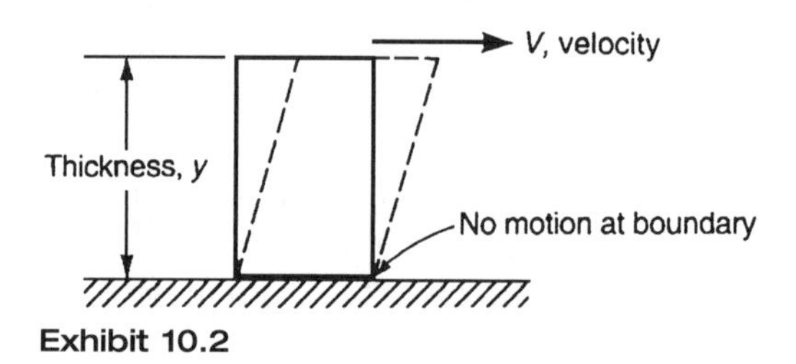

Exhibit 10.2

10.3 **c.** The internal pressure must equal the local external pressure. Externally, the pressure is

$$p = \gamma h = (\text{SG})\ (\gamma_{\text{water}})\ (\text{h})$$

$$p = (1.03)\left(9.81 \frac{\text{kN}}{\text{m}^3}\right)(50\ \text{m}) = 505\ \text{kPa (gage)}$$

10.4 **b.** Since the situation is static, gage pressure at the base is 1.5 m of water. In the air chamber it is 1.5 m of water, less 1 m of water less 1.2 m of oil. $p = \gamma h$

$$p = 1.5\ (9810) - 1\ (9810) - 1.2\ (0.8)\ (9810) = -4513\ \text{Pa}$$

10.5 **c.** Pressure $= \gamma h = 9810\ (100) = 981{,}000$ Pa $= 981$ kPa

10.6 **d.** The density of air can be calculated from the ideal gas law using

$$R = 0.286 \frac{\text{kN}\bullet\text{m}}{\text{kg}\bullet\text{K}}$$

$$\rho = \frac{p}{RT} = \frac{100\ \text{kN/m}^3}{0.286 \frac{\text{kN}\bullet\text{m}}{\text{kg}\bullet\text{K}} (20+273)\text{K}} = 1.19 \frac{\text{kg}}{\text{m}^3}$$

The specific weight of air $\gamma = 1.19 \frac{\text{kg}}{\text{m}^3} \bullet \left(\frac{9.81 \frac{\text{m}}{\text{s}^2}}{1 \frac{\text{kg} \bullet \text{m}}{\text{N}\bullet\text{s}^2}} \right) = 11.7 \frac{\text{N}}{\text{m}^3}$

$$p = \gamma h \text{ or } h = \frac{p}{\gamma} \text{ and } h = \frac{15{,}000 \frac{\text{N}}{\text{m}^2}}{11.7\ \text{N/m}^3} = 1282\ \text{m}$$

10.7 **a.** Assuming atmospheric pressure at sea level at 101.3 kPa and a constant specific weight of air at 11.7 N/m^3 (as previously calculated).

$$p = p_{\text{SL}} - \gamma h = 101.3 - \frac{11.7(1200)}{1000} = 87.26\ \text{kPa}$$

10.8 **b.**

$$\text{Force} = PA = \gamma hA = 9.81 \frac{\text{kN}}{\text{m}^3} \times 3\ \text{m} \times 500\ \text{cm}^2 \times \left(\frac{1\ \text{m}}{100\ \text{cm}} \right)^2 = 1.47\ \text{kN}$$

10.9 **d.** The pressure at the tank base $= p = \gamma_W h = (9810)(2.5) = 24{,}325\ \text{N/m}^3$

Area of tank base, $A = \frac{\pi}{4}(3)^2 = 7.07\,\text{m}^2$

Force on tank base $= pA = 24{,}325\ (7.07) = 171{,}978$ N

10.10 **c.** The average pressure exerted on one side is the pressure that exists at the centroid of the side times the area of the side.

$$F = \gamma h_c A$$

where h_c = the depth in meters from the fluid-free surface to the centroid of the area, and A = area. Since the sides are square, h_c = 1.5/2 = 0.75 m.

$$F = 9.81(0.75)(1.5 \times 1.5) = 16.55\ \text{kN}$$

10.11 **d.** The static force at $A = \left(700 \frac{\text{kN}}{\text{m}^2} \right) \left[\frac{\pi}{4} (0.1\ \text{m})^2 \right] = 5.50$ kN tension on the bolts at A.

The static force at $B = \left(700 \frac{\text{kN}}{\text{m}^2} \right) \left[\frac{\pi}{4} (0.05\ \text{m})^2 \right] = 1.37$ kN tension on the bolts at B.

The end restraint by the pipes opposes a net force of 5.50 – 1.37 = 4.13 kN to the right on the reducing section.

10.12 d. Area of port $= \frac{\pi}{4} D^2 = \frac{\pi}{4}(0.6)^2 = 0.283\,\text{m}^2$

$$\text{Pressure} = 100 \text{ kPa} = 100{,}000 \frac{\text{N}}{\text{m}^2}$$

$$F = pA = 100{,}000\ (0.283) = 28{,}300 \text{ N}$$

10.13 b. The volume of the bubble equals the volume of the displaced seawater, which equals

$$V_D = \frac{4}{3}\pi r^3 = \frac{4}{3}\pi(0.0125)^3 = 8.18 \times 10^{-6}\,\text{m}^3$$

Since the specific weight of seawater is

$$(\text{SG})(\gamma_W) = 1.03\left(9810 \frac{\text{N}}{\text{m}^2}\right) = 10{,}104 \frac{\text{N}}{\text{m}^2}$$

The buoyant force, B, is

$$B = \gamma V_D (10{,}104)(8.18 \times 10^{-6}) = 0.0826 \text{ N}$$

10.14 d. A buoyant force is equal to the weight of fluid displaced. At equilibrium, or floating, the weight downward is equal to the buoyant force.

Let V_1 = total volume of the iceberg in m^3. Its weight is $V_1(9810)(0.922) = 9045(V_1)$ N.

Let V_2 = immersed volume of the iceberg, which equals the volume of seawater displaced. The weight of seawater displaced is then $V_2(9810)(1.03) = 10{,}104(V_2)$ N.

Hence $\frac{V_2}{V_1} = \frac{9045}{10{,}104} = 0.895$ is the volume fraction of the iceberg immersed, and the volume fraction exposed is $1 - 0.895 = 0.105 =$ 10.5%.

10.15 b. Archimedes' principle applies equally well to gases. Thus a body located in any fluid, whether liquid or gaseous, is buoyed up by a force equal to the weight of the fluid displaced. A balloon filled with a gas lighter than air readily demonstrates the buoyant force.

Thus the weight of the cork is equal to the weight of water displaced plus the weight of air displaced. When the air within the vessel is removed, the cork is no longer provided a buoyant force equal to the weight of air displaced. For equilibrium, the cork will sink somewhat in the water.

10.16 c. $$V_D = \frac{W}{\gamma} = \frac{9.32\,\text{N}}{9810\,\frac{\text{N}}{\text{m}^2}} = 0.00095\,\text{m}^3 = 950\ \text{cm}^3$$

The change in total volume, ΔV, beneath the water surface equals the area of the cylindrical container, A, times the change in water level, dh, or $dV = A\,dh$. The depth of the water will increase

$$dh = \frac{dV}{A} = \frac{950\,\text{cm}^3}{\frac{\pi}{4}(20)^2} = 3.02\ \text{cm}$$

10.17 d. Let x = height of wood block, W = width of wood block, L = length of wood block, and γ = specific weight of water, N/m^3. The weight of the block is equal to the weight of the liquid displaced.

Weight of the block in water = $(x - 15)WL\gamma(1.0)$
Weight of the block in alcohol = $(x - 10)WL\gamma(0.82)$

Since the weight of the block is constant,

$$(x-15)WL\gamma = 0.82(x-10)WL\gamma$$
$$x-15 = 0.82x - 8.2$$
$$x = \frac{6.8}{0.18} = 37.8\,\text{cm}$$

The specific gravity of the wood block is, by definition,

$$\frac{\text{Volume of water displaced}}{\text{Total volume}} = \frac{(x-15)WL}{xWL} = \frac{37.8-15}{37.8} = 0.603$$

10.18 d. Continuity requires that

$$Q = A_1V_1 = A_2V_2 = A_1(3) = \frac{A_1}{4}V_2, \qquad V_2 = 12\,\text{m/s}$$

10.19 c. Continuity requires that the mass flow rate be the same at all sections in steady flow. Calculate $\dot{m}$ at Section 1, where the velocity is given, using $\dot{m} = \rho AV$. This will also be the mass flow rate at Section 2.

Cross-sectional area at Section 1:

$$\frac{\pi}{4}(0.10)^2 = .00785\ \text{m}^2$$

$$\dot{m} = \rho AV\left(1000\frac{\text{kg}}{\text{m}^3}\right)(.00785\,\text{m}^2)(10\,\text{m/s}) = 78.5\,\text{kg/s}$$

10.20 a. The mass flow rate $\dot{m} = \rho Q = \rho AV$. The density of air at 200 kPa and 150°C (423°K) is obtained from the ideal gas law:

$$\frac{p}{\rho} = RT$$

Use $R = 286 \frac{\text{N•m}}{\text{kg•°K}}$ for air.

$$\rho = \frac{P}{RT} = \frac{200{,}000 \frac{\text{N}}{\text{m}^2}}{286 \frac{\text{N•m}}{\text{kg•°K}} \bullet 423\text{°K}} = 1.65 \frac{\text{kg}}{\text{m}^3}$$

The cross-sectional area = $A = \frac{\pi}{4} D^2 = 0.785(.025 \text{ m})^2 =$ $490 \times 10^{-6} \text{ m}^2$, and

$$\dot{m} = \rho A v = \left(1.65 \frac{\text{kg}}{\text{m}^3}\right)(490 \times 10^{-6} \text{ m}^2)\left(10 \frac{\text{m}}{\text{s}}\right) = 0.00809 \frac{\text{kg}}{\text{s}}$$

10.21 b.

$$Q = 10 \frac{\text{L}}{\text{min}} \times \frac{\text{m}^3}{1000 \text{ L}} \times \frac{\text{min}}{60 \text{ s}} = 167 \times 10^{-6} \frac{\text{m}^3}{\text{s}}$$

The cross-sectional area $A = \frac{\pi}{4} D^2 = 0.785(0.01)^2 = 78.5 \times 10^{-6} \text{ m}^2$

$$V = \frac{Q}{A} = \frac{167 \times 10^{-6}}{78.5 \times 10^{-6}} = 2.13 \text{ m/s}$$

10.22 a. The pressure (flow) energy change is 45 – 9 = 36 N•m/N, or

$$\gamma = \frac{g}{g_c} \rho = \frac{9.81}{1.0}(800) = 7848 \text{ N/m}^3$$

$$\frac{\Delta p}{\gamma} = 36, \qquad \Delta p = 36\gamma = 36(7848) = 282{,}500 \text{ Pa} = 282.5 \text{ kPa}$$

10.23 a. The volume flow rate is

$$55 \frac{\text{L}}{\text{min}} \bullet \frac{\text{m}^3}{1000 \text{ L}} \bullet \frac{\text{min}}{60 \text{ s}} = 917 \times 10^{-6} \text{ m}^3\text{/s}.$$

Ignoring the head loss, h_L, from friction, the required energy input is $52.5 - 12 = 40.5 \frac{\text{N•m}}{\text{N}}$.

$$\text{Power} = \gamma Q h_A \left(7848 \frac{\text{N}}{\text{m}^3}\right)(917 \times 10^{-6} \text{m}^3\text{/s})\left(40.5 \frac{\text{N} - \text{m}}{\text{N}}\right)$$
$$= 291 \text{W} = 0.291 \text{ kW}$$

10.24 d. In a venturi throat, the increased velocity required by continuity results in a *KE* (velocity) increase that occurs at the expense of pressure (flow) energy. Since the system is horizontal, no change in potential energy has occurred. At B the pressure (flow energy) decreases and *KE* increases. For a well-designed venturi, the conditions existent at A are essentially restored at C.

10.25 b. Apply an energy balance of the fluid flowing: Total energy in = Total energy out + Energy losses – Energy inputs. Thus, 135 = 130 + h_L – 0. The head loss h_L = 5 N•m/m, or 5 meters.

10.26 b. Ignoring frictional losses, pump inefficiency, and noting that any changes in KE or pressure are essentially 0, pumping power is equal to the increase in potential energy between the reservoirs.

The potential energy increase per lb_m is Z or h = 120 meters or $\frac{\text{N}\bullet\text{m}}{\text{N}}$

The volume flow rate, Q, is

$$400\frac{\text{L}}{\text{min}}\bullet\frac{\text{m}^3}{1000\text{ L}}\bullet\frac{\text{min}}{60\text{ s}}=6.67\times10^{-3}\text{ m}^3/\text{s}$$

The power required is

$$P=\gamma Qh_a=9.81\frac{\text{kN}}{\text{m}^3}\bullet6.67\times10^{-3}/\text{s}\bullet120\text{ m}=7.85\text{ kW}$$

10.27 c.

$$h=\frac{V^2}{2g}\quad\text{or}\quad V=(2gh)^{1/2}=(2\times9.81\times10)^{1/2}=14\text{ m/s}$$

10.28 a. The head is

$$h=\frac{V^2}{2g}=\frac{10^2}{2(9.81)}=5.10\text{ m}$$

10.29 c. A **piezometer tube** indicates static pressure and is equivalent to a static pressure gage.

Stagnation pressure is an increased pressure developed at the entrance to a pitot tube when the velocity locally becomes zero.

The **hydraulic grade line** is a flow energy or pressure head in m, which can be plotted vertically above the pipe centerline along the pipe.

The **energy grade line** is the total mechanical energy (flow energy or pressure head, plus kinetic energy or dynamic head, plus potential energy or height above datum) in m, which may be plotted vertically above the datum along the pipe.

The **friction head** is the head loss h_f in m caused by fluid friction.

The **critical depth** above the channel floor in open channels is the depth for minimum potential and kinetic energy for the given discharge.
Tranquil flow (low *KE* and high *PE*) exists when the actual flow is above critical depth, and rapid flow (high *KE* and low *PE*) exists when the actual flow is below critical depth.

10.30 b. The steady impulse-momentum equation is $F_{net} = \dot{m}(\overline{v}_2 - \overline{v}_1)$ if the pressure in the fluid stream is zero at each end of the "U." Then

$$F = 30 \text{ kg/s}(-6-6)\text{ m/s} = 360 \frac{\text{kg}\bullet\text{m}}{\text{s}^2} = 360 \text{ N}$$

Since the original velocity was 6 m/s, the final reversed velocity is –6 m/s. This force from impulse-momentum is the force *on* the fluid to achieve the velocity change. In reaction, the fluid exerts an equal and opposite force, 360 N, to the right on the fitting.

10.31 a. The impulse-momentum equation is $F = pQ(V_2 - V_1)$. Here $\rho Q = 1$ kg/s, the final velocity of the exhaust is $V_2 = 150$ m/s, and $V_1 = 0$. Hence

$$F = \left(\frac{1 \text{ kg}}{\text{s}}\right)(150-0)\frac{\text{m}}{\text{s}} = 150 \frac{\text{kg}\bullet\text{m}}{\text{s}^2} = 150 \text{ N}$$

10.32 d. There is a drop in the energy line from the 25-cm main to the 30-cm main. This head loss must be equal in both 15-cm lines, or

$$h_{f300} = h_{f450}$$

The Darcy equation is

$$h_f = f \frac{L}{d} \frac{V^2}{2g}$$

where h_f = head loss in meters, f = friction factor, L = length of pipeline in meters, d = diameter of pipe in meters, and $g = 9.81$ m/s^2. Thus, in this situation,

$$f \frac{300}{0.15} \frac{V_{300}^2}{2(9.81)} = f \frac{450}{0.15} \frac{V_{450}^2}{2(9.81)}$$

which reduces to

$$300V_{300}^2 = 450(3)^2$$

$$V_{300} = \sqrt{\frac{4050}{300}} = 3.67 \text{ m/s}$$

10.33 c. Laminar (streamline, viscous) flow is compared with turbulent flow in the following table:

	Laminar Flow	Turbulent Flow
Motion of fluid particles	Parallel to stream velocity. Paths of particles do not cross.	Particle paths cross and move in all directions.
Energy loss, h_f	$h_f = f\left(\frac{L}{D}\right)\left(\frac{V^2}{2g}\right)$ f is independent of surface roughness and decreases with Re. $f = \frac{64}{\text{Re}}$	$h_f = f\left(\frac{L}{D}\right)\left(\frac{V^2}{2g}\right)$ f varies with surface roughness, decreases with Re to a constant value. See Moody diagram.
Velocity distribution in pipe	Average is ½ of maximum at centerline. parabolic distribution. Zero at wall.	Essentially same throughout, except for thin boundary layer at wall. Follows 1/7 power law.
Reynolds number $\text{Re} = DV\rho/\mu$	Less than 2300	Greater than 2300

Newton's law of viscosity defines μ on the basis of shear stress and the rate of fluid angular deformation. The Reynolds number contains μ as a contributing parameter. Very viscous liquids usually move in laminar flow.

On the basis of the above data, select (c). Do not confuse the energy loss, h_f, with the friction factor, f.

10.34 d. In turbulent flow the Reynolds number is *greater* than 2300.

10.35 a. Turbulent flow is highly agitated flow with individual particles crossing paths and colliding; critical flow is a point at which some property of the fluid—or some parameter related to it—changes; dynamic flow is redundant; uniform flow is of constant rate.

10.36 b. Choice (a) is mass flow rate, $\dot{m}$, in kg/s.

Choice (b) has these dimensions:

$$pAV = \left(\frac{\text{N}}{\text{m}^2}\right)(\text{m}^2)\left(\frac{\text{m}}{\text{s}}\right) = \frac{\text{N}\bullet\text{m}}{\text{s}} = \text{W}$$

Choice (c) is the Reynolds number, Re; it is the dimensionless ratio of inertial force to viscous force.

Choice (d) is the Euler number, Eu; it is the dimensionless ratio of inertial force to pressure force.

10.37 **a.** Below the critical velocity (Re < 2300) flow is laminar, and $f = 64/\text{Re}$. Substitution of this term into the Darcy equation for friction loss in pipes, $h_f = f\dfrac{L}{D}\dfrac{V^2}{2g}$, yields the Hagen-Poiseuille equation,

$$h_f = \frac{32\mu LV}{\gamma d^2}.$$

10.38 **a.** Line *D* applies to all roughness ratios in laminar flow (Re < 2300) because the boundary layer at the wall makes the friction factor independent of roughness ratio:

$$f = \frac{64}{\text{Re}}$$

In turbulent flow (Re > 2300), increasing roughness is represented by *A* for a smooth pipe to *C* for a very rough pipe; moreover, only the thinnest boundary layer exists in a turbulent flow, so the friction factor is very dependent on surface roughness.

10.39 **a.** Use an energy balance to determine h_f. Upon substituting the given data, the resulting equation is

$$Z_A + \frac{V_A^2}{2g} + \frac{P_A}{\gamma} = Z_B + \frac{V_B^2}{2g} + \frac{P_B}{\gamma} + h_f$$

Since the pipe diameter is unchanged, continuity requires that *V* be the same at both points. Thus the kinetic energy terms can be deleted from both sides of the equation.

$$Z_A + \frac{P_A}{\gamma} = Z_B + \frac{P_B}{\gamma} + h_f$$

$$h_f = \frac{200 - 230}{9.81} + 50 - 40 = 6.94\,\text{m}$$

10.40 **a.** Typical head losses for the above items are expressed as an empirical average constant, *K* or *C* times the kinetic energy, $V^2/2g$:

$$h_f = K\frac{V^2}{2g}$$

10.41 **b.** The discharge coefficient is $C_D = C_c\,C_v$, where C_c = coefficient of contraction = (area of v.c.)/(area of orifice) and C_v = coefficient of velocity = (actual velocity at v.c.)/(theoretical velocity at v.c.) ignoring losses.

The theoretical velocity at the v.c. is

$$V = \sqrt{2gh} = \sqrt{2(9.81)(5)} = 9.9 \text{ m/s}$$

$$C_v = \frac{9.0}{9.9} = 0.909$$

The area of the v.c. is

$$A = \frac{Q}{V} = \frac{.015}{9} = 0.00167\,\text{m}^2$$

The area of the orifice $= \frac{\pi}{4}D^2 = \frac{\pi}{4}(.05)^2 = 0.00196\,\text{m}^2$. Thus,

$$C_c = \frac{0.00167}{0.00196} = 0.852 \quad \text{and} \quad C_D = C_c C_v = (0.852)(0.909) = 0.774$$

10.42 d. The maximum height to which a fluid may be siphoned is determined when the pressure of the fluid column plus its vapor pressure equals the external pressure. The minimum pressure at the highest point is 0 kPa plus vapor pressure.

Ignoring the vapor pressure (small),

$$p = \gamma h \quad \text{or} \quad h = \frac{P}{\gamma}$$

$$h = \frac{-101.3\ \text{kN/m}^2}{(0.8)9.81\ \text{kN/m}^3} = 12.91\ \text{m in depth (or height)}$$

10.43 b. A pitot tube generates a stagnation pressure as fluid kinetic energy is converted to pressure head. Hence

$$V = \sqrt{2g\left(\frac{\Delta p}{\gamma}\right)}$$

$$V = \sqrt{2(9.81)\frac{\text{m}}{\text{s}^2}\left(\frac{7000\ \text{N/m}^3}{9810\ \text{N/m}^3}\right)} = 3.74\ \text{m/s}$$

$$Q = AV = \frac{\pi}{4}D^2V = \frac{\pi}{4}(0.15)^2(3.74) = 0.0661\ \text{m}^2\text{/s}$$

10.44 a. For a static head orifice discharging freely into the atmosphere

$$Q = CA\sqrt{2gh}$$

10.45 c. Hydraulic radius, R, is defined as cross-sectional area, divided by wetted perimeter.

10.46 d. Choices (a) and (c) are not dimensionless, as required for a Reynolds number. Since the hydraulic radius R = (cross sectional area)/(wetted perimeter), for a circular cross section,

$$R = \frac{\frac{\pi D^2}{4}}{\pi D} = \frac{D}{4} \quad \text{or} \quad D = 4R$$

Therefore,

$$\text{Re} = \frac{4RV\rho}{\mu} = \frac{4RV}{\nu}$$

CHAPTER 11

Thermodynamics

Robert Michel

OUTLINE

This chapter reviews the basic knowledge and working tools required to pass the thermodynamics portion of the Fundamentals of Engineering (FE/EIT) exam. It excludes areas of thermodyamics not normally included on the exam.

Typical solved problems are grouped at the end of this chapter and are referred to in the text section of the review material.

THE FIRST LAW OF THERMODYNAMICS

Except for nuclear physics, which involves the conversion of mass into energy, mass is conserved in a process. In thermodynamics, the mass in a closed system (such as a piston–cylinder) is constant. In an open system, where mass is flowing in and out, the sum of the mass flowing into the system equals the mass flowing out if there is no accumulation of mass within the system. This is called conservation of mass.

Energy is also conserved and must be accounted for. The types of energy that are important in thermodynamics are the following:

- Internal energy
- Flow energy
- Kinetic energy
- Potential energy
- Heat
- Work

The first four energy types are a function of the state, or condition, of the substance. Heat and work are forms of energy that cross the boundary of systems and are not a function of the state.

The first law of thermodynamics is a bookkeeping system to keep track of these energies.

Closed System

For a typical closed thermodynamic system (Fig. 11.1), the kinetic and potential energy are not important, and since there is no flow, the bookkeeping is simple and reduces to

$$u_1 + q = u_2 + w$$

Open System

For an open system (Fig. 11.2), the bookkeeping system yields

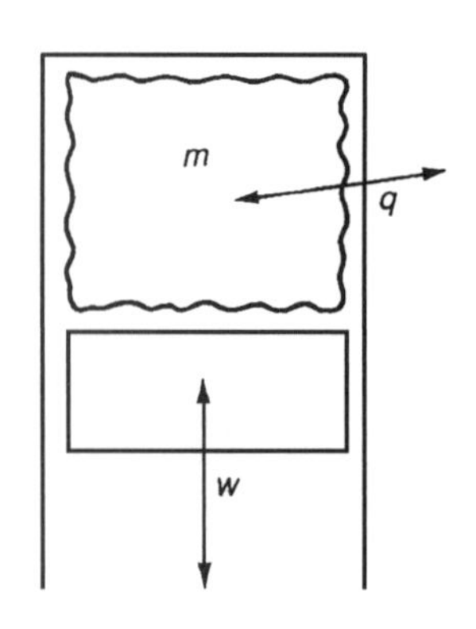

Figure 11.1

$$u_1 + p_1v_1 + \frac{V_1^2}{2g_c} + \frac{Z_1g}{g_c} + q = u_2 + p_2v_2 + \frac{V_2^2}{2g_c} + \frac{Z_2g}{g_c} + w$$

In many common open-system processes, the kinetic energy ($V^2/2g_c$) and the potential energy (Zg/g_c) are not important. For convenience, u and pv are combined into h (enthalpy), and the equation then reduces to

$$h_1 + q = h_z + w$$

The thermodynamic sign convention for heat and work is that q_{in} is positive and w_{out} is positive. This is the normal flow of heat and work in an engine or power plant.

Note also that a "change in" a property or state refers to the second point value minus the first point value.

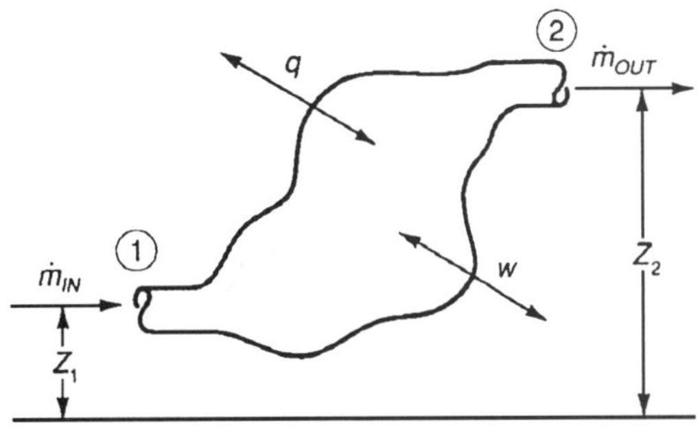

Figure 11.2

Care must be taken to keep the units consistent. For example, the normal units for internal energy, heat, and work are kJ/kg. In the SI system, $V^2/2g_c$ and Zg/g_c must be divided by 1000. Remember that g_c in the SI system has a value of 1 but is 32.2 in the U.S. system.

The first law is the most popular one by far in solving thermodynamics problems!

Example **11.1**

A piston–cylinder contains 5 kg of air. During a compression process 100 kJ of heat is removed while 250 kJ of work is done on the air. Find the change in internal energy of the air.

Solution

The system is a closed system since the mass is fixed. So

$$u_1 + q = u_2 + w$$

Here, $q = -100$ kJ, since heat leaving a system is negative, and $w = -250$ kJ, since work done *on* a system is negative. Thus

$$u_2 - u_1 = -100 - (-250) = 150 \text{ kJ}, \qquad \frac{150 \text{ kJ}}{5 \text{ kg}} = 30 \text{ kJ/kg}$$

Example **11.2**

Heated air enters a turbine at a flow rate of 5 kg/s. The entering and leaving conditions are shown in the following table. The heat loss from the turbine is 50 kW. Find the power produced.

	Inlet	**Exit**
Pressure, kPa	1000	100
Temperature, °K	800	500
Velocity, m/s	100	200
Specific internal energy, kJ/kg	137	85
Specific volume, m^3/kg	0.23	1.44
Elevation, m	3	10

Solution

The system has mass flowing across the boundaries, so it is an open system.

$$u_1 + p_1 v_1 + \frac{Z_1 g}{g_c} + \frac{V_1^2}{2g_c} + q = u_2 + p_2 v_2 + \frac{Z_2 g}{g_c} + \frac{V_2^2}{2g_c} + w$$

$$w = (u_1 - u_2) + (p_1 v_1 - p_2 v_2) + \frac{(Z_1 - Z_2)g}{g_c} + \frac{V_1^2 - V_2^2}{2g_c} + q$$

$$= (137 - 85) + (1000 \times 0.23 - 100 \times 1.44)$$

$$+ \frac{(3-10)}{1000} \times \frac{9.81}{1.0} + \frac{100^2 - 200^2}{2 \times 1 \times 1000} + \left(-\frac{50}{5}\right)$$

$$= 52 + 86 - 0.069 - 15 - 10 = 112.9 \ \frac{\text{kJ}}{\text{kg}}$$

$$\dot{w} = w \times \dot{m} = 112.9 \times 5 = 564.5 \text{ kW}$$

PROPERTIES OF PURE SUBSTANCES

Use of Steam Tables

Property tables for substances that go through a phase change during normal thermodynamic processes, such as H_2O, R-12, and NH_3, are divided into three groups:

- Saturated Tables. These show the properties (v, u, h, s) of the saturated liquid and saturated vapor. For convenience, there are usually two sets: one using T as the entering argument and one using P. The highest temperature/pressure entry is usually the critical point. The quality x is needed to define properties in the mixture region.
- Superheated Tables. These show the properties (v, u, h, s) as a function of T and P in the superheated area to the right of the saturated vapor curve and are usually grouped by pressure. *Any two* properties (v, u, h, s, T, P) may be used to define the state. The saturated state is usually noted. At moderate pressures and temperatures well away from saturation, perfect gas relationships may be used as an approximation.
- Subcooled or Compressed Liquid Tables. These tables show properties (v, u, h, and s) as a function of T and P in the area to the left of the saturated liquid curve and are usually grouped by pressure. As with the Superheated Tables, any two properties may be used to define the state, and the saturated state is usually noted. For points in the region that are below the tabulated pressures, the properties are approximated as those of the saturated liquid at the same temperature (v, u, h, and s are weak functions of pressure).

The procedure for finding the state-point properties for given data where the condition of the substances is not defined is best done in a structured manner:

1. Always look at the saturation tables first to determine whether the state point is liquid, vapor, or "wet" (Fig. 11.3).
2. When one is given T or P and h, v, or u:

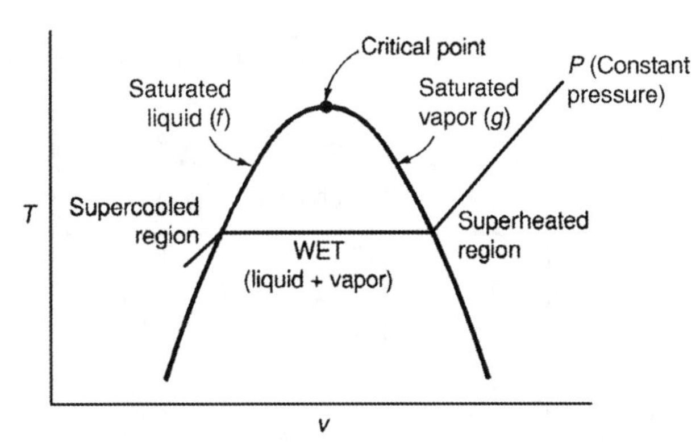

Figure 11.3

a. If h, v, or u is between f and g, the point is in the wet region (inside the dome). Calculate the quality x using saturation properties and given properties. For example,

$$x = \frac{u - u_f}{u_{fg}}$$

b. If h, v, or u is greater than the saturated-vapor value, the point is in the superheated region. Locate the properties in the superheated tables.

c. If h, v, or u is less than the saturated-liquid value, the point is in the subcooled liquid region. Locate the properties in the liquid tables and/or calculate from the saturated-liquid properties.

3. Given T or P and x:

   Go directly to the saturated tables, since an intermediate value of x implies it is in the "wet" region.

4. Given T and P:

   Compare the saturation temperature with the given temperature. The saturation temperature is read at the given pressure.

   If T is greater than T_{sat}, the point is superheated vapor.
   If T is less than T_{sat}, the point is subcooled liquid.

Example 11.3

Fill in the following table for steam (water) using the extracted tabulations.

	T, °C	P, kPa	x, %	h, kJ/kg	u, kJ/kg	v, m^3/kg
a)	200	—	—	852.38	—	—
b)	—	143.3	—	—	1000	—
c)	300	800	—	—	—	—
d)	200	5000	—	—	—	—
e)	—	300	—	—	—	0.85
f)	300	—	80	—	—	—

Superheated Water Tables

T Temp. °C	v m³/kg	u kJ/kg	h kJ/kg	s kJ/(kg·K)	v m³/kg	u kJ/kg	h kJ/kg	s kJ/(kg·K)
	P = 0.01 MPa (45.81°C)				P = 0.05 MPa (81.33°C)			
Sat.	14.674	2437.9	2584.7	8.1502	3.240	2483.9	2645.9	7.5939
50	14.869	2443.9	2592.6	8.1749				
100	17.196	2515.5	2687.5	8.4479	3.418	2511.6	2682.5	7.6947
150	19.512	2587.9	2783.0	8.6882	3.889	2585.6	2780.1	7.9401
200	**21.825**	**2661.3**	**2879.5**	**8.9038**	**4.356**	**2659.9**	**2877.7**	**8.1580**
250	24.136	2736.0	2977.3	9.1002	4.820	2735.0	2976.0	8.3556
300	26.445	2812.1	3076.5	9.2813	5.284	2811.3	3075.5	8.5373
400	31.063	2968.9	3279.6	9.6077	6.209	2968.5	3278.9	8.8642
500	35.679	3132.3	3489.1	9.8978	7.134	3132.0	3488.7	9.1546
600	**40.295**	**3302.5**	**3705.4**	**10.1608**	**8.057**	**3302.2**	**3705.1**	**9.4178**
700	44.911	3479.6	3928.7	10.4028	8.981	3479.4	3928.5	9.6599
800	49.526	3663.8	4159.0	10.6281	9.904	3663.6	4158.9	9.8852
900	54.141	3855.0	4396.4	10.8396	10.828	3854.9	4396.3	10.0967
1000	58.757	4053.0	4640.6	11.0393	11.751	4052.9	4640.5	10.2964
1100	**63.372**	**4257.5**	**4891.2**	**11.2287**	**12.674**	**4257.4**	**4891.1**	**10.4859**
1200	67.987	4467.9	5147.8	11.4091	13.597	4467.8	5147.7	10.6662
1300	72.602	4683.7	5409.7	11.5811	14.521	4683.6	5409.6	10.8382
	P = 0.10 MPa (99.63°C)				P = 0.20 MPa (120.23°C)			
Sat.	1.6940	2506.1	2675.5	7.3594	0.8857	2529.5	2706.7	7.1272
100	1.6958	2506.7	2676.2	7.3614				
150	1.9364	2582.8	2776.4	7.6134	0.9596	2576.9	2768.8	7.2795
200	2.172	2658.1	2875.3	7.8343	1.0803	2654.4	2870.5	7.5066
250	**2.406**	**2733.7**	**2974.3**	**8.0333**	**1.1988**	**2731.2**	**2971.0**	**7.7086**
300	2.639	2810.4	3074.3	8.2158	1.3162	2808.6	3071.8	7.8926
400	3.103	2967.9	3278.2	8.5435	1.5493	2966.7	3276.6	8.2218
500	3.565	3131.6	3488.1	8.8342	1.7814	3130.8	3487.1	8.5133
600	4.028	3301.9	3704.4	9.0976	2.013	3301.4	3704.0	8.7770
700	**4.490**	**3479.2**	**3928.2**	**9.3398**	**2.244**	**3478.8**	**3927.6**	**9.0194**
800	4.952	3663.5	4158.6	9.5652	2.475	3663.1	4158.2	9.2449
900	5.414	3854.8	4396.1	9.7767	2.705	3854.5	4395.8	9.4566
1000	5.875	4052.8	4640.3	9.9764	2.937	4052.5	4640.0	9.6563
1100	6.337	4257.3	4891.0	10.1659	3.168	4257.0	4890.7	9.8458
1200	**6.799**	**4467.7**	**5147.6**	**10.3463**	**3.399**	**4467.5**	**5147.5**	**10.0262**
1300	7.260	4683.5	5409.5	10.5183	3.630	4683.2	5409.3	10.1982
	P = 0.40 MPa (143.63°C)				P = 0.60 MPa (158.85°C)			
Sat.	0.4625	2553.6	2738.6	6.8959	0.3157	2567.4	2756.8	6.7600
150	0.4708	2564.5	2752.8	6.9299				
200	0.5342	2646.8	2860.5	7.1706	0.3520	2638.9	2850.1	6.9665
250	0.5951	2726.1	2964.2	7.3789	0.3938	2720.9	2957.2	7.1816
300	**0.6548**	**2804.8**	**3066.8**	**7.5662**	**0.4344**	**2801.0**	**3061.6**	**7.3724**
350					0.4742	2881.2	3165.7	7.5464
400	0.7726	2964.4	3273.4	7.8985	0.5137	2962.1	3270.3	7.7079
500	0.8893	3129.2	3484.9	8.1913	0.5920	3127.6	3482.8	8.0021
600	1.0055	3300.2	3702.4	8.4558	0.6697	3299.1	3700.9	8.2674
700	**1.1215**	**3477.9**	**3926.5**	**8.6987**	**0.7472**	**3477.0**	**3925.3**	**8.5107**
800	1.2372	3662.4	4157.3	8.9244	0.8245	3661.8	4156.5	8.7367
900	1.3529	3853.9	4395.1	9.1362	0.9017	3853.4	4394.4	8.9486
1000	1.4685	4052.0	4639.4	9.3360	0.9788	4051.5	4638.8	9.1485
1100	1.5840	4256.5	4890.2	9.5256	1.0559	4256.1	4889.6	9.3381
1200	**1.6996**	**4467.0**	**5146.8**	**9.7060**	**1.1330**	**4466.5**	**5146.3**	**9.5185**
1300	1.8151	4682.8	5408.8	9.8780	1.2101	4682.3	5408.3	9.6906
	P = 0.80 MPa (170.43°C)				P = 1.00 MPa (179.91°C)			
Sat.	0.2404	2576.8	2769.1	6.6628	0.194 44	2583.6	2778.1	6.5865
200	0.2608	2630.6	2839.3	6.8158	0.2060	2621.9	2827.9	6.6940
250	0.2931	2715.5	2950.0	7.0384	0.2327	2709.9	2942.6	6.9247
300	0.3241	2797.2	3056.5	7.2328	0.2579	2793.2	3051.2	7.1229
350	**0.3544**	**2878.2**	**3161.7**	**7.4089**	**0.2825**	**2875.2**	**3157.7**	**7.3011**
400	0.3843	2959.7	3267.1	7.5716	0.3066	2957.3	3263.9	7.4651
500	0.4433	3126.0	3480.6	7.8673	0.3541	3124.4	3478.5	7.7622
600	0.5018	3297.9	3699.4	8.1333	0.4011	3296.8	3697.9	8.0290
700	0.5601	3476.2	3924.2	8.3770	0.4478	3475.3	3923.1	8.2731
800	**0.6181**	**3661.1**	**4155.6**	**8.6033**	**0.4943**	**3660.4**	**4154.7**	**8.4996**
900	0.6761	3852.8	4393.7	8.8153	0.5407	3852.2	4392.9	8.7118
1000	0.7340	4051.0	4638.2	9.0153	0.5871	4050.5	4637.6	8.9119
1100	0.7919	4255.6	4889.1	9.2050	0.6335	4255.1	4888.6	9.1017
1200	0.8497	4466.1	5145.9	9.3855	0.6798	4465.6	5145.4	9.2822
1300	**0.9076**	**4681.8**	**5407.9**	**9.5575**	**0.7261**	**4681.3**	**5407.4**	**9.4543**

Saturated Water - Temperature Table

Temp. °C T	Sat. Press. kPa P_{sat}	Specific Volume m^3/kg		Internal Energy kJ/kg			Enthalpy kJ/kg			Entropy kJ/(kg·K)		
		Sat. liquid v_f	Sat. vapor v_g	Sat. liquid u_f	Evap. u_{fg}	Sat. vapor u_g	Sat. liquid h_f	Evap. h_{fg}	Sat. vapor h_g	Sat. liquid s_f	Evap. s_{fg}	Sat. vapor s_g
0.01	0.6113	0.001 000	206.14	0.00	2375.3	2375.3	0.01	2501.3	2501.4	0.0000	9.1562	9.1562
5	0.8721	0.001 000	147.12	20.97	2361.3	2382.3	20.98	2489.6	2510.6	0.0761	8.9496	9.0257
10	1.2276	0.001 000	106.38	42.00	2347.2	2389.2	42.01	2477.7	2519.8	0.1510	8.7498	8.9008
15	1.7051	0.001 001	77.93	62.99	2333.1	2396.1	62.99	2465.9	2528.9	0.2245	8.5569	8.7814
20	**2.339**	**0.001 002**	**57.79**	**83.95**	**2319.0**	**2402.9**	**83.96**	**2454.1**	**2538.1**	**0.2966**	**8.3706**	**8.6672**
25	3.169	0.001 003	43.36	104.88	2304.9	2409.8	104.89	2442.3	2547.2	0.3674	8.1905	8.5580
30	4.246	0.001 004	32.89	125.78	2290.8	2416.6	125.79	2430.5	2556.3	0.4369	8.0164	8.4533
35	5.628	0.001 006	25.22	146.67	2276.7	2423.4	146.68	2418.6	2565.3	0.5053	7.8478	8.3531
40	7.384	0.001 008	19.52	167.56	2262.6	2430.1	167.57	2406.7	2574.3	0.5725	7.6845	8.2570
45	**9.593**	**0.001 010**	**15.26**	**188.44**	**2248.4**	**2436.8**	**188.45**	**2394.8**	**2583.2**	**0.6387**	**7.5261**	**8.1648**
50	12.349	0.001 012	12.03	209.32	2234.2	2443.5	209.33	2382.7	2592.1	0.7038	7.3725	8.0763
55	15.758	0.001 015	9.568	230.21	2219.9	2450.1	230.23	2370.7	2600.9	0.7679	7.2234	7.9913
60	19.940	0.001 017	7.671	251.11	2205.5	2456.6	251.13	2358.5	2609.6	0.8312	7.0784	7.9096
65	25.03	0.001 020	6.197	272.02	2191.1	2463.1	272.06	2346.2	2618.3	0.8935	6.9375	7.8310
70	**31.19**	**0.001 023**	**5.042**	**292.95**	**2176.6**	**2569.6**	**292.98**	**2333.8**	**2626.8**	**0.9549**	**6.8004**	**7.7553**
75	38.58	0.001 026	4.131	313.90	2162.0	2475.9	313.93	2321.4	2635.3	1.0155	6.6669	7.6824
80	47.39	0.001 029	3.407	334.86	2147.4	2482.2	334.91	2308.8	2643.7	1.0753	6.5369	7.6122
85	57.83	0.001 033	2.828	355.84	2132.6	2488.4	355.90	2296.0	2651.9	1.1343	6.4102	7.5445
90	70.14	0.001 036	2.361	376.85	2117.7	2494.5	376.92	2283.2	2660.1	1.1925	6.2866	7.4791
95	**84.55**	**0.001 040**	**1.982**	**397.88**	**2102.7**	**2500.6**	**397.96**	**2270.2**	**2668.1**	**1.2500**	**6.1659**	**7.4159**
	MPa											
100	0.101 35	0.001 044	1.6729	418.94	2087.6	2506.5	419.04	2257.0	2676.1	1.3069	6.0480	7.3549
105	0.120 82	0.001 048	1.4194	440.02	2072.3	2512.4	440.15	2243.7	2683.8	1.3630	5.9328	7.2958
110	0.143 27	0.001 052	1.2102	461.14	2057.0	2518.1	461.30	2230.2	2691.5	1.4185	5.8202	7.2387
115	0.169 06	0.001 056	1.0366	482.30	2041.4	2523.7	482.48	2216.5	2699.0	1.4734	5.7100	7.1833
120	**0.198 53**	**0.001 060**	**0.8919**	**503.50**	**2025.8**	**2529.3**	**503.71**	**2202.6**	**2706.3**	**1.5276**	**5.6020**	**7.1296**
125	0.2321	0.001 065	0.7706	524.74	2009.9	2534.6	524.99	2188.5	2713.5	1.5813	5.4962	7.0775
130	0.2701	0.001 070	0.6685	546.02	1993.9	2539.9	546.31	2174.2	2720.5	1.6344	5.3925	7.0269
135	0.3130	0.001 075	0.5822	567.35	1977.7	2545.0	567.69	2159.6	2727.3	1.6870	5.2907	6.9777
140	0.3613	0.001 080	0.5089	588.74	1961.3	2550.0	589.13	2144.7	2733.9	1.7391	5.1908	6.9299
145	**0.4154**	**0.001 085**	**0.4463**	**610.18**	**1944.7**	**2554.9**	**610.63**	**2129.6**	**2740.3**	**1.7907**	**5.0926**	**6.8833**
150	0.4758	0.001 091	0.3928	631.68	1927.9	2559.5	632.20	2114.3	2746.5	1.8418	4.9960	6.8379
155	0.5431	0.001 096	0.3468	653.24	1910.8	2564.1	653.84	2098.6	2752.4	1.8925	4.9010	6.7935
160	0.6178	0.001 102	0.3071	674.87	1893.5	2568.4	675.55	2082.6	2758.1	1.9427	4.8075	6.7502
165	0.7005	0.001 108	0.2727	696.56	1876.0	2572.5	697.34	2066.2	2763.5	1.9925	4.7153	6.7078
170	**0.7917**	**0.001 114**	**0.2428**	**718.33**	**1858.1**	**2576.5**	**719.21**	**2049.5**	**2768.7**	**2.0419**	**4.6244**	**6.6663**
175	0.8920	0.001 121	0.2168	740.17	1840.0	2580.2	741.17	2032.4	2773.6	2.0909	4.5347	6.6256
180	1.0021	0.001 127	0.194 05	762.09	1821.6	2583.7	763.22	2015.0	2778.2	2.1396	4.4461	6.5857
185	1.1227	0.001 134	0.174 09	784.10	1802.9	2587.0	785.37	1997.1	2782.4	2.1879	4.3586	6.5465
190	1.2544	0.001 141	0.156 54	806.19	1783.8	2590.0	807.62	1978.8	2786.4	2.2359	4.2720	6.5079
195	**1.3978**	**0.001 149**	**0.141 05**	**828.37**	**1764.4**	**2592.8**	**829.98**	**1960.0**	**2790.0**	**2.2835**	**4.1863**	**6.4698**
200	1.5538	0.001 157	0.127 36	850.65	1744.7	2595.3	852.45	1940.7	2793.2	2.3309	4.1014	6.4323
205	1.7230	0.001 164	0.115 21	873.04	1724.5	2597.5	875.04	1921.0	2796.0	2.3780	4.0172	6.3952
210	1.9062	0.001 173	0.104 41	895.53	1703.9	2599.5	897.76	1900.7	2798.5	2.4248	3.9337	6.3585
215	2.104	0.001 181	0.094 79	918.14	1682.9	2601.1	920.62	1879.9	2800.5	2.4714	3.8507	6.3221
220	**2.318**	**0.001 190**	**0.086 19**	**940.87**	**1661.5**	**2602.4**	**943.62**	**1858.5**	**2802.1**	**2.5178**	**3.7683**	**6.2861**
225	2.548	0.001 199	0.078 49	963.73	1639.6	2603.3	966.78	1836.5	2803.3	2.5639	3.6863	6.2503
230	2.795	0.001 209	0.071 58	986.74	1617.2	2603.9	990.12	1813.8	2804.0	2.6099	3.6047	6.2146
235	3.060	0.001 219	0.065 37	1009.89	1594.2	2604.1	1013.62	1790.5	2804.2	2.6558	3.5233	6.1791
240	3.344	0.001 229	0.059 76	1033.21	1570.8	2604.0	1037.32	1766.5	2803.8	2.7015	3.4422	6.1437
245	**3.648**	**0.001 240**	**0.054 71**	**1056.71**	**1546.7**	**2603.4**	**1061.23**	**1741.7**	**2803.0**	**2.7472**	**3.3612**	**6.1083**
250	3.973	0.001 251	0.050 13	1080.39	1522.0	2602.4	1085.36	1716.2	2801.5	2.7927	3.2802	6.0730
255	4.319	0.001 263	0.045 98	1104.28	1596.7	2600.9	1109.73	1689.8	2799.5	2.8383	3.1992	6.0375
260	4.688	0.001 276	0.042 21	1128.39	1470.6	2599.0	1134.37	1662.5	2796.9	2.8838	3.1181	6.0019
265	5.081	0.001 289	0.038 77	1152.74	1443.9	2596.6	1159.28	1634.4	2793.6	2.9294	3.0368	5.9662
270	**5.499**	**0.001 302**	**0.035 64**	**1177.36**	**1416.3**	**2593.7**	**1184.51**	**1605.2**	**2789.7**	**2.9751**	**2.9551**	**5.9301**
275	5.942	0.001 317	0.032 79	1202.25	1387.9	2590.2	1210.07	1574.9	2785.0	3.0208	2.8730	5.8938
280	6.412	0.001 332	0.030 17	1227.46	1358.7	2586.1	1235.99	1543.6	2779.6	3.0668	2.7903	5.8571
285	6.909	0.001 348	0.027 77	1253.00	1328.4	2581.4	1262.31	1511.0	2773.3	3.1130	2.7070	5.8199
290	7.436	0.001 366	0.025 57	1278.92	1297.1	2576.0	1289.07	1477.1	2766.2	3.1594	2.6227	5.7821
295	**7.993**	**0.001 384**	**0.023 54**	**1305.2**	**1264.7**	**2569.9**	**1316.3**	**1441.8**	**2758.1**	**3.2062**	**2.5375**	**5.7437**
300	8.581	0.001 404	0.021 67	1332.0	1231.0	2563.0	1344.0	1404.9	2749.0	3.2534	2.4511	5.7045
305	9.202	0.001 425	0.019 948	1359.3	1195.9	2555.2	1372.4	1366.4	2738.7	3.3010	2.3633	5.6643
310	9.856	0.001 447	0.018 350	1387.1	1159.4	2546.4	1401.3	1326.0	2727.3	3.3493	2.2737	5.6230
315	10.547	0.001 472	0.016 867	1415.5	1121.1	2536.6	1431.0	1283.5	2714.5	3.3982	2.1821	5.5804
320	**11.274**	**0.001 499**	**0.015 488**	**1444.6**	**1080.9**	**2525.5**	**1461.5**	**1238.6**	**2700.1**	**3.4480**	**2.0882**	**5.5362**
330	12.845	0.001 561	0.012 996	1505.3	993.7	2498.9	1525.3	1140.6	2665.9	3.5507	1.8909	5.4417
340	14.586	0.001 638	0.010 797	1570.3	894.3	2464.6	1594.2	1027.9	2622.0	3.6594	1.6763	5.3357
350	16.513	0.001 740	0.008 813	1641.9	776.6	2418.4	1670.6	893.4	2563.9	3.7777	1.4335	5.2112
360	18.651	0.001 893	0.006 945	1725.2	626.3	2351.5	1760.5	720.3	2481.0	3.9147	1.1379	5.0526
370	**21.03**	**0.002 213**	**0.004 925**	**1844.0**	**384.5**	**2228.5**	**1890.5**	**441.6**	**2332.1**	**4.1106**	**0.6865**	**4.7971**
374.14	22.09	0.003 155	0.003 155	2029.6	0	2029.6	2099.3	0	2099.3	4.4298	0	4.4298

Solution

See the explanations in the list below the table.

	T, °C	P, kPa	x, %	h, kJ/kg	u, kJ/kg	v, m^3/kg
a)	**200**	1553.8	0	**852.45**	850.65	0.0012
b)	110	**143.3**	26.2	1046	**1000**	0.318
c)	**300**	**800**	—	3056.5	2797.2	0.3241
d)	**200**	**5000**	—	852	851	0.0012
e)	466	**400**	—	3413	3075	**0.85**
f)	**300**	8581	**80**	2467.9	2316.8	0.0176

a) This state falls directly on the saturated liquid line.

b) The given value for internal energy falls between u_f and u_g, so the state is inside the vapor dome. Calculate x and use x to find h and v.

c) The temperature of 300° C is higher than $T_{sat} = 170.4$° C, so the state is superheated.

d) The temperature of 200° C is less than $T_{sat} = 264$° C, so the state is subcooled. Find the properties in the saturation tables at 200° C (good approximation).

e) The specific volume is greater than $v_g = 0.6058$ at the saturation pressure of 300 kPa, so it is superheated (interpolation required).

f) A quality value (x) is specified, so the state is obviously in the vapor dome. Use x to find h, u, and v.

Use of R-134a Tables, NH_3 Tables, and *P-h* Diagrams

Typically, the four working fluids used in thermodynamic texts and on the EIT exam are H_2O, R-134a, NH_3, and air. The refrigerant R-22 has characteristics similar to R-134a, so it is not customarily tabled. The refrigerants R-134a and NH_3 exhibit the same phase-diagram characteristics as water, so they have the same type of tabled data, that is, subcooled liquid, saturated mixture, and superheated vapor. The guidelines given for finding one's way around the steam tables apply equally well to R-134a and NH_3 tables.

By tradition and for convenience, refrigerant properties are shown in a *P-h* diagram; this is especially useful when analyzing a vapor compression refrigeration cycle, since it operates at two basic pressure levels. A skeleton *P-h* chart is shown in Fig. 11.4. A larger *P-h* diagram on page 381 will be used in the next example and in the section on the vapor compression refrigeration cycle.

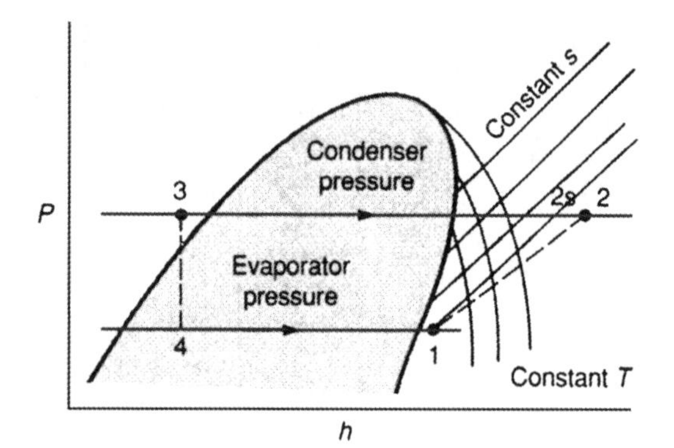

Figure 11.4

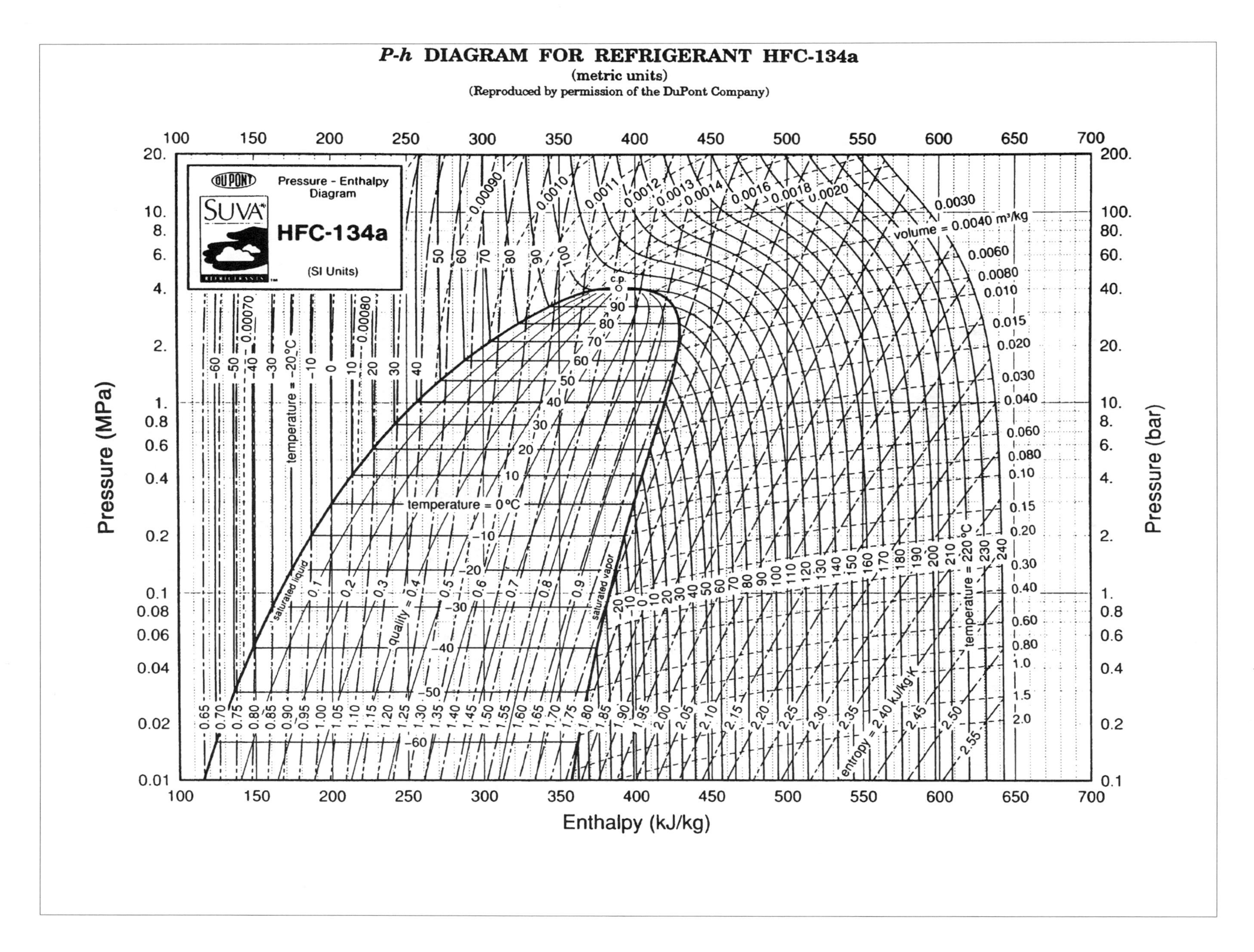
P-h DIAGRAM FOR REFRIGERANT HFC-134a
(metric units)
(Reproduced by permission of the DuPont Company)
DUPONT
Pressure - Enthalpy Diagram
SUVA
HFC-134a
(SI Units)
Pressure (MPa)
Pressure (bar)
Enthalpy (kJ/kg)
saturated liquid
saturated vapor
temperature = 0°C
temperature = -20°C
temperature = 220°C
volume = 0.0040 m³/kg
entropy = 2.40 kJ/kg·K
quality = 0.4
c.p.

Example 11.4

Saturated liquid R-134a at P = 2 bars is heated to 100°C at constant pressure. Find the original and final state point properties.

Solution

At P = 2 bars (.2 MPa) on the sat. liquid line,

$$v_f = .00075 \text{ m}^3/\text{kg}$$
$$h_f = 186 \text{ kJ/kg}$$
$$T_f = -10 \text{ °C}$$
$$s_f = 0.95 \text{ kJ/kg} \bullet \text{°K}$$

At P = 2 bars and 100°C, the R-134a is superheated.

$$v = .148 \text{ m}^3/\text{kg}$$
$$h = 491 \text{ kJ/kg}$$
$$s = 2.0484 \text{ kJ/kg} \bullet \text{°K}$$

IDEAL GASES

Equations of State

There are many relationships that relate the state-point properties of gases with varying degrees of accuracy. In order of increasing accuracy they are as follows:

1. Ideal gas
2. Van der Waals
3. Beattie-Bridgeman
4. Generalized compressibility
5. Property tables such as H_2O, R-134a, NH_3

Tables for these relationships are more conveniently simulated with software programs.

When one is dealing with a wide range of temperature and pressure for gases that are far from their saturated vapor state, the ideal gas relationship is accurate for most engineering work. Some working expressions are

$$Pv = \text{R}T$$

$$PV = m\text{R}T = n\overline{\text{R}}T, \qquad \text{R} = \frac{\overline{\text{R}}}{\text{m.w.}}$$

where

v = specific volume, m^3/kg
V = total volume, m^3
R = gas constant for the specific gas kJ/(kg • °K)
$\overline{\text{R}}$ = universal gas constant
n = number of moles
m = mass, kg
m.w. = molecular weight

Example **11.5**

Calculate and compare the specific volumes of water (H_2O) at P = 10 MPa and T = 400°C. The gas (steam) table value is v = 0.02641 m^3/kg. For an ideal gas,

$$R = \frac{8.31}{\text{m.w.}} = \frac{8.31}{18} = 0.462 \ \frac{\text{kJ}}{\text{kJ/kg} \bullet \text{°K}}$$

$$v = \frac{RT}{P} = \frac{0.462(673)}{10{,}000} = 0.0311 \text{ m}^3/\text{kg}$$

The steam table value is the most accurate, and the ideal gas value is the least accurate.

Enthalpy and Internal Energy Changes

The other condition which is usually considered to be part of the definition of ideal gas is that the specific heats (c_p and c_v) are constant. This allows a number of working relationships to be developed:

$$h_2 - h_1 = c_p\,(T_2 - T_1) \qquad c_p = \frac{kR}{k-1}$$

$$u_2 - u_1 = c_v\,(T_2 - T_1) \qquad c_v = \frac{R}{k-1}$$

$$c_p = c_v + R \quad \text{(always true)}$$

$$k = \frac{c_p}{c_v} = \text{constant}$$

PROCESSES

Thermodynamic processes usually involve a "working fluid" such as a pure substance (like water) or a gas (like air), so tables of properties or ideal gas relationships are used. For the process path to be known, the process must be reversible. If the process involves friction or turbulence and is irreversible, then only the first law of thermodynamics applies.

So, generally, the processes are considered to be reversible (no friction or turbulence) and are one of the following:

- constant pressure (Isobaric)
- constant volume (Isometric)
- constant temperature (Isothermal)
- no heat flow (adiabatic) (Isentropic)

Table 11.1 shows the applicable relationships for the various processes.

Table 11.1 First and Second Thermodynamics Law Formulas for Reversible Processes of an Ideal Gas*

Process	Closed System (non-flow)	Open System (steady flow)
General $(Pv = RT)$ $\frac{p_1v_1}{T_1} = \frac{p_2v_2}{T_2}$	$q = c_v(T_2 - T_1) + w$ $w = \int_1^2 P\,dv$ $s_2 - s_1 = c_p \ln \frac{T_2}{T_1} - R \ln \frac{p_2}{p_1}$	$q = c_p(T_2 - T_1) + \Delta KE - \Delta PE + w$ $w = -\int_1^2 v\,dP - \Delta KE - \Delta PE$ $s_2 - s_1 = c_v \ln \frac{T_2}{T_1} - R \ln \frac{v_2}{v_1}$
	($s_2 - s_1$ for closed or open systems)	
Polytropic $Pv^n = \text{constant}$ $\frac{p_2}{p_1} = \left(\frac{T_2}{T_1}\right)^{n/(n-1)} = \left(\frac{v_1}{v_2}\right)^n$ $\frac{T_2}{T_1} = \left(\frac{p_2}{p_1}\right)^{(n-1)/n} = \left(\frac{v_1}{v_2}\right)^{n-1}$ $\frac{v_2}{v_1} = \left(\frac{p_2}{p_1}\right)^{1/n} = \left(\frac{T_1}{T_2}\right)^{1/n}$	$q = \frac{k-n}{1-n} c_v(T_2 - T_1)$ $w = \frac{k-1}{1-n} c_v(T_2 - T_1)$ $s_2 - s_1 = c_p \ln \frac{T_2}{T_1} - R \ln \frac{p_2}{p_1}$ $s_2 - s_1 = \phi_2 - \phi_1 - R \ln \frac{p_2}{p_1} = \phi_2 - \phi_1 + R \ln \frac{v_2}{v_1}$	$q = \frac{k-n}{1-n} c_v(T_2 - T_1)$ $w = n \frac{k-1}{1-n} c_v(T_2 - T_1) - \Delta KE - \Delta PE$ $s_2 - s_1 = c_v \ln \frac{T_2}{T_1} - R \ln \frac{v_2}{v_1}$
Constant with volume (isometric) $v_2 = v_1, \quad n = \infty$ $\frac{p_2}{T_2} = \frac{p_1}{T_1}$	$w = 0$ $s_2 - s_1 = c_v \ln(T_2/T_1)$	$q = c_v(T_2 - T_1) \quad q = c_v(T_2 - T_1)$ $w = -v(p_2 - p_1) - \Delta KE - \Delta PE$ $s_2 - s_1 = c_v \ln(T_2/T_1)$
Constant pressure (isobaric) $p_2 = p_1, \quad n = 0$ $\frac{v_2}{T_2} = \frac{v_1}{T_1}$	$w = p(v_2 - v_1)$ $w = R(T_2 - T_1)$ $s_2 - s_1 = c_p \ln(T_2/T_1)$	$q = c_p(T_2 - T_1) \quad q = c_p(T_2 - T_1)$ $w = -\Delta KE - \Delta PE$ $s_2 - s_1 = c_p \ln(T_2/T_1)$
Constant temperature (isothermal) $T_2 = T_1, \quad n = 1$ $p_2 v_2 = p_1 v_1$	$q = w = T(s_2 - s_1)$ $q = w = RT \ln \frac{v_2}{v_1}$ or $\frac{p_1}{p_2}$ $s_2 - s_1 = R \ln \frac{v_2}{v_1}$ or $\frac{p_1}{p_2}$	$q = T(s_2 - s_1) = RT \ln \frac{v_2}{v_1}$ or $\frac{p_1}{p_2}$ $w = RT \ln \frac{v_2}{v_1} - \Delta KE - \Delta PE$ or $w = RT \ln \frac{p_1}{p_2} - \Delta KE - \Delta PE$ $s_2 - s_1 = R \ln \frac{v_2}{v_1} = R \ln \frac{p_1}{p_2}$
Adiabatic (isentropic) $n = k$ $s_2 = s_1$	$q = 0$ $w = c_v(T_1 - T_2)$ $w = \frac{p_1v_1 - p_2v_2}{k-1}$ $w = R(T_1 - T_2)/(k-1)$ $s_2 - s_1 = 0$	$q = 0$ $w = c_p(T_1 - T_2) - \Delta KE - \Delta PE$ $w = \frac{k(p_1v_1 - p_2v_2)}{k-1}\ \Delta KE - \Delta PE$ $w = kR(T_1 - T_2)/(k-1) - \Delta KE - \Delta PE$ $s_2 - s_1 = 0$

*Per-unit mass basis and constant (average) specific heats (c_v, c_p) assumed. $R = c_p - c_v$, $k = c_p/c_v$, $c_p = kR/(k-1)$, $C_v = R/(k-1)$.
$\Delta u = u_2 - u_1 = c_v(T_2 - T_1)$, $\Delta h = h_2 - h_1 = c_p(T_2 - T_1)$

$$\Delta KE\ (\text{S.I.}) = \frac{v_2^2 - v_1^2}{2000 \times g_c}, \quad \Delta PE\ (\text{S.I.}) = \frac{g(z_2 - z_1)}{1000 \times g_c}$$

ΔKE and ΔKE may be negligible for many open systems.

Example 11.6

Air at 27°C is heated to 927°C. Find the change in enthalpy and internal energy, treating air as a perfect gas (c_p and c_v constant).

Solution

For air at room temperature, $c_p = 1.00$ and $c_v = 0.718$ kJ/kg, so

$$h_2 - h_1 = (927 - 27) \times 1.00 = 900.0 \text{ kJ/kg}$$
$$u_2 - u_1 = (927 - 27) \times 0.718 = 646.2 \text{ kJ/kg}$$

THE SECOND LAW OF THERMODYNAMICS

In the morning section of the FE examination, there will typically be several problems that involve the application of one of the statements of the second law or the Carnot cycle. Useful statements of the second law are

a. Whenever energy is transferred, some energy is reduced to a lower level.

b. No heat cycle is possible without the rejection of some heat.

c. A Carnot cycle converts the maximum amount of heat into work; it has the highest thermal efficiency.

d. All Carnot cycles operating between two temperature reservoirs have the same efficiency.

e. A Carnot machine's efficiency, or coefficient of performance (COP), is a function of only the two reservoir temperatures.

A Carnot cycle consists of the following four processes:

4–1 Reversible adiabatic compression
1–2 Reversible adiabatic constant temperature heat addition
2–3 Reversible adiabatic expansion
3–4 Reversible constant temperature heat rejection

The normal property diagrams are shown in Figs. 11.5 and 11.6. If the processes proceed in a clockwise direction, the Carnot engine operates as a power-producing engine; if in a counterclockwise direction, the engine is a refrigerator or a heat pump.

The efficiencies, or COPs, are

$$\eta = \frac{W_{\text{out}}}{Q_{\text{in}}} = 1 - \frac{T_C}{T_H}$$

$$\text{COP}_{\text{REFR}} = \frac{Q_{\text{in}}}{W_{\text{in}}} = \frac{T_C}{T_H - T_C}$$

$$\text{COP}_{\text{HEATPUMP}} = \frac{Q_{\text{out}}}{W_{\text{in}}} = \frac{T_H}{T_H - T_C}$$

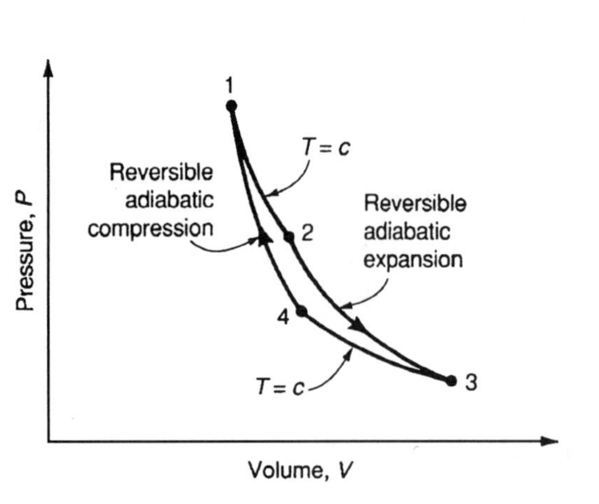

Figure 11.5

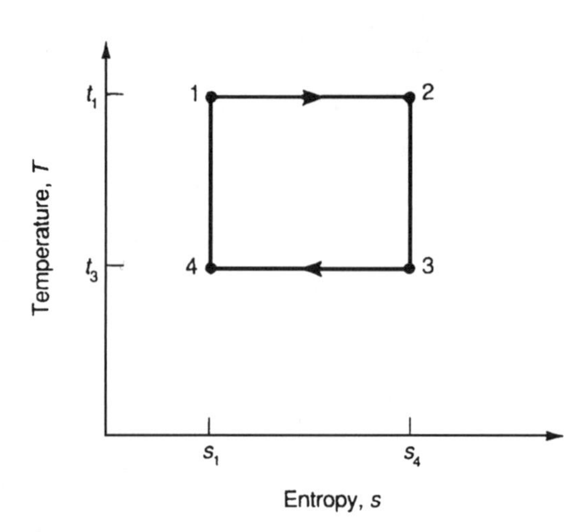

Figure 11.6

Example 11.7

A Carnot machine operates between a hot reservoir at 200°C and a cold reservoir at 20°C. (a) When operated as an engine, it receives 1000 kJ/kg; find the work output. (b) Find the COP when the machine is operated as a refrigerator and as a heat pump.

Solution

(a)

$$\eta = \frac{T_H - T_C}{T_H} = \frac{200 - 20}{473} = 0.381$$

$$W = \eta \times Q_{\text{in}} = 0.381 \times 1000 = 381 \text{ kJ/kg}$$

(b)

$$\text{COP}_{\text{REFR}} = \frac{T_C}{T_H - T_C} = \frac{293}{200 - 20} = 1.63$$

$$\text{COP}_{\text{HEATPUMP}} = \frac{T_H}{T_H - T_C} = \text{COP}_{\text{REFR}} + 1.0 = 2.63$$

ENTROPY

Entropy is another thermodynamic property that is useful in the evaluation of thermodynamic systems and processes. The following statements are useful in solving problems:

a. Natural processes (which typically involve friction) result in an increase in entropy.

b. Entroy will always *increase* when heat is added.

c. Entropy will remain *constant* when processes are reversible and adiabatic.

d. Entropy can *decrease* only when heat is removed.

For reversible processes,

$$ds = \frac{dq}{T}$$

$$T\,ds = du + P\,dv = dh - v\,dP$$

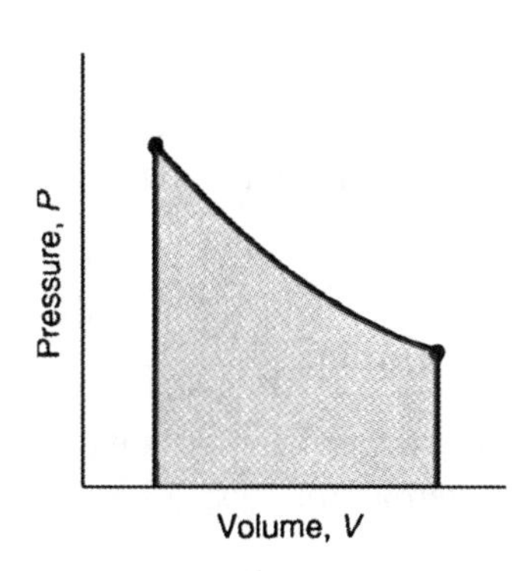

Figure 11.7

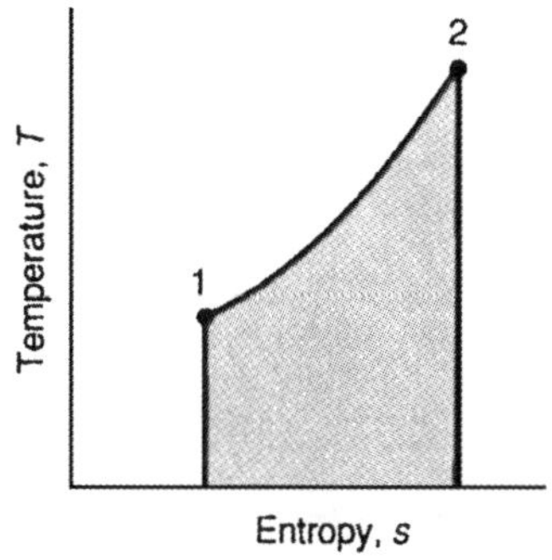

For an ideal gas,

$$
\begin{aligned}
s_2 - s_1 &= c_v \ln\frac{T_2}{T_1} + \mathrm{R}\ln\frac{v_2}{v_1} \\
&= c_p \ln\frac{T_2}{T_1} - \mathrm{R}\ln\frac{p_2}{p_1}
\end{aligned}
$$

Just as work for a closed system,

$$W = \int_1^2 P\,dv$$

can be shown as an area on a *P-V* diagram (Fig. 11.7), so can heat,

$$Q = \int_1^2 T\,ds$$

Figure 11.8

be shown as an area on a *T-s* diagram (Fig. 11.8), and the area *enclosed* by the process lines on a *T-s* diagram shows the *net* heat flow in a cycle. This, of course, is equal to the *net* work.

An **isentropic process** is defined as one that is reversible and adiabatic. Of course, on a property diagram showing entropy (s), the process would appear as a straight line. For several important thermodynamic devices, the isentropic process is a standard of comparison and is used in the calculation of the component efficiency (turbine, compressor, pump, nozzle).

Since all natural processes produce an increase in entropy, the ideal (isentropic) and the actual processes can be compared, as shown in Figs. 11.9 and 11.10.

$$
\begin{aligned}
\eta_{\text{turbine nozzle}} &= \frac{h_1 - h_3}{h_1 - h_2} \\
\eta_{\text{compr., pump}} &= \frac{h_2 - h_1}{h_3 - h_1}
\end{aligned}
$$

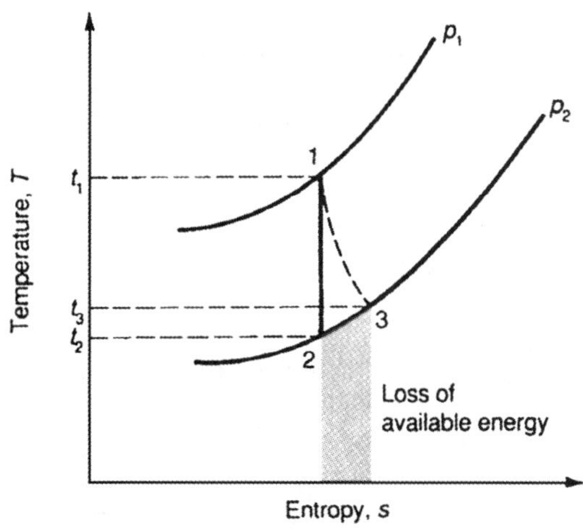

Figure 11.9

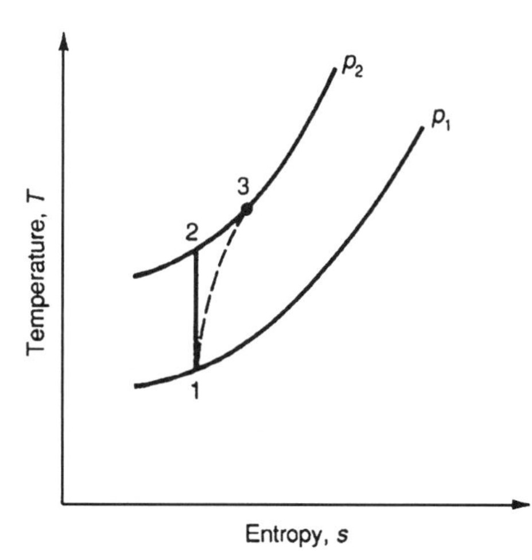

Figure 11.10

CYCLES

The FE afternoon problems will usually include a Rankine cycle problem and a vapor compression problem. Typically, there will be others involving a knowledge of the Brayton, Otto, or Diesel cycles.

Rankine Cycle (Steam)

An ideal Rankine cycle with superheated steam flowing into the turbine is shown in Fig. 11.11. The four open-system components are analyzed as follows:

a. Boiler

$$q_{in} = h_2 - h_1 \left(\frac{\text{kJ}}{\text{kg}}\right)$$
$$\dot{Q}_{in} = \dot{m}_{stm}(h_2 - h_1)(\text{kW})$$

b. Turbine

$$w_T = h_3 - h_2 \text{ (kJ/kg)}$$
$$\dot{W}_T = \dot{m}_{stm}(h_3 - h_2) \text{ (kW)}$$

c. Condenser

$$q_{out} = h_3 - h_4 = c_p \Delta T \text{ (kJ/kg)}$$
$$\dot{Q}_{out} = \dot{m}_{stm}(h_3 - h_4) = \dot{m}_{cw} c_p \Delta T_{cw} \text{ (kW)}$$

d. Pump

$$w_p = h_1 - h_4 = v_4(p_1 - p_4) \text{ (kJ/kg)}$$
$$\dot{W}_p = \dot{m}_{stm}(h_1 - h_4) \text{ (kW)}$$

State points are found in property (steam) tables. State 2 is usually given by pressure and temperature and can be either saturated or (normally) superheated. State 2 can be found in the steam tables. State 3 is found by using an isentropic process, so that entropy (State 2) and pressure (condensing) are known. State 4 is saturated liquid found in the tables at the condensing pressure. Also, State 1—

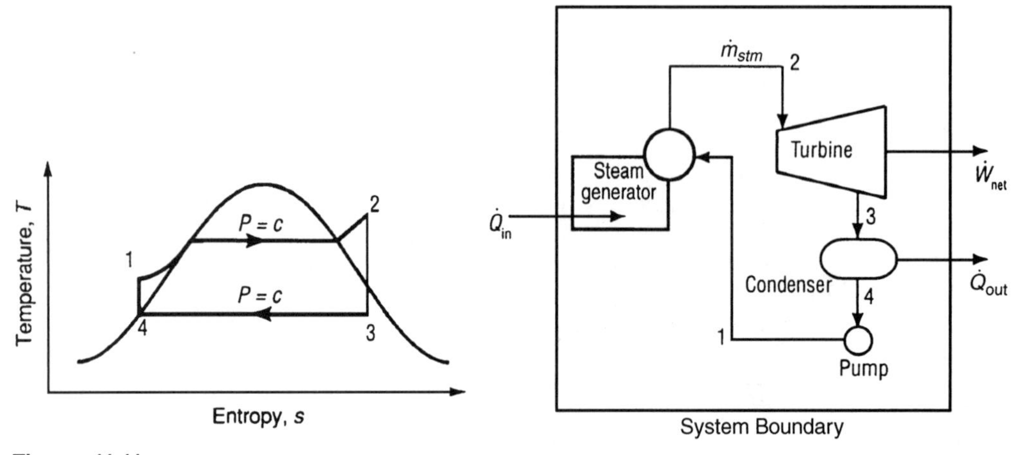

Figure 11.11

using an isentropic process—can be found by the tables but is more easily found by

$$h_1 = h_4 + v_4(p_1 - p_4)$$

The thermal efficiency is

$$\eta = \frac{W_T - W_P}{Q_{in}} = \frac{h_2 - h_3 - (h_1 - h_4)}{h_2 - h_1}$$

Adding $h_1 - h_4$ (the pump work) to both the numerator and denominator results in

$$\eta_{approx.} = \frac{h_2 - h_3}{h_2 - h_4}$$

Example 11.8

A Rankine cycle using steam has turbine inlet conditions of $P = .4$ MPa, $T = 300$°C, and a condenser temperature of 70°C. The turbine efficiency is 90%, and the pump efficiency is 80%.

For both the ideal cycle and the cycle considering the component efficiencies, find (a) the thermal efficiency (η), (b) the turbine discharge quality (x), and (c) the steam flow rate ($\dot{m}$) for 1 MW of net power.

Properties of water (SI units): superheated-vapor table

Temp., °C	v	u	h	s
		0.4 MPa (T_{sat} = 143.6 °C)		
300	.6548	2805	3067	7.566

v, m^3/kg; u, kJ/kg; h, kJ/kg; s, kJ/(kg•°K)

Properties of saturated water (SI units): pressure table

		Specific volume		Internal energy		Enthalpy			Entropy	
Press. Bars kPa	Temp. °C T	Sat. Liquid vf	Sat. Vapor v_g	Sat. Liquid u_f	Sat. Vapor u_g	Sat. Liquid h_f	Evap. h_{fg}	Sat. Vapor h_g	Sat. Liquid s_f	Sat. Vapor s_g
31.2	70	.00102	5.04	293	2570	293	2334	2627	.9549	7.7553

v, m^3/kg; u and h, kJ/kg; s, kJ/(kg • °K)

Solution

Starting with State 2 (turbine inlet), the properties h and s can be found with the steam tables.

$$h_2 = 3067\frac{\text{kJ}}{\text{kg}} \qquad s_2 = 7.566\frac{\text{kJ}}{\text{kg}\bullet{}^\circ\text{K}}$$

The ideal turbine discharge (State 3) is found at $T_3 = 70$°C and $s_3 = 7.566$:

$$x = \frac{s - s_f}{s_{fg}} = \frac{7.566 - .9549}{7.7553 - .9549} = 0.972$$

$$h_3 = h_f + xh_{fg} = 293 + .972 \times 2334 = 2562 \text{ kJ/kg}$$

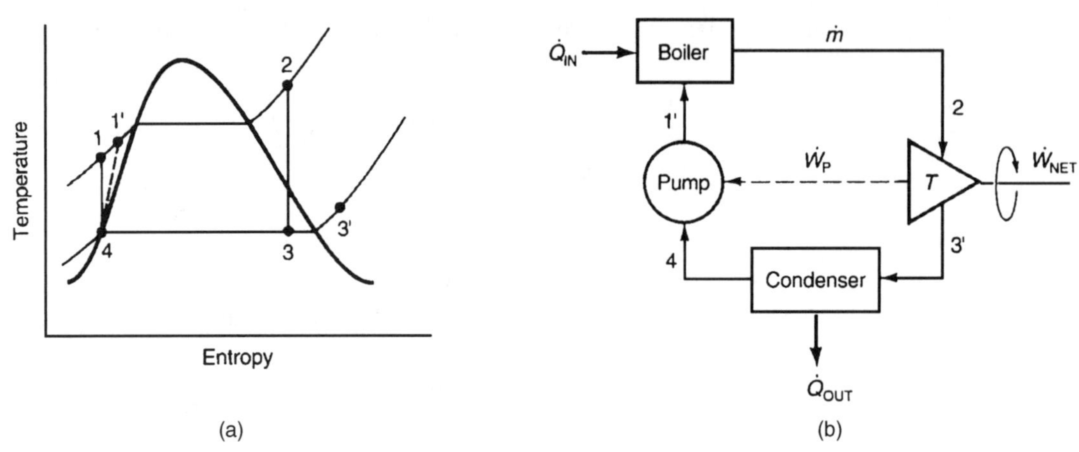

Figure 11.12

The saturated liquid state leaving the condenser (State 4) is read from the tables:

$$h_4 = 293\,\text{kJ/kg}$$
$$s_4 = .9549\frac{\text{kJ}}{\text{kg}\bullet{}^\circ\text{K}}$$
$$v_4 = 0.00102\,\text{m}^3/\text{kg}$$

The compressed (subcooled) liquid leaving the pump (State 1) can be found in the tables if values are available for the condition. The usual approximation is to calculate

$$\begin{aligned} h_1 &= h_4 + v_4(p_4 - p_1) \\ &= 293 + .00102(400 - 31.2) \\ &= 293 + .4 = 293.4\frac{\text{kJ}}{\text{kg}} \end{aligned}$$

For the ideal cycle,

a) $$\eta = \frac{W_{\text{net}}}{Q_{\text{in}}} = \frac{W_T - |W_P|}{Q_{\text{in}}} = \frac{(3067 - 2562) - .4}{3067 - 293.4}$$
$$= \frac{505 - .4}{2773.6} = 0.182$$

b) $x = 0.972$

c) $$\dot{m} = \frac{1000\text{ kW}}{W_{\text{net}}} = \frac{1000}{505} = 2\text{ kg/s}$$

For the cycle considering the component efficiencies (Fig. 11.12),

$$\eta_{\text{turb.}} = \frac{h_2 - h_{3'}}{h_2 - h_3}$$

$h_{3'} = 3067 - .9(3067 - 2562) = 2613$ kJ/kg, and $P = 31.2$ kPa.

$$h_{1'} = h_4 + \frac{h_1 - h_4}{\eta_p} = 293 + \frac{.4}{.8} = 293.5 \ \frac{\text{kJ}}{\text{kg}}$$

$$x = \frac{h_{3'} - h_f}{h_{fg}} = \frac{2613 - 293}{2334} = 0.99$$

$$\eta = \frac{W_T - W_P}{Q_{\text{in}}} = \frac{h_2 - h_{3'} - h_{1'} - h_4}{h_2 - h_{1'}}$$

$$= \frac{(3067 - 2613) - \frac{.4}{.8}}{3067 - \left(293 + \frac{.4}{.8}\right)} = \frac{453.5}{2773.5} = .164$$

$$\dot{m} = \frac{1000}{453.5} = 2.2 \ \frac{\text{kg}}{\text{s}}$$

The pump work makes little numerical difference and can usually be ignored in the calculation of thermal efficiency and steam flow rate.

Vapor Compression Cycle (Refrigeration)

An ideal vapor compression cycle is shown in Fig. 11.13. The four open-system components are analyzed below:

a. Compressor

$$W_{\text{in}} = h_2 - h_1$$
$$\dot{W}_{\text{in}} = \dot{m}\,(h_2 - h_1)$$

b. Condenser

$$Q_{\text{out}} = h_2 - h_3$$
$$\dot{Q}_{\text{out}} = \dot{m}(h_2 - h_3)$$

c. Expansion valve

$$h_3 = h_4$$

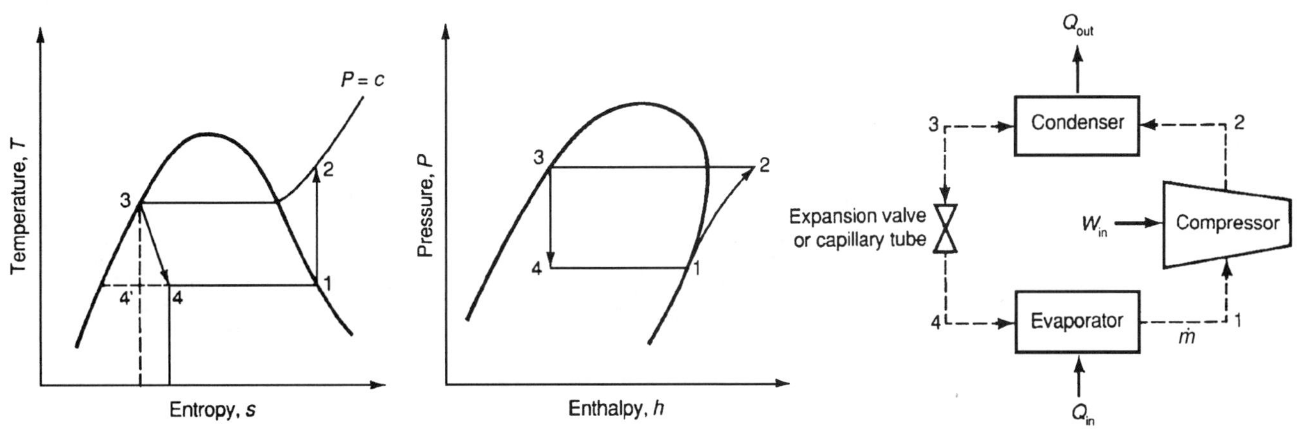

Figure 11.13

d. Evaporator

$$Q_{in} = h_1 - h_4$$
$$\dot{Q}_{in} = \dot{m}(h_1 - h_4)$$

The coefficient of performance (COP), if the cycle is used as a refrigerator, is

$$COP_{REFR} = \frac{Q_{in}}{W_{in}} = \frac{h_1 - h_4}{h_2 - h_1}$$

If the cycle is used as a heat pump,

$$COP_{HEATPUMP} = \frac{Q_{out}}{W_{in}} = \frac{h_2 - h_3}{h_2 - h_1} = COP_{REFR} + 1.0$$

The state points are found in property tables and/or property diagrams (*P-h*). State 1 is usually given as saturated vapor at a given pressure or temperature. State 2 is found by assuming an isentropic process so that the entropy and pressure (or corresponding condensing temperature) are known. This is best done on a *P-h* diagram. State 3 is saturated liquid at the given condensing pressure. State 4 is found at the same enthalpy as State 3 and at the evaporating pressure.

Example 11.9

An ideal vapor compression refrigeration cycle using R-134a operates between 100 kPa and 1000 kPa. Find the COP and the mass flow rate required for 100 kW of cooling.

Solution

Refer to the *P-h* diagram on page 387 and a schematic of the components (Exhibit 1). State 1 (saturated vapor at 100 kPa), from *P-h* diagram or from property tables:

$$h_1 = 383 \text{ kJ/kg}$$
$$s_1 = 1.746 \text{ kJ/kg} \bullet {}^\circ\text{K}$$
$$T_1 = -26.31^\circ\text{C}$$

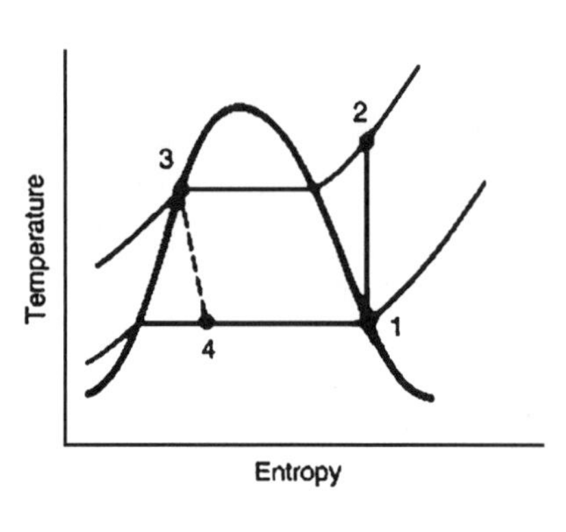

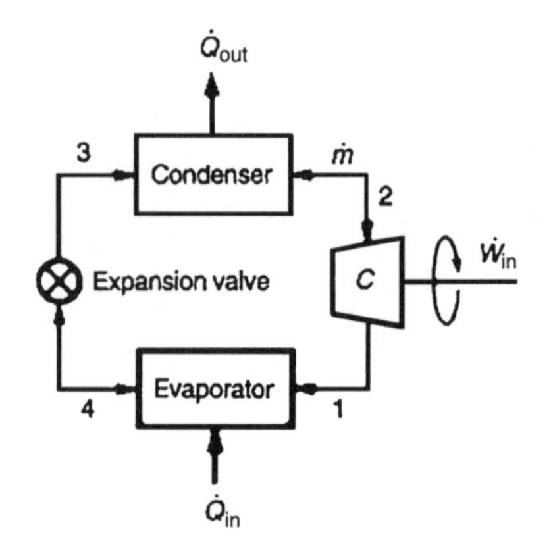

Exhibit 1

State 2 ($P_2 = 1000$ kPa)

$$s_1 = s_2 = 1.746 \text{ kJ/kg} \bullet {}^\circ\text{K}$$
$$h_2 = 431 \text{ kJ/kg}$$
$$T_2 = 49.8^\circ\text{C}$$

State 3 (saturated liquid)

$$h_3 = 255.6 \text{ kJ/kg}$$
$$T_3 = 39.3^\circ\text{C}$$

State 4 (liquid + vapor)

$$h_3 = h_4 = 225.6 \text{ kJ/kg}$$
$$\text{COP}_\text{R} = \frac{\dot{q}_\text{evap}}{w_c} = \frac{h_1 - h_4}{h_2 - h_1} = \frac{383 - 255.6}{431 - 383} = \frac{127.4}{48} = 2.65$$
$$\dot{q}_\text{evap} = \dot{m}(h_1 - h_4), \qquad \dot{m} = \frac{100}{127.4} = 0.478 \frac{\text{kg}}{\text{s}}$$

Otto Cycle (Gasoline Engine)

An ideal Otto cycle is shown in Fig. 11.14. It consists of the following four processes:

1. An isentropic compression for 1 to 2.
2. A constant volume heat addition from 2 to 3.
3. An isentropic expansion from 3 to 4.
4. A constant volume heat rejection from 4 to 1.

The Otto cycle is an air standard cycle, that is, a cycle that uses ideal air as the working media and has ideal processes. An equipment sketch consists of only a piston and cylinder, since it is a closed-system cycle using a fixed quantity of mass. The four closed-system processes reduce to

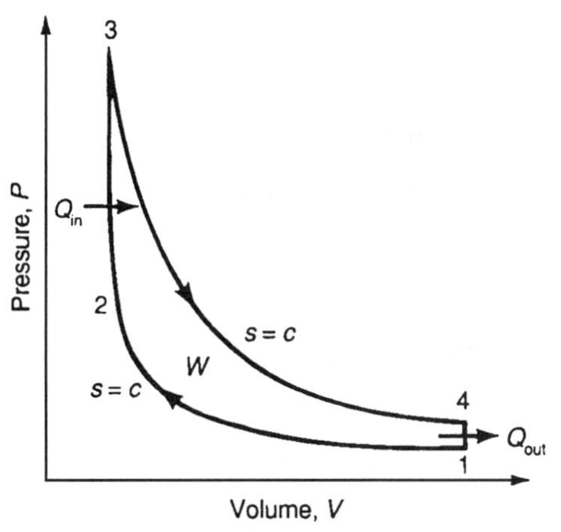

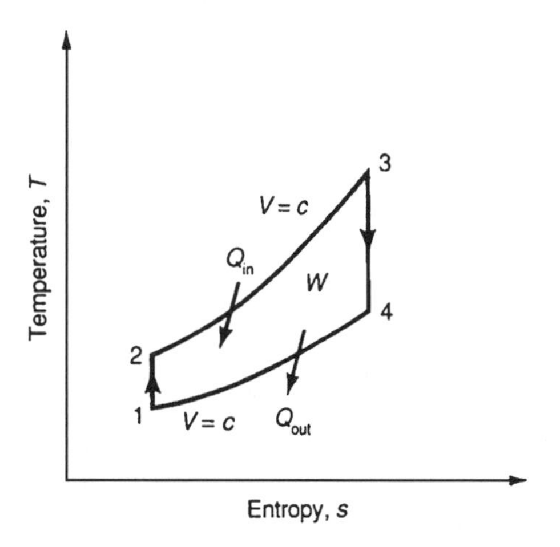

Figure 11.14

1. Isentropic compression

$$u_1 + q = u_2 - W$$
$$q = 0$$
$$w_{\text{comp}} = u_2 - u_1 = c_v(T_2 - T_1)$$

2. Heat addition

$$u_2 + q_{\text{in}} = u_3 + w$$
$$w = 0$$
$$q_{\text{in}} = u_3 - u_2 = c_v(T_3 - T_2)$$

3. Isentropic expansion

$$u_3 + q = u_4 + w$$
$$q = 0$$
$$w_{\text{exp}} = u_3 - u_4 = c_v(T_3 - T_4)$$

4. Heat rejection

$$u_4 + q_{\text{out}} = u_1 + w$$
$$w = 0$$
$$q_{\text{out}} = u_1 - u_4$$

The thermal efficiency is

$$\eta = \frac{W_{\text{net}}}{Q_{\text{in}}} = \frac{W_{\text{exp}} - W_{\text{comp}}}{Q_{\text{in}}} = \frac{u_3 - u_4 - (u_2 - u_1)}{u_3 - u_2} = \frac{T_3 - T_4 - (T_2 - T_1)}{T_3 - T_2} = 1 - \frac{1}{r_c^{k-1}}$$

Note that r_c is the compression ratio, a ratio of the *volume* at the bottom of the piston stroke (bottom dead center) to the *volume* of the top of the stroke (top dead center). This is also equal to v_1/v_2.

The state points are found by ideal gas laws or air tables: State 1 is usually given by T and P. State 2 is found by using an isentropic process; for an ideal gas,

$$\frac{T_2}{T_1} = \left(\frac{v_1}{v_2}\right)^{k-1} = r_c^{k-1}$$

If air tables are used,

$$\frac{v_{r_2}}{v_{r_1}} = \frac{v_2}{v_1} = \frac{1}{r_c}$$

State 3 is usually found by knowing the heat addition:

$$Q_{\text{in}} = u_3 - u_2 = c_v(T_3 - T_2)$$
$$T_3 = \frac{q}{c_v} + T_2$$

State 4 is found by an isentropic process; for an ideal gas

$$\frac{T_3}{T_4} = \left(\frac{V_4}{V_3}\right)^{k-1} = r_c^{k-1}$$
$$T_4 = \frac{T_3}{r_c^{k-1}}$$

Example 11.10

An engine operates on an air standard Otto cycle with a temperature and pressure of 27°C and 100 kPa at the beginning of compression. The compression ratio is 8.0, and the heat added is 1840 kJ/kg. Find the state point properties and the thermal efficiency. The properties for ideal air are $c_p = 1.008$ kJ/kg • °K, $c_v = 0.718$ kJ/kg • °K, $R = 0.287$ kJ/kg • °K, and $k = 1.4$.

Solution

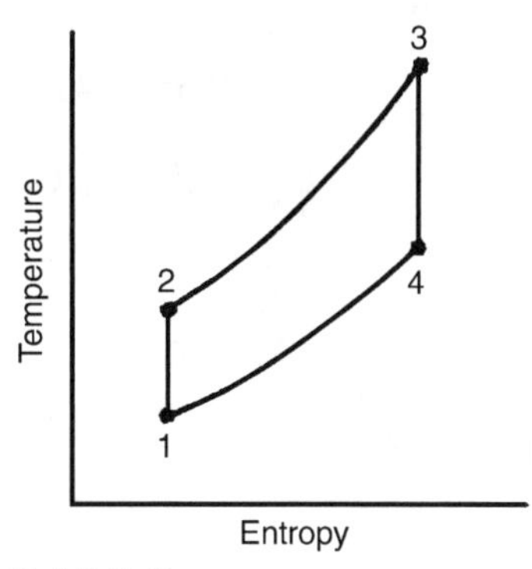

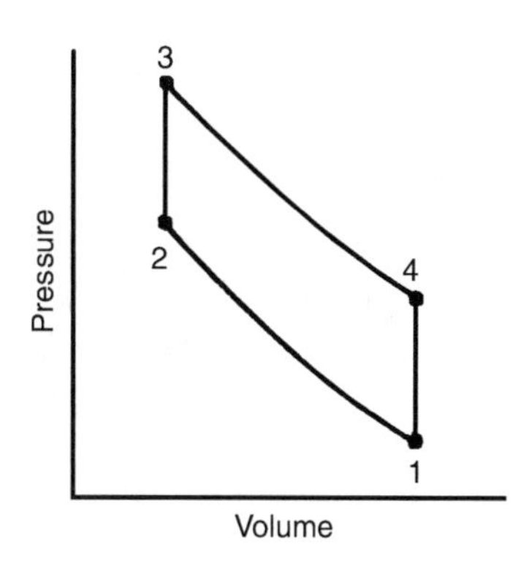

Exhibit 2

Refer to the *T-s* diagram and *P-V* diagram in Exhibit 2.

State 1

$$P_1 = 100 \text{ kPa}$$
$$T_1 = 300\ °\text{K}$$
$$v_1 = \frac{RT_1}{P_1} = 0.287 \times \frac{300}{100} = 0.861 \text{ m}^3/\text{kg}$$

State 2

$$v_2 = \frac{v_1}{8} = 0.1076 \text{ m}^3/\text{kg}$$
$$T_2 = T_1\left(\frac{v_1}{v_2}\right)^{k-1} = 300 \times (8.0)^{0.4} = 689°\text{K}$$
$$P_2 = P_1\left(\frac{v_1}{v_2}\right)^{k} = 100 \times (8.0)^{1.4} = 1838 \text{ kPa}$$

State 3

$$v_3 = v_2 = 0.1076 \text{ m}^3/\text{kg}$$
$$u_3 = u_2 + q_{\text{in}}$$
$$q_{\text{in}} = u_3 - u_2 = c_v(T_3 - T_2)$$
$$T_3 = \frac{1840}{0.718} + 689 = 3255°\text{K}$$
$$P_3 = P_2\left(\frac{T_3}{T_2}\right) = 1838 \times \frac{3255}{689} = 8683 \text{ kPa}$$

State 4

$$v_4 = v_1 = 0.861 \text{ m}^3/\text{kg}$$

$$T_4 = \frac{T_3}{r_c^{k-1}} = \frac{3255}{(8.0)^{0.4}} = 1417°\text{K}$$

$$P_4 = \frac{P_3}{r_c^k} = \frac{8676}{(8.0)^{1.4}} = 472 \text{ kPa}$$

$$\eta_{\text{TH}} = \frac{W_{\text{net}}}{Q_{\text{in}}} = \frac{W_{3-4} - W_{1-2}}{Q_{\text{in}}} = \frac{c_v(T_3 - T_4) - c_v(T_2 - T_1)}{1840}$$

$$= \frac{0.718(3255 - 1416) - 0.718(689 - 300)}{1840}$$

$$= \frac{1318 - 279}{1840} = \frac{1039}{1840} = 0.565$$

Check:

$$\eta_{\text{TH}} = \frac{Q_{\text{in}} - Q_{\text{out}}}{Q_{\text{in}}} = \frac{1840 - c_v(T_4 - T_1)}{1840} = \frac{1840 - 800.9}{1840} = 0.565$$

$$\eta_{\text{TH}} = 1 - \frac{1}{r_c^{k-1}} = 1 - \frac{1}{(8.0)^{0.4}} = 0.565$$

MISCELLANEOUS

Mixture of Gases

The composition of a closed mixture of gases may be expressed in terms of volume (mol) fractions or mass fractions. These are related through the component molecular weight (m.w.) and are best shown in tabular form. If volume fractions are given, convert to mass fraction:

Gas	Volume Fraction	m.w.	Volume Fraction × m.w.	Mass Fraction
O_2	0.2	32	6.4	$\frac{6.4}{38.4} = 0.167$
N_2	0.2	28	5.6	$\frac{5.6}{38.4} = 0.146$
CO_2	0.6	44	26.4	$\frac{26.4}{38.4} = 0.687$
			38.4	1.000

If mass fraction is given, convert to volume fractions:

Gas	Mass fraction	m.w.	$\frac{\text{Mass Fraction}}{\text{m.w.}}$	Volume Fraction
O_2	0.1	32	0.00313	$\frac{0.00313}{0.03135} = 0.100$
N_2	0.6	28	0.0214	$\frac{0.0214}{0.03135} = 0.683$
CO_2	0.3	44	0.00682	$\frac{0.00682}{0.03135} = 0.217$
			0.03135	1.000

The mass fraction is sometimes called the gravimetric fraction. Component pressure and molecular weight are volume fraction functions; u, h, c_p, c_v, and R are mass fraction functions.

Heat Transfer

The three modes of heat transfer are conduction, convection, and radiation. The heat transfer "laws" are based on both empirical observations and theory but are consistent with the first and second laws of thermodynamics. That is, energy is conserved and heat flows from hot to cold.

Conduction

Conduction occurs in all phases of materials (Fig. 11.15). The equation for one-dimensional, planar, steady-state conduction heat transfer is

$$\dot{Q} = kA\frac{T_H - T_C}{x} \quad \text{(watts)}$$

The **conductivity**, k, is a property of the material and is evaluated at the average temperature of the material. The **heat flow rate**, q, is sometimes expressed as a heat **flux** $\dot{Q}/A$.

For multiple layers of different materials (Fig. 11.16), as in composite structures, it is usually best to use an electrical analogy:

$$\dot{Q}_1 = \frac{T_H - T_{x_1}}{R_1} \qquad R_1 = \frac{x_1}{A_1 k_1}$$

$$\dot{Q} = \dot{Q}_1 = \dot{Q}_2 = \dot{Q}_3$$

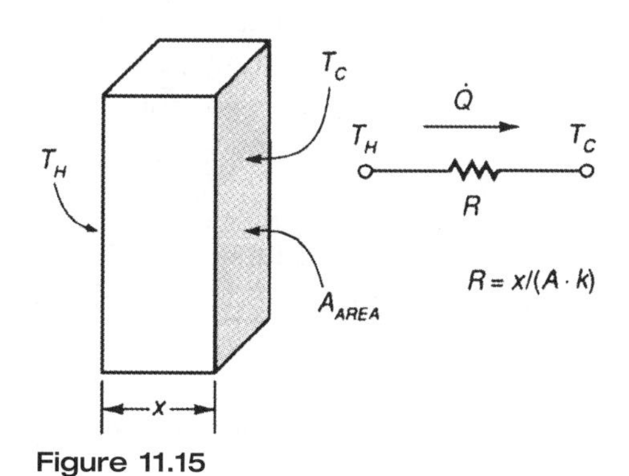

Figure 11.15

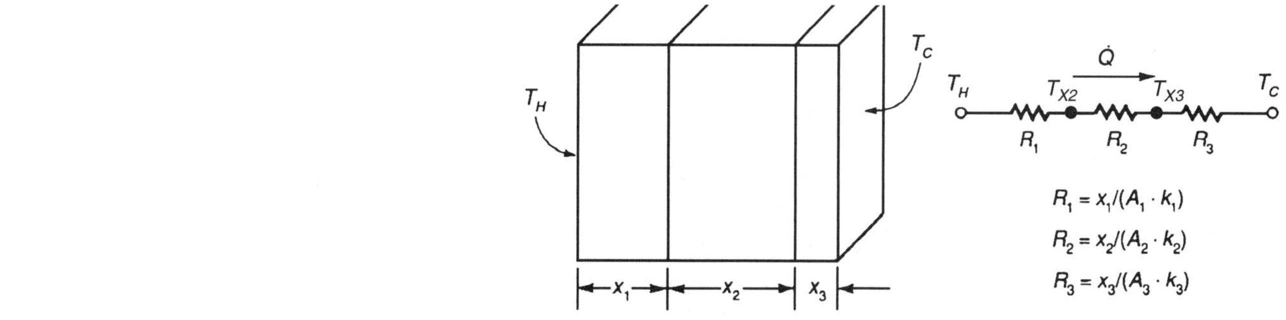

Figure 11.16

$$\dot{Q}_2 = \frac{T_{x_1} - T_{x_2}}{R_2} \qquad R_2 = \frac{x_2}{A_2 k_2}$$
$$\dot{Q}_3 = \frac{T_{x_2} - T_C}{R_3} \qquad R_3 = \frac{x_3}{A_3 k_3}$$
$$\dot{Q}_T = \frac{T_H - T_C}{R_1 + R_2 + R_3}$$

Example **11.11**

A plane wall is 2 m high by 3 m wide and is 20 cm thick. It is made of material that has a thermal conductivity of 0.5 W/(m • °K). A temperature difference of 60°C is imposed on the two large faces. Find the heat flow, the heat flux, and the conductive resistance.

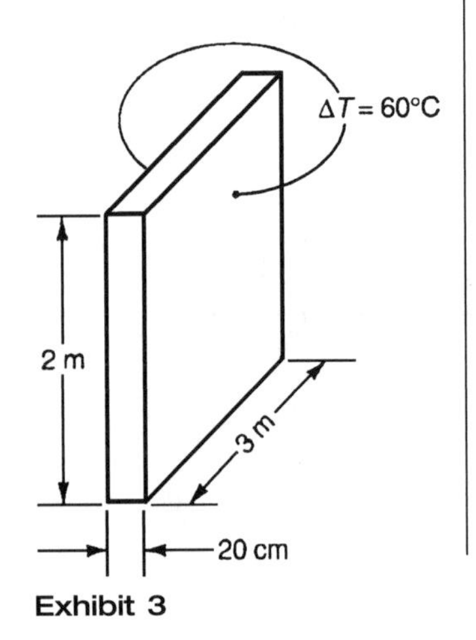

Exhibit 3

Solution

Refer to Exhibit 3.

$$\dot{Q} = \frac{kA(T_H - T_C)}{x} = \frac{0.5 \times 3 \times 2 \times 60}{0.20} = 900 \text{ W}$$
$$\frac{\dot{Q}}{A} = \frac{900}{3 \times 2} = 150 \frac{\text{W}}{\text{m}^2}$$
$$R = \frac{x}{kA} = \frac{0.2}{0.5 \times 3 \times 2} = 0.0667 \frac{°\text{K}}{\text{W}}$$

Convection

Convection occurs at the boundary of a solid and a fluid (liquid or gas) when there is a temperature difference. The mechanism is complex and can be evaluated analytically only for a few simple cases; most situations are evaluated empirically. The equation for convective heat transfer is

$$q = hA(T_{\text{surface}} - T_{\text{fluid}})$$

The evaluation of h, the heat transfer coefficient, normally involves use of data correlated in the form of dimensionless parameters: for example, Nussult number, Reynolds number, Prandtl number.

The conduction and convection mechanisms can be combined as shown in Fig. 11.17. So the temperature of the surface, T_s, is dependent on the relative magnitude of the two resistances.

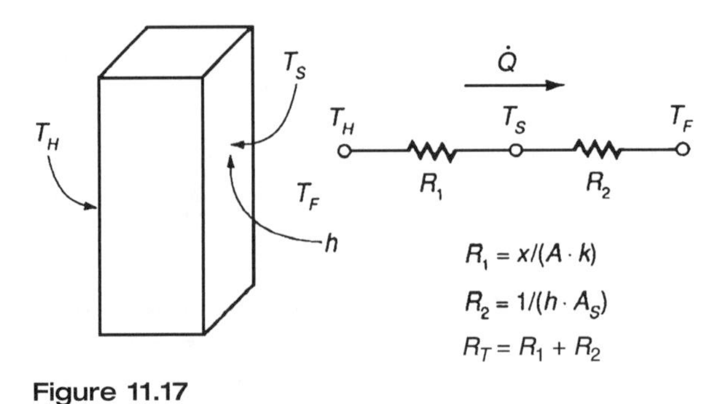

Figure 11.17

Example 11.12

Water at an average temperature of 20°C flows through a 5-cm-diameter pipe that is 2 m long. The pipe wall is heated by steam and is held at 100°C. The convective heat transfer coefficient is 2.2 × 10^4 W/m^2 • °K). Find the heat flow, the heat flux, and the convective resistance.

Solution

Refer to Exhibit 4.

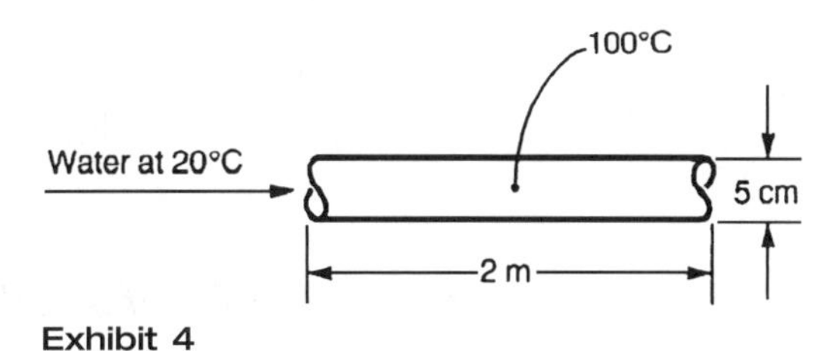

Exhibit 4

$$\dot{Q} = hA(T_H - T_C) = 2.2 \times 10^4 \times (\pi \times 0.05 \times 2) \times (100 - 20) = 5.53 \times 10^5 \text{ W}$$

$$\frac{\dot{Q}}{A} = \frac{5.53 \times 10^5}{\pi \times 0.05 \times 2} = 1.76 \frac{\text{MW}}{\text{m}^2}$$

$$R = \frac{1}{hA} = \frac{1}{2.2 \times 10^4 \times \pi \times 0.05 \times 2} = 1.45 \times 10^{-4} \frac{°\text{K}}{\text{W}}$$

Radiation

Radiation heat transfer occurs between two surfaces via electromagnetic waves and *does not* require an intervening medium to permit the energy flow. In fact, it travels best through a vacuum as radiant energy does from the sun. The equation for radiation energy exchange between two surfaces is

$$q = \sigma A_1 F_e F_s \left(T_1^4 - T_2^4\right)$$

where the Stefan-Boltzmann constant is $\sigma = 5.67 \times 10^{-8}$ W/(m^2 • °K^4) F_e is a factor that is a function of the emissivity of the two surfaces with a value from 0 to 1.0, and F_s is a modulus that is a function of the relative geometries of the two surfaces with a value from 0 to 1.0.

Note that the heat flow is not proportional to the linear temperature difference but is a function of the temperature of the surfaces to the fourth power.

The simplest, and by far the most common, case of radiation energy exchange occurs in the case of a small surface radiating to large surroundings. In this case, the equation simplifies to

$$q = \sigma A_1 \varepsilon_1 \left(T_1^4 - T_2^4\right)$$

where ε_1 is the emissivity of the radiating surface.

Example 11.13

A steam pipe with a surface area of 5 m^2 and a surface temperature of 600°C radiates into a large room (which acts as a black body), the surfaces of which are at 25°C. The pipe gray-body surface emissivity is 0.6. Find the heat flow and heat flux from the surface to the room.

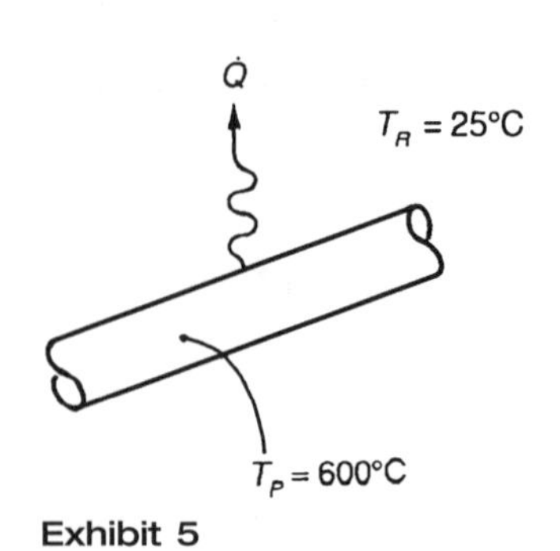

Exhibit 5

Solution

Refer to Exhibit 5.

$$\dot{Q}_{1-2} = \sigma A F_e F_s \left(T_1^4 - T_2^4\right)$$

For a gray body radiating to a black-body enclosure,

$$F_e F_s = \varepsilon_1$$
$$\dot{Q}_{1-2} = 5.67 \times 10^{-8} \times 5 \times 0.6 \times (873^4 - 298^4) = 9.75 \times 10^4 \text{ W}$$
$$\frac{\dot{Q}}{A} = \frac{9.75 \times 10^4}{5} = 1.95 \times 10^4 \ \frac{\text{W}}{\text{m}^2}$$

SELECTED SYMBOLS AND ABBREVIATIONS

Symbol or Abbreviation	Description
c	specific heat
h	enthalpy, heat transfer coefficient
k	thermal conductivity
m	mass
P, p	total pressure, partial pressure
P_r	relative pressure
Q	heat taken in or given off
q	heat (energy transfer), emitted radiation
r_c	compression ratio
R	gas constant
$\overline{\text{R}}$	universal gas constant
R	thermal resistance
s	entropy
T, t	temperature
T_H	high or hot temperature
T_L, T_C	low temperature, cold temperature
u	internal energy
V_r	relative volume
V, v	volume
v	specific volume
W, w	work
Z	compressibility factor
h	thermal efficiency of a heat engine
ε	emissivity

PROBLEMS

11.1 Equations of state for a single component can be any of the following, **except**

a. the ideal gas law, $Pv = RT$
b. the ideal gas law modified by insertion of a compressibility factor, $Pv = ZRT$
c. any relationship interrelating three or more state functions
d. a mathematical expression defining a path between states

11.2 Given the following data for a fluid, what is its state at 40°C and 3 kPa?

Saturated Property Table

T, °C	P, kPa	v_f, m³/kg	v_g, m³/kg	h_f, kJ/kg	h_g, kJ/kg	s_f, kJ/(kg • °K)	s_g, kJ/(kg • °K)
40	7.38	.001008	19.52	167.57	2574.3	.5725	8.257
80	47.39	.001029	3.407	334.9	2643.7	1.1343	7.5445
120	198.5	.001060	.8919	503.71	2706.3	1.5276	7.1296

a. saturated liquid
b. superheated vapor
c. compressed liquid
d. saturated vapor

11.3 Using the data table in 11.2, what is the fluid's entropy in kJ/(kg • °K) at 120°C and 80% quality?

a. 1.53
b. 6.009
c. 7.13
d. 28.8

11.4 Using the refrigerant data table in 11.2, what is its latent heat (heat of vaporization) in kJ/kg at 80°C?

a. 198.5
b. 2706
c. 1306
d. 2308.8

11.5 A nonflow (closed) system contains 1 kg of an ideal gas ($c_p = 1.0$, $c_v = .713$). The gas temperature is increased by 10°C while 5 kJ of work are done by the gas. What is the heat transfer in kJ?

a. −3.3
b. −2.6
c. +12.1
d. +7.4

11.6 Shaft work of −15 kJ/kg and heat transfer of −10 kJ/kg change the enthalpy of a system by

a. −25 kJ/kg
b. −15 kJ/kg
c. −10 kJ/kg
d. +5 kJ/kg

11.7 A quantity of 55 cubic meters of water passes through a heat exchanger and absorbs 2,8000,000 kJ. The exit temperature is 95°C. The entrance water temperature in °C is nearest to

a. 49
b. 56
c. 68
d. 83

11.8 A fluid at 690 kPa has a specific volume of .25 m^3/kg and enters an apparatus with a velocity of 150 m/s. Heat radiation losses in the apparatus are equal to 25 kJ/kg of fluid supplied. The fluid leaves the apparatus at 135 kPa with a specific volume of .9 m^3/kg and a velocity of 300 m/s. In the apparatus, the shaft work done by the fluid is equal to 900 kJ/kg. Does the internal energy of the fluid increase or decrease, and how much is the change?

a. 858 kJ/kg (increase) c. 908 kJ/kg (increase)
b. 858 kJ/kg (decrease) d. 908 kJ/kg (decrease)

11.9 Exhaust steam from a turbine exhausts into a surface condenser at a mass flow rate of 4000 kJ/hr, 9.59 kPa, and 92% quality. Cooling water enters the condenser at 15°C and leaves at the steam inlet temperature.

Properties of Saturated Water (US units): Temperature Table

		Specific Volume		Internal Energy		Enthalpy			Entropy	
Temp. °C T	Press. kPa P	Sat. Liquid v_f	Sat. Vapor v_g	Sat. Liquid u_f	Sat. Vapor u_g	Sat. Liquid h_f	Evap. h_{fg}	Sat. Vapor h_g	Sat. Liquid s_f	Sat. Vapor s_g
15	1.705	.001	77.9	62.99	2396	62.99	2466	2529	.2245	8.781

v, m^3/kg; u and h, kJ/kg; s, kJ/(kg • °K)

The cooling water mass flow rate in kg/hr is closest to

a. 157,200 c. 95,000
b. 70,200 d. 88,000

11.10 The mass flow rate of a Freon refrigerant through a heat exchanger is 5 kg/min. The enthalpy of entry Freon is 238 kJ/kg, and the enthalpy of exit Freon is 60.6 kJ/kg. Water coolant is allowed to rise 6°C. The water flow rate in kg/min is

a. 24 c. 83
b. 35 d. 112

11.11 The maximum thermal efficiency that can be obtained in an ideal reversible heat engine operating between 833°C and 170°C is closest to

a. 100% c. 78%
b. 60% d. 40%

11.12 A 2.2-kW refrigerator or heat pump operates between −17°C and 38°C. The maximum theoretical heat that can be transferred from the cold reservoir is nearest to

a. 7.6 kW c. 15.6 kW
b. 4.7 kW d. 10.2 kW

11.13 In any non-quasistatic thermodynamic process, the overall entropy of an isolated system will

a. increase and then decrease c. stay the same
b. decrease and then increase d. increase only

11.14 For spontaneously occurring natural processes in an isolated system, which expression best expresses ds?

a. $ds = \dfrac{dq}{T}$ c. $ds > 0$
b. $ds = 0$ d. $ds < 0$

11.15 Which of the following statements about entropy is **false**?

a. The entropy of a mixture is greater than that of its components under the same conditions.
b. An irreversible process increases the entropy of the universe.
c. The entropy of a crystal at 0°C is zero.
d. The net entropy change in any closed cycle is zero.

11.16 A high-velocity flow of gas at 250 m/s possesses kinetic energy nearest to which of the following?

a. 3.13 kJ/kg c. 31,300 kJ/kg
b. 313 kJ/kg d. 31.3 kJ/kg

11.17 $(u + Pv)$ is a quantity called

a. flow energy c. entropy
b. shaft work d. enthalpy

11.18 In flow process, neglecting KE and PE changes, $-\int v\, dP$ represents which item below?

a. heat transfer c. closed system work
b. shaft work d. flow energy

11.19 Power may be expressed in units of

a. joules c. kJ
b. watts d. newtons

11.20 The temperature-entropy diagram in Exhibit 11.20 represents a

a. Rankine cycle with superheated vapor
b. Carnot cycle
c. diesel cycle
d. refrigeration cycle

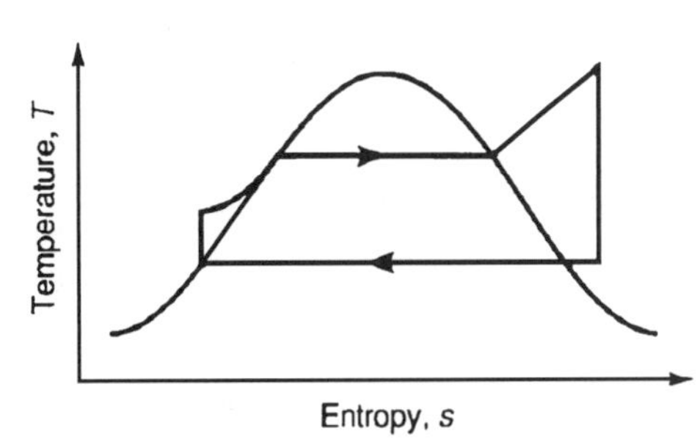

Exhibit 11.20

11.21 Entropy is the measure of
a. the change in enthalpy of a system
b. the internal energy of a gas
c. the heat capacity of a substance
d. randomness or disorder

11.22 A Carnot heat engine cycle is represented on the T-s and P-V diagrams in Exhibit 11.22. Which of the several areas bounded by numbers or letters represents the amount of heat rejected by the fluid during one cycle?
a. Area 1–2–6–5 c. Area 3–4–5–6
b. area B–C–H–G d. Area D–A–E–F

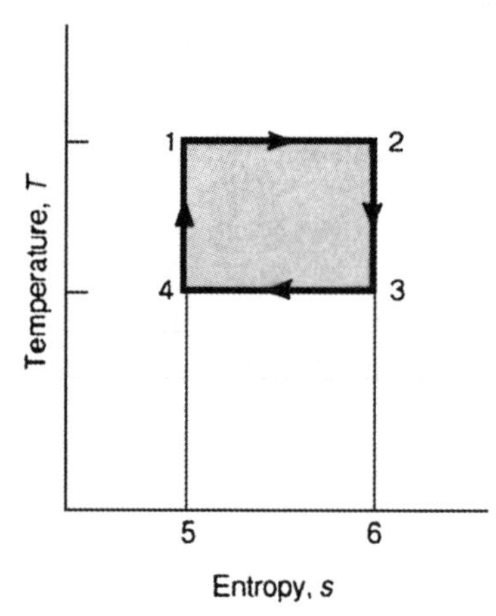

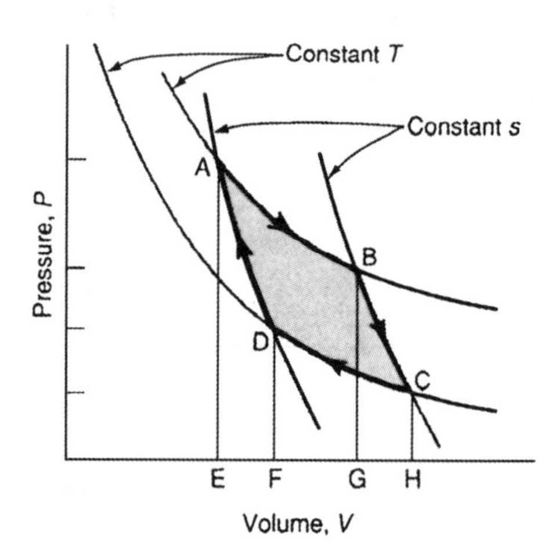

Exhibit 11.22

11.23 A Carnot engine operating between 70°C and 2000°C is modified solely by raising the high temperature by 150°C and raising the low temperature by 100°C. Which of the following statements is **false**?
a. The thermodynamic efficiency is increased.
b. More work is done during the isothermal expansion.
c. More work is done during the isentropic compression.
d. More work is done during the reversible adiabatic expansion.

11.24 In the ideal heat pump system represented in Exhibit 11.24, the expansion valve 4–1 performs the process that is located on the T-s diagram between points
a. A and B c. C and D
b. B and C d. E and A

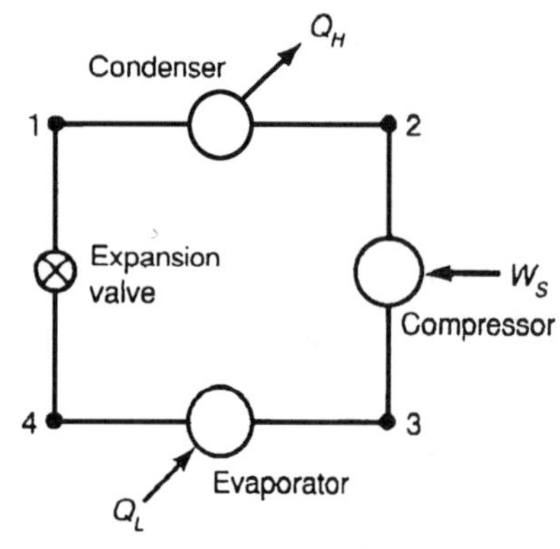

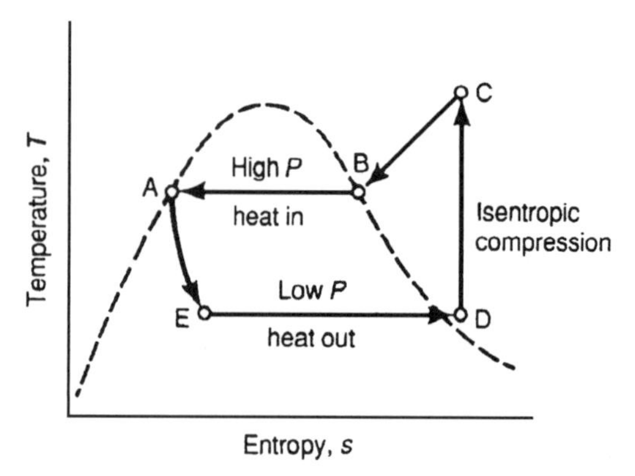

Exhibit 11.24

11.25 Data in the following table describe two states of a working fluid that exist at two locations in a piece of hardware:

	P, kPa	*v*, m³/kg	*T*, °C	*h*, kJ/kg	*s*, kJ/kg•k
State 1	25	0.011	20	19.2	0.0424
State 2	125	0.823	180	203.7	0.3649

Which of the following statements about the path from State 1 to 2 is **false**?

a. The path results in an expansion.
b. The path determines the amount of work done.
c. The path is indeterminate from these data.
d. The path is reversible and adiabatic.

11.26 Name the process that has no heat transfer.

a. Isentropic c. Quasistatic
b. Isothermal d. Reversible

11.27 In a closed system with a moving boundary, which of the following represents work done during an isothermal process?

a. $W = P(V_2 - V_1)$

b. $W = 0$

c. $W = P_1V_1 \ln\left(\frac{P_1}{P_2}\right) = P_1V_1 \ln\left(\frac{V_2}{V_1}\right) = mRT \ln\left(\frac{P_1}{P_2}\right)$

d. $W = \frac{P_2V_2 - P_1V_1}{1-k} = \frac{mR(T_2 - T_1)}{1-k}$

11.28 The work of a polytropic ($n = 1.21$) compression of air ($c_p/c_v = 1.40$) in a system with moving boundary from $P_1 = 15$ kPa, $V_1 = 1.0$ m^3 to $P_2 = 150$ kPa, $V_2 = 0.15$ m^3 is

a. −35.7 kJ c. 1080 kJ
b. −324 kJ d. 5150 kJ

11.29 The isentropic compression of 1 m^3 of air, $c_p/c_v = 1.40$, from 20 kPa to a pressure of 100 kPa gives a final volume of

a. 0.16 m^3 c. 0.32 m^3
b. 0.20 m^3 d. 0.40 m^3

11.30 An ideal gas at a pressure of 500 kPa and a temperature of 75°C is contained in a cylinder with a volume of 700 m^3. Some of the gas is released so that the pressure in the cylinder drops to 250 kPa. The expansion of the gas is isentropic. The specific heat ratio is 1.40, and the gas constant is .287 kJ/kgK. The mass of the gas (in kg) remaining in the cylinder is nearest to

a. 900 c. 1500
b. 1300 d. 2140

11.31 The theoretical power required for the isothermal compression of 800 m^3/min of air from 100 to 900 kPa is closest to

a. 70 c. 130
b. 90 d. 290

11.32 Which of the following statements is **false** concerning the deviations of real gases from ideal gas behavior?

a. Molecular attraction interactions are compensated for in the ideal gas law.
b. Deviations from ideal gas behavior become large near the saturation curve.
c. Deviations from ideal gas behavior become significant at pressures above the critical point.
d. Molecular volume becomes significant as specific volume is decreased.

11.33 There are 3 kg of air in a rigid container at 250 kPa and 50°C. The gas constant for air is .287 kJ/kg•°K. The volume of the container, in m^3, is nearest to

a. 2.2 c. 2.8
b. 1.1 d. 3.1

11.34 A mixture at 100 kPa and 20°C that is 30% by weight CO_2 (m.w. = 44) and 70% by weight N_2 (m.w. = 28) has a partial pressure of CO_2 in kPa that is nearest to

a. 21.4 c. 68.3
b. 31.5 d. 78.6

11.35 Dry air has an average molecular weight of 28.79, consisting of 21 mole-percent O_2, 78 mole-percent N_2, and 1 mole-percent Ar (and traces of CO_2). The weight-percent of O_2 is nearest to

a. 21.0 c. 23.2
b. 22.4 d. 24.6

11.36 The temperature difference between the two sides of a solid rectangular slab of area A and thickness L, as shown in Exhibit 11.36, is ΔT. The heat transferred through the slab by conduction in time, t, is proportional to

a. $AL\Delta Tt$ c. $AL\frac{t}{\Delta T}$

b. $AL\frac{\Delta T}{t}$ d. $\frac{A\Delta Tt}{L}$

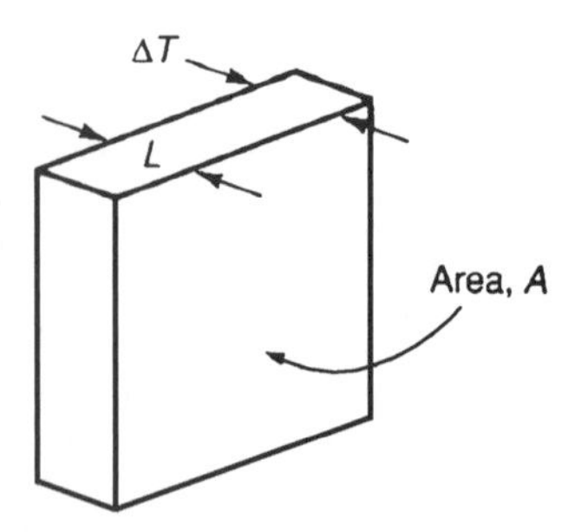

Exhibit 11.36

11.37 The composite wall in Exhibit 11.37 has an outer temperature $T_1 = 20°C$ and an inner temperature $T_4 = 70°C$. The temperature T_3, in °C, is nearest to

a. 38 c. 58
b. 46 d. 69

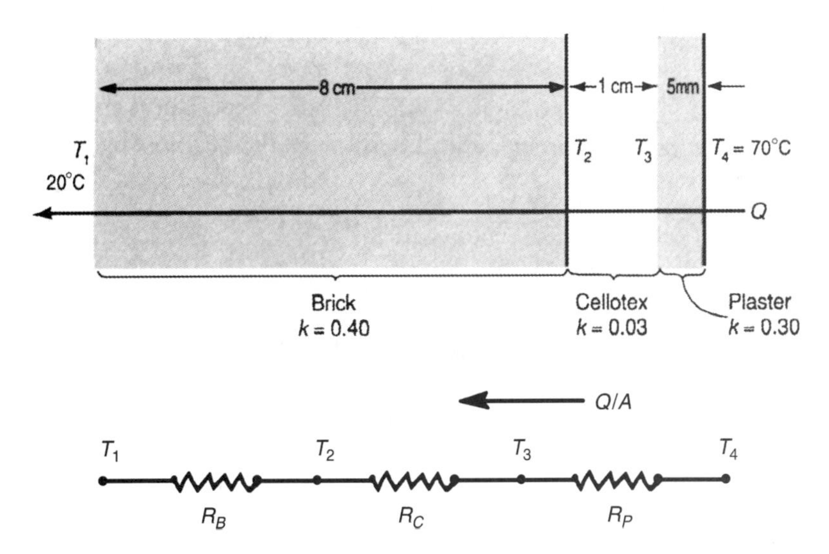

Exhibit 11.37

11.38 In Exhibit 11.38, the inner wall is at 30°C, and the outer wall is exposed to ambient wind and surroundings at 10°C. The film coefficient, h, for convective heat transfer in a 7-m/s wind is about 20 W/m^2 • °C. Ignoring any radiation losses, an overall coefficient (in the same units) for the conduction and convection losses is most nearly

a. 1.4 c. 12.5
b. 2.6 d. 7.1

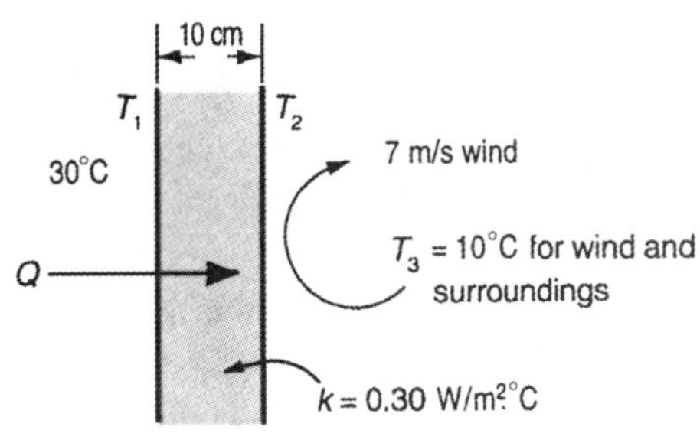

Exhibit 11.38

11.39 Heat is transferred by conduction from left to right through the composite wall shown in Exhibit 11.39. Assume the three materials are in good thermal contact and that no significant thermal resistance exists at any of the interfaces. The overall coefficient U in W/m^2 • °C is most nearly

a. 0.04 c. 0.35
b. 0.20 d. 0.91

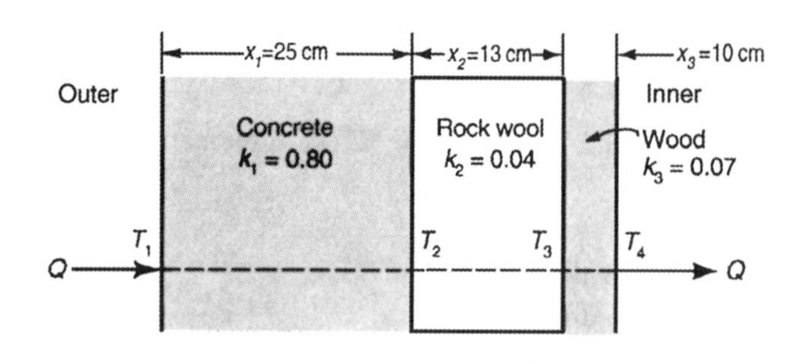

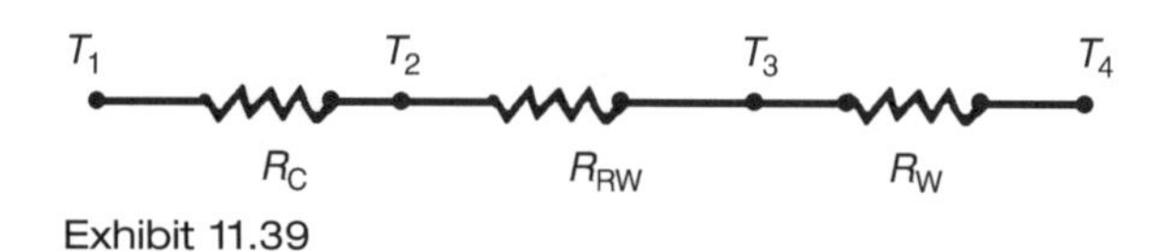

Exhibit 11.39

11.40 The heat loss per hour through 1 m^2 of furnace wall that is 40 cm thick is 520 W. The inside wall temperature is 1000°C, and its average thermal conductivity is 0.61 W/m^2 • °C.

The outside surface temperature of the wall is nearest to

a. 100°C
b. 300°C
c. 700°C
d. 1000°C

11.41 Which of the following is the usual expression for the power/unit-area Stefan-Boltzmann constant for black-body radiation?

a. 1.36×10^{-12} cal/(s•cm^2•°K^4)
b. 5.67×10^{-5} ergs/(s•cm^2•°K^4)
c. 5.67×10^{-8} watts/(m^2•°K^4)
d. 5.67×10^{-8} coulombs/(s•m^2•°K^4)

SOLUTIONS

11.1 **d.** All *except* (d) are correct. The ideal gas law is the simplest equation of state; it is often applied to real gases by using a compressibility factor Z. Any relationships that interrelate thermodynamic state function data are equations of state. Answer (d) expresses the path of a process between states rather than a relationship between variables at a single point or state.

11.2 **b.** At 40°C, equilibrium between liquid and gas exists at 7.38 kPa. Below 7.38 kPa superheated vapor exists, and above 7.38 kPa only pressurized liquid exists.

11.3 **b.** At 120°C, s_f = 1.5276 and s_{fg} = 7.1296 − 1.5276 = 5.602. Here s_f is saturated liquid at 0% quality and s_g is saturated vapor of 100% quality. Thus s at 80% quality = s_f + (0.80) s_{fg} = 1.5276 + .8 × 5.602 = 6.009 kJ/(kg • °K).

11.4 **d.** Here, $h_{fg} = h_g - h_f$ = 2643.7 − 334.9 = 2308.8 kJ/kg.

11.5 **c.** The thermodynamic sign convention is + for heat in and + for work out of a system. Apply the first law for a closed system and an ideal gas working fluid:

$$\Delta U = mc_v \Delta T = q - w$$

$$.713(10) = q - (+5), \qquad 7.13 = q - 5, \qquad q = 12.13$$

11.6 **d.** The first law applied to a flow system is $h = q - w_s = -10 - (-15) = +5$.

11.7 **d.** For liquid water, $c_p = 4.18$ kJ/(kg • °C):

$$Q = mc_p \Delta T = mc_P(T_2 - T_1)$$

$$2{,}800{,}000 = (55\,\text{m}^3)\left(\frac{1000\,\text{kg}}{\text{m}^3}\right)(4.18)(95 - T_1)$$

$$12.2 = 95 - T_1 \qquad T_1 = 82.8°\text{C}$$

11.8 **d.** The basis of the calculation will be 1 kg. Use the thermodynamic sign convention that heat in and work out are positive. The first-law energy balance for the flow system is $h_2 + KE_2 - h_1 - KE_1 = Q - W_s$. Since the working fluid is unspecified and the internal energy change is desired, use the definition $h = u + Pv$. Then

$$u_2 + P_2v_2 + KE_2 - u_1 - P_1v_1 - KE_1 = Q - W_s \quad \text{or}$$

$$u_2 - u_1 = Q -$$

$$W_s + P_1v_1 + KE_1 - P_2v_2 - KE_2$$

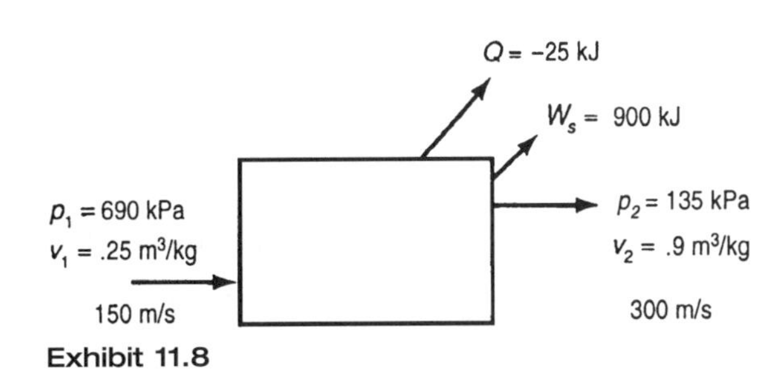

Exhibit 11.8

Now calculate numerical values for all terms except $u_2 - u_1$:

$$P_2v_2 = 135 \times .9 = 121.5 \text{ kJ/kg} \qquad P_1v_1 = 690 \times .25 = 172.5 \text{ kJ/kg}$$

$$KE_2 = \frac{V^2}{2gJ} = \frac{300^2}{2000} = 45 \text{ kJ/kg}$$

$$KE_1 = \frac{V^2}{2gJ} = \frac{(150)^2}{2000} = 11.3 \text{ kJ/kg}$$

$$W_s = +900 \text{ kJ/kg}$$

Therefore,

$$u_2 - u_1 = -25 - 900 + 172.5 + 11.3 - 121.5 - 45 = -907.7 \text{ kJ/kg}$$

11.9 **b.** Saturated steam table data at 9.59 kPA are

T, °C	h_f, kJ/kg	h_{fg}, kJ/kg	h_g, kJ/kg
45	188.45	2394.8	2583.2

The enthalpy of steam at 92% quality = $h_1 = h_f + 0.92h_{fg}$ = 188.45 + 92 × 2394.8 = 2391.7. The enthalpy of liquid water at 45°C = h_2 = 188.45 kJ/kg. The enthalpy of liquid water at 15°C = h_3 = 62.99 kJ/kg above reference of 0 °C.

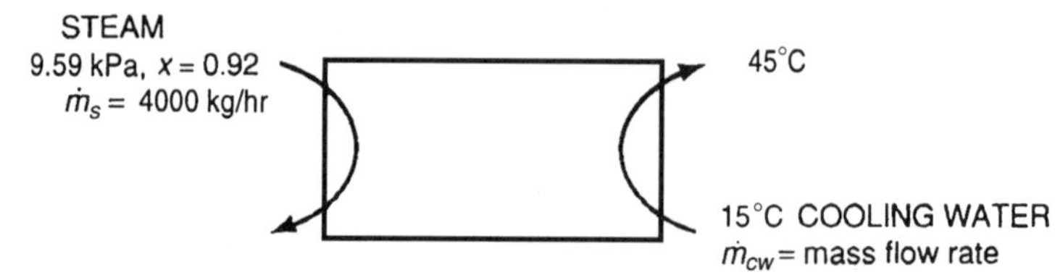

Exhibit 11.9

In the absence of data, assume that the steam condensate leaves at 45°C; if a heat balance is written over a 1-hour period, then the heat from steam = heat to cooling water, or

$$\dot{m}_s(h_1 - h_2) = \dot{m}_{cw}(h_2 - h_3)$$
$$4000(2391.7 - 188.45) = \dot{m}_{cw}(188.45 - 62.99)$$
$$\dot{m}_{cw} = 70{,}245 \text{ kg/hr}$$

11.10 **b.** Over a 1-minute period, the heat gain by water equals heat loss by Freon

$$\dot{m}_{CW} c_p \Delta T = \dot{m}_F(h_1 - h_2)$$
$$\dot{m}_{CW} \times 4.2 \times 6 = 5(238 - 60.6)$$
$$\dot{m}_{CW} = \frac{887}{25.2} = 35.2 \text{kg/min}$$

11.11 **b.** Maximum efficiency is achieved with a Carnot engine.

T_L = 170 + 273 = 443°K $\quad$ T_H = 833 + 273 = 1106°K

$$\eta_{\text{TH}} = \frac{w}{q_H} = \frac{q_H - q_L}{q_H}$$
$$= 1 - \frac{Q_L}{Q_H} = 1 - \frac{T_L}{T_H}$$
$$= 1 - \frac{443}{1106} = 1 - 0.40 = 0.60 = 60\%$$

11.12 **d.** The coefficient of performance of a Carnot refrigerator or heat pump is

$$\text{COP} = \frac{T_L}{T_H - T_L} = \frac{256°\text{K}}{311°\text{K} - 256°\text{K}} = 4.65 = \frac{q_L}{w} = 4.65$$

$$\text{COP} = \frac{q_L}{q_H - q_L} = \frac{q_L}{w} = \frac{q_L}{2.2}; \qquad q_L = 4.65 \times 2.2 = 10.2 \text{ kW}$$

11.13 **d.** Quasistatic means infinitely slow, lossless, hypothetical, by differential increments. The overall entropy will increase for an isolated system or for the system plus surroundings.

11.14 **c.** (a) $ds = \frac{dq_{\text{rev}}}{T}$ only. The reversible requirement is necessary to generate the exact height vs. rectangular area equivalence on the Carnot cycle *T-s* diagram.

(b) Only a reversible adiabatic process is isentropic by definition.

(c) All naturally occurring spontaneous processes are irreversible and result in an entropy increase.

(d) An energy input from the surroundings is required to reduce the entropy.

11.15 **c.** All are true except (c). The entropy of a perfect crystal at absolute zero (0°K or 0°R) is zero. This is the third law of thermodynamics. There is presumably no randomness at this temperature in a crystal without flaws, impurities, or dislocations.

11.16 **d.** Per 1 kg of flowing fluid,

$$KE = \frac{V^2}{2g_c} \qquad \text{where } V \text{ is in m/s, and } g_c = 1.0$$

Use 1000 to convert J to kJ:

$$KE = \frac{250^2}{2 \times 1000} = 31.3 \text{ kJ/kg}$$

11.17 **d.** Flow energy is Pv. Shaft work, W_s, is $-\int v\,dP$. Entropy is s. Internal energy is u. Enthalpy h is defined as $u + Pv$, the sum of internal energy plus flow energy.

11.18 **b.** Shaft work is work or mechanical energy crossing the fixed boundary (control volume) of a flow (open) system. Shaft work W_s is defined, in the absence of *PE* and *KE* changes, by $dh = T\,ds + v\,dP$, where $T\,ds = dq_{\text{rev}}$ and $-v\,dP$ is dW_s. In integrated form, $\Delta h = \int T\,ds + \int v\,dP = q_{\text{rev}} - W_s$, where W_s is represented by $-\int v\,dP$. Closed system work W is defined by $du = T\,ds - P\,dv$, or $\Delta u = \int T\,ds - \int P\,dv = q_{\text{rev}} - W$. Thus, closed system work is $+\int P\,dv$. Flow energy is the Pv term, and enthalpy change is ΔH.

11.19 **b.** Power is energy per unit time. The usual power units are watts.

11.20 **a.**

11.21 **d.**

11.22 **c.** The table below gives the significance of each area of the diagrams:

Process	*T-s* Diagram: Area Representing Heat	*P-V* Diagram: Area Representing Work
Isothermal expansion, 1–2 and A–B	1–2–6–5 = heat in from high-temp. reservoir	A–B–G–E = work done by fluid
Isentropic expansion, 2–3 and B–C	2–3–6 = 0 heat transfer	B–C–H–G = work done by fluid
Isothermal compression, 3–4 and C–D	3–4–5–6 = heat out to low-temp. reservoir	C–D–F–H = work done on fluid
Isentropic compression, 4–1 and D–A	4–1–5 = 0 heat transfer	D–A–E–F = work done on fluid
Net result of process	1–2–3–4 = net heat converted to work	A–B–C–D = net work done by process

11.23 **a.** The Carnot cycle efficiency is originally

$$\eta = \frac{T_H - T_L}{T_H} = \frac{2273 - 343}{2273} = 0.849$$

After the change,

$$\eta = \frac{2423 - 443}{2423} = .817 \qquad \text{(efficiency is reduced)}$$

On the *T-s* and *P-V* diagrams in Exhibit 11.23, the original cycle is shown as ABCD, and the modified cycle is shown as A′B′C′D.

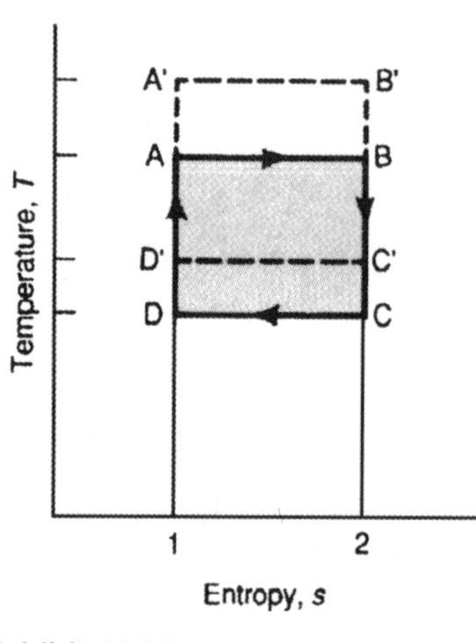

Exhibit 11.23

Compare the work done during the isothermal expansion (A to B vs. A′ to B′):
Original: area A–B–8–4
Modified: area A′–B′–7–3 is larger
Compare the work done during the isentropic compression (D to A vs. D′ to A′):
Original: area D–A–4–6
Modified: area D′–A′–3–5 is larger
Compare the work during the reversible (isentropic) expansion (B to C vs. B′ to C′):
Original: area B–C–10–8
Modified: area B′–C′–9–7 is larger

Compare the work during the isothermal compression (C to D vs. C′ to D′):

Original: area C–D–6–10

Modified: area C′–D′–5–9 is larger

Statements (b), (c), (d), and (e) are correct.

11.24 d. The vapor compression–reversed Rankine cycle is conducted counterclockwise on both the schematic and the *T-s* diagram. Numbers on the schematic and letters on the *T-s* diagram are related: 1 = A, 2 = B, 3 = D, and 4 = E. Process C–B–A occurs in the condenser between 2 and 1. The expansion process A–E occurs between 1 and 4.

11.25 d. The large volume and entropy changes indicate a change from a condensed phase to a vapor phase. Temperature, pressure, and enthalpy increases require an energy input. The path from 1 to 2 is indeterminate because no information on intermediate states is given. Work is always path dependent. The entropy increase means the process cannot be reversible and adiabatic (isentropic).

11.26 a. An *isentropic* process is reversible and adiabatic. An *adiabatic* process has no heat exchange with its surroundings. An *isothermal* process is conducted at constant temperature. A *quasistatic* (almost static) process departs only infinitesimally from an equilibrium state. A *reversible* process can have its initial state restored without any change (energy gain or loss) taking place in the surroundings.

11.27 c. For a closed system (piston-cylinder type, nonrepetitious) the work done is $W = \int P\,dV$. The equations in the problem are valid for ideal gases in the following processes, respectively:

a. constant pressure　　c. isothermal process

b. constant volume　　d. isentropic process

11.28 a. The work of a closed system (moving boundary) polytropic process for an ideal gas is

$$W = \frac{P_2V_2 - P_1V_1}{1-n} = \frac{[150(0.15) - 15(1.0)]}{1-1.21} = -35.7 \text{ kJ}$$

which is work done on the gas.

11.29 c. An isentropic process for an ideal gas follows the path

$$PV^k = P_1V_1^k = P_2V_2^k = \text{constant} \qquad \text{where } k = c_p/c_v$$

$$20(1)^{1.4} = 100(V_2)^{1.4}; \qquad V_2^{1.4} = 0.20; \qquad \text{hence, } V_2 = 0.317 \text{ m}^3$$

11.30 d. Given:

$k = c_p/c_v = 1.40$　　$R = .287 \dfrac{\text{kJ}}{\text{kg} \bullet {}^\circ\text{K}}$

$P_1 = 500$ kPa　　$P_2 = 250$ kPa

$V_1 = 700 \text{ m}^3$　　$V_2 = 700 \text{ m}^3$

$T_1 = 75^\circ\text{C} + 273 = 348\ {}^\circ\text{K}$　　$T_2 = ?$

$w_2 = ?$

Basis: The ideal gas law may be written $PV = m\text{R}T$, and the basic equation for reversible adiabatic (isentropic) expansion is

$$\frac{T_2}{T_1} = \left(\frac{P_2}{P_1}\right)^{(k-1)/k}$$

The gas remaining in the tank cools as it expands; the new temperature is

$$T_2 = T_1\left(\frac{P_2}{P_1}\right)^{(k-1)/k} = 348\left(\frac{250}{500}\right)^{(1.4-1)/1.4} = 348\left(\frac{1}{2}\right)^{0.2857} = 285°\text{K}$$

Now apply the gas law at State 2, $P_2V_2 = m_2\text{R}T_2$:

$$250 \times 700 = m_2 \times .287 \times 285$$

$$m_2 = \frac{(250)(700)}{(.287)(285)} = 2139 \text{ kg}$$

11.31 **d.** Since a volume flow rate is specified, the process is a flow process. The work of isothermal compression of an ideal gas is numerically the same in a steady flow process as in a closed system:

$$PV = \text{constant} = P_1V_1 = P_2V_2 = m\text{R}T$$

In a closed system,

$$W = \int_{V_1}^{V_2} P\,dV = P_1V_1 \ln\frac{V_2}{V_1} = P_1V_1 \ln\frac{P_1}{P_2} = m\text{R}T\ln\frac{V_2}{V_1} = m\text{R}T\ln\frac{P_1}{P_2}$$

In a flow system,

$$W_s = \int_{P_1}^{P_2} V\,dP = -P_1V_1 \ln\frac{P_2}{P_1} = P_1V_1 \ln\frac{P_1}{P_2} = P_1V_1 \ln\frac{V_2}{V_1} = m\text{R}T\ln\frac{V_2}{V_1} = m\text{R}T\ln\frac{P_1}{P_2}$$

Over a 1-minute interval,

$$W_s = P_1V_1 \ln\frac{P_1}{P_2} = (100)(800)\left(\frac{100}{900}\right) = -17{,}580 \text{ kJ/min}$$

$$W_s\left(\frac{-17{,}580}{60}\right) = -293 \text{ kW}$$

11.32 **a.** All statements except (a) are true. The ideal gas law does not consider the volume of the molecules or any interaction other than elastic collisions.

11.33 **b.** The ideal gas law is $PV = m\text{R}T$. Here $P = 250$ kPa and $T_1 = 50°\text{C} + 273 = 323°\text{K}$. Hence,

$$250V = 3 \times .287 \times 323$$

$$V = \frac{(3)(.287)(323)}{250} = 1.11 \text{ m}^3$$

11.34 **a.** The calculation is based on 1 kg of mixed gases. (1) Calculate the weight of each component and the number of moles of each that is present. (2) Compute the mole fraction of each, and apportion the total pressure in proportion to the mole fraction. The computations are in the following table:

Component	Weight, kg	Number of kg-mol	Mole Fraction	Partial Pressure, kPa
CO_2	0.30	$\frac{0.30}{44} = 0.00682$	$\frac{0.00682}{0.03182} = 0.214$	21.4
N_2	0.70	$\frac{0.70}{28} = 0.0250$	$\frac{0.0250}{0.03182} = 0.786$	78.6
Total	1.00	0.03182	1.000	100

Since the mole fraction of a gas is the same as the volume fraction, the composition of the mixture is 21.4% vol. CO_2 and 78.6% vol. N_2. From the table, the correct partial pressure of CO_2 is 21.4 kPa.

11.35 **c.** The calculation will be based on 1 kg-mol of dry air and arranged in the following table:

Component	m.w.	Mole Fraction	Weight, kg	Weight, %
O_2	32.0	0.21	6.72	23.2
N_2	28.0	0.78	21.80	75.4
Ar	40.0	0.01	0.40	1.4
Totals		1.00	28.92	100.0

11.36 **d.** The heat transfer rate through the slab by conduction is governed by the equation

$$Q = kA\Delta T/L$$

In time t the amount of heat transfer is proportional to

$$A\frac{\Delta T}{L}t$$

The symbol k is the coefficient of thermal conductivity of the material; hence, the heat transfer in a given material is proportional to the other variables.

11.37 **d.** At steady state the same Q flows across each material, and the temperatures descend in direct proportion to the thermal resistances (reciprocal of conductivity).

$$\text{Resistance of brick} = \frac{x}{k} = \frac{.08\text{ m}}{.4\frac{\text{W}}{\text{m}\bullet{}^\circ\text{C}}} = .2\frac{\text{m}^2}{\text{W}\bullet{}^\circ\text{C}}$$

$$\text{Resistance of Cellotex} = \frac{x}{k} = \frac{.01}{.03} = .333\frac{\text{m}^2}{\text{W}\bullet{}^\circ\text{C}}$$

$$\text{Resistance of plaster} = \frac{x}{k} = \frac{.005}{.3} = .017\frac{\text{m}^2}{\text{W} \bullet {}^\circ\text{C}}$$

$$\text{Total resistance} = .2 + .333 + .017 = .55\ \text{m}^2/\text{W} \bullet {}^\circ\text{C}$$

$$Q/A = \frac{\Delta T_{\text{total}}}{\text{total resistance}} = \left(\frac{\Delta T}{x/k}\right)_{\text{layer}}$$

Hence,

$$Q/A = \frac{50}{.55} = \frac{T_4 - T_3}{.017} = \frac{T_3 - T_2}{.333} = \frac{T_2 - T_1}{.2} = 90.9\ \text{W/m}^2$$

$T_4 - T_3 = 1.5°\text{C}$, since $T_4 = 70°\text{C}$, $T_3 = 68.5°\text{C}$.
$T_3 - T_2 = 30.3°\text{C}$, since $T_3 = 68.5°\text{C}$, $T_2 = 38.2°\text{C}$.
$T_2 - T_1 = 18.2°\text{C}$, since $T_2 = 38.2°\text{C}$, $T_1 = 20°\text{C}$
(in agreement with given data)

11.38 **b.** Since conduction and convection are based on ΔT, absolute temperatures are not required. For steady state, the heat conducted through a wall must equal the heat lost by convection:

$$Q = \frac{kA(T_1 - T_2)}{x} = hA(T_2 - T_3) \qquad \textbf{(11.1)}$$

In a similar way, Q can be expressed by an overall coefficient

$$Q = UA(T_1 - T_3) \qquad \textbf{(11.2)}$$

Here, U is calculated in a manner analogous to that used for thermal conductivities in series:

$$U = \frac{1}{\frac{1}{h_1} + \frac{x_1}{k_1}} \qquad \textbf{(11.3)}$$

In this case,

$$U = \frac{1}{\frac{1}{20} + \frac{.1}{.3}} = \frac{1}{.05 + .333} = 2.61\ \text{W/m}^2 \bullet {}^\circ\text{C}$$

11.39 **b.** The overall coefficient U, the thermal conductivity k/x, and the film coefficient h are the reciprocals of their thermal resistances. Thermal resistances in series are handled analogously to series electrical resistances; hence

$$U = \left(\frac{1}{\sum_i \frac{x_i}{k_i}}\right)^{-1} = \frac{1}{R_T} = \frac{1}{R_1 + R_2 + R_3}$$

The overall coefficient U is then used in the simplified conduction equation $Q = UADT$.

In this problem

$$U = \frac{1}{\frac{.25}{.80} + \frac{.13}{.04} + \frac{.10}{.07}} = \frac{1}{.313 + 3.25 + 1.43} = 0.20 \text{ W/m}^2 \bullet {}^\circ\text{C}$$

11.40 **c.** The heat conduction equation is

$$Q = k\frac{A}{L}(T_1 - T_2)$$

where $T_1 = 1000°\text{C}$, T_2 = outside temperature, $k = 0.61$ W/m^2• °C, $Q/A = 520$ W/m^2, and $L = .4$ m.

Solving for T_2, one has

$$T_2 = -\frac{Q}{A}\frac{L}{k} + T_1 = -520 \times \frac{.4}{.61} + 1000 = 659°\text{C}$$

11.41 **c.** All are numerically correct conversions of the constant in terms of power per unit area. The units of watts/m^2• °K^4, however, are normally used in heat transfer.

CHAPTER 12

Electrical Circuits

Lincoln D. Jones

OUTLINE

This chapter assumes the reader completed an introductory electrical engineering or circuits course but needs to review points of theory and application. For instance, the solution for a single voltage source in a series circuit might easily come to mind, but if one changed the circuit by inserting another source (multiple source network) that may be acting as a load, the method of solution may not be readily apparent without a short review.

This chapter presents the portions of theory that are relevant to the EIT exam. An attempt is made to reduce the more detailed theory to the level for the expected problem and to leave out the more abstract formal development of a solution. For example, rather than setting up a three-phase ac problem for a complete solution—including possibly an unbalanced network—this review will assume the more likely case of all three legs of the circuit being part of a balanced network and proceed with a much simpler method of solution.

The objective of this presentation will be to jog the memory so that the reader can feel comfortable with the simpler type of "quick solution" problems expected on the examination. It should be noted that multiple choice answers for any such problems are significantly different, which allows one to quickly locate the correct answer if the simplest possible solution can be found. As an example, in a series ac circuits problem one can frequently plot a phasor solution with ruler and protractor that is accurate enough to pick the correct answer. One needs to be careful to not solve for more than what is asked for.

ELECTRICAL QUANTITIES AND DEFINITIONS

Electrostatics

The subject of electric fields is sometimes omitted in a formal course in electrical engineering (the knowledge is frequently assumed from a prerequisite physics course). Thus, one needs to review electric fields and flux densities attributable to electrical charges. These fields, forces, and flux densities require three-dimensional vector notation. However, most problems will reduce in complexity to two dimensions; this reduction will permit quick graphical solutions.

Whereas the forces exerted between charges depend on whether the charges are in motion, one may start with stationary electric charges that produce electric fields. These fields may be defined in terms of the forces they produce on one another.

The smallest amount of charge that can exist is the charge of one electron, which is 1.602×10^{-19} coulombs (C). One coulomb of charge is thus equivalent to 6.24×10^{18} electrons. This is the amount of charge that is necessary to develop a force of one newton in an electric field of one volt per meter. An electric field, **E** (bold face for vector quantity), is the amount of force (F) that would be exerted on a *positive charge* (assuming it is concentrated at a point) *of one coulomb* if it were placed in that field

$$\mathbf{F} = Q\mathbf{E}$$

The electric field is not thought of as a point but rather as being distributed throughout a small region. This is a vector quantity; the direction of force on the one coulomb would be toward the point source of the field if that point were a negative charge. The units of measurement are newtons per unit of charge; alternately, it may be given in terms of volts/meter since

$$\frac{\text{force}}{\text{charge}} = \frac{\text{force} \times \text{distance}}{\text{charge} \times \text{distance}} = \frac{\text{energy}}{\text{charge} \times \text{distance}} = \frac{\text{voltage}}{\text{distance}}$$

For point charges, the force is directly proportional to the product of the two charges and inversely proportional to the square of the distance between them (similar to the laws of gravity).

$$\mathbf{F}_2 = \frac{Q_1 Q_2}{4\pi\varepsilon r^2}\mathbf{a}_{r12}, \quad \text{where} \tag{12.1a}$$

$\mathbf{F}_2$ = the force on charge 2 due to charge 1,

Q_i = the ith point charge,

r = the distance between charges 1 and 2,

$\mathbf{a}_{r12}$ = a unit vector directed from 1 to 2, and

ε = the permittivity (or dielectric constant) of the medium.

The constant of proportionality depends on the medium between the two charges. This constant is $1/(4\pi\varepsilon)$ or approximately 9×10^9 for free space. If the medium is not free space, the 9×10^9 is merely divided by the relative permittivity to give the simplified "in line" form as

$$F = (9 \times 10^9/\varepsilon_{rel})Q_1Q_2/r^2 \text{ newtons} \qquad \textbf{(12.1b)}$$

$$\varepsilon = \varepsilon_{rel}\varepsilon_0 \text{ F/m} \qquad \textbf{(12.2)}$$

and ε_0 = permittivity of free space = 8.85×10^{-12} farads/meter. Permittivity of air is approximately the same as that of free space or a vacuum (that is, $\varepsilon_{rel} = 1.0006$), but if the medium happens to be caster oil, ε_{rel} is almost 5.

A convenient way of thinking of these fields is to imagine flux lines radiating either away from or toward a point source, as shown in Fig. 12.1. If one imagines a positive point source charge at the center of a sphere and arrows pointed away from the center, these arrows would be the flux lines (the bigger the charge, the more arrows).

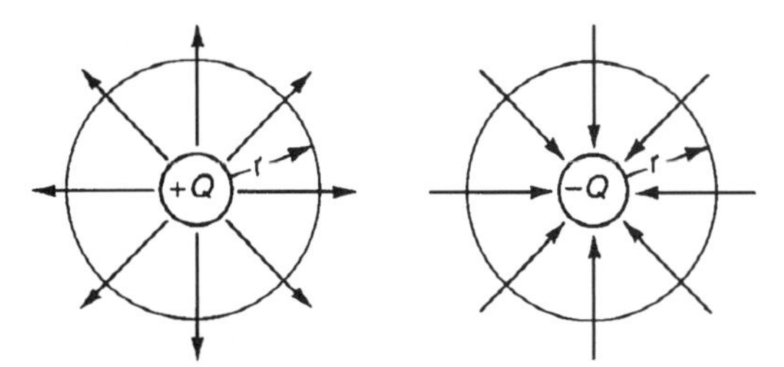

Figure 12.1 Electric flux lines around a charge

The flux density on the surface of the sphere (whose center is located at the point charge) would then be the number of arrows through a unit area on the sphere.

$$D = Q/A \qquad \textbf{(12.3a)}$$

$$Q = \mathbf{D} \bullet \mathbf{A} \qquad \textbf{(12.3b)}$$

where A is the area of the sphere ($4\pi r^2$) and Q is the quantity of charge in coulombs (assuming the area is normal to the flux). Equation (12.3b) is presented for those familiar with the dot and cross product vector notation (this notation guarantees the portion of the surface being considered is normal to the flux). If one were to divide the area of the sphere into very small areas, dA's, the sum of areas times the amount of flux through each area would be the amount of charge at the center of sphere. More formally, as the size of each area approaches zero, and as one integrates over the entire area of the sphere, the total charge enclosed is Q; this is known as Gauss's Law, Eqs. (12.4a) and (12.4b). But, by using the dot product notation, one is not limited to a spherical shape with the charge at the center; the formal law is then given as Eq. (12.4b) or (12.4c):

$$Q = \sum D\, dA \text{ (for entire surface)} \qquad \textbf{(12.4a)}$$

$$Q = \oint_s \mathbf{D} \bullet dA \qquad \textbf{(12.4b)}$$

$$Q/\varepsilon = \oint_s \mathbf{E} \bullet dA \qquad \textbf{(12.4c)}$$

The field strength, **E**, at the sphere's surface is then proportional to D,

$$\mathrm{E} = (D/\varepsilon)\mathbf{a}_r \qquad \textbf{(12.5)}$$

where $\mathbf{a}_r$ is a unit radial vector direction. If there are a number of charges throughout a region, it is usually easier to use the flux density concept in solving problems. In this chapter, unit vectors will be denoted as **a** so that *e* can be reserved for other purposes.

Example 12.1

Assume three point charges, A, B, and C as shown in Exhibit 1. Points A and B are 2 meters apart; Point C is on a perpendicular bisector between A and B and is 1 meter lower. Point A has 4×10^{-6} coulombs of negative charge, Point B has 10×10^{-6} coulombs of positive charge, and Point C has no charge yet. Determine the force and the field on B (in this case, due only to the charge on A) and the flux density at Point C.

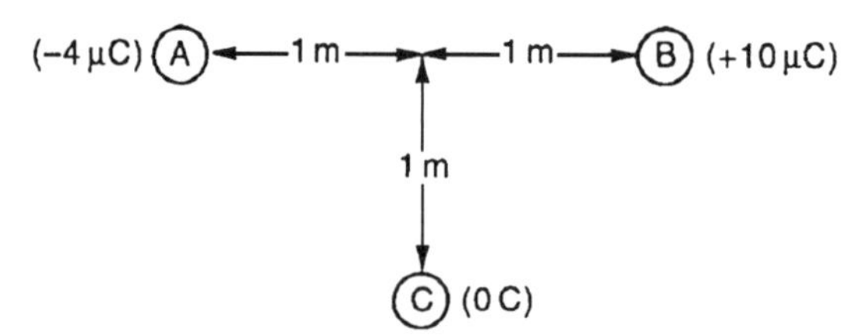

Exhibit 1 Point charge locations

Solution

The force on B due to A is directly proportional to the charges on A and B and inversely proportional to the square of the distance of separation. Using Eq. (12.1),

$$\mathbf{F} = (9 \times 10^9)(-4 \times 10^{-6})(10 \times 10^{-6})/(2^2) = -9 \times 10^{-2} \text{ newtons}$$

with the direction of the force toward each other.

$$\mathbf{E} \text{ (at B)} = \mathbf{F}/Q_B = 9 \times 10^{-2}/10^{-5} = 9 \times 10^3 \text{ V/m}$$

(If the medium happened to be a special oil with relative permittivity 5.0, both **F** and **E** would only be one-fifth as large.) A more orderly solution (especially if several charges are involved—or none at C in this case) is to use the flux density relationship to find the individual Ds, then convert to Es. All that is now necessary to find the net **E** is to use vector summation of the flux densities and divide by ε. To find the flux density at C attributable to charge A, imagine a sphere passing through C with its center at A; repeat for B.

$$D_{CA} \text{ (at C attributable to A)} = Q_A/(4\pi r^2)$$
$$\mathbf{E}_{CA} = (D/\varepsilon)\mathbf{a}_{rCA} = (Q_A/r^2)(1/4\pi\varepsilon)\mathbf{a}_{rCA} \text{ V/m}$$

Recall that $1/(4\pi\varepsilon)$ for free space is 9×10^9, then

$$\mathbf{E}_{CA} = (4\times10^{-6})/\left[\left(\sqrt{2^2}\right)(9\times10^9)\right]\mathbf{a}_{rCA} = 1.8\times10^4\mathbf{a}_{rCA} \text{ V/m}$$

$$\mathbf{E}_{CB} = (10\times10^{-6})/\left[\left(\sqrt{2^2}\right)(9\times10^9)\right]\mathbf{a}_{rCA} = 4.5\times10^4\mathbf{a}_{rCB} \text{ V/m}$$

To find the E_C net one may use the more formal procedure of finding the rectangular components of each of the field vectors since only two dimensions are involved. Then, summing the horizontal and vertical components, the net field vector is the square root of the sum of the squares. If one assumes the reference vector, $\mathbf{a}_r$, has the horizontal component a and the vertical component ja (the j implies the 90° or vertical axis), then

$$\mathbf{E}_{CA} = |\mathbf{E}_{CA}|(\cos\theta + j\sin\theta) = 1.8 \times 10^4 (\cos 135° + j\sin 135°)$$
$$= (-1.27 + j1.27) \times 10^4 \text{ V/m}$$

$$\mathbf{E}_{CB} = 4.5 \times 10^4 (\cos 225° + j\sin 225°) = (-3.18 - j3.18) \times 10^4 \text{ V/m}$$

$$\mathbf{E}_{C\text{net}} = (-4.45 - j1.91) \times 10^4 = 4.84 \times 10^4 \angle 203.2° \text{ V/m}$$

where the angle θ is the angle between the vector $\mathbf{E}$ and a horizontal reference line. However, solving for the vector field at C may be done faster by using a graphical solution, assuming reasonable accuracy. The points A, B, and C are set up (to their own scale); then, working with Point C, one places the vectors (using any suitable scale) that represent the two other fields. The reference direction and magnitude is then pictured as shown in Exhibit 2.

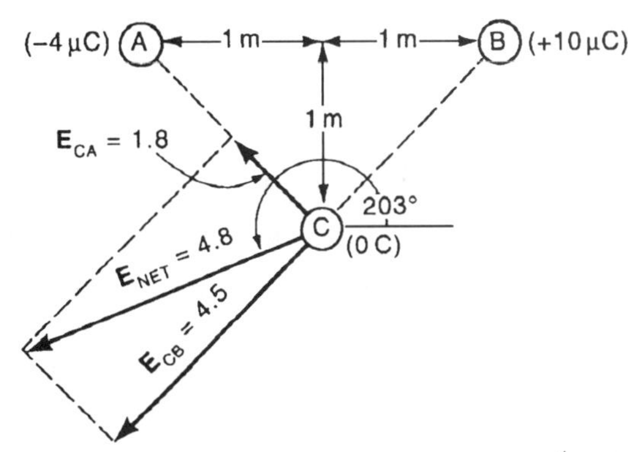

Exhibit 2 Sample problem vector fields (×10^4) at C

Now assume the charge at C is $+5 \times 10^{-6}$ coulombs; find the net vector field at Point B. To find this field at Point B, merely remove the charge at B.

$$\mathbf{E}_{BA} = (Q_A/r^2)(9 \times 10^9)\mathbf{a}_{rBA} = 0.90 \times 10^4\ \mathbf{a}_{rBA} \text{ V/m}$$
$$\mathbf{E}_{BC} = (Q_C/r^2)(9 \times 10^9)\mathbf{a}_{rBC} = 2.25 \times 10^4\ \mathbf{a}_{rBC} \text{ V/m}$$

Again, solve graphically for the solution (Exhibit 3).

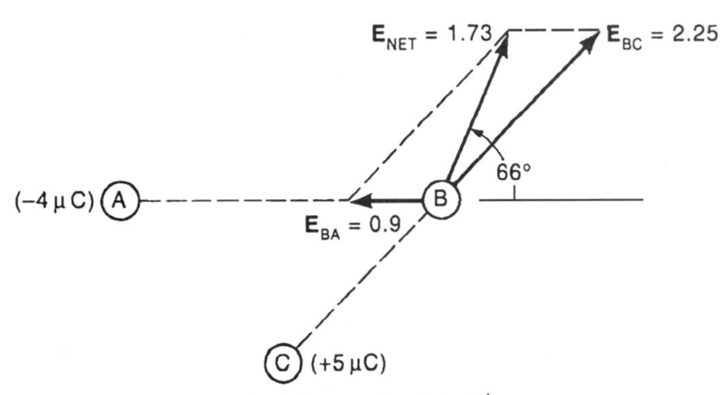

Exhibit 3 The electric field at point B (×10^4)

The actual solution, using vector notation, is found to be $\mathbf{E}$ (at B) $= 1.73 \times 10^4$ $\angle 66.5°$ volts/meter.

A more complex problem involving the x, y, and z axes is no more complicated except for the bookkeeping difficulty of solving a three-dimensional problem.

Before leaving the subject of static electric fields, the subject of energy storage in a capacitor (a pair of parallel plates separated by a distance, d) will be introduced. Capacitance is the ratio of the total stored charge uniformly distributed on the plates to the voltage difference between the plates. The separation distance, d, between plates will be considered to be small compared to the plate area (so that fringing may be neglected), then the field and flux lines will be perpendicular to the plates. The capacitance is

$$C = q_c(t)/v_c(t) \text{ farads (coulombs/volt)} \tag{12.6a}$$

and for non–time varying quantities

$$C = Q/(Ed) = AD/(Ed) = A(E\varepsilon)/(Ed) = A\varepsilon/d \tag{12.6b}$$

Permittivity may now be defined in a slightly different manner, where ε equals the charge induced on one square meter of the capacitor plates by an electric field intensity of one volt/meter. Thus, for a free space separation of the plates, 8.85×10^{-12} coulombs is induced on one square meter of a conducting plate by an electric field having an intensity of one volt per meter.

Magnetic Quantities and Definitions

Magnetic effects are related to the motion of charges, or currents. From the previous section on electrostatics, a force of one newton is produced by a charge of one coulomb in an electric field intensity of one volt per meter. Electric current, on the other hand, can be thought of as moving charges. **Current** is the time rate of change of the electric charge passing through a surface area. This definition is expressed as

$$i = dq/dt \tag{12.7}$$

The unit of current is the **ampere**, A, which equals one coulomb per second.

From the concept of moving charges, one can begin to understand magnetic fields. For permanent magnets (due to "static" magnetic fields) one recalls from physics that for some materials the molecular structure has the electron orbits of the atoms aligned. These tiny moving charges of electrons produce tiny currents. This alignment results in magnetic fields; actually ferromagnetic materials can be thought of as a large number of magnetic domains, with the domains being mostly aligned. However, when these (magnetic) domains are in disarray or randomly aligned, the material is unmagnetized.

Oersted, in 1819, observed that a magnetic flux existed about a wire carrying an electric current. (Flux lines can almost be visualized by observing the pattern of sprinkled iron filings on a piece of paper held over one pole of a magnet.) A few years after Oersted observed the effect of magnetic flux, Ampere found that wire coils carrying a current acted in the same manner as magnets. Simply stated, a coil of several turns of wire produced a stronger magnetic flux than only one turn of wire for the same current. And, if there were a ferromagnetic material to carry (or to provide a path for) the magnetic flux, the flux strength would be much greater.

Consider a toroidal ferromagnetic ring wrapped with a coil of several turns of wire, with the coil connected to a variable current source. Assume the current is zero and the material is not magnetized to begin with, then with an increase of current, an increase in flux results. One tends to think that the relationship would be linear; but the relationship is actually rather complex for a ferromagnetic material. As the current is further increased, the flux tends to level off. The leveling off is caused by most of the magnetic domains in the material aligning themselves in the same direction; consequently, increasing the current beyond the "knee of the bend" does not significantly increase the flux. If the current is decreased, most (but not all) of the magnetic domains return to random directions for zero current. The "going up" path is not necessarily the same as the "going down" path; these paths are known as the hysteresis curve for a particular material. When the current reaches zero on the "coming down" side, the material is left partially magnetized; this is called the residual magnetism. Again, assuming the material is unmagnetized to begin with, the flux is caused by the current through the number of turns of wire. This product of the current and the number of turns of wire, or amp-turns (A • *t*), is frequently called the magneto-motive force, (mmf) and indicated with a script *F*. The mmf is the source that causes the magnetic flux (ϕ). A plot of these two quantities is given in Fig. 12.2(b); this plot is referred to as the hysteresis curve.

If the width of the hysteresis curve were small (that is, the residual magnetic flux is small), only the first quadrant of the hysteresis curve is needed and can be drawn as a single line; then, only one curve per kind of magnetic material is needed.

Before a numerical problem can be solved, one needs to quantify and further define several terms. Rather than using flux, ϕ (units of webers), it is more appropriate to use flux density, *B* (units of tesla or webers/unit area). Flux density is the amount of flux passing through a unit area (normal to the flux). Also, rather than using the straight magneto-motive force designation, a more useful one is mmf/length, called the magnetic field intensity, *H*. These standardized quantities will allow a plot of *B* vs. *H*; this plot is always given for solid magnetic materials (that is, before any air gap might be cut into them).

One needs to define further the direction of magnetic flux and fields; this is difficult to do without the concept of magnetic poles. To show this effect, consider

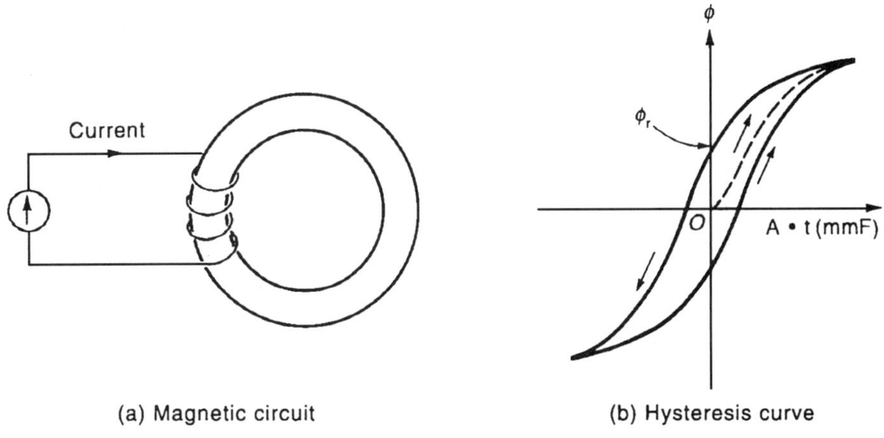

Figure 12.2 Magnetic characteristics

an air gap cut into the ferromagnetic material. However, analogous to electrical charges being positive or negative, one can think of magnetic poles (they always occur in pairs) as north and south (recall, like poles repel and unlike ones attract). As an example, consider a freely suspended magnet (say, a magnetized needle in a compass) in the earth's magnetic field; one end of the magnet points toward the geographical north. By common usage, this end of the magnet is referred to as the "north-seeking pole." For the toroidal ring with the air gap cut in it, if the positive direction of the current enters the top end of the coil, the direction of the flux in the ring will be clockwise, and the top of the air gap is the north pole, whereas the bottom is the south pole.

In the previous example, the magnetic flux in the air gap is concentrated and may be very high. The lines of magnetic flux produced in the material and the air gap are the same and continuous when there is a current in the wire. Without the ferromagnetic material in the path, the relationship between the current (causing the magnetic flux) and the resulting magnetic field would be linear but very weak. The linear constant of proportionality is called the permeability of free space, μ_0 (as in the air gap). This constant is

$$\mu_0 = 4\pi \times 10^{-7} \text{ F/m} \tag{12.8a}$$

For ferromagnetic materials, the relative permeability is nonlinear; for approximate calculations, it is sometimes linearized in the region of the curve before magnetic saturation is reached. The slope of the curve in the saturation region approaches that of free space or air, thus

$$\mu_r \mu_0 \rightarrow \mu_0 \tag{12.8b}$$

The relationship between B and H may then be expressed as the slope of a B vs. H curve

$$\mu = \mu_r \mu_0 = B/H \tag{12.8c}$$

when B and H are normal to each other.

As stated previously, flux lines are continuous (that is, the lines of flux in the ring and air gap are the same), whereas the flux density in the ring may or may not be the same as that in the air gap—frequently they are considered the same by neglecting the fringing effect in the gap. Also, the cause of the flux (that is, the amp-turns) is thought of as being distributed along the whole ring, thus it is appropriate to use field intensity, H (amp-turns/meter or mmf/m). The length for this example is merely the mean circumference of the ring (the width of the air gap usually being negligible). Kirchhoff's voltage law in electrical series circuits states that the net voltage drop in a loop equals zero; one may use the same analogy for series magnetic circuits. That is, the net mmf in a magnetic series loop equals zero. Stated another way, the mmf source must equal the sum of the mmf drops (or losses) in the series circuit. Thus, summing these mmf drops in the iron and the air gap is all that is needed to find the mmf for the source.

Although the previous example involved a coil of wire, a more fundamental problem is one involving a long straight wire.

For a wire on the z-axis carrying a current

$$\boldsymbol{H} = \frac{\boldsymbol{B}}{\mu} = \frac{I\boldsymbol{a}_\theta}{2\pi r} \tag{12.9a}$$

where

$\boldsymbol{H}$ = the magnetic field strength (amps/meter)

$\boldsymbol{B}$ = the magnetic flux density (tesla)

$\boldsymbol{a}_\theta$ = the unit vector in positive θ direction in cylindrical coordinates

I = the current

μ = the permeability of the medium (for air, $\mu = \mu_0 = 4\pi \times 10^{-7}\,\text{H/m}$)

The flux density at some radial distance, r, from the center line of the wire is given in Eq. (12.9b).

$$B = (\mu i)/(2\pi r) \qquad \textbf{(12.9b)}$$

For the direction of the current shown in Fig. 12.3, the flux density direction is shown with the head of an arrow as (⊙) and the tail of an arrow as (⊗); this notation will be the same for current.

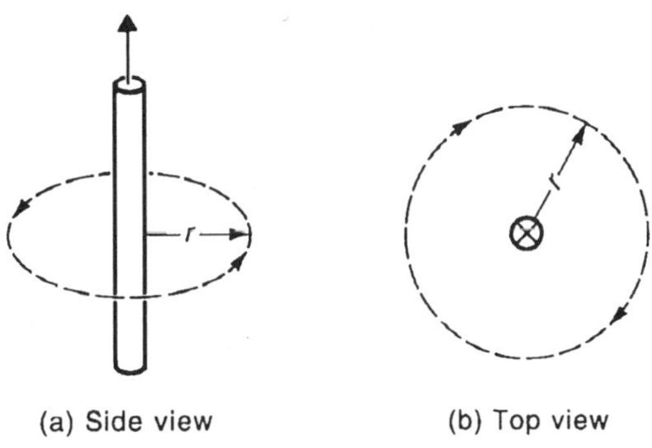

Figure 12.3 The flux density around a wire

The direction of the flux may be remembered by using the right-hand rule: place the thumb in the direction of the current; the partially closed fingers will point in the direction of the magnetic flux. Now, if the long wire were formed into a circular coil of radius r, one could use calculus to consider a differential length of the wire and integrate around the closed loop to find the flux density within the loop. If there were several turns, N, for the loop, the equation for the flux density would be

$$B = (\mu N i)/r \qquad \textbf{(12.10)}$$

The right-hand rule may also be used to find the direction that a current would flow if a voltage is induced (or generated) in a wire by the relative motion of a magnetic field and a wire (see Fig. 12.4). Consider a straight wire being moved within a magnetic field (assume the wire and motion is normal to the magnetic field). A voltage induced in the wire will be proportional to the strength of the field, the length (l) of wire within the field, and the velocity, v, of the wire.

$$e \text{ (induced voltage)} = Blv \qquad \textbf{(12.11a)}$$

For a coil of N turns enclosing flux ϕ

$$e = -N\, d\phi/dt \qquad \textbf{(12.11b)}$$

where ϕ = the flux (webers) enclosed by the N conductor turns and

$$\phi = \int_A B \cdot dA$$

The direction of the flux is from the north pole face to the south pole face. It will help to think of the wire within the field as being a voltage generator (the induced voltage); then if there is a closed path (perhaps through an external resistive load) so that a current could flow, the polarity of the "generator" must be plus where the current (thumb direction) leaves the magnetic field and minus where it enters the field.

In this section, vector notation involving three-dimensional space has been kept to a minimum; the FE examination will probably not include this added complication for magnetic fields. Should such a problem occur, one could still attempt to tailor the problem using vector components. As an example, assume one is asked to find an induced voltage for the motion of a wire in a magnetic field as depicted in Fig. 12.4 using Eq. (12.11a). However, assume also that the wire velocity direction is not normal to the magnetic field but at an angle θ from the horizontal. The solution and the equation would be the same, except that one must use the vertical component of the velocity rather than the magnitude of the entire velocity.

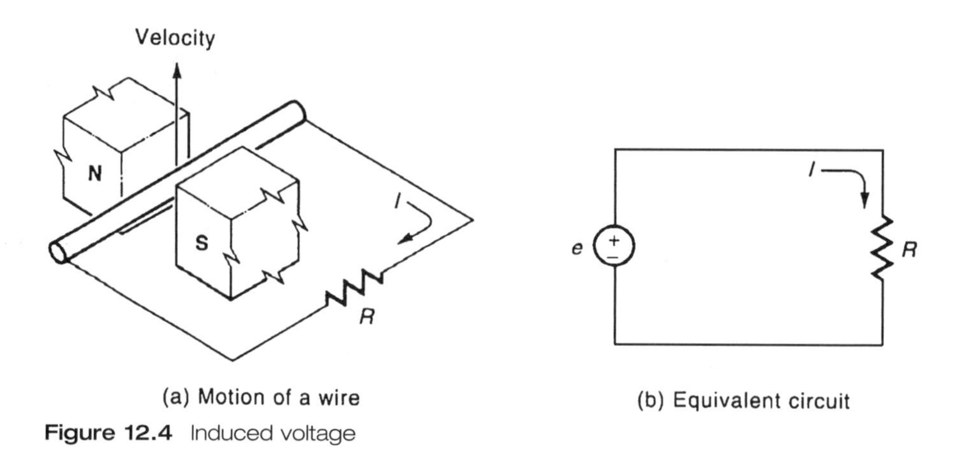

Figure 12.4 Induced voltage

Power, Energy, and Measurements

Before the discussion of power, energy, and measurements is undertaken, three items need to be reviewed. The first is notation, the second is the calculation of resistance, and the third is a more complete definition of voltage.

Notation

In this section, the notation for time-dependent quantities will normally be given by lowercase, italic letters (for example, for voltage, current, and such, $v(t)$, $i(t)$, where the (t) may or may not be included). A constant value, such as a battery voltage, will be given in uppercase, italic letters (for example, V and I); the effective or rms (root mean square) values and other quantities that have a non-varying value (such as the average power) will also be expressed as an uppercase letter. Where confusion is possible, a subscript is normally used. Further discussion of notation will be presented where appropriate.

Resistance

Most reviews assume that one is familiar with simple wire resistance. Although the actual resistance of a wire conductor is usually considered to be negligible in circuit analysis, it is easily calculated. The parameters are shown in Fig. 12.5.

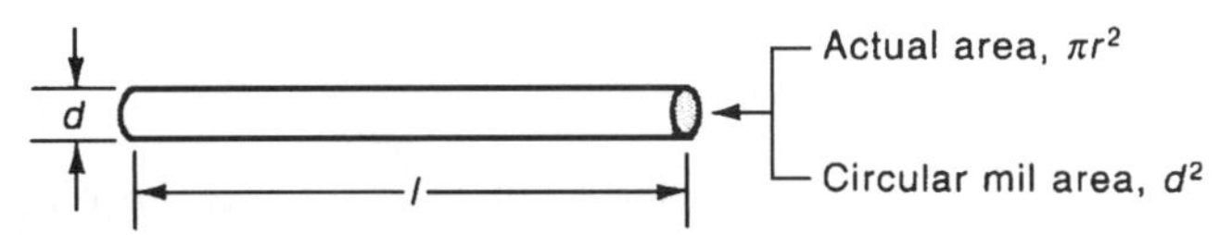

Figure 12.5 Wire resistance parameters

The resistance is proportional to its length and inversely proportional to its cross-sectional area; the constant of proportionality is the resistivity, ρ:

$$R = \frac{\rho L}{A} \tag{12.12}$$

The resistivity of a particular conductor is normally given for a standard temperature of 20°C. However, the units depend on the length and area. For the MKS system of units, the value of ρ for copper is $1.7 \times 10^{-8}\ \Omega \bullet \text{m}$ (for aluminum, it is almost twice this value).

Example 12.2

As an example, for a 5-meter length of 12-gauge (approximate diameter of 2 mm) copper wire at room temperature (near 20°C) find the resistance, R.

Solution

$$R = (1.7 \times 10^{-8})(5/[\pi(0.001)^2]) = 0.027\ \ \Omega$$

On the other hand, for another temperature, the resistivity is modified by the temperature coefficient, α, to be

$$\rho = \rho_0[1 + \alpha(\mathrm{T} - \mathrm{T}_0)] \tag{12.13a}$$

or

$$R = R_0[1 + \alpha(\mathrm{T} - \mathrm{T}_0)] \tag{12.13b}$$

where $(\mathrm{T} - \mathrm{T}_0)$ is the change of temperature from 20°C.

Generally, these two equations will solve most resistive type problems. However here is one note of caution: unfortunately, the inch, foot, and circular mil units for area are still in use. These units can be confusing since the area of the wire may be given in circular mils; this area omits π in the true area, πr^2, computation and, instead, includes it in the resistivity constant. In this case, the resistivity for copper wire is 10.4 Ω circ-mils/foot. Actually, the computation is made easier; assume a length of 5 feet for a wire diameter of 0.03 inches (or 30 mils):

$$R = (10.4)(5)/(30)^2 = 0.0578\ \Omega$$

The key to this type of problem is to check carefully by dimensional analysis to determine what units are being used. Wire tables, listing these parameters, should be furnished with the EIT examination when they are needed.

Voltage

Whereas a simple definition of **voltage** is the potential difference that will cause a current of one ampere to flow through a resistance of one ohm, a more formal definition is "a charge of one coulomb receives or delivers an energy of one joule moving through a voltage of one volt." The instantaneous voltage is defined by Eq. (12.14).

$$v = dw/dq \qquad \textbf{(12.14)}$$

In other words, if a unit quantity of electricity (coulomb) gives up energy equal to one joule as it proceeds from one point to another, the difference in potential is one volt. The joule is sometimes called the coulomb-volt. Likewise, the energy acquired by an electron when it is raised through a difference of potential of one volt is called an electron-volt. The definition of voltage leads directly into the subject of power and energy.

Power and Energy

Power (instantaneous, i.e., lowercase italic notation) is the rate of change of energy, *dw/dt*, or

$$p(t) = dw/dt = (dw/dq)(dq/dt) = v(t)(i) \qquad \textbf{(12.15)}$$

The total energy is then given in Eq. (12.16). (The function of time, *t*, is implied and will be dropped for simplicity.)

$$w = \int_0^T p\,dt = \int_0^T vi\,dt \qquad \textbf{(12.16)}$$

Energy Storage

Energy may be stored in some electrical circuit elements. For electrical circuits, there are three basic elements: resistance, inductance, and capacitance. For the resistor, electrical energy cannot be stored since it is turned into heat; however, both the inductor and capacitor are capable of energy storage (see Fig. 12.6).

The inductor, *L* (measured in units of henries, H), usually a coil of wire that can produce a magnetic field, stores energy in the magnetic field; the capacitor stores energy within its electrical field (i.e., within its dielectric medium). The circuit

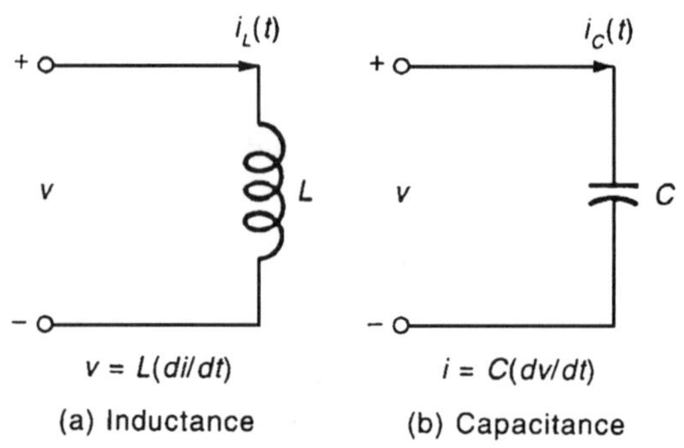

Figure 12.6 Energy storage elements

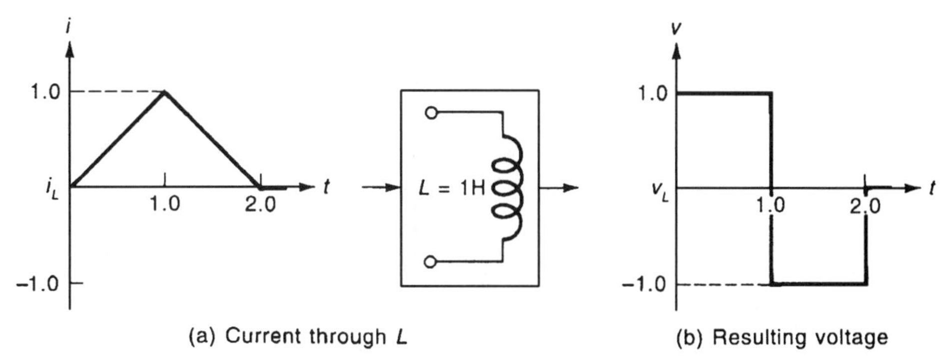

Figure 12.7 Inductor current-voltage relationship

symbols and voltage-current relationships are given in Fig. 12.7. Eqs. (12.17) and (12.18) give the energy stored as

$$v_L = L(di_L/dt) \quad \textbf{(12.17a)}$$

$$i_C = C(dv_C/dt) \quad \textbf{(12.17b)}$$

$$w_L = (1/2)Li_L^2 \quad \textbf{(12.18a)}$$

$$w_C = (1/2)Cv_C^2 \quad \textbf{(12.18b)}$$

The voltage across an inductor is a function of the rate of change of current. If there is no rate of change (or direct current), then the voltage must be zero, and the inductor acts like a short circuit or zero resistance. Actually, there is always some wire resistance, but usually this is negligible. For current to flow in a capacitor, there must be a rate of change of voltage. If there is no rate of change, there is no current, and the capacitor acts like an open circuit (this will be discussed in more detail later). As an example, assume the current (from a variable current source) through an inductor is as given in Fig. 12.7(a); then the voltage across the inductor must be as shown in Fig. 12.7(b).

As shown in Fig. 12.8, the power "taken" (absorbed or stored) by the inductor is a function of time until $t = 1$ second and is positive; for the current with a negative slope, power is "given up" (or returned to the circuit source) from $t = 1$ second until $t = 2$ seconds. On the other hand, the energy stored (also a function of time) follows the square law given directly in Eq. (12.18a). The resulting curves are found by splitting Eq. (12.16):

$$w = \int_0^2 p\,dt = \int_0^1 vi\,dt + \int_1^2 vi\,dt$$

Average Voltage and Current

All dc quantities, being constant values, are represented by uppercase letters. However, for changing quantities one must be careful with notation. The formal definition involves the time period of interest, *T*. The average voltage (or current) is given by

$$V_{\text{avg}} = (1/T)\int_0^T v\,dt \quad \textbf{(12.19)}$$

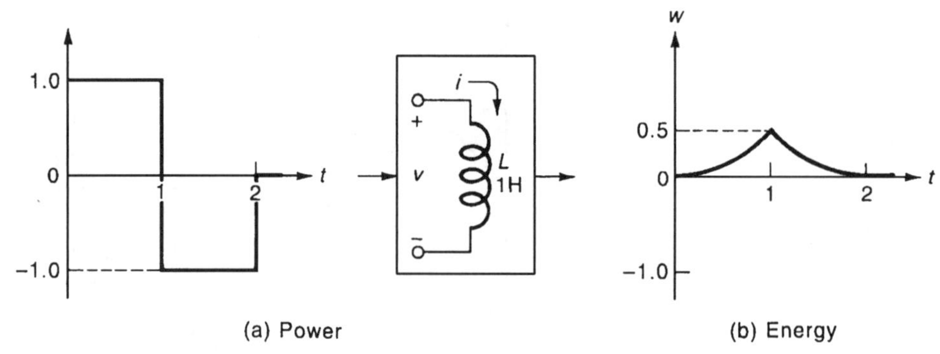

Figure 12.8 Power and energy plot for the current/voltage curves of Fig. 12.7

or the area under the voltage curve divided by the period of time. Of course, the same holds for current.

If the positive and negative areas balance over a specific period of time, then the net area is zero. It should then be obvious that the average voltage from a sine wave generator over a full period, T (or 2π), is zero:

$$V_{avg} = (1/T)\int_0^T V_{max} \sin \omega t \, dt = 0$$

Over a *half period*, $(1/2)T$, the average voltage is

$$V_{avg} = (2/T)\int_0^{T/2} V_{max} \sin \omega t \, dt = 2V_{max}/\pi$$

(*Caution:* a half wave over a full period has an average value of V_{max}/π.)

Average Power

Since power is defined as the product of voltage and current, then for simple dc voltage or current sources, the average power is simply the product of the current and voltage. This is written as uppercase, italic P (without subscript) as $P = VI$. (Generally, use this equation only for dc quantities, as ac or other types of wave forms may have a phase shift. This will be discussed later.)

Effective Values

For most wave forms, other than straight dc, one must define an effective or rms value of current and voltage since an average value of current or voltage could be zero for a sinusoidal signal. Another way of stating this is the following: "What equivalent value of current will cause the same heating power in a resistor as would a dc current?" The effective (or rms) value of current or voltage is defined as

$$I_{eff} = I_{rms} = I = \sqrt{(1/T)\int_0^T i^2 dt} \qquad \textbf{(12.20)}$$

If $i = I_{max} \sin \omega t$ or $i = I_{max} \cos \omega t$, then

$$I = 0.707 I_{max} = I_{max}/\sqrt{2} \qquad \textbf{(12.21a)}$$

$$V = 0.707 V_{max} = V_{max}/\sqrt{2} \qquad \textbf{(12.21b)}$$

Although Eq. (12.21) and Fig. 12.9 are for sinusoids, the effective or rms values may be found for any periodic wave form.

For sinusoids, the average power may involve phase angles (more on this later); however, one is safe in finding the power developed as heat in a resistor by using

$$P_{avg} = P = I^2R = \left(I_m/\sqrt{2}\right)^2 R \text{ watts} \tag{12.22}$$

For a short example, assume a known sinusoidal current of $i = 14.1 \cos \omega t$ flows through a resistor of 25 ohms. The effective, or rms, current is easily computed from Eq. (12.20) or (12.21) and the power by Eq. (12.22):

$$P = \left(14.1/\sqrt{2}\right)^2 (25) = 2500 \text{ watts}$$

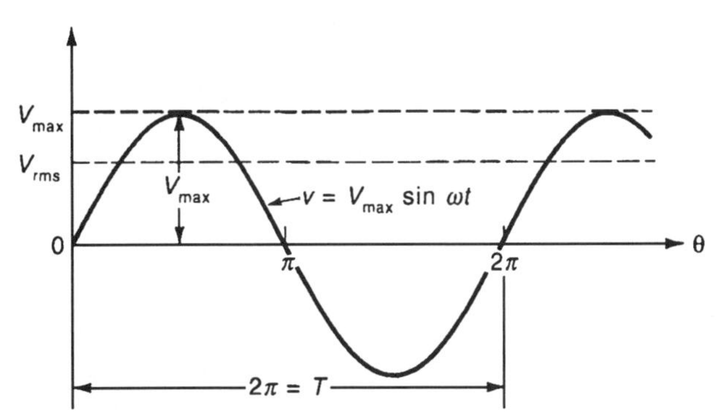

Figure 12.9 Effective or rms value for a sinusoid

If the sinusoid is replaced with another kind of periodic wave form, Eq. (12.20) would, of course, still be used. However, sometimes a table of "standard wave forms" with a list of effective or rms values is available.

DC CIRCUITS AND RESISTANCE

Ohm's and Kirchhoff's Laws for a Single Source Network

The solutions of all circuits problems whether they are dc, ac, single source, or multiple source, involve the use of these laws. The first is Kirchhoff's voltage law (KVL):

For any closed loop in a circuit, the voltage algebraically sums to zero. $\Sigma v = 0$	or	The voltage rises equal the voltage drops in a closed loop. $\Sigma V_{rises} = \Sigma V_{drops}$

Another way of stating this relationship is to say the sum of the voltage rises is equal to the sum of the voltage drops in any loop. Both of these statements are true of both dc and instantaneous values. However, one needs to be careful when dealing with ac analysis with possibly different phase angles; this relationship will be discussed later.

The second is Kirchhoff's current law (KCL); this law involves any junction, or node, in an electrical circuit:

For any node in an electrical circuit, the net current algebraically sums to zero. $\Sigma i = 0$	or	The total current entering a junction equals the total current leaving. $\Sigma I_{in} = \Sigma I_{out}$

Restated, the sum of the currents entering a junction must equal the sum of the currents leaving the junction. Again, this is true of both dc and instantaneous values; however, for ac circuit analysis care is required in its application.

Ohm's law for resistance ($R = v/i$) is well known and is the same for dc, instantaneous quantitities, and ac values. For a series circuit (where the current is the same in every element) the equivalent resistance merely becomes

$$R_T = R_1 + R_2 + \cdots + R_n = \sum R \qquad \textbf{(12.23a)}$$

For parallel circuits (where the voltage across every element is the same) the equivalent resistance, R_T, is

$$1/R_T = 1/R_1 + 1/R_2 + \cdots + 1/R_n = \sum G \qquad \textbf{(12.23b)}$$

The conductance G is simply the inverse of the resistance (whose units are siemens or mhos, $1/\Omega$). From Eq. (12.23b), one can quickly determine that the equivalent resistance for two resistors (or impedances) in parallel is $R_T = R_1R_2/(R_1 + R_2)$.

Example 12.3

A simple example follows; for this type of circuit problem, it is usually quicker to convert the parallel resistors to an equivalent value. The simpler series circuit is easily solved: by finding the voltage drop across any element, the voltage across the equivalent resistance is found. Assume the question is to find the power dissipated in the 6-Ω resistor in Exhibit 4.

Solution

The equivalent parallel resistance is found to be 3/2 Ω. This value yields a total series resistance of 4 Ω as shown in Exhibit 4b. The series current is easily found to be 3 A; the voltage drop across the parallel resistance is $V_{eq} = I(R_T) = (3)(3/2) = 9/2$ volts. The current through the 6-Ω resistor is $I = V_{eq}/R_6 = (9/2)/6 = 3/4$ A. The power dissipated in the 6-Ω resistor is $P_{6\Omega} = I^2R = (3/4)^2 6 = 27/8 = 3.375$ watts.

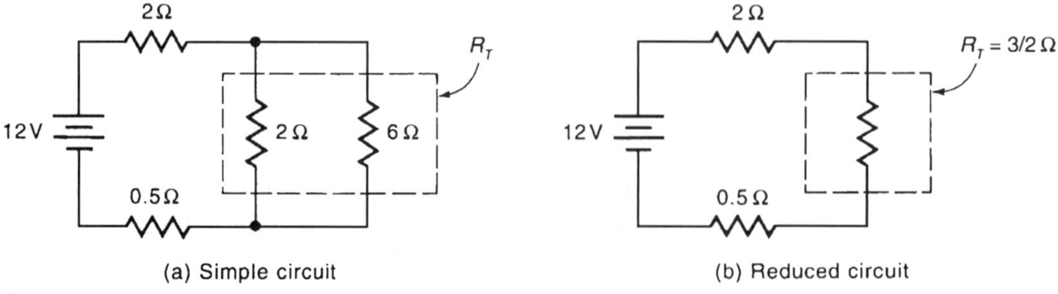

Exhibit 4 A simple voltage source circuit

Here, the power from the source equals the sum of the power dissipated in each of the resistors. It is instructive to compute the sum of the power lost; this is

$$P_{\text{source}} = V_{\text{source}}\, I = (12)(3) = 36 \text{ watts}$$

$$\sum P_{\text{lost}} = I^2\left(\sum R_{\text{ser}}\right) = 9(2 + 3/2 + 1/2) = 36 \text{ watts}$$

For this problem, a somewhat more formal method is to use the voltage dividing equation after R_T is found (all elements are now in series). The voltage across R_T may be calculated to be $V_T = (R_T/\Sigma R)V_{\text{bat}}$; then, the current through R_6 is determined as before.

Example 12.4

Consider another example with a current source: Here, the series portion of the circuit should be reduced to one resistor; with all of the resistors now in parallel, the voltage across this group is easily found. Now the question is to find the amount of power dissipated in the 2-ohm resistor in Exhibit 5.

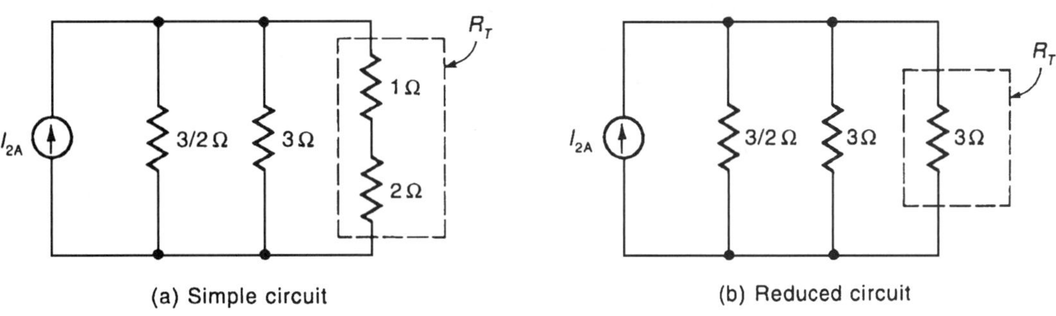

Exhibit 5 A simple single current source circuit

Solution

After finding the one series resistor, R_s, that is equivalent to the two resistances in series, all resistances are now in parallel. Then using Eq. (12.23b), or merely taking two resistors at a time, one easily computes the total equivalent one resistance to be 3/4 Ω. The voltage across the parallel circuit is $RI = (3/4)(2) = 3/2$ volts. The current through the series equivalent branch is $I_{se} = V_{par}/R_{se} = (3/2)/3 = 1/2$ amperes; the power dissipated in the 2-Ω resistor is $P = I^2R = (1/2)^2 2 = 0.5$ watts.

Again, a quick check is to find if the power taken from the source ($P_s = VI = (3/2)2 = 3$ watts) and the power dissipated in the resistors are equal:

$$\sum P_{\text{lost}} = V^2(1/R_{3/2} + 1/R_3 + 1/R_{se}) = (3/2)^2(2/3 + 1/3 + 1/3) = 3 \text{ watts}$$

For this problem, as before, there is a somewhat more formal method of finding the current through R_{se} directly; $I_{se} = (G_{se}/\Sigma G)I_{tot}$. Knowing the branch current, the power is easily found.

Multiple Source Networks and Theorems

Although the term multiple source circuits implies multiple power output, this implication may be incorrect as the voltage or current source may actually absorb power. If the current is flowing into a voltage "source"—from plus to minus—the

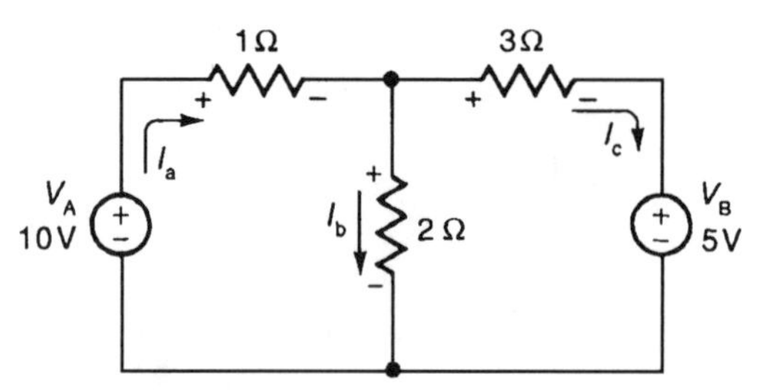

Figure 12.10 A two-loop network

"source" is actually a "load" taking power (that is, a battery being charged). Therefore, it is very important to determine the direction of current (or voltage polarity as the case may be). The procedure is to assume current directions and carefully label these on a circuit diagram; then let the mathematics determine the actual direction. One way to do this is to assume a current through each element (of course series elements will have the same current) of a circuit. This is in contrast to assuming loop or mesh currents; however, both methods amount to the same thing. A short example (see Fig. 12.10) will show the method. (To keep the circuit diagrams uniform and to later allow for time varying voltage sources, a circle is used to indicate the voltage source rather than showing a battery.)

After labeling the assumed current directions, place a plus sign where the current enters the resistor and a negative sign where it leaves. Then, by Kirchhoff's voltage law, sum the voltages around each loop:

$$V_A - R_1 I_a - R_2 I_b = 0 \qquad \textbf{(i)}$$
$$V_B + R_3 I_c - R_2 I_b = 0 \qquad \textbf{(ii)}$$

Since three variables are present, another equation is needed; Kirchhoff's current law will produce the third equation (actually J − 1 node junction equations are needed; here J = 2). Summing currents at one junction (say, the upper middle node) gives $I_a = I_b + I_c$.

Rewriting the equations for loops one and two (here, replacing all I_a), yields.

$$V_A = R_1(I_b + I_c) + R_2 I_b, \qquad 10 = 3I_b + I_c \qquad \textbf{(i)}$$
$$V_B = R_2 I_b - R_3 I_c, \qquad 5 = 2I_b - 3I_c \qquad \textbf{(ii)}$$

Solving for the currents, the results are

$$I_a = 3.64\text{ A}, \quad I_b = 3.18\text{ A}, \quad I_c = 0.455\text{ A}$$

Since all of the currents are positive, the assumed directions of the currents are correct; therefore, the voltage "source" B is really a load (perhaps a battery being charged). As a check, the sum of the voltages around the outside loop should equal zero (starting at the upper left hand corner): does 1(3.64) + 3(0.455) + 5 − 10 = 0? Or, 3.64 + 1.365 + 5 − 10 = 0.005 (acceptable).

Another way of analyzing the circuit is to use the "mesh" technique of analysis. Notice in the previous two-loop problem, the current through the middle resistance that was identified as I_b is nothing more than $I_a - I_c$. If we now consider going around the entire left loop as I_a and around the entire right loop as I_c, the currents mesh at the middle resistance. Because voltages are summed around each loop, the net voltage across the middle resistor (when going around the left loop) is $RI_a - RI_c$; the loop currents are in opposite directions. The same kind of equation is written for the right hand loop. Thus one current variable (I_b) is eliminated. There are still other ways of analyzing the circuit, as will be pointed out later.

Example 12.5

Another example that involves mixed sources of current and voltages is shown in Exhibit 6. Find the voltage across the current source for this circuit configuration.

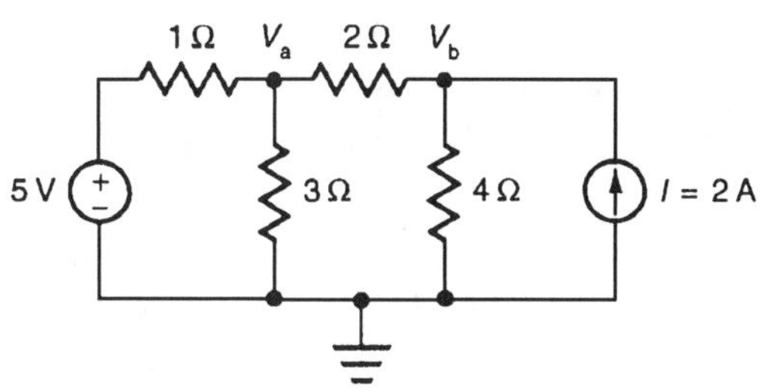

Exhibit 6 Mixed source circuit

Solution

Here, the node-voltage method of analysis will be used. This method finds voltages with respect to some common reference point called a ground node (see Exhibit 6). By wisely selecting this common point to include as many junctions as possible, the number of simultaneous equations may be reduced. The node voltage will be considered as a voltage rise from the common reference node (that is, current directions may then be assumed away from the node while writing the equations for that particular node).

$$\text{At } V_a, \sum I \qquad 0 = (V_a - 5)/1 + (V_a - V_b)/2 + (V_a - 0)/3$$

$$\text{At } V_b, \sum I \qquad 0 = (V_b - V_a)/2 + (V_b - 0)/4 + (-2)$$

Simplifying (multiplying through by 6 and 4, respectively) and collecting terms,

$$30 = 11V_a - 3V_b$$
$$8 = -2V_a + 3V_b$$

Solving these equations for the node voltages gives $V_a = 4.22$ V and $V_b = 5.48$ V. From these voltages, any desired currents or powers may be found. To check the solution, find the sum of the power for the current and voltage sources and compare this result with the sum of the power dissipated by the resistors ($P_{\text{sources}} = 14.9$ watts $= \sum P_{\text{resistors}}$). Either of the apparent sources could actually be a load by taking power from the other source. If the voltage source is a battery and it is found that the actual current flows through the battery from plus to minus, then the battery is being charged and acts as a load.

Network Reduction, Thevenin and Norton Equivalents

Most circuits may be greatly reduced in complexity and thus yield a simpler solution. This is especially so if a variable load is involved and requires a number of repeated calculations. Consider the circuit shown in Fig. 12.11; many calculations would be involved if one were asked to compute the power dissipated in R_L

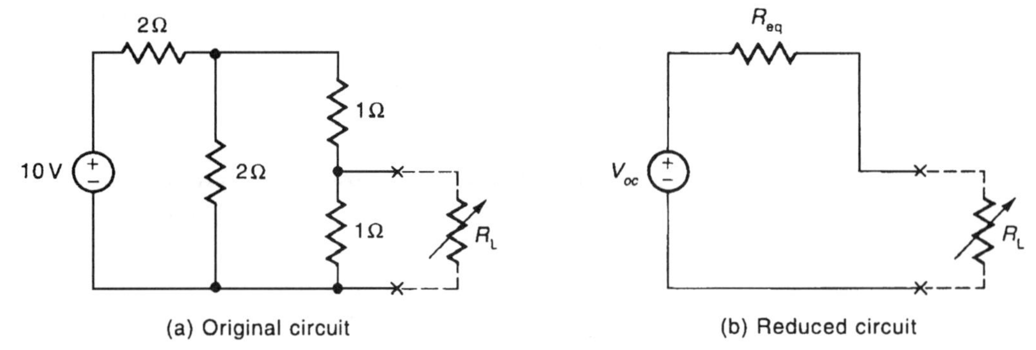

Figure 12.11 Circuit for Thevenin's reduction

for several different values of R_L. If one could reduce the network (all of that circuitry to the left of × – ×) to one simple voltage source in series with one resistance [see Fig. 12.11(b)], then calculating the current to R_L would be simple and easy. Reducing the network is done with Thevenin's theorem. One can find the Thevenin voltage by merely opening the network at the point of interest (that is, temporarily removing R_L) and measuring the open circuit voltage at × – ×; the open circuit voltage is the Thevenin voltage. Next, place a short circuit (an ammeter, assumed to have zero resistance) across this section and measure the short circuit current. The Thevenin resistance is given as

$$R_{eq} = V_{oc}I_{sc}$$

To find a Thevenin equivalent circuit on an examination, one cannot measure the various voltages or currents but must calculate them. Calculating the open circuit voltage does not present a problem, but more needs to be said about calculating the resistance.

There are at least two ways to calculate the Thevenin resistance. One is to calculate the short circuit current and proceed as before. The other method is to replace all voltage sources with their internal resistance (usually zero) and any current source with its internal resistance (usually infinity) and then calculate the resistance one would see looking back into the circuit (to the left of × – × in Fig. 12.11). However, caution is needed if any dependent sources are present. For the values given in Fig. 12.11, the calculated open circuit voltage is 5/3 volts, and the short circuit current is 5/2 amperes; thus the Thevenin equivalent circuit is as given in Fig. 12.12(a).

The current source that is equivalent to the Thevenin circuit is Norton's equivalent circuit. Norton's circuit is the current source equivalency for Thevenin's voltage source [see Fig. 12.12(b)]. The value of the resistance is the same except that it is usually stated in terms of conductance ($1/R$) and the units are mho (old)

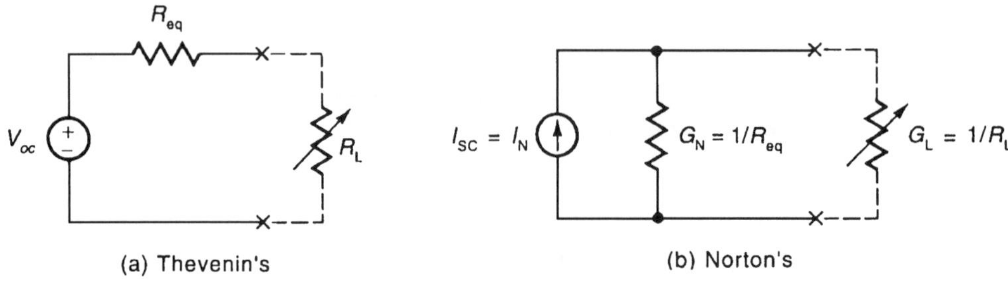

Figure 12.12 Equivalent circuits

or siemens (new). The current source may be found directly from Thevenin's circuit by finding the short-circuit current directly from Thevenin's circuit. If one is asked for a Norton equivalent circuit, one can first find Thevenin's circuit then convert to Norton's circuit.

Assume one wishes to determine the maximum power that could be dissipated in a variable load resistor that is nested in a multiple source circuit. First, it is necessary to know the size of the resistor. The **maximum power transfer theorem** (for dc circuits) states that maximum power is extracted from a circuit when the circuit is converted to Thevenin's equivalent circuit and the load resistance is equal to Thevenin's resistance. A consequence of this theorem is that half of the power is dissipated in the load and the other half in Thevenin's equivalent resistance. Consider the circuit of Fig. 12.13; all that is necessary is to isolate the load resistor from being "buried" in the circuit. This is done by rearranging the circuit as shown in Fig. 12.13(b). One way is to move the load resistor to the right side of the circuit and then find Thevenin's equivalent circuit for the portion on the left.

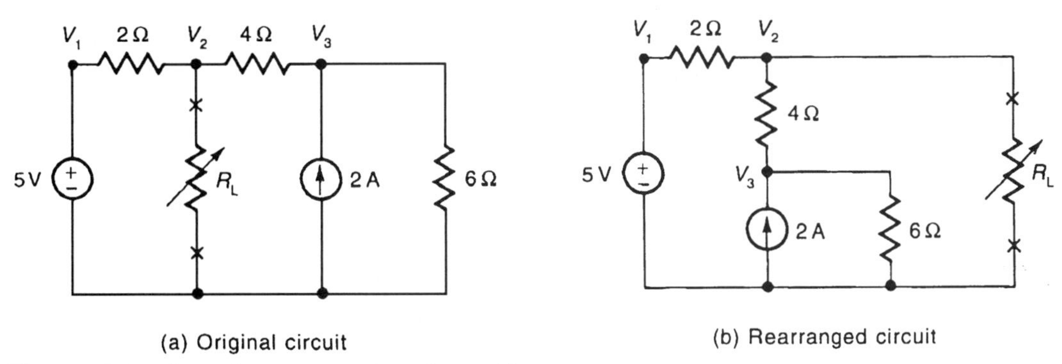

Figure 12.13 A maximum power transfer problem

For the rearranged circuit of Fig. 12.13, the open circuit voltage (that is, with R_L temporarily disconnected) will be found by the node method of analysis:

$$\text{At } V_2, \sum I \quad (V_2 - V_1)/2 + (V_2 - V_3)/4 = 0$$

$$\text{At } V_3, \sum I \quad (V_3 - V_2)/4 - 2 + V_3/6 = 0$$

Solving (with $V_1 = 5$ volts) for $V_2 = V_T$ yields 6.116 volts. Thevenin's resistance can now be found by using the simpler method of merely replacing all independent voltage sources with zero and all independent current sources with an open circuit; then, looking back into the circuit × – ×, the equivalent resistance can now be found. This resistance yields $R_{oc} = 1.667\ \Omega$. From the maximum power relationship, $R_L = R_{oc}$. The power dissipated in $R_L = 5.61$ watts, see Fig. 12.14.

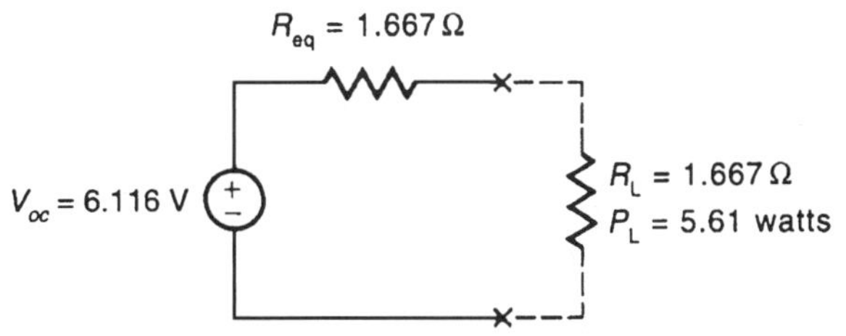

Figure 12.14 Final load resistance using Thevenin's circuit

Transient Response for a Single Energy Storage Element

If an energy storage element, such as a capacitor or inductor, is present in a circuit with one or more resistors, the problem solution will be a function of time. The voltage across a capacitor or current through an inductor cannot change instantaneously but takes time (unless infinite pulses are involved). It is assumed that the circuit is connected to either a voltage or current source or has some initial values.

Capacitive Circuits

If a capacitor (assume initially uncharged) is being charged from a voltage source through a resistor as shown in Fig. 12.15(a), then it takes time for it to reach the charging voltage. The voltage vs. time solution is plotted in Fig. 12.15(b). Here it is obvious that the amount of time for the capacitor voltage to reach its final value is infinite. On the other hand, if the capacitor already has an initial charge (or voltage) and is being discharged through a resistor, it again takes time for it to discharge [refer to Fig. 12.16(b)]; this amount of time is also infinite.

Rather than use infinity as the changing time, it is more convenient to define a time constant, τ, to obtain a practical result. The time constant is some value of t that makes the exponent, x, of e^x, equal to -1. As an example, if $x = -t/RC$, then

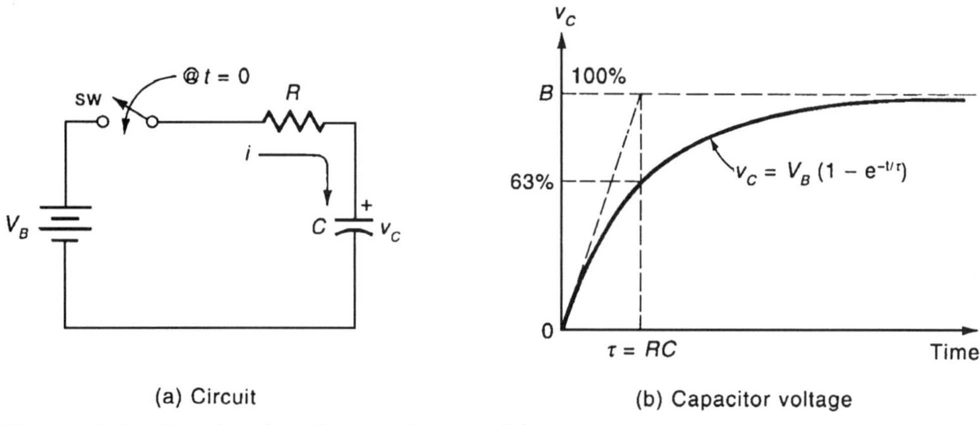

Figure 12.15 The charging of a capacitor, $\tau = RC$

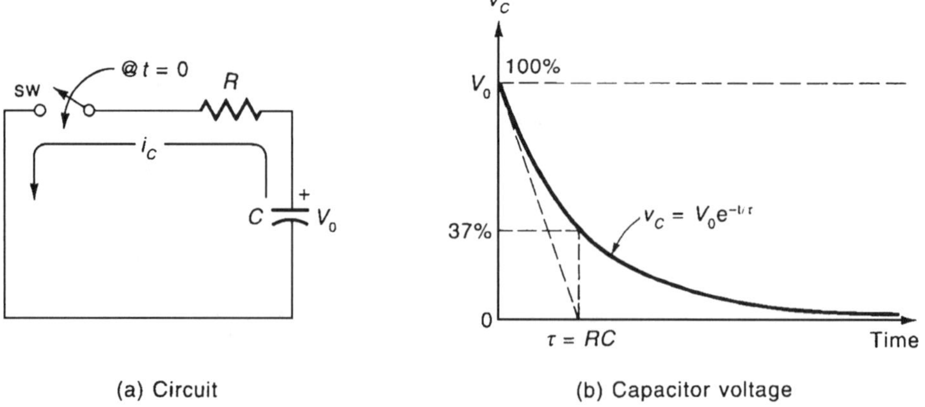

Figure 12.16 The discharging of a capacitor, $\tau = RC$

the time constant is equal to the value of *RC*. This value of e^{-1} is 0.3679, or approximately 37%; it is so often used in electrical transient analysis that it probably should be committed to memory. In Fig. 12.15, V_B is the final value of the voltage, and at $t = \tau$ the parenthetical quantity is approximately 63% so the variable is within 37% of its final value. Also note that in Fig. 12.16, the voltage decreases from the initial value, V_o, by 63% (or is within 37% of the final value) at $t = \tau$. If the initial slope of the curves in both Figures 12.15 and 12.16, were extrapolated forward, the intersection of the asymptotes with the horizontal axis would occur at $t = \tau$. To be realistic, "within 37%" is not good enough for engineering purposes, and a more practical approach is to consider the voltage at "almost" its final value in five time constants (well within 1%).

The equations in Fig. 12.15 and Fig. 12.16 are the solutions of the integral-differential equations that describe the circuit(s). Kirchhoff's voltage law states that sums of voltages around any closed loop must be equal to zero at *any* instant of time. In Fig. 12.15, the switch closes at $t = 0$; the voltage around the loop (for $t > 0$) is $V_B = Ri + v_c$, where v_c is q/C. Since current is defined as dq/dt, v_c is the integral of the present expression. If the initial voltage (or charge) on the capacitor was zero at the instant before the switch was closed, then $V_B = Ri + q/C = Ri + (I/C)\int_0^t i\,dt$.

The solution to this equation is

$$v_c(t) = v_c(0)e^{-t/RC} + V(I - e^{-t/RC}) \qquad \textbf{(12.24a)}$$

and Eq. (12.24a) frequently reduces to Eq. (12.24b)

$$v_C = v_B(1 - e^{-t/\tau}) \qquad \textbf{(12.24b)}$$

where $\tau = RC$. If t goes to infinity, then $v_C \rightarrow V_B$. If the initial value was not zero, as in Fig. 12.16, it must also be considered. The differential equation used to obtain the solution for Fig.12.16(b) may be found by the node voltage method; use the lower junction of *RC* as the reference node, and the currents to the top voltage node are given by $i_C + i_R = 0 = C(dv/dt) + v/R$. Knowing the initial voltage, $V_C(0^-)$, the voltage across the capacitor just before the switch was closed, and that $V_C(0^-) = V_C(0) = V_C(0^+)$, the differential equation is easily solved. The solution, where $\tau = RC$, is $v_C = V_C(0)e^{-t/\tau}$.

A somewhat more comprehensive example involving a multiple time constant circuit is presented in Fig. 12.17. The switch is assumed to be open for a long time (so long that any previous voltage across the capacitor has been reduced to zero because of R_3). The switch is then moved to the middle position for five seconds ($0 < t < 5$); it will then be switched to the lower position and remain there; this lower position will have a time, t', that starts at the time of switching to this new position. It is desired to know the current through the capacitor at $t =$ 9.8 s ($t' = 4.8$ s). Although this problem may seem more comprehensive than most, the reader should follow the details of the solution to note simplification techniques. First, it should be obvious that the problem may be broken into two parts. The first part spans the first 5 seconds, and the second runs from $t = 5$ (or $t' > 0$) seconds. When the switch is in the middle position, one needs to make a Thevenin equivalent circuit (refer to Fig. 12.18) to the left of × – × and then solve for the capacitor voltage at $t = 5$s as done on the previous problem. This voltage at $t = 5$ s is easily computed from the capacitor charge equation as $v_C(t = 5) = 7.5(1 - e^{-5/3}) = 6.083$ volts.

One could almost obtain this value from a sketch, as at one time constant (3 seconds) the voltage would be 63% of 7.5 volts (4.72 volts), and at two time constants (6 seconds) the voltage would be 87% of 7.5 volts (6.49 volts). Thus at 5/3 time constant, a plot would yield about 6.1 volts. Of course the final voltage of Fig. 12.18 is the initial voltage when the switch is moved from the middle to the bottom position in Fig. 12.19. The voltage at $t' = 4.8$ ($t = 9.8$) seconds is 37% of the initial value, $0.37 \times 6.08 = 2.25$ volts. The current at this instant of time is $v_C/R_{Eq} = 1.87\ \mu$A.

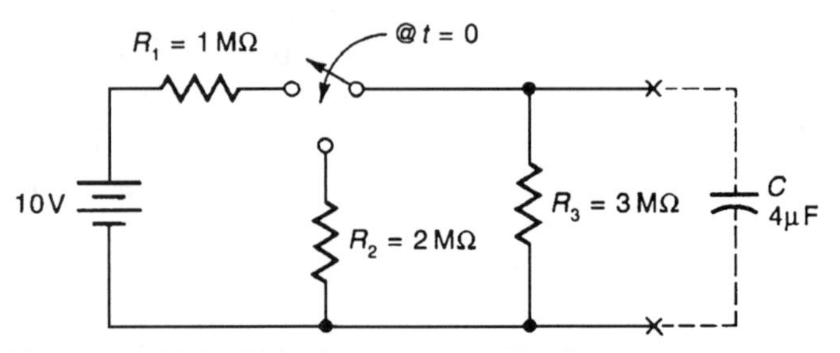

Figure 12.17 Multiple time constant circuit

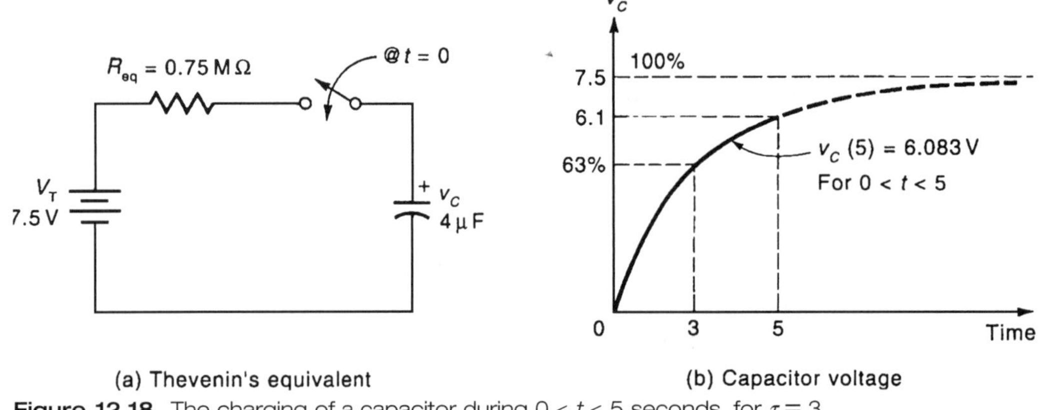

Figure 12.18 The charging of a capacitor during $0 < t < 5$ seconds, for $\tau = 3$

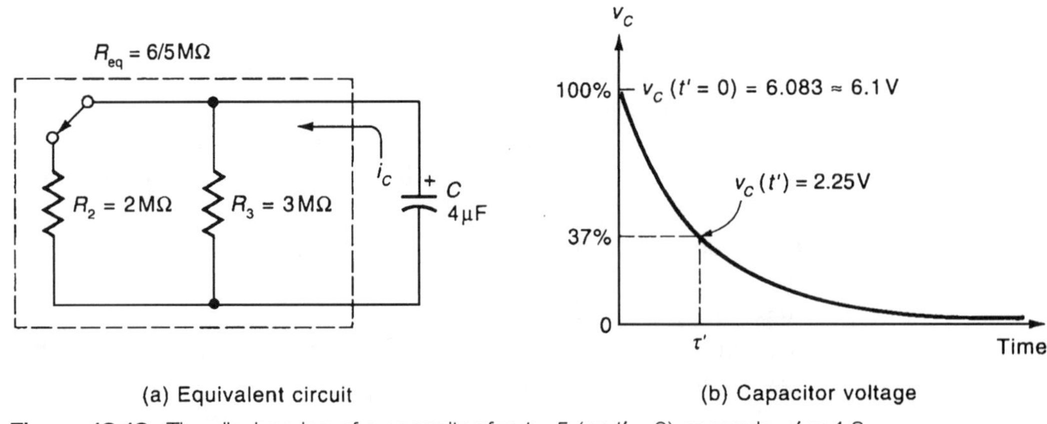

Figure 12.19 The discharging of a capacitor for $t > 5$ (or $t' > 0$) seconds, $\tau' = 4.8$

Inductive Circuits

When an inductor (instead of a capacitor) is in a circuit, the current through the inductor takes time to change (like a voltage across a capacitor) and the circuit may be treated much like the previous voltage-capacitor relationship,

$$i(t) = i(0)e^{-Rt/L} + V/R(1 - e^{-Rt/L}) \quad \textbf{(12.25)}$$

An example *RL* circuit problem is presented in Fig. 12.20. Here, the time constant, τ, is L/R. Assume zero initial current in the inductor. A voltage will be induced across an inductor equal to $L(di/dt)$ because of the rate of change of magnetic flux through a coil or around a wire. The polarity of this voltage depends on whether the flux is increasing or decreasing. Unlike a capacitor, where the capacitor acts as an open circuit (that is, no current flow) after the *voltage* stabilizes, the inductor acts like a short circuit after the *current* stabilizes. Thus the final value of the current in Fig. 12.20 is the applied voltage divided by the series resistance.

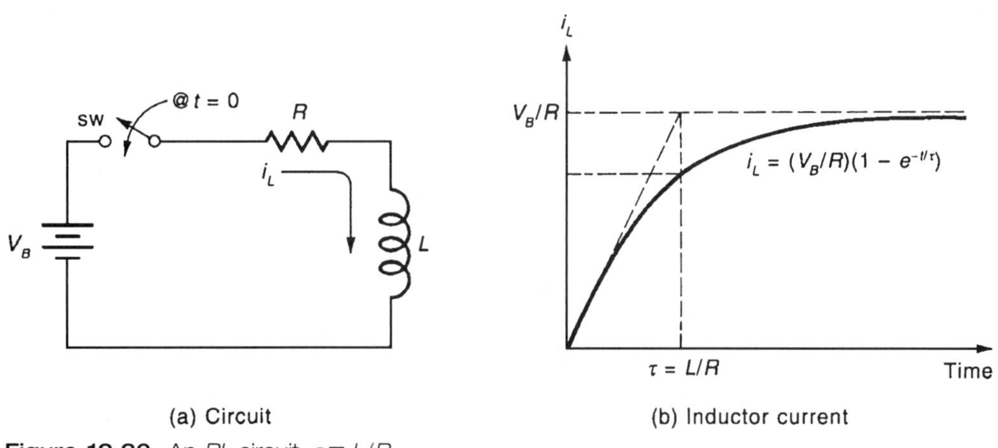

Figure 12.20 An *RL* circuit, $\tau = L/R$

Example 12.6

An inductor has an initial current caused by the switch in Exhibit 7 being connected to a voltage source for a long period of time and then suddenly switched to the open position at $t = 0$. Find the current at $t = 0.1$ seconds and the voltage across the inductor at the same time.

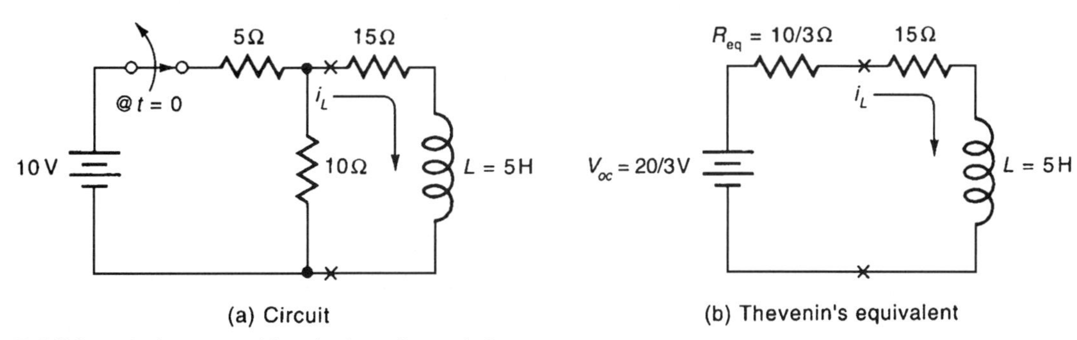

Exhibit 7 Inductor problem before the switch opens

Solution

Before the switch is open, the easy way to solve the problem is to make a Thevenin equivalent circuit of that portion of the circuit to the left of × – ×. While the switch is connected to the voltage source, the polarity of the voltage across the inductor is positive at the upper terminal of the inductor (while the voltage source is providing an increasing current in the inductor). Just before the switch opens the current in the inductor is $I(0^-)$. Because current cannot change instantaneously, $I_0(0^+)$ is the same current immediately after the switch is opened. However, the voltage polarity switches immediately as the energy stored in the inductor causes the current to continue to flow in the same direction (that is, $L(di/dt)$ acts like a voltage source that is strong enough to keep the current flowing). The current plot is given in Exhibit 8. The voltage is applied for a long period of time, so any rates of change, $L(di/dt)$, are completed; the inductor acts like a short circuit current and is easily found to be 0.3636. At $t(0)$, this current is the initial current $I(0) = I_0$ for the equation in Exhibit 8(c). The current at $t = 0.1$ second is found directly from the exponential equation where the time constant, $\tau = L/R$, has a resistance equal to the sum of the two resistors in series, $R = 25\ \Omega$, so $\tau = (5/25) = 0.2$ seconds. This current is $i = I_0 e^{-t/\tau} = 0.3636e^{-(0.1/0.2)} = 0.220$ A.

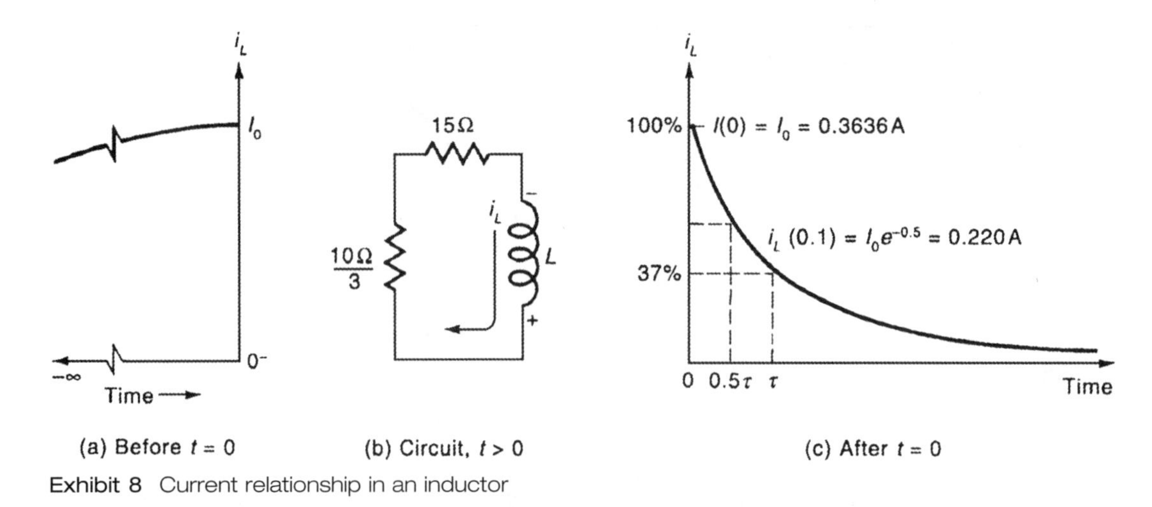

(a) Before $t = 0$ (b) Circuit, $t > 0$ (c) After $t = 0$

Exhibit 8 Current relationship in an inductor

The voltage across $L(di/dt)$ must be the same as the voltage drop across the resistors, since the loop is closed. Thus at $t = 0.2$ seconds, the voltage is $iR = 5.50$ volts; the polarity of this voltage is with the plus sign at the bottom of the inductance.

For an *RLC* circuit that involves two energy storage elements, *L* and *C*, the solution is complex. The governing equation is a second order differential equation whose solution may involve complex conjugate parameters.

AC CIRCUITS

In this review, the emphasis is on easily visualized graphical solutions rather than formal mathematical techniques. The review is based on assumptions that should be understood before the material is presented. The assumptions are the following:

- All sources are sinusoidal and are of the same frequency in any circuit unless specifically noted. For certain kinds of circuits, such as filters and resonant circuits, frequency changes will be made and will be considered in steady state and of the same frequency.

- Voltages and/or currents are considered at steady state. Any transients that may have resulted from the sudden closing of a switch are not considered. This assumption greatly simplifies the analysis; differential equation solutions will be bypassed and are replaced with phase-shifting operators. Although switch closing may be part of a circuit, it is assumed that one waits a short period of time (actually five time constants or more) for any transient or nonsinusoidal effects to die out before analyzing the circuit.
- Current or voltage sources are single phase unless otherwise noted (such as in three-phase circuit analysis).
- Circuits are linear. If a sinusoidal signal is applied to a circuit, then a current or voltage measured anyplace in the circuit will also be sinusoidal (with no harmonics). The only difference between an input and an output sinusoid will be a possible change in magnitude and phase.

An oversimplified pictorial (see Fig. 12.21) will help one visualize a sinusoidal voltage source along with some of the above assumptions. The oversimplification is that of showing only one loop of wire in a magnetic field producing a pure sine wave. The frequency of the sine wave is directly proportional to shaft speed. And, the time for one revolution for a single pair of poles (called the period, T) is the inverse of the frequency. The wave is periodic and continuous and may be represented with a phasor.

Consider how one could draw a sine wave rather accurately by placing the tip of a pencil at the center of an X–Y plot (start with it laying horizontally to right of the origin) and then allowing it to rotate counterclockwise through 360°. If one were to view the eraser from afar to the left and then to project horizontally the tip of the eraser onto the vertical axis of a time plot, the projection would trace out a pure sine wave (see Fig. 12.22); the pencil could be thought of as a phasor, a two-dimensional vector. This phasor (length) represents the maximum value (height) of the sine wave (it is usually more convenient to substitute the rms value rather than the maximum value for the length). This maximum value will, at first, be used to demonstrate how the rotating phasor is visualized. When using these phasors, one's real concern is a time "snapshot" of the relative positions of the phasors. Also, rather than plotting the voltage vs. θ as in Fig. 12.21 and 12.22, it is more convenient to plot it against ωt or $2\pi ft$. The angle the phasor passes through, with respect to some reference position or another phasor, represents a phase angle.

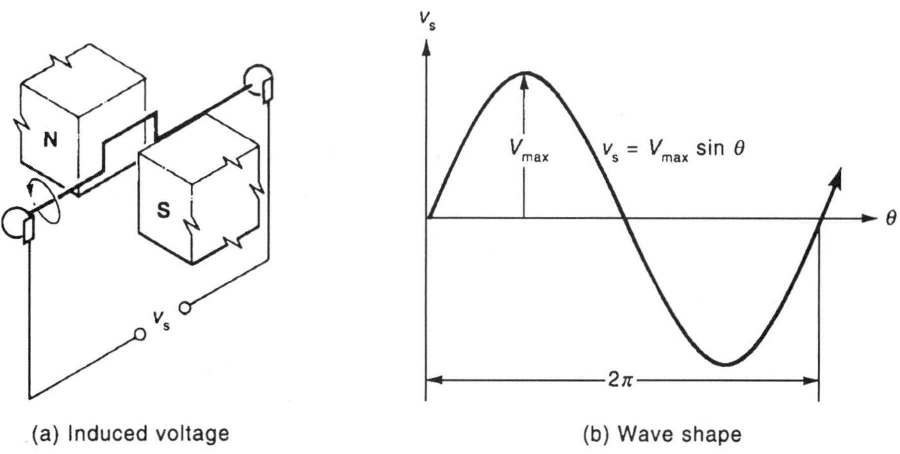

Figure 12.21 Simplified ac voltage source

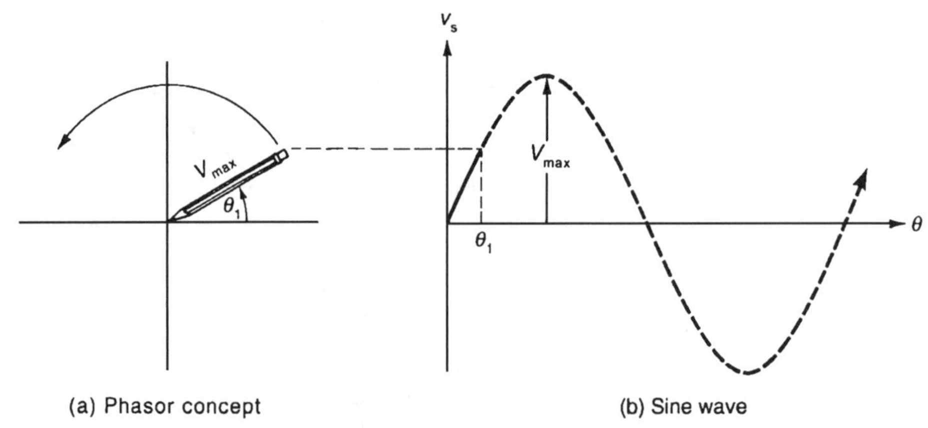

(a) Phasor concept (b) Sine wave

Figure 12.22 Graphical production of a sine wave

As an example of the simplicity of ac circuit analysis when using this phasor representation, consider summing two ac voltage sources graphically as in Fig. 12.23(b). Here, one would sum the two voltages and continually draw the sum of the two waves; the result is

$$v_{\text{total}} = V_{\text{max1}} \sin \omega t + V_{\text{max2}} \sin(\omega t + \alpha) = V_{\text{total}} \sin(\omega t - \theta)$$

This graphic summing produces the desired total voltage and also displays the phase relationship with respect to either of the original sine waves. However, in Fig. 12.23(c), the parallelogram formed from the two phasors yields the resulting phasor (the length is $v_{\text{max total}}$) much easier than summing the two sine waves. Furthermore, it gives the phase angle directly. The result is the same, in terms of relative magnitude and phase angle, for any instant of time. A formal mathematical description is

$$v_{\text{total}} = Im[V_{\text{max total}} e^{j(\omega t - \theta)}] \tag{12.26}$$

The notation "*Im*" implies the imaginary part of the expression that follows. A much simpler descriptive notation that most engineers use is to describe the representative phasor by a magnitude and its phase angle as $\mathbf{V}_{\text{max total}} \angle{-\theta}$. In electrical engineering, a positive rotation of the phasor is considered to be counterclockwise. Thus, in Fig. 12.23, the parallelogram resultant phasor may be listed as being θ degrees behind (or lagging) the phasor V_1 or as ϕ degrees ahead (or leading) V_2.

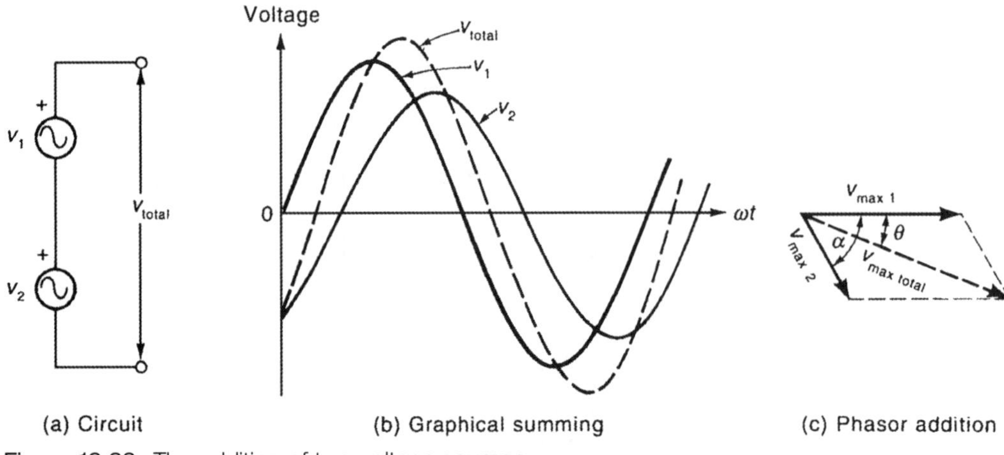

(a) Circuit (b) Graphical summing (c) Phasor addition

Figure 12.23 The addition of two voltage sources

Phasor Manipulation

Although adding two phasors is easily accomplished with the parallelogram, adding more than two might better be done by breaking each phasor into its x and y components and then summing the components. Multiplying phasors is done by multiplying the magnitudes and summing the angles, and division is done by dividing the magnitudes and subtracting the angles. As an example, two phasors $\mathbf{A} = 10\angle 0°$ and $\mathbf{B} = 5\angle 90°$ sum to a resultant of $\mathbf{R} = 11.2\angle 26.6°$. And $\mathbf{AB} = 50\angle 90°$; or $\mathbf{A/B} = 2\angle -90°$. The concept of rotation is important. Assigning phasor **B** to be **B** $10\angle 180°$, then $\mathbf{B} = -\mathbf{A}$; one could say that **B** is the same as **A** if **A** were rotated through 180°. Here, multiplying phasor **A** by –1 is considered as operating on **A** to rotate it through 180°. One could then say that multiplying a phasor by the square root of –1 is also an operator that rotates the phasor through 90°. This imaginary value, the square root of –1, is referred to as j. The operator j then rotates the phasor by 90°. As an example, the only difference between a sine wave and a cosine wave (for the same amplitude) is that a sine wave is shifted through 90° to become a cosine wave,

$$\cos\theta = \sin(\theta + 90°) = j \sin\theta \qquad \textbf{(12.27)}$$

Because of the effect of Eq. (12.27), it will be more convenient to use the real (x-axis) and imaginary j (y-axis) notation when dealing with phasors.

For most ac voltages or currents, one is usually interested in rms values rather than peak or maximum quantities. It was previously shown that the rms (or in Eqs. (12.20) and (12.21) "what a meter would read") value of a sine wave is its maximum value divided by the square root of two (or $V_{rms} = 0.707V_{max}$). Thus, one usually begins with rms magnitudes when using phasors. This makes voltage, current, and (especially) power calculations much simpler; the notation is even easier (the rms subscripts are dropped). In ac circuits there are really only three important devices, resistors (energy dissipating element), capacitors (energy storage element), and inductors (also an energy storage element).

> Since all phasors are relative, plot phasor lengths as rms values rather than peak values.

Resistors

As in dc circuits, this resistive element dissipates power equal to the current squared times the resistance. This is not the only equation for power, but *it is the safest equation to use in ac circuits* unless one is very careful with notation and is experienced in ac circuit theory. Also, the voltage across a resistor is always in phase with the current through the resistor. The resistor stores no energy but dissipates the power in terms of heat.

> Current and voltage are in phase for a resistor. $\mathbf{V}_R = R\mathbf{I}$

CAPACITORS

The capacitor stores energy for half a cycle and gives it up on the other half; the net (real) energy is zero. However, it is sometimes convenient to refer to the product of voltage and current as imaginary power (using the symbol Q and calling it reactive power). The result of this energy interchange is that the current through the capacitor has a 90° phase difference for the two phasors; the current is ahead of the voltage (or leads the voltage). Whereas the resistance, R, impedes the flow of current, the impending quantity for a capacitor is the reactance, called X_C. Not only does X_C impede the current, but it also causes a phase shift of –90° so that the voltage across the capacitor is lagging the current by 90°; these descriptive words may be replaced by the symbol $-j$.

> The voltage drop across a capacitor is 90° behind the current.
>
> $$\mathbf{V}_C = -jX_C\mathbf{I}, \quad \text{where } X_C = \frac{1}{2\pi f C}$$

The instantaneous voltage across a capacitor for a given current is

$$v = 1/C \int i\, dt \quad \textbf{(12.28a)}$$

and if the given current is $i = I_{max} \sin \omega t$ then

$$v = 1/C \int I_{max} \sin \omega t\, dt \quad \textbf{(12.28b)}$$

and the solution is

$$v = [-1/(\omega C)] I_{max} \cos \omega t = [1/(\omega C)] I_{max} (-j) \sin \omega t \quad \textbf{(12.28c)}$$

Using phasor notation, one can show that Eq. (12.28c) is equivalent to

$$\mathbf{V}_C = -jX_C\mathbf{I}, \quad \text{where } X_C = 1/(\omega C). \quad \textbf{(12.29)}$$

It is obvious that a numerical value of X_C is a function of frequency. In fact, if the frequency is infinite, X_C is zero as in a short circuit; and if the frequency is zero, X_C is infinity as in an open circuit. This relationship between current and voltage is shown graphically in Fig. 12.24.

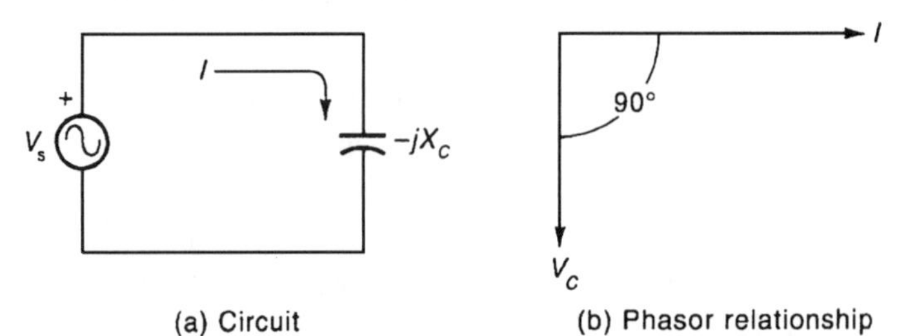

(a) Circuit (b) Phasor relationship

Figure 12.24 Voltage-current relationship for a capacitor. (The voltage and current are not to the same scale.)

Example 12.7

Consider a series R–C circuit whose measured current is 5 amperes (rms). Determine the voltage source required to produce this current. Assume that the voltage source is known to have a frequency of 60 Hz, with $R = 10$ ohms and $C = (0.1/377)$ farads. The factor $2\pi f$ for 60 Hz is 377.

Solution

The effect of the capacitor is to create a 90° phase shift between the current and voltage while the resistor produces no phase shift. The instantaneous Kirchhoff voltage equation is

$$v_{\text{total}} = Ri + v_C = Ri + (1/C)\int i\,dt \qquad \textbf{(12.30a)}$$

Rather than solving the above equation, phasors will be used here in the form

$$\mathbf{V}_s = \mathbf{V}_R + \mathbf{V}_C = R\mathbf{I} - jX_C\mathbf{I} \qquad \textbf{(12.30b)}$$

Now R and jX_C should be viewed as operators (they operate on the current to produce a voltage—and the j is to produce a ninety degree phase shift). Then X_C is calculated from $1/(2\pi fC)$ to be 10 ohms.

$$\mathbf{V}_s = 10\mathbf{I} - j10\mathbf{I} \qquad \textbf{(12.30c)}$$

Remember, the $-j10$ operator multiplies the current phasor by ten and shifts it through −90°. With the two voltage phasors, one only needs the parallelogram construction to get the voltage source phasor along with its angular relationship to the current, as shown in Exhibit 9.

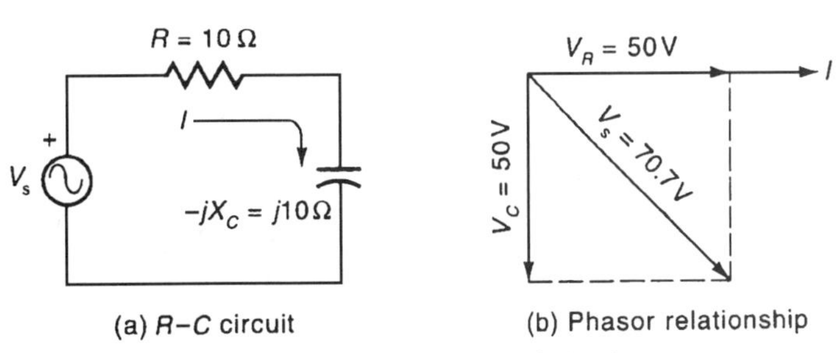

Exhibit 9 Voltage relationship for an *R–C* circuit

The voltage source is 70.7 volts with the voltage phasor lagging 45° behind the current. Or if the voltage source is considered to be the reference, one could say that the current leads the voltage by 45°. Observe that all of the (real) power from the source is dissipated by heat loss in the resistor (no real power is dissipated in the capacitor):

$$P_R = I^2R = 5^2(10) = 250 \text{ watts}$$

The power from the source, P_s, is the product of voltage and current times the power factor, which is defined as the cosine of the angle between the voltage and current. Thus

$$P_s = VI\cos 45° = 70.7 \times 5 \times 0.707 = 250 \text{ watts}$$

The reactive power, Q, is calculated as follows:

$$P_{\text{reactive}} = Q = VI\sin\theta = 70.7 \times 5 \times 0.707 = 250 \text{ VARs}$$

(*Note:* P_s and Q match only because θ is 45°.)

Inductors

Like the capacitor, the inductor stores energy for half a cycle and returns it to the circuit for the next half cycle. In circuit analysis, the inductor behaves in a similar manner as a capacitor except the sign is reversed; here the term is $+jX_L$ where X_L is directly proportional to the frequency and is equal to $\omega_L = 2\pi fL$. For the inductor the voltage leads the current by 90°, or the current lags the voltage.

> A voltage drop across an inductor leads the current by 90°.
>
> $$\mathbf{V}_L = jX_L\mathbf{I}, \quad \text{where } X_L = 2\pi fL$$

Also, at zero frequency, the reactance is zero; and for an infinite frequency, the reactance is infinite. The phase relationship is shown in Fig. 12.25.

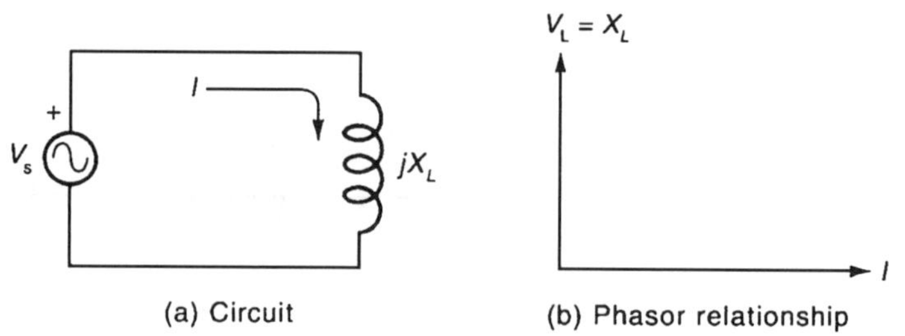

Figure 12.25 Voltage-current relationship for an inductor

Example 12.8

Consider the circuit of Exhibit 10, where the current is, as before, 5 amperes, R is 10 Ω, and $L = 10/377$ henries.

Solution

Here X_L is ohms (Exhibit 10). The answers are unchanged from Example 6 except for the phase relationships.

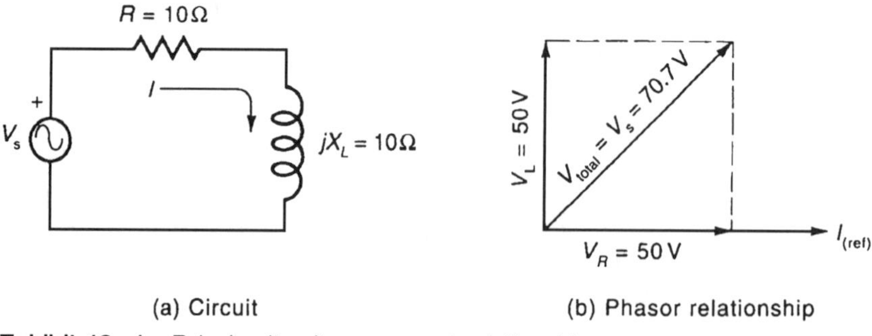

Exhibit 10 An *R-L* circuit voltage-current relationship

Impedance Relationship

In general, ac voltages and currents are related by a quantity called impedance (Z). It is usually complex, and it is sometimes referred to as Ohm's law for ac circuits. This association is easily explained by reference to either of the two

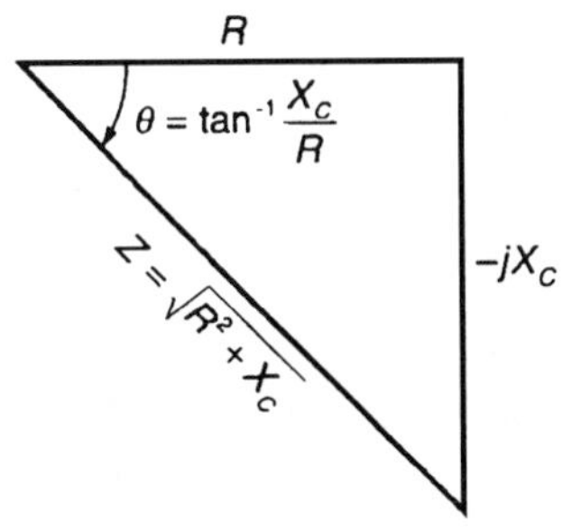

Figure 12.26 The impedance relationship

previous example problems. For instance, in Exhibit 9 and Eq. (12.30), the current was assumed to be known. If the current were unknown and the source voltage was given as 70.7 volts, then $\mathbf{V}_s = 70.7\angle 0°$ volts can be the known reference phasor. From Eq. (12.30b), the unknown current is easily factored out of the expression to give $\mathbf{V}_s = (R - jX_C)\mathbf{I}$.

The parenthetical quantity—or the operators—for this problem, as in Eq. (30c), is the complex impedance. It has a right-triangle relationship as in Fig. 12.26 and is given by $\mathbf{Z} = R - jX_C = 10 - j10 = 10\sqrt{2}\ \angle -45°$.

The $10 - j10$ is known as the rectangular form, and the $10\sqrt{2}\ \angle -45°$ is known as the polar form. The rectangular form is usually used when adding or subtracting is involved, and the polar form is used when one is multiplying or dividing. To finish the problem, the current is found as $\mathbf{I} = \mathbf{V}_s/\mathbf{Z} = (70.7\angle 0°)/(10\sqrt{2}\ \angle -45°) = \angle +45°$ A.

Example 12.9

Another series circuit problem will expand on the complex impedance concept; see Exhibit 11. Here, all of the resistances and reactances and the source voltage are given for a particular frequency. It is desired to find the power dissipated in the 20-ohm load resistance and also to find the total power from the source voltage. Since the circuit is a series one, the current is common for all elements; therefore the current should be found. The voltage source is 100 volts.

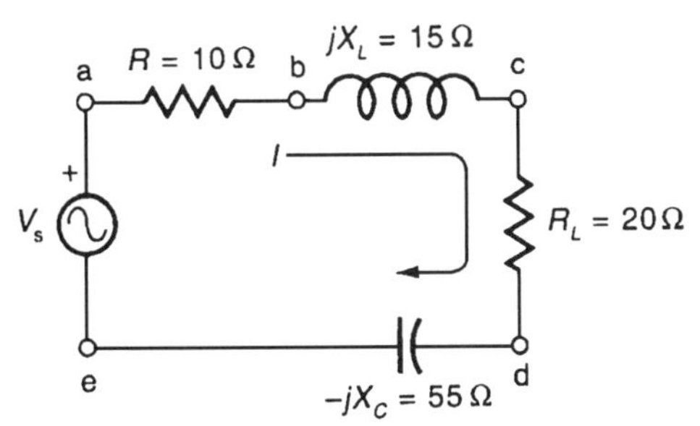

Exhibit 11 A series circuit with known parameters

Solution

Find the total impedance to find the current (see Exhibit 12): $Z = 10 + j15 + 20 - j55 = (10 + 20) + j(15 - 55) = 30 - j40 = 50\ \angle -53.1°\ \Omega$.

From this complex impedance, current and power can be found directly:

$$\mathbf{I} = \mathbf{V}/\mathbf{Z} = (100\angle 0°)/(50\angle -53.1°) = 2\ \angle +53.1° \text{ amperes}$$
$$P_{RL} = I^2R_L = 2^2(20) = 80 \text{ watts}$$
$$P_s = V_sI\cos\theta = 100 \times 2\cos(-53.1°) = 120 \text{ watts}$$

(*Check:* $P_s = \Sigma I^2R = 2^2(10 + 20) = 120$ watts.)

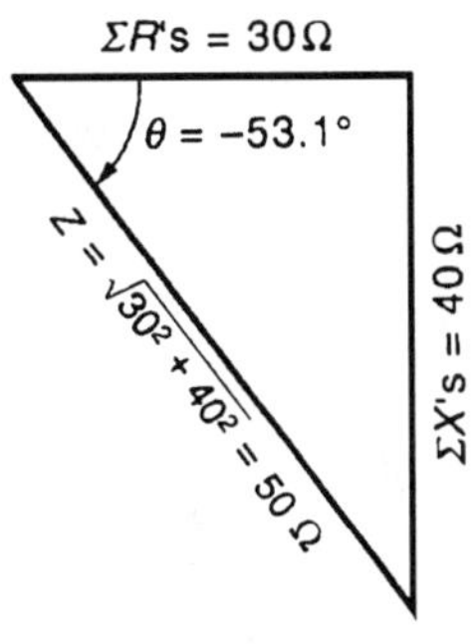

Exhibit 12 Impedance triangle

Example 12.10

Suppose, in Example 9, one is asked to find the magnitude and phase of the voltage across two points—say the voltage from point b to point d.

Solution

The phasor plot readily yields this answer. Since the current is common to all elements, the current should be made the reference phasor. Again, the currents and voltages are not necessarily plotted to the same scale. Rather than having all phasors emanating from the origin, it is convenient to plot the voltages in a cumulative fashion, from head to tail (see Exhibit 13).

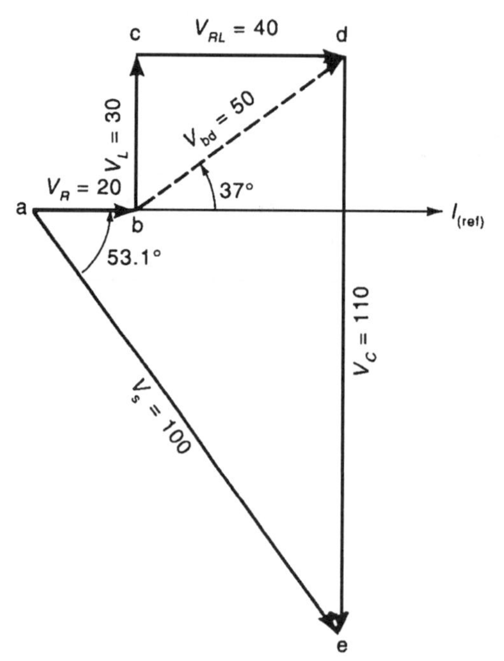

Exhibit 13 Plot of voltages vs. current

The voltage, V_{bd}, may be read directly from the plot to be 50 volts at an angle of 37° ahead of the current. An interesting aspect of this problem is brought out if the frequency of the source is doubled. Doubling the frequency causes no change in the resistive elements but dramatic changes in the reactive ones. For the inductor, X_L is 15 × 2 = 30 ohms and for the capacitor X_C = 0.5 × 55 = 27.5 ohms. The impedance then becomes $\mathbf{Z} = 10 + j30 + 20 - j27.5 = 30 + j2.5 = 30.1\angle 4.76°$ ohms. The current is $\mathbf{I} = (100\angle 0°)/30.1\angle 4.76° = 3.32\angle{-4.76°}$ A, and the power dissipated in the 20-ohm resistor is

$$P_{RL} = I^2 R_L = 3.32^2 \times 20 = 220 \text{ watts}$$

At this point, several practical observations are in order. For approximate answers, if either the real or imaginary part is more than ten times the other, the hypotenuse is almost equal to the longest leg, and the angle is within a few degrees of being zero. Using this approximation for impedance, the approximate current is 3.3 amperes and P_{RL} is 222 watts. Another observation is that by changing the frequency, the power output has gone from only 80 watts to well over 200. There is a specific frequency to get maximum current and power; this is called the resonant frequency.

Resonant Frequency

For series circuits, the resonant frequency occurs when all of the reactive components cancel so that $X_L = X_C$, $\omega L = 1/\omega C$. For this particular frequency, the subscript notation may be given as "res", "0", or "n".

Series Resonant Frequency

$$\omega_0 = 2\pi f_0 = \frac{1}{\sqrt{LC}}$$

The series resonant frequency concept leads directly to the maximum power transfer theorem, which should now be obvious. Maximum power occurs when the reactive components in an ac series circuit cancel.

Parallel ac circuits are no more difficult than series ones but may require slightly more bookkeeping; also, since many reciprocals are involved, additional definitions may be used to identify these reciprocals. As an example, the reciprocal of impedance, Z, is admittance, Y; the units of Y are siemens (old notation mhos) abbreviated S. The admittance may be further expanded to have a real and imaginary part. The real part is called the conductance, G, and the imaginary part is called the susceptance, B; both parts have the units of siemens. However, for simple problems, these definitions are optional; consider the example shown in Fig. 12.27.

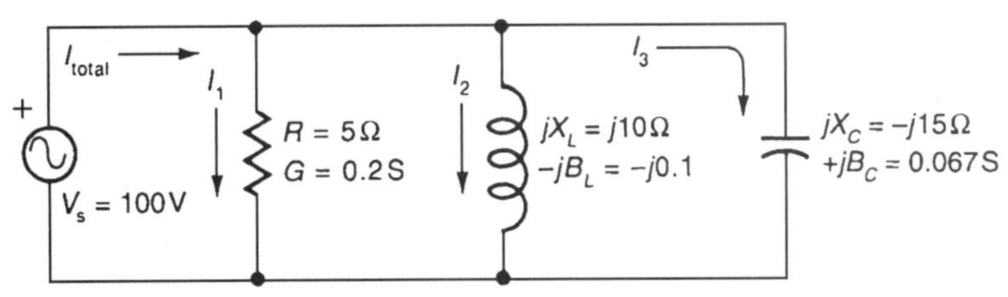

Figure 12.27 A parallel circuit

Example **12.11**

Solution

The equivalent impedance for this problem may be found by either dividing the source voltage by the total current (see Exhibit 14) or by taking the reciprocal of the total admittance.

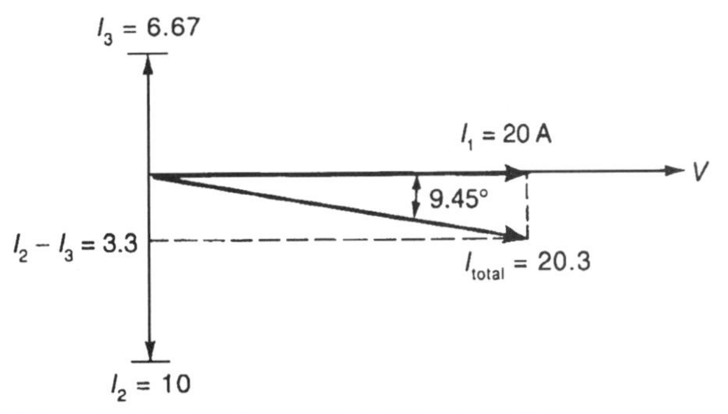

Exhibit 14 The phasor relationship for Fig. 12.27

$$\mathbf{I}_{\text{total}} = \mathbf{I}_1 + \mathbf{I}_2 + \mathbf{I}_3$$

where

$\mathbf{I}_1 = \mathbf{V}/R = (100\angle 0°)/5 = 20\angle 0°$ A;
$\mathbf{I}_2 = \mathbf{V}/(jX_L) = (100\angle 0°)/(10\angle 90°) = 10\angle -90°$ A;
$\mathbf{I}_3 = \mathbf{V}/(-jX_C) = (100\angle 0°)/(15\angle -90°) = 6.67\angle 90°$ A;
$\mathbf{I}_{\text{total}} = 20 - j10 + j6.67 = 20 - j3.33 = 20.28\angle -9.46°$

$$\mathbf{Z} = \mathbf{V}/\mathbf{I} = (100\angle 0°)/(20.28\angle -9.46°) = 4.93\angle 9.46° \text{ ohms}$$

However, using admittances, Y may be found directly:

$$G = 0.20 \text{ S}, \quad -jB_L = -j0.10 \text{ S}, \quad jB_C = j0.066 \text{ S}$$
$$\mathbf{Y} = G - jX_L + jX_C = 0.20 - j0.10 + j0.066 = 0.20 - j0.03 = 0.203\angle -9.45° \text{ S}$$
$$\mathbf{Z} = 1/\mathbf{Y} = 1/(0.203\angle -9.45°) = 4.93\angle 9.45° \text{ ohms}$$

Example 12.12

A final example of a combined series and parallel circuit is given in Exhibit 15. For this kind of problem, the reader is urged to make two separate plots to find the individual branch currents, then add the two current phasors on a third plot to find the total current. Find the power taken from the source.

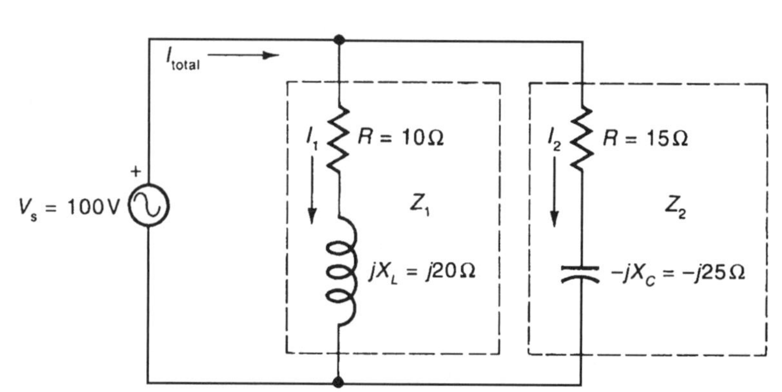

Exhibit 15 Combined series and parallel circuit

Solution

First, find the currents:

$$\mathbf{I}_1 = \mathbf{V}/\mathbf{Z}_1 = (100\angle 0°)/(10 + j20) = 4.47\angle -63.4° = 2.00 - j4.00$$
$$\mathbf{I}_2 = \mathbf{V}/\mathbf{Z}_2 = (100\angle 0°)/(15 - j25) = 3.43\angle 59.0° = 1.76 + j2.94$$
$$\mathbf{I}_{\text{total}} = \mathbf{I}_1 + \mathbf{I}_2 = (2.00 + 1.76) + j(-4.00 + 2.94) = 3.76 - j1.06 = 3.91\angle -15.7°$$

See Exhibit 16 for the phasor diagram. The power from the source is now found to be

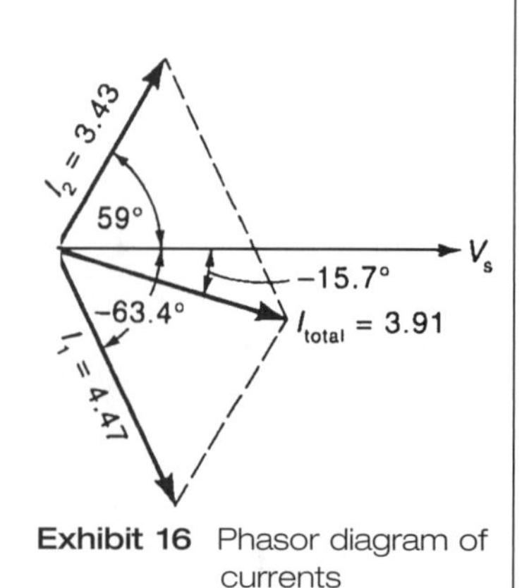

Exhibit 16 Phasor diagram of currents

$$P_s = \mathbf{V}\mathbf{I}_{\text{total}} \cos\theta = 100 \times 3.91 \times \cos(-15.7) = 376 \text{ watts}$$

(*Check:* $P_{\text{total}} = (\mathbf{I}_1)^2R_1 + (\mathbf{I}_2)^2R_2 = 200 + 176 = 376$ watts.)

Although calculations are straightforward, the reader is urged to sketch the phasor diagram. Again, for a multiple choice problem, the graphical solution may be sufficiently accurate for one to select the correct answer.

Quality Factor

For parallel resonance as for series resonance circuits, Z still equals R at resonant frequency. For the series circuit the inductive and capacitive reactances cancel; and for parallel circuits, the inductive and capacitive susceptances (in siemens) cancel. In describing the behavior of these resonant circuits another descriptive quantity is frequently used—especially so for parallel circuits—which is called the *quality factor*, Q, of the circuit. For a series circuit Q is a measure of the energy stored in an inductor compared to the energy dissipated in the resistance; this ratio may also be given as

$$Q_{\text{series}} = \omega_o L/R = 1/(\omega_o CR) \tag{12.31}$$

where Q is dimensionless and may be quite high for a sharp resonant peak as a narrow bandwidth circuit. On the other hand, for a parallel circuit, Q may be defined (see Fig. 12.27) as

$$Q_{\text{parallel}} = 1/Q_{\text{series}} = \omega_o RC = R/(\omega_o L) \tag{12.32}$$

and the bandwidth, BW, may easily be found as

$$\text{BW} = \omega_o/Q_{\text{parallel}} \text{ (rad/s)} \tag{12.33}$$

ELECTRONIC CIRCUITS

A thorough review of electronics is not undertaken here. But two topics—the solid state diode and the operational amplifier have been selected as deserving attention. The complexity of certain aspects of both solid state theory and integrated circuit theory is well beyond the scope of this review, so only the idealized devices will be dealt with.

Diodes

The solid state theory of diode operation depends on the particular kind of diode being considered. For example, for the junction diode, knowledge of solid state theory and the behavior of majority and minority carriers in the presence of an electric field is desirable. But when an idealized diode is treated, the full theory is not essential. For a p–n junction diode, a very simplistic explanation of the operation at the junction between a "p" and an "n" type semiconductor is that in each type of material there are many free charges available; these are holes (p, positive) and electrons (n, negative), respectively. If the diode is biased in favor of forcing positive charges near the boundary (the positive voltage being connected to the p material and the negative terminal to the n material), a rapid

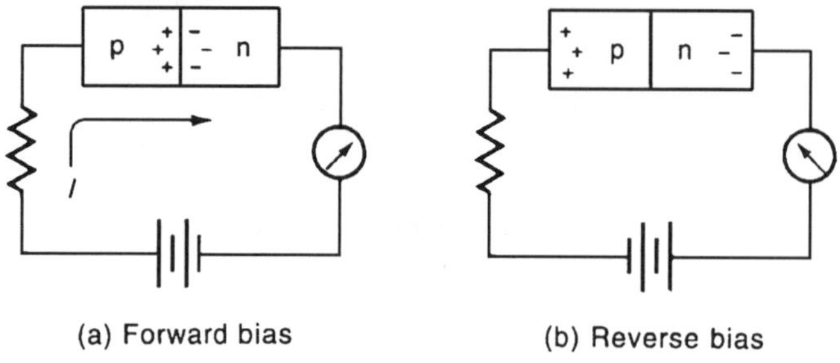

Figure 12.28 Charges inside a junction diode

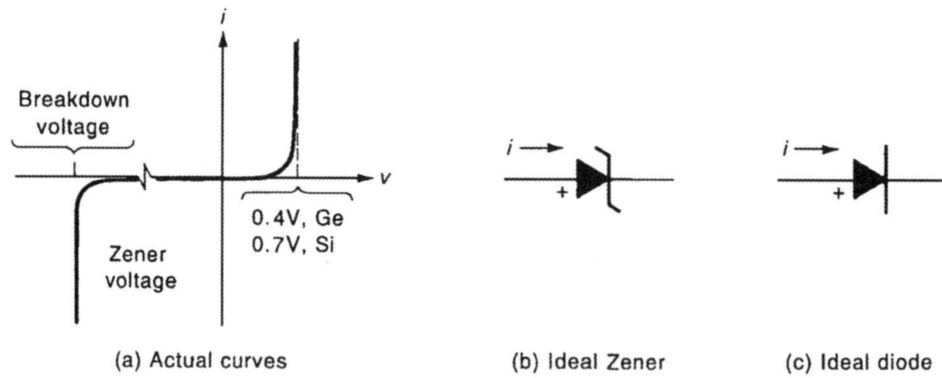

Figure 12.29 Junction diode voltage–current characteristics

recombining of the charges takes place [see Fig. 12.28(a)]; this is the direction of easy current flow. On the other hand, if the diode is reverse biased [Fig. 12.28(b)], the free charges are attracted toward their bias polarities, leaving a dearth of charges at the junction for very little recombination and almost no current flow. When the bias is in the forward direction, the voltage necessary to cause the charges to recombine is fairly small but still could be significant (see Fig. 12.29). The voltage is near half of one volt (approximately 0.4 V for germanium and near 0.7 for silicon) and is nonlinear. In the reverse direction, the current is essentially zero until breakdown voltage is reached; this breakdown is referred to as Zener voltage, which ranges from a few volts to several hundred volts. When designing circuits that use these diodes, care must be taken to work well within the reverse peak breakdown voltage. If one ignores the nonlinearity by assuming that the half volt is negligible for the forward direction, then the symbol shown in Fig. 12.29(b) represents the ideal diode (the arrowhead side is the anode, and the other side is the cathode). These diodes, of course, have many applications; in this review, the applications will be limited to the ideal case in pure rectifying circuits.

Rectifying Circuits

Most diode rectifying circuits are used to convert ac voltages to "pulsed dc" voltages (see Fig. 12.30). The diode is used to convert an ac voltage to a half-wave sinusoid so that the meter measurement found is the average current. Rectification is not limited to sinusoid; the following table (see Table 12.1) shows a sample of various kinds of signals and kinds of rectification, either full- or half-wave, relationships.

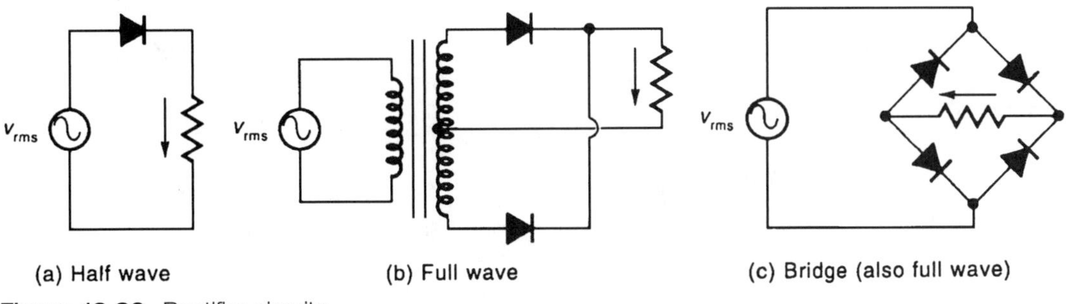

(a) Half wave (b) Full wave (c) Bridge (also full wave)

Figure 12.30 Rectifier circuits

Table 12.1 Rectified values for three signals

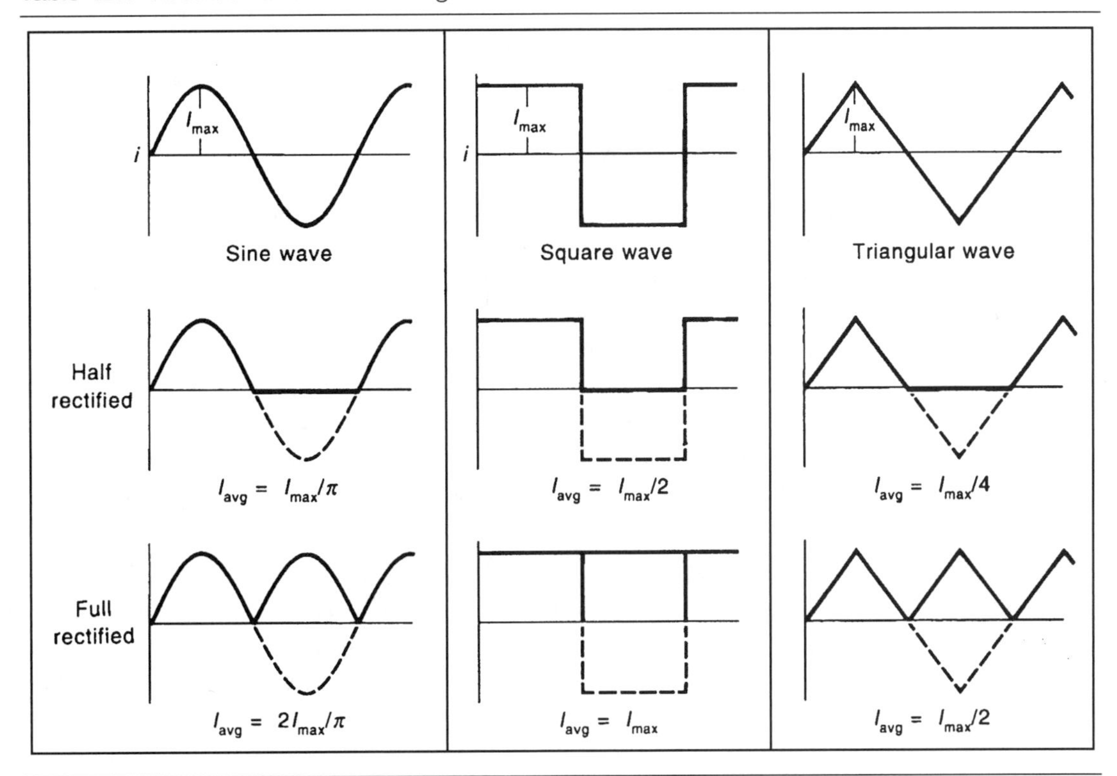

Example 12.13

A triangular wave form with a peak value of 50 volts is rectified by a half-wave rectifier and goes to a load resistor of 100 ohms (see Exhibit 17); the ammeter in the circuit is a dc type that measures average current. What is this current?

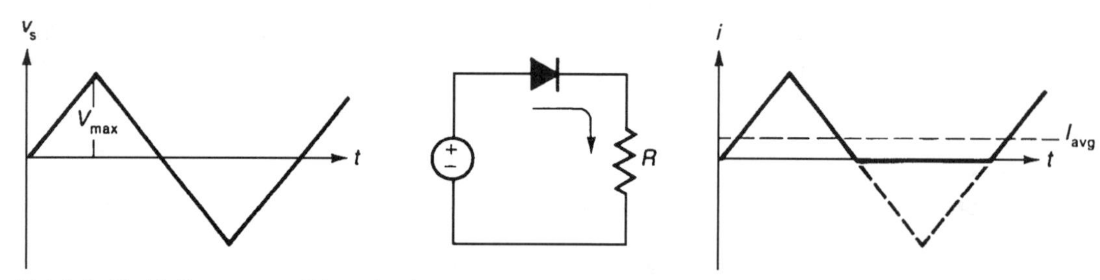

Exhibit 17 Half-wave rectifying circuit

Solution

The current is v/*R* and is found from the equation for average values as

$$I_{avg} = 1/R\left(1/T\int_0^T v\,dt\right) = 1/R\left[1/T\int_0^{T/2} v\,dt + 1/T\int_{T/2}^{T} 0\,dt\right] \qquad (12.34)$$

The integral is the area of the triangle over a half period divided by a full period, *T*.

$$I_{avg} = [(V_{max}/R)(0.5)(0.5T)]/T = (50/100)(1/4) = 0.125 \text{ A}$$

Or, the current can be found by reading directly from Table 12.1; the average reading is one-quarter of the peak (or maximum) value of the current.

Operational Amplifiers

The operational amplifier is a high gain differential amplifier circuit (see following list) that has been highly developed over the years. Since the cost has gone from a few hundred dollars (old vacuum tube era) to a few cents for highly developed integrated circuits, the applications for this device cover almost all areas of engineering instrumentation. This discussion focuses on an ideal operational amplifier that is treated externally as a black box. This operational amplifier (referred to as an "op-amp") differs from a normal amplifier. For example, a home hi-fi audio amplifier's frequency response is considered good if it amplifies voltages over a frequency range from approximately 20 Hz to 20,000 Hz. An op-amp has much higher amplification. It also has several other significant characteristics:

- The amplification is usually of the order 100,000 or more. It is based on the input voltage being the difference between two very small voltages, $v_d = v_p - v_n = v_1 - v_2$. These small voltages are designated as positive and negative; however, the actual polarity of the applied voltages could be either. This implies that the output voltage is zero if the two small input voltages are equal.
- The amplification is flat over a frequency range from zero Hz (that is, "down to dc") to some very high frequency (high compared to the highest frequency being amplified). Before the low-drift transistor, great care and clever circuit design was required to obtain a "dc amplification."
- The currents to the actual input pins (both positive and negative) are very small and may usually be neglected; this means that the input resistance (or impedance) is very high. Also, the output resistance (or impedance) is very low; again, "low" means relatively small compared with the external circuitry values.
- The device is linear over a known range, which means that the superposition theorem applies over a range of voltages (for example, if the input voltage were doubled, the output voltage would double).

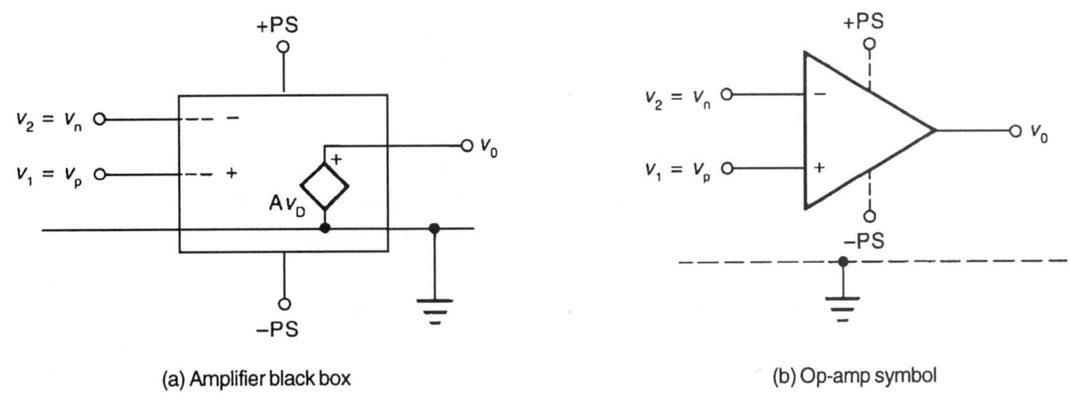

(a) Amplifier black box (b) Op-amp symbol

Figure 12.31 The operational amplifier

One may visualize an op-amp as the equivalent circuit shown in Fig. 12.31(a). In this figure, the positive and negative power supply (+PS, –PS) connections are shown, but, as in Fig. 12.31(b), they may not be shown. Since input voltages may be positive or negative, the power supply voltages are both relative to some common point, usually referred to as ground (signal ground). The power supply voltages must be larger than the largest expected output voltage within the limits of the op-amp. These constraints are normally assumed and are not usually stated or even shown on a diagram [see Fig. 12.31(b)]; the triangular symbol may have the vertical side shown as slightly rounded in some diagrams.

The op-amp is used in circuits designed for use in either the inverting or the noninverting mode. Consider the inverting mode circuit (see Fig. 12.32); here, since the two voltage input pins are assumed to go to an open circuit, the input currents for both the plus and minus pins are zero, and the voltage difference between these two pins is essentially zero. Since the gain is so high and output voltages are in the several-volts range, the input voltages are in the microvolt range. This makes the circuit analysis especially easy because if one knows one input voltage, the other is essentially the same (actually only a few microvolts different). So, if one input happens to be grounded [as in Fig. 12.32(a)], the other input is almost zero. The equations for the output voltage given in Fig. 12.32(a) are easily obtained by realizing that the current into the op-amp itself is zero. By summing currents at Node 1, i_a must equal $-i_f$ and the voltage at the junction is near zero volts. Since the node voltage is essentially zero, i_1 is v_a/R_1 and i_f is v_0/R_f. Thus, the circuit amplifies (by a factor of $-R_f/R_1$) as it inverts and has a relatively low current input at v_a (to the external circuit if R_1 is high) and whose output acts almost like an ideal voltage source. For more than one input, one simply sums the currents at Node 1 to yield the equation in Fig. 12.32(b).

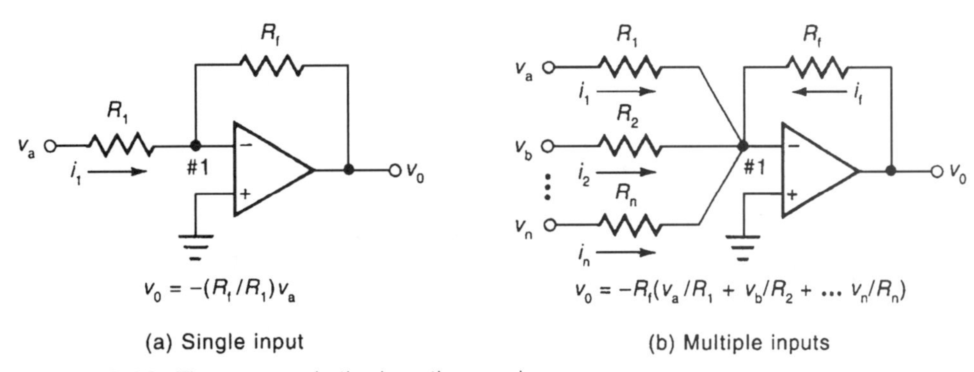

(a) Single input (b) Multiple inputs

Figure 12.32 The op-amp in the inverting mode

As an example problem [for Fig. 12.32(b)], four different transducers produce a possible maximum ac voltage of 0.1, 1.0, 5.0, and 10 volts, respectively. Each transducer is to be recorded on a one channel recorder whose desirable input signal level is one volt but whose input impedance will not allow a direct connection to the transducers. Since only one input at a time may be recorded, it is necessary only to determine each resistance ratio for a summing op-amp circuit. Summing is not required, but multiplying by a constant is. The ratios are easily found; in changing the level of the first voltage from 0.1 to 1, the ratio is ten. The common feedback resistor is arbitrarily chosen to be 100k ohms (the typical range of values runs from a few kilo-ohms to several megohms), then R_1 must be 10k ohms. Calculate the numerical values for the other resistor ratios shown in the circuit of Fig. 12.33. When using the op-amp as a summer (all inputs connected at the same time), the instantaneous algebraic sum of the input voltages multiplied by their resistor ratio values will be summed together. They should not exceed the specified linear output voltage.

The other standard circuit configuration for the op-amp is the noninverting mode. For this, the negative input pin is normally fed from the output [perhaps through a resistive network as in Fig. 12.34(b)], and the plus pin terminal is connected as an input. For the direct-feedback design [see Fig. 12.34(a)], the output voltage—being also the minus terminal input—must be within a few microvolts of the plus terminal. Thus, this circuit has an output that is essentially the same as the input, and this circuit has a gain of unity with no change in polarity. The circuit is usually referred to as a buffer or voltage follower; the advantages are that it takes almost no power from an input source (that is, it does not "load" the input), and the output is almost as though it were an ideal voltage source. For the circuit shown in Fig. 12.34(b), the minus input voltage must follow a certain percentage of the output and is considered a noninverting amplifier whose gain is $v_0/v_{in} = \text{Gain} = 1 + R_2/R_1$. As an example, if a nonverting op-amp with a gain of three is desired, the resistor ratio must be two. If R_2 is 10 kΩ, then R_1 must be 5 kΩ.

The key to solving these operational amplifier circuits is that the current input to the op-amp itself is considered to be zero. And, if either the plus or minus input voltage is known (or is a ratio of some other voltage), the other input voltage is considered to be the same. The application of Kirchhoff's laws to the rest of the external circuit will usually yield the correct answer.

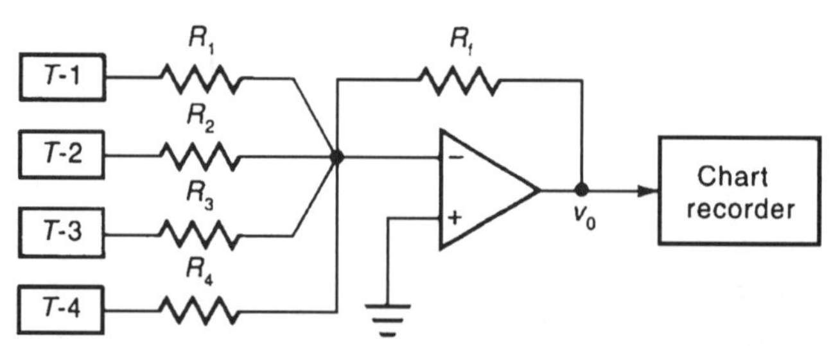

Figure 12.33 An op-amp circuit for matching voltage levels of various transducers. (*Caution:* only one switch to be closed at any one time.)

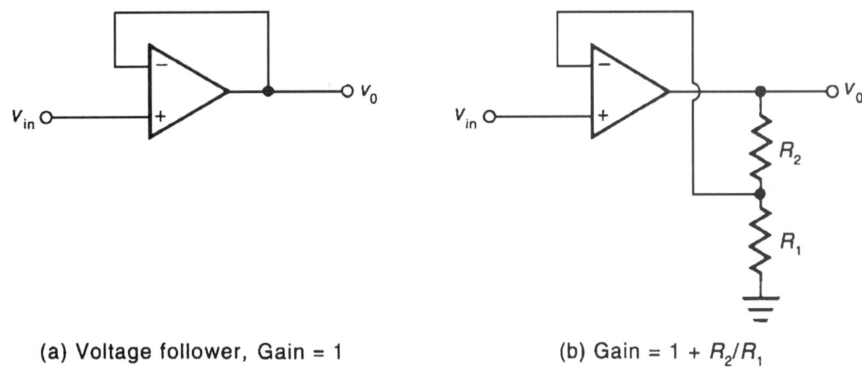

(a) Voltage follower, Gain = 1 (b) Gain = 1 + R_2/R_1

Figure 12.34 Noninverting op-amp circuits

TRANSFORMERS

The transformer is principally used in ac circuits to convert voltages from one level to another through the medium of magnetic fields. There are many other uses for the transformer, but they are specialized (such as pulse transformers). For ac circuits (implying sinusoidal wave forms), the ideal transformer is considered as lossless with 100% efficiency; actually, the efficiency of a typical transformer is greater than 90%. For these ideal devices, the product of the input volt–amps equals the product of output volt–amps (for larger power transformers, $kVA_{in} = kVA_{out}$). The product is the apparent power (VA) rather than real power (watts). The nameplate rating of the transformer is important. The manufacturer gives the normal operating conditions; these nameplate ratings include the frequency, the voltage and the voltage ratio, and the kVA rating. The voltage ratios are the same as the turns ratio, *a*, and the current ratios are inversely related to the turns ratio. Whether a voltage is stepped up or down depends on which side one considers as primary and as secondary. For this discussion, assume the left side is primary (1) and the right side is secondary (2).

Example 12.14

For example, consider a transformer with a nameplate rating of 5 k*VA*, 60 Hz, and 880 : 220 V. The primary side might come from an 880-volt source, and the secondary would be at 220 volts.

Solution

The voltage/current rating is always given in rms values, and the turns ratio, of course, is 4:1. The current (or load) on the secondary side could be as high as $I_2 = VA/V_2 = 5000/220 = 22.7$ amperes, whereas the primary side would be $I_1 = 5000/880 = 5.68$ amperes, or 1/4 of 22.7 amperes. Note (from Exhibit 18) that a resistive load on the secondary side is $R_L = V/I = 220/22.7 = 9.69$ ohms, whereas it would be 880/5.68 = 155 ohms on the primary side. This equivalent resistance on the left side is a^2 times R_L:

$$R'_L = a^2R_L = (4)^2 \times 9.69 = 155\ \Omega$$

Here, $kVA_{in} = kVA_{out}$ and $kW_{in} = kW_{out}$ because the load is a pure resistive one (the power factor of the load is unity). If the load were $5 + j5$ (or $Z_L = \sqrt{2} \times 5\angle 45°$), the current would be

$$I_2 = V_2/Z_L = 220/\left(\sqrt{2} \times 5\angle 45°\right) = 31.1\angle -45°\ A$$

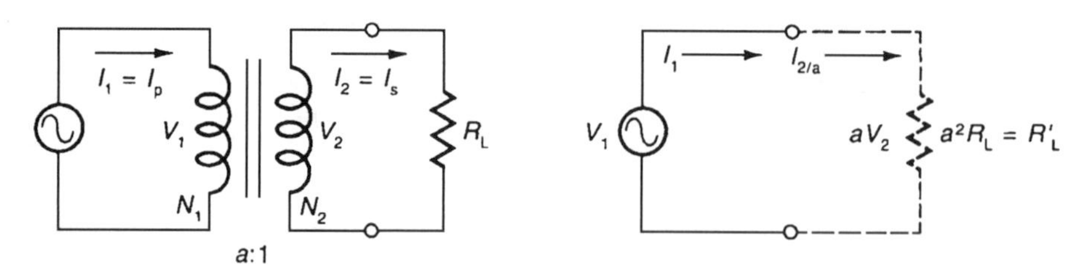

(a) Circuit with load (b) Circuit with equivalent load

Exhibit 18 A typical two-winding, loaded transformer

The current on the left side is $I_1 = I_2/a = 7.78$ amperes. These exceed the name plate values, and a larger transformer would have to be selected. The equivalent impedance, Z'_L, of the load if reflected to the left side is

$$Z'_L = a^2 Z_L = (4)^2 5 + j(4)^2 5 = 80 + j80\ \Omega$$

Caution is needed here because the power-in/power-out relationship is misleading, $P = VI \cos\theta = 220 \times 31.1 \times 0.707 = 4837$ W, or less than 5 kW; the transformer rating is 40 kVA, not necessarily 40 kW.

This chapter has reviewed selected electrical fundamentals. The depth of coverage has been limited. The reader who has more study time should select a text written about electrical engineering for all engineers. Of the many books available, several stand out. Any edition of the following should be available at a library or book store:

- Carlson and Gisser, *Electrical Engineering Concepts and Applications*, Addison-Wesley.
- Clement and Johnson, *Electrical Engineering Science*, McGraw-Hill.
- Smith, *Circuits, Devices and Systems*, Wiley.

SELECTED SYMBOLS AND ABBREVIATIONS

Symbol or Abbreviation	Definition
A	area
A	amperes
A • t	amp-turns
ac	alternating current
B	magnetic flux density; susceptance
C	coulomb
C	capacitance
dc	direct current
E	electric field
e	induced voltage
ε	permittivity
F	farad
F	force

(Continued)

(Continued)

Symbol or Abbreviation	Definition
G	conductance
H	henry
H	magnetic field intensity
I_c	core loss
I_e	exciting current
I_f	field current
I_i	current
I_m	magnetizing current
K	dielectric constant
KCL	Kirchhoff's Current Law
KVL	Kirchhoff's Voltage Law
L	inductance
l	length
ma	milliamp
mv	millivolts
mmf	magneto-motive force
N	number of turns
N_{ag}	Newton air gap
n	speed
n_s	synchronous speed
P_i, p	power
ϕ	magnetic flux
Q	point change, reactive power
R	resistance
R_{eq}	Thevenin's resistance
rms	root mean square
S	siemens
s	slip
T	tesla
T	period
T_d	developed torque
θ	angle
V, v	voltage
v	velocity
V_i, V_{oc}	terminal voltage, Thevenin's voltage
W	work
w	total energy
Y	admittance
Z, z	impedance
N_P	number of pole pairs

PROBLEMS

12.1 For a parallel plate capacitor separated by an air gap of 1 cm and with an applied dc voltage across the plates of 500 volts, determine the force on an electron mass of 18.2×10^{-31} kg inserted in the space. The mass of an electron is 9.1×10^{-31} kg.

a. 3.2×10^{-14} N c. 9.1×10^{-31} N
b. 1.6×10^{-14} N d. 1.6×10^{-19} N

12.2 Assume a point charge of 0.3×10^{-3} C at an origin. What is the magnitude of the electric field intensity at a point located 2 meters in the x-direction, 3 meters in the y-direction, and 4 meters in the z-direction from the origin?

a. 500 kV/m c. 93 kV/m
b. 5 kV/m d. 9.3 MV/m

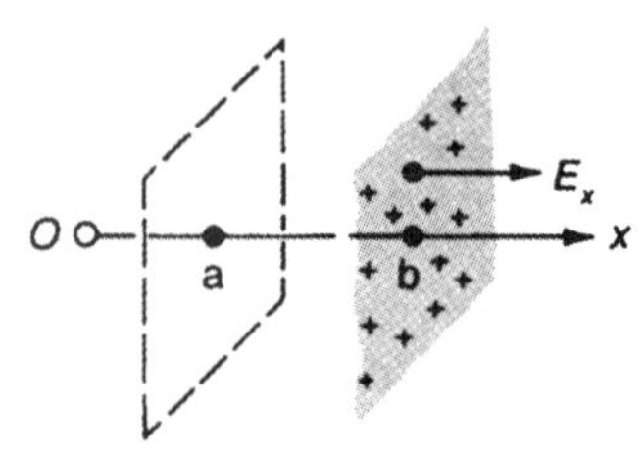

Exhibit 12.3 Charge placement

12.3 An infinite sheet of charge with a positive charge density, σ, has an electric field of

$$\mathbf{E} = [\sigma/(2\varepsilon_0)]_{ax} \quad \text{for } x > \text{b}$$

If a second sheet of charge with a charge density of $-\sigma$ is then placed (see Exhibit 12.3), what is the electric field for $a < x < b$?

a. $(\sigma/\varepsilon_0)_{ax}$ c. $(-\sigma/\varepsilon_0)_{ax}$
b. 0 d. $(\sigma/2\varepsilon_0)_{ax}$

12.4 Two equal charges of 10 μC are located one meter apart on a horizontal line, and another charge of 5 μC is placed one meter below the first charge (forming a right triangle). What is the magnitude of the force on the 5 μC charge?

a. 0.09×10^{6} N c. 6.39×10^{4} N
b. 12.6×10^{4} N d. 63×10^{-2} N

12.5 For a coil of 100 turns wound around a toroidal core of iron with a relative permeability of 1000, find the current needed to produce a magnetic flux density of 0.5 tesla in the core. The dimensions of the core are given in Exhibit 12.5.

a. 390 A c. 1.2 A
b. 39 A d. 12.2 A

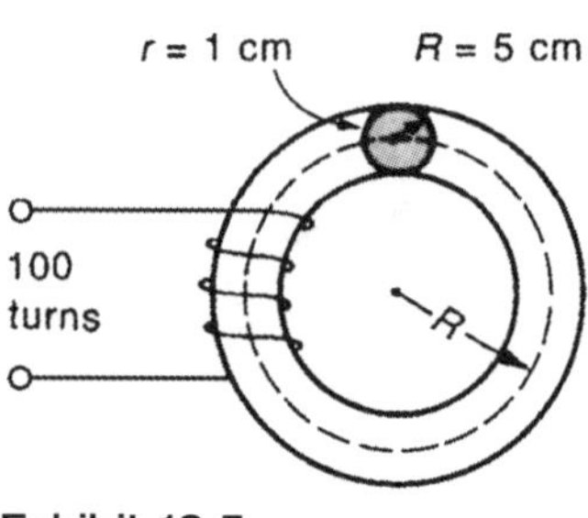

Exhibit 12.5

12.6 Two long straight wires, bundled together, have a magnetic flux density around them. One wire carries a current of 5 amperes, and the other carries a current of 1 ampere in the opposite direction. Determine the magnitude of the flux density at a point 0.2 meters away (i.e., normal to the wires).

a. $2\pi \times 10^{-6}$ T c. 4×10^{-6} T
b. $4\pi \times 10^{-6}$ T d. $16\pi \times 10^{-6}$ T

12.7 The root mean square of $i(t)$ for Exhibit 12.7 is

a. 1.6 A
b. 3.0 A
c. 2.0 A
d. 0.0 A

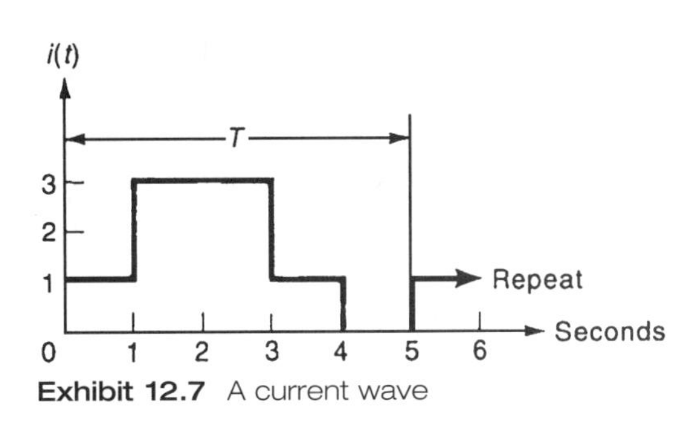

Exhibit 12.7 A current wave

12.8 A current phases through a 0.2-henry inductor. The current increases in a linear fashion from a value of zero at $t = 0$ to 20 A at $t = 20$ seconds. What is the amount of energy stored in the inductor at 10 seconds?

a. 0 J c. 40 J
b. 10 J d. 1 J

12.9 A sine wave of 10 volts (rms) is applied to a 10-ohm resistor through a half-wave rectifier. What is the average value of the current (that is, what would a dc meter read)?

a. 0.32 A c. 1.0 A
b. 0.45 A d. 0.9 A

12.10 Two resistors of 2 ohms each are connected in parallel, another resistor of 1 ohm is connected in series with the parallel combination, and the resistive combination is connected to a 2-volt source. How much power is dissipated in either one of the parallel resistors?

a. 0.25 W c. 1.0 W
b. 0.5 W d. 1.5 W

12.11 A current source of 2 amperes is connected to four resistors, all in parallel. The resistors have values of 1, 2, 3, and 4 ohms, respectively. How much power is dissipated in the 2-ohm resistor?

a. 8.0 W c. 2.0 W
b. 4.0 W d. 0.46 W

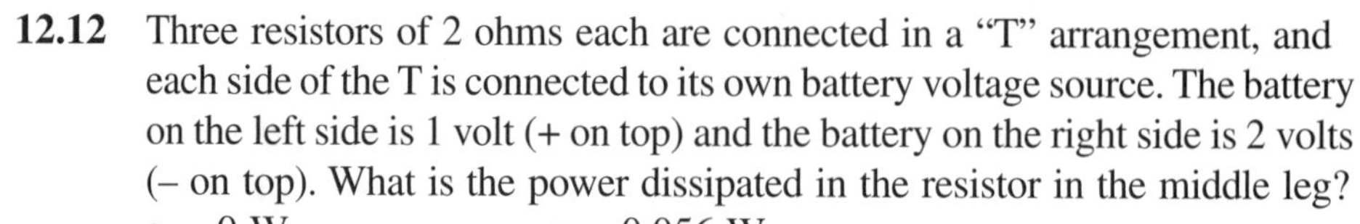

12.12 Three resistors of 2 ohms each are connected in a "T" arrangement, and each side of the T is connected to its own battery voltage source. The battery on the left side is 1 volt (+ on top) and the battery on the right side is 2 volts (– on top). What is the power dissipated in the resistor in the middle leg?

a. 0 W
b. 4.5 W
c. 0.056 W
d. 0.89 W

Exhibit 12.13 Original circuit

12.13 For the circuit in Exhibit 12.13, the load resistor, R_L, might "see" a Thevenin equivalent circuit in its place. What are the values of the equivalent circuit?

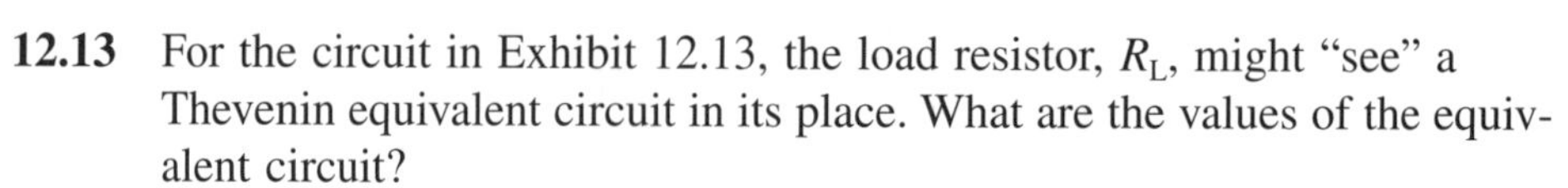

a. $V_{oc} = 5$ V, $R_{eq} = 2\ \Omega$
b. $V_{oc} = 5$ V, $R_{eq} = 1\ \Omega$
c. $V_{oc} = 4$ V, $R_{eq} = 2\ \Omega$
d. $V_{oc} = 1$ V, $R_{eq} = 2\ \Omega$

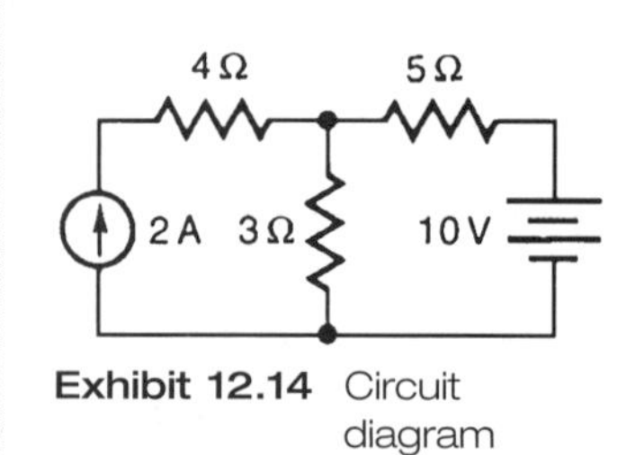

Exhibit 12.14 Circuit diagram

12.14 For the circuit shown in Exhibit 12.14, determine the magnitude of the voltage across the current source.

a. 16 V
b. 30 V
c. 20 V
d. 23 V

12.15 A charged 100-pF capacitor has initial voltage across it of 10 volts; a 100 kΩ resistor is suddenly connected across the capacitor (at $t = 0$) to discharge it. What is the magnitude of the instantaneous current at 20 microseconds?

a. 13.5 μA
b. 36.8 μA
c. 13.5 pA
d. 36.8 pA

12.16 Three 2-ohm resistors are arranged in "T" configuration. Connected to the left side is a 10-volt battery and switch; to the right is a 1-farad capacitor. The switch closes at $t = 0$. Find the time for the capacitor to reach 63% of its final voltage.

a. 0.2 s
b. 2.0 s
c. 0.3 s
d. 3.0 s

12.17 Three 2-ohm resistors are arranged in a "T" configuration. Connected to the left side is a 10-volt battery; to the right side is connected a switch and a 1-henry inductor. Assume the switch closes at $t = 0$. Find the amount of time for the inductor to reach 63% of its final current.

a. 0.2 s
b. 2.0 s
c. 0.3 s
d. 3.0 s

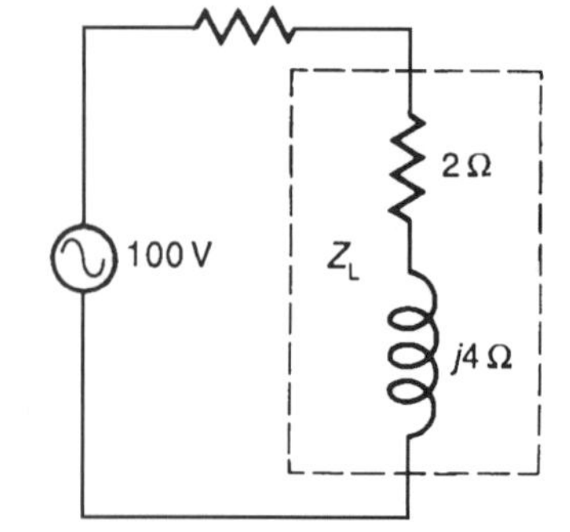

Exhibit 12.18 An ac circuit

12.18 For the circuit in the Exhibit 12.18, determine the power dissipated in the impedance, Z_L.

a. 1250 W
b. 625 W
c. 312 W
d. 1.7 kW

12.19 A resistance of 10 Ω, an inductor of 10/377 henries, and a capacitor of 20/377 farads are all connected in series to a 60-hertz, 100-volt source. Determine the magnitude of the current.

a. 3.3 A c. 10 A
b. 0.38 A d. $10\sqrt{2}$ A

12.20 If in Problem 12.19, all parameters were the same except that the frequency of the voltage source were doubled to 120 Hz, what would be the magnitude of the current?

a. 3.3 A c. 10 A
b. 0.38 A d. 5.5 A

12.21 A 10-ohm resistor and a capacitor with a capacitive reactance of $-j10$ Ω are connected in parallel. The parallel combination is connected in series through an inductive reactance of $+j10$ Ω to a 100-volt ac source. Determine the magnitude of the current from the source.

a. 6.7 A c. 10 A
b. 7.1 A d. $10\sqrt{2}$ A

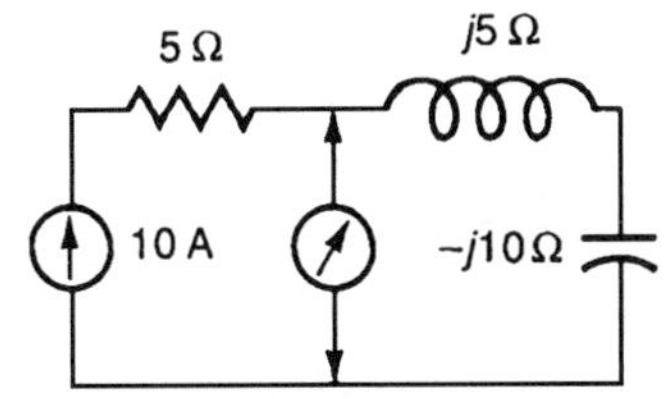

Exhibit 12.22 An ac circuit

12.22 An ac current source of 10 amperes is connected to a series circuit as shown in Exhibit 12.22. What would a voltmeter measure if the meter were connected across the inductor and capacitor combination?

a. 200 V c. 70 V
b. 150 V d. 50 V

12.23 For an *R–L–C* series circuit with $R = 10$ Ω, $L = 0.1$ henry, and $C = 0.1$ farad, all connected to an ac voltage source of 100 volts whose frequency could be varied, determine the frequency if maximum power is to be dissipated in *R*.

a. 10 Hz c. 1.6 Hz
b. 0 Hz d. 16 Hz

12.24 A low frequency ac voltage of 5 volts (rms) is applied to the input of the op-amp circuit in Exhibit 12.24. Determine the magnitude of the output voltage (rms).

a. 5 V c. 10 V
b. 2.5 V d. 20 V

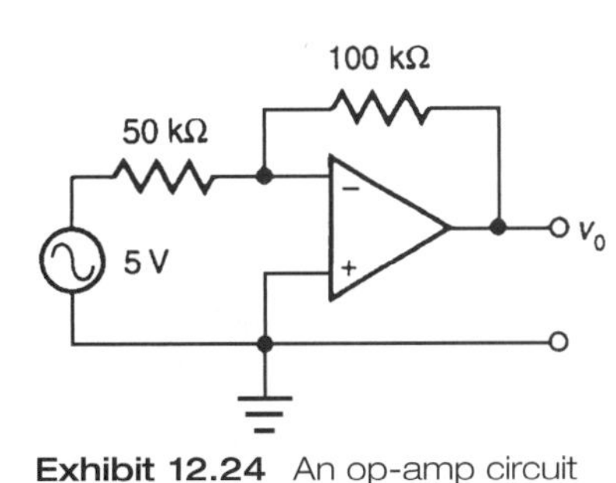

Exhibit 12.24 An op-amp circuit

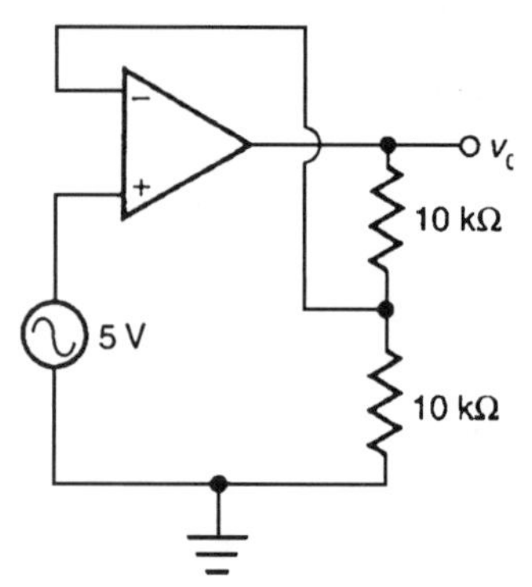

Exhibit 12.25 An op-amp circuit

12.25 A low frequency ac voltage of 5 volts (rms) is applied to the input of the op-amp circuit in Exhibit 12.25. Determine the magnitude of the output voltage (rms).

a. 5 V c. 10 V
b. 2.5 V d. 20 V

12.26 The inputs for the op-amp circuits are given in Exhibit 12.26. What is the output voltage from the last op-amp?

a. +1 V c. +3 V
b. −1 V d. −3 V

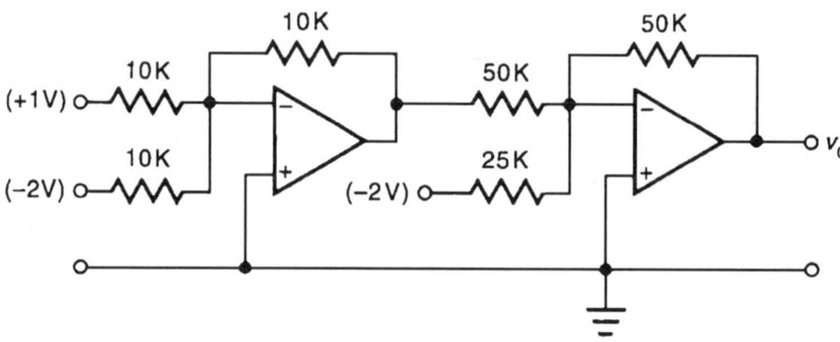

Exhibit 12.26 Summing circuits

12.27 A square wave of 1 V (or 2 V peak-to-peak) is applied to a bridge rectifier circuit where a 1-ohm resistor is connected to the output. What is the average current through the resistor?

a. 0.5 A c. π A
b. $\pi/2$ A d. 1.0 A

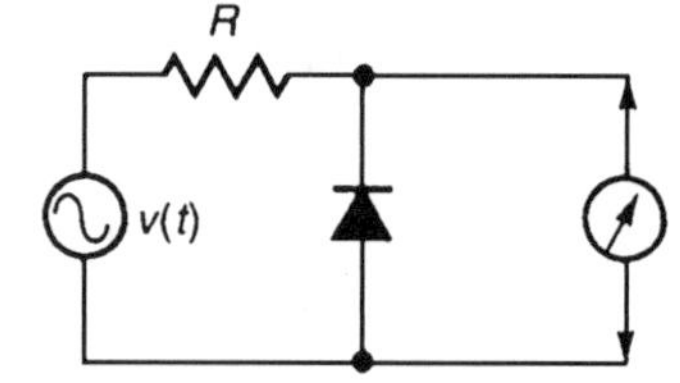

Exhibit 12.28 Clipping circuit

12.28 A sine wave voltage $v(t) = 14 \sin 2\pi ft$ is applied to the clipping circuit of Exhibit 12.28. A dc voltmeter (average reading instrument) is connected to the output. What does the meter read?

a. 2 V c. 7.5 V
b. 4.5 V d. 9 V

12.29 For an ideal, two-winding transformer whose name plate reads: 5 kVA, 400:200 V, and 60 Hz, determine the magnitude of a load that could be connected to the low voltage side for rated conditions. Assume the load impedance is $\mathbf{Z}_L = Z\angle 45°$, where Z is

a. 4 Ω c. 16 Ω
b. 8 Ω d. 25 Ω

SOLUTIONS

12.1 **b.** The "mass" of 2 electrons has a charge $Q = 3.2 \times 10^{-19}$ C, thus the electric field is $\mathbf{E} = 500$ V/0.01 m $= 50 \times 10^3$ V/m. the force is then $\mathbf{F} = Q\mathbf{E} = (3.2 \times 10^{-19}) \times (50 \times 10^3) = 1.6 \times 10^{-14}$ N.

12.2 **c.** The magnitude of the length of the resultant vector, R, in the x, y, z plane is

$$R = \sqrt{2^2 + 3^2 + 4^2} = \sqrt{29}$$

The magnitude of the electric field, **E,** is $Q/(4\pi\varepsilon R^2) = (0.3 \times 10^{-3})/(4\pi \times 8.85 \times 10^{-12} \times 29) = 93{,}000$ V/m.

12.3 **c.** On the b plane (for that plane alone), $\mathbf{E}^+ = (-\sigma/2\varepsilon)_{\mathbf{a}x}$, and for the negatively charged plane at the a plane (again for that plane alone, but acting to the right of a), is the same as before, therefore:

$$\mathbf{E} = \mathbf{E}^+ + \mathbf{E}^- = (-\sigma/\varepsilon_0)_{\mathbf{a}x} \text{ V/m}$$

12.4 **d.** From a sketch of the vectors in Exhibit 12.4, the value of $\mathbf{E}_{\text{net}}$ is 12.6×10^4 V/m. **F** is then found to be

$$\mathbf{E} = \frac{10 \times 10^{-6}}{4\pi(8.85 \times 10^{-12})\, r^2}$$

$$\mathbf{F} = Q_C\mathbf{E}_{\text{net}} = 5 \times 10^{-6} \times 12.6 \times 10^4 = 63 \times 10^{-2} \text{ N}$$

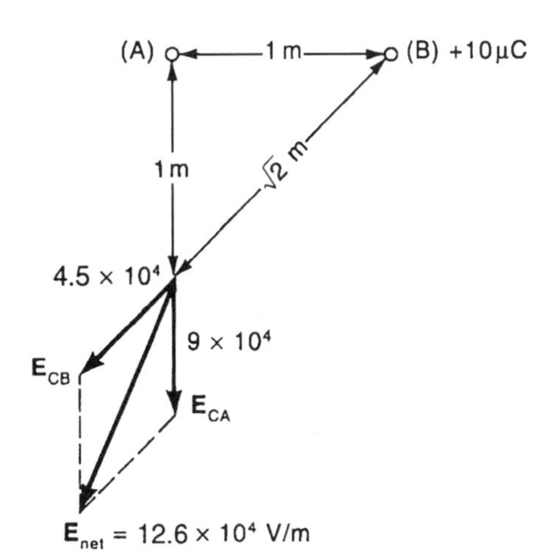

Exhibit 12.4 Vector field

12.5 **c.** The cross-sectional area of the iron core is $\pi r^2 = \pi \times 10^{-4}$ m^2. The path length is $l = 2\pi R = 2\pi \bullet 5 \times 10^{-2} = 0.1\pi$ m. The permeability is $\mu = \mu_0\mu_r = 4\pi \bullet 10^{-7} \times 10^3 = 4\pi \bullet 10^{-4}$.

Hence, $H = B/\mu = 0.5/(4\pi \bullet 10^{-4}) = 390$ A • t/m, and since $H =$ mmf/length, mmf $= H \times$ length $= 390 \times 0.1\pi = 122.5$ A • t. Thus, $I =$ mmf/turns $= 122.5/100 = 1.225$ A.

12.6 **c.** Assume the two wires are bundled close together (with respect to the 0.2-meter position); then the net current to the right is $I_{net} = 5 - 1 = 4$ A. The flux density is given by $B = (\mu I)/(2\pi r) = (4\pi \times 10^{-7})(4)/(2\pi \times 0.2) = 4 \times 10^{-6}$ tesla.

12.7 **c.**

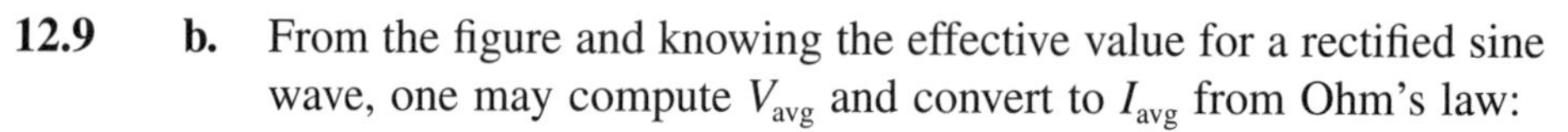

$$I_{rms} = \sqrt{(1/T)\int_0^T i^2 dt} = \sqrt{(1/5)[(1)^2 \times 1 + (3)^2 \times 2 + (1)^2 \times 1 + 0]} = 2 \text{ A}$$

12.8 **b.** The current at 10 seconds is 10 A. And from the equation for energy storage in an inductor, it is found that $W = (1/2)\,Li^2 = (1/2)(0.2)(10)^2 = 10$ J.

12.9 **b.** From the figure and knowing the effective value for a rectified sine wave, one may compute V_{avg} and convert to I_{avg} from Ohm's law:

$$V_{max} = \sqrt{2}V_{rms} = \sqrt{2} \times 10 \text{ V}, \quad V_{avg} = V_{max}/\pi = \sqrt{2} \times 10/\pi = 4.49$$

$$I_{avg} = V/R = 4.5/10 = 0.45 \text{ A}$$

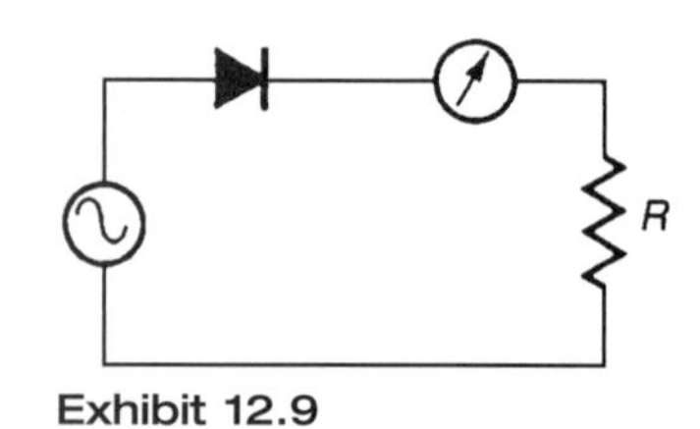

Exhibit 12.9

12.10 **b.** In the Exhibit 12.10, the resistance of the parallel combination is 1 Ω; then the series combination is 2 Ω. The current I_1 is $V/R = 2/2 = 1$ A, and the current for the parallel branch is

$$I_2 = \left[G_2/\left(\sum G\right)\right]I_1 = (1/2)/[(1/2)(1/2)]1$$

$$= 0.5 \text{ A}, \ P_2 = I_2^2 R = (0.5)^2\, 2 = 0.5 \text{ watts}$$

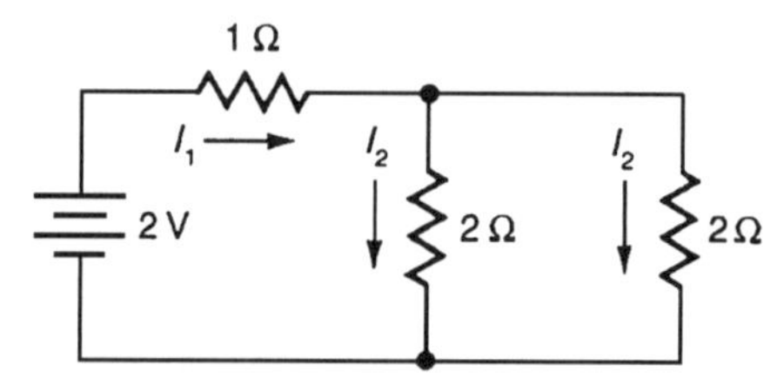

Exhibit 12.10 Circuit configuration

12.11 **d.** Refer to Exhibit 12.11. The current is given by $I_2 = [G_2/\Sigma G)]I_{total} = [(1/2)/(1 + 1/2 + 1/3 + 1/4)](2) = 0.48$ A, $P_2 = I^2R = (0.48)^2(2) = 0.46$ watts.

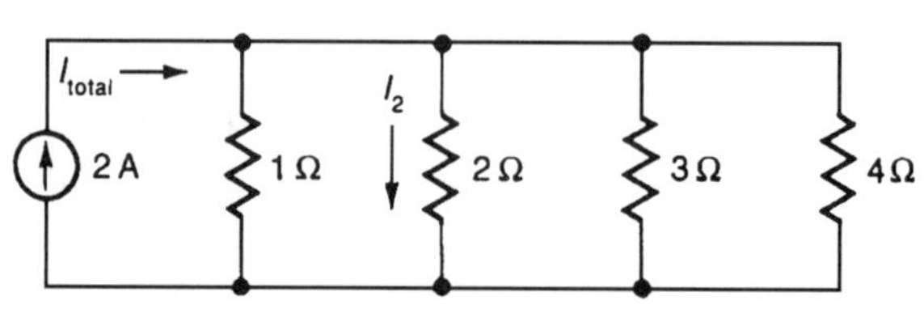

Exhibit 12.11 Circuit configuration

12.12 **c.** From the Exhibit 12.12, set up mesh currents, I_1, and I_2, then sum voltages around each loop:

Loop 1: $1 = 2I_1 + 2(I_1 - I_2)$;
Loop 2: $2 = 2(I_2 - I_1) + 2I_2$

$$I_1 = \frac{\begin{vmatrix} 1 & -2 \\ 2 & 4 \end{vmatrix}}{\begin{vmatrix} 4 & -2 \\ -2 & 4 \end{vmatrix}} = 2/3\ A \qquad I_2 = \frac{\begin{vmatrix} 4 & 1 \\ -2 & 2 \end{vmatrix}}{\begin{vmatrix} 4 & -2 \\ -2 & 4 \end{vmatrix}} = 5/6\ A$$

$$I_{\text{mid-leg}} = I_1 - I_2 = -1/6\ \text{A}; \qquad P_{\text{mid-leg}} = I^2R = (-1/6)^2 2 = 0.056\ \text{W}$$

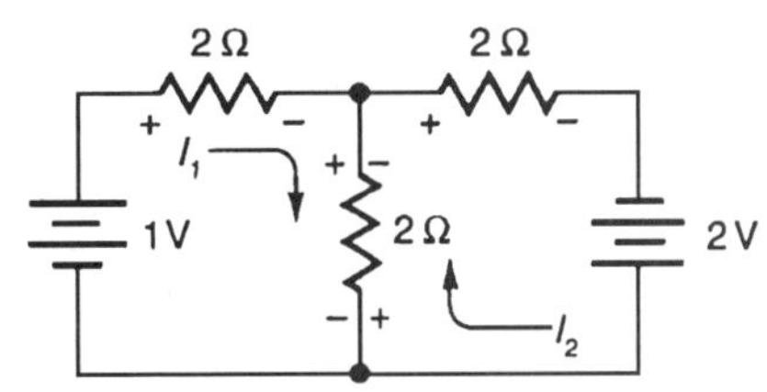

Exhibit 12.12 Circuit configuration

12.13 **a.** To find the Thevenin voltage, find the voltage across the 4-ohm resistor with R_L removed, which yields 5 volts. Thevenin resistance may be found by either of two methods. One method is to look back into the circuit with all active sources replaced with their own impedances and to calculate this output impedance; the other method is to short out the terminals of interest. For this second method, the impedance is

$$R_{eq} = V_{oc}/I_{sc} = 5/(10/4) = 2\ \Omega$$

12.14 **a.** First find the node voltage, V_1, at the intersection of the three resistors by summing the currents (start by assuming all currents flow away from the node). Assume that I_1 is to the left, I_3 to the right, and I_2 down.

$$\sum I = 0: (-2) + V_1/3 + (V_1 - 10)/5 = 0, \quad 8V_1 = 60, \quad V_1 = 7.5\ \text{V};$$
$$V_A = V_1 - I_1R_4 = 7.5 - (-2)4 = 15.5\ \text{V}$$

12.15 **a.** Refer to Exhibit 12.15. The time constant, τ, for an R–C circuit is RC. The voltage across the capacitor will discharge at a rate given by

$$v_C = V_C e^{-t/\tau} = 10\ e^{-20/10} = 1.35\ \text{V}, \quad i = v/R = 1.35/10^5 = 13.5 \times 10^{-6}\ \text{A}$$

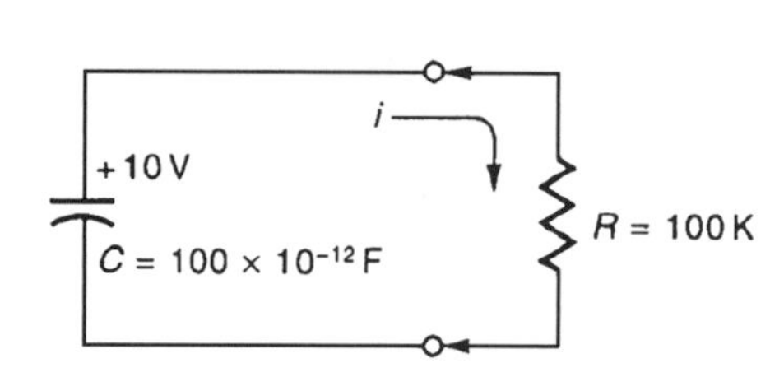

(a) R–C circuit

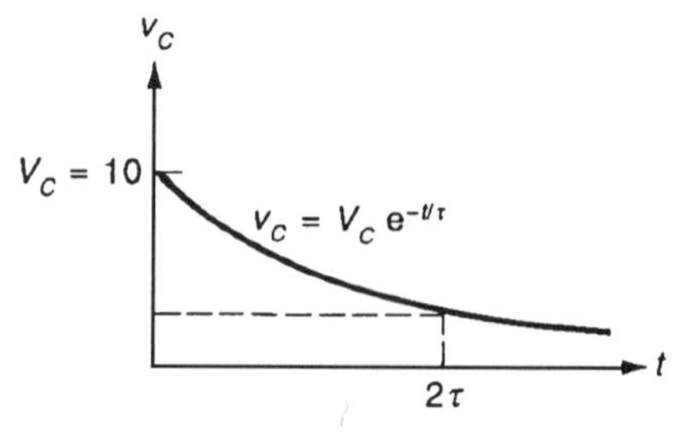

(b) Voltage discharge curve

Exhibit 12.15 R–C circuit

12.16 **d.** First construct Thevenin's circuit so that all elements will be in series; this makes the R of the R–C circuit especially easy to determine. Over one time constant, the voltage will build up to 63% of its final value.

$$V_{oc} = 5 \text{ V}, \quad R_{eq} = 5\ \Omega, \quad \tau = R_{ser}\ C = 3 \times 1 = 3 \text{ s}.$$

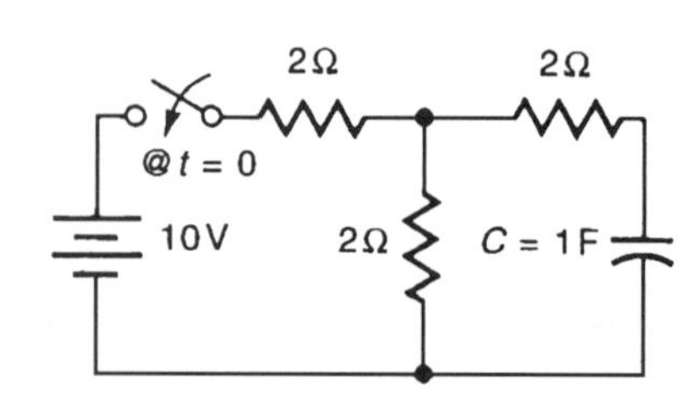

(a) Original circuit

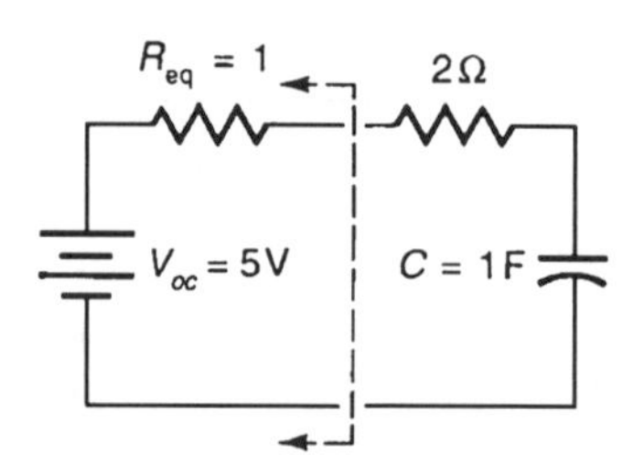

(b) Thevenin's circuit

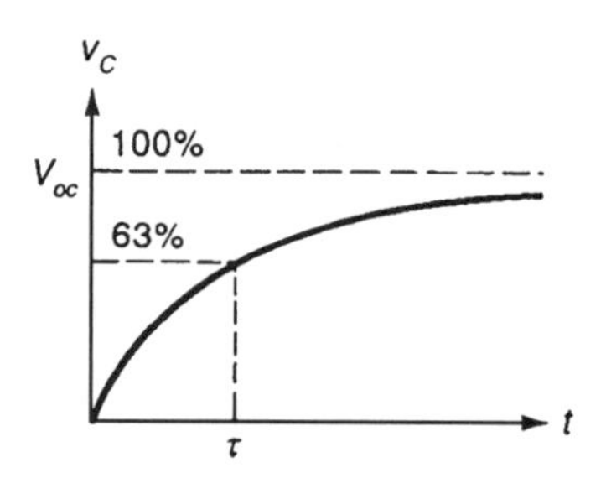

(c) Charging curve

Exhibit 12.16 Simplifying an R–C circuit

12.17 **c.** Make a Thevenin's equivalent circuit for finding the series resistance. The time constant for an L–R circuit is $\tau = L/R = 1/3$ s. For a charging circuit, in one time constant, the current builds up to 63% of its final value.

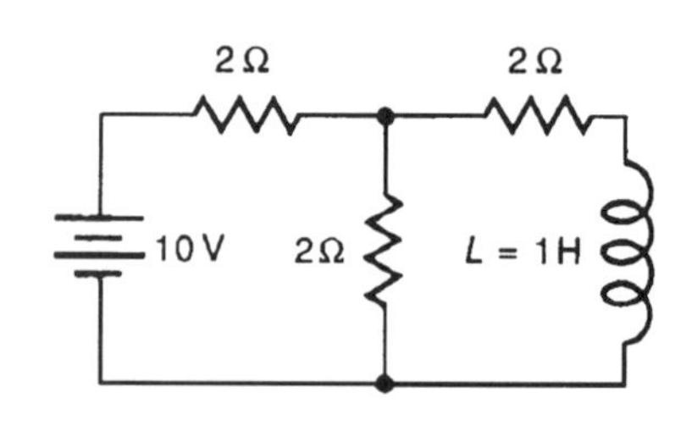

(a) Original circuit

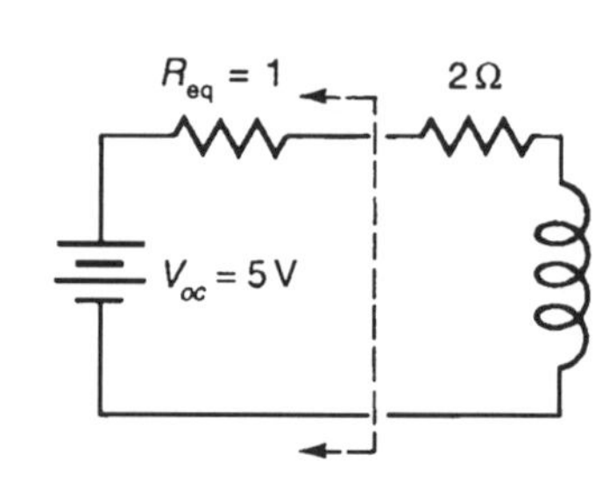

(b) Thevenin's circuit

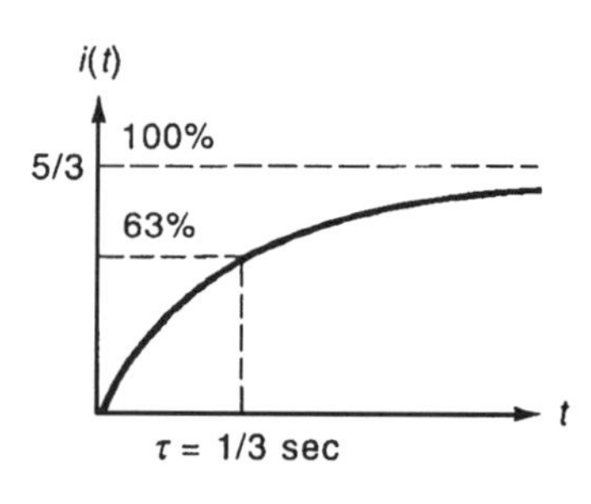

(c) Charging curve

Exhibit 12.17 Simplifying an R–L circuit

12.18 **b.** For a series circuit, $\mathbf{Z}_{total} = R_2 + \mathbf{Z}_L$, then

$$\mathbf{I} = \mathbf{V}/\mathbf{Z} = 100\angle 0°/(4 + j4) = 100\angle 0°/\left(\sqrt{2} \times 4\angle 45°\right) = 17.7\ \angle -45°$$

$$P_2 = I^2R = (17.7)^2(2) = 625 \text{ watts}$$

12.19 **c.** First it is necessary to compute the reactances (recall that $2\pi f = 377$ for 60 Hz): $X_L = 2\pi fL = (377)(10/377) = 10\ \Omega$, and $X_C = 1/(2\pi fC) = 10\ \Omega$. Thus

$$\mathbf{Z} = 10 + j10 - j10 = 10\angle 0°$$
$$\mathbf{I} = \mathbf{V/Z} = 100\angle 0°/10\angle 0° = 10\text{ A}$$

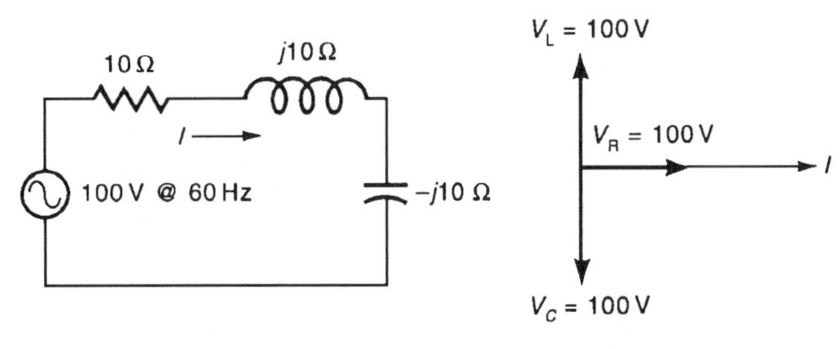

(a) Circuit diagram (b) Phasor diagram

Exhibit 12.19 An *R–L–C* circuit

12.20 **d.** If the frequency is doubled, the reactances will change to $X_L = 2 \times 10 = 20\ \Omega$, $X_C = 0.5 \times 10 = 5\ \Omega$; hence, $Z = 10 + j20 - j5 = 18.2\angle 56.3°$. $|I| = |V/Z|$ (answer only requires magnitude) $= 100/18.2 = 5.5$ A.

12.21 **d.** First find the series equivalent to the parallel portion of the circuit:

$$Z_P = (10)(-j10)/(10 - j10) = \left(10/\sqrt{2}\right)\angle -45° = 5j5\,\Omega;$$

$$Z_{\text{total}} = j10 + 5 - j5 = 5\sqrt{2}\ \angle 45°\ I = V/Z = 100/\left(5\sqrt{2}\right) = 10\sqrt{2}\text{ A.}$$

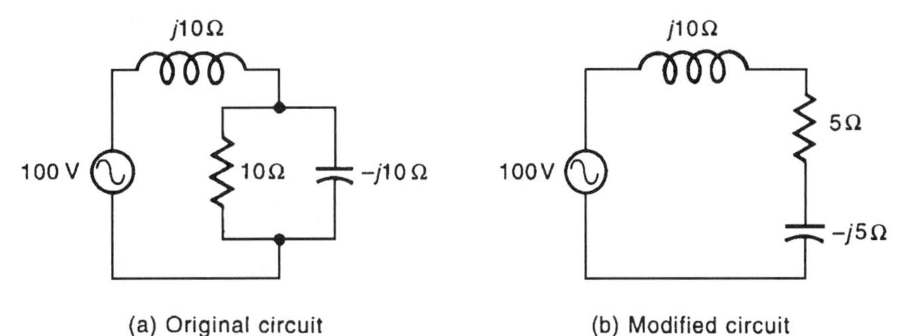

(a) Original circuit (b) Modified circuit

Exhibit 12.21 Reduced circuit

12.22 **d.** The impedance for the *L–C* portion of the circuit is $\mathbf{Z}_{L\text{-}C} = j5 - j10 = -j5 = 5\angle -90°$; $V_{L\text{-}C} = IZ_{L\text{-}C} = (10)(5) = 50$ V.

12.23 **c.** Maximum current occurs if X_L and X_C just cancel and Z is left with only resistance: $X_L = 2\pi fL$, $X_C = 1/(2\pi fC)$; $f_{\text{res}} = 1/(2\pi\sqrt{LC}) = 1.6$ Hz.

12.24 **c.** The op-amp circuit produces a gain of $-(R_f/R_1)$ for an inverting amplifier; and here, only the magnitude is requested so the sign is not important.

$$G = (100k/50k) = 2; \quad V_0 = 2V_{in}, \quad V_0 = 2 \times 5 = 10 \text{ V}.$$

12.25 **c.** This op-amp circuit produces a gain of $1 + (R_2/R_1)$ for a noninverting amplifier.

$$G = 1 + (10k/10k) = 2; \quad V_0 = 2 \times 5 = 10 \text{ V}.$$

12.26 **c.** The output from the first summing op-amp is $v_0 = -[(R_f/R_1)v_1 + (R_f/R_2)v_2] = -[(10k/10k)1 + (10k/10k)(-2)] = +1$ V. The output from the second summing op-amp is $v_0 = -[(R_f/R_3)v_3 + (R_f/R_4)v_4] = -[(50k/50k)1 + (50k/25k)(-2)] = +3$ V.

12.27 **d.** Refer to Exhibit 12.27. The bridge rectifier is a full-wave type, and, for a square wave, the result of taking the average is the maximum of the input itself. $I = V/R = 1/1 = 1$ A (average).

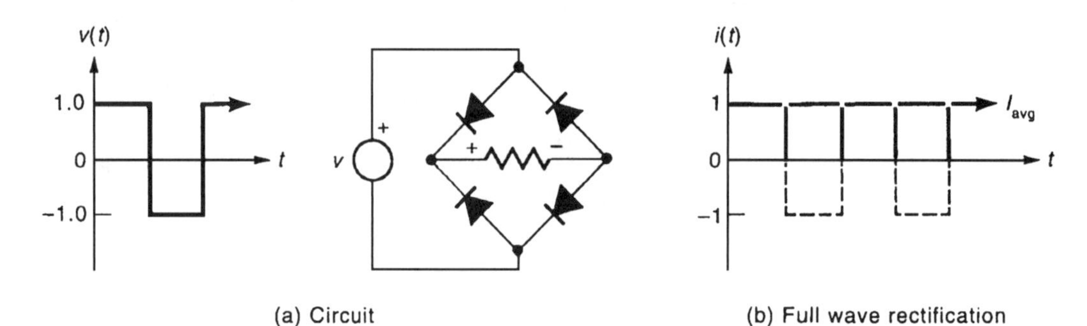

(a) Circuit (b) Full wave rectification

Exhibit 12.27 A full-wave bridge rectifier circuit

12.28 **b.** Since the output is unloaded (that is, the meter resistance is assumed to be ≫ R), the output is "shorted out" by the ideal diode for the negative half cycle; thus the output voltage appears as an average value. For a half-wave rectifier, the voltage is $V_{avg} = V_{max}/\pi = 14/\pi = 4.5$ V.

12.29 **b.** Assume the load side is V_2, then under full load conditions (rated current at rated voltage), the current is $I_2 = 5000$ VA/200 V = 25 A; $Z_L = 200$ V/25 A = 8 Ω regardless of phase.

Materials Engineering

Lawrence H. Van Vlack
with updates by Brian Flinn

OUTLINE

All engineering products are made of materials. Thus, engineers become directly involved with materials, whether they be designs engineers, production engineers, or applications engineers. Their familiarity with a wide spectrum of materials becomes particularly important as they advance through management and into administration, where they must oversee the activities of additional engineers on their technical staffs.

The way that an engineering product performs in service is a consequence of the combination of the components of the product. Thus, a cellular phone must have the diodes, resistors, capacitors, and other components that function together to meet its design requirements. Likewise, a competitively produced car must possess a carefully designed engine with its numerous parts, as well as safety features and operating characteristics that meet customer approval. Materials are pertinent to each and every design consideration.

Just as it is to be expected that the internal circuitry of a four-function hand calculator will differ from the internal circuitry of its multifunctional scientific counterpart, the internal structure of a steel gear differs from that of the sheet steel to be used in an automotive fender. Their roles, and therefore their properties, are designed to be different.

The variations in the internal structures of materials that lead to property differences include variations in atomic coordination and electronic energies, differences in internal geometries (microstructures), and the incorporation of larger structures, sometimes called macrostructures. Each of these is considered in the following sections, along with procedures for obtaining desired structures and properties.

ATOMIC ORDER IN ENGINEERING MATERIALS

Atoms, Ions, and Electrons

There is an order within atoms. Each atom has an integer number of protons. That number is called the **atomic number**. The natural elements possess, progressively, 1 to 92 of these protons, which carry a positive charge. A neutral atom has a number of electrons equal to the number of protons. Each electron is negative with a charge of 1.6×10^{-17} coulombs. Electrons are only allowed in given orbitals (called shells) that correspond to specific allowed energy levels.

With the exception of the principal isotope of hydrogen, each atom possesses neutrons. While these are charge-neutral, they add to the mass of each atom. The protons and neutrons reside in the nucleus of the atom. Figure 13.1 shows the Bohr model of a sodium atom. The lighter elements contain approximately equal numbers of protons and neutrons; however, in heavier elements the number of neutrons exceeds the number of protons. Furthermore, the number of neutrons per atom is not fixed. Thus, we encounter several **isotopes** for most atoms.

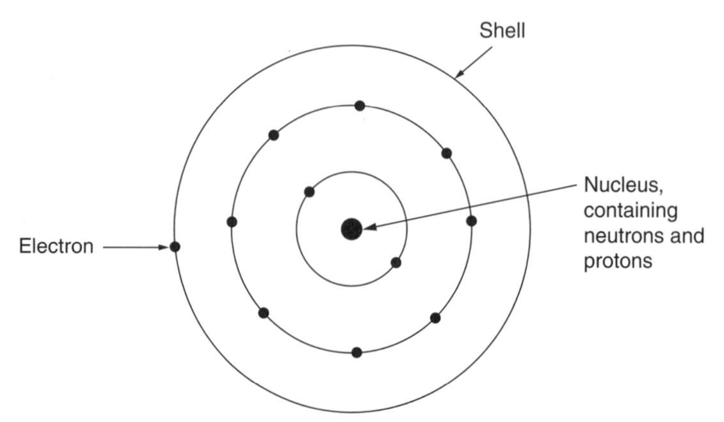

Figure 13.1 Bohr model of a sodium atom

Table 13.1 Data for selected atoms

Element	Protons Electrons Atomic No.	Neutrons (in natural isotopes)	Atomic Mass Units	Grams per Avogadro's Number*
Hydrogen	1	0 or 1	1.008	1.008
Carbon	6	6 or 7	12.011	12.011
Oxygen	8	8, 9, or 10	15.995	15.995
Chlorine	17	18 or 20	35.453	35.453
Iron	26	28, 30, 31, or 32	55.847	55.847
Gold	79	118	196.97	196.97
Uranium	92	142, 143, or 146	238.03	238.03

*Avogadro's number = 6.022×10^{23}

Since the mass of an electron is appreciably less than one percent of that of protons and neutrons, the mass of an atom is directly related to the combined number of the latter two. By definition, an **atomic mass unit** (amu) is one-twelfth of the mass of a carbon isotope that has six protons and six neutrons, C^{12}. The **atomic mass** of an element is equal to the number of these atomic mass units. (Selected values are listed in Table 13.1.) Thus while there are integer numbers of neutrons, protons, and electrons in each atom, the mass generally is not an integer, because more than one isotope is typically present.

A limited number of electrons may be accepted or released by an atom, thus introducing a charge on the atom (due the difference in the number of protons and electrons). A charged atom is an **ion**. Negative ions have accepted extra electrons; and are called **anions**; electrons have been released by positive-ions, which are called **cations**. Because they are charged, ions respond to electric fields. These fields may involve macroscopic dimensions (in electroplating baths); or they may involve interatomic distances (in molecules). Unlike charges attract; like charges repel.

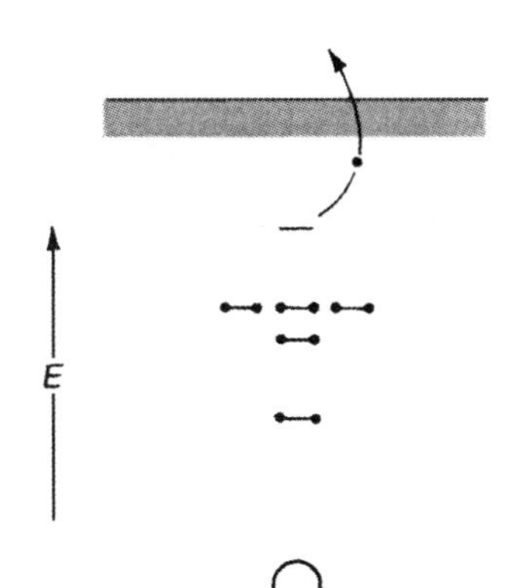

Figure 13.2 Ionization energy (schematic for sodium). Electrons reside at specific energy **states** (levels). Energy must be supplied to remove an electron from an atom, producing a positive ion. Two electrons (of opposite magnetic spins) may reside in each state.

Energy is required to remove an electron from a neutral atom. Figure 13.2 shows this schematically for a sodium atom. Conversely, fixed quantities of energy are released when electrons are captured by a positive ion. These energy levels (**states**) associated with an atom are fixed. Furthermore each state may accept only two electrons, and these must have opposite magnetic characteristics. Electronic, magnetic and optical properties of materials must be interpreted accordingly.

Molecules

Atoms can join to one another; this is called **bonding**. Strong attractive forces can develop between atoms by three mechanisms; (1) coulombic attraction between oppositely charged ions, forming ionic bonds; (2) sharing of electrons to fill outer shells, creating covalent bonds; and (3) formation of ion cores surrounded by valence electrons that have been excited above the Fermi level and have become free electrons, forming metallic bonds. More detailed examples of these three primary types of bonds follows.

Ionic bonds form between metallic and nonmetallic atoms. The metallic atoms release their valence electrons to become cations (which are positively charged) and the nonmetallic atoms accept them to become anions. Ionic bonds are nondirectional. For example, a sodium ion that has lost an electron (Na^+) associates with as many negatively charged chlorine ions (Cl^-) as space will allow. And each Cl^- ion will become *coordinated* with as many Na^+ ions as necessary to balance the charge. The resulting structure will continue to grow in three dimensions until all available ions are positioned. Energy is released with each added ion.

Covalent bonds form between atoms that share valence electrons in order to fill their outer shells. In the simplest case, two hydrogen atoms release energy as they combine to produce a hydrogen molecule:

$$2H \rightarrow H_2 \quad \text{or} \quad 2H \rightarrow H{-}H \tag{13.1}$$

The bond between the two involves a pair of shared electrons. This mechanism is common among many atomic pairs. In this case only one pair of atoms is involved. The covalent bonds of molecules are stereospecific, that is, they are between specific atoms and are therefore directional bonds.

In polymers a string or network of thousands or millions of atoms are bonded together. Examples include polyethylene, which has the structure shown in Fig. 13.3(a), in which there is a backbone of carbon atoms that are covalently bonded; that is, they share pairs of valence electrons. Since each carbon atom has four valence electrons, it can form four covalent bonds, thus adding two hydrogens at the side of the chain in addition to the two bonds along the chain. Polyvinyl chloride, Fig. 13.3(b), is related but has one of the four hydrogen atoms replaced by a chlorine atom.

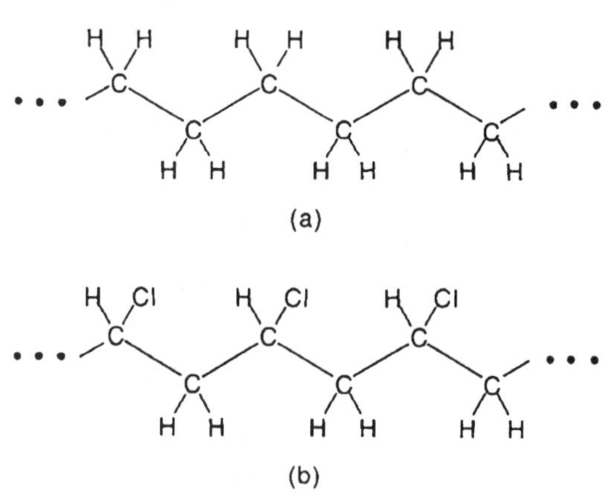

Figure 13.3 Covalent bonds. (a) Polyethylene; (b) Polyvinyl chloride

Metallic bonds form between the elements on the left side of the periodic table known as metals. Metallic bonds can form between atoms of the same element, to form a pure metal such as gold (Au), or between different metal atoms to form an alloy, such as brass, a mixture of copper (Cu) and zinc (Zn). The basis for the metallic bond is the formation of ion cores created when the metal's valence electrons are no longer associated with a specific atom. These electrons (called free electrons) move freely around surrounding metal ion cores, shielding them from each other. These freely moving electrons are often referred to as a "sea of electrons". Metallic bonds are nondirectional.

The type and strength of bonding between atoms and molecules determines many properties of materials. As a general rule (and with other things being equal), the stronger the atomic bonding, the higher the melting temperature hardness, and elastic modulus of materials. Ionically bonded materials are usually electrical and thermal insulators, while materials with metallic bonds have high electrical and thermal conductivities.

Crystallinity

The repetition of atomic coordinations in three dimensions produces a periodic structure that is called a **crystal**. The basic building block of a crystal is called a unit cell. Figure 13.4 illustrates the unit cell structure of iron. Each atom in this metal has eight nearest neighbors, which are symmetrically coordinated to give a **body-centered cubic** (BCC) crystal. About 30 percent of the metals have this structure. The atoms of aluminum and copper, among another 30 percent of the metals, become coordinated with 12 nearest neighbors with the result that **face-centered cubic** (FCC) crystal lattices are formed, as shown in Fig. 13.5. A third group of metals form **hexagonal crystals** with **close packing** (HCP), as shown in Fig. 13.6. As in the FCC crystals, each atom is coordinated with 12 nearest neighbors in HCP crystals.

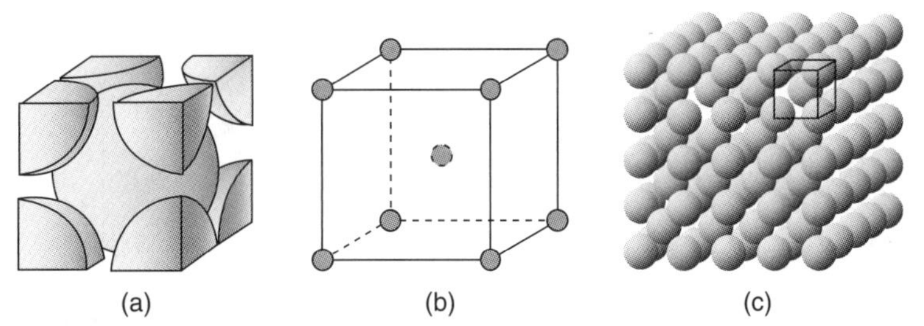

Figure 13.4 Body-centered cubic crystal strucutre. (a) a hard-sphere unit cell representation; (b) a reduced-sphere unit cell; (c) an aggregate of many atoms

It is possible to have a very high degree of perfection in crystals. For example, the repetition dimension (**lattice constant**) of the FCC lattice of pure copper is constant to the fifth significant figure (and to the sixth if the thermal expansion is factored in). This high degree of ordering provides a quantitative base for anticipating properties. Included are density calculations, certain thermal properties, and some of the effects of alloying.

Directions and Planes

Many properties are **anisotropic**, that is, they differ with direction and orientation. We can identify crystal directions by selecting any zero location (the origin) and determining the x, y, and z coordinates for any point along the direction ray. A corner of the unit cell is often used as the origin. The unit length is the edge length of the unit cell. The direction of easy magnetization in iron is parallel to one of the crystal axes. This is labeled the [100] direction, because that direction passes from the origin through a point that is one unit along the x-axis and zero

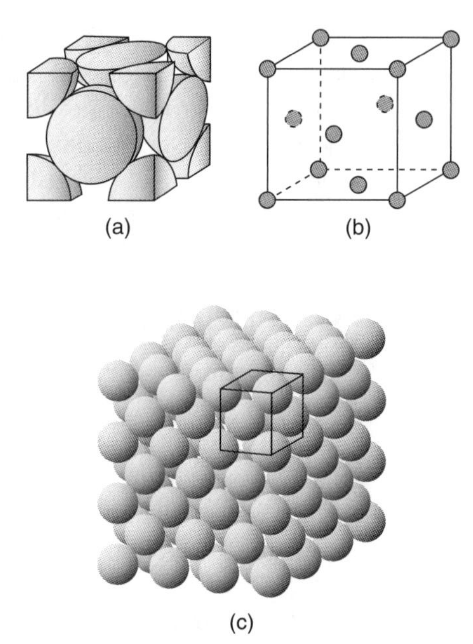

Figure 13.5 Face-centered cubic crystal structure. (a) a hard-sphere unit cell representation; (b) a reduced-sphere unit cell; (c) an aggregate of many atoms

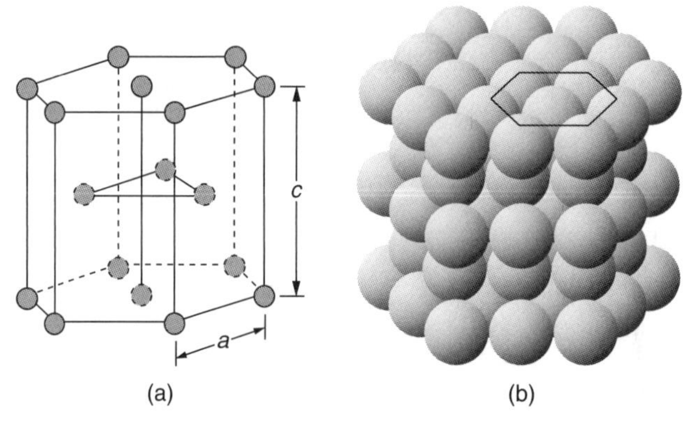

Figure 13.6 Hexagonal close-packed crystal structure. (a) a reduced-sphere unit cell (*a* and *c* represent the short and long edge lengths, respectively); (b) an aggregate of many atoms

units along the other two axes. Figure 13.7(a) shows a ray in the [120] direction and a ray in the $[1\bar{1}0]$ direction, where the overbar indicates a negative direction. Parallel directions carry the same label. We use square brackets, [ ], for closures for direction rays.

In cubic crystals, each of the four directions that are diagonal through the cube are identical (because the three axes are identical). We label these four directions as a *family* with pointed arrows for closures, <111>.

Figure 13.7(b) identifies the three shaded planes as (001), (210), and (111). Here the labeling procedure (known as indexing) is somewhat more complicated than for directions (but leads to simplified mathematics for complex calculations). As an example we will draw the (210) plane. We first invert the three indices,

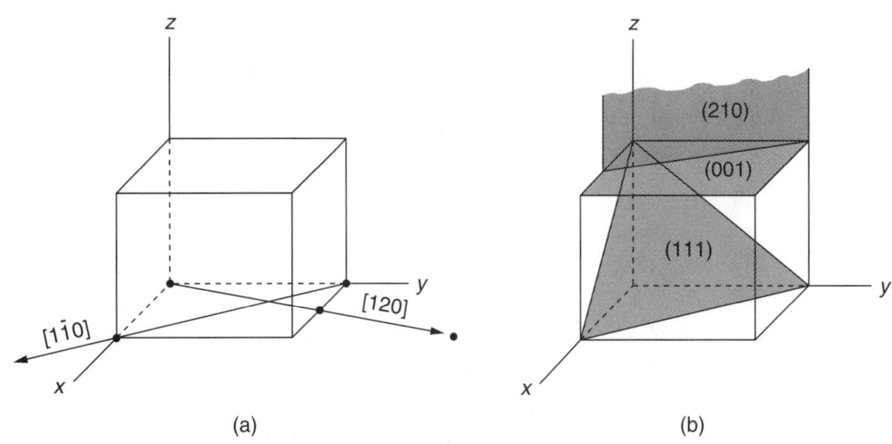

Figure 13.7 Crystal notation. (a) Directions, [120] and [1$\bar{1}$0]. The *x*, *y*, and *z* coordinates are used for crystal directions. Square brackets are used as closures. Negative coordinates are indicated with an overbar. (b) Planes, (001), (210), and (111). The reciprocals of the axial intercepts are used for crystal planes. Parentheses are used as closures. Any point may be selected as an arbitrary origin.

1/**2**, 1/**1**, 1/**0**. These are the intercept dimensions of the plane across the three axes; specifically, 0.5 on the *x*-axis, 1.0 on the *y*-axis, and infinity along the *z*-axis. An adjacent parallel plane with intercepts of 1, 2, and ∞, carries the same (210) index. We use parentheses, ( ), as closures for the indices of individual planes, and braces, { }, as closures for a family of comparable planes. To index an unknown plane, the procedure is reversed: (1) Choose an origin that the plane does not pass through. (2) Determine the intercepts of the plane on the *x*, *y*, and *z* axes. (3) Take reciprocals of these intercepts. (4) Clear fractions and enclose in parentheses ( ).

Characteristics of Ordered Solids

There are several useful properties of unit cells that can be determined through geometric relations and 3-D visualization. These include:

- Number of atoms per unit cell
- Number of nearest-neighbor atoms (coordination number)
- Lattice parameter (spacing of atoms)
- Distance of nearest approach of atoms
- Atomic packing factor (the volume of atoms per unit volume of the solid)
- Density

Many of these relationships are given in Table 13.2 and in the NCEES *FE Supplied-Reference Handbook* and therefore do not need to be memorized, but it is useful to see how these are determined. The following examples will illustrate these relationships.

Table 13.2 Characteristics of selected crystal structures

Unit Cell	Number of Atoms per Unit Cell	Coordination Number	Lattice Parameter	Packing Factor
BCC	2	8	$a = 4R/\sqrt{3}$	0.68
FCC	4	12	$a = 2R/\sqrt{2}$	0.72
HCP		12		0.72

Figure 13.4 showed a unit cell of α iron with an atom at its center. Inasmuch as each of the eight corner atoms is shared by the eight adjacent unit cells, we can note that there are (1 + 8/8 = 2) atoms per unit cell. From Table 13.1, each iron atom has a mass of 55.85 g per 0.6022×10^{24} atoms. Since iron forms a body-centered cubic crystal, its unit cell has a mass of 1.855×10^{-26} g. X-ray diffraction techniques give a lattice constant value of 2.866×10^{-18} m. As a result, the mass per unit volume, that is, the **density**, may be calculated to be nearly 7.88 g/cm^3. Careful density measurements give a value of slightly more than 7.87 g/cm^3 at ambient temperatures. The close agreement for the simple property of density implies that the concept of the crystal structure is valid.

The relationship between the size of the atoms (atomic radius, r) and lattice parameter (a) for BCC crystals is demonstrated in Fig. 13.8. Note that the atoms touch along the body diagonal. By inspection this gives a known length of the body diagonal of $4r$.

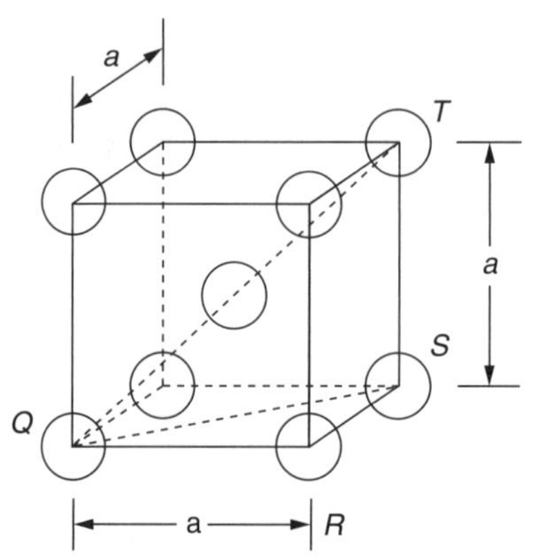

Figure 13.8 BCC unit cell

Using the triangle QRS

$$\| \overline{QS}^2 \| = a^2 + a^2 = 2a^2$$

and for triangle QST

$$\| \overline{QT}^2 \| = \| \overline{TS}^2 \| + \| \overline{QS}^2 \|$$

But $\overline{QT} = 4r$, r being the atomic radius. Also, $\overline{TS} = a$. Therefore,

$$(4r)^2 = a^2 + 2a^2$$

or

$$a = \frac{4r}{\sqrt{3}}$$

The relationship between r and a for FCC crystals can be similarly determined by noting that the atoms touch along the face diagonal as shown in Fig. 13.9.

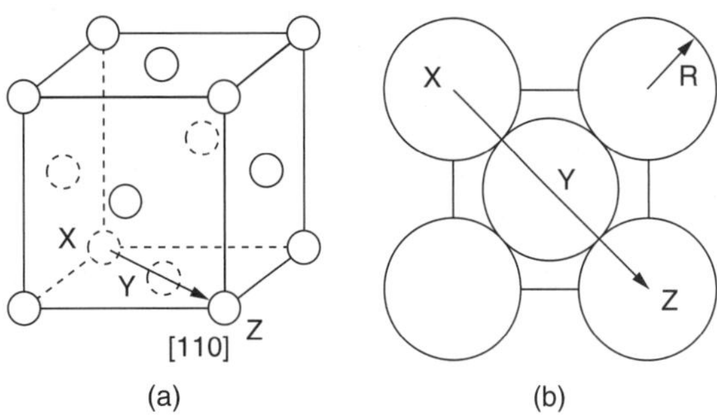

Figure 13.9 (a) Reduced-sphere FCC unit cell with the [110] direction indicated. (b) The bottom face-plane of the FCC unit cell in (a) on which is shown the atomic spacing in the [110] direction, through atoms labeled X, Y, and Z.

The plastic deformation of these solids is also related to crystal structure. To illustrate, in Fig. 13.10 the {111} plane of an FCC structure is shown. The {111} planes of aluminum and other FCC metals are the most densely packed planes; each atom of those planes is surrounded by six other atoms in the same plane. There is no arrangement where more atoms could have been included. This is not true for other planes within the FCC crystal. Since there is a fixed number of atoms per unit volume, it is apparent that the interplanar spacings between parallel (111) planes through the centers of atoms must be greater than between planes of other orientations. It is thus not surprising that sliding (**slip**) occurs there at lower stresses than on other planes.

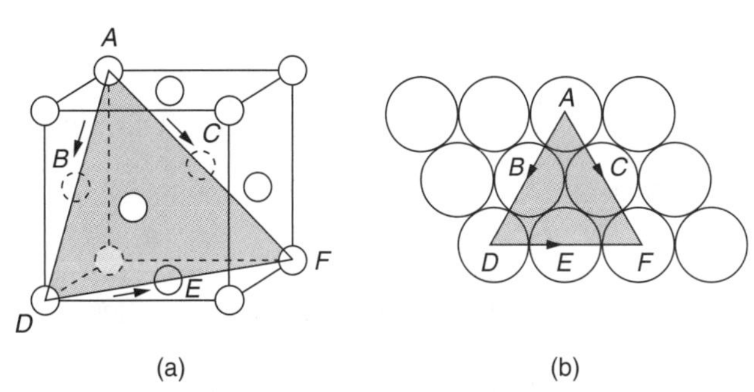

Figure 13.10 (a) A {111}<110> slip system shown within an FCC unit cell. (b) The (111) plane from (a) and three <110> slip directions (as indicated by arrows) within that plane comprise possible slip systems.

Not only is the required shear stress for slip low on the {111} planes, the <110> directions on those planes require less stress for slip than other directions. This is because the step distance between like crystal positions is the shortest, specifically, $2r$ where r is the atomic radius. We speak of the <110>{111} **slip system** for FCC metals. In a BCC metal, <111> {110} is the prominent slip system. Note especially that the {111} planes and the <110> directions are operative in FCC metals, whereas it is the {110} planes and the <111> directions in BCC metals. Hexagonal metals such as Mg, Zn, and Ti do not deform as readily as BCC and FCC metals because there are fewer combinations of directions and planes for slip by shear stresses.

Example 13.1

Carbon (12.011 amu) contains C^{12} and C^{13} isotopes with masses of 12.00000 amu and 13.00335 amu, respectively. What are the percentages of each?

Solution

$$12.011 = x(12.00000) + (1 - x)(13.00335)$$
$$1.00335\,x = 0.99335$$
$$x = 98.9\%$$

Carbon is 98.9% C^{12} and 1.1% C^{13}.

Example 13.2

Aluminum has a face-centered cubic unit cell, that is, an atom at each corner of the unit cell and an atom at the center of each face (see Fig. 13.5). The Al–Al distance (=$2r$) is 0.2863 nm. Calculate the density of aluminum. (The mass of an aluminum atom is 26.98 amu.)

Solution

$$\text{Volume} = [2(0.2863 \times 10^{-9}\,\text{m})/\sqrt{2}]^3 = 6.638 \times 10^{-29}\ \text{m}^3$$
$$\text{Mass} = (8/8 + 6/2\ \text{atoms})(26.98\ \text{g}/6.022 \times 10^{23}\ \text{atoms}) = 1.792 \times 10^{-22}\ \text{g}$$
$$\text{Density} = (1.792 \times 10^{-22}\ \text{g})/(6.638 \times 10^{-29}\ \text{m}^3) = 2.700 \times 10^{6}\ \text{g/m}^3 = 2.700\ \text{g/cm}^3$$
$$\text{Actual density} = 2.699\ \text{g/cm}^3$$

Example 13.3

What is the repeat distance along a <211> direction of a copper crystal that is face-centered cubic and has a unit cell dimension (lattice constant) of 0.3615 nm?

Solution

Select the center of any atom as the origin. Make a sketch of a cubic unit cell with that origin arbitrarily set at the lower left rear corner, as shown in Exhibit 1. One of the <211> directions, with coefficients of 2, 1, and 1, exits the first unit cell through the center of its front face, where another atom is centered (with no other intervening atoms).

$$d^2 = a^2 + \left(\frac{a}{2}\right)^2 + \left(\frac{a}{2}\right)^2 = \left(\frac{6}{4}\right)(0.3615\ \text{nm})^2;\ d = 0.4427\ \text{nm}$$

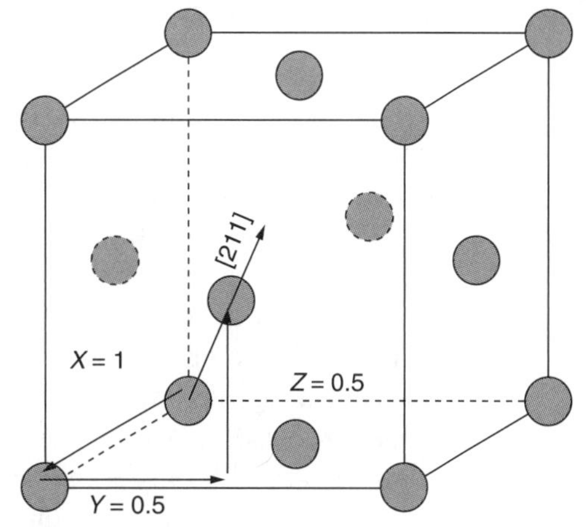

Exhibit 1 FCC unit cell

Example **13.4**

Assuming spherical atoms, calculate the packing factor of a BCC metal.

Solution

The packing factor is the volume of atoms per unit volume of the solid. Based on Fig. 13.4, there are (8/8 + 1) atoms/unit cell.

$$\text{Volume of atoms} = 2 \times (4\pi/3)r^3 = 8.38r^3$$

Since the cube diagonal is $4r$,

$$\text{Volume of unit cell} = a^3 = \left(\frac{4r}{\sqrt{3}}\right)^3 = 12.32r^3$$

$$\text{Packing factor} = \frac{8.38r^3}{12.32r^3} = 68\%$$

Example **13.5**

How many atoms are there per mm^2 on one of the {110} planes of copper (FCC)?

Solution

From Example 3, $a_{Cu} = 0.3615$ nm. A {110} plane lies diagonally through the unit cell and is parallel to one of the axes. There are (4/4 + 2/2 atoms) in an area measuring a by $a\sqrt{2}$. The number of atoms is 2 atoms/[$(0.3615 \times 10^{-6}$ mm$)^2 \sqrt{2}$] = 10.8×10^{12}/mm^2.

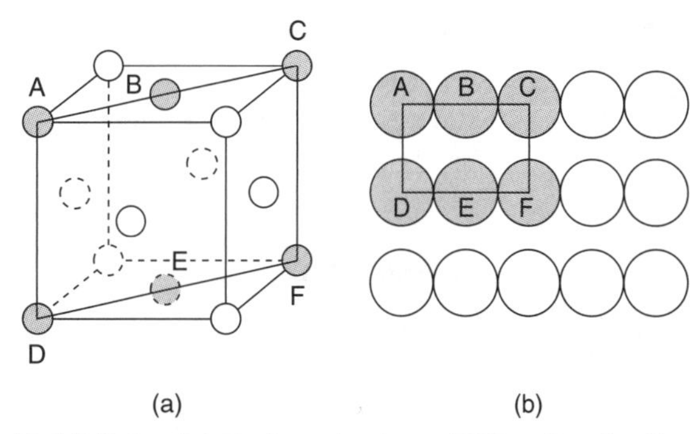

Exhibit 2 (a) Reduced-sphere FCC unit cell with (110) plane. (b) Atomic packing of an FCC (110) plane. Corresponding atom positions from (a) are indicated.

ATOMIC DISORDER IN SOLIDS

In the previous section, we paid attention to the orderly combinations that can exist in engineering materials. A variety of properties and behaviors are closely related to that ordering. Examples that were cited included density, slip systems, and molecular melting.

However, no solid has perfect order. There are always imperfections present, and these may be highly significant. A few missing potassium atoms in a compound such as KBr do not detectably affect the density; however, their absence introduces color. Likewise, the absence of a partial plane of atoms in a metal significantly modifies the shear stress required by a slip system for plastic deformation. Also, a rubber is vulcanized by the joining (**crosslinking**) of adjacent molecules with only a minor compositional change (sulfur addition). As a final example, the thermal conductivity is doubled in diamond if the one percent of naturally present C^{13} has been removed.

Crystal defects can be characterized as point, line (one-dimensional), plane (two-dimensional), or volume (three-dimensional) defects.

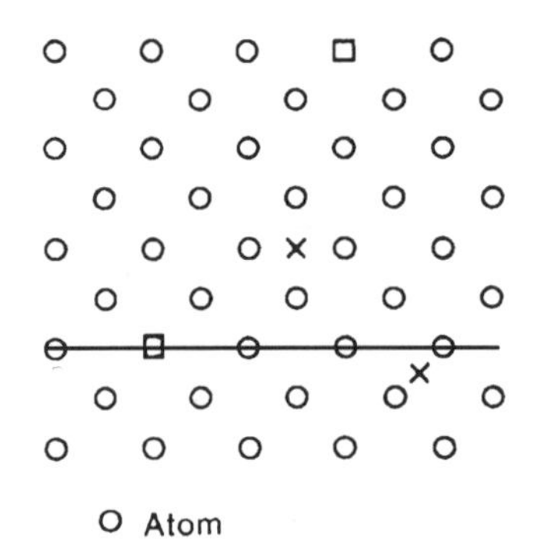

Figure 13.11 Point imperfections. These defects can originate from imperfect crystallization, or through a relocation of energized atoms.

Point Imperfections

Imperfections may be atomically local in nature. Missing atoms (**vacancies**), extra atoms (**interstitials**) **displaced** atoms, and impurity atoms are called **point imperfections** (Fig. 13.11). Their existence facilitates the transport of atoms (**diffusion**), thus becoming important in materials processing. In service, the presence of point imperfections scatters internal waves and thus reduces energy transport. These include elastic waves for thermal conductivity, light waves for optical transparency, and electron waves for electrical conduction.

Lineal Imperfections

The principal defect of this type is a **dislocation**. It is most readily visualized (Fig. 13.12) as a partial displacement along a slip system or an extra half-plane of atoms. Dislocations facilitate plastic deformation by slip; however, increased numbers of dislocations lead to dislocation tangles or "traffic snarls" and therefore to interference of slip. Thus, ductility decreases and strength increases. Dislocations may develop during initial crystal growth as well as from plastic deformation.

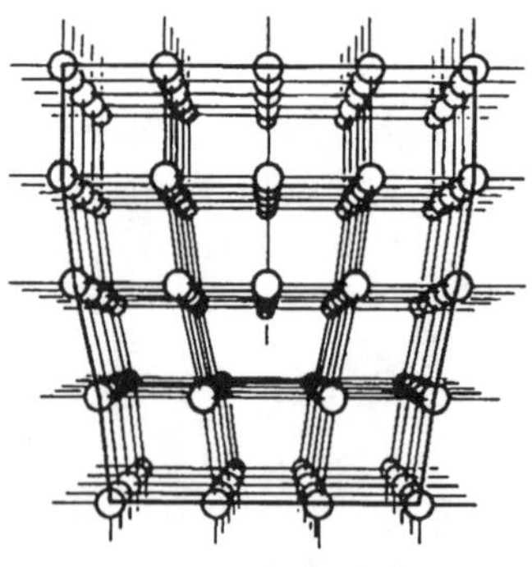

Figure 13.12 Dislocation, ⊥ (schematic). These lineal imperfections facilitate slip within crystals. Excessive numbers of defects, however, lead to their entanglement and a resistance to deformation. The resulting increase in strength is called **work hardening**.

Boundaries

No liquid or solid is infinite; each has a **surface**. The resulting two-dimensional boundary has a different structure and bonding than that encountered in the underlying material. Since atoms are absent from one side of the boundary, the atoms at these surfaces possess additional energy, and subsurface distortions are introduced.

Grain Boundaries

Boundaries also occur where two growing crystals meet. These are called grain boundaries. There are atoms on each side of the boundary; however, any misorientation between the two crystal grains leads to local inefficiencies in atomic packing. As a result some atom-to-atom distances are compressed; others are

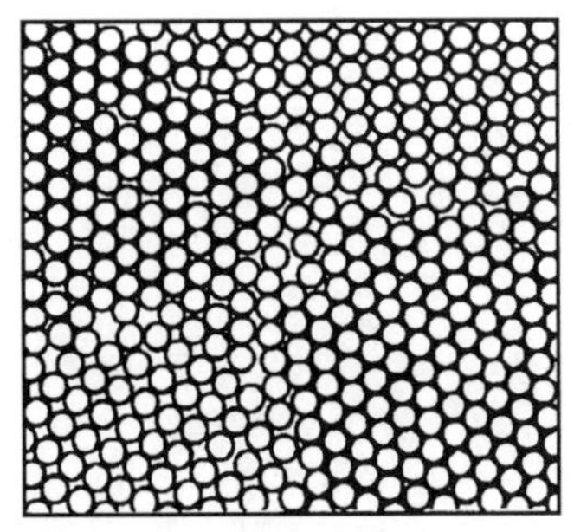

Figure 13.13 Grain boundaries (schematic). Most materials contain a multitude of grains, each of which is a separate crystal. The boundary between grains is a zone of mismatch. Atoms along grain boundaries possess added energy because they are not as efficiently coordinated with their neighbors.

stretched. Both distortions increase the energy of the atoms along the grain boundaries. This **grain-boundary energy** introduces reactive sites for structural modification during processing and in service. The imperfect grain boundary is also an "avenue" for atomic diffusion within solid materials (Fig. 13.13). Further, the mismatch at a grain boundary blocks slip that might otherwise occur, particularly at ambient temperatures where atom-by-atom mobility is limited.

Solutions

Both sugar and salt dissolve in water, each producing a **solution** (commonly called a syrup and a brine, respectively). There are many familiar solutions. Lower melting temperatures, increased conductivity, and altered viscosities commonly result.

Solid Solutions

Impurities, both unwanted and intentional, may also dissolve into a solid. A crystal cannot be perfect when foreign atoms are present. Common brass (70Cu–30Zn) is a familiar example. Zinc atoms simply substitute for copper atoms to produce a **substitutional solid solution**. It has the face-centered cubic crystal structure of pure copper; however, with approximately one-third of the copper atoms replaced by zinc atoms, we can anticipate certain changes. First, the size of the unit cell is increased because zinc atoms are approximately four percent larger than copper atoms. (Fifteen percent mismatch is the practical limit for extensive substitutional solid solution.) Second, charge transport is greatly reduced because the electrons are scattered as they travel toward the positive electrode through locally varying electrical fields. Thus, the electrical and thermal **conductivities** are decreased. Also, atoms of a different size immobilize dislocation movements and in turn produce **solution hardening**.

There are some important situations in which small atoms can be positioned among larger atoms. The most widely encountered example is the solution of carbon in face-centered-cubic iron at elevated temperatures. The result is an **interstitial solid solution**. Iron changes to body-centered-cubic at ambient temperatures. The BCC iron cannot accommodate many carbon atoms in its interstices. This loss of interstitial solubility plays a major role in the various heat treatments of steel.

Amorphous Solids

Materials lose their crystallinity when they melt. The long-range order of the crystals is not maintained. As a liquid, the material is amorphous (literally, without form). Those materials that are closely packed—for example, metals—will expand on melting [Fig. 13.14(a)]. Being thermally agitated, the atoms do not maintain close coordination with neighboring atoms (or molecules). A few materials, such as ice, which has stereospecific bonds, lose volume on melting and become more dense [Fig. 13.14(b)].

In certain cases, crystallization may be avoided during cooling. An amorphous solid can result. Two common examples are window glass and the candy part of peanut brittle. In neither case is there time for the relatively complex crystalline structures to form. Very rapid cooling rates are required to avoid crystallization when the crystalline structures are less complex. For example, metals must be

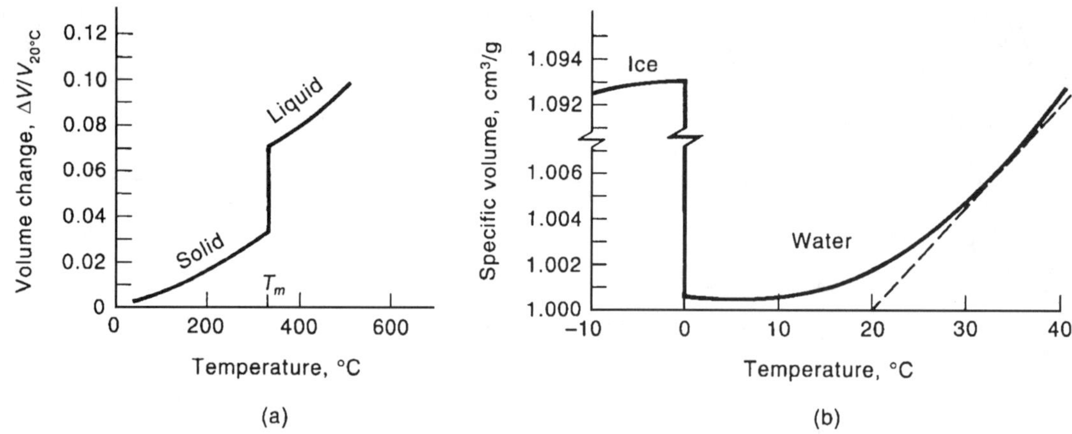

Figure 13.14 Volume changes during heating. (a) Lead (FCC); (b) H_2O. Melting destroys the efficient packing of metallic atoms within solids, so most metals *expand* when melted. The crystalline structure of ice, silicon, and a number of related materials with stereospecific bonds have low, inefficient atomic packing within solids. Therefore, they lose volume when melted.

quenched a thousand degrees in milliseconds to avoid crystallization. The amorphous materials that result are considered to be **vitreous**, or glass-like.

Example 13.6

Sterling silver contains 92.5% silver and 7.5% copper by weight. What percentage of the atoms on a (111) plane are silver? Copper?

Solution

The alloy is a random solid solution; therefore the percentage of atoms on the (111) plane or any other plane will be the same as throughout the alloy. Change weight percent to atom percent. The atomic masses are 107.87 and 63.54 amu, respectively.

Basis: 1000 amu = 925 amu Ag + 75 amu Cu
Ag: 925 amu/(107.87 amu/Ag) = 8.58 Ag atoms
Cu: 75 amu/(63.54 amu/Cu) = 1.18 Cu atoms
Total: = 9.76 atoms

Ag atoms = 8.58/9.76 = 87.9%; Cu atoms = 12.1%

MICROSTRUCTURES OF SOLID MATERIALS

The atomic coordination within solids is on the nanometer scale. The resulting structures involve either crystalline solids, or amorphous solids such as the glasses. As discussed, certain properties arise from these atom-to-atom relationships. Other properties arise from longer-range structures, generally with micrometer to millimeter dimensions, called **microstructures**.

Atomic Movements in Materials

Our initial examination of crystals implied that an atom becomes permanently coordinated with adjacent atoms and remains fixed in position. This is not entirely true. In the first place, there is thermal vibration of the atoms within the crystal. Thus, while the lattice constant and the mean interatomic distances are fixed to several significant figures, the instantaneous interatomic distances vary. The amplitude of vibration increases with temperature. At the melting temperature, the crystal is literally "shaken apart." As the melting point is approached, a measurable fraction of atoms jump out of their crystalline positions. They may return, or they may move to other sites, producing the vacancies and interstitials discussed in the previous section.

Diffusion

Within a single-component material, such as pure copper, there is equal probability that like numbers of copper atoms will jump in each of the coordinate directions. Thus, there is no net change.

Imagine, however, one location, x_1, in nickel containing 2000 atoms of copper for every mm^3 of nickel, whereas one mm to the right at x_2 there are 1000 atoms of copper for every mm^3. Although all copper atoms have the same probability for jumping in either direction, there is a net movement of copper atoms to the right simply because there are unequal numbers of copper atoms in the two locations. There is a copper **concentration gradient**, $\Delta C/\Delta x$. In this case

$$\Delta C/\Delta x = (C_2 - C_1 \text{ Cu/mm}^3)/(x_2 - x_1 \text{ mm}) = -(1000 \text{ Cu/mm}^3)/\text{mm} \quad \textbf{(13.2)}$$

The rate of diffusion, called the **flux**, J, is proportional to the concentration gradient

$$J = -D\frac{dC}{dx} \quad \textbf{(13.3)}$$

where D is the *diffusivity*, also called the **diffusion coefficient**. (Its units are m^2s^{-1} since the units for flux and concentration gradient are $m^{-2}s^{-1}$ and m^{-4}, respectively.)

Among the various factors that affect the diffusivity are (1) the size of the diffusing atom, (2) the crystal structure of the matrix, (3) bond strength, and (4) temperature. Comparisons are made in Table 13.3. The diffusivity of the C in Fe_{FCC} is higher than for the Fe in Fe_{FCC} because the diffusing carbon atom is smaller than the iron atom. The diffusivity of the C in Fe_{BCC} is higher than for the C in Fe_{FCC} because the latter contains more iron atoms per m^3. The FCC packing factor is higher than the BCC form. The diffusivity for Fe in Fe_{FCC} is lower than for Cu in Cu_{FCC} because the iron atoms are more strongly bonded than the copper atoms. (The melting temperatures provide evidence of this.) In each case, higher diffusion coefficients accompany higher temperatures.

The process engineer obtains many of the properties required of a material by heat treating procedures. Diffusion plays the predominant role in achieving the required microstructures.

Table 13.3 Selected diffusion coefficients

Diffusing Atom	Host Structure	Diffusion Coefficient, D, m^2/s	
		500°C (930°F)	1000°C (1830°F)
Fe	FCC Fe	2×10^{-23}	2×10^{-16}
C	FCC Fe	5×10^{-15}	3×10^{-11}
C	BCC Fe	1×10^{-12}	2×10^{-9}
Cu	FCC Cu	1×10^{-18}	2×10^{-13}

Structures of Single-Phase Solids

Many materials possess only one structure. Examples include copper wire, transparent polystyrene cups, and Al_2O_3 substrates for electronic circuits. The wire contains only face-centered-cubic crystals. The polystyrene is an amorphous solid with minimal crystallinity. The substrates have numerous crystals, all with the same crystalline structure. We speak of single phases because none of these materials contains a second structure.

Grains

Each of the individual crystals in a copper wire is called a **grain**. Recall from the previous section that adjacent crystals may be misoriented with respect to each other, and that there is a boundary between them. This is shown schematically in Fig. 13.15.

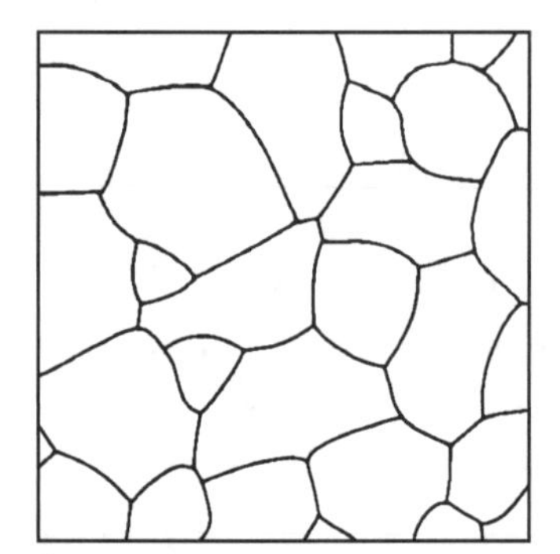

Figure 13.15 Grains (schematic). Each grain is a separate crystal. There is a surface of mismatch between grains because adjacent grains have unlike orientations. The **grain boundary area** is inversely related to grain size.

The **grain size** is an important structural parameter, because the grain boundary area varies inversely with the grain size. Diffusion is faster along grain boundaries because there is less-perfect packing of the atoms and, consequently, a more open structure. At ambient temperatures, grain boundaries interfere with plastic deformation, thus increasing the strength. At elevated temperatures the grain boundaries contribute to creep and therefore are to be minimized. Grain boundaries also serve as locations that initiate structural changes within a solid.

Grain growth may occur in a single-phase material. The driving force is the fact that the atoms at the grain boundary possess extra energy. Grain growth reduces this excess energy by minimizing the boundary area. Higher temperatures increase the rate of grain growth because the atoms migrate faster. However, since the growth rate is inversely related to grain size, we see a decrease in the rate of growth with time.

The texture of a single-phase solid can also depend on **grain shape** and **orientation**. Even a noncrystalline material may possess a structure. For example, the molecular chains within a nylon fiber have been aligned during processing to provide greater tensile strength.

Phase Diagrams

Many materials possess more than one phase. A simple and obvious example is a cup containing both ice and water. While ice and water have the same composition, the structures of the two phases are different. There is a **phase boundary** between the two phases. A less obvious, but equally important material is the steel used as a bridge beam. The steel in the beam contains two phases: nearly pure iron (body-centered cubic), and an iron carbide, Fe_3C.

In these two examples, we have a mixture of two phases. Solutions are phases with more than one component. In the previous section, we encountered brass, a solid solution with copper and zinc as its components.

Although the steel beam just cited contains a mixture of two phases at ambient temperatures, it contained only one phase when it was red-hot during the shaping process. At that temperature, the iron was face-centered cubic and was thereby able to dissolve all of the carbon into its interstices. No Fe_3C remained. The phases within a material can be displayed on a **phase diagram** as a function of temperature and composition.

Phase diagrams are useful to the engineer because they indicate the temperature and composition requirements for attaining the required internal structures and accompanying service properties. The phase diagram shows us (1) *what* phases to expect, (2) the *composition* of each phase that is present, and (3) the *quantity* of each phase within a mixture of phases.

What Phases?

Sterling silver contains 92.5Ag–7.5Cu (weight percent). Using the Ag–Cu phase diagram of Fig. 13.16, we observe that this composition is liquid above 910°C; below 740°C, it contains a mixture of the α and β solid structures. The former is a solid solution of silver plus a limited amount of copper; the latter is a solid solution of copper plus a limited amount of silver. Between 740 and 810°C, only one phase is present, α. In that temperature range, all of the copper can be dissolved in the α solid solution. From 810 to 910°C, the alloy changes from no liquid to all liquid.

Phase Compositions?

Pure silver has a face-centered-cubic structure. Copper forms the same crystalline structure. Not surprisingly, copper atoms can be substituted for silver. But the solid solution in α is limited, because the silver atom is 13 percent larger in diameter than is copper. However, the solubility increases with temperature to 8.8 weight-percent copper at 780°C. This is shown by the boundary of the silver-rich shaded area of Fig. 13.16. Within that shaded area, we have a single phase called α.

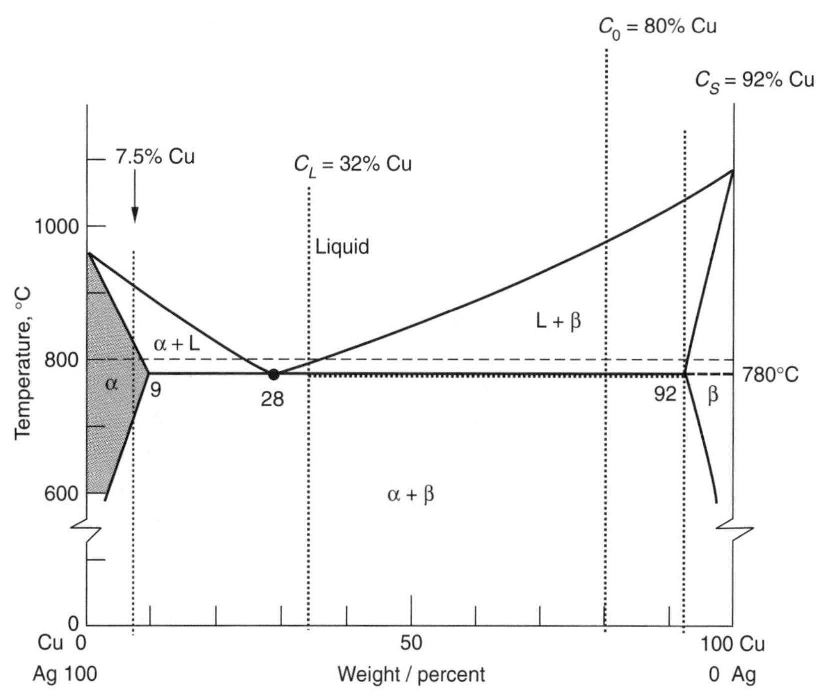

Figure 13.16 Silver-copper phase diagram. Sterling silver contains 7.5% copper; therefore, it has a single phase, α, in the 740–810°C range.

Conversely, the FCC structure of copper can dissolve silver atoms in solid solution. At room temperature the solubility is very small. Again, as the temperature is raised the solubility limit increases to 8 weight-percent silver (92 wt. % Cu) at 780°C. This copper-rich phase is called β.

Silver and copper are mutually soluble in a liquid solution. Above 1100°C there is no limit to the solubility. As the temperature is reduced below the melting point of copper (1084°C), the copper solubility limit decreases from 100% to 28%. Excess copper produces the β phase. Likewise, below the melting point of silver (962°C), the silver solubility limit decreases from 100% to 72%. Excess silver produces the α phase. The two solubility curves cross at approximately 72Ag–28Cu and 780°C. We call this low-melting liquid **eutectic**.

How Much of Each Phase in a Mixture?

At 800°C, a 72Ag–28Cu alloy is entirely liquid. At the same temperature, but with added copper, the solubility limit is reached at 32% Cu. Beyond that limit, still at 800°C, any additional copper precipitates as β. Halfway across the two-phase field of (L + β), there will be equal quantities of the two phases. Additional copper in the alloy increases the amount of β. Within this two-phase region the liquid remains saturated with 32% Cu, and the solid β is saturated with 8% Ag (thus, 92% Cu).

The *composition* of each phase is dictated by the solubility limits. (In a one-phase field, the solubility limits are not factors; the alloy composition and the phase composition are identical.) The *amount* of each phase in a two-phase field is determined by interpolation between the solubility limits using what is called the lever law. For example, at 800°C, an alloy consisting of 80% Cu and 20% Ag would consist of a liquid phase containing 32% Cu and a solid phase of 92% Cu. The weight fraction of liquid phase, W_L, and solid phase, W_s, is determined by

$$W_L = \frac{C_S - C_0}{C_S - C_L} \quad \text{and} \quad W_S = \frac{C_0 - C_L}{C_s - C_L}$$

In addition $W_L + W_S = 1$.

Reaction Rates in Solids

A phase diagram is normally an equilibrium diagram; that is, all reactions have been completed. The time required to reach equilibrium generally increases with decreasing temperature. Thus, a material may not always possess the expected phases with predicted amounts or compositions. Even so, the phase diagram is valuable. For example, sterling silver (92.5Ag–7.5Cu) contains only one phase when equilibrated at 775°C (Fig. 13.16). Slow cooling to room temperature precipitates β as a minor second phase, as would be expected when plenty of time is available. Rapid cooling, however, traps the copper atoms within the α solid solution. This situation is used to advantage, because the solid solution is stronger than the (α + β) combination. Also, a single-phase alloy corrodes less readily. This explains why the "impure" sterling silver is commonly preferred over pure silver.

The selection of compositions and processing treatments is generally based on a knowledge of equilibrium diagrams plus a knowledge of how equilibrium is circumvented.

Microstructures of Multiphase Solids

The microstructure of a single-phase, crystalline solid includes the *size, shape*, and *orientation* of the grains. Variations in these properties are also found in multiphase solids. In addition, the microstructure of a multiphase solid may also vary in the *amount* of each phase and the *distribution* of the phases. In an equilibrated microstructure, the amounts of the phases may be predicted directly from the phase diagrams using the lever law. From Fig. 13.17, a 1080 steel (primarily iron, with 0.80 percent carbon) will have twice as much Fe_3C (W_{Fe3C}) at room temperature as a 1040 steel (with 0.40 percent carbon). The Fe_3C is a hard phase. Therefore, with all other factors equal, we expect a 1080 steel to be harder than a 1040 steel, and it is.

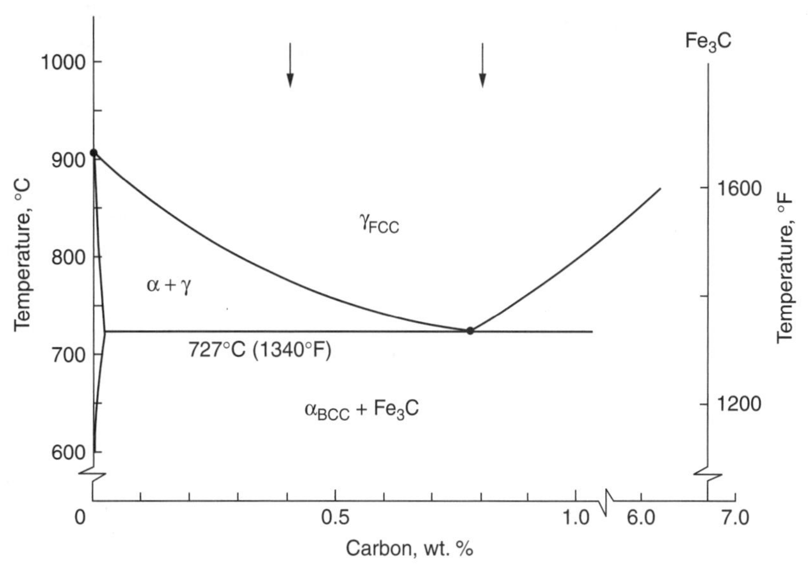

Figure 13.17 Portion of the Fe–Fe_3C phase diagram. Most steels are heat treated by initially forming austenite, γ, which is Fe_{FCC}. It changes to a mixture of ferrite, α, and carbide at lower temperatures.

$$W_{Fe_3C} = \frac{C_0 - C_\alpha}{C_{Fe_3C} - C_\alpha}$$

The distribution of phases within microstructures is more difficult to quantify, so only descriptive examples will be cited. Similar to sterling silver, aluminum will dissolve several percent of copper in solid solution at 550°C. If it is cooled slowly, the copper precipitates as a minor, hard, brittle compound ($CuAl_2$) along grain boundaries of the aluminum. The alloy is weak and brittle and has little practical use. If the same alloy is cooled rapidly from 550°C, trapping the copper atoms within the aluminum grains as shown in Fig. 13.18(a), the quenched solid solution is stronger and more ductile and has commercial uses. If the quenched alloy is reheated to 100°C, the $CuAl_2$ precipitates, as expected from the phase diagram shown in Fig. 13.18(b) and (c). In this case, however, the precipitate is very finely dispersed within the grains of aluminum. The alloy retains its toughness, because the brittle $CuAl_2$ does not form a network for fracture paths. In addition, the strength is increased because the submicroscopic hard particles interfere with the deformation of the aluminum along slip planes. Alloys of this type are used in airplane construction because they are light, strong, and tough. Observe that all of the examples in this paragraph are for the same alloy. The properties have been varied through microstructural control.

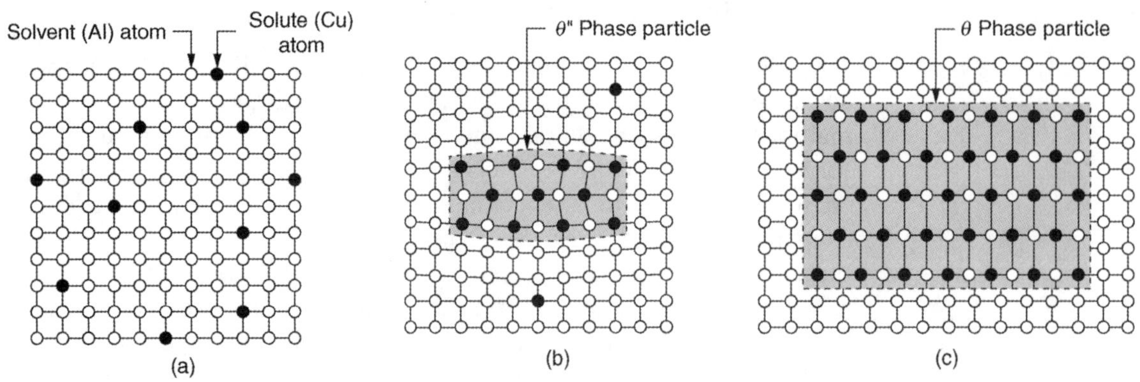

Figure 13.18 Schematic depiction of several stages in the formation of the equilibrium precipitate (θ) phase. (a) A supersaturated α solid solution; (b) a transition, θ'', precipitate phase; (c) the equilibrium θ phase, within the α-matrix phase. Actual phase particle sizes are much larger than shown here.

Two distinct microstructures may be produced in a majority of steels. In one, called **spheroidite**, the hard Fe_3C is present as rounded particles in a matrix of ductile ferrite (α). In the other, called **pearlite** Fe_3C and ferrite form fine alternating layers, or lamellae. Spheroidite is softer but tougher; pearlite is harder and less ductile (Fig. 13.19). The mechanical properties are controlled by the spacing of the Fe_3C, because the hard Fe_3C phase stops dislocation motion. The closer the spacing between Fe_3C particles, the greater the strength or hardness, but the lower the ductility. Note the differing magnifications of the spheroidite and pearlite photographs.

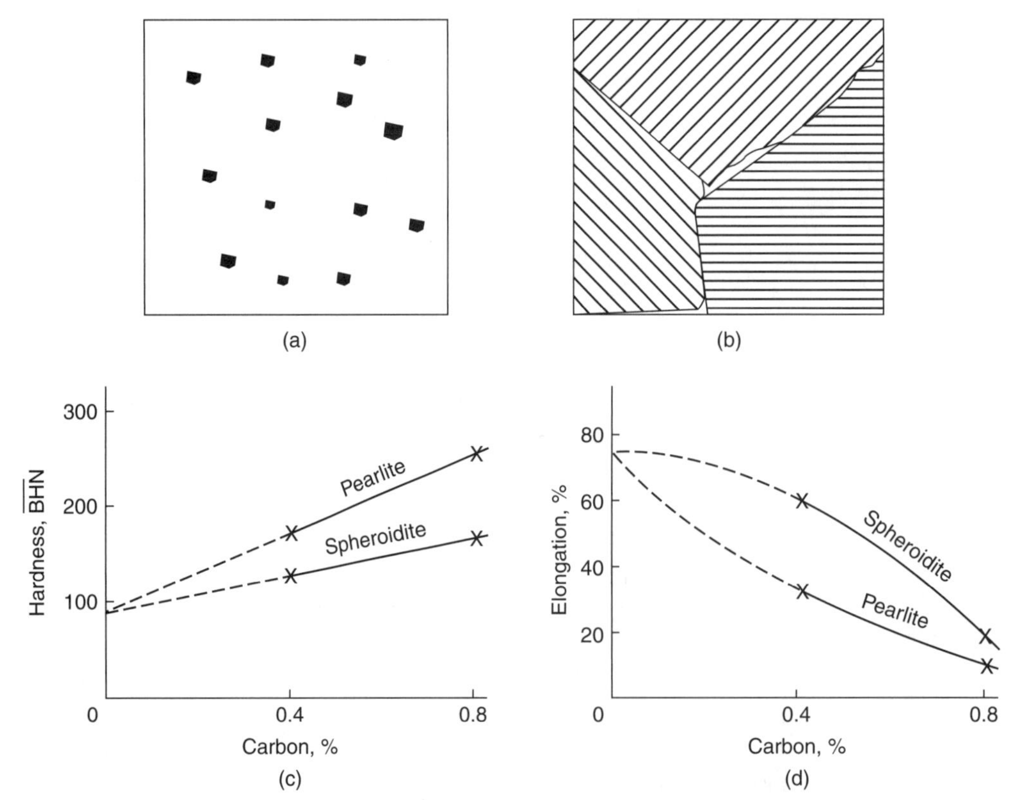

Figure 13.19 Phase distributions (schematic). The two samples are of the same steel, both containing ferrite (white) and carbide (black) but heat treated differently. Spheroidite is shown in (a), and pearlite in (b). The hardness of the two structures is shown in (c), and ductility is shown in (d).

Example 13.7

At 780°C, an Ag–Cu liquid of eutectic composition (Fig. 13.16) solidifies to solid α and solid β. (a) What are the compositions of the two solid phases just below the eutectic temperature? (b) How much of each of these two phases exists per 100 grams of alloy?

Solution

(a) From Fig. 13.16, the α phase is 91Ag–9Cu, and the β phase is 8Ag–92Cu.

(b) Interpolation along the 780 °C tie-line between the solubility limits yields

$$\beta\text{: (using Cu) } [(28 - 9)/(92 - 9)](100 \text{ g}) = 23 \text{ g}$$

$$\alpha\text{: (using Cu) } [(92 - 28)/(92 - 9)](100 \text{ g}) = 77 \text{ g}$$

or

$$\alpha\text{: (using Ag) } [(72 - 8)/(92 - 9)](100 \text{ g}) = 77 \text{ g}$$

Example 13.8

The phases and microstructures of an SAE 1040 steel may be related to the Fe-Fe_3C phase diagram (Fig. 13.17). (a) Assume equilibrium. What are the phases and their weight percents at 728°C? At 726°C? (b) Pearlite has the eutectoid composition of 0.77% C. What percent of the steel will *not* be contained in the pearlite?

Solution

(a) At 728°:

$$\alpha\text{ (0.02\% C): } (0.77 - 0.40)/(0.77 - 0.02) = 49\%\ \alpha$$
$$\gamma\text{ (0.77\% C): } (0.40 - 0.02)/(0.77 - 0.02) = 51\%\ \gamma$$
$$Fe_3C\text{ contains } 12/[12 + 3(55.85)] = 6.7\%\text{ C}$$

At 726°:

$$\alpha\text{ (0.02\% C): } (6.7 - 0.40)/(6.7 - 0.02) = 94\%\ \alpha$$
$$Fe_3C\text{: } (0.40 - 0.02)/(6.7 - 0.02) = 6\%\ Fe_3C$$

(b) All of the pearlite, $\alpha + Fe_3C$, comes from the γ. This FCC phase changes to pearlite at 727°C. The α that was present at 728°C remains unchanged. Therefore, the answer is $P = 51\%$; and unchanged $\alpha = 49\%$.

MATERIALS PROCESSING

For many engineers, materials are first encountered in terms of handbook data or stockroom inventories. However, materials always have a prior history. They must be obtained from natural sources, then subjected to compositional modifications and complex shaping processes. Finally, specific microstructures are developed to achieve the properties that are necessary for extended service.

Extraction and Compositional Adjustments

Few materials are used in their natural form. Wood is cut and reshaped into lumber, chip-board, or plywood; most metals must be extracted from their ores; rubber latex is useless unless it is vulcanized.

Extraction from Ores

Most ores are oxides or sulfides that require chemical reduction. Commonly oxide ores are chemically reduced with carbon- or CO-containing gas. For example,

$$Fe_2O_3 + 3\ CO \rightarrow 2\ Fe + 3\ CO_2 \qquad \textbf{(13.4)}$$

Elevated temperatures are used to speed up the reactions and more completely reduce the metal. If the metal is melted, it is more readily separated from the accompanying gangue materials. The reduced product is a carbon-saturated metallic iron. As such, it has only limited applications. Normally further processing is required.

Refining

Dissolved impurities must be removed. Even if the above ore were of the highest quality iron oxide, it would be necessary to refine it, because the metallic product is saturated with carbon. In practice, small but undesirable quantities of silicon and other species are also reduced and dissolved in the iron. They are removed by closely controlled **reoxidation** at chemically appropriate temperatures (followed by **deoxidation**). The product is a **steel**. Alloying additions are made as specified to create different types of steel.

Chemicals from which a variety of plastics are produced are refinined from petroleum. The principal step of petroleum refinement involves selective distillation of liquid petroleum. Lightweight fractions are removed first. Controlled temperatures and pressures distill the molecular fractions that serve as precursors for polymers. Residual fractions are directed to other products.

Polymerization makes macromolecules out of the smaller molecules that are the product of distillation. **Addition polymerization** involving a C=C→C—C reaction is encountered in the polymerization of ethylene, $H_2C{=}CH_2$; vinyl chloride, $H_2C{=}CHCl$; and styrene, $H_2C{=}CH(C_6H_5)$.

$$C{=}C \rightarrow C{-}C{-} \qquad \textbf{(13.5)}$$

Shaping Processes

The earliest cultural ages of human activities produced artifacts that had been shaped from stone, bronze, or iron. In modern technology, we speak of casting, deformation, cutting, and joining in addition to more specialized shaping procedures.

Casting

The concept of casting is straightforward. A liquid is solidified within a mold of the required shape. For metal casting, attention must be given to volume changes; in most cases there is shrinkage. In order to avoid porosity, provision must be made for feeding molten metal from a **riser**. **Segregation** may occur at the solidifying front because of compositional differences in the $(\alpha + L)$ range.

(See the earlier discussion of phase diagrams and the discussion of annealing below.)

A number of ceramic products are made by **slip casting**. The slip is a slurry of fine powders suspended in a fluid, usually water. The mold is typically of gypsum plaster with a porosity that absorbs water from the adjacent suspension. When the shell forming inside the mold is sufficiently thick, the remaining slip is drained. Subsequent processing steps are **drying** and **firing**. The latter high-temperature step bonds the powder into a coherent product.

The casting process is also used in forming polymeric products. Here the solidification is accomplished by polymerization. There is a chemical reaction between the small precursor molecules of the liquid to produce macromolecules and a resulting solid.

Deformation Processes

Deformation processes include forging, rolling, extrusion, and drawing, plus a number of variants. In each case, a force is applied, and a dimensional change results. **Forging** involves shape change by impact. **Rolling** may be used for sheet products as well as for products with constant cross sections, for example, structural beams. **Extrusion** is accomplished through open or closed dies. The former requires that the product be of uniform cross section, such as plastic pipe or siderails for aluminum ladders. Closed dies are molds into which the material is forced. These forming processes can be done at ambient temperature (cold working) or elevated temperatures ($T > 0.4T_m$ in Kelvin; hot working). The forming temperature has a strong effect on mechanical properties. Products formed at ambient temperature have high dislocation densities and hence greater strength and hardness but lower ductility than hot-worked products.

In general, ceramics do not lend themselves to the above deformation processes because they lack ductility. Major exceptions are the glasses, which deform not by crystalline slip but by viscous flow.

Cutting

Chiseling and sawing are cutting processes that predate history. Current technology includes **machining** in which a cutting tool and the product move with respect to each other. Depth and rates of cut are adjustable to meet requirements. **Grinding** is a variant of machining that is used for surface removal.

Joining

The process of **welding** produces a joint along which the abutting material has been melted and filled with matching metal. **Soldering** and **brazing** processes use fillers that have a lower melting point than the adjoining materials, which remain solid. There are glass solders as well as metallic ones. Adhesives have long been used for joining wood and plastic components, and many have now been developed to join metals and ceramics.

Annealing Processes

Annealing processes involve reheating a material sufficiently that internal adjustments may be made between atoms or between molecules. The temperature of annealing varies with (1) the material, (2) the amount of time available, and (3) the structural changes that are desired.

Homogenization

The dynamics of processing will produce segregation. For example, when an 80Cu–20Ni alloy starts to solidify at 1200°C, the first solid contains 30% Ni. When solidification is completed, the final liquid has only 12% Ni. Uniformity can be obtained if the alloy is reheated to a temperature at which the atoms can relocate by diffusion. There is a time-temperature relationship (log t vs. $1/T$). In this case an increase in temperature from 500 to 550°C reduces the necessary annealing time by a factor of eight.

Recrystallization

Networks of dislocations are introduced when most metals are plastically deformed at ambient temperatures (cold working). The result is a work hardening and loss of ductility. Whereas the resulting increase in strength is often desired, the property changes resulting from dislocations make further deformation processing more difficult. Annealing will remove the dislocations and restore the initial workable characteristics by forming new, strain-free crystals.

A one-hour heat treatment is a common shop practice because it allows for temperature equalization as well as scheduling requirements. For that time frame, it is necessary to heat a metal to approximately 40 percent of its melting temperature (on the absolute scale). Thus, copper which melts at 1085°C (1358 K) may be expected to recrystallize in the hour at 270°C (545 K). The recrystallization time is shorter at higher temperatures and longer at lower temperatures.

Grain Growth

Extended annealing, beyond that required for recrystallization, produces grain growth and therefore coarser grains. Normally, this is to be avoided. However, grain growth has merit in certain applications because grain boundaries hinder magnetic domain boundary movements, reducing creep. So, coarse grains are preferred in the sheet steel used in transformers, for example. At high temperatures, grain boundaries permit creep to occur under applied stresses, producing changes in dimensions.

Residual Stresses

Expansions and contractions occur within materials during heat-treating operations. These are isotropic for many materials; or they may vary with crystal orientation. Also, differential expansions exist between the two or more phases in a multiphase material. The latter, plus the presence of thermal gradients, introduce internal stresses, which can lead to delayed fracture if not removed.

It is generally desirable to eliminate these residual stresses by an annealing process called **stress relief**. The required temperature is less than that for recrystallization because atomic diffusion is generally not necessary; rather, adjustments are made through the local movement of dislocations. Stress relief is performed on metals before the final machining or grinding operation. Annealing is always performed on glass products, because any residual surface tension easily activates cracks in this nonductile material.

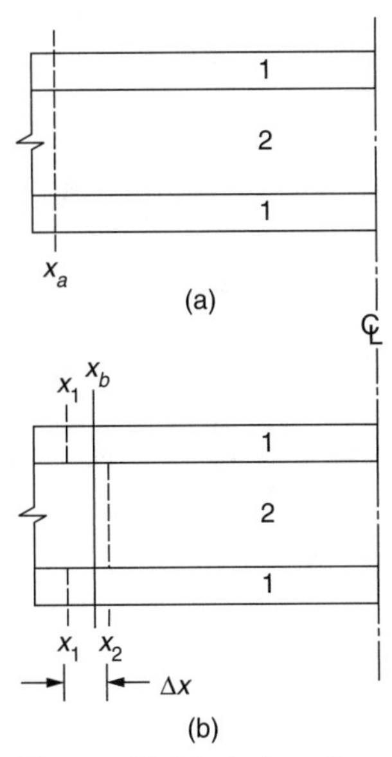

Figure 13.20 Induced stresses (sandwich glass). (a) The previously bonded composite containing glasses 1 and 2 is annealed. There are no internal stresses at x_a. (b) If the three layers were separate, the outer layers which have a lower thermal coefficient, would contract less during cooling than the center glass (x_1 vs. x_2). Since the layers are bonded together, the restricted contraction of the center (x_a to x_b rather than x_a to x_2) induces compression within the surface layers (x_1 to x_b).

Induced Stresses

In apparent contradiction to the last statement, residual compressive stresses may be prescribed for certain glasses, since glass like most nonductile materials is strong in **compression** but weak in **tension**. As an example, a familiar dinnerware product is made from a "sandwich" glass sheet in which the "bread" layers have a lower thermal expansion than does the "meat" layer. The processing involves heating the dinnerware to relieve all stresses (annealing). As the dinnerware pieces are cooled, the center layer tries to contract more than the surface layers, placing the surfaces under compression (Fig. 13.20). Any tension encountered in service must overcome the residual compression before a crack can propagate.

Time-Dependent Processes

We have seen in a previous section that sterling silver is solution treated to dissolve all of the copper within the silver-rich α phase. The single phase is preserved by rapidly cooling the alloy. This avoids the precipitation of the copper-rich β phase, as required for equilibrium. The cooling rate need not be drastic for sterling silver because it takes a minute or more to nucleate and grow the precipitate, β.

Even more time is available for the production of a **silicate glass**—a supercooled liquid. The necessary bond breaking and rearrangements are very slow. The available processing times can approach an hour or more. However, in order to produce a **metallic glass**, the cooling rate must approach 1000 °C per *millisecond*. Otherwise, individual metal atoms rapidly order themselves into one of the crystalline patterns described earlier.

Martensitic Reactions

Most steel processing treatments initially heat the steel to provide a single-phase microstructure of γ or **austenite** (Fe_{FCC}). This is face-centered cubic and dissolves all of the carbon that is present. Normal cooling produces **pearlite**, a lamellar microstructure containing layers of α or **ferrite** (Fe_{BCC}) and **carbide**, (Fe_3C), as shown previously on the Fe–Fe_3C phase diagram (Fig. 13.17).

$$Fe_{FCC} \xrightarrow{\text{cooling}} Fe_{BCC} + Fe_3C \tag{13.6}$$

If **quenched**, a different structure forms:

$$Fe_{FCC} \xrightarrow{\text{quenching}} \text{Martensite} \tag{13.7}$$

Martensite is a transition phase. It offers an interesting possibility for many applications because it is much harder than pearlite. Unfortunately, martensite is also very brittle, and its usefulness in steel is severely limited. Martensite will exist almost indefinitely at ambient temperatures, but reheating the steel provides an opportunity for the completion of the reaction of Eq. (13.6).

$$\text{Martensite} \xrightarrow{\text{tempering}} Fe_3C \tag{13.8}$$

The reheating process is called **tempering**. The product, which has a microstructure of finely dispersed carbide particles in a ferrite matrix, is both hard and tough. It is widely used and is called **tempered martensite**.

Hardenability

Tempered martensite is the preferred microstructure of many high-strength steels. Processing requires a sufficiently rapid quench to obtain the intermediate martensite, Eq. (13.7), followed by tempering, Eq. (13.8). This means severe quenching for products of Fe–C steels that are larger than needles or razor blades. Even then, martensite forms only at the quenched surface. The subsurface metal transforms directly to the ferrite and carbide, Eq. (13.6).

The reaction rate of Eq. (13.6) can be decreased by the presence of various alloying elements in the steel. Thus, gears and similar products commonly contain fractional percentages of nickel, chromium, molybdenum, or other metals. Quenching severity can be reduced, and larger components can be hardened throughout. These alloying elements delay the formation of carbide because, not only must be small carbon atoms relocate, it is also necessary for the larger metal atoms to choose between residence in the ferrite or in the carbide.

Surface Modification

Products may be treated so that their surfaces are modified and therefore possess different properties than the original material. Chrome plating is a familiar example.

Carburizing

Strength and hardness of a steel increase with carbon content. Concurrently, ductility and toughness decrease. With these variables, the engineer must consider trade-offs when specifying steels for mechanical applications. An alternate possibility is to alter the surface zone. A common example is the choice of a low-carbon, tough steel that has had carbon diffused into the subsurface (<1 mm) to produce hard carbide particles. Wear resistance is developed without decreasing bulk toughness.

Nitriding

Results similar to carburizing are possible for a steel containing small amounts of aluminum. Nitrogen can be diffused through the surface to form a subscale containing particles of aluminum nitride. Since AlN has structure and properties that are related to diamond, wear resistance is increased for the steel.

Shot Peening

Superficial deformation occurs when a ductile material is impacted by sand or by hardened shot. A process employing shot peening places the surface zone in local compression and therefore lowers the probability for fracture initiation during tensile loading.

Example 13.9

Assume a single spherical shrinkage cavity forms inside a 2-kg lead casting as it is solidified at 327°C. What is the initial diameter of the cavity after solidification? (The greatest density of molten lead is 10.6 g/cm^3).

Solution

Refer to Fig. 13.14(a). Based on a unit volume at 20°C, lead shrinks from 1.07 to 1.035 during solidification.

The volume of molten lead is (2000 g)/(10.6 g/cm^3) = 188.7 cm^3

$$\text{Volume of solid lead at 327°C: } (188.7\text{ cm}^3)(1.035/1.07) = 182.5\text{ cm}^3$$
$$\text{Shrinkage: } 188.7\text{ cm}^3 - 182.5\text{ cm}^3 = \pi d^3/6;\ d = 2.28\text{ cm}$$

Example **13.10**

How much energy is involved in polymerizing one gram of ethylene (C_2H_4) into polyethylene, Eq. (13.5). The double carbon bond possesses 162 kcal/mole, and the single bond has 88 kcal/mole.

Solution

As shown in Eq. (13.5), one double carbon bond changes to two single bonds/mer:

$$(1\text{ g})/(24 + 4\text{ g/mole}) = 0.0357\text{ moles}$$

The energy required to break 0.0357 moles of double bonds is

$$(1)(0.0357)(162\text{ kcal}) = 5.79\text{ kcal}$$

The energy released in joining twice as many single bonds is

$$(2)(0.0357)(-88\text{ kcal}) = -6.29\text{ kcal}$$

The net energy change is –6.29 + 5.79 kcal = 500 cal released/g.

Example **13.11**

There are 36 equiaxed grains per mm^2 observed at a magnification of 100 in a selected area of copper. The copper is heated to double the average grain diameter. (a) How many grains exist per mm^2? (b) What will be the percentage (increase, decrease) in grain boundary area?

Solution

(a) Doubling a lineal dimension decreases the number of grains by a factor of four, so there are 9 grains per mm^2.

(b) Surface area is a function of the lineal dimension squared:

$$a_1/a_2 \propto d_1^2/d_2^2$$

Thus, $a_1/a_2 = 0.25$, or a 75% decrease in grain boundary area.

Example **13.12**

The two surface layers of glass in Fig. 13.20 are 1 mm thick and have an expansion coefficient of 3×10^{-6}/°C. The interior layer is 4 mm thick and has an expansion coefficient of 6×10^{-6}/°C. Assume no stresses at 340°C and an elastic modulus for each of 70 GPa. What stresses develop during cooling to 40°C?

Solution

The forces in the two layers must be equal in magnitude and opposite in direction, or $-F_1 = F_2$. Since changes in length must match,

$$\Delta x_1 = \Delta x_2 \text{ or } \alpha_1 \Delta T + s_1/E_1 = \alpha_2 \Delta T + s_2/E_2$$

Rearranging,

$$s_1 - s_2 = (6 - 3)(10^{-6}/°C)(40 - 340\,°C)(70{,}000 \text{ MPa}) = -63 \text{ MPa}$$

$$-63 \text{ MPa} = F_1/(2 \times 0.001w \text{ m}^2) - F_2/(0.004w \text{ m}^2) = 3.0\, F_1/0.004w \text{ m}^2$$

Let $w = 1$; $F_1 = (-63 \text{ MPa})(0.004w \text{ m}^2)/3 = -0.084$ MN, and

$$s_1 = (-0.084 \text{ MN})/(0.002w \text{ m}^2) = -42 \text{ MPa};\ s_2 = 21 \text{ MPa}.$$

MATERIALS IN SERVICE

Products of engineering are made to be used. Conditions that are encountered in service most often vary tremendously from those present in the stockroom: static and dynamic loads, elevated and subambient temperatures, solar and nuclear radiation exposure, and many other reactions with the surrounding environment. All of these situations can lead to deterioration and even to failure. The design engineer should be able to anticipate the conditions of failure.

Mechanical Failure

Under ambient conditions, excessive loads can lead to bending or to cracking, the principal modes of mechanical failure. The former depends upon geometry and the stress level (**stress** is defined as load/cross sectional area, $s = p/a$). Cracking (and succeeding fracture) includes those two considerations, plus the loading rate.

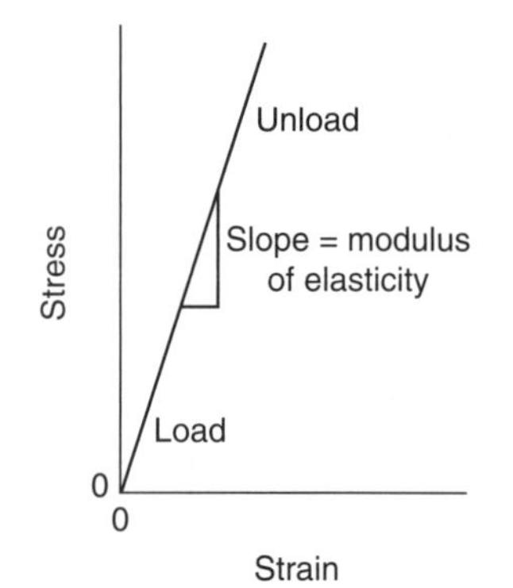

Figure 13.21 Schematic stress-strain diagram showing linear elastic deformation for loading and unloading cycles

Yield Strength

When solids are stressed, strain occurs (strain is change in length/original length $e = \Delta L/L_0$). The ratio of stress to strain (called the **elastic modulus**) is initially constant in most solids as shown in Fig. 13.21. The interatomic spacings are altered as the load is increased. Initial slip starts at a threshold level called the **yield strength**, S_y, shown in Fig. 13.22. Higher stresses will produce a permanent distortion, which will be called failure if the product was designed to maintain its initial shape. The toughness of a material is related to the energy or work required for fracture, a product of both strength and ductility. In a tensile test, the energy is the area under the stress-strain curve. Ductile fracture involves significant plastic deformation and hence absorbs much more energy than brittle fracture, which has little or no plastic deformation. A complete stress-strain curve is shown in Fig. 13.23.

Fracture

Breakage always starts at a location of **stress concentration**. This may be at a "flaw" of microscopic size, such as an abrasion scratch produced while cleaning eyeglasses, or it may be of larger dimensions, such as a hatchway on a ship.

With a crack, the **stress intensity factor**, K_I, is a function of the applied stress, s, and of the square root of the crack depth, c:

$$K_I = Ys\sqrt{\pi c} \qquad \textbf{(13.9)}$$

The proportionality constant, Y, relates to cross section dimensions and is generally near unity.

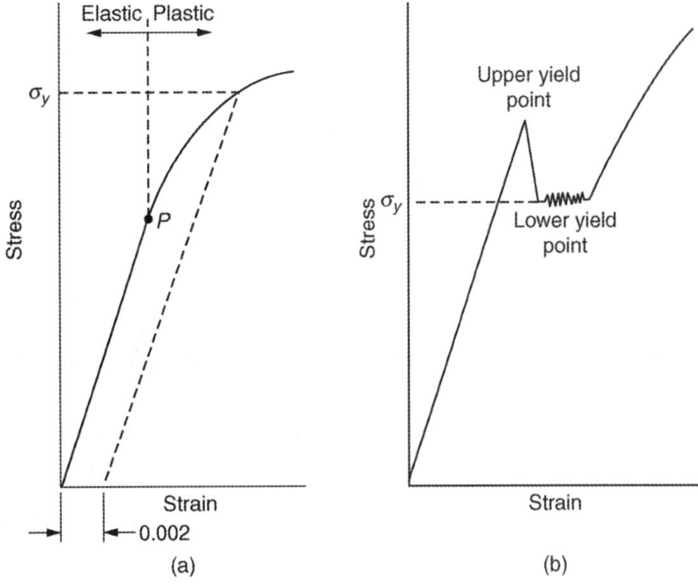

Figure 13.22 (a) Typical stress-strain behavior for a metal, showing elastic and plastic deformations, the proportional limit P, and the yield strength σ_y, as determined using the 0.002 strain offset method; (b) representative stress-strain behavior found for some steels, demonstrating the yield point phenomenon

Fracture toughness is the **critical** stress intensity factor, K_{Ic}, to propagate fracture and is a property of the material. This corresponds to the yield strength, S_y, being the critical stress to initiate slip. However, stronger materials generally have lower fracture toughness and *vice versa*.

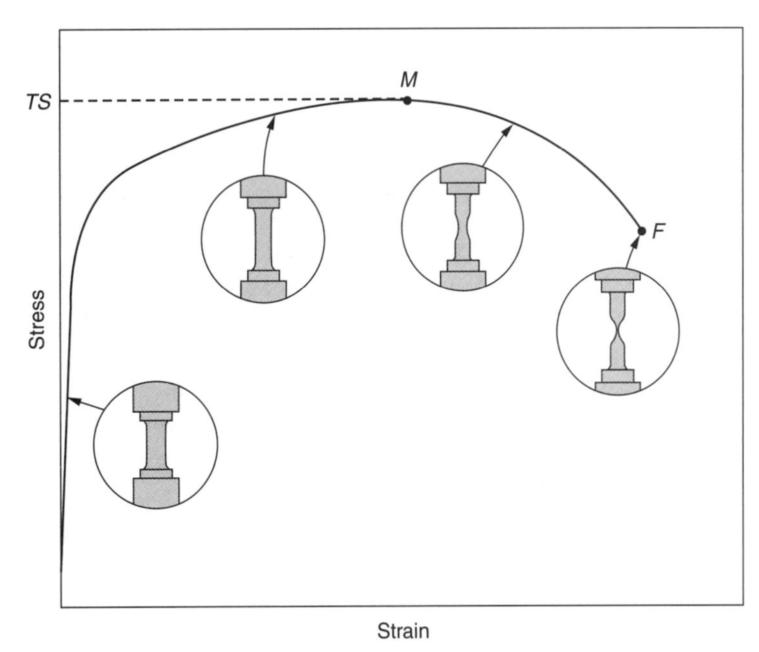

Figure 13.23 Typical engineering stress-strain behavior to fracture, point F. The tensile strength TS is indicated at point M. The circular insets represent the geometry of the deformed specimen at various points along the curve.

To illustrate the relationship between strength and toughness, consider a steel that has a yield strength, S_y, of 1200 MPa, and a critical stress intensity factor, K_{Ic}, of 90 MPa•$m^{1/2}$.

In the presence of a 2-mm crack, a stress of 1135 MPa would be required to propagate a fracture, according to Eq. (13.9). This is below the yield stress. If the value of K_{Ic} had been 100 MPa• $m^{1/2}$, fracture would not occur; rather, the metal would deform at 1200 MPa. A 2.2-mm crack would be required to initiate fracture without yielding.

Fatigue

Cyclic loading reduces permissible design stresses, as illustrated in Fig. 13.24. Minute structural changes that introduce cracks occur during each stress cycle. The crack extends as the cycles accumulate, leading to eventual fracture. When this delayed behavior was first observed, it was assumed that the material got tired; hence the term *fatigue*. Steels and certain other materials possess an *endurance limit*, below which unlimited cycling can be tolerated.

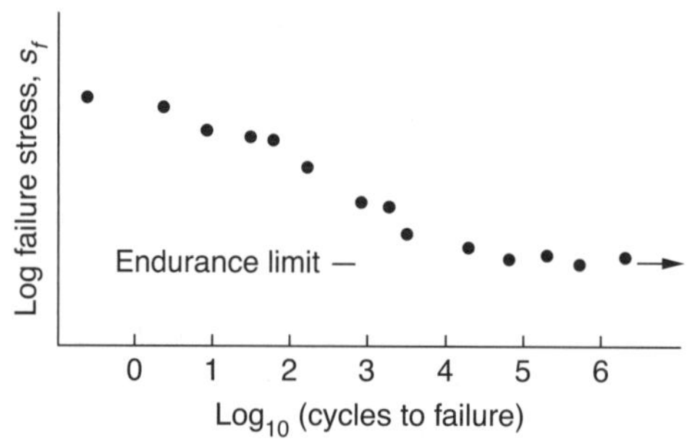

Figure 13.24 Cyclic fatigue. The stress for failure s_f, decreases as the number of loading cycles is increased. Most steels exhibit an endurance limit, a stress level for unlimited cycling. (A static tensile test is only one-fourth of a cycle.)

Thermal Failure

Melting is the most obvious change in a material at elevated temperatures. Overheating, short of melting, can also introduce microstructural changes. For example, the tempered martensite of a tool steel is processed so that it has a very fine dispersion of carbide particles in a tough ferrite matrix. It is both hard and tough and serves well for machining purposes. However, an excessive cutting speed raises the temperature of the cutting edge, causing the carbide particles to grow and softening the steel; if heating continues, failure eventually occurs by melting at the cutting edge. “High-speed” tools incorporate alloy additions, such as vanadium and chromium, that form carbides that are more stable than iron carbide. Thus, they can tolerate the higher temperatures that accompany faster cutting speeds.

Creep

As the name implies, **creep** describes a slow (<0.001%/hr) dimensional change within a material. It becomes important in long-term service of months or years. Slow viscous flow is commonly encountered in plastic materials. Refractories (temperature-resistant ceramics) will slowly slump when small amounts of liquid accumulate.

In metals, creep occurs when atoms become sufficiently mobile to migrate from compressive regions of the microstructure into tensile regions. Grain boundary areas are heavily involved. For this reason, coarser grained metals are advantageous for high-temperature applications. Three stages of creep are identified in Fig. 13.25. Following the initial elastic strain, Stage 1 of creep is fairly rapid as stress variations are equalized. In stage 2, the *creep rate*, dL/dt, is essentially constant. Design considerations are focused on this stage. Stage 3, which accelerates when the cross-sectional area starts to be reduced, leads to eventual rupture.

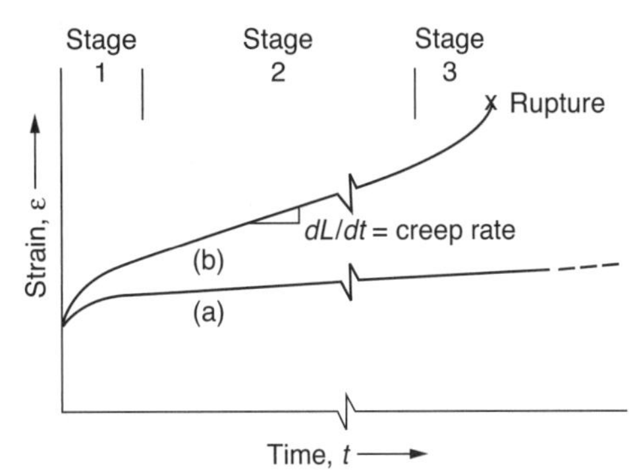

Figure 13.25 Creep. (a) Low stresses and/or low temperatures; (b) high stresses, high temperatures. The initial strain is elastic, followed by rapid strain adjustments (Stage 1). Design calculations are commonly based on the steady-state strain rate (Stage 2). Strain accelerates (Stage 3) when the area reduction becomes significant. Rupture occurs at ×.

Spalling

Spalling is thermal cracking. It is the result of stress caused by differential volume changes during processing or service. As discussed earlier, stresses can be introduced into a material (1) by thermal gradients, (2) by anisotropic volume changes, or (3) by differences in expansion coefficients in multiphase materials. Cyclic heating and cooling lead to **thermal fatigue** when the differential stresses produce localized cracking.

The spalling resistance index (SRI) of a material is increased by higher thermal conductivities, k, and greater strengths, S; it is reduced with greater values of the thermal expansion coefficients, α, and higher elastic moduli. In functional form

$$\mathrm{SRI} = f(kS/\alpha E) \tag{13.10}$$

Low-Temperature Embrittlement

Many materials display an abrupt drop in ductility and toughness as the temperature is lowered. In glass and other amorphous materials, this change is at the temperature below which atoms or molecules cannot relocate in response to the applied stresses. This is the **glass-transition temperature**, T_g. Metals are crystalline and do not have a glass transition. However, steels and a number of other metals have a **ductility-transition temperature** below which fracture is nonductile. The impact energy required for fracture can drop by an order of magnitude at this transition temperature. Thus it becomes a very significant consideration in design for structural applications. The **ductile-to-brittle transition temperature** (DBTT) is often measured using the Charpy Impact test, with samples soaked at different temperatures immediately prior to testing. Representative curves and the influence of carbon content on the DBTT in steel are shown in Fig. 13.26.

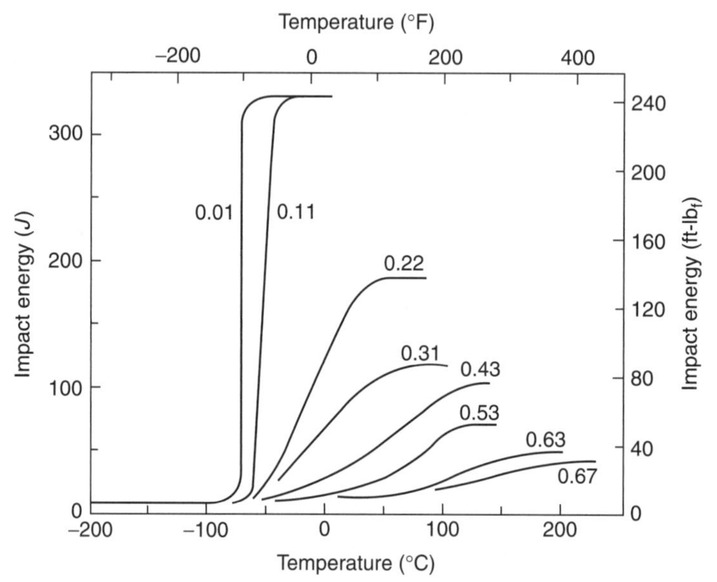

Figure 13.26 Influence of carbon content on the Charpy V-notch energy-versus-temperature behavior for steel

Radiation Damage

Unlike heat, which energizes all of the atoms and molecules within a structure, radiation introduces energy at "pinpoints" called **thermal spikes**. Individual bonds are broken, specific atoms are dislodged, molecules are ruptured, and electrons are energized. Each of these actions disorders the structure and alters the properties of the material. As expected, slip and deformation are resisted. Therefore, while strength and hardness increase, ductility and toughness decrease. Electrical and thermal conductivities drop within metals, because there is more scattering of the electrons as they move along the voltage or thermal gradient. These property changes, among others, are considered to be damaging changes, especially when they contradict carefully considered design requirements.

Damage Recovery

Partial correction of radiation damage is possible by annealing. Reheating a material allows internal readjustments, since atoms that have been displaced are able to relocate into a more ordered structure. It is similar to recrystallization after work hardening, where dislocations involving lineal imperfections composed of many atoms must be removed, except that radiation damage involves "pinpoint" imperfections, which means that it can be removed at somewhat lower temperatures than those required for recrystallization.

Chemical Alteration

Oxidation

Materials can be damaged by reacting chemically with their environments. All metals except gold oxidize in ambient air. Admittedly, some oxidize very slowly at ordinary temperatures. Others, such as aluminum, form a protective oxide surface that inhibits further oxidation. However, all metals—including gold—will oxidize significantly at elevated temperatures or in chemical environments that consume the protective oxidation.

Several actions are required for oxide scale to accumulate on a metal surface. Using iron as an example, the iron atom must be ionized to Fe^{2+} before it or the electrons move through the scale to produce oxygen ions, O^{2-}, at the outer surface. There, FeO or Fe_3O_4 accumulates. As the scale thickens, the oxidation rate decreases. However, exceptions exist. For example, the volume of MgO is less than the volume of the original magnesium metal. Therefore, the scale cracks and admits oxygen directly to the underlying metal. Also, an Al_2O_3 scale is insulating so the ionization steps are precluded.

Moisture

Moisture can produce chemical **hydration**. As examples, MgO reacts with water to produce $Mg(OH)_2$, and Fe_2O_3 can be hydrated to form $Fe(OH)_3$.

Water can be absorbed into materials. Small H_2O molecules are able to diffuse among certain large polymeric molecules. Consequently, polymers such as the aramids, which we normally consider to be very strong, are weakened—a fact that the design engineer must consider in specifications.

Corrosion

Metallic corrosion is familiar to every reader who owns a car, since rust—$Fe(OH)_3$—is the most obvious product of corrosion. Oxidation produces positive ions and electrons:

$$M \rightarrow M^{n+} + ne^- \qquad \textbf{(13.11)}$$

The reaction stops unless the electrons are removed. Oxygen accompanied by water (Fig. 13.27) is a common consumer of the electrons:

$$O_2 + 2\ H_2O + 4e^- \rightarrow 4\ (OH)^- \qquad \textbf{(13.12)}$$

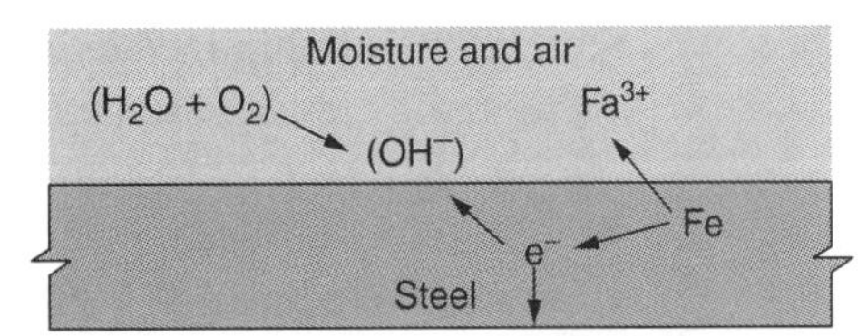

Figure 13.27 Rust formation. Electrons are removed from iron atoms and react with water and oxygen to produce $Fe(OH)_3$—rust.

Alternatively, if ions of a metal with a low oxidation potential are present, they can be reduced, consuming electrons from the preceding corrosion reaction, Eq. (13.11). Copper is cited as a common example:

$$Cu^{2+} + 2e^- \rightarrow Cu^0 \qquad \textbf{(13.13)}$$

Electroplating uses this reaction advantageously to deposit metals from a solution by the addition of electrons. The relative reactivity of metals with respect to standard electrodes is represented by the electromotive force (emf) series given in Table 13.4. However in real environments such as sea water the galvanic series as shown in Table 13.5, is more commonly used to determine the likelihood of corrosion.

Table 13.4 The standard emf series

	Electrode Reaction	Standard Electrode Potential V^0(V)
↑	$Au^{3+} + 3e^- \rightarrow Au$	+1.420
	$O_2 + 4H^- + 4e^- \rightarrow 2H_2O$	+1.229
	$Pl^{2+} + 2e^- \rightarrow Pl$	+1.2
	$Ag^+ + e^- \rightarrow Ag$	+0.800
Increasingly	$Fe^{3+} + e^- \rightarrow Fe^{2+}$	+0.771
inert (cathodic)	$O_2 + 2H_2O + 4e^- \rightarrow 4(OH^-)$	+0.401
	$Cu^{2+} + 2e^- \rightarrow Cu$	+0.340
	$2H^+ + 2e^- \rightarrow H_2$	0.000
	$Pb^{2+} + 2e^- \rightarrow Pb$	–0.126
	$Sn^{2+} + 2e^- \rightarrow Sn$	–0.136
	$Ni^{2+} + 2e^- \rightarrow Ni$	–0.250
	$Co^{2+} + 2e^- \rightarrow Co$	–0.277
	$Cd^{2+} + 2e^- \rightarrow Cd$	–0.403
	$Fe^{2+} + 2e^- \rightarrow Fe$	–0.440
Increasingly	$Cr^{3+} + 3e^- \rightarrow Cr$	–0.744
active (anodic)	$Zn^{2+} + 2e^- \rightarrow Zn$	–0.763
	$Al^{2+} + 3e^- \rightarrow Al$	–1.662
	$Mg^{2+} + 2e^- \rightarrow Mg$	–2.363
↓	$Na^+ + e^- \rightarrow Na$	–2.714
	$K^+ + e^- \rightarrow K$	–2.924

Corrosion Control

Corrosion is minimized by a variety of means. Some involve the avoidance of one or more of the above reactions. Feed water for steam-power boilers is deaerated; surfaces are painted to limit the access of water and air; junctions between unlike metals are electrically insulated. Other control procedures induce a reverse reaction: sacrificial anodes such as magnesium are attached to the side of a ship; iron sheet is galvanized with zinc. Each corrodes preferentially to the underlying steel and forces the iron to assume the role of Eq. (13.13). Corrosion may also be restricted by an impressed voltage. Natural gas lines utilize this procedure by connecting a negative dc voltage to the pipe.

Table 13.5 The galvanic series

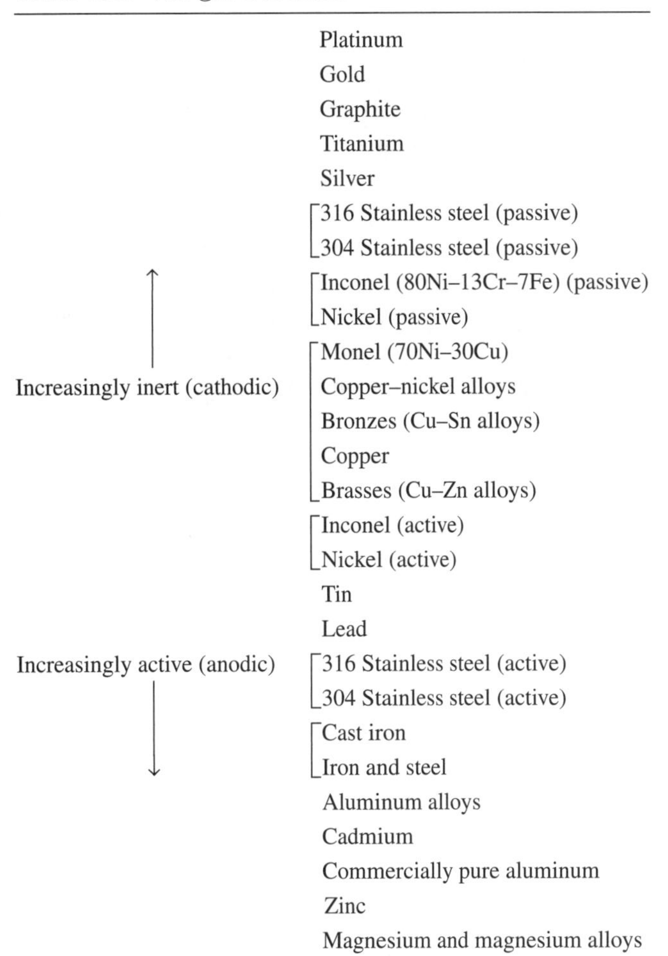

	Platinum
	Gold
	Graphite
	Titanium
	Silver
	[316 Stainless steel (passive)
	[304 Stainless steel (passive)
↑	[Inconel (80Ni–13Cr–7Fe) (passive)
	[Nickel (passive)
	[Monel (70Ni–30Cu)
Increasingly inert (cathodic)	Copper–nickel alloys
	Bronzes (Cu–Sn alloys)
	Copper
	Brasses (Cu–Zn alloys)]
	[Inconel (active)
	Nickel (active)]
	Tin
	Lead
Increasingly active (anodic)	[316 Stainless steel (active)
	304 Stainless steel (active)]
	[Cast iron
↓	Iron and steel]
	Aluminum alloys
	Cadmium
	Commercially pure aluminum
	Zinc
	Magnesium and magnesium alloys

Example 13.13

(a) What is the maximum static force that can be supported without permanent deformation by a 2-mm-diameter wire that has a yield strength of 1225 MPa?

(b) The elastic deformation at this threshold stress is 0.015 m/m. What is its elastic modulus?

Solution

(a) $S_y = F/A$; $F = \pi(2\ \text{mm}/2)^2(1{,}225\ \text{MPa}) = 3{,}800\ \text{N}$
(b) $E = s/e = 1{,}225\ \text{MPa}/\text{-}0.015\ \text{m/m} = 82{,}000\ \text{MPa}$

Example 13.14

The value of K_{Ic} for steel is 186 MPa•m$^{1/2}$. What is the maximum tolerable crack length, c, when the steel carries a nominal stress of 800 MPa? (Assume 1.1 as the proportionality constant.)

Solution

$$186\ \text{MPa}\bullet\text{m}^{1/2} = 800\ \text{MPa}\ (1.1)(\pi c)^{1/2};\ c = 0.014\ \text{m} = 14\ \text{mm}.$$

COMPOSITES

Composites are not new. Straw in brick and steel reinforcing rods in concrete have been used for a long time. But there is current interest in the development of new composites by designing materials appropriately. It is possible to benefit from the positive features of each of the contributors in order to optimize the properties of the composite.

The internal structures of composites may be viewed as enlarged polycomponent micro-structures, which were previously discussed. Attention is given to size, shape, amounts, and distribution of the contributing materials. However, there is commonly a significant difference in processing composites. Typically, the internal structure of a composite is a function of mechanical processing steps—mixing, emplacement, surface deposition, and so on—rather than thermal processing. These processing differences suggest a different approach in examining property-structure relationships.

Reinforced Materials

In familiar composites, such as reinforced concrete, the steel carries the tensile load. Also, there must be bonding between the reinforcement and the matrix. In a "glass" fishing rod, the glass fibers are all oriented longitudinally within the polymer matrix. If it is fractured, the break exhibits a splintered appearance with a noticeable amount of fiber pull-out.

Steel reinforcing bars (rebar) are used to reinforce concrete to meet strength and cost factors. Positioning is dictated by stress calculations. Rebar surfaces are commonly merloned (ridged circumferentially) for better anchorage.

Glass is widely used in fiber-reinforced plastics (FRP). The positioning of the fibers varies with the product. Continuous fibers are used in such structures as fuel storage tanks or rocket casings. However, chopped fibers are required when FRP are processed within molds. A matrix-to-glass bond must be achieved through chemical coatings at the interfaces.

Silicon carbide (SiC) and alumina (Al_2O_3) fibers are used increasingly in high-temperature composites with either metallic or ceramic matrices.

Inert fillers such as silica flour and wood flour serve multiple purposes in many polymers. They add strength, rigidity, and hardness to the product. In addition, they generally are less expensive than the matrix polymer.

Reinforcing bars or fibers are expected to carry the bulk of the tensile load. To be effective, the reinforcement must have a higher elastic modulus, E_r, than does the matrix because the reinforcement and the matrix undergo the same tensile strain when loaded ($e_r = e_m$). Therefore, $(E/s)_r = (E/s)_m$; and

$$s_r/s_m = E_r/E_m \tag{13.14}$$

Mixture Rules

We commonly study and analyze properties of composites in terms of the properties of the contributing materials. Mixture rules can then be formulated in which the properties are a function of the amounts and geometric distributions of each of the contributors. The simplest mixture rules are based on volume fraction, f. For example, the density of the mixture, ρ_m, is the volume-weighted average:

$$\rho_m = f_1\rho_1 + f_2\rho_2 + \cdots \tag{13.15a}$$

or simply

$$\rho_m = \sum f_i \rho_i \quad \textbf{(13.15b)}$$

Likewise, for *heat capacity*, *c*:

$$c_m = \sum f_i c_i \quad \textbf{(13.16)}$$

Directional Properties

Many composites are anisotropic. Laminates of two or more layers are bidirectional, as is a matted FRP that uses chopped fibers. The previously cited "glass" fishing rod has a structure with uniaxial anisotropy.

When considering conductivity, elastic moduli, strength, or other properties that are directional, attention must be given to the anisotropy of the composite. Consider the electrical resistance across a laminate that contains alternate layers of two materials with different resistances, R_1 and R_2. In a direction perpendicular to the laminate, the plies are in series. Thus the relationship is

$$R_\perp = L_1 R_1 + L_2 R_2 + \cdots \quad \textbf{(13.17)}$$

where L_i is the thickness of each ply. For unit dimensions, volume fractions, f_i, and the resistivities, ρ_i, are applicable:

$$\rho_\perp = f_1 \rho_1 + f_2 \rho_2 + \cdots \quad \textbf{(13.18)}$$

In contrast, for directions parallel to the laminate, resistivities follow the physics analog for parallel circuits:

$$\frac{1}{\rho_\parallel} = \frac{f_1}{\rho_1} + \frac{f_2}{\rho_2} + \cdots \quad \textbf{(13.19)}$$

The mixture rules for conductivities, either thermal, *k*, or electrical, σ, are inverted to those for resistivities:

$$\frac{1}{k_\perp} = \frac{f_1}{k_1} + \frac{f_2}{k_2} + \cdots \quad \textbf{(13.20)}$$

and

$$k_\parallel = f_1 k_1 + f_2 k_2 + \cdots \quad \textbf{(13.21)}$$

More elaborate mixture rules must be used when the structure of the composite is geometrically more complex, such as the use of particulate fillers or chopped fibers. In general, however, the property of the composite falls between the calculated values using the parallel and series versions of the preceding equations.

Preparation

Composites receive their name from the fact that two or more distinct starting materials are combined into a unified material. The application of a protective coating to metal or wood is one form of composite processing. The process is not as simple as it first appears because priming treatments are commonly required after the surface has been initially cleared of contaminants and moisture; otherwise, peeling and other forms of deterioration may develop. Electroplating, Eq. (13.13), commonly requires several intermediate surface preparations so that the final plated layer meets life requirements.

Wood paneling combines wood and polymeric materials into large sheets. **Plywood** not only has dimensional merit, it transforms the longitudinal anisotropy into a more desirable two-dimensional material. Related products include chipboard and similar composites. Composite panels include those with veneer surfaces where appearance and technical properties are valued.

The concept of uniformly mixing particulate **fillers** into a composite is simple. The resulting product is isotropic.

Several considerations are required for the use of **fibers** in a composite. Must the fiber be continuous? What are the directions of loading? What is the shear strength between the fiber and the matrix? Chopped fibers provide more reinforcement than do particles, and at the same time permit molding operations. The **aspect ratio** (L/D) must be relatively low for die molding. Higher ratios, and therefore more reinforcement, are used in sheet molding. (Sheet molded products are used where strength is not critical in the third dimension.) **Continuous fibers** maximize the mechanical properties of FRP composites. Their uses, however, are generally limited to products that permit parallel layments.

Example **13.15**

A rod contains 40 volume percent longitudinally aligned glass fibers within a plastic matrix. The glass has an elastic modulus of 70,000 MPa; the plastic, 3500 MPa. What fraction of a 700-N tensile load is carried by the glass?

Solution

Based on Eq. (13.14)

$$(F_{gl}/0.4A)/(F_p/0.6A) = (70{,}000\text{ MPa}/3500\text{ MPa}) = 20$$
$$F_{gl} = 20(0.4/0.6)\,F_p = 13.3\,F_p$$
$$F_{gl}/(F_{gl} + F_p) = (13.3\,F_p)/(13.3\,F_p + F_p) = 93\%$$

Example **13.16**

An electric highline cable contains one cold-drawn steel wire and six annealed aluminum wires, all with a 2-mm diameter. (The steel provides the required strength; the aluminum, the conductivity.) Using the following data, calculate (a) the resistivity and (b) the elastic modulus of the composite wire.

$$\text{Steel: } \rho = 17 \times 10^{-6}\,\Omega\bullet\text{cm},\ E = 205{,}000\text{ MPa}$$
$$\text{Aluminum: } \rho = 3 \times 10^{-6}\,\Omega\bullet\text{cm},\ E = 70{,}000\text{ MPa}$$

Solution

(a) From Eq. (13.19)

$$1/\rho_{\parallel} = (1/7)/(17 \times 10^{-6}\ \Omega\bullet\text{cm}) + (6/7)/(3 \times 10^{-6}\ \Omega\bullet\text{cm}) = 0.294 \times 10^{6}\ \Omega^{-1}\bullet\text{cm}^{-1}$$
$$\rho_{\parallel} = 3.4 \times 10^{-6}\ \Omega\bullet\text{cm}$$

(b) We must write a mixture rule for the elastic modulus of a composite in parallel. Let A be the area of one wire.

Since $e_C = e_{Al} = e_{St}$, $[(F/A)/E]_C = [(F/A)/E]_{Al} = [(F/A)/E]_{St}$
Also, $F_C = F_{Al} + F_{St} = F_C(f_{Al}A_C/A_C)(E_{St}/E_C) + F_C(f_{St}A_C/A_C)(E_{St}/E_C)$
Canceling,

$$E_C = f_{Al}E_{Al} + f_{St}E_{St} = (6/7)(70{,}000 \text{ MPa}) + (1/7)(205{,}000 \text{ MPa}) = 89{,}000 \text{ MPa}$$

The apparent modulus will be lower because there will also be cable extension by the straightening of the cable wire.

SELECTED SYMBOLS AND ABBREVIATIONS

Symbol or Abbreviation	Description
amu	atomic mass unit
E	elastic modulus
e	engineering strain
F	volume fraction
K_I	stress intensity factor
nm	nanometer
σ	conductivity
s	stress
SRI	Spalling Resistance Index
S_y	yield strength
Y	proportionality constant

REFERENCE

Elements of Materials Science and Engineering, 6th ed., L. Van Vlack, 1989, Addison-Wesley

PROBLEMS

13.1 For a neutral atom,

a. the atomic mass equals the mass of the neutrons plus the mass of the protons
b. the atomic number equals the atomic mass
c. the number of protons equals the atomic number
d. the number of electrons equals the number of neutrons

13.2 All isotopes of a given element have

a. the same number of protons
b. the same number of neutrons
c. equal numbers of protons and neutrons
d. the same number of atomic mass units

13.3 Which of the following statements is *not* true?

a. An anion has more electrons than protons.
b. Energy is released when water is solidified to ice.
c. Energy is required to remove an electron from a neutral atom.
d. Energy is released when a H_2 molecule is separated into two hydrogen atoms.

13.4 Select the correct statement.

a. Crystals possess long-range order.
b. Within a crystal, like ions attract and unlike ions repel.
c. A body-centered cubic metallic crystal (for example, iron) has nine atoms per unit cell.
d. A face-centered cubic metallic crystal (for example, copper) has fourteen atoms per unit cell.

13.5 The (110) plane of diamond contains all of the following directions *except*

a. [1^{1}1]
b. [110]
c. [$\bar{2}$21]
d. [1^{1}2]

13.6 In a cubic crystal, a is the edge of a unit cell. The shortest repeat distance in the [111] direction of a body-centered cubic crystal is

a. $a\sqrt{2}$
b. $2a$
c. $a\sqrt{3}/2$
d. $a\sqrt{3}/4$

13.7 The line of intersection between the (101) and (110) planes lies along what direction?

a. [11$\bar{1}$]
b. [211]
c. [^{1}11]
d. [111]

13.8 All but which of the following data are required to calculate the density of aluminum in g/m^3?

a. Avogadro's number, which is 6.0×10^{23}
b. atomic number of Al, which is 13
c. crystal structure of Al, which is face-centered cubic
d. atomic mass of Al, which is 27 amu

13.9 The atomic packing factor of gold, an FCC metal, is

a. $(4\pi r^3/3)/(4r/\sqrt{3})^3$

b. $4(4\pi r^3/3)/(4r/\sqrt{2})^3$

c. $4(4\pi r^3/3)/(r/\sqrt{2})^3$

d. $4(2r/\sqrt{2})^3/(4\pi r^3/3)$

13.10 The lattice constant for a unit cell of FCC nickel is 0.3525 nm. How many atoms/nm^2 are there on the (011) plane?
a. $(2/2 + 4/8)/(0.3525\ \text{nm})^2$
b. $(1.5)/[(0.3525\ \text{nm})^2\ \sqrt{2}]$
c. $2/(0.3525\ \text{nm})^2$
d. $\sqrt{2}/(0.3525\ \text{nm})^2$

13.11 The lattice constant for a unit cell of BCC chromium is 0.288 nm. How many atoms/nm are there in the [011] direction of chromium?
a. $\sqrt{2}\,/(0.288\ \text{nm})$ c. $1/(0.288\ \text{nm})$
b. $2\sqrt{2}/(0.288\ \text{nm})$ d. $1/(0.288\ \text{nm}\sqrt{2})$

13.12 Ethylene is C_2H_4. To meet bonding requirements, how many bonds are present?
a. 6 single c. 1 double and 4 single
b. 4 single and 2 double d. 12 single

13.13 When comparing ethylene and vinyl chloride, each of the following statements is true except
a. they have the same molecular weight
b. they have the same bonding changes during polymerization
c. they have the same number of single bonds
d. they have the same number of atoms

13.14 MgO is cubic. Every Mg^{2+} and O^{2-} ion has six neighbors. Their radii are r_{Mg} and R_O. Which of the following is *not* true?
a. the lattice constant, a, equals $2(r_{Mg} + R_O)$
b. there are eight ions per unit cell
c. the body diagonal of the unit cell equals $a\sqrt{3}$
d. the face diagonal of the unit cell is 4 R_O

13.15 Each of the following groups of plastics is thermoplastic except
a. polyvinyl chloride (PVC) and a polyvinyl acetate
b. phenolics, melamine, and epoxy
c. polyethylene, polypropylene, and polystyrene
d. acrylic (Lucite) and polyamide (Nylon)

13.16 Gold is FCC and has a density of 19.3 g/cm^3. Its atomic mass is 197 amu. Its atomic radius, r, may be calculated using which of the following?
a. $19.3\ \text{g/cm}^3 = (197)(6.02 \times 10^{23})/[(4r/\sqrt{2})^3]$
b. $19.3\ \text{g/cm}^3 = 2\ (197/6.02 \times 10^{23})/[(4r/\sqrt{2})^3]$
c. $19.3\ \text{g/cm}^3 = 4\ (197/6.02 \times 10^{23})/[(4r/\sqrt{2})^3]$
d. $19.3\ \text{g/cm}^3 = 6\ (197)(6.02 \times 10^{23})/[(4r/\sqrt{2})^3]$

13.17 X-ray diffraction shows that sodium is BCC and has an atomic radius of 0.186 nm. Its density is 0.97 g/cm^3 (= 0.97 Mg/m^3). From which of the following can one calculate its atomic mass?

a. $0.97\ \text{Mg/m}^3 = [4(\text{amu}_{\text{Na}})/6.02 \times 10^{23}]/[4r/\sqrt{3}]^{1/3}$
b. $0.97\ \text{Mg/m}^3 = [(\text{amu}_{\text{Na}})/6.02 \times 10^{23}]/[4r/\sqrt{3}]^{1/3}$
c. $0.97\ \text{Mg/m}^3 = [(\text{amu}_{\text{Na}})/6.02 \times 10^{23}]/[4r/\sqrt{3}]^{3}$
d. $0.97\ \text{Mg/m}^3 = [2(\text{amu}_{\text{Na}})/6.02 \times 10^{23}]/[4r/\sqrt{3}]^{3}$

13.18 How is the distance between adjacent (111) planes in FCC aluminum related to the distance between adjacent (110) planes?

a. the same
b. greater than
c. (1+1+1)/(1+1+0) times as great
d. (1+1+0)/(1+1+1) times as great

13.19 How does the distance compare between adjacent (100) planes and between adjacent (110) planes in BCC tungsten.

a. they are the same
b. (100) distance is greater
c. (100) distance is (1+0+0)/(1+1+0) times as great
d. (100) distance is less

13.20 The repeat distance in the [011] direction of nickel (FCC) is 0.2492 nm. Therefore, the atomic radius is how many nm?

a. $(0.2492)(\sqrt{2})$
b. $(0.2492/2)/(\sqrt{3})$
c. $(0.2492)/\sqrt{2})$
d. 0.1246

13.21 The repeat distance in the [110] direction of iron (BCC metal) is 0.4052 nm. The atomic radius is how many nm?

a. $(0.4052)(\sqrt{2})$
b. 0.2026
c. $[(0.4052)/(\sqrt{2})][(\sqrt{3})/4]$
d. 0.4052

13.22 The average molecular mass of PVC molecules in a plastic is 50,000 amu. How many mers in each? (Atomic masses are from Table 13.1.)

a. 50,000/[2(12) + 3(1) + 35.5]
b. (50,000)[2(12) + 3(1) + 35.5]
c. 50,000/[2(12) + 2(1) + 2(35.5)]
d. 50,000 (24 + 2.71)

13.23 Styrene resembles vinyl chloride, C_2H_3C1, except that the chlorine is replaced by a benzene ring. The mass of each mer is

a. 8(12) + 9(1) amu
b. 26 + 78 amu
c. 27 + 6(12) + 6(1) amu
d. 2(12) + 3(1) + 77 amu

13.24 The <1$\bar{1}$0> family of directions in a cubic crystal include all but which of the following? (An overbar is a negative coefficient):

a. [110] c. [101]

b. [0$\bar{1}$1] d. [1$\bar{1}$1]

13.25 The {112} family of planes in a cubic crystal includes all but which of the following directions?

a. (212) c. (1$\bar{1}$2)

b. (211) d. (121)

13.26 Crystal imperfections include all but which of the following?

a. dislocations c. interstitials

b. displaced atoms d. dispersions

13.27 A dislocation may be described as a

a. displaced atom
b. shift in the lattice constant
c. a slip plane
d. lineal imperfection

13.28 A grain within a microstructure is

a. a particle the size of a grain of sand
b. the nucleus of solidification
c. a particle the size of a grain of rice
d. an individual crystal

13.29 Which of the following does *not* apply to a typical brass?

a. an alloy of copper and zinc
b. a single-phase alloy
c. an interstitial alloy of copper and zinc
d. a substitutional solid solution

13.30 Which of the following is a crystalline material having long-range order?

a. a plate glass window
b. a supercooled syrup
c. a pendant on a chandelier
d. a random solid solution of copper in silver

13.31 Sterling silver, as normally sold,

a. is pure silver
b. is a supersaturated solid solution of 7.5% copper in silver
c. is 24-carat silver
d. has higher conductivity than pure silver

13.32 Atomic diffusion in solids matches all but which of the following generalities?

a. Diffusion is faster in FCC metals than in BCC metals.
b. Smaller atoms diffuse faster than do larger atoms.
c. Diffusion is faster at elevated temperatures.
d. Diffusion flux is proportional to the concentration gradient.

13.33 Grain growth involves all but which of the following?
a. reduced growth rates with increased time
b. an increase in grain boundary area per unit volume
c. atom movements across grain boundaries
d. a decrease in the number of grains per unit volume

13.34 Imperfections within metallic crystal structures may be any but which of the following?
a. lattice vacancies and extra interstitial atoms
b. displacements of atoms to interstitial site (Frenkel defects)
c. lineal defects or slippage dislocations caused by shear
d. ion pairs missing in ionic crystals (Shottky imperfections)

13.35 All but which of the following statements about solid solutions are correct?
a. In metallic solid solutions, larger solute atoms occupy the interstitial space among solvent atoms in the lattice sites.
b. Solid solutions may result from the substitution of one atomic species for another, provided radii and electronic structures are compatible.
c. Defect structures exist in solid solutions of ionic compounds when there are differences in the oxidation state of the solute and solvent ions, because vacancies are required to maintain an overall charge balance.
d. Order-to-disorder transitions that occur at increased temperatures in solid solutions result from thermal agitation that dislodges atoms from their preferred neighbors.

13.36 In ferrous oxide, $Fe_{1-x}O$, two percent of the cation sites are vacant. What is the Fe^{3+}/Fe^{2+} ratio?
a. 2/98
b. 0.04/0.94
c. 0.04/0.96
d. 0.06/0.94

13.37 A solid solution of MgO and FeO contains 25 atomic percent Mg^{2+} and 25 atomic percent Fe^{2+}. What is the weight fraction of MgO? (Mg: 24; Fe: 56; and O: 16 amu)
a. 40/(40 + 72)
b. 24/(24 + 56), or 4/80
c. 25/(25 + 25)
d. (25 + 25)/(50 + 50)

13.38 The boundary between two metal grains provides all but which of the following?
a. an impediment to dislocation movements
b. a basis for an increase in the elastic modulus
c. a site for the nucleation of a new phase
d. interference to slip

13.39 If 5 percent copper is added to silver,
a. the hardness is decreased
b. the strength is decreased
c. the thermal conductivity is decreased
d. the electrical resistivity is decreased

13.40 An interstitial site in FCC iron is at the center of the edge of the unit cell. The radius, r, of an iron atom is 0.127 nm. What is the diameter, d, of the interstitial site?

a. 2(0.127)
b. $(0.127)\sqrt{2} - (0.127)$
c. $(\sqrt{2} - 1)(2)(0.127)$
d. $[4(0.127)/\sqrt{3}] - 2(0.127)$

13.41 Stoichiometric zinc oxide is ZnO. Normally a few Zn^{+} ions are present. In order to maintain a neutral charge, that presence of Zn^{+} in the crystal could be accommodated by any but which of the following?

a. interstitial zinc ions
b. anion vacancies
c. cation vacancies
d. the introduction of Zn^{3+} ions

13.42 All but which of the following statements about diffusion and grain growth are correct?

a. Atoms can diffuse both within grains and across grain (crystal) boundaries.
b. The activation energy for diffusion through solids is inversely proportional to the atomic packing factor of the lattice.
c. Grain growth results from local diffusion and minimizes total grain boundary area. Large grains grow at the expense of small ones, and grain boundaries move toward their centers of curvature.
d. Net diffusion requires an activation energy and is irreversible. Its rate increases exponentially with temperature. It follows the diffusion equation, in which flux equals the product of diffusivity and the concentration gradient.

13.43 Refer to the accompanying Mg-Zn phase diagram, Exhibit 13.43. Select an alloy of composition C (71Mg–29Zn) and raise it to 575 °C so that only liquid is present. Change the composition to 60Mg–40Zn by adding zinc. When this new liquid is cooled, what will be the first solid to separate?

a. a solid intermetallic compound
b. a mixture of solid intermetallic compound and solid eutectic C (71Mg–29Zn)
c. a solid eutectic C (71Mg–29Zn)
d. a solid solution containing less than 1% intermetallic compound dissolved in Mg

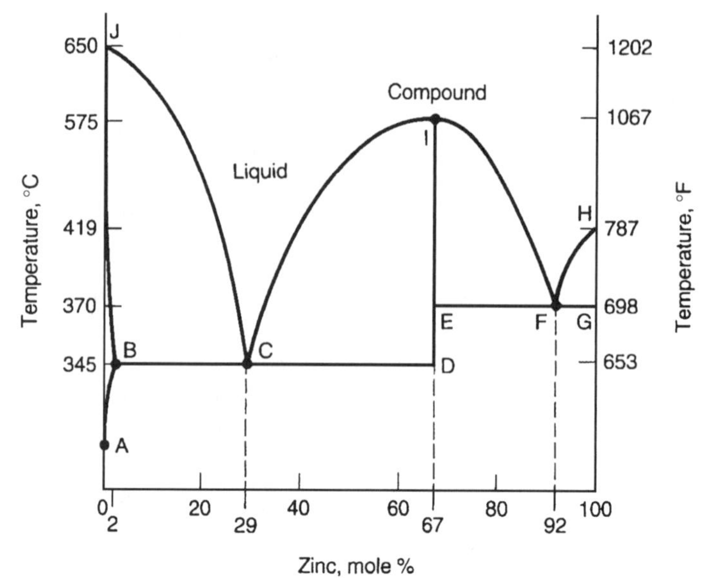

Exhibit 13.43 Magnesium-zinc phase diagram

13.44 Refer to the Mg-Zn phase diagram of Exhibit 13.43. Which of the following compounds is present?

a. Mg_3Zn_2 c. MgZn
b. Mg_2Zn_3 d. $MgZn_2$

13.45 Refer to Exhibit 13.45, a schematic sketch of the Fe-Fe_3C phase diagram. All but which of the following statements are true?

a. A eutectoid reaction occurs at location C, 727°C (1340°F).
b. The eutectic composition is 99.2 weight percent Fe and 0.8 weight percent C.
c. A peritectic reaction occurs at K, 1500°C (2732°F).
d. A eutectic reaction occurs at G, 1130°C (2202°F).

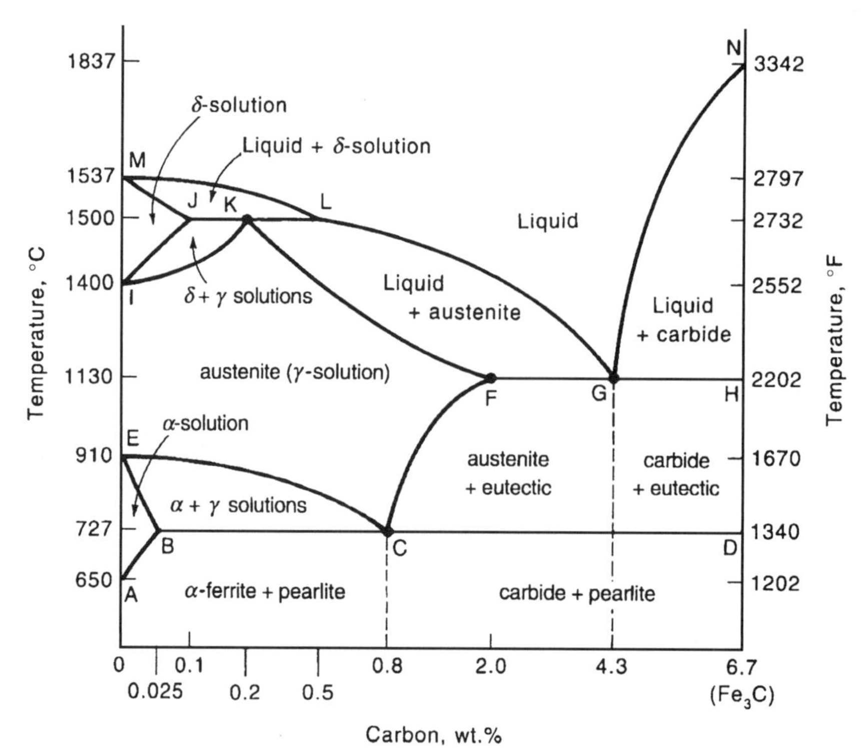

Exhibit 13.45 Iron-iron carbide phase diagram (schematic)

13.46 Refer to Exhibit 13.45, the Fe-Fe_3C phase diagram. Pearlite contains ferrite (α) and carbide (Fe_3C). The weight fraction of carbide in pearlite is

a. 0.8%
b. BC/CD
c. CD/BD
d. BC/BD

13.47 Consider the Ag-Cu phase diagram (Fig. 13.16). Silver-copper alloys can contain *approximately* half liquid and half solid at all but which of the following situations?

a. 40Ag–60Cu at 781°C
b. 81Ag–19Cu at 781°C
c. 20Ag–80Cu at 910°C
d. 50Ag–50Cu at 779°C

13.48 Refer to the Ag-Cu phase diagram of Fig. 13.16. Which of the following statements is wrong?

a. The solubility limit of copper in the liquid at 900°C is approximately 62%.
b. The solubility limit of silver in β at 700°C is approximately 5%.
c. The solubility limit of silver in β at 900°C is approximately 7%.
d. The solubility limit of copper in β at 800°C is approximately 92%.

13.49 Other factors being equal, diffusion flux is facilitated by all but which of the following?

a. smaller grain sizes
b. smaller solute (diffusing) atoms
c. lower concentration gradients
d. lower-melting solvent (host structure)

13.50 Consider copper. All but which of the following statements are applicable for grain growth?

a. Atoms jump the boundary from large grains to small grains.
b. Grain size varies inversely with boundary area.
c. Grain growth occurs because the boundary atoms possess higher energy than interior atoms.
d. Grain growth occurs because larger grains have less boundary area.

13.51 When equilibrated, single-phase microstructures may possess which of the following variations in the grains?

a. size and composition
b. conductivity and size
c. density and shape
d. shape and orientation

13.52 A phase diagram can provide answers for all but which of the following questions?

a. What are the direction of the planes at a given temperature?
b. What phases are present at a given temperature?
c. What are the phase compositions at a given temperature?
d. How much of each phase is present at a given temperature?

13.53 In the Ag-Cu system (Fig. 13.16), three equilibrated phases may be present at

a. 930°C
b. 880°C
c. 830°C
d. 780°C

13.54 The microstructures of multiphase materials may involve all but which of the following variables?

a. phase shape and phase density
b. phase amounts and distribution
c. phase quantities and shape
d. phase composition and shape

13.55 During heating, a 72Ag–28Cu alloy (Fig. 13.16) may have any but which one of the following equilibrium relationships?

a. ~25% β at 600°C, where β contains ~2% Ag
b. less than 50% β (95Cu–5Ag) at 700°C
c. ~74% α(93Ag–7Cu) at 750°C
d. α(26% Cu) and β(32% Cu) at 800°C

13.56 During cooling, a (20Ag–80Cu) alloy (Fig. 13.16) has all but which of the following equilibrium situations?

a. the first solid forms at 980°C
b. the first solid contains 5% Ag
c. the second solid appears at 780°C
d. the last liquid contains 32% Cu

13.57 Add copper to 100 g of silver at 781°C (Fig. 13.16). Assume equilibrium. Which statement is correct?

a. Liquid first appears with the addition of exactly 9 g of copper.
b. The last α disappears when approximately 39 g of copper has been added.
c. The solubility limit of copper in solid silver (α) is 28% copper.
d. Solid β first appears with the addition of 92 g of copper.

13.58 All but which of the following statements about strain hardening is correct?

a. Strain hardening is produced by cold working.
b. Strain hardening is relieved during annealing above the recrystallization temperature.
c. With more strain hardening, more time-temperature exposure is required for relief.
d. Strain hardening is relieved during recrystallization. Recrystallization produces less strained and more ordered structures.

13.59 Which process is used for the high-temperature shaping of many materials?

a. reduction
b. recrystallization
c. polymerization
d. extrusion

13.60 All but which of the following processes strengthens metals?

a. precipitation processes that produce submicroscopic particles during a low-temperature heat treatment
b. increasing the carbon content of low-carbon steels
c. annealing above the recrystallization temperature
d. mechanical deformation below the recrystallization temperature (cold working)

13.61 All but which of the following statements about the austentite-martensite-bainite transformations is correct?

a. Pearlite is a stable lamellar mixture consisting of BCC ferrite (α) plus carbide (Fe_3C). It forms through eutectoid decomposition during slow cooling of austenite. Most alloying elements in steel retard this transformation.

b. Martensite has a body-centered structure of iron that is tetragonal and is supersaturated with carbon. It forms by shear during the rapid quenching of austenite (FCC iron).

c. Tempering of martensite is accomplished by reheating martensite to precipitate fine particles of carbide within a ferrite matrix, thus producing a tough, strong structure.

d. Bainite and tempered martensite have distinctly different microstructures.

13.62 Steel can be strengthened by all but which of the following practices?

a. annealing
b. quenching and tempering
c. age or precipitation hardening
d. work hardening

13.63 Residual stresses can produce any but which of the following?

a. warpage
b. distortion in machined metal parts
c. cracking of glass
d. reduced melting temperatures

13.64 The reaction ($\gamma \rightarrow \alpha + Fe_3C$) is most rapid at

a. the eutectoid temperature
b. 10°C above the eutectoid temperature
c. the eutectic temperature
d. 100°C below the eutectoid temperature

13.65 If 1080 steel (0.80 wt.% carbon) is annealed by very slow cooling from 800°C to ambient temperature, its microstructure will consist almost entirely of

a. bainite
b. austenite
c. martensite
d. pearlite

13.66 Grain growth, which reduces boundary area, may be expected to

a. decrease the thermal conductivity of ceramics
b. increase the hardness of a solid
c. decrease the creep rate of a metal
d. increase the recrystallization rate

13.67 Crystallization from a liquid is slowest for

a. soda-lime glasses
b. pure silica glasses
c. metallic alloys, such as iron with additions of C, P, or Si
d. pure metals

13.68 Crystallization from a liquid is fastest for
a. soda-lime glasses
b. pure silica glasses
c. metallic alloys, such as iron with additions of C, P, or Si
d. pure metals

13.69 Rapid cooling can produce which one of the following in a material such as sterling silver?
a. homogenization
b. phase separation
c. grain boundary contraction
d. supersaturation

13.70 Martensite, which may be obtained in steel, is a
a. supersaturated solid solution of carbon in iron
b. supercooled iron carbide, Fe_3C
c. undercooled FCC structure of austenite
d. superconductor with zero resistivity at low temperature

13.71 Alloying elements produce all but which one of the following effects in steels?
a. They alter the number of atoms in a unit cell of austenite.
b. They increase the depth of hardening in quenched steel.
c. They increase the hardness of ferrite in pearlite.
d. They retard the decomposition of austenite.

13.72 Hardenability may be defined as
a. resistance to indentation
b. the hardness attained for a specified cooling rate
c. another measure of strength
d. rate of increased hardness

13.73 Hardenability tests are used for steel alloys. Why not for copper alloys?
a. The copper-rich phase is stable at all temperatures below the melting temperature.
b. Copper alloys never are as hard as any of the steels.
c. Copper-rich phases are FCC.
d. Copper alloys form substitutional and not interstitial phases.

13.74 The linear portion of the stress-strain diagram of steel is known as the
a. irreversible range
b. scant modulus
c. modulus of elasticity
d. elastic range

13.75 The ultimate (tensile) strength of a material is calculated from
a. the applied force divided by the true area at fracture
b. the applied force times the true area at fracture
c. the tensile force at the initiation of slip
d. the applied force and the original area

13.76 All but which one of the following statements about slip are correct?

a. Slip occurs most readily along crystal planes that are least densely populated.

b. Slip, or shear along crystal planes, results in an irreversible plastic deformation or permanent set.

c. Ease of slippage is directly related to the number of low-energy slip planes within the lattice structure.

d. Slip is impeded by solution hardening, with odd-sized solute atoms serving as anchor points around which slippage does not occur.

13.77 When a metal is cold worked more severely, all but which one of the following generally occur?

a. the recrystallization temperature decreases

b. the tensile strength increases

c. grains become equiaxed

d. slip and/or twinning occur

13.78 All but which of the following statements about the rusting of iron are correct?

a. Contact with water or oxygen is required for rusting to occur.

b. Halides aggravate rusting, a process which involves electrochemical oxidation-reduction reactions.

c. Contact with a more electropositive metal restricts rusting.

d. Corrosion occurs in oxygen-rich areas.

13.79 All but which of the following statements about mechanical and thermal failure is true?

a. Creep is time-dependent, plastic deformation that accelerates at increased temperatures. Stress rupture is the failure following creep.

b. Ductile fracture is characterized by significant amounts of energy absorption and plastic deformation (evidenced by elongation and reduction in cross-sectional area).

c. Fatigue failure from cyclic stresses is frequency-dependent.

d. Brittle fracture occurs with little plastic deformation and relatively low energy absorption.

13.80 The stress intensity factor is calculated from

a. yield stress and crack depth

b. applied stress and crack depth

c. tensile stress and strain rate

d. crack depth and strain rate

13.81 Service failure from applied loads can occur in all but which of the following cases?

a. cyclic loading, tension to compression

b. glide normal to the slip plane

c. cyclic loading, low tension to higher tension

d. Stage 2 creep

13.82 Brittle failure becomes more common when

a. the endurance limit is increased
b. the glass-transition temperature is decreased
c. the critical stress intensity factor is increased
d. the ductility-transition temperature is increased

13.83 Where applicable, all but which of the following procedures may reduce corrosion?

a. avoidance of bimetallic contacts
b. sacrificial anodes
c. aeration of feed water
d. impressed voltages

13.84 A fiber-reinforced rod contains 50 volume percent glass fibers (E = 70 GPa, S_y = 700 MPa) and 50 volume percent plastic (E = 7 GPa). The glass carries what part of a 5000-N tensile load?

a. [(7000 MPa)(0.5)]/[70,000 MPa)(0.5)] = 0.1; F_{gl} = (0.1)(5000 N) = 500 N
b. [(70,000)(0.5) + 7000(0.5)] = 5000/x; x = 0.0002
c. [(F_{gl}/0.5A)/(F_p/0.5A)] = (70/7) = 10; $F_{gl}/(F_p + F_{gl}) = 10F_p/(10F_p + F_p)$ = 0.91
d. (70)/[70 + 2(7)] = 0.83

SOLUTIONS

13.1 **c.** Each step through the periodic table introduces an additional proton and electron to a neutral atom.

13.2 **a.** The number of protons are fixed for an individual element. If the number of protons (and electrons) were varied, the chemical properties would be affected.

13.3 **d.** To separate H_2 into hydrogen atoms, the H-to-H bond would have to be broken, thus requiring energy.

13.4 **a.** Unlike ions attract. FCC metals possess four atoms per unit cell; BCC metals have two.

13.5 **b.** The (110) plane lies diagonally through the unit cell, parallel to the c-axis. The [110] direction is perpendicular to the c-axis.

13.6 **c.** The [111] direction passes diagonally through the unit cell. The distance is $a\sqrt{3}$, which equals two repeat distances.

13.7 **c.** $[\bar{1}11]$ lies in both planes. (This may be checked with a sketch or by verifying that both dot products are equal to zero.)

13.8 **b.** The mass is determined from 27 amu per 6.0×10^{23} atoms. Each cell of four atoms has a volume of $(4r/\sqrt{2})^3$.

13.9 **b.** Assuming spherical atoms, there are four atoms of radius r per unit cell. The cube edge is $4r/\sqrt{2}$.

13.10 **d.** Within the unit cell, the (011) plane contains (2/2 + 4/4) atoms in an area of $a^2\sqrt{2}$.

13.11 **d.** In BCC atoms the [011] repeat distance is $a\sqrt{2}$.

13.12 **c.** There is a double bond between the two carbons. Each hydrogen is held with a single bond.

13.13 **a.** The two are equal except that one hydrogen is replaced with a chlorine.

13.14 **d.** The unlike ions make contact. The oxygen anions avoid contact with each other. The face diagonal is $2(r_{Mg} + R_O)\sqrt{2}$.

13.15 **b.** Thermoplastic materials are polymerized but soften for molding at elevated temperatures. The polymeric molecules are linear. Thus they include the ethylene-type compounds that are bifunctional (two reaction sites per mer).

Thermosetting materials develop three-dimensional structures that become rigid during processing. For example, phenol is trifunctional and thus forms a network structure. Reheating does not soften them.

13.16 **c.** Density is mass/volume. The mass per FCC unit cell is 4 Au × (197 g/6.02 × 10^{23} Au). The volume per FCC unit cell of a metal is (face diagonal/$\sqrt{2}$)^3 or $(4r/\sqrt{2})^3$.

13.17 **d.** Density is mass/volume. With 2 Na per cell, the mass per unit cell is (2Na)[g/(6.0 × 10^{23} Na)]. Volume per BCC unit cell is (body diagonal/$\sqrt{3}$)^3 or $(4r/\sqrt{3})^3$.

13.18 **b.** Make a sketch. There are three (111) interplanar spacings per cube body diagonal in an FCC metal unit cell. There are four (110) interplanar spacings per face diagonal in an FCC metal unit cell.

13.19 **d.** Make a sketch. There are two (100) interplanar spacings per cube edge in a BCC metal unit cell. There are two (110) interplanar spacings per face diagonal in a BCC metal unit cell.

13.20 **d.** In FCC metals, the atoms have contact in the <110> directions. Therefore, the repeat distance is $2r$.

13.21 **c.** In a BCC metal, the repeat distance in the [110] is $a\sqrt{2}$. The atoms have contact in the <111> directions. Make a sketch.

13.22 **a.** Polyvinyl chloride (PVC) resembles polyethylene (PE) except that one of the four hydrogens is replaced by a chlorine, $-(C_2H_3Cl)-$ vs. $-(C_2H_4)-$. The mass of each mer is [2(12) + 3(1) + 35.5 amu].

13.23 **d.** Benzene is C_6H_6; however, in styrene, one hydrogen is absent at the connection to the C_2H_3–base.

13.24 **d.** Since a cubic crystal has interchangeable *x*-, *y*-, and *z*-axes, the <110> family includes all directions with permutations of 1, 1, and 0 (either + or –). (This is not necessarily true for noncubic crystals.)

13.25 **a.** Since a cubic crystal has interchangeable *x*-, *y*-, and *z*-axes, the {112} family includes all planes with index permutations of 1, 1, and 2 (either + or –). (This is not necessarily true for noncubic crystals.)

13.26 **d.** (b) and (c) involve individual atoms (point imperfections). (a) is a lineal imperfection. Boundaries result from a two-dimensional mismatch of crystal structures.

13.27 **d.** There are two types of dislocations: (1) An edge dislocation may be described as an edge of a missing half-plane of atoms; (2) A screw dislocation is the core of a helix.

13.28 **d.** Unless special efforts are made to grow single crystals, many crystals are nucleated and grow until they encounter neighboring crystals. Each grain is individually oriented.

13.29 **c.** Zinc is sufficiently near copper in size and electrical behavior to proxy for copper in the crystal structure. It is too big for the interstices.

13.30 **d.** Some glass products are called "crystal" because they can be shaped to give refracted colors, as do transparent gem crystals. Actually, glasses are amorphous, supercooled liquids.

13.31 **b.** The 7.5% copper replaces silver atoms. If it is cooled rapidly, the copper is retained in solid solution. The copper atoms interfere with electron movements within the silver.

13.32 **a.** FCC metals have a higher packing factor than do BCC metals; therefore, with other factors equal, diffusion is reduced.

13.33 **b.** As the grains grow, their volume increases by the third power. Their surface area increases by the square.

13.34 **d.** Metallic crystals are not ionic and do not have discrete ions.

13.35 **a.** The interstitial sites are smaller than the atoms in metals.

13.36 **b.** To balance the charge, each missing Fe^{2+} ion must be compensated by two Fe^{3+} ions. Therefore, out of 100 cation sites, two are vacant, four are Fe^{3+}, and thus 94 are Fe^{2+}.

13.37 **a.** Using a computational basis of four atoms, $(1\ Mg^{2+} + 1\ O^{2-}) + (1\ Fe^{2+} + 1\ O^{2-}) = (24 + 16) + (56 + 16) = (40 + 72)$.

13.38 **b.** The elastic strains between atoms along the boundary follow the same relationships as the strains among atoms within the grains.

13.39 **c.** Solid solution increases strength (solution hardening). It also decreases conductivity (and increases resistivity). Sterling silver is 92.5Ag–7.5Cu.

13.40 **c.** The face diagonal of an FCC metal is $4r$; therefore, the edge of the unit cell is $2r(\sqrt{2})$, and also $(2r + d)$.

13.41 **c.** Charges can be balanced by $Zn^{+}F^{-}$ for $Zn^{2+}O^{2-}$, by $Zn^{+}Zn^{3+}$ for $2\ Zn^{2+}$, or by an anion vacancy ($Zn^{+}Zn^{+}O^{2-}$ for $2\ Zn^{+}O^{2-}$), but not by a cation vacancy. The cation interstitial ($Zn^{+}Zn_{i}^{+}$ for $1\ Zn^{2+}$) is the principal adjustment in pure zinc oxide.

13.42 **b.** When atoms are moved from one site to another, bonds are broken and reconstituted. During transition, an activation energy is required to distort the lattice. Small solute atoms, low-melting-point solvents, and lower atomic packing factors in a lattice all require a lower activation energy. Hence activation energy for diffusion is *directly* proportional to the packing factor.

13.43 **a.** On cooling, curve CI is encountered at approximately 420°C (790°F). That curve is the solubility limit of Zn in that liquid. Zinc in excess of the solubility limit separates as the intermetallic compound, $MgZn_2$, which is plotted as the vertical line EI.

13.44 **d.** A ratio of 67 Zn atoms to 33 Mg atoms is 2-to-1; therefore, $MgZn_2$.

13.45 **b.** The eutectic composition is that of a low-melting liquid saturated with two solids. The 0.8 weight percent composition is a solid, not a liquid, at 727°C.

13.46 **d.** The (ferrite + carbide) area extends across the lower part of the phase diagram from nil carbon to 6.7 carbon. At the left side there is no Fe_3C; at the right side there is only carbide (Fe_3C contains 6.7% carbon). The amount of carbide between the two extremes may be determined by linear interpolation.

13.47 **d.** The eutectic temperature is 780°C (1445°F). That is the lowest temperature at which liquid can exist of equilibrium. At 779°C there are approximately equal amounts of α and β, the two solid solutions.

13.48 **d.** The curves of a phase diagram are solubility limits for the phases within the single-phase regions. Since copper is the solvent for the β structure, β has no upper limit of copper solubility (other than 100%).

13.49 **c.** The diffusion flux is proportional to the concentration gradient. (The other choices cited reduce the activation energy needed for diffusion.)

13.50 **a.** Boundary atoms possess higher energy. Therefore, the boundary is reduced and the grain size is increased at temperatures where the atoms can move. The net movement of the atoms is to the larger grains (with less boundary area).

13.51 **d.** Size, shape, and orientation are variables in single-phase microstructures.

13.52 **a.** Phase diagrams cannot be used to predict crystalline properties.

13.53 **d.** Above the eutectic temperature, (α + L) can be present concurrently, as can (L + β), but not ($\alpha + \beta$). Below the eutectic temperature, ($\alpha + \beta$) may be present, but no liquid. As the eutectic temperature is passed during cooling or heating, all three phases may coexist.

13.54 **a.** Microstructure is a geometric characterization. Density is independent of size, shape, and location.

13.55 **d.** At 800°C, a 72Ag–28Cu alloy is fully liquid, which therefore has 72–28 composition.

13.56 **d.** During equilibrium cooling, the final liquid for this alloy does not disappear until the eutectic temperature is reached at 780°C. At that temperature, the liquid composition is 72Ag–28Cu.

13.57 **b.** All α disappears on the right side of the (α + L) field, where the composition is 72Ag–28Cu, or 100g Ag to 39 g Cu.

13.58 **c.** As the temperature is increased, the atoms gain additional energy and can relocate, eliminating the strain energy that accompanies dislocations. *Less time* is required at higher temperature. *Less time* is also required for a highly cold-worked material because there is additional stored energy present.

13.59 **d.** Reduction and polymerization involve chemical reactions. Tempering and recrystallization involve reheating but no shape change.

13.60 **c.** Strength and hardness are increased at the expense of ductility and toughness (opposite of brittleness). The increase is facilitated by microstructures that interfere with dislocation movements. These include a high density of dislocations from plastic deformation, and the presence of many fine, hard particles. Annealing removes dislocations and permits the agglomeration of particles into fewer large particles.

13.61 **d.** The production of tempered martensite is indicated in (c) above. Bainite is formed by isothermally decomposing austenite directly to a microstructure of fine carbide particles within a ferrite matrix. Although the processing differs, the resulting microstructure and properties are nearly identical.

13.62 **a.** Annealing removes the hardness that was introduced by cold work. Quenched and tempered steels are harder with higher carbon contents, because more hard carbide particles are present. Alloying elements perform several hardening functions: They solution-harden the ferrite matrix; they slow down grain growth; and they delay the formation of pearlite, thus permitting more martensite with slower cooling rates (in turn, more tempered martensite may be realized farther below the quenched surface).

13.63 **d.** Stresses will relax below the melting temperature. Tensile stresses facilitate the cracking of brittle materials; compression limits cracking.

13.64 **d.** The reaction occurs only below the eutectoid temperature.

13.65 **d.** At equilibrium at 800°C a 1080 steel is fully austenitic. During slow cooling past the eutectoid temperature to ambient, austenite decomposes to pearlite (ferrite plus lamellar carbides). Rapid cooling (quenching) yields metastable martensite. On tempering, the martensite yields fine particles of carbide in a ferrite matrix.

13.66 **c.** While grain boundaries interfere with slip at low temperatures, they facilitate creep at elevated temperatures.

13.67 **b.** Both silica glass and crystallized silica have the same strong Si—O bonds. Therefore, only a very small (long-range) energy change is available to drive the crystallization process. Na^+ and Ca^{2+} ions modify the silica network to permit easier atom rearrangements and crystallization.

13.68 **d.** All first-neighbor bonds are identical in pure metals. Therefore only nearest-neighbor relocations are necessary.

13.69 **d.** The processing step of rapid cooling, such as quenching, retains the structures that existed at higher temperatures, even though a solubility limit is exceeded.

13.70 **a.** Martensite is a transition phase between austenite and ferrite, which retains carbon interstitially. Given an opportunity, the carbon separates as Fe_3C.

13.71 **a.** Alloying elements can dissolve substitutionally in austenite, which remains FCC.

13.72 **b.** Hardenability may be described as the ability, or the ease, by which martensitic hardness is obtained at various cooling rates (as located on an end-quenched, or Jominy, test).

13.73 **a.** Hardenability is the result of phase transitions that produce new structures.

13.74 **d.** The ratio of stress-to-strain is defined as the elastic modulus.

13.75 **d.** $S_u = F/A_0$.

13.76 **a.** Slip occurs most readily in directions that have the shortest steps, and along planes that are farthest apart. The latter are automatically the planes that are most densely populated.

13.77 **c.** Cold working—such as rolling, forging, drawing, or extrusion—deforms the material at temperatures below the recrystallization temperature. Strain hardening occurs, increasing both the yield and ultimate strength. Internal strains and minute cracks are introduced as slip or twinning occur. Ductility, elongation, and the recrystallization temperature are decreased. A preferred grain orientation is introduced in the direction of elongation, and the grains are flattened.

13.78 **d.** Since oxygen is required for rust formation, oxygen-depleted areas become the anode and are corroded. This may lead to pitting, particularly if rust or other corrosion products are accumulated locally to prevent the access of oxygen. Iron and other metals may be protected from corrosion by the presence of a more electropositive metal such as magnesium or zinc. This is the reason for coating steel with zinc to produce galvanized sheet.

13.79 **c.** Although fatigue strength is not sensitive to temperature or loading rates, it is very sensitive to surface imperfections from which cracks originate and propagate.

13.80 **b.** The stress intensity factor is proportional to the applied stress and the square root of the crack depth.

13.81 **b.** Glide occurs parallel to the slip plane.

13.82 **d.** Brittleness exists below T_{DT}.

13.83 **c.** Corrosion commonly occurs in the combined presence of oxygen and water. Protection may be obtained by making a cathode out of the critical part or by avoiding air.

13.84 **c.** With equal strains, $(s_{gl}/s_p) = E_{gl}/E_p = 10$. Likewise with equal areas, $F_{gl} = 10F_p$, and $F_{gl}/(F_{gl} + F_p) = 10/(10 + 1)$.

CHAPTER 14

Chemistry

E. Vernon Ballou

OUTLINE

This chapter is organized into 13 sections, starting with a general description of states of matter. As an understanding of the periodic table of the elements is helpful to an organized approach to inorganic chemistry, this subject is treated next. Periodicity is followed by the section on nomenclature so that the references to names of inorganic compounds will be more meaningful. The **shorthand** of chemical reaction representation is treated next. The numerical relationships of reacting elements is then treated more fully in the section on stoichiometry, and electronic changes in chemical reactions are explored in sections devoted to oxidation and reduction and electrochemistry.

Descriptive material on general classes of elements and compounds is presented in the sections on acids and bases and metals and nonmetals. These are followed by three sections that are more generally oriented toward physical chemistry, and its studies of equilibrium, solutions, and kinetics. The final section deals with organic chemistry, since the numerous and diverse organic chemical species are better introduced separately, and they are often treated as a separate category of chemical knowledge. The selection of 13 chemistry sections for this chapter follows the 13 chemical subdivision covered in the NCEES Sample Examination. The subdivisions have been reordered here to give a more logical flow to the subject matter.

The Periodic Table of the Elements on pages 540 and 541 has been reprinted, with permission, from a commercial chemical catalogue and gives four alternate methods of numbering the groups of the periodic table. The method referred to in the text is that noted in the chart as **Former Chemical Abstract Service**. In this method, the elements of the first three periods of the periodic table, and the elements that fall beneath them in other periods, are designated from IA to VIIIA. Transition metals and other metals in Periods 4 through 7 are designated as Groups IB through VIIIB. This method of designation is consistent with textbook discussions of the elements.

STATES OF MATTER

All elements and compounds exist in states ranging from solid to gaseous, depending on temperature, pressure, and the bonding between atoms and molecules.

The most tenuous, and in some respects the simplest and most amenable to theoretical understanding, is the gaseous state. Here matter is separated into distinct units, usually atoms or small molecules. Helium (He), argon (Ar), nitrogen (N_2), oxygen (O_2), chlorine (Cl_2), carbon dioxide (CO_2), and water vapor (H_2O) are found in the gaseous state at sufficiently high temperature and low pressure, corresponding to ambient conditions at the surface of the earth.

Gas molecules are (a) relatively far apart compared to their own dimensions, (b) traveling at high speed, and (c) frequently colliding with each other. The physical relationships of gaseous systems is treated by the **kinetic theory of gases**, in which their properties are quantified and related to the pressure, temperature, and volume of the system. A basic relationship, and one of great utility in chemical calculations involving gases, is the **perfect gas law**:

$$\frac{PV}{T} = k \text{ (constant)}$$

This relationship shows that, for a confined gas, the pressure and volume are inversely proportional at constant temperature, and either pressure or volume are directly proportional to the temperature, T, when T is expressed in units of absolute temperature. This law is stated more specifically as:

$$PV = n\mathrm{R}T$$

where n refers to the number of moles (molecular weight expressed in grams) of the gaseous species, and R is a constant known as the **gas constant**. To use this equation, all the parameters must be expressed in consistent units. One of the common expression is for P in atmospheres, V in liters, n in moles, and T in Kelvin. In this case the gas constant, R, is 0.08206, and has units of liter • atmospheres/mole •Kelvin.

Example 14.1

One mole of an ideal gas at 0°C and 1.00 atm pressure contains 6.022×10^{23} molecules (Avogadro's number) and occupies a volume of 22.414 L. A good quality, bakeable fitting can be used at pressures of 10^{-8} torr and some vacuum valves are good to 10^{-11} torr. What is the number of molecules in 1 cc of a vacuum system at 0°C and (a) 10^{-8} torr and (b) 10^{-11} torr?

$$\text{Note: } PV = nRT, \text{ where } R = 0.08206 \frac{\text{liter} \bullet \text{atm}}{°\text{K} \bullet \text{mol}}$$

$$1 \text{ atm} = 759.9 \text{ torr}; \ t_K = t_C + 273.15$$

Solution

$$\text{Solve } PV = nRT \text{ for } n = \frac{PV}{RT}$$

(a)

$$P = 1\times10^{-8} \text{ torr} \times \frac{1 \text{ atm}}{759.9 \text{ torr}} = 1.316\times10^{-11} \text{ atm}$$

$$V = 1 \text{ cm}^3 \times \frac{1 \text{ L}}{10^3 \text{ cm}} = 1\times10^{-3} \text{ L}$$

$$T = 0 + 273.15 = 273.15 \text{ K}$$

$$n = \frac{PV}{RT} = \frac{(1.316\times10^{-11} \text{ atm})\times(1\times10^{-3} \text{ L})}{0.08206\times273.15 \text{ K}} = 5.871\times10^{-16} \text{ mol}$$

$$\text{Number of molecules} = (5.871\times10^{-16} \text{ mol}) \times \left(6.022\times10^{23} \frac{\text{molecules}}{\text{mol}}\right)$$

$$= 3.535\times10^{8} \text{ molecules in 1 cc at } 10^{-8} \text{ torr}$$

(b)

$$P = 1\times10^{-11} \text{ torr} \times \frac{1 \text{ atm}}{759.9 \text{ torr}} = 1.316\times10^{-14} \text{ atm}$$

$$n = \frac{PV}{RT} = \frac{(1.316\times10^{-14} \text{ atm})\times(1\times10^{-3} \text{ L})}{0.08206\times273.15 \text{ K}} = 5.871\times10^{-19} \text{ mol}$$

$$\text{Number of molecules} = (5.871\times10^{-19} \text{ mol})\times\left(6.022\times10^{23} \frac{\text{molecules}}{\text{mol}}\right)$$

$$= 3.535\times10^{5} \text{ molecules in 1 cc at } 10^{-11} \text{ torr}$$

There are more refined equations for gas behavior, one of which is the **Van der Waals** equation,

$$\left(P + \frac{a}{V^2}\right)(V - b) = RT$$

where a is a constant for the interionic attraction between the molecules and b is a constant for the actual volume occupied by the molecules.

The kinetic theory of gases also allows us to relate the mass and temperature of a molecule to its linear velocity. The equation is

$$PV = \frac{1}{3} Nmu^2 = nRT$$

where N is the number of molecules, m is the mass of a single molecule of the gas, n is the number of moles of gas, and u is the molecular velocity. For one mole of gas ($n = 1$), N is Avogadro's number (6.022×10^{23}) and Nm = the molecular weight of the gas, M g/mol. The equation can be solved for the root-mean-square (rms) velocity, u, when T is the temperature in degrees absolute (K) and R = 8.314×10^3 g•m ^{2}/(s^2 •K) •mol, to give the velocity in m/s. The speed for an average molecule of air is about 500 m/s. It can be calculated that the mean free path of a molecule in air is only about 6×10^{-8} m and the number of collisions an air molecule encounters is about 6×10^9 molecules/s.

When the pressure is increased and the temperature is decreased, gas molecules will, at some point, condense to the liquid phase. The higher pressure beings the molecules closer together and the lower temperature reduces the velocity and kinetic energy of the molecules until the point is reached at which the attractive forces between molecules are great enough to keep molecules adjacent to each other. Although adjacent, the molecules still are in motion relative to each other. In fact, the molecules slide past each other rather easily, giving rise to the observed liquid property of flow. There is some evidence of a pattern or **structure** of molecules, especially near the freezing point, but the outstanding property of liquids is flow. Under the influence of gravity or a centripetal force they fill their container to a level consistent with the total volume of the molecules.

At the appropriate temperature and pressure for condensation the gaseous molecules do not all go into the liquid phase. Instead, a sufficient number of molecules are left in the gaseous phase to account for the full properties of a gas at the appropriate temperature and pressure. This is a gas-liquid equilibrium. At a given temperature, the pressure in the gas phase of the molecules of the species forming the liquid is a fundamental constant of the system—called the **vapor pressure**. If the temperature is raised, the concentration, or pressure, of the molecules in the gaseous phase goes up; if the temperature is lowered, the concentration in the gaseous phase goes down. This is all at equilibrium, meaning that the liquid and gaseous phase are identical in temperature and sufficient time has elapsed after a change in conditions to allow the situation representing the new condition to stabilize. The change of vapor pressure with temperature for a given species is given by the equation

$$\log 2.3\ \frac{p_2}{p_1} = \frac{\Delta H}{\mathrm{R}}\left(\frac{1}{T_1} - \frac{1}{T_2}\right)$$

where p_1 and p_2 are the vapor pressures under the two sets of conditions, and T_1 and T_2 are the corresponding temperatures. A new parameter introduced here is ΔH, which is the heat of vaporization. When ΔH is in units of calories per mole, R, the gas constant, is 1.987 calories per mole, and T is absolute temperature in Kelvin. When the heat of vaporization is known, the vapor pressure at a new temperature can be calculated from that at the original temperature. The units of vapor pressure are not critical, as long as both pressures are in the same units, as the ratio of pressures appears in the equation.

The molecules of the liquid phase are held together by molecular interactions, of which the most pervasive is a diffuse type of interaction known as **Van der Waals forces**. These forces are largely proportional to the bulk of the individual molecules. More specific attractive interactions occur between molecules in which the charge centers are not symmetrical with the mass centers and these are called **dipole interactions**. As a polar molecule may induce a charge in a nonpolar molecule, another class of forces is dipole-induced dipole interactions. A small number of liquid molecular species are held in association with their neighbors by a **hydrogen bond** force, in which the shared proton acts as an attractive bridge between two molecules. This is a very significant interaction because it is strong in the most common liquid and solvent encountered in chemistry—H_2O, or water.

Liquids are known for their lack of compressibility, which means that the liquid phase is very small compared to the gas phase. They also have a property called **viscosity**, which is not unique to the liquid phase, but is often referred to in the liquid phase. Viscosity represents the frictional drag of layers of liquid sliding by each other. It can be measured by a variety of methods. The flow of liquid through a capillary tube is one method, in which the viscosity can be directly related to the rate of flow and the capillary dimensions by Poiseuille's equation:

$$\eta = \frac{(P_1 - P_2)r^4 t}{8Vl}$$

where η is the viscosity; $P_1 - P_2$ is the pressure drop in the capillary tube; r is the radius, and l the length, of the capillary tube; and the volume of liquid V flows the length of the capillary in time t.

Liquids also are known in chemistry for their solvent properties. Water is the principal solvent for all ionic species—salts, acids, and bases—that may react chemically with each other or take part in electrochemical reactions at an electrode in an electrochemical cell. The reason water is such an excellent solvent is its dipole moment, as the molecule is not linear, as H—O—H, but has an approximately 108° bond angle, as

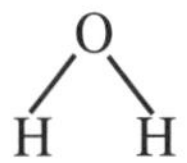

so that positive and negative charges are separated in the molecule. The resulting polar molecules can cluster around charged ionic species by having the opposite charged end of the molecule face the ion. This brings about good solvent properties for ionic materials. Nonpolar liquids, like hexane and benzene, are good solvents for nonionic species, such as small organic molecules. Some of the organic liquid species, such as acetone and alcohol, have polarity and dipole moment due to their oxygen groups. These are good solvents for organic species with a polar group attached.

When liquids are cooled the molecular motion decreases and the intermolecular forces tend not only to keep molecules together, but also to keep the molecules in a pattern or arrangement relative to each other. Such a pattern is usually a **crystal**, a three-dimensional structure with the molecules in precise fixed positions. In crystals the molecules still vibrate with thermal energy, but they no longer slide past each other as they did in the liquid state. This is the solid state, and the transition from the liquid state is accomplished by the loss of the **heat of fusion**.

In the solid state the assembled molecules have a form and do not **flow** from minor forces such as gravitation, but they do yield under pressure. Solids are also relatively good heat conductors, and monatomic solids have a heat capacity of close to 6 cal $\cdot K^{-1} \cdot mol^{-1}$. Compared to gases and liquids, solids are less reactive. However, the surface of the solid may readily react with a component in the gas phase, and a reaction can proceed quickly at the surface of a finely divided solid. This can also be the case when a solid and a liquid are in contact, sometimes involving the dissolution of the solid in the liquid. There can also be a reaction between two solid phases in physical contact. However, the rate of such a reaction is controlled by the rate of diffusion in the solid phases and is slow.

The borderline of solid, liquid, and gas is not always clear. This is evident with solids, as a noncrystalline material can also be quite solid, but its transition to the liquid phase is not sharp. One class of compounds of this type is the thermoplastic polymers, such as polystyrene, which deform and flow under pressure as the temperature is raised. Another class which behaves in this manner is the inorganic glasses. The solid glass has an amorphous, rather than crystalline, structure. It softens at elevated temperatures, and the transition to liquid phase is not sharp. We should also consider the **mesophases**, such as the **smetic phase**. Examples are liquids used in **liquid crystal displays**. These liquids respond to an applied electric charge in a way that brings about some order in the spatial relationship of the molecules.

The gas and liquid phases can be also have an area of coexistence in which properties of both phases are evident. This is seen in processes in which the gas is pressurized but kept above its critical temperature. The critical temperature is the temperature that the gas must be cooled to before it can become liquid, no matter what the pressure. At high pressure, and temperatures above the critical temperature, a gas has some liquid-like properties. It can be used as a solvent, but is not a true liquid. Water and carbon dioxide are two species that can be used in this state. The pressurized water, for example, exhibits some liquid properties, but it acts more like a nonpolar solvent than the polar solvent known as liquid water. Carbon dioxide is primarily thought of as either a solid or a gas, but it can also be pressurized to have liquid-like properties.

The above discussion applies to three-dimensional states. There is also a world of two-dimensional states, or the relationships of matter adsorbed on a surface. The primary evidence for this has been from the deposition of monolayers of polar organic compounds on an aqueous surface. This was the technique of the Langmuir-Blodgett balance, in which fatty acids, for example, could be deposited on water in a trough with a barrier. Measurements of the force-area relationships showed a two-dimensional analog to three-dimensional gas relationships.

PERIODICITY

The periodicity of the properties of the elements is based on the electronic structure of the atoms and reflected in their order and position in the **periodic table**. Similarities of elemental properties, and their periodicity, was observed long before atomic structure was understood. Definitions of atomic structure, and the understanding thereof, came largely from spectroscopic observations and their mathematical interpretation.

The periodic table now has seven horizontal rows or **periods** and eight, or more, vertical columns or **groups**. The properties of the elements in each period vary from metallic to nonmetallic going from left to right, ending with the nonreactive noble gases. The groups from I to VII (with noble gases as Group VIII), have metals or nonmetals of somewhat similar chemical and physical properties. However, the heavier elements, going down a column in the table, tend to be more metallic in character, as particularly noted on the nonmetal side of the table.

The first period contains only the two elements hydrogen (H) and helium (He), but the second and third periods contain eight elements each, from lithium (Li) through neon (Ne) for the second period and from sodium (Na) through argon (Ar) for the third period. In the fourth period, the number of groups has to be considerably expanded, as 10 elements come in order of atomic numbers between calcium (Ca) in the second group and gallium (Ga) in the third group. These are all metals and are called the first transition series.

The fifth period is similar to the fourth, in that 10 metallic elements—the second transition series—occur in order of atomic number between strontium (Sr) in the second group and indium (In) in the third group. A third transition series occurs in the sixth period, between barium (Ba) in the second group and thallium (T1) in the third group.

In order to handle the periods with 18 elements, the modern periodic table divides the elemental groups into A groups and B groups. The eight groups consistent with the elements listed in the second and third periods are IA, IIA, IIIA, IVA, VA, VIA, VIIA and VIIIA. The transition series elements are divided into groups IB, IIB, IIIB, IVB, VB, VIB, VIIB, and VIIIB.

A further complication of the periodic table is the occurrence of 14 elements in the sixth period between lanthanum (La) in Group IIIB and hafnium (Hf) in Group IVB. These are known as the lanthanide series or **rare earth** elements, and are not listed in the main sequence of the periodic table. Again in the seventh period, 14 elements occur between actinium (Ac) and the unknown element 104, and are known as the actinide series. As in the case of the transition elements, their occurrence at this place in the sequence of atomic numbers can be explained by the theory of atomic structure.

Another comment on the periodic table concerns Groups IIIA, IVA, VA, and VIA. These groups all start as nonmetals in the second period, namely with boron (B), carbon (C), nitrogen (N), and oxygen (O) respectively. However, going down the columns corresponding to these groups, the elements change to metallic properties. This occurs in the third, fourth, fifth, and sixth period, respectively, for Groups IIIA–VIA; the first metallic elements are aluminum (A1), germanium (Ge), antimony (Sb), and polonium (Po) in each of these groups.

With the overall structure of the periodic table in mind, let us look at some properties that show periodicity. One is atomic radius, which is a maximum for each Group IA element (the alkali metals—Li, Na, K, Rb, and Cs). The atomic radius decreases going across the period to the halogen elements of Group VIIA, complicated by minima in the transition elements and the lanthanide-actinide series. The ionization potential, or first ionization energy, represents the energy needed to remove one electron from an atom. It is a minimum for the Group IA metals and reaches a maximum for the Group VIII-rare gases.

NOMENCLATURE

Chemical nomenclature is concerned with the prefixes, suffixes, and other modifications of the root word to denote the state of the elements in a compound. We shall be concerned here primarily with inorganic compounds, as the nomenclature of organic compounds is a topic of its own and will be considered in the section on organic chemistry.

When compounds of metals and nonmetals occur, the compound is considered a salt; and the naming of the compound uses the elemental name of the metal first, followed by the root of the nonmetal element name and the ending **ide**. Sodium brom*ide* and calcium flur*ide* are examples. The nomenclature becomes slightly more complicated when the metal may have more than one valence state is its compound with the nonmetal. An example is iron oxide, which is known more precisely as either ferrous oxide or ferric oxide, depending on the valence state of the iron. Also, these compounds would be known as oxides, rather than salts, as the compound would not dissociate to oxide ions when dissolved in aqueous solution.

When the metal ion can have both a lower valence and a higher valence, the compounds of the lower valence have the root word with the suffix **ous**, while the compounds of the higher valence have the root word with the suffix **ic**. In the case of the iron oxide mentioned above, the elemental root is not iron but **ferr**. The root still designates the same element but comes form a different linguistic source.

The compounds of hydrogen and nonmetals maintain the salt type of nomenclature, with hydrogen fluoride, hydrogen chloride, hydrogen bromide, and hydrogen iodide as designations for the halogen compounds in the unionized gaseous phase. The nomenclature is consistent with other hydrogen compounds such as

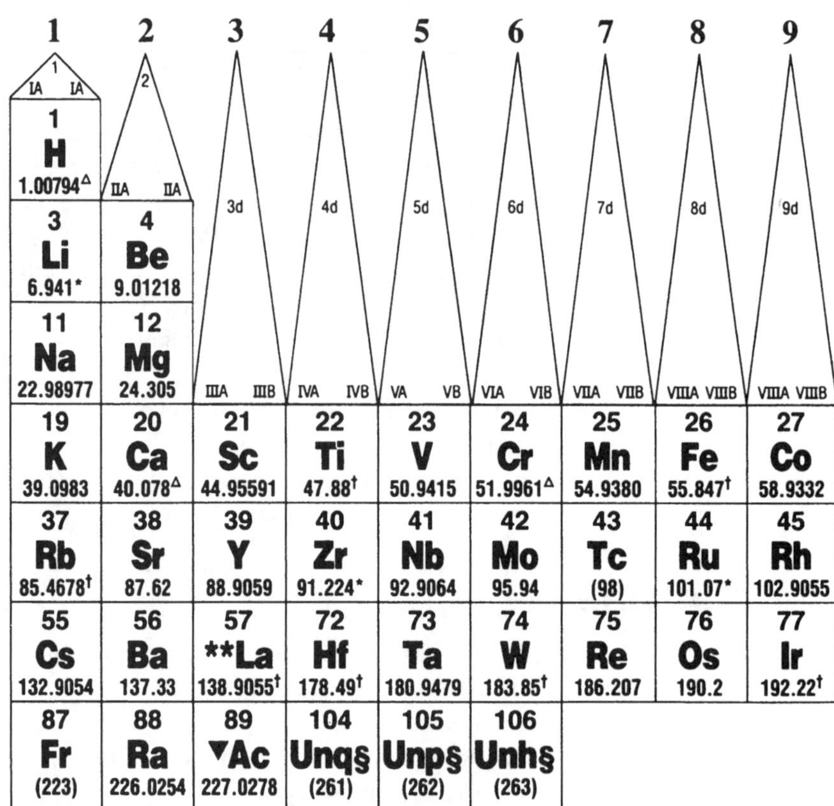

Fisher Scientific PERIODIC CHART OF THE ELEMENTS

1 (IA IA)	2 (IIA IIA)	3 (3d; IIIA IIIB)	4 (4d; IVA IVB)	5 (5d; VA VB)	6 (6d; VIA VIB)	7 (7d; VIIA VIIB)	8 (8d; VIIIA VIIIB)	9 (9d; VIIIA VIIIB)
1 H 1.00794Δ								
3 Li 6.941*	4 Be 9.01218							
11 Na 22.98977	12 Mg 24.305							
19 K 39.0983	20 Ca 40.078Δ	21 Sc 44.95591	22 Ti 47.88†	23 V 50.9415	24 Cr 51.9961Δ	25 Mn 54.9380	26 Fe 55.847†	27 Co 58.9332
37 Rb 85.4678†	38 Sr 87.62	39 Y 88.9059	40 Zr 91.224*	41 Nb 92.9064	42 Mo 95.94	43 Tc (98)	44 Ru 101.07*	45 Rh 102.9055
55 Cs 132.9054	56 Ba 137.33	57 **La 138.9055†	72 Hf 178.49†	73 Ta 180.9479	74 W 183.85†	75 Re 186.207	76 Os 190.2	77 Ir 192.22†
87 Fr (223)	88 Ra 226.0254	89 ▼Ac 227.0278	104 Unq§ (261)	105 Unp§ (262)	106 Unh§ (263)			

**Lanthanides

58	59	60	61	62
Ce	Pr	Nd	Pm	Sm
140.12	140.9077	144.24†	(145)	150.36†

▼Actinides

90	91	92	93	94
Th	Pa	U	Np	Pu
232.0381	231.0359	238.0289	237.0482	(244)

FISHER SCIENTIFIC
CAT NO. 05-702-10

§The International Union of Pure and Applied Chemistry (IUPAC) has not adopted official names or symbols for these elements.

*These weights are considered reliable to ±2 in the last place.

†These weights are considered reliable to ±3 in the last place.

ΔThese weights are considered reliable in the last place, as follows: Calcium and Gallium ±4; Boron ±5; Chromium and Sulfur ±6; Hydrogen and Tin ±7.

All other weights are reliable to ±1 in the last place. All reliabilities are based on an uncertainty scale of ±1 to 9.

Atomic weights corrected to conform to the most recent values of the Commission on Atomic Weights. Column nomenclature conforms to IUPAC system and data in this chart have been checked by the National Bureau of Standards' Office of Standard Reference Data.

hydrogen sulfide (H_2S) and hydrogen selenide (H_2Se). However, hydrogen oxide (H_2O) is known universally by its more common name, **water**.

When the above hydrogen compounds are in aqueous solution, they become ionized to inorganic acids. The inorganic acids always contain hydrogen as the cation and the cation root is designated as **hydro**—leading to hydrofluoric (HF), hydrochloric (HCl), hydrobromic (HBr), hydroiodic (HI), hydrosulfuric (H_2S), and hydroselenic (H_2Se) acids. Hydrogen oxide, as the solvent phase, remains designated as water.

Perhaps the most confusing of the designation nomenclature is that of the oxyacids. These are acids containing various amounts of oxygen in the anion. The compounds with sulfur in the anion are illustrative. Here the designation for hydrogen as the cation is inherent in the name **acid**, so it is not explicitly given. The most common oxyanion of sulfur is the sulfate ion, SO_4^{2-}, where the sulfur has an oxidation number of +6. As the acid, the name becomes sulfuric—with the root **sulfur** and the suffix **ic**. When the anion has one less oxygen, SO_3^{2-}, the name of the acid has the **sulfur** root and the suffix **ous**. Sulfur*ous* acid is H_2SO_3, where the sulfur has an oxidation number of +4.

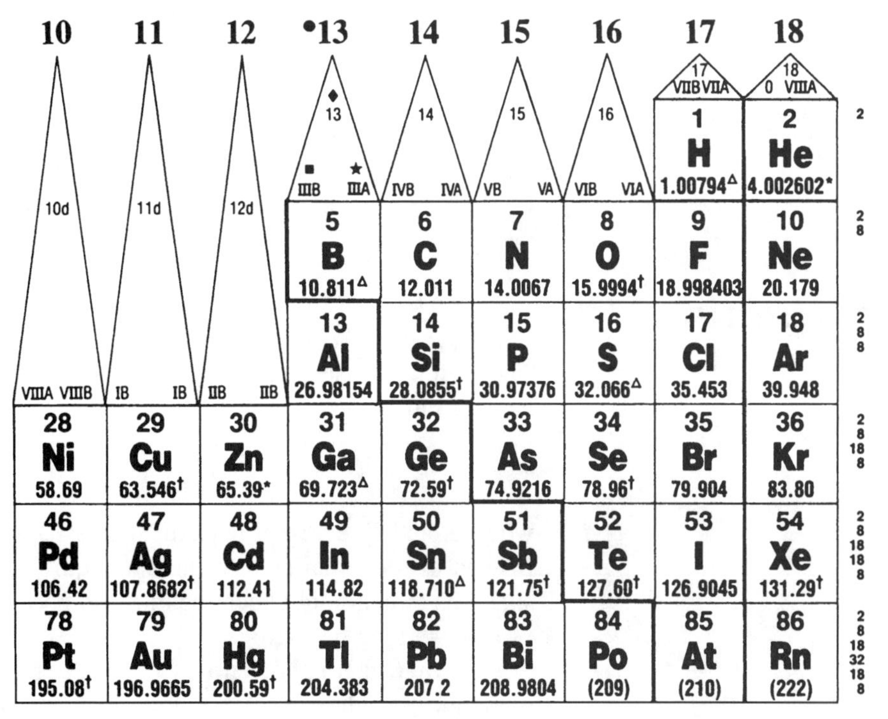

Example 14.2

The oxyacids of sulfur and phosphorous have the ending **ic** for the element in the most stable oxidation state and **ous** for the element in the next lower oxidation state. What are the names of the corresponding salts of these acids?

Solution

The *sulfuric* acid salts have the endings **ate** for the salt in which both hydrogens are replaced and **hydrogen ate** when only one hydrogen is replaced—as in sodium sulf*ate* and sodium *hydrogen* sulf*ate*. The latter salt is also known as sodium *bi*sulf*ate*. The *sulfurous* acid salts have similar names, except the *ate* ending is replaced by *ite*. The comparable names for the *ous* acid salts are sodium sulf*ite* and sodium hydrogen sulf*ite* or sodium bisulf*ite*.

The *phosphoric* acid salts also have the ending **ate** for the salt when all three hydrogens are replaced—as in sodium phosph*ate*. However, the salt is the **dihydrogen ate** when only one hydrogen is replaced and the **hydrogen ate** when two of the three hydrogens are replaced, as in sodium dihydrogen phosphate and sodium hydrogen phosphate. The *phosphorous* acid salts have similar names, except that the *ate* ending is replaced by *ite*.

The **ic** and **ous** suffix designation carries over to the oxidation state of cations in salt-type compounds with nonmetals. Thus iron in the +3 state forms ferr*ic* chloride, $FeCl_3$, and ferr*ic* oxide, Fe_2O_3; iron in the +2 state forms ferr*ous* chloride, $FeCl_2$, and ferr*ous* oxide, FeO. The same designation is used with tin compounds, with the root designated by **stann**. Since the oxidation states of tin are +4 and +2, $SnCl_4$ is stann*ic* chloride and $SnCl_2$ is stann*ous* chloride.

EQUATIONS

The reaction of one chemical species with another is defined by a chemical equation of the form

$$A + B = C + D$$

where A and B represent reactant species and C and D represent product species. Instead of the equal sign to indicate equilibrium, an arrow (→) is often used to show the direction in which the reaction goes to completion. Also, a double arrow (↔ or ⇌) may be used.

In a **metathetical** reaction, two species react to form two other species without oxidation or reduction. A common example would be an acid-base reaction (in solution), as

$$HCl + NaOH = NaCl + H_2O$$

The simplest form of chemical equation is a unimolecular reaction, such as a dissociation:

$$HgO\ (+\ heat) = Hg + O_2$$

Here we encounter another requirement of chemical equations—that they be **balanced**. A balanced equation has the same number of atoms of each species on the left side of the equation as on the right side of the equation. The above reaction should, therefore, be

$$2\ HgO = 2\ Hg + O_2$$

The balanced equation is very informative in a quantitative manner; the amounts of each species involved in the reaction are given by the relative number of atoms or molecules in the equation. Therefore, the gram-atoms or gram-moles of each species can be calculated. In the case of gases, the volume of the gas at standard conditions can also be calculated, and modified for the volume at non-standard conditions.

Example 14.3

A reaction that makes gas is that of the metal zinc with hydrochloric acid:

$$Zn + HCl = ZnCl_2 + H_2$$

In a laboratory experiment, 1.00 g Zn is added to a solution of 100 mL of 1.00 M HCl. How many cc of H_2 is produced in the reaction at room temperature (25°C) and one atmosphere pressure? (a.w. Zn = 65.4, Cl = 35.5, H = 1.01; 1.00 mol ideal gas at standard conditions = 22.4 L)

Solution

The coefficients to balance the equation must be added. A causal inspection shows that two molecules of HCl are required to supply two atoms of Cl for $ZnCl_2$ and two atoms of H for H_2:

$$Zn + 2\ HCl = ZnCl_2 + H_2$$

This balanced equation shows that one mole of zinc (65.4 g) reacts with two moles of HCl (73.0 g) to yield one mole of zinc chloride (136.4 g) and one mole (2.0 g) of hydrogen. Any ratio of these amounts can be used to fulfill the reaction predicted by the balanced equation.

In the experiment, a small amount of zinc, 1 g, is added to the solution. The number of moles of Zn is

$$\frac{1.00\text{ g}}{65.4\text{ g/mol}} = 0.015\text{ mol}$$

The number of moles of HCl in solution is

$$0.100\text{ L} \times \frac{1.00\text{ mol}}{\text{L}} = 0.100\text{ mol}$$

The balanced equation shows that 0.015 mol Zn requires 2 × 0.015 = 0.030 mol HCl to completely react, so there is an excess of HCl and the amount of Zn limits the amount of H_2 formed. The amount of H_2 formed will then be the same as the number of moles of Zn, or 0.015 mol. The volume of H_2 can be calculated from the physical constant for an ideal gas at standard conditions (22.4 L/mol):

$$0.015\text{ mol} \times \frac{22.4\text{ L}}{\text{mol}} = 0.336\text{ L at standard conditions}$$

25°C = 298 K, so

$$0.336\text{ L} \times \frac{298}{273} = 0.367\text{ L} = 367\text{ mL}$$

at 25°C and one atmosphere.

STOICHIOMETRY

The calculation of weights, volumes, and elemental proportions in a chemical reaction can be carried out from basic premises and information. One premise is that the combination of atoms to form a molecule takes place with whole numbers of atoms, and the ratio of the atoms to each other is a ratio of integers and is characteristic of the particular molecular species. Thus, if two atomic species, A and B, join to form a molecule, they are in a fixed ratio and the molecular formula is written with subscript numbers denoting the number of atoms of each atomic species in the molecular formula. A molecular formula may be of the form AB, AB_2, A_2B, A_2B_3 or other whole number subscripts, depending on the valence or bonding capacity of the species A and B. Examples are NaCl, $CaCl_2$, K_2O, and Fe_2O_3.

Chemical compounds can be characterized by both the **molecular formula** and an **empirical formula**. The molecular formula gives the whole numbers of atoms actually joined together by chemical bonding to assemble the molecular species in question. An empirical formula, on the other hand, gives the correct atomic ratio by the subscript numbers, but may be less than the actual number of atoms in the molecule.

For example, the empirical formula for benzene is CH, whereas the molecular formula is C_6H_6, as there are actually six atoms of carbon and six atoms of hydrogen in a molecule of benzene. The empirical formula, or atomic ratio, is the simplest whole number atomic ratio and may be calculated from experimental data, such as a combustion analysis. Either prior knowledge of the chemical species involved or some other experimental data, such as molecular weight determination, is needed to derive the molecular formula.

A second premise is that the equation for the reaction is known and can be presented in balanced form by use of the proper stoichiometric coefficients. The **stoichiometric coefficient** is the number that each molecular species is multiplied by so that the same number of each atom is found on both the left-hand side and the right-hand side of the equation. By convention, the stoichiometric coefficients, are usually whole numbers. When fractional values appear both sides of the equation are multiplied by the lowest number that will make all coefficients whole numbers. Examples of balanced equations are

$$2\ KClO_3 + \text{heat} = 2\ KCl + 3\ O_2$$
$$2\ C_2H_6 + 7\ O_2 = 4\ CO_2 + 6\ H_2O$$
$$K_2Cr_2O_7 + 9\ KI + 14\ HCl = 3\ KI(I_2) + 2\ CrCl_3 + 8\ KCl + 7\ H_2O$$

and

$$2\ FeCr_2O_4 + 7\ Na_2O_2 = 2\ NaFeO_2 + 4\ Na_2CrO_4 + 2\ Na_2O$$

Given the whole number ratios of atoms in the molecule and knowledge of a reaction as a balanced equation, the atomic and molecular formulas can be related to the physical world in which quantities of chemicals are measured. This is done by the concept of a quantity called a mole, which is defined in the metric system as the molecular weight of the species expressed in grams, also known as the gram molecular weight. The molecular weight of a molecular species is the sum of the atomic weights and the atomic weight of an element is the number representing the weight of an atom of the element relative to carbon-12, or ^{12}C. For various reasons the atomic weights are not necessarily whole numbers, primarily because a sampling of a given element in the earth environment may include two or more isotopes of different atomic weight. However, the isotope ratio is, for the purpose of chemical reactions, very constant.

The mole, which has the abbreviation **mol** when used after a number, is a quantity which has, by definition, the same number of **elementary entities** as there are carbon atoms in 12 g of carbon-12. This number is known as the **Avogadro constant** and has been experimentally determined to be 6.022×10^{23}. Since all atomic weights are related to that of carbon-12, a mole of any species has the same number of molecules—for example, there are the same number of molecule in 2.016 g of H_2, 71.0 g of Cl_2, and 58.5 g of NaCl. The great importance of this concept is that the molecular formulas and chemical equations are written for discrete numbers of atoms and molecules, but since all moles have the same number of molecules, the molecular formulas and chemical equations can be related to the measured weights of chemicals in the physical world.

When one of the reactants or products of a chemical reaction is a gas, the usefulness of the mole concept becomes further evident. This is because one mole of gas of any species still stands for a certain number of molecules (6×10^{23}), but in the case of gases this number accounts for a certain volume of gas at a given temperature and pressure. The number that is useful in relating equations to real physical measurements is that one mole of a gas—any gas—occupies 22.4 liters at so-called **standard** conditions. Standard conditions are one atmosphere and 0°C or 273 K. A common ambient temperature is 25°C or 298 K, and the ideal gas law correction indicates that a mole of gas at 25°C occupies 24.4 liters.

Example 14.4

A simple equation involving a gaseous product was previously presented, the decomposition of $KClO_3$ to yield oxygen gas:

$$2\ KClO_3 + heat = 2\ KCl + 3\ O_2$$

How much oxygen could we obtain by heating 100 g of $KClO_3$?

Solution

A mole of $KClO_3$ is 122.6 g (its molecular weight in grams) and the equation shows that 2 moles of $KClO_3$ can yield 3 moles of O_2. However, we only have 100/122.6 = 0.82 mol $KClO_3$. The equation shows that the number of moles of O_2 yielded will be $3/2 \times 0.82 = 1.23$ and the volume at one atmosphere and 25°C will be $1.23 \times 24.4 = 30.0$ liters.

OXIDATION AND REDUCTION

Oxidation refers to the loss of electrons by an atom, which is the case when the metallic elements react with oxygen. The element that gains the electrons has been reduced, reduction has taken place. Thus, in the reaction of magnesium (Mg) with oxygen (O_2), there is a transfer of two electrons from the Mg, which then exists as the positive ion, whereas oxygen becomes O^{2-}, the oxide ion:

$$2\ Mg + O_2 = 2\ MgO$$

The product, MgO, is an ionic lattice, and there is ionic bonding in the lattice between the Mg^{2+} ions and the O^{2-} ions.

Although the reaction with elemental oxygen is a common example of oxidation, there are other oxidation reactions in which oxygen is not involved. For example, the burning of hydrogen (H_2) in chlorine (Cl_2) is an example of oxidation of the H_2, as it loses electrons to the chlorine in the process:

$$H_2 + Cl_2 = 2\ HCl$$

Oxidation, or reduction, can also take place when an element changes in oxidation state, or the number of electrons in the outer shell, in a reaction. For example, in the reaction of sulfur dioxide (as a solution of sulfurous acid) with iodine in solution

$$SO_2 + 2\ I + H_2O = H_2SO_4 + 2\ HI$$

the sulfur atom, S changes from an oxidation state of +4 to +6. Therefore it loses electrons in its outer shell and is oxidized. At the same time the iodine atom, I, has gained an electron and changed from a state of no charge to a reduced state of −1. The iodine therefore has gained an electron and is reduced. The **oxidation state** is also known as the **oxidation number**, and is a useful concept in balancing equations in which oxidation and reduction take place.

Example 14.5

A standard solution of sodium thiosulfate, $Na_2S_2O_3$, is made by dissolving 25 g of the hydrated salt, $Na_2S_2O_3 \bullet 5\ H_2O$ in water, and diluting to 1.0 L of solution. The solution is standardized with iodine, I_2, according to the reaction

$$I_2 + 2\ Na_2S_2O_3 = 2\ NaI + Na_2S_4O_6$$

where the iodine, I, is reduced to iodide, I^-, and the thiouslfate, $S_2O_3^{2-}$, is oxidized to tetrathionate, $S_4O_6^{2-}$. If 0.500 g of pure iodine is dissolved for the standardization and requires 39.40 ml of solution to reach the end point (disappearance of the iodine color), what is the normality of the thiosulfate solution in this reaction?

Solution

From the equation, two moles of thiosulfate react with one mole of iodine (I_2). The 0.500 g of pure iodine is

$$\frac{0.500\ \text{g}}{(2 \times 126.90\ \text{g/mol})} = 0.00197\ \text{mol}\ I_2$$

and 0.00197 mol I_2 reacts with (2 × 0.00197) = 0.00394 mol thiosulfate. The morality of thiosulfate solution is

$$\frac{0.00394\ \text{mol}}{0.0394\ \text{L}} = 0.100\ \text{mol/L}$$

Since there is change of one transferred electron per thiosulfate molecule in the oxidation $2\ S_2O_3^{2-} - 2e^- = S_4O_6^{2-}$, the equivalent weight of thiosulfate is the same as the molecular weight and the solution is also 0.100 N. The original 25 g of sodium thiosulfate weighed out would be

$$M = \frac{\text{g}}{\text{m.w}} = \frac{25.0}{(2 \times 23.00) + (2 \times 32.07) + (3 \times 15.00) + (5 \times 18.02)}$$
$$= \frac{25.0}{248.24} = 0.101\ \text{M}$$

so the titration showed that the hydrated salt was slightly impure.

Although all oxidation reactions involve reduction, and a large class of reactions are called oxidation-reduction or **redox** reactions, there are also reactions that are primarily thought of as reduction reactions. Some of these are related to the production of metals from ores. For example, one blast furnace reaction is the reduction of iron sulfide:

$$FeS + CaO + CO = Fe + CaS + CO_2$$

Here, the Fe of the FeS is reduced by CO, and CO is oxidized in the process to CO_2.

Oxidation and reduction reactions can be considered quantitatively from the viewpoint of an electrochemical cell. Such a cell has two electrodes and oxidation takes place at the negative pole (electrons are given up to the electrode by the species being oxidized) and reduction takes place at the positive pole (electrons are taken from the electrode by the species being reduced). The total cell potential is the sum of the two reactions:

$$E_{cell} = E_{ox} + E_{red}$$

The potential of the half-cell $H^+ + 2e^- = H_2$ is assigned the value 0.0 volts at standard conditions. If this reaction takes place at one of the electrodes, then the measured cell potential when another reaction takes place at the other electrode is the half-cell or single electrode potential for the reaction of interest. This is at standard conditions, which means unit **activity** or approximately unit concentration (that is, one mole per liter solution concentration or one atmosphere gas pressure) and 25°C.

Tables are available for reduction potentials, which is the cell potential when the reduction of the species being considered takes place at the positive electrode and the other electrode is the hydrogen electrode reaction with 0 volts half-cell potential (by definition). When the reduction takes place spontaneously the resultant half-cell potential is positive. All the positive potential half-reactions take place spontaneously when the other electrode is the hydrogen electrode, and each reaction at a higher potential takes place more readily than one at a lower potential. Therefore, if a reduction reaction in the table is linked in a cell to an electrode with a lower reduction potential, the initially considered reaction should still take place (at standard conditions).

Below the $2H^+ + 2e^- = H_2$ reaction in the table the potential values are negative, indicating that the reduction of the species shown takes place less readily than that for the hydrogen electrode reaction. The values for reduction potential continue negative to the bottom of the table, where the most difficult species to reduce are listed. As in the top part of the table, each species is more easily reduced than the ones below it (more negative reduction potential) and presumably would take place when coupled in a cell to a reaction with a more negative reduction potential.

Although the half-cell potentials may be measured, in principle, as indicated above, this is not always experimentally feasible. Therefore some data on single electrode potentials is calculated from other thermodynamic data.

A partial list of species and reactions in order of their electrochemical single electrode reduction potential is shown in Table 14.1.

ELECTROCHEMISTRY

A chemical reaction that involves an electron transfer can, when properly connected in a cell, be a source of an electric potential and an electric current. However, a single chemical reaction will only function as a **half-cell**, so that two

Table 14.1 Partial electrochemical series

Oxidizing Agents	Reaction	Voltage	Reducing Agents
(strong oxidizing)			(weak reducing)
F_2	$F_2 + 2e^- = 2F^-$	2.9	F^-
Au^+	$Au^+ + e^- = Au$	1.7	Au
MnO_4^-	$MnO_4^- + 8H^+ + 5e^- = Mn^{2+} + 4H_2O$	1.5	Mn^{2+}
N_2O_4	$N_2O_4 + 4H^+ + 4e^- = 2NO + H_2O$	1.0	NO
I_2	$I_2 + 2e^- = 2I^-$	0.5	I^-
Cu^{2+}	$Cu^{2+} + 2e^- = Cu$	0.3	Cu
H^+	$2H^+ + 2e^- = H_2$	0.0	H_2
Ni^{2+}	$Ni^{2+} + 2e^- = Ni$	–0.2	Ni
S	$S + 2e^- = S^{2-}$	–0.5	S^{2-}
Be^{2+}	$Be^{2+} + 2e^- = Be$	–1.7	Be
Cs^+	$Cs^+ + e^- = Cs$	–2.9	Cs
(weak oxidizing)			(strong reducing)

chemical reactions must be coupled to complete an electrochemical cell. In order to be electrically coupled, the chemical reaction at the electrode must have (1) one species that is conducting and connected to an external circuit, or (2) electrical contact between the reacting species and a conducting phase that can be connected to the external circuit. There must also be a continuity of the reacting solution between the electrodes, which can be provided by a **salt bridge**, porous barrier, or even an interface between two solutions of difference densities.

In order to have data that can be used to compare a number of combinations of chemical reactions in cells, it has become conventional to list the open circuit potential of reactions versus the potential of a hydrogen electrode. A table of this type was shown in the previous section on oxidation and reduction. The half-cell potential of the reaction

$$H_2 - 2e^- = 2H^+$$

is taken to be 0 volts, and half-cell potentials of all other combinations giving electrode reactions are measured against this standard. As the concentration of the species involved also affects the emf, the concentrations considered are, by convention, that of gas phases at one atmosphere and solution phases at a concentration of one mole per liter.

Example 14.6

$Zn - 2e^- = Zn^{2+}$ standard electrode potential = +0.76 volts, and $Cl_2 + 2e^- = 2Cl^-$ standard electrode potential = +1.36 volts. What is the open circuit potential at standard conditions when the two half-cells are combined?

Solution

When the two half-cells are combined as an electrochemical cell, the open circuit emf will be 0.76 + 1.36 = 2.12 volts. This is for standard concentration conditions, which in this case would be 1 mole/liter Z^{2+}+ and 1 atm Cl_2.

The effect of the concentration of reactant species on the emf of the cell can be calculated from the Nernst equation, which, for the reaction $A + B = C + D$, is

$$E = E_0 - \frac{RT}{nF} \ln \frac{[C][D]}{[A][B]}$$

where

$$\frac{[C][D]}{[A][B]}$$

is the **reaction quotient**, sometimes designated by Q. The equation is a thermodynamic derivation and includes the gas constant, R; the Faraday constant, F; the absolute temperature, T; and the number of transferred elections, n. At 25°C, with $n = 1$, and converting to $\log_{10}$, the factor RT/nF in the equation is 0.059.

Example **14.7**

For the Zn and Cl_2 reaction

$$Zn + Cl_2(g) = Zn^{2+} + 2\,Cl^-$$

$$E = E_0 - \frac{0.059}{2}\log\frac{[Zn^{2+}][Cl^-]^2}{[Zn][Cl_2]}$$

If the $ZnCl_2$ is 0.1 molar and the Cl_2 gas is 0.01 atmosphere, what is the electrochemical potential of the cell?

Solution

$$\begin{aligned} E &= 2.12 - \frac{0.059}{2}\log\frac{(0.1)(0.2)^2}{(1)(0.01)} \\ &= 2.12 - 0.030\log\frac{0.004}{0.01} \\ &= 2.12 - 0.030\log 0.4 \\ &= 2.12 + 0.012 = 2.14 \text{ volts} \end{aligned}$$

ACIDS AND BASES

Acids are a species of chemical compounds that function as proton donors or, in some cases, electron acceptors in reactions with other chemical species. The species with which acids react are bases, and are, therefore, compounds that function in a reaction as proton acceptors or electron donors.

The most common applications of the terms acid and base are in aqueous solution chemistry, where the dissolved acidic species **dissociates** to yield a hydrated proton, or hydronium ion, H_3O^+. The remainder of the dissociated species will be the negative ion, or **anion**. A basic species will dissociate in aqueous solution to yield a hydroxyl ion, OH^-, and a positive ion, or **cation**.

The law controlling the behavior of both acids and bases in aqueous solution arises from that fact that the solvent itself, water (H_2O), dissociates to form both the hydronium ion, H_3O^+, and the hydroxyl ion, OH^-. Furthermore, the amount of either hydronium ion or hydroxyl ion in aqueous solution, including that contributed by added acid or base, is controlled by the equilibrium constant, K_w, for the dissociation of H_2O.

At a given temperature, the product of the concentration of hydronium and hydroxyl ions is fixed. The number is very small, and varies with temperature. At 25°C, with concentrations expressed in moles/liter $[H_3O^+] \times [OH^-] = 1.0 \times 10^{-14}$. Any addition of acid reduces the OH^- concentration and any addition of base reduces the H_3O^+ concentration, maintaining the product of the concentration of the two ions in the 10^{-14} range.

When both acid and base are added to an aqueous solution the 1.0×10^{-14} value for the concentration product of $[H_3O^+]$ and $[OH^-]$ is exceeded, and the two species rapidly combine to form undissociated H_2O until the equilibrium value of the product of the concentrations is reestablished. The acid and base have then **neutralized** each other, with accompanying heat liberation and, possibly, a vigorous reaction.

Acids react with bases according to the **metathetical** reaction $HA + BOH = BA + H_2O$, where A is an anion and B is a cation. The driving force for this reaction is the formation of water, which will be in equilibrium with an ion product of 1×10^{-14} moles/liter of hydronium $[H_3O^+]$ and hydroxyl $[OH^-]$ ions. An example of a strong acid–strong base reaction is

$$HCl + NaOH = NaCl + H_2O$$

where both reactants and products are completely ionized, except for the H_2O. Weak acids and weak bases react to the full extent of their equilibrium concentrations, even though the reactant dissociates only to a small extent. This is because the reactant species continues to ionize as the dissociated hydrogen ion or hydroxyl ion is used up by reaction with the other species until the new equilibrium appropriate to the hydrolysis constant of the salt formed in the reaction is reached.

The measurement and characterization of the degree of acidity or basicity of a solution is done on the pH scale. The exact degree of acidity or basicity of a solution is given by the concentration of the hydrogen ions (hydrated to hydronium ions) or hydroxyl ions in solution. The product of the two (at 25°C) is 1×10^{-14} moles/liter, so that their concentration in a neutral solution is 1×10^{-7} moles/liter of each species. Numbers of this magnitude can be awkward in specifications and producers, and the so-called **pH value** is commonly used, where

$$pH = -\log_{10} [H^+] \quad \text{and} \quad [H^+] = 10^{-pH} \text{ moles/liter}$$

The number in brackets, [ ], is a concentration in the solution. Acid solutions then have a pH below 7.0 and basic solutions a pH above 7.0. Sometimes the pOH value is used, which has comparable definitions in terms of OH^- concentration.

Example 14.8

Calculate the $[H^+]$ and $[OH^-]$ concentrations, and the pH, for a solution of 0.080 M HNO_3.

Solution

HNO_3 will be completely ionized at this concentration, so

$$[H^+] = 0.080 \text{ M (mol/L)}$$

Since $K_w = [H^+]\,[OH^-] = 1.0 \times 10^{-14}$,

$$[OH^-] = \frac{1.0 \times 10^{-14}}{8.0 \times 10^{-2}}$$

$$= 0.125 \times 10^{-12} = 1.25 \times 10^{-13} \text{ M}$$

$$pH = -\log_{10}[H^+] = -\log_{10}(8.0 \times 10^{-2}) = -(-2 + 0.903) = 1.1$$

METALS AND NONMETALS

The two general classes of elements are metals and nonmetals. The division between metal and nonmetal is not always sharp, and the nonreactive elements (noble gases) do not fit into either category.

Nevertheless, there is a similarity in the physical aspects and reactivity of the large group known as metals and in the large group known as nonmetals, so that it is convenient and logical to consider the elements as part of each group. Also, the electronic structure of the atoms of the elements can be fundamentally grouped into electron donors, or metals, and electron acceptors, or nonmetals, when participating in chemical reactions. However, there are cases when the metal-nonmetal classification breaks down from this point of view, as, for example, (a) for many elements the loss or gain of electrons in reactions depends on reaction conditions as well as on the other element participating in the reaction, and (b) compound formation may result in the sharing of available electrons rather than an identifiable loss or gain of electrons by the elements participating in the reaction.

Referring to the periodic table of the elements, the period and row in which an element is located is closely related to its metallic or nonmetallic properties. Generally the elements on the left side of the periodic table are metals and those on the right side are nonmetals. Elements in a particular group become more metallic toward the bottom of the table (heavier elements) and the elements in the middle of the table, including the transition series elements, have metallic properties.

Metals are generally known for their physical properties, which include the characteristic metallic luster, cohesive strength, ductility, and high electrical and thermal conductivity. These properties are well explained by the theory of metals that views metal atoms as points in a crystalline structure surrounded by their **electron cloud**. The outer electrons travel freely through the crystalline structure, bonding the atoms together and providing for flow of electrical charge. The nonmetals, on the other hand, are often gases, or colored solids with no luster and low electrical and thermal conductivity.

The most active metals are those of Group IA of the periodic table, the alkali metals. These are lithium, sodium, potassium, rubidium, and cesium—all with a valence of +1. The next most active metals are the Group IIA alkaline earth metals, beryllium, magnesium, calcium, strontium, barium, and radium—all with a valence of +2. Group IB metals—copper, silver, and gold—are distinctly metallic, with valences from +1 to +3. The Group IIB elements—zinc, cadmium, and mercury—have +1 or +2 valence.

The Group IIIA metals, aluminum, gallium, indium, and thallium, all have a +3 valence. The Group IIIB metallic elements include scandium, yttrium, lanthanum, and actinium. Groups IVB and VB include titanium, zirconium, hafnium, vanadium, niobium, and tantalum, with valences in the +2 to +5 range. The Group VIB and VIIB elements—chromium, molybdenum, tungsten, manganese, technetium, and rhenium—also have variable positive valences. The Group IVB, VB, VIB, and VIIB elements (as well as IB, IIIB, and VIIIB) are **transition metals**.

The nonmetals are on the right side of the periodic table of the elements—or, more specifically, on the upper right side of the table, as elements in Group IIIA through Group VIA become metallic toward the bottom of the table. We include as nonmetals the Group VIIIA elements, also known as the noble gases. These elements, helium (He), neon (Ne), argon (Ar), krypton (Kr), xenon (Xe), and radon (Rn), are certainly nonmetallic, but do not quite fit into the concept of nonmetals because they are almost totally unreactive. However, a limited number of noble gas compounds have been prepared, including the xenon compounds XeF_2, XeF_4, and XeF_6.

The next group of nonmetals to consider are the Group VIIA halogens: fluorine (F), chlorine (Cl), bromine (Br), iodine (I), and astatine (At). These elements are characterized electronically by an unfilled *p* orbital in the outer electron shell, and a great tendency to fill the orbital and assume the same electron configuration as the rare gases. As elements, this configuration is achieved by the diatomic molecule, x_2, where x stands for any halogen, and the diatomic state shows covalent sharing of an electron pair.

The physical properties of the halogens vary with atomic number and weight. Under most ambient conditions, fluorine is a pale yellow gas, chlorine a greenish-yellow gas, bromine a brownish liquid, and iodine a violet solid. The reactivity of the halogens is outstanding because they react with almost all metals, with hydrogen, with phosphorous, with water, and with each other. Although all halogens are reactive, fluorine and chlorine are more reactive than iodine and bromine. The compounds of the halogens with Group IA and IIA metals are generally ionic, but other metallic compounds, such as $TiCl_4$, are nonionic. Fluorides tend to have the most ionic character and iodides the least ionic character.

The nonmetals of Group VIA are oxygen (O), sulfur (S), selenium (Se), and tellurium (Te). The lightest of these, oxygen, exists in elemental form as the diatomic molecule, O_2, but also can from the triatomic ozone molecule, O_3. Oxygen is very abundant on the earth's surface, not only in the atmosphere, but also as the major weight component of H_2O and in the silicates, oxides, sulfates, and carbonates of the earth's crust.

Elemental oxygen, O_2, is, next to fluorine, the most reactive of the nonmetals; but the reactions may require elevated temperatures to activate and break the O—O bond. All metals, except the noble metals like silver, gold, and platinum, react with O_2, forming oxides or superoxides.

Oxygen also reacts with all nonmetals, except for the halogens and the noble gases. Carbon (C), sulfur (S), and nitrogen (N), all react to form oxides, with the oxide formulation being sensitive to the relative amounts of reactants and the reaction conditions. Oxygen also reacts with compounds, such as metal sulfides and hydrocarbons, to form oxides of each element of the original compound. This is illustrated in the **roasting** of sulfide ores, for example:

$$2\ ZnS + 3\ O_2 = 2\ ZnO + 2\ SO_2$$

S, Se, and Te all form hydrogen compounds of the formula H_2S, where S can also be Se or Te. Many metal sulfides are insoluble, accounting for the use of H_2S as an analytical reagent in tests for metal ions. Metal ions forming insoluble sulfides by reaction with H_2S include bismuth (Bi^{3+}) cadmium (Cd^{2+}), copper (Cu^{2+}), mercury (Hg^{2+}), antimony (Sb^{3+}), iron (Fe^{2+}), manganese (Mn^{2+}), nickel (Ni^{2+}), and zinc (Zn^{2+}).

The Group VA nonmetals are limited to nitrogen (N), phosphorus (P), and arsenic (As), as antimony (Sb) and bismuth (Bi) are considered metals.

Elemental nitrogen is unreactive at ambient temperature, primarily from the strength of the N—N bond in the molecule. At high temperature nitrogen reacts with a number of metals, forming compounds that are either ionic, interstitial, or covalent. Interstitial means the N atoms are in the interstices of the metal lattice, as in vanadium nitride, VN, and titanium nitride, TiN. These latter materials are high melting, metallic appearing, and relatively inert.

Phosphorus reacts with a number of metals, forming a variety of compounds with Group IA and IIA metals, Group IIIA elements, and the transition metals. Compounds of the Group IA and IIA metals hydrolyze to form phosphine, as in

$$Ca_3P_2 + 6\ H_2O = 3\ Ca(OH)_2 + PH_3$$

The hydrides and halides of the Group VA nonmetals are known. The best known nitrogen hydride is ammonia, NH_3, but hydrazine, N_2H_2, and hydrazoic acid, HN_3, can also be prepared. The trifluoride of nitrogen, NF_3, is a stable gas, but the other trihalides, NX_3, are unstable. The oxides of nitrogen include nitrous oxide, N_2O, which is stable and has anesthetic properties (laughing gas), as well as nitric oxide, NO, which is formed in lightning storms or electric arcs and which reacts with oxygen at ambient temperature. Nitrogen dioxide, NO_2, is a brown gas that exists in equilibrium with colorless dinitrogen tetroxide, N_2O_4: $2\ NO_2 = N_2O_4$.

Carbon (C) and silicon (Si) are the two nonmetals in Group IV, and each have two filled *s* orbitals and two filled *p* orbitals in the outer electron shell, so that four electrons are needed to complete an octet. This is generally done by covalent bonding.

The outstanding property of carbon is the ability to form long chains or rings of carbon atoms joined together. Also the carbon atoms may join together in various types of bonds, involving single, double, or triple electron pairs. Elemental carbon occurs in two primary crystalline forms—the three-dimensional tetrahedral structure known as the diamond, and the two-dimensional lattice structure known as graphite. Carbon also joins to other elements, such as O, N, and S, with multiple bonds. This is delineated more fully in the section on organic chemistry. Carbon forms compounds known as **carbides** with a number of metals.

The two most prevalent oxides of carbon are the monoxide, CO, and the dioxide, CO_2. CO is relatively active, as it burns, reacts with halogens and sulfur vapor, and reduces metal oxides,

$$FeO + CO = Fe + CO_2$$

CO_2 is relatively inert and does not burn or support combustion.

Carbon also reacts with other elements to form liquid products that are useful as solvents, such as carbon disulfide, CS_2, and carbon tetrachloride, CCl_4. The latter is one of a series of halocarbon solvents with various commercial names such as **Freon** or **UCON**, followed by a number designating the particular halocarbon. Other important carbon compounds are the carbon-hydrogen compounds, which start with methane, CH_4, and are called **hydrocarbons**. These and compounds containing oxygen, nitrogen, and sulfur are considered further in the section on organic chemistry.

Silicon (Si), the other nonmetallic element of Group IVA, has many similarities to carbon. Elemental silicon is a grey, lustrous solid with a structure similar to the carbon diamond structure. The electronic energy level of elemental silicon is critical to its extensive use in semiconductors. Unlike carbon the energy gap between the valence band for electrons and the conduction band for electrons is small enough that heating can promote electrons into the conduction band, making the element an intrinsic semiconductor.

The addition of a **dopant** such as arsenic (As) or gallium (Ga) to Si decreases the energy gap and allows conduction at lower temperatures. Arsenic has one more electron in the outer shell than Si, and this electron is promoted into the conduction band to create an *n*-type semiconductor. Gallium has one less electron than Si in the outer shell and can accept electrons from the occupied level, leaving **holes** and resulting in a *p*-type semiconductor.

Boron is generally found in the earth's crust as a borate, such as borax, $Na_2B_4O_7 \bullet 8H_2O$. It can be prepared in elemental form (black, lustrous crystals) by the reduction of boron tribromide (BBr_3) with hydrogen (H_2).

EQUILIBRIUM

Equilibrium is the condition at which the steady state concentration of chemical reactant species and products does not change with time, as reactions are proceeding at equal rates in both directions. The condition of equilibrium is characterized by the equilibrium constant, K, which is given for a certain temperature. Both reactants and products must be at this temperature for the equilibrium constant to be valid. In general terms, if the reaction is: $A + B = C + D$, the equilibrium constant is

$$K = \frac{[C]\,[D]}{[A]\,[B]}$$

A, B, C, and D appear in brackets as the values of the concentrations which, in the appropriate units, are inserted into the equation. Thus the concentration of any reactant or product is a function of the concentration of the other reactants or products, and the concentration of a single species can be affected by addition or removal of a different species by external means.

Example 14.9

Hydrogen (H_2) and carbon dioxide (CO_2) react at 100°C to form water (H_2O) and carbon monoxide (CO):

$$CO_2(g) + H_2(g) = CO(g) + H_2O(g)$$

Assume the reactants and products are at equilibrium and the number of moles of each gas in a 1.00-L reaction vessel is 0.050, 0.025, 0.040, and 0.055 for CO_2, H_2, CO, and H_2O, respectively. Calculate the equilibrium constant, K, for the reaction at this temperature.

Solution

$$K = \frac{[0.040]\,[0.055]}{[0.050]\,[0.025]} = 1.8$$

The units of each term are mol/liter; units cancel.

When the balanced equation involves molecular quantities with a coefficient other than one, as in

$$A + 2\,B = AB_2$$

the equilibrium constant equation has the molecular coefficients as concentration exponents, as

$$K = \frac{[AB_2]}{[A]\,[B]^2}$$

This would be the case of the gaseous equilibrium

$$N_2 + 3\ H_2 = 2\ NH_3$$

where

$$K = \frac{[NH_3]^2}{[N_2]\,[H_2]^3}$$

so that the concentration of ammonia is squared, and that of hydrogen is cubed in the equation. In addition, this reaction, in which both the reactants and the product are gaseous, shows the effect of pressure on the reaction equilibrium. Since four molecules of reactants become two molecules of product, this is the same (Dalton's Law) as four volumes of reactants becoming two volumes of products. Increased pressure will shift the reaction to the right (Le Chatelier's Principle). The position of the equilibrium is, therefore, pressure dependent.

Equilibrium of solid, liquid, and gas phases is also relevant to chemical reactions. For example, the solubility of a salt such as barium sulfate, $BaSO_4$, is characterized by the solubility product constant of the ions in aqueous solution. In this case: $K_{sp} = [Ba^{2+}]\ [SO^{2-}]\ 1.1 \times 10^{-10}$, showing that we have a material of very low solubility. This low solubility is the driving force for the reaction of barium chloride and sulfuric acid:

$$[Ba^{2+}] + 2\ [Cl^-] + 2\ [H^+] + [SO_4^{2-}] = [BaSO_4] + 2\ [H^+] + 2\ [Cl^-]$$

The $BaSO_4$ is a precipitate, but it is in equilibrium with the concentration of Ba^{2+} ions and SO_4^{2-} ions in solution whose concentration product is equal to K_{sp}.

A gas may also be in equilibrium with the same species in aqueous solution, and the undissociated species in solution may be in equilibrium with the ionic dissociated form. This is the case for ammonia, NH_3, where the equilibrium is

$$NH_3(g) = NH_3(aq) + H_2O = NH_4^+ + OH^-$$

There are two equilibria involved in the above equation:

$$K_1 = \frac{[NH_3]_{aq}}{[NH_3]_g} \quad \text{and} \quad K_2 = \frac{[NH_4^+]\,[OH^-]}{[NH_3]_{aq}}$$

Since the concentration of dissolved NH_3 will be a function of NH_3 pressure in the gas phase, it follows that the basicity of the solution (the OH^- concentration) will also be a function of the NH_3 gas pressure.

SOLUTIONS

Solution chemistry usually means aqueous solution chemistry, although there is also chemistry for species in other solvents, such as liquid ammonia or propylene glycol. However, the polar nature of the water molecule makes water an excellent solvent for all the ionic species, such as acids, bases, and salts.

One characteristic often measured for solutions is the electrical conductivity, generally measured with an AC field to avoid polarization at the electrodes. The conductivity of all ionic solutions increases as the amount of dissolved species increases. In order to compare different electrolytes, the equivalent conductivity, λ, may be used, where

$$\lambda = \frac{1000}{C} k$$

The specific conductance, k, is measured in a cell with a known cell constant, and has the units $ohm^{-1} \bullet cm^{-1}$. C is the concentration of the solution in equivalents/liter, or equivalents per 1000 cc. The units of λ are then

$$\frac{1}{\text{ohm}} \bullet \frac{1}{\text{cm}} \bullet \frac{\text{cm}^3}{\text{equivalents}} = \text{cm}^2 \bullet \text{equiv}^{-1} \bullet \text{ohm}^{-1}$$

Values of equivalent conductance of 120 to 140 are characteristic of salts, whether monovalent or divalent. However, acids are much more conductive, HCl having an equivalent conductance value of over 400, and bases are also more conductive than salts, with a value of 250 for NaOH. The calculated equivalent conductance of the strongly ionized species decreases slowly with increasing concentration, due to **interionic attraction**. The case of weak electrolytes like acetic acid (CH_3COOH) and ammonium hydroxyide (NH_4OH) is quite different, as the equivalent conductance falls rapidly with increased concentration. The equivalent conductance is a function of the degree of dissociation of the electrolyte. The assumption is that the observed equivalent conductance divided by the calculated equivalent conductance at infinite dilution is equal to the degree of dissociation.

Besides conductivity, other important properties of solutions are their so-called **colligative** properties. These are properties that are dependent on the number of particulate species—molecules or ions—dissolved in the solvent. The fundamental property affected is vapor pressure, and a law known as **Raoult's law** states that the relative lowering of the vapor pressure of the solvent is equal to the mole fraction of the solute in solution (where mole fraction of solute is the number of moles of solute divided by the total number of moles of solute plus solvent in a given volume of solution). With a small amount of rearrangement of terms, this relationship is most useful as

$$p = x_1 \, p^0$$

where x_1 is the mole fraction of the solvent, p^0 is the vapor pressure of the pure solvent, and p is the vapor pressure of the solvent as part of the solution. It may be noted that this relationship assumes that the solution is relatively dilute (approximating so-called **ideal solutions**). The solvent is not limited to water and the solute itself may be a substance that, in the pure state, has some volatility. For example, a solution of benzaldehyde in ether follows the law fairly well. Also, since the effect is related to the number of moles of solute, values for the lowering of the vapor pressure can be used to calculate the molecular weight of a solute of known weight concentration.

A more readily measurable colligative property is the elevation of the boiling point of a solvent from the addition of a solute. The amount that the boiling point of a solvent is raised by a solute is a fundamental constant of that solvent and is called the **molal boiling point elevation**. The values vary with the solvent, ranging from 0.51 for water to 5.0 for carbon tetrachloride (CCl_4). The number indicates the amount that the boiling point would be raised by a one molal solution (one mole solute per 1000 grams solvent), but the application must be to much more dilute solutions to meet the criterion of approaching **ideal** behavior. The molecular weight of a dissolved solute can then the calculated from the relation

$$M_2 = K_b \frac{1000 w_2}{\Delta T\, w_1}$$

where K_b is the molal boiling point elevation constant for the solvent, w_2 is weight of dissolved solute, w_1 is the weight of solvent and ΔT is the boiling point elevation in °C.

The same type of relationship holds for the depression of the freezing point by a dissolved solute. Molal freezing point depressions range from 1.86 degrees for water to 20.2 for cyclohexane. The same form of equation as that used in the case of the boiling point elevation can be used to determine the molecular weight of a solute from freezing point depression.

The relationships for colligative properties change if the system is one in which the dissolved solid becomes ionized, the usual case for acids, bases, and salts dissolved in water. In this case the observed change in the vapor pressure, boiling point, or freezing point is up to several times that expected from solutes of known molecular weights.

Another colligative property related to those discussed above is the osmotic pressure of solutions. This is the pressure that needs to be exerted on the solution behind a semipermeable membrane (permeable to the solvent but not to the solute) to keep the solvent from entering the solution and diluting it. A simplified formula can be derived as

$$P = cRT$$

where P is the osmotic pressure in atmospheres, c is the solute concentration in moles liter^{-1} (a simplification for dilute solutions), R is the gas constant (0.082 liter atmospheres degree^{-1}), and T is the absolute temperature in Kelvin. The relationship is similar in form to that of the ideal gas laws.

If the solute is an ionized salt, and the solvent is water, the equation should also include the Van't Hoff factor, i. A calculation will show that the osmotic pressure effect can involve considerable pressures.

Example **14.10**

Sea water has been found to contain 3.26% by weight of the nine major elements, chiefly sodium and chlorine. Tables show that this concentration of sea water salts has an effect on colligative properties equivalent to a 0.513 M NaCl solution and that such an NaCl solution has the osmotic effect of a 0.944 m unionized solute. (a) What is the value of the Van't Hoff factor, i, for such an NaCl solution? (b) What is the pressure in psi that must be applied to such a sea water solution at 13.0°C in a reverse osmosis purification apparatus to start a flow of purified water? (a.w. Na = 29.99, Cl = 35.45; density 0.513 M NaCl = 1.019 g/mL; 1.00 atm = 14.69 psi)

Solution

The Van't Hoff factor, i, is the ratio of the apparent molality, m, of a solute as determined from its colligative effect, to its actual molality. A small correction should be made to the value of M (molarity), to get the value of m (molality):

$$1.00 \text{ L solution} \times 1.019 = 1.019 \text{ kg} = 1019 \text{ g}$$

$$0.513 \text{ M NaCl} \times 58.44 \text{ g/mol} = 30.0 \text{ g NaCl}$$
$$1019 \text{ g solution} - 30 \text{ g NaCl} = 989 \text{ g } H_2O = 0.989 \text{ kg } H_2O$$

$$m = \frac{0.513 \text{ mol NaCl}}{0.989 \text{ kg } H_2O} = 0.519 \text{ m solution}$$

$$\frac{0.944 \text{ m}}{0.519 \text{ m}} = 1.82 = i$$

The osmotic pressure can be calculated from the formula

$$p = icRT$$

where p is the pressure in atmospheres, i is the Van't Hoff factor, c is the solute concentration in mol/L, R is the gas constant in the appropriate units, and T is the absolute temperature.

$$p = 1.82 \times \frac{0.513 \text{ mol}}{\text{L}} \times \frac{0.08206}{\text{K} \bullet \text{mol}} \text{L} \bullet \text{ atm} \times (273.2 + 13.0) \text{ K} = 21.9 \text{ atm}$$
$$21.9 \text{ atm} \times 14.69 = 322 \text{ psi}$$

This is the pressure, on the basis of the above data, that must be applied to the reverse osmosis purification apparatus to balance the osmotic pressure. Any flow of purified water requires a greater pressure.

KINETICS

Kinetics is the study of the rate at which chemical reactions take place. For a reaction to take place, the reacting species must come into close physical approximation to each other, generally of the order of atomic dimensions. This is true whether the reaction is taking place in the gas, liquid, or solid phase, and generally involves some type of diffusion of a reacting species to either contact the other species or go to a **site** on a catalyst or other surface where reaction can take place. The mixing or contact of reacting species can also be helped by external stirring, although diffusion may still have a role in allowing the species to come close enough to react.

When reactions take place in the gas phase, they can be unimolecular, bimolecular, or, more rarely, termolecular. The term signifies whether the reacting molecule is alone, as in a decomposition, or reacts in the presence of other species. The kinetics of the reaction is then known as first order, second order, or third order. In first order kinetics, the reaction rate R, is a function only of the species reacting as:

$$R = k_1 [A]$$

In a second order reaction, the rate is a function of the concentrations of two reacting species, as in:

$$R = k_2 [A] [B]$$

The constants k_1 and k_2 are first- and second order reaction rate constants, respectively.

Example 14.11

The following set of data has been accumulated for the gaseous reaction

$$2\,A + 3\,B \rightarrow 2\,C + D$$

Initial Conc. of *A*, M	Initial Conc. of *B*, M	Initial Rate of Appearance of *C*, mol/hr
(1) 0.10	0.10	0.40×10^{-6}
(2) 0.10	0.30	1.2×10^{-6}
(3) 0.20	0.30	4.8×10^{-6}
(4) 0.20	0.10	1.6×10^{-6}
(5) 0.30	0.30	1.08×10^{-5}

The rate can be expressed as $R = k\,[A]^m[B]^n$. What is the order of the reaction with respect to reactant A?

Solution

Compare two tests in which the initial concentration of reactant A varies, but that of reactant B stays the same.

Comparing test (3) with test (2)

$$\frac{R_3}{R_2} = \frac{k\,[A]_3^m\,[B]_3^n}{k\,[A]_2^m\,[B]_2^n} \quad \text{or} \quad \frac{4.8\times10^{-6}}{1.2\times10^{-6}} = \frac{k\,(0.20)^m(0.30)^n}{k\,(0.10)^m(0.30)^n}$$

Dividing and canceling, as appropriate, yields

$$4 = 2^m \quad \text{or} \quad m = 2$$

Therefore, the reaction is second order with respect to A, as the exponent, m, = 2.

Reaction rates in solutions, when they involve ionic species, are generally rapid, depending only on the rate of diffusion or mixing of the species involved. However, when the reacting molecules are large and complex, the reaction rate in solution must account for steric effects as well as the appropriate activation energy. Reaction rates in nonaqueous liquids can also be fast but often are not ionic reactions and, therefore, are inhibited by steric and other factors. Reaction rates in melts will be slowed by diffusion if the melt viscosity is high. Solid reactions will be the slowest because of the low rate of diffusion and will only become appreciable at higher temperatures.

The role of a catalyst in chemical kinetics is to increase the rate of the reaction, while not affecting the thermodynamic equilibrium of reactants and products. The catalyst for gas phase is usually a solid material with a high surface area. It can be a metal or a **supported** metal that acts as a site for the reaction to take place at a lower activation energy barrier than in the gas phase. The catalyst is not a part of an intermediate compound in the reaction, although there is clearly an interaction with the reacting species.

Catalysts may be part of the phase in which the reaction takes place, and are then referred to as **homogeneous** catalysts. This is often the case in solution phase reactions and can be very important in biochemical reactions where the catalyst is an **enzyme**.

ORGANIC CHEMISTRY

Organic compounds are far greater in number than inorganic compounds, even though they are all based on one element, carbon. The chemistry of organic compounds is known as **organic chemistry** and is a distinct discipline. Its importance stems not only from the great number of organic compounds, but also from the fact that all life forms are based on organic compounds. In addition, the importance of petroleum and petroleum products to civilized society has greatly enhanced the interest in organic chemistry.

The unique structure of the carbon atom, with atomic number 6 and atomic weight 12, is the foundation of its ability to form complex compounds. The four bonds are normally directed toward the corners of a tetrahedron, with the carbon nucleus at its center. This allows carbon atoms to form chains of almost unlimited length, as well as side chains, while the remainder of the bonds are directed to hydrogen atoms and other species, such as halogens, oxygen, nitrogen, and sulfur. Another important aspect of carbon bonding is its ability to form double bonds or triple bonds with another carbon and with other elements. The double bonds may alternate between carbon atoms, and this leads to the ability to form rings in which extra bonding strength is provided by **delocalized** electrons around the ring. The most stable ring of this type contains six carbons and is known as the benzene ring. It is the foundation of a large class of ring compounds known as **aromatic compounds**.

We may start in organic chemistry by considering the lower molecular weight compounds, as much of their chemistry is consistent with that of higher molecular weight (longer chain) compounds. The simplest organic compounds are those of only carbon and hydrogen, and are known as **hydrocarbons**. One carbon bonded to four hydrogen atoms is CH_4, or methane. Successive carbons may be added to form the compounds ethane (C_2H_6), propane (C_3H_8), butane (C_4H_{10}), pentane (C_5H_{12}), hexane (C_6H_{14}), heptane (C_7H_{16}), octane (C_8H_{18}), and on to any number—although individual names are lacking above about 30 carbons in a chain. The aforementioned hydrocarbons are known as **alkanes**.

Combustion is a reaction that all organic compounds are subject to, but the hydrocarbons, as mentioned, are often used for that purpose. Carbon dioxide and water are formed, as in the reaction of hexane:

$$2\ C_6H_{14} + 19\ O_2 \rightarrow 12\ CO_2 + 14\ H_2O$$

Another reaction of hydrocarbons is halogenation, such as

$$C_3H_8 + Cl_2 \rightarrow C_3H_7Cl + HCl$$

Example 14.12

A compound has an empirical formula of C_2H_3Br and a molecular weight of 213.9 g/mol (a.w. C = 12.01, H = 1.008, Br = 79.90). What is its molecular formula and a possible structural formula with name?

Solution

The empirical formula weight is $(2 \times 12.01) + (3 \times 1.008) + 79.90 = 106.94$ g/empirical formula.

$$\text{empirical formulas/mol} = \frac{213.9\ \text{g/mol}}{106.94\ \text{g/empirical formula}} = 2.0$$

Therefore, the molecular formula is $C_4H_6Br_2$. A number of structural formulas can be devised that correspond to this molecular formula. Two compounds with this formula are

1,4-dibromo-2-butene, or $Br—CH_2—CH{=}CH—CH_2—Br$

and

3,4-dibromo-1-butene, or $CH_2{=}CH—CHBr—CH_2Br$

The carbon compounds with a double bond are called **unsaturated hydrocarbons** or **unsaturates**. They are notably more reactive than the alkanes—one reaction being to hydrogenate (add H_2 with heat, pressure, and possibly catalyst) back to form the **saturated** or alkane compound. They also react readily with halogens, as in

$$H_2C{=}CH_2 + Br_2 \rightarrow H_2BrC—CBrH_2$$

The product of the reaction is dibromoethane or, more correctly, 1,2-dibromoethaneor ethane, 1,2-dibromo. The compound is also known as ethylene dibromide, where, it may be noted, the **ene** suffix is appropriate because of the reactant material from which the compound was made.

The hydrocarbon compounds may also contain the triple bond, or three pair of electrons bonding two carbon atoms. The best known example of a compound with this type of bond is acetylene, C_2H_2, which may be written as $HC \equiv CH$. The triple bonded, or acetylene type, hydrocarbons are noted for their reactivity, and the instability of the triple bond. Acetylene is used as a fuel to reach high temperatures with the oxyacetylene torch.

As has been shown, the hydrocarbons may have halogen atoms substituted for the hydrogen, thus giving the generic class of halocarbons. Other groups may also be substituted for one or more hydrogens. These include the OH group, leading to a class of organic compounds called alcohols; the NH_2 group, leading to a class of compounds called amines; the NO_2 group, leading to the nitro compounds; the CN group, leading to a class of compounds called organic cyanides; the O atom in various groups, leading to classes of compound called aldehydes, ketones, acids, and oxides; and the S atom, leading to classes of compounds called mercaptans and sulfides. There is also a major class of compounds with C—Si bonds, generally known as silanes, as well as compounds of phosphorous, arsenic, and numerous metallo-organics.

The **aromatic** compounds form a large class and can contain substituent groups, just as the alkanes can. When an OH group is substituted on a benzene ring (six carbons in a ring with alternating double and single bonds in resonance), the product is phenol, C_6H_5OH. Unlike the alcohols, phenol is an acid, ionizing in water to yield the proton H^+—or the hydrated proton, hydronium ion, H_3O^+—and the phenylate anion, $C_6H_5O^-$. When two OH groups are on the benzene ring, the compound is resorcinol, and the compound with three OH groups is catechol. Although they have some reactions in common, the aromatic OH compounds, or phenols, are generally considered as a separate chemical group from the alcohols.

The amines, such as methylamine, CH_3NH_2, ethylamine, $C_2H_5NH_2$, and propylamine, $C_3H_7NH_2$), are odoriferous liquids, the NH_2 groups of which are quite reactive.

The nitro compounds, such as nitromethane, NO_2CH_3, nitroethane, $NO_2C_2H_5$, and nitropropane, $NO_2C_3H_7$, are all reactive. Aromatic nitro compounds, such as nitrobenzene ($C_6H_5NO_2$) can be reduced to primary aromatic amines, which are a starting point for many organic synthesis experiments.

Some of the sulfur compounds are analogous to the alcohols, as the S may be substituted for O so that the hydrocarbon is attached to an SH group. These compounds are known as mercaptans, for example, methyl mercaptan, CH_3SH, and ethyl mercaptan, C_2H_5SH. They are odoriferous and reactive.

Sulfur can also be bonded between two carbon atoms, as in the sulfides such as dimethylsulfide, CH_3SCH_3, and diethylsulfide, $C_2H_5SC_2H_5$. The sulfides are reactive.

Besides the alcohols, there are other large classes of organic compounds with oxygen in the substituent group. Among these are aldehydes, ketones, esters, and acids. The characteristic group for aldehydes is HC═O (with a single bond below the C); for ketones, C═O (with single bonds above and below the C); for ethers, —O—; for esters, —O—C═O (with a single bond below the C); and for acids, HO—C═O (with a single bond below the C). The double bond is shown to emphasize the fact that in four of these cases the oxygen atom is doubly bonded to the carbon atom. In the other case, ethers, the oxygen atom is singly bonded to each of two carbon atoms.

Aldehydes and ketones are similar in many respects, as both contain the carboxyl group, C═O. In the aldehydes the C═O group is at the end of the carbon chain, with the carboxyl carbon atom also bonded to a hydrogen atom. In ketones, the C═O group is somewhere in the middle of the chain, bonded to carbon atoms on both sides.

SELECTED SYMBOLS AND ABBREVIATIONS

Symbol or Abbreviation	Description
α	absorptivity
E	electrode potential
F	Faraday constant
H	heat of vaporization
h	enthalpy
h	Planck constant
k	Boltzman reaction constant
k	rate constant
K	equilibrium constant
m	molality
M	molarity
n	number of moles
N	normality
P	gas pressure
R	gas constant
R	reaction rate
torr	equivalent to 1 mm/Hg
T	absolute temperature
V	gas volume

PROBLEMS

14.1 In a Boyle's law experiment (perfect gas law at constant temperature), the following data were taken:

Pressure (torr)	Volume (mL)
904	27.0
827	29.5
760	32.1
650	38.5
552	44.2

a. The volume value which was recorded incorrectly is most nearly:

a. 27.0 c. 32.1
b. 29.5 d. 38.5

b. The predicted volume of that pressure is most nearly:

a. 26.0 c. 31.1
b. 28.5 d. 37.5

14.2 The volume of one mole of gaseous water at 100°C and 1 atmosphere pressure is

a. 18.0 mL c. 22.4 L
b. 100 mL d. 30.6 L

14.3 First ionization energy refers to

a. removal of an electron from a gas atom
b. energy to form the most probable ion
c. trapping an ion in a lattice structure
d. formation of a −1 anion

14.4 From the periodic table predict the molecular formula of silicon (Si) oxide (O).

a. SiO c. SiO_4
b. Si_2O d. SiO_2

14.5 Two metals that each commonly form +1 and +2 ions in solution are

a. Cu and Hg c. Fe and Cu
b. Au and Ag d. Zn and Cd

14.6 The formula for potassium aluminum sulfate is (not including water of hydration)

a. $KAlSO_4$ c. K_2AlSO_4
b. $KAl(SO_4)_2$ d. $KAl(SO_4)_3$

14.7 To indicate a compound is pentahydrate, you would write as part of the formula

a. • 5 (H^+) c. • 5 (OH^-)
b. • 5 (H^-) d. • 5 H_2O

14.8 Hydroiodic acid contains how many elements?

a. 1 c. 3
b. 2 d. 4

14.9 The common name of the oxide of nitrogen with the formula N_2O is

a. nitrogen dioxide c. nitric oxide
b. nitrous oxide d. dinitrogen oxide

14.10 The name of $(NH_4)_2Cr_2O_7$ is

a. diammonium chromate c. ammonium(II) chromate
b. ammonium chromate d. ammonium dichromate

14.11 A sample of 1.38 moles of manganese (IV) oxide contains

a. how many moles of manganese ions?

a. 1.00 c. 1.38
b. 2.76 d. 4.00

b. how many moles of oxygen ions?

a. 1.38 c. 2.00
b. 2.76 d. 1.76

c. how many grams of material? (a.w. Mn = 54.9, O = 16.0)

a. 1.38 c. 2.76
b. 86.9 d. 120

14.12 A student in a chemistry laboratory wants to weigh out 2.00 moles of calcium carbonate ($CaCO_3$) but picks up sodium chloride (NaCl) by mistake. How many moles of NaCl will the student weigh out? (m.w. $CaCO_3$ = 40.1 + 12.0 + (3 × 16.0) = 100.1; NaCl = 23.0 + 35.5 = 58.5)

a. 3.42 c. 1.16
b. 1.87 d. 2.00

14.13 A protein molecule is known to bind one molecule of oxygen (O_2) per molecule of protein. If 12.2 g of protein bind 9.8 mg of O_2, what is the molecular weight of the protein? (m.w. $O_2 = 2 \times 16.0 = 32.0$)

a. 25,700 c. 38,900
b. 79,600 d. 39,800

14.14 An impure sample of $FeSO_4 \cdot 7H_2O$ weighing 1.285 g is analyzed for Fe(II) content by titration with 0.03820 M $KMnO_4$ in acid solution. The endpoint is 21.83 ml. What is the weight % of Fe(II) in the impure sample? (a.w. Fe = 55.85, K = 39.10, S = 32.07, Mn = 54.94, O = 16.0, H = 1.008)

a. 18.12% c. 9.060%
b. 36.76% d. 20.09%

14.15 A student wishes to check the concentration of a bottle of hydrogen peroxide, H_2O_2. To do so, the student carried out a redox titration of the H_2O_2 with $KMnO_4$, as shown in the reaction

$$5\ H_2O_2(aq) + 2\ KMnO_4(aq) + 3\ H_2SO_4(aq) \rightarrow$$

$$K_2SO_4(aq) + 2\ MnSO_4(aq) + 8\ H_2O(1) + 5\ O_2(g)$$

It required 39.7 ml of 0.0103 M $KMnO_4(aq)$ to react with 1.546 g of the hydrogen peroxide solution. What was the percent of H_2O_2 by mass in the solution? (a.w K = 39.10, H = 1.008, O = 16.00, Mn = 54.94, S = 32.07)

a. 1.80%
c. 1.65%
b. 2.70%
d. 2.24%

14.16 What weight in g of $SnCl_2$ is needed to react with 40 mL of a 0.10 normal I_2 solution as follows?

$Sn^{2+} + I_2 \rightarrow Sn^{4+} + 2\ I^-$ (a.w. Sn = 118.70, I = 126.90, C1 = 35.45)

a. 0.38 g
c. 3.8 g
b. 0.76 g
d. 0.19 g

14.17 A solution is prepared by dissolving 5.88 g $K_2Cr_2O_7$ in dilute acid and diluting with water to 1.000 L.

Calculate the normality of the solution assuming that the half-reaction for the $Cr_2O_7^2$ in solution is

$$14\ H^+ + Cr_2O_7^{2-} \rightarrow 2\ Cr^{3+} + 7\ H_2O$$

a. 0.163 N
c. 0.120 N
b. 0.060 N
d. 0.050 N

14.18 Given the following standard electrode potentials:

$$Al^{3+} + 3\ e^- \rightarrow Al;\ E^0 = -1.66\ V$$

$$Cd^{2+} + 2\ e^- \rightarrow Cd;\ E^0 = -0.40\ V$$

Determine the standard cell potential for the reaction

$$Al + 3\ Cd^{2+} \rightarrow 2\ Al^{3+} + 3\ Cd$$

a. $E^0 = 2.06$ V
c. $E^0 = 1.23$ V
b. $E^0 = -1.26$ V
d. $E^0 = 1.26$ V

14.19 Calculate the standard cell potential, E^0, for the reactions below, using the correct half-reactions and the corresponding standard electrode potentials, E^0. State whether each reaction is spontaneous or nonspontaneous at standard conditions.

a. $Fe(s) + 2\ H^+ \rightarrow Fe^{2+} + H_2(g)$

a. –0.41 V (nonspontaneous)
c. 0.41 V (spontaneous)
b. 0.20 V (spontaneous)
d. 0.20 V (nonspontaneous)

Table of standard reduction potentitals

Half-Reaction	E^0
$Fe^{2+} + 2e^- = Fe(s)$	–0.41 V
$Sn^{2+} + 2e^- = Sn(s)$	–0.14 V
$2H^+ + 2e^- = H_2(g)$	0.00 V
$Sn^{4+} + 2e^- = Sn^{2+}$	0.15 V
$Cu^{2+} + e^- = Cu^+$	0.16 V
$Fe^{3+} + 2e^- = Cu(s)$	0.34 V
$Fe^{3+} + e^- = Fe^{2+}$	0.77 V

b. $Cu(s) + 2H^+ \rightarrow Cu^{2+} + H_2(g)$

a. 0.34 V (nonspontaneous)
b. 0.17 V (spontaneous)
c. –0.34 V (spontaneous)
d. –0.34 V (nonspontaneous)

c. $Sn^{2+} + 2Fe^{3+} \rightarrow Sn^{4+} + 2Fe^{2+}$

a. 0.62 V (spontaneous)
b. 0.92 V (spontaneous)
c. 0.62 V (nonspontaneous)
d. –0.62 V (spontaneous)

14.20 Given the cell $Cu|Cu^{2+} \| Fe^{3+}, Fe^{2+}|$ Pt and the standard reduction potentials

$$Cu^{2+} + 2e^- \rightarrow Cu(s); \quad E^0 = 0.34\text{ V}$$
$$Fe^{3+} + e^- \rightarrow Fe^{2+}; \quad E^0 = 0.77\text{ V}$$

calculate the cell potential (emf) of the cell at 25°C when $[Fe^{2+}] = 0.040$ M, $[Fe^{3+}] = 0.0020$ M, and $[Cu^{2+}] = 0.050$ M.

a. 0.43 V
b. 0.35 V
c. 1.07 V
d. 0.39 V

14.21 A chromium salt–containing solution is electrolyzed using a current of 2.00 A (amperes) for 2.00 hr (hours). The metallic chromium deposited weighs 2.59 g (grams). Determine a factor for moles of electrons needed per mole of chromium deposited and deduce the oxidation state of the chromium in the salt. For this problem, F (Faraday constant) = 96,485 C (coulomb) • mol^{-1}. (a.w. Cr = 52.00)

a. 0.33, species in solution is Cr^{3+}
b. 3.0, species in solution is Cr^{3+}
c. 2.0, species in solution is $(Cr_2O_7)^{2-}$
d. 2.0, species in solution is $(CrO_4)^{2-}$

14.22 Using a current of 3.75 A (amperes), how long (in seconds) would it take to electroplate 6.19 g metallic chromium from a solution in which the chromium was in the +3 oxidation state (Cr^{3+}). (a.w. Cr = 52.00, Faraday constant (F) = 96,485 C • mol^{-1})

a. 2.70×10^5 s
b. 1.02×10^3 s
c. 1.29×10^5 s
d. 9.18×10^3 s

14.23 A solution of nickel sulfate ($NiSO_4$) was electrolyzed for 0.75 hr between inert electrodes. If 17.5 g of nickel metal was deposited, what was the average current? (a.w. Ni = 58.69, Faraday constant, F = 96,485 C • mol^{-1})

a. 1.1×10^1
b. 1.3×10^3
c. 2.1×10^1
d. 1.6×10^1

14.24 A spoon, with a surface of 45 cm^2, is suspended in a cell filed with a 0.10 M solution of gold (III) chloride, $AuCl_3$. A current of 0.52 A has been passed through the cell, until a coating of gold 0.10 mm thick, has plated on the spoon. How long did the current run? (density Au = 19.3 g/cm^3, a.w. Au = 196.97, Faraday constant, F = 96,485)

a. 1.3×10^4 sec (3.6 hr) c. 4.1×10^4 sec (11 hr)
b. 2.5×10^4 sec (6.9 hr) d. 8.2×10^3 sec (2.3 hr)

14.25 Assuming 100% dissociation, calculate the molarity of the H^+ (H_3O^+) ions in a solution made by diluting 10.0 mL of 0.10 m HNO_3(aq) to 500.0 mL.

a. 0.0010 M c. 0.020 M
b. 0.010 M d. 0.0020 M

14.26 How many mL of water need to be added to 30.0 mL of a 12.0 molar (M) solution of HCl to give a 3.00 M solution?

a. 120 mL c. 90 mL
b. 133 mL d. 60 mL

14.27 A stock solution is prepared by dissolving 30.0 g of NaOH in water and diluting to a final volume of 500 mL. How many mL of this stock solution are necessary to prepare 1.00 L of 0.100 M NaOH? (a.w. Na = 22.99, O = 16.00, H = 1.008)

a. 66.7 mL c. 150 mL
b. 37.5 mL d. 167 mL

14.28 In the laser cutting of polyvinyl chloride, hydrogen chloride gas is formed. In a certain laser-cutting process for polyvinyl chloride, the hydrogen chloride formed was dissolved in water and titrated with 0.0100 M sodium hydroxide (NaOH) solution. The titration required 37.68 mL of NaOH to reach the endpoint as indicated by a color change of bromothymol blue indicator. How many milligrams of hydrogen chloride were formed in this laser-cutting process? (a.w. H = 1.008, Cl = 35.45)

a. 1.03 mg c. 103 mg
b. 1.37 mg d. 13.7 mg

14.29 Calculate the volume of 0.250 N $Ca(OH)_2$ needed to completely neutralized 75.0 mL of 0.150 N H_3PO_4.

a. 45.0 mL c. 22.5 mL
b. 125 mL d. 41.7 mL

14.30 Metallic elements are found where in the periodic table?

a. in the far left-hand and far right-hand groups
b. in the middle of the table and Group VIIIA
c. in the left-hand and middle groups
d. only in Groups IA and IIA

14.31 Metals have:

a. both high electrical and high thermal conductivity
b. high electrical but low thermal conductivity
c. low cohesive strength and high luster
d. high luster and low ductility

14.32 Nonmetals are
a. malleable but not ductile
b. very reactive with acids
c. good conductors of electricity
d. able to form halides, which react with water to give an oxyacid

14.33 The following metals are in a single group in the periodic table:
a. lithium, sodium, potassium, strontium, and cesium
b. iron, cobalt, nickel, platinum, and gold
c. boron, aluminum, gallium, indium, and thallium
d. beryllium, magnesium, calcium, barium, and radium

14.34 The halogens:
a. will not react with each other
b. are strong electron donors
c. form strong oxyacids of the formula HOX_3
d. form strong covalent bonds with Group 1A metals

14.35 The dissolution of sulfur dioxide (SO_2) in water produces
a. a weak solution of sulfuric acid, H_2SO_4
b. a weak solution of pyrosulfuric acid (disulfuric acid), $H_2S_2O_7$
c. a solution used as an analytical reagent to precipitate metal cations
d. a weak solution of sulfurous acid, H_2SO_3

14.36 Nitrogen N_2, forms several oxides. Two that exist in equilibrium at ambient temperature are
a. NO_2 and N_2O_4, which exist in the equilibrium $2\ NO_2 = N_2O_4$
b. NO and NO_2, which exist in the equilibrium $2\ NO + O_2 = 2NO_2$
c. NO and N_2O, which exist in the equilibrium $3\ NO = N_2O + NO_2$
d. N_2O_5 and N_2O_4, which exist in the equilibrium $2\ N_2O_5 = 2\ N_2O_4 + O_2$

14.37 A number of carbon compounds with metals hydrolyze to release
a. carbon dioxide (CO_2)
b. carbon monoxide (CO)
c. a mixture of carbon dioxide and carbon monoxide
d. either acetylene (C_2H_2) or methane (CH_4)

14.38 Silicon (Si) is important in the semiconductor industry because
a. of its outer electron configuration of two electrons in *s* orbitals and two electrons in *p* orbitals
b. the energy band gap between valence electrons and conductance electrons in the crystal is relatively small
c. its melting point is low enough that it can be melted and cast into **chips**
d. it is the most dense of the Group IVA elements and this allows a high electron density in devices

14.39 The substitution of Al^{3+} for Si in the compounds with Si—O—Si is important to
a. the use of Si in semiconductor chips
b. the production of silicone polymers
c. the formation of clays and zeolites
d. the production of specialty glasses

14.40 The reaction below is carred out in a 5.00-L reaction vessel at 600 K.

$$CO(g) + H_2O(g) = CO_2(g) + H_2(g)$$

At equilibrium it is found that 0.020 mol CO, 0.0215 mol H_2O, 0.070 mol CO_2, and 2.00 mol H_2 are present. Evaluate K_c for this reaction.

a. 65.1 c. 236
b. 0.00307 d. 326

14.41 Consider the following reaction at equilibrium:

$$N_2O_4(g) = 2\ NO_2(g)$$

If a 5.00-L reaction vessel, held at constant temperature, is initially filled with 10.0 mol pure $N_2O_4(g)$, and if 3.5 mol $NO_2(g)$ are found in the vessel once equilibrium has been established, what is the value of the equilibrium constant, K_c, for this reaction (at the temperature of the experiment)?

a. 0.297 M c. 0.424 M
b. 1.48 M d. 0.0594 M

14.42 At a given temperature the equilibrium concentrations in a reactor were found to be $PCl_3 = 0.025$ M, $Cl_2 = 0.25$ M, and $PCl_5 = 0.125$ M. Calculate the equilibrium constant at the given temperature for the reaction $PCl_5 = PCl_3 + Cl_2$.

a. 0.125 M c. 0.50 M
b. 2.0 M d. 0.25 M

14.43 What is the molarity of a solution prepared by dissolving 15.0 g of $La(NO_3)_4$ in 800 mL of water? (a.w. La = 138.91, N = 14.01, O = 16.00)

a. 0.093 M c. 0.048 M
b. 0.031 M d. 0.039 M

14.44 If 15.0 g of 26% LiBr solution is diluted to a volume of 80.0 ml, what is the molarity of the resultant solution? (a.w. Li = 6.9, Br = 79.9)

a. 0.71 M c. 0.44 M
b. 0.36 M d. 0.56 M

14.45 The vapor pressure of pure ethyl alcohol (C_2H_5OH) at 30°C is 71.2 torr and that of pure carbon tetrachloride (CCl_4) is 121.6 torr. Calculate the pressure above a solution containing 20.0 g of C_2H_5OH and 60.0 g of CCl_4 assuming ideal behavior. (a.w. C = 12.01, H = 1.008, O = 16.00, Cl = 35.45)

a. 109 torr c. 75.2 torr
b. 97.8 torr d. 95.0 torr

14.46 A quantity of a covalent substance of molecular weight 73 is dissolved in 425 g water. If the resulting solution has a boiling point of 100.31°C at 1.00 atm, how many grams of this substance were added? ($K_b = 0.512$ °C • m^{-1} for water)

a. 104 g c. 5.8 g
b. 2.7 g d. 19 g

14.47 The following initial rate data were obtained for the reaction: $2\ B \rightarrow C$.

[B] M	$r = \dfrac{d[B]}{dt}\ \mathrm{M \bullet sec^{-1}}$
(1) 0.245	2.92×10^{-4}
(2) 0.490	4.13×10^{-4}

Find the exponential coefficient of the rate law $r = k\,[B]^n$ for this reaction.

a. $n = -0.5$ c. $n = 1.0$
b. $n = 0.5$ d. $n = 2.0$

14.48 Consider the following gas phase reaction:

$$CH_3CHO \rightarrow CH_4 + CO$$

Can the order of this reaction be determined from the above balanced equation? If so, determine the order and explain; if not, explain.

a. Yes—the reaction is second order because each molecule of reactant must collide with a second molecule of reactant to decompose
b. Yes—the reaction is third order because each molecule of reactant collides with other reactant species as well as with each of two species of product molecules
c. No—the reaction order can't be determined from balanced equation and reaction coefficients alone; experimental data is reuired
d. Yes—it is a first order reaction because the only reactant is CH_3CHO, which can only react with itself

14.49 Consider the following set of data:

Set	Rate, $mol^3/L^3 \bullet s$	[A], mol/L	[B], mol/L
(1)	0.020	0.10	0.20
(2)	0.080	0.10	0.40
(3)	0.040	0.20	0.20
(4)	0.060	0.30	0.20

Write the rate equation (R = rate), the overall order of the reaction, and the specific rate constant, k, in proper units.

a. $R = k\,[A]\,[B]^2$, order = 3, $k = 5.0\ s^{-1}$
b. $R = k\,[A]^2\,[B]$, order = 3, $k = 10\ s^{-1}$
c. $R = k\,[A]^2\,[B]^2$, order = 4, $k = 50\ L \bullet mol^{-1} \bullet s^{-1}$
d. $R = k\,[A]\,[B]^0$, order = 1, $k = 0.20\ mol^2 \bullet L^{-2} \bullet s^{-1}$

14.50 An analysis of a 2.147-g sample of a hydrocarbon produced 7.260 g of carbon dioxide and 1.485 g of water. (a.w. C = 12.01, O = 16.00, H = 1.008)

Calculate the percentage composition of the hydrocarbon.

a. 97.27% C, 2.73% H c. 93.73% C, 6.27% H
b. 92.27% C, 7.73% H d. 91.27% C, 8.73% H

14.51 Cyclopropane is 85.7% carbon. Cyclohexane has

a. somewhat more than 85.7% carbon
b. somewhat less than 85.7% carbon
c. exactly 85.7% carbon
d. exactly half the carbon percentage of cyclopropane

14.52 An organic compound has an empirical formula of C_3H_8O. This formula can represent

a. three alcohols
b. one organic acid
c. two alcohols and one ether
d. two ethers and one alcohol

14.53 Name the *type* of compound shown in each of the following structural formulas:

a. $CH_3—CH_2—OH$

a. aldehyde
b. alcohol
c. organic acid
d. ketone

b.

$$\begin{array}{c} \quad\;\; H \\ \quad\;\; | \\ CH_3—C{=}O \end{array}$$

a. alcohol
b. ester
c. alkyne
d. aldehyde

c.

$$\begin{array}{c} O \\ \| \\ CH_3—C—CH_3 \end{array}$$

a. ester
b. aldehyde
c. organic acid
d. glycol

d. $CH_3—CH_2—NH_2$

a. nitrile
b. alkyne
c. amide
d. amine

e. $CH_3—C{\equiv}N$

a. amine
b. nitrile
c. amide
d. azide

f.

$$\begin{array}{c} O \\ \| \\ CH_3—C—CH_3 \end{array}$$

a. aldehyde
b. alkane
c. ketone
d. ether

SOLUTIONS

14.1

a. **d.** Boyle's law and the perfect gas law state that PV should be constant at constant temperature. The PV product of the above data are 24408, 24396, 24396, 25025, and 24398. The fourth value, 25025, is inconsistent with the others. Therefore the volume value 38.5 is incorrect.

b. **d.** The average PV product for the four good readings is 24400. Then 24400/650 = 37.5, so the value of the volume at pressure 650 torr should have been 37.5 mL.

14.2 **d.** The volume of one mole of any gas is 22.4 L at STP conditions (1 atm pressure and 0°C). The conditions are 1 atm so there is no pressure correction. However the temperature correction is done by Charles' law (perfect gas law at constant pressure), after the temperature is converted to K (Kelvin):

$$\frac{P_1V_1}{T_1} = \frac{P_2V_2}{T_2}$$

At constant pressure

$$\frac{V_1}{T_1} = \frac{V_2}{T_2}$$

$$V_2 = V_1 \times \frac{T_2}{T_1}$$

$$V_2 = 22.4 \text{ L} \times \frac{(100 + 273) \text{ K}}{273 \text{ K}} = 30.6 \text{ L}$$

14.3 **a.** Choice (a) is the definition of ionization energy.
Choice (b) is incorrect since the second and third ionization energy may be pertinent to forming the most probable ion.
Choice (c) is incorrect since the ion formed does not have to be trapped in a lattice structure, and the energy is unrelated.
Choice (d) is incorrect since a +1 cation forms.

14.4 **d.** Oxygen (atomic no. 8) is in the second period and Group VIA of the periodic table. Silicon (atomic no. 14) is in the third period and Group IVA of the periodic table. The most probable electronic structure for oxygen in a compound is to pick up two electrons to fill its outermost shell, giving it the electron configuration of neon (atomic no. 10). Silicon, with four electrons in the outer shell, could either gain four electrons or lose four electrons to form a complete octet in the outer shell. However, since O wants to gain electrons, it is more likely that Si will lose them, making it also isoelectronic with neon. The appropriate number of Si and O atoms must then combine so that the number of electrons gained by the O atoms equals the number of electron lost by the Si atoms. Therefore, there must be twice as many O atoms as Si atoms in the molecular formula, and the simplest molecular formula is SiO_2.

14.5 **a.** Choice (a) is correct because Cu forms Cu^+ and Cu^{2+} and Hg forms Hg^+ and Hg^{2+}, although the monovalent Hg ion is considered to be Hg_2^{2+}. Au forms Au^+ and Au^{3+}, Ag forms Ag^+, Fe forms Fe^{2+} and Fe^{3+}, Zn forms Zn^{2+}, and Cd forms Cd^{2+}.

14.6 **b.** In this mixed salt (alum), the K has a valence of +1, the Al has a valence of +3, and (SO_4) is a group with a valence of −2. The formula is $KAl(SO_4)_2$. Alum also includes 12 H_2O of hydration in the crystal.

14.7 **d.** **Penta** means 5, **hydrate** means water. This is one of the few times a coefficient shows up inside a formula. Some prefer to write salt$(H_2O)_5$.

14.8 **b.** Any acid beginning with **hydro** contains hydrogen and only one other element.

14.9 **b.** There are six oxides of nitrogen with oxidation numbers of the nitrogen ranging from +1 to +5. The formulas are N_2O, NO, N_2O_3, NO_2, and N_2O_5. Their common names are nitrous oxide, nitric oxide, dinitrogen trioxide, nitrogen dioxide, and dinitrogen pentaoxide (or pentoxide), respectively. N_2O_4 is the dimmer of NO_2 and N is +4 in both species. The systematic name of N_2O is dinitrogen oxide and of NO is nitrogen monoxide. Nitric oxide is the correct common name for N_2O. Although (d) is a correct name for N_2O, it is a systematic name, and the question calls for the common name.

14.10 **d.** Chromium forms two oxyanions with Cr in the +6 oxidation state. These are the chromate ion, $(CrO_4)^{2-}$, and the dichromate ion, $(Cr_2O_7)^{2-}$. Choices (a), (b), and (c) are chromates, which is incorrect. Choice (d) is systematically named but ignores the fact that the name of an ionic compound is built from the names of the ions present.

14.11

a. **c.** 1 mol MnO_2 = 1 mol Mn^{4+} + 2 mol O^{2-}; 1.38 mol MnO_2 × 1.0 mol Mn^{4+}/1.0 mol MnO_2 = 1.38 mol Mn^{4+}.

b. **b.** 1.38 mol MnO_2 × 2.0 mol O^{2-}/1.0 mol MnO_2 = 2.76 mol O^{2-}.

c. **d.** m.w. MnO_2 = 54.9 + (2 × 16.0) = 86.9; 1.38 mol MnO_2 × 86.9 g MnO_2/1 mol MnO_2 = 120 g MnO_2.

14.12 **a.** Thinking he or she has $CaCO_3$, the student will weigh out 2.00 mol × (100.1 g $CaCO_3$/mol) = 200.2 g. But since it was NaCl that was weighed, the number of moles of NaCl is 200.2 g × (1 mol NaCl/58.5 g) = 3.42 NaCl.

14.13

$$9.8 \text{ mg } O_2 \times 10^{-3} \text{ g/1 mg} = 9.8 \times 10^{-3} \text{ g } O_2$$

$$1 \text{ molecule } O_2/1 \text{ molecule protein} = 1 \text{ mol } O_2/1 \text{ mol protein}$$

$$1 \text{ mol } O_2 = 32.0 \text{ g } O_2$$

$$\frac{12.2 \text{ g protein}}{9.8\times10^{-3} \text{ g } O_2}\times\frac{32.0 \text{ g } O_2}{1 \text{ mol } O_2}\times\frac{1 \text{ mol } O_2}{1 \text{ mol protein}}=\frac{39.8\times10^3 \text{ g protein}}{1 \text{ mol protein}}$$

$$\text{m.w. protein} = 39.8 \times 10^3 = 39{,}800$$

or

$$\frac{9.8\times10^{-3} \text{ g } O_2}{32.0 \text{ g/1 mol } O_2}=0.3062\times10^{-3} \text{ mol } O_2 = 0362\times10^{-3} \text{ mol protein}$$

$$\text{g} = \text{mol} \times \text{m.w.}$$

$$\text{m.w.} = \frac{\text{g}}{\text{mol}}$$

$$\frac{12.2 \text{ g protein}}{0.3062\times10^{-3} \text{ mol protein}}=39.8\times10^3 = 39{,}800 \text{ m.w. protein}$$

14.14 **a.** First calculate moles of $KMnO_4$ to reach the endpoint:

$$21.83 \text{ mL } KMnO_4\times\frac{1 \text{ L}}{10^3 \text{ mL}}\times\frac{0.03820 \text{ mol}}{1 \text{ L}}=8.339\times10^{-4} \text{ mol } KMnO_4$$

We need the balanced equation to relate moles of $KMnO_4$ to moles of Fe(II):

$$5 \text{ Fe}^{2+} + MnO_4^- + 8 \text{ H}^+ \rightarrow 5 \text{ Fe}^{3+} + Mn^{2+} + 4 \text{ H}_2O.$$

since of MnO_4^- = moles of $KMnO_4$,

$$8.399\times10^{-4} \text{ mol } MnO_4^-\times\frac{5 \text{ mol Fe}^{2+}}{1 \text{ mol } MnO_4^-}=4.170\times10^{-3} \text{ mol Fe}^{2+}$$

$$4.170 \times 10^{-3} \text{ mol Fe}^{2+} \times \frac{5{,}585 \text{ g}}{\text{mol Fe}} = 0.239 \text{ g Fe}^{2+} \text{ in sample}$$

$$\frac{0.2329 \text{ g Fe}^{2+}}{1.285 \text{ g sample}}\times100=18.12\% \text{ Fe}^{2+} \text{ in sample}$$

14.15 **d.** Percent H_2O_2 by mass is equal to

$$\frac{\text{g } H_2O_2}{\text{g sample}}\times100$$

so we must find g H_2O_2. This information is supplied by the titration and the balanced equation

$$\text{moles } KMnO_4=\frac{0.0103 \text{ ml}}{1 \text{ L}}\times\frac{0.0397 \text{ L}}{\text{titre}}=4.09\times10^{-4} \text{ mol}$$

From the balanced equation the reacting ratio is

$$\frac{5 \text{ mol } H_2O_2}{2 \text{ mol } KMnO_4}$$

$$\frac{5 \text{ mol } H_2O_2}{2 \text{ mol } KMnO_4}\times\frac{4.09\times10^{-4} \text{ mol } KMnO_4}{\text{sample}}=1.02\times10^{-3} \text{ mol } H_2O_2$$

$$\text{g } H_2O_2 = \text{mol} \times \text{m.w.} = 1.02\times10^{-3} \text{ mol} \times\frac{34.02 \text{ g}}{1 \text{ mol}}=3.47\times10^{-2}$$

$$\text{percent } H_2O_2=\frac{0.0347 \text{ g } H_2O_2}{1.546 \text{ g solution}}\times100=2.24\% \text{ } H_2O_2$$

14.16 **a.** First calculate the number of equivalent of I_2:

$$\text{Equivalents} = \text{volume (L)} \times \text{normality (N)}$$
$$= 0.040 \text{ L} \times 0.10 = 0.0040 \text{ equiv. } I_2$$

Since the values of normality and equivalents are taken with respect to the particular reaction being considered,

$$\text{equivalents } SnCl_2 = \text{equivalents } I_2 = 0.040$$

The half-reaction for $Sn^{2}+$ is

$$Sn^{2+} \rightarrow Sn^{4+} + 2\,e^{-}$$

As the oxidation number of Sn changes from +2 to +4, a change of two, the equivalent weight of $SnCl_2$ is

$$\frac{\text{formula weight}}{2} = \frac{189.60}{2} = 94.80$$

$$\text{weight } SnCl_2 = \frac{94.80 \text{ g}}{\text{equivalent}} \times 0.0040 \text{ equiv.} = 0.38 \text{ g}$$

14.17 **c.** In the half-cell reaction for $Cr_2O_7^{2-}$, the oxidation state of Cr changes from +6 to +3. Since two Cr atoms are in the $Cr_2O_7^{2-}$ formula, the change for the species is six and the equivalent weight is the molecular weight divided by six. This can also be seen by balancing the half-cell charges by the addition of six e^{-} to the left side of the equation.

$$\text{equivalent weight } K_2Cr_2O_7 = \frac{294.20}{6} = 49.03$$

(Note: Since we are relating this figure to the weight of dissolved $K_2Cr_2O_7$, it is the molecular weight—or formula weight—of the K salt that is used in the calculation.)

$$\text{equivalents } K_2Cr_2O_7 = \frac{5.88 \text{ g}}{49.03}$$

$$\frac{\text{g equivalent}}{\text{g}} = 0.120 \text{ equivalent}$$

$$\text{normality} = \frac{\text{equivalents}}{\text{liter}} = \frac{0.120 \text{ equivalent}}{1.00 \text{ L}} = 0.120 \text{ N}$$

14.18 **d.** The cell reaction shows Al as oxidized (loss of electrons). Therefore, the equation for the standard electrode potential must be rewritten as an oxidation reaction, and the sign of E^0 will be positive:

$$Al \rightarrow Al^{3+} + 3\,e^{-};\ E^0 = 1.66 \text{ V}$$

The cell reaction for Cd is written correctly, as Cd^{2+} is reduced (gains electrons). However the equations must be multiplied by the appropriate factors so that each has the same number of electrons. The factor is 2 for the Al equation and 3 for the Cd equation, to give

$$2\ Al \rightarrow 2\ Al^{3+} + 6\ e^-;\ E^0 = 1.66\ V$$
$$3\ Cd^{2+} + 6\ e^- \rightarrow 3\ Cd;\ E^0 = -0.40\ V$$

The equations are added to give:

$$2\ Al + 3\ Cd^{2+} = 2\ Al^{3+} + 3\ Cd \text{ (the electrons cancel)};\ E^0 = 1.26\ V$$

Note that the values of E^0 are not multiplied by the coefficients used to balance the electron transfer in the equation.

14.19

a. **c.** $$Fe(s) \rightarrow Fe^{2+} + 2\ e^+:$$

+0.41 V; (oxidation half-reaction, reversed from reduction half-reaction table, E^0 sign reversed)

$$2\ H^+ + 2\ e^- \rightarrow H_2(g):$$

0.00 V; (reduction half-reaction from table, same sign of E^0)

$$Fe(s) + 2\ H^+ = Fe^{2+} + H_2(g):$$

0.41 V; (add half-reactions, electrons cancel, E^0 additive)
Reaction is spontaneous at standard conditions when E^0 positive.

b. **d.** $Cu(s) \rightarrow Cu^{2+} + 2\ e^-$:

–0.34 V; (oxidation half-reaction, reversed from reduction half-reaction table, E^0 sign reversed)

$$2\ H^+ + 2\ e^- \rightarrow H_2(g):$$

0.00 V; (reduction half-reaction, from table, same sign of E^0)

$$Cu(s) + 2\ H^+ \rightarrow Cu^{2+} + H_2(g):$$

–0.34 V; (add half-reactions, electrons cancel, E^0 additive)
Reaction is nonspontaneous at standard conditions when E^0 negative.

c. **a.** $Sn^{2+} \rightarrow Sn^{4+} + 2\ e^-$:

–0.15 V; (oxidation half-reaction, reversed from reduction half-reaction table, E^0 sign reversed)

$$Fe^{3+} + e^- \rightarrow Fe^{2+}:$$

0.77 V; (reduction half-reaction from table, same sign of E^0)

$$2\ Fe^{3+} + 2\ e^- \rightarrow 2\ Fe^{2+}:$$

0.77 V; (double equation so electrons will cancel; note E^0 does not double)

$$Sn^{2+} + 2\ Fe^{3+} \rightarrow Sn^{4+} + 2\ Fe^{2+}:$$

0.62 V; (add half-reactions, electrons cancel, E^0 additive)
Reaction is spontaneous at standard conditions when E^0 positive.

14.20 **d.**

$$Cu(s) \rightarrow Cu^{2+} + 2\,e^-:$$

$E^0 = -0.34$ V (anode half-reaction; reverse equation above and change sign of E^0 to give oxidation potential)

$$2\,Fe^{3+} + 2\,e^- \rightarrow 2\,Fe^{2+}:$$

$E^0 = 0.77$ V (cathode half-reaction; double above equation so equal number electrons in each half-reaction; E^0, reduction potential, remains the same)

$$Cu(s) + 2\,Fe^{3+} \rightarrow Cu^{2+} + 2\,Fe^{2+}:$$

$E^0 = 0.43$ V (add half-reactions, electrons cancel, cell E^0 is E^0_{ox} + E^0_{red})

Cell potential at concentrations given is calculated using the Nernst equation:

$$\begin{aligned} E &= E^0 - \frac{0.0592}{n}\log\frac{[Cu^{2+}]\,[Fe^{2+}]^2}{[Fe^{3+}]^2} \\ &= E^0 - \frac{0.0592}{n}\log\frac{[0.050]\,[0.040]^2}{[0.0020]^2} \\ &= 0.43 - \frac{0.0592}{n}\log 20 = 0.43 - (0.0296)(1.301) \\ &= 0.43 - 0.038 = 0.39\text{ V} \end{aligned}$$

14.21 **b.** Time of electrolysis = 2.00 hr × 3600 s/hr = 7200 s
Coulombs = amperes × seconds = 2.00 amp × 7200 s = 14,400 C

$$14{,}400\text{ C} \times \frac{1\text{ mol}}{96{,}485\text{ C}} = 0.1492\text{ mol } e^-$$

$$2.59\text{ g Cr} \times \frac{1\text{ mol}}{52.00\text{ g}} = 0.04981\text{ mol Cr}$$

$$\frac{1\text{ mol Cr}}{x\text{ mol } e^-} = \frac{0.04981\text{ mol Cr}}{0.1492\text{ mol } e^-}$$

$$x = \frac{0.1492}{0.04981} = 2.995 = 3.0$$

$$\text{factor} = \frac{1\text{ mol Cr}}{3\text{ mol } e^-}$$

The oxidation state of the Cr in solution should be +3, so that $Cr^{3+}(aq) - 3\,e^- \rightarrow Cr(s)$, (3 F, or 3 mol e^- for 1 mol Cr deposited).

14.22 **d.** The reaction for electroplating Cr from Cr^{3+} in solution is $Cr^{3+}(aq) + 3\,e^- \rightarrow Cr(s)$.

$$\text{factor} = \frac{1\text{ mol Cr}}{3\text{ mol } e^-}$$

$$6.19\text{ g Cr} \times \frac{1\text{ mol Cr}}{52.00\text{ g Cr}} = 0.1190\text{ mol Cr}$$

$$0.1190\text{molCr} \times \frac{3 \text{ mol } e^-}{1 \text{ mol Cr}} = 0.3570\text{mol } e^- \text{ used to electroplate } 6.19\text{gCr}$$

$$0.357 \text{ mol } e^- \times \frac{96,465 \text{ C}}{1 \text{ mol } e^-} = 34,445 \text{ C}$$

$$\text{Coulombs} = \text{amperes} \times \text{seconds} \qquad \text{seconds} = \frac{\text{C}}{\text{amperes}} = \frac{34445 \text{ C}}{3.75 \text{ A}}$$

$$= 9185 \text{ s} = 9.18 \times 10^3 \text{ s}$$

14.23

c. $$\text{Moles of Ni} = 17.5 \text{ g} \times \frac{1 \text{ mol}}{58.69 \text{ g}} = 2.98 \times 10^{-1} \text{ mol}$$

Reaction is $Ni^{2+}(aq) + 2\,e^- = Ni(s)$

$$\text{Factor is } \frac{1 \text{ mol Ni}}{2 \text{ mol } e^-} = \frac{1 \text{ mol Ni}}{2 \text{ F}}$$

$$\text{Coulombs } = 2.98 \times 10^{-1} \text{mol Ni} \times \frac{2 \text{ F}}{1 \text{ mol Ni}} \times \frac{96,485}{1 \text{ F}} = 5.75 \times 10^4$$

C = amp × sec

Number of seconds = 0.75 hr × 3600 = 2.70×10^3 s

$$\text{amperes} = \frac{5.75 \times 10^4 \text{C}}{2.70 \times 10^3 \text{sec}} = 2.1 \times 10^1 \text{ A}$$

14.24 **b.** The electrode reaction is $Au^{3+}(aq) + 3\,e^- = Au(s)$

$$\text{Factor is } \frac{1 \text{ mol Au}}{3 \text{ mol } e^-}$$

$$\text{The volume of Au deposited} = 45 \text{ cm}^2 \times 0.10 \text{ mm} \times \frac{1.00 \text{cm}}{10 \text{ mm}} = 0.45 \text{cm}^3$$

$$\text{The weight of Au deposited is} = 0.45 \text{ cm}^3 \times \frac{19.3 \text{ g}}{1 \text{ cm}^3} = 8.68 \text{ g Au}$$

$$\text{The coulombs passed} = 8.68 \text{ g Au} \times \frac{1 \text{ mol Au}}{196.97 \text{ g Au}} \times \frac{3 \text{ mol } e^-}{1 \text{ mol Au}} \times \frac{96485 \text{ C}}{\text{mol } e^-}$$

$$= 1.28 \times 10^4 \text{ C}$$

$$\text{C} = \text{amp} \times \text{sec} \quad \text{sec} = \frac{\text{C}}{\text{amp}}$$

$$\text{Time of run} = \frac{1.28 \times 10^4 \text{ C}}{0.52 \text{ amp}} = 2.46 \times 10^4 \text{ sec} = 2.5 \times 10^4 \text{ sec } (=6.9 \text{ hr})$$

14.25 **d.** $$\text{Molarity (M)} = \frac{\text{moles solute}}{\text{liters (L) solution}}$$

Therefore, moles solute = M × L solution

moles solute = 0.10 × 0.010 L = 0.0010 mol HNO_3

A quantity of 0.0010 mol HNO_3 is diluted to 500 mL. The molarity (M) of the solution is given by

$$M = \frac{0.0010 \text{ moles } HNO_3}{0.500 \text{ L}} = 0.0020$$

Since $HNO_3(aq)$ is 100% dissociated into ions, $HNO_3 \rightarrow H^+ + NO_3^-$, the 0.0020 M solution will contain 0.0020 M $H^+(H_3O^+)$ and 0.0020 M (NO_3^-).

14.26 **c.** Since the number of moles of HCl remains the same, that is, moles HCl = moles HCl, we can use the relationship $(\text{volume}_1) \times (\text{molarity}_1) = (\text{volume}_2) \times (\text{molarity}_2)$.

(30.0 ml) × (12.0 M) = (x mL) × (3.00 M)

x = 120 mL for volume of solution_2

Since the original volume (solution_1) is 30.0 mL, the volume of water to be added is 120 – 30 = 90 mL.

14.27 **a.** First calculate the number of moles of NaOH dissolved, and the molarity (M) of the stock solution.

$$\frac{30.0 \text{ g NaOH}}{40.0 \text{ g/mol}} = 0.750 \text{ mol NaOH}$$

$$M = \frac{\text{moles}}{\text{liter}} = \frac{0.750 \text{ mol}}{0.500 \text{ L}} = 1.50$$

In 1.00 L of 0.100 M NaOH, there are

$$\frac{0.100 \text{ mol}}{\text{L}} \times 1.00 \text{ L} = 0.100 \text{ mol NaOH}$$

Now calculate the volume of 150 M NaOH solution which contains 0.100 mol NaOH.

$$M = \frac{\text{moles}}{\text{L}}$$

$$L = \frac{0.100}{1.50} = 0.0667 \text{ L} = 66.7 \text{ mL}$$

14.28 **d.** First, calculate how many moles of NaOH were used in the titration:

$$\frac{0.0100 \text{ mol NaOH}}{1 \text{ L}} \times (37.68 \text{ mL NaOH}) \times \frac{1 \text{ L}}{1000 \text{ mL}}$$
$$= 3.77 \times 10^{-4} \text{ mol NaOH}$$

The balanced equation for the reaction is

$$NaOH(aq) + HCl(aq) \rightarrow NaCl(aq) + H_2O$$

showing that 1 mol of NaOH neutralized 1 mol HCl. Then, the weight of HCl can be calculated from

$$(3.77 \times 10^{-4} \text{mol NaOH}) \times \frac{1 \text{ mol HCl}}{1 \text{ mol NaOH}} \times \frac{36.46 \text{ g HCl}}{1 \text{ mol HCl}} \times \frac{100 \text{ mg HCl}}{1 \text{ g HCl}}$$
$$= 13.7 \text{ mg HCl}$$

14.29 **a.** By definition one equivalent of acid is equal to one equivalent of base in a reaction. The number of equivalents of $Ca(OH)_2$ at the neutralization point must be equal to the number of equivalents of H_3PO_4 at the neutralization point. In a two-step solution find the number of equivalents of H_3PO_4 first, from equivalents = volume in L × normality:

$$\text{equiv.} = \text{vol(L)} \times \text{N} = 0.0750 \times 0.150 = 0.01125$$

Since the same number of equivalents of $Ca(OH)_2$ were used in the reaction,

$$0.01125 = x(\text{L}) \times 0.250$$
$$x = 0.0450 \text{ L } Ca(OH)_2 \text{ solution}$$

A one-step solution is

$$\text{equivalents } Ca(OH)_2 = \text{equivalents } H_3PO_4$$

$$\text{Since normality (N)} = \frac{\text{equivalents}}{\text{volume solution (L)}}$$

$$N_1 \times \text{vol(L)}_1 = N_2 \times \text{vol(L)}_2$$
$$(0.250) \text{ vol(L)}_1 = (0.150)\ (0.0750 \text{ L})$$
$$\text{vol(L)}_1 = 0.0450 \text{ L} = 45.0 \text{ mL } Ca(OH)_2$$

It is also convenient to multiply $N_1 \times \text{vol(L)}_1 = N_2 \times \text{vol(L)}_2$ by 1000, so that $N_1 \times \text{vol(mL)}_1 = N_2 \times \text{vol(mL)}_2$ and:

$$\text{vol(mL)}_1 = \frac{(0.150)\ (75.0)}{0.250} = 45.0 \text{ mL } Ca(OH)_2$$

14.30 **c.** The metallic elements are found primarily in Groups IA and IIA at the left of the table and Groups IIIB-VIIIB and IB-IIB in the middle. They are also in the higher-period elements of Groups IIIA–VIA. There are no metals in the far right Groups and VIIIA, ruling out (a) and (b). There are metals in a number of groups besides IA and IIA, ruling out (d).

14.31 **a.** The mobile electron cloud of outer electrons gives metals both high electrical and high thermal conductivity. The mobile electrons and accompanying crystal structure also give metals high luster, cohesive strength, and high ductility. The chemical reactivity of metals varies from high (Li, Na) to low (Pt, Au).

14.32 **d.** The nonmetals are generally neither malleable nor ductile. They do not react with acids and are poor conductors of electricity. Often they have low melting points and volatile. However they do form covalent halides, which react with water to give oxyacids.

14.33 **d.** Strontium is a member of Group IIA, not IA. Gold is in Group IB, while the other metals in choice (b) are in Group VIIIB. All the elements in choice (c) are in IIIA, but boron is not considered a metal. The elements in choice (d) are all metals in Group IIA.

14.34 **c.** Halogens form a number of interhalogen compounds, such as ClF and ICl. They are noted as strong electron acceptors. They do form strong oxyacids of the formula HOX_3, as well as HOX_4. Halogens form ionic bonds in compounds with Group IA metals. The halogens bond with some nonmetals, such as sulfur and phosphorus, as well as other halogens.

14.35 **d.** Sulfuric acid forms from the reaction of SO_3 with water. Pyrosulfuric acid is formed from SO_3 and concentrated sulfuric acid. SO_2 solution is used as a reducing agent, but it is H_2S that is used in analytical determination of metal cations. SO_2 can be converted to SO_3 by oxidation in the presence of a catalyst. A weak and unstable solution of sulfurous acid is formed upon dissolution in water.

14.36 **a.** Dark brown NO_2 and colorless N_2O_4 exist in equilibrium at ambient temperature. NO reacts rapidly with O_2 to form NO_2 at ambient temperature. NO is not normally in equilibrium with N_2O and NO_2 but can decompose to yield these species when pressurized at moderate temperatures. N_2O_5 decomposes at moderate temperatures to yield N_2O_4 or NO_2. N_2O_3 decomposes when vaporized to $NO + NO_2$. Although all the reactions above take place under appropriate conditions, (a) is the only one that could be considered in equilibrium at ambient temperature.

14.37 **d.** Carbides of some metals (for example copper, silver, and calcium) are believed to contain the acetylide ion, C_2^{2-}, which hydrolyzes to acetylene. Carbides of other metals (for example Al) may contain a methanide ion, C^{4-}, and hydrolyze to methane.

14.38 **b.** The relatively small energy band gap makes silicon an intrinsic semiconductor and leads to its industrial use.

14.39 **c.** Semiconductor silicon is very pure Si until it is **doped** with a 3+ or 5+ element. Si—O—Si may occur in surface coatings. Silicone polymers have Si—C bonds as well as Si—O bonds. Clays and zeolites often contain trivalent Al in the Si—O—Si structure and a counterion such as Na^+ or K^+ to balance the charge. Glasses are supercooled liquids with oxides of boron, lead, sodium, and calcium mixed with silica. Quartz, with a high melting point, is essentially pure silica.

14.40 **d.** The chemical equation is balanced so:

$$K_c = \frac{[CO_2(g)][H_2(g)]}{[CO_2(g)][H_2O(g)]}$$

$$\frac{\left[\frac{0.070 \text{ mol } CO_2}{5 \text{ L}}\right]\left[\frac{2.00 \text{ mol } H_2}{5 \text{ L}}\right]}{\left[\frac{0.020 \text{ mol CO}}{5 \text{ L}}\right]\left[\frac{0.0215 \text{ mol } H_2O}{5 \text{ L}}\right]} = 326$$

Note that the units cancel, leaving K_c dimensionless.

14.41 **a.** A good general approach to equilibrium problem is (1) write the balanced equilibrium reaction equation, (2) write the equilibrium constant expression in terms of the balanced equation, and (3) identify in a simple tabular form what is known about the concentrations of species initially and at equilibrium, and what change(s) occur. For the above problem, these procedures give

(1) (as given) $N_2O_4(g) = 2\ NO_2(g)$

(2) $K_c = \frac{[NO_2]^2}{[N_2O_4]}$

(3)

Species:	**mol N_2O_4**	**mol NO_2**
Initial amt.	10.0	0
Change	(−1.75)	(+3.50)
Equilibrium amt.	(8.25)	3.50

The values *not* in parentheses are given with the problem and are filled in first. The blanks are filled in with the values in parentheses from the calculations below.

In order to produce 3.50 mol NO_2, the equation shows us that half this amount of N_2O_4 has decomposed, or:

$$3.50 \text{ mol } NO_2 \times \frac{1 \text{ mol } N_2O_4}{2 \text{ mol } NO_2} = 1.75 \text{ mol } N_2O_4$$

Therefore, at equilibrium, 10.0 − 1.75 = 8.25 mol N_2O_4 remain. The equation for the equilibrium expression is then

$$K_c = \frac{\left[\frac{3.50 \text{ mol NO}_2}{5.00 \text{ L}}\right]^2}{\left[\frac{8.25 \text{ mol N}_2\text{O}_4}{5.00 \text{ L}}\right]} = \frac{[0.70 \text{ M}]^2}{[1.65 \text{ M}]} = 0.297 \text{ M}$$

Note that the presence of the square term dictates the equilibrium constant units (mol/L = M). Note also, that the vessel size (5.00 L) must be included in the calculations.

14.42 **c.** The expression for the equilibrium constant is

$$K_c = \frac{[PCl_3][Cl_2]}{[PCl_5]} = \frac{[0.25 \text{ M}][0.25 \text{ M}]}{[0.125 \text{ M}]} = \frac{[0.0625 \text{ M}]^2}{[0.125 \text{ M}]} = 0.50 \text{ M}$$

14.43 **c.** First, calculate the molecular weigh of $La(NO_3)_4$

$$1 \times (\text{a. w. La}) + 4 \times (\text{a.w. N}) + 12 \times (\text{a.w. O})$$
$$138.91 + (4 \times 14.01) + (12 \times 16.00) = 386.9$$

$$\text{Number of moles in 15.0 g La} = \frac{15.0 \text{ g}}{386.9 \text{ g/mol}} = 0.0388 \text{ mol}$$

$$\text{Molarity} = \text{M} \frac{0.0388 \text{ mol}}{0.800 \text{ L}} = 0.048 \text{ M}$$

14.44 **d.** The moles of solute before dilution equal the moles of solute after dilution.

$$\text{moles} = \frac{\text{g}}{\text{m.w.}} = \text{volume (L)} \times \text{ molarity (M)}$$

$$\frac{0.26 \times 15.0}{86.8} = 0.080 \text{ L} \times \text{M}$$

$$\text{M} = 0.56$$

14.45 **d.** The total pressure above an ideal solution is the sum of the partial pressures of each component multiplied by the mole fraction of each component:

$$P = X_1 P_1^0 + X_2 P_2^0$$

$$n_{C_2H_5OH} = \frac{20.0 \text{ g}}{46.07 \text{ g/mol}} = 0.434 \text{ mol}$$

$$n_{CCl_4} = \frac{60.0 \text{ g}}{153.8 \text{ g/mol}} = 0.390 \text{ mol}$$

$$n_{total} = 0.434 + 0.390 = 0.824 \text{ mol}$$

$$X_1 = \frac{0.434}{0.824} = 0.527 \text{ C}_2\text{H}_5\text{OH}$$

$$X_2 = \frac{0.390}{0.824} = 0.473 \text{ CCl}_4$$

$$P = (0.527)\ (71.2 \text{ torr}) + (0.473)\ (121.6 \text{ torr}) = 37.5 + 57.5 = 95.0 \text{ torr}$$

14.46 **d.** $\Delta T = K_b m$. Rearrange to

$$m = \frac{\Delta T}{K_b} = \frac{0.31°C}{0.512°C \bullet m^{-1}} = 0.606 \text{ mol} \bullet \text{kg}^{-1}_{\text{solvent}}$$

$$\text{m.w.} = 73 \text{ g/mole}$$

$$0.606 \text{ mol kg}^{-1}_{\text{solvent}} \times 73 \text{ g} \bullet \text{mol}^{-1} = 44.2 \text{ g} \bullet \text{kg}^{-1}_{\text{solvent}}$$

$$425 \text{ g water} = 0.425 \text{ kg solvent}$$

$$44.2 \text{ g} \bullet \text{kg}^{-1}_{\text{solvent}} \times 0.425 \text{ kg solvent} = 19 \text{ g}$$

14.47 **b.** Divide equation for data (1) by equation for data (2) to find a single value of n.

$$\frac{r_1}{r_2} = \frac{k\,[B]_1^n}{k\,[B]_2^n} \quad \text{or} \quad \frac{r_1}{r_2} = \frac{[B]_1^n}{[B]_2^n} \quad (k \text{ cancels})$$

Take the log of both sides of the equation:

$$\log \frac{r_1}{r_2} = \log \frac{[B]_1^n}{[B]_2^n} = n \log \frac{[B]_1}{[B]_2}$$

Solve for n:

$$n = \frac{\log \frac{r_1}{r_2}}{\log \frac{[B]_1}{[B]_2}} = \frac{\log \frac{2.92 \times 10^{-4} \text{ M} \bullet \text{s}^{-1}}{4.13 \times 10^{-4} \text{ M} \bullet \text{s}^{-1}}}{\log \frac{0.245 \text{ M}}{0.490 \text{ M}}} = \frac{-0.151}{-0.301} = 0.5 \qquad r = \text{k}\,[B]^{0.5}$$

14.48 **c.** The reaction order can be obtained for a reaction by a balanced stoichiometric equation only by considering adequate experimental data. Such data would come from experiments where the concentration of reactants is varied one at a time and compared with the rate of disappearance of a reactant or appearance of a product.

There is possible confusion because the reaction written as a stoichiometric equation can derive from one or more steps called **elementary reactions**. The elementary reactions can be unimolecular, bimolecular, or termolecular. When an elementary reaction step is established, the reaction order of that particular elementary step follows from its equation.

14.49 **a.** From set (1) and set (2) of data, when the concentration of $[B]$ doubles, the rate quadruples. Therefore, the reaction is second order in $[B]$. From set (1) and set (3) of data, when the concentration of $[A]$ doubles, the rate doubles. Therefore, the reaction is first order in $[A]$.

$$R = k\,[A]\,[B]^2$$

Overall order is 1 + 2 = 3.

The specific rate constant, k, can be calculated from any set of data:

Set	Rate, $mol^3/L^3 \cdot s$	[*A*], mol/L	[*B*], mol/L	*k*, s^{-1}
(1)	0.020	0.10	0.20	5.0
(2)	0.080	0.10	0.40	5.0
(3)	0.040	0.20	0.20	5.0
(4)	0.060	0.30	0.20	5.0

14.50 **b.** All the carbon in the carbon dioxide and all the hydrogen in the water come form the hydrocarbon.

$$\% \text{ C in } CO_2: \quad \frac{12.01 \text{ g C/mol}}{44.01 \text{ g } CO_2\text{/mol}} \times 100 = 27.29\% \text{ C}$$

$$\text{g C} = 7.260 \text{ g } CO_2 \times 0.2729 = 1.981 \text{ g C}$$

$$\% \text{ H in } H_2O: \quad \frac{2.016 \text{ g H/mol}}{18.016 \text{ g } H_2O\text{/mol}} \times 100 = 11.19\% \text{ H}$$

$$\text{g H} = 1.485 \text{ g } H_2O \times 0.1119 = 0.166 \text{ g H}$$

Calculate elemental percentages:

$$\% \text{ C} = \frac{1.981 \text{ g C}}{2.147 \text{ g compound}} \times 100 = 92.27\% \text{ C}$$

$$\% \text{ H} = \frac{0.166 \text{ g H}}{2.147 \text{ compound}} \times 100 = 7.73\% \text{ H}$$

Note: Since the compound is defined as a hydrocarbon, the percent of each element calculated independently should add to 100%.

14.51 **c.** Both compounds contain only carbon and hydrogen. Cyclopropane is a ring structure of three carbons singly bonded to each other and two hydrogens attached to each carbon. The formula is C_3H_6. The ratio of carbon to hydrogen is 1:2. Cyclohexane is also a ring structure with six carbons singly bonded to each other and two hydrogens attached to each carbon. The formula is C_6H_{12}. The ratio of carbon to hydrogen in cyclohexane is 1:2. Therefore cyclohexane must have the same percentage of carbon as cyclopropane.

Check: $6 \times 12.01 = 72.06$ g C per moel cyclohexane; (84.16 m.w. cyclohexane)

$12 \times 1.008 = 12.096$ g H per mole cyclohexane

$$\frac{72.1}{84.1} = 0.857 \times 100 = 85.7\% \text{ C in cyclohexane}$$

14.52 **c.** Only alcohols (R—OH) and ethers (R—O—R′) fit the general formula C_3H_8O. Three structures are possible:

the alcohols

```
CH3CH2CH2   and   CH3CHCH3
        |              |
        OH             OH
```

and the ether CH_3—O—CH_2CH_3

14.53

a. **b.** The compound is an alcohol, characterized by the OH group bonded to a carbon.

b. **d.** The compound is an aldehyde, characterized by the CHO group bonded to a carbon.

c. **a.** The compound is an ester, characterized by the COO in which the characteristic carbon is double bonded to an O atom and single bonded to another C, and an O atom. In the ester, the second O atom is also bonded to another C.

d. **d.** The compound is an amine, as the carbon chain ends in a carbon bond to the N of an NH_2 group.

e. **b.** The compound is a nitrile, characterized by the triple bond between the end C atom and an N atom.

f. **c.** The compound is a ketone, characterized by the C—O group in which the C is bonded to each of two other C atoms.

Engineering Economics

Donald G. Newnan

OUTLINE

This is a review of the field known as **engineering economics, engineering economy**, or **engineering economic analysis**. Since engineering economics is straightforward and logical, even people who have not had a formal course should be able to gain sufficient knowledge from this chapter to successfully solve most engineering economics problems.

There are 29 example problems throughout this review. These examples are an integral part of the review and should be examined as you come to them.

The field of engineering economics uses mathematical and economic techniques to systematically analyze situations that pose alternative courses of action. The initial step in engineering economics problems is to resolve a situation, or each possible alternative in a given situation, into its favorable and unfavorable consequences or factors. These are then measured in some common unit, usually money. Factors that cannot readily be equated to money are called **intangible** or **irreducible** factors. Such factors are considered in conjunction with the monetary analysis when making the final decision on proposed courses of action.

CASH FLOW

A cash flow table shows the "money consequences" of a situation and its timing. For example, a simple problem might be to list the year-by-year consequences of purchasing and owning a used car:

Year	Cash Flow	
Beginning of first year, Year 0	–$4500	Car purchased "now" for $4500 cash. The minus sign indicates a disbursement.
End of Year 1	–350	Maintenance costs are $350 per year.
End of Year 2	–350	
End of Year 3	–350	
End of Year 4	+2000	The car is sold at the end of the 4th year for $2000. The plus sign represents a receipt of money.

This same cash flow may be represented graphically, as shown in Fig. 15.1. The upward arrow represents a receipt of money, and the downward arrows represent disbursements. The horizontal axis represents the passage of time.

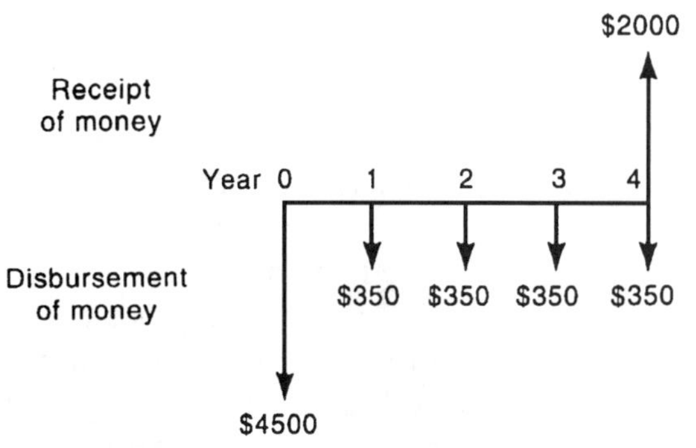

Figure 15.1

Example 15.1

In January 1993 a firm purchased a used typewriter for $500. Repairs cost nothing in 1993 or 1994. Repairs are $85 in 1995, $130 in 1996, and $140 in 1997. The machine is sold in 1997 for $300. Compute the cash flow table.

Solution

Unless otherwise stated, the customary assumption is beginning-of-year purchase, followed by end-of-year receipts or disbursements, and an end-of-year resale or salvage value. Thus the typewriter repairs and the typewriter sale are assumed to occur at the end of the year. Letting a minus sign represent a disbursement of money, and a plus sign a receipt of money, we are able to set up the cash flow table:

Year	Cash Flow
Beginning of 1993	–$500
End of 1993	0
End of 1994	0
End of 1995	–85
End of 1996	–130
End of 1997	+160

Notice that at the end of 1997 the cash flow table shows +160, which is the net sum of –140 and +300. If we define Year 0 as the beginning of 1993, the cash flow table becomes

Year	Cash Flow
0	–$500
1	0
2	0
3	–85
4	–130
5	+160

From this cash flow table, the definitions of Year 0 and Year 1 become clear. Year 0 is defined as the *beginning* of Year 1. Year 1 is the *end* of Year 1, and so forth.

TIME VALUE OF MONEY

When the money consequences of an alternative occur in a short period of time—say, less than one year—we might simply add up the various sums of money and obtain the net result. But we cannot treat money this way over longer periods of time. This is because money today does not have the same value as money at some future time.

Consider this question: Which would you prefer, \$100 today or the assurance of receiving \$100 a year from now? Clearly, you would prefer the \$100 today. If you had the money today, rather than a year from now, you could use it for the year. And if you had no use for it, you could lend it to someone who would pay interest for the privilege of using your money for the year.

EQUIVALENCE

In the preceding section we saw that money at different points in time (for example, \$100 today or \$100 one year hence) may be equal in the sense that they both are \$100, but \$100 a year hence is *not* an acceptable substitute for \$100 today. When we have acceptable substitutes, we say they are *equivalent* to each other. Thus at 8% interest, \$108 a year hence is equivalent to \$100 today.

Example 15.2

At a 10% anual interest rate, \$500 now is *equivalent* to how much three years hence?

Solution

\$500 now will increase by 10% in each of the three years.

$$\begin{aligned} \text{Now} &= \$500.00 \\ \text{End of 1st year} &= 500 + 10\%\ (500) = 550.00 \\ \text{End of 2nd year} &= 550 + 10\%\ (550) = 605.00 \\ \text{End of 3rd year} &= 605 + 10\%\ (605) = 665.50 \end{aligned}$$

Thus \$500 now is *equivalent* to \$665.50 at the end of three years. Note that interest is charged each year on the original \$500 *plus* the unpaid interest.

Equivalence is an essential factor in engineering economics. Suppose we wish to select the better of two alternatives. First, we must compute their cash flows. For example

	Alternative	
Year	**A**	**B**
0	−\$2000	−\$2800
1	+8000	+1100
2	+800	+1100
3	+800	+1100

The larger investment in alternative B results in larger subsequent benefits, but we have no direct way of knowing whether it is better than alternative A. To make a decision, we must resolve the alternatives into *equivalent* sums so they may be accurately compared.

COMPOUND INTEREST

To facilitate equivalence computations, a series of compound interest factors will be defined here, and their use will be illustrated in examples.

Symbols and Functional Notation

i = effective interest rate per interest period. In equations, the interest rate is stated as a decimal (that is, 8% interest is 0.08).

n = number of interest periods. Usually the interest period is one year, but it could be another length of time.

P = a present sum of money.

F = a future sum of money. The future sum F is an amount that is equivalent to P at interest rate i interest periods from the present n.

A = an end-of-period cash receipt or disbursement in a uniform series continuing for n periods. The entire series is equivalent to P or F at interest rate i.

G = uniform period-by-period increase in cash flows; the uniform gradient.

r = nominal annual interest rate.

From Table 15.1 we see that the functional notation scheme is based on writing (to find/given,i,n). Thus, if we wish to find the future sum F, given a uniform series of receipts A, the proper compound interest factor to use would be denoted $(F/A,i,n)$.

Table 15.1 Periodic compounding: functional notation and formulas

Factor	To Find	Given	Functional Notation	Formula
• ***Single payment***				
Compound Amount	F	P	$(F/P,i,n)$	$F = P(1+i)^n$
Present worth	P	F	$(P/F,i,n)$	$P = F(1+i)^{-n}$
• ***Uniform payment series***				
Sinking fund	A	F	$(A/F,i,n)$	$A = F\left[\frac{i}{(1+i)^n - 1}\right]$
Capital recovery	A	P	$(A/P,i,n)$	$A = P\left[\frac{i(1+i)^n}{(1+i)^n - 1}\right]$
Compound amount	F	A	$(F/A,i,n)$	$F = A\left[\frac{(1+i)^n - 1}{i}\right]$
Present worth	P	A	$(P/A,i,n)$	$P = A\left[\frac{(1+i)^n - 1}{i(1+i)^n}\right]$
• ***Uniform g***				
Gradient present worth	P	G	$(P/G,i,n)$	$P = G\left[\frac{(1+i)^n - 1}{i^2(1+i)^n} - \frac{n}{i(1+i)^n}\right]$
Gradient future worth	F	G	$(F/G,i,n)$	$F = G\left[\frac{(1+i)^n - 1}{i^2} - \frac{n}{i}\right]$
Gradient uniform series	A	G	$(A/G,i,n)$	$A = G\left[\frac{1}{i} - \frac{n}{(1+i)^n - 1}\right]$

Single Payment Formulas

Suppose a present sum of money P is invested for one year at interest rate i. At the end of the year, the initial investment P is received together with interest equal to Pi for a total amount $P + Pi$. Factoring P, the sum at the end of one year is $P(1 + i)$. If the investment is allowed to remain for subsequent years, the progression is as follows:

	Amount at Beginning of Period	+	Interest for the Period	=	Amount at End of the Period
1st year	P	+	Pi	=	$P(1+i)$
2nd year	$P(1+i)$	+	$Pi(1+i)$	=	$P(1+i)^2$
3rd year	$P(1+i)^2$	+	$Pi(1+i)^2$	=	$P(1+i)^3$
nth year	$P(1+i)^{n-1}$	+	$Pi(1+i)^{n-1}$	=	$P(1+i)^n$

The present sum P increases in n periods to $P(1 + i)^n$. This gives a relation between a present sum P and its equivalent future sum F:

$$\text{future sum} = (\text{present sum})(1 + i)^n$$
$$F = P(1 + i)^n$$

This is the **single payment compound amount formula**. In functional notation it is written

$$F = P(F/P,i,n)$$

The relationship may be rewritten as

$$\text{present sum} = (\text{future sum})(1 + i)^{-n}$$
$$P = F(1 + i)^{-n}$$

This is the **single payment present worth formula**. It is written in functional notation as

$$P = F(P/F,i,n)$$

Example 15.3

At a 10% per year interest rate, \$500 now is *equivalent* to how much three years hence?

Solution

This problem was solved in Example 15.2. Now it can be solved using a single payment formula, with P = \$500, n = 3 years, i = 10%, and F = unknown:

$$F = P(1 + i)^n = 500(1 + 0.10)^3 = \$665.50$$

This problem also may be solved using a compound interest table. In functional notation, the problem is

$$F = P(F/P,i,n) = 500(F/P,10\%,3)$$

From the 10% compound interest table, read $(F/P,10\%,3) = 1.331$.

$$F = 500(F/P,10\%,3) = 500(1.331) = \$665.50$$

Example 15.4

To raise money for a new business, a man asks you to lend him some money. He offers to pay you \$3000 at the end of four years. How much should you give him now if you want 12% interest per year?

Solution

Using the single payment present worth formula with P = unknown, F = \$3000, n = 4 years, and i = 12%:

$$P = F(1 + i)^{-n} = 3000(1 + 0.12)^{-4} = \$1906.55$$

The alternate computation using a compound interest table:

$$P = F(P/F,i,n) = 3000(P/F,12\%,4) = 3000(0.6355) = \$1906.50$$

Note that the solution based on the compound interest table is slightly different from the exact solution using a hand-held calculator. In engineering economics the compound interest tables are always considered to be sufficiently accurate.

Uniform Payment Series Formulas

Consider the situation shown in Fig. 15.2. Using the single payment compound amount factor, we can write an equation for F in terms of A:

$$F = A + A(1 + i) + A(1 + i)^2 \qquad \text{(i)}$$

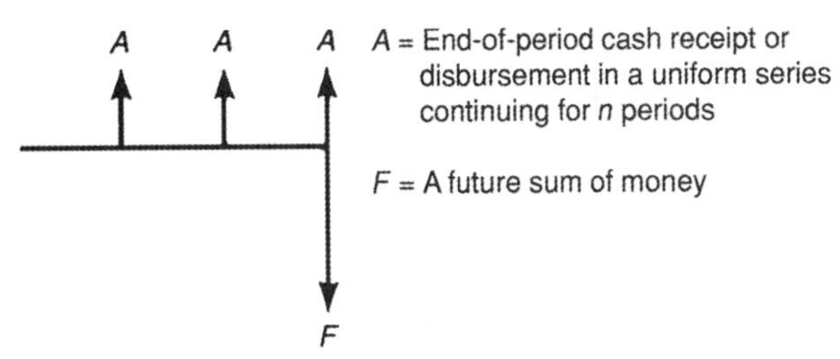

Figure 15.2

In this situation, with n = 3, Eq. (i) may be written in a more general form:

$$F = A + A(1 + i) + A(1 + i)^{n-1} \qquad \text{(ii)}$$

Multiply Eq. (ii) by $(1 + i)$: $(1 + i)F = A(1 + i) + A(1 + i)^{n-1} + A(1 + i)^n$ **(iii)**

Subtract Eq. (ii): $-F = A + A(1 + i) + A(1 + i)^{n-1}$ **(ii)**

(15.iii) – (ii): $iF = -A + A(1 + i)^n$

This produces the **uniform series compound amount formula**:

$$F = A\left(\frac{(1+i)^n - 1}{i}\right)$$

Solving this equation for A produces the **uniform series sinking fund formula**:

$$A = F\left(\frac{i}{(1+i)^n - 1}\right)$$

Since $F = P(1 + i)^n$, we can substitute this expression for F in the equation and obtain the **uniform series capital recovery formula**:

$$A = P\left(\frac{i(1+i)^n}{(1+i)^n - 1}\right)$$

Solving the equation for P produces the **uniform series present worth formula**:

$$P = A\left(\frac{(1+i)^n - 1}{i(1+i)^n}\right)$$

In functional notation, the uniform series factors are

Compound amount: $(F/A, i, n)$

Sinking fund: $(A/F, i, n)$

Capital recovery: $(A/P, i, n)$

Present worth: $(P/A, i, n)$

Example **15.5**

If $100 is deposited at the end of each year in a savings account that pays 6% interest per year, how much money will be in the account at the end of five years?

Solution

$A = \$100$, $F =$ unknown, $n = 5$ years, and $i = 6\%$:

$$F = A(F/A, i, n) = 100(F/A, 6\%, 5) = 100(5.637) = \$563.70$$

Example **15.6**

A fund established to produce a desired amount at the end of a given period, by means of a series of payments throughout the period, is called a **sinking fund**. A sinking fund is to be established to accumulate money to replace a $10,000 machine. If the machine is to be replaced at the end of 12 years, how much should be deposited in the sinking fund each year? Assume the fund earns 10% annual interest.

Solution

$$\begin{aligned}\text{Annual sinking fund deposit } A &= 10{,}000(A/F, 10\%, 12) \\ &= 10{,}000(0.0468) = \$468\end{aligned}$$

Example **15.7**

An individual is considering the purchase of a used automobile. The total price is $6200. With $1240 as a down payment, and the balance paid in 48 equal monthly payments at an interest rate of 1% per month, compute the monthly payment. The payments are due at the end of each month.

Solution

The amount to be repaid by the 48 monthly payments is the cost of the automobile *minus* the \$1240 down payment.

$P = \$4960$, $A =$ unknown, $n = 48$ monthly payments, and $i = 1\%$ per month:

$$A = P(A/P,1\%,48) = 4960(0.0263) = \$130.45$$

Example 15.8

A couple sells their home. In addition to cash, they accept a mortgage that will be paid off by monthly payments of \$450 for 50 months. The couple decides to sell the mortgage to a local bank. The bank will buy the mortgage, but it requires a 1% per month interest rate on their investment. How much will the bank pay for the mortgage?

Solution

$A = \$450$, $n = 50$ months, $i = 1\%$ per month, and $P =$ unknown:

$$P = A(P/A,i,n) = 450(P/A,1\%,50) = 450(39.196) = \$17,638$$

Uniform Gradient

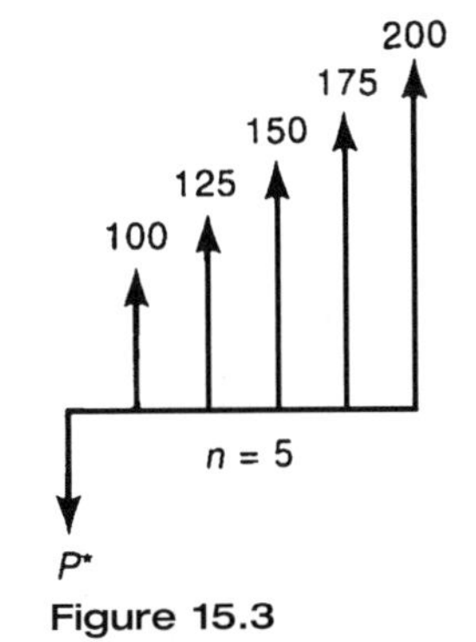

Figure 15.3

In some cases the cash flow series is not a constant amount A. Instead, it is an increasing series. The cash flow shown in Fig. 15.3 may be resolved into two components (Fig. 15.4). The value of P^* is the sum of P' and P. We already have the equation for P': $P' = A(P/A,i,n)$.

The value for the other component of P^*, P, is:

$$P = G\left[\frac{(1+i)^n - 1}{i^2(1+i)^n} - \frac{n}{i(1+i)^n}\right]$$

This is the **uniform gradient present worth formula**. In functional notation, the relationship is $P = G(P/G,i,n)$.

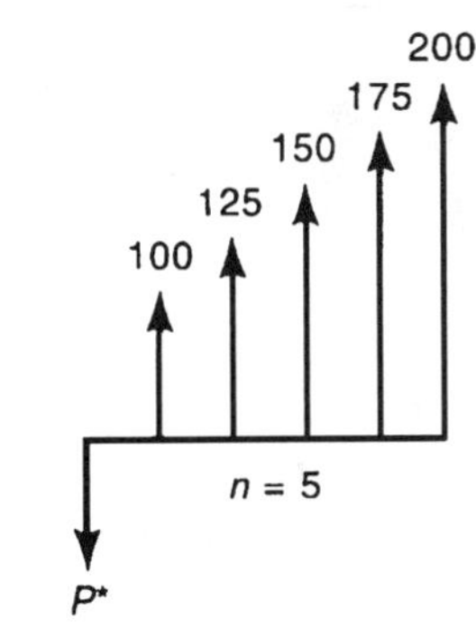

=
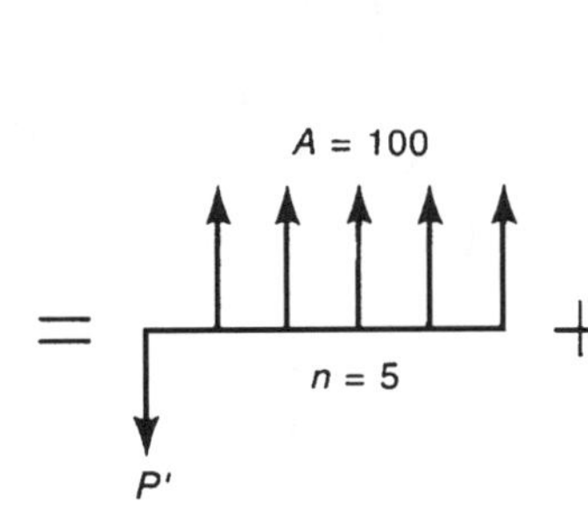

+
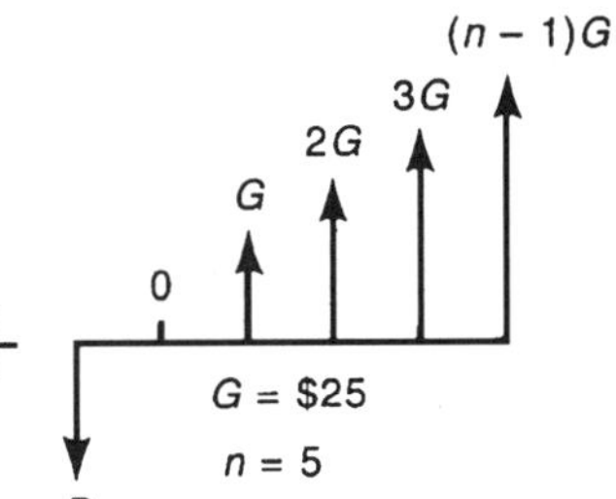

Figure 15.4

Example 15.9

The maintenance on a machine is expected to be \$155 at the end of the first year, and it is expected to increase \$35 each year for the following seven years (Exhibit 1). What sum of money should be set aside now to pay the maintenance for the eight-year period? Assume 6% interest.

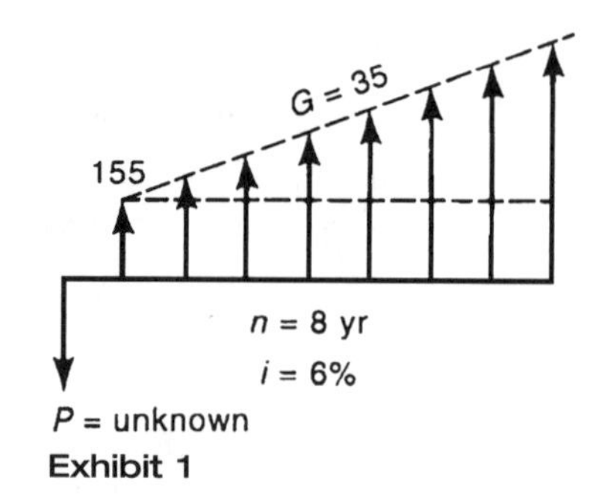

Exhibit 1

Solution

$$P = 155(P/A,6\%,8) + 35(P/G,6\%,8)$$
$$= 155(6.210) + 35(19.841) = \$1656.99$$

In the gradient series, if—instead of the present sum P—an equivalent uniform series A is desired, the problem might appear as shown in Fig. 15.5. The relationship between A' and G in the right-hand diagram is:

$$A' = G\left[\frac{1}{i} - \frac{n}{(1+i)^n - 1)}\right]$$

In functional notation, the uniform gradient (to) uniform series factor is $A' = G(A/G,i,n)$.

The **uniform gradient uniform series factor** may be read from the compound interest tables directly, or computed as

$$(A/G,i,n) = \frac{1 - n(A/F,i,n)}{i}$$

Figure 15.5

Note carefully the diagrams for the uniform gradient factors. The first term in the uniform gradient is zero and the last term is $(n - 1)G$. But we use n in the equations and functional notation. The derivations (not shown here) were done on this basis, and the uniform gradient compound interest tables are computed this way.

Example 15.10

For the situation in Example 15.9, compute the uniform annual maintenance cost. Compute an equivalent A for the maintenance costs.

Solution

Refer to Exhibit 2. The equivalent uniform annual maintenance cost is

$$A = 155 + 35(A/G,6\%,8) = 155 + 35(3.195) = \$266.83$$

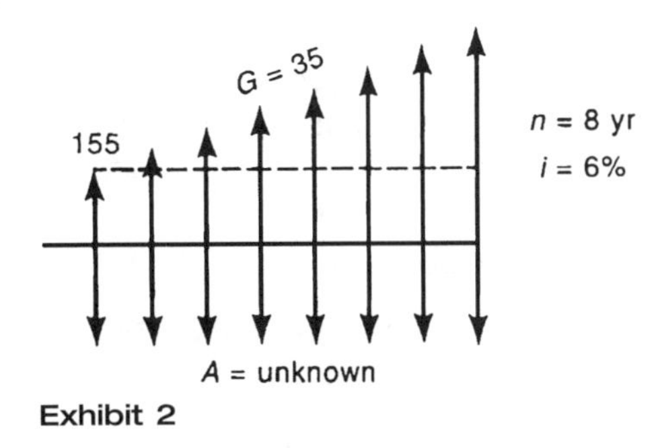

Exhibit 2

Standard compound interest tables give values for eight interest factors: two single payment factors, four uniform-payment series factors, and two uniform gradient factors. The tables do *not* given the uniform gradient future worth factor, $(F/G,i,n)$. If it is needed, it may be computed from tabulated factors:

$$(F/G,i,n) = \frac{(F/A,i,n) - n}{i} = (F/A,i,n) \times (A/G,i,n)$$

For example, if i = 10% and n = 12 years, then

$$\begin{aligned}(F/G,10\%,12) &= (F/A,10\%,12) \times (A/G,10\%,12) \\ &= 21.384 \times 4.388 = 93.833\end{aligned}$$

A second method of computing the uniform gradient future worth factor is

$$(F/G,i,n) = \frac{(F/A,i,n) - n}{i}$$

Other tabulated factors could be similarly computed. For example,

$$(P/G,i,n) = (P/A,i,n) \times (A/G,i,n)$$

Continuous Compounding

Table 15.2 Continuous compounding: functional notation and formulas

Factor	To Find	Given	Functional Notation	Formula
• ***Single payment***				
Compound amount	F	P	$[F/P,r,n]$	$F = P[e^{rn}]$
Present worth	P	F	$[P/F,r,n]$	$P = F[e^{-rn}]$
• ***Uniform payment series***				
Sinking fund	A	F	$[A/F,r,n]$	$A = F\left[\frac{e^r - 1}{e^{rn} - 1}\right]$
Capital recovery	A	P	$[A/P,r,n]$	$A = P\left[\frac{e^r - 1}{1 - e^{-rn}}\right]$
Compound amount	F	A	$[F/A,r,n]$	$F = A\left[\frac{e^{rn} - 1}{e^r - 1}\right]$
Present worth	P	A	$[P/A,r,n]$	$P = A\left[\frac{1 - e^{-rn}}{e^r - 1}\right]$

r = nominal annual interest rate; n = number of years

Example 15.11

Five hundred dollars is deposited each year into a savings bank account that pays 5% nominal interest, compounded continuously. How much will be in the account at the end of five years?

Solution

$A = \$500, \quad r = 0.05, \quad n = 5$ years

$$F = A[F/A, r, n] = A\left[\frac{e^{rn} - 1}{e^{r} - 1}\right] = 500\left[\frac{e^{0.05(5)} - 1}{e^{0.05} - 1}\right] = \$2769.84$$

NOMINAL AND EFFECTIVE INTEREST

Nominal interest is the annual interest rate without considering the effect of any compounding. **Effective interest** is the annual interest rate taking into account the effect of any compounding during the year.

Non-Annual Compounding

Frequently an interest rate is described as an annual rate, even though the interest period may be something other than one year. A bank may pay 1% interest on the amount in a savings account every three months. The *nominal* interest rate in this situation is $4 \times 1\% = 4\%$. But if you deposited \$1000 in such an account, would you have $104\%(1000) = \$1040$ in the account at the end of one year? The answer is no; you would have more. The amount in the account would increase as follows:

Amount in Account

At beginning of year = \$1000.00

End of 3 months: 1000.00 + 1%(1000.00) = 1010.00

End of 6 months: 1010.00 + 1%(1010.00) = 1020.10

End of 9 months: 1020.10 + 1%(1020.10) = 1030.30

End of one year: 1030.30 + 1%(1030.30) = 1040.60

At the end of one year, the interest of \$40.60, divided by the original \$1000, gives a rate of 4.06%. This is the *effective* interest rate.

$$\text{Effective annual interest rate} = i_e = (1 + r/m)^m - 1$$

where

r = nominal annual interest rate
m = number of compound periods per year
r/m = effective interest rate per period

Example 15.12

A bank charges 1.5% interest per month on the unpaid balance for purchases made on its credit card. What nominal interest rate is it charging? What effective interest rate?

Solution

The nominal interest rate is simply the annual interest ignoring compounding, or 12(1.5%) = 18%.

$$\text{Effective interest rate} = (1 + 0.015)^{12} - 1 = 0.1956 = 19.56\%$$

SOLVING ENGINEERING ECONOMICS PROBLEMS

The techniques presented so far illustrate how to convert single amounts of money, and uniform or gradient series of payments, into some equivalent sum of money at another point in time. These compound interest computations are an essential part of engineering economics problems.

The typical engineering economics problem describes a number of alternatives and poses the question, which alternative should we select? The customary method of solution is to express each alternative in some common form and then choose the best, taking both the monetary and intangible factors into account. In most computations an interest rate must be used. It is often called the **minimum attractive rate of return (MARR)** to indicate that this is the smallest interest rate, or rate of return, at which one is willing to invest money.

Criteria

Engineering economics problems inevitably fall into one of three categories:

1. Fixed input. The amount of money or other input resources is fixed. *Example*: A project engineer has a budget of $450,000 to overhaul a plant.
2. Fixed output. There is a fixed task, or other output to be accomplished. *Example*: A mechanical contractor has been awarded a fixed-price contract to air-condition a building.
3. Neither input nor output fixed. This is the general situation, in which neither the amount of money (or other inputs) nor the amount of benefits (or other outputs) is fixed. *Example*: A consulting engineering firm has more work available than it can handle. It is considering paying the staff to work evenings to increase the amount of design work it can perform.

There are five major methods of comparing alternatives: present worth, future worth, annual cost, rate of return, and benefit-cost analysis. These are presented in the sections that follow.

Present Worth

Present-worth analysis converts all of the money consequences of an alternative into an equivalent present sum. The criteria are listed in the following table:

Category	Present-Worth Criterion
Fixed input	Maximize the present worth of benefits or other outputs
Fixed output	Minimize the present worth of costs or other inputs
Neither input nor output fixed	Maximize the present worth of benefits minus present worth of costs, or maximize net present worth

Appropriate Problems

Present-worth analysis is most frequently used to determine the present value of future money receipts and disbursements. We might want to know, for example, the present worth of an income-producing property, such as an oil well. This should provide an estimate of the price at which the property could be bought or sold.

An important restriction in the use of present-worth calculations is that there must be a common analysis period when comparing alternatives. It would be incorrect, for example, to compare the present worth (PW) of the cost of pump *A*, expected to last 6 years, with the PW of the cost of pump *B*, expected to last 12 years (Fig. 15.6). In situations like this, the solution is either to use some other analysis technique (generally the annual cost method is suitable in these situations) or to restructure the problem so that there is a common analysis period.

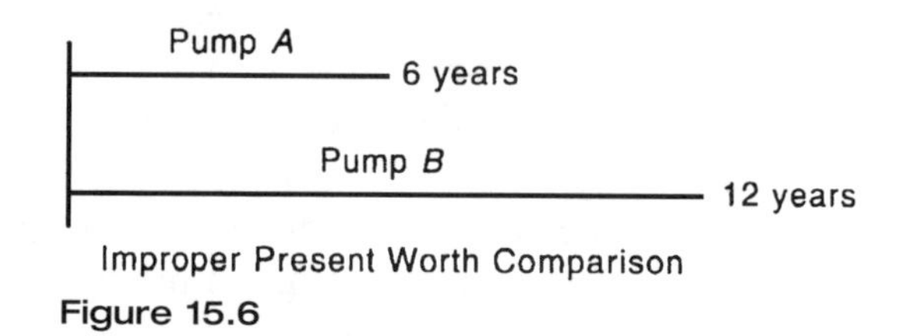

Figure 15.6

In the example above, a customary assumption would be that a pump is needed for 12 years and that pump *A* will be replaced by an identical pump *A* at the end of 6 years. This gives a 12-year common analysis period (Fig. 15.7). This approach is easy to use when the different lives of the alternatives have a practical least common multiple life. When this is not true (for example, life of *J* equals 7 years and the life of *K* equals 11 years), some assumptions must be made to select a suitable common analysis period, or the present-worth method should not be used.

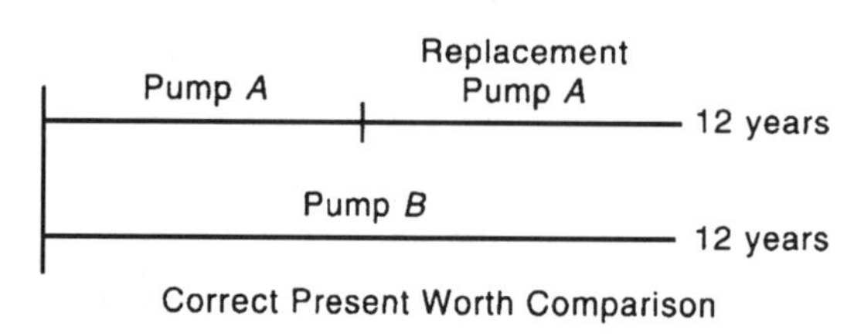

Figure 15.7

Example 15.13

Machine *X* has an initial cost \$10,000, an annual maintenance of \$500 per year, and no salvage value at the end of its 4-year useful life. Machine *Y* costs \$20,000, and the first year there is no maintenance cost. Maintenance is \$100 the second year, and it increases \$100 per year thereafter. Machine *Y* has an anticipated \$5000 salvage value at the end of its 12-year useful life. If the minimum attractive rate of return is 8%, which machine should be selected?

Solution

The analysis period is not stated in the problem. Therefore, we select the least common multiple of the lives, or 12 years, as the analysis period.

Present worth of cost of 12 years of machine *X*

$$= 10{,}000 + 10{,}000(P/F,8\%,4) + 10{,}000(P/F,8\%8) + 500(P/A,8\%,12)$$
$$= 10{,}000 + 10{,}000(0.7350) + 10{,}000(0.5403) + 500(7.536) = \$26{,}521$$

Present worth of cost of 12 years of machine Y

$$= 20{,}000 + 100(P/G,8\%,12) - 5000(P/F,8\%,12)$$
$$= 20{,}000 + 100(34.634) - 5000(0.3971) = \$21{,}478$$

Choose machine Y with its smaller PW of cost.

Example 15.14

Two alternatives have the following cash flows:

	Alternative	
Year	**A**	**B**
0	–$2000	–$2800
1	+800	+1100
2	+800	+1100
3	+800	+1100

At a 4% interest rate, which alternative should be selected?

Solution

Compute the net present worth (NPW) of each alternative.

$$\text{NPW} = \text{PW of benefits} - \text{PW of cost}$$
$$\text{NPW}_A = 800(P/A,4\%,3) - 2000 = 800(2.775) - 2000 = \$220.00$$
$$\text{NPW}_B = 1100(P/A,4\%,3) - 2800 = 1100)(2.775) - 2800 = \$252.50$$

To maximize NPW, choose alternative B.

Infinite Life and Capitalized Cost

In the special situation where the analysis period is infinite ($n = \infty$), an analysis of the present worth of cost is called **capitalized cost**. There are a few public projects where the analysis period is infinity. Other examples are permanent endowments and cemetery perpetual care.

When n equals infinity, a present sum P will accrue interest of Pi for every future interest period. For the principal sum P to continue undiminished (an essential requirement for $n = \infty$), the end-of-period sum A that can be disbursed is Pi (Fig. 15.8).

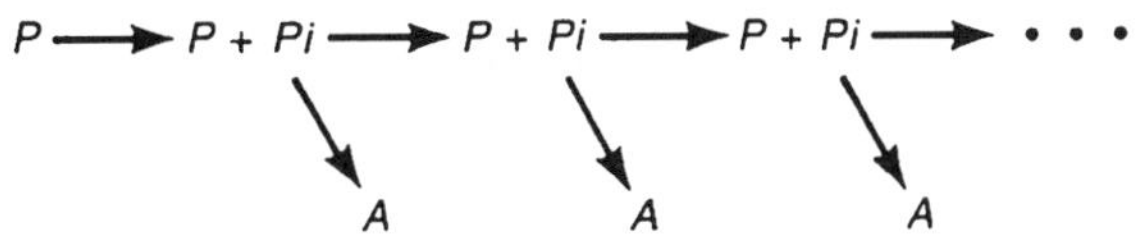

Figure 15.8

When $n = \infty$, the fundamental relationship is

$$A = Pi$$

Some form of this equation is used whenever there is a problem with an infinite analysis period. Thus

$$\text{Capitalized Cost } P = \frac{A}{i}$$

Example 15.15

In his will, a man wishes to establish a perpetual trust to provide for the maintenance of a small local park. If the annual maintenance is $7500 per year and the trust account can earn 5% interest, how much money must be set aside in the trust?

Solution

When $n = \infty$, $A = Pi$ or $P = A/i$. The capitalized cost is $P = A/i = \$7500/.05 = \$150{,}000$.

Future Worth or Value

In present-worth analysis, alternatives are compared in terms of the equivalent present costs and benefits. However, the analysis need not be made in terms of the present—it can be made in terms of a past, present, or future time. Although the numerical calculations may look different, the decision is unaffected by the selected point in time. Often we want to know what the future situation will be if we take some particular couse of action now. An analysis based on some future point in time is called **future-worth analysis**.

Category	Future-Worth Criterion
Fixed input	Maximize the future worth of benefits or other outputs
Fixed output	Minimize the future worth of costs or other inputs
Neither input nor output fixed	maximize future worth of benefits minus future worth of costs, or maximize net future worth

Example 15.16

Two alternatives have the following cash flows:

	Alternative	
Year	A	B
0	–$2000	–$2800
1	+800	+1100
2	+800	+1100
3	+800	+1100

At a 4% interest rate, which alternative should be selected?

Solution

In Example 14, this problem was solved by present-worth analysis at Year 0. Here it will be solved by future-worth analysis at the end of Year 3.

$$\text{NFW} = \text{FW of benefits} - \text{FW of cost}$$

$$\text{NFW}_A = 800(F/A,4\%,3) - 2000(F/P,4\%3) = 800(3.122) - 2000(1.125) = \$247.60$$

$$\text{NFW}_B = 1100(F/A,4\%,3) - 2800(F/P,4\%,3) = 1100(3.122) - 2800(1.125) = \$284.20$$

To maximize NFW, choose alternative B.

Annual Cost

The annual cost method is more accurately described as the method of equivalent uniform annual cost (EUAC). When the computation is of benefits, it is called the method of equivalent uniform annual benefits (EUAB).

Criteria

For each of the three possible categories of problems, there is an annual cost criterion for economic efficiency.

Category	Annual Cost Criterion
Fixed input	Maximize the equivalent uniform annual benefits (EUAB)
Fixed output	Minimize the equivalent uniform annual cost (EUAC)
Neither input nor output fixed	Maximize EUAB − EUAC

Application of Annual Cost Analysis

In the section on present worth, we pointed out that the present worth method requires a common analysis period for all alternatives. This restriction does not apply in all annual cost calculations, but it is important to understand the circumstances that justify comparing alternatives with different service lives.

Frequently, an analysis is done to provide for a more-or-less continuing requirement. For example, one might need to pump water from a well on a continuing basis. Regardless of whether the pump has a useful service life of 6 years of 12 years, we would select the one whose annual cost is a minimum. This would still be the case if the pumps' useful lives were the more troublesome 7 and 11 years. Thus, if we can assume a continuing need for an item, an annual cost comparison among alternatives of differing service lives is valid. This is because the underlying assumption made in these situations is that the shorter-lived alternative can be replaced with an identical item with identical costs when it has reached the end of its useful life. This means the EUAC of the initial alternative is equal to the EUAC for the continuing series of replacements.

On the other hand, if there is a specific requirement to pump water for 10 years, then each pump must be evaluated to see what costs will be incurred during the analysis period and what salvage value, if any, may be recovered at the end of the analysis period. The annual cost comparison needs to consider the actual circumstances of the situation.

Examination problems are often readily solved by the annual cost method. Also, the underlying "continuing requirement" is usually present, so an annual cost comparison of unequal-lived alternatives is an appropriate method of analysis.

Example 15.17

Consider the following alternatives:

	A	B
First cost	$5000	$10,000
Annual maintenance	500	200
End-of-useful-life salvage value	600	1000
Useful life	5 years	15 years

Based on an 8% interest rate, which alternative should be selected?

Solution

Assuming both alternatives perform the same task and there is a continuing requirement, the goal is to minimize EUAC.

Alternative A:

$$\text{EUAC} = 5000(A/P,8\%,5) + 500 - 600(A/F,8\%,5)$$
$$= 5000(0.2505) + 500 - 600(0.1705) = \$1650$$

Alternative B:

$$\text{EUAC} = 10{,}000(A/P,8\%,15) + 200 - 1000(A/F,8\%,15)$$
$$= 10{,}000(0.1168) + 200 - 1000(0.0368) = \$1331$$

To minimize EUAC, select alternative B.

Rate-of-Return Analysis

The rate-of-return analysis method is typically used for problems that can be stated as a cash flow representing the costs and benefits. The rate of return may be defined as the interest rate where PW of cost = PW of benefits, EUAC = EUAB, or PW of cost = PW of benefits. The minimum attractive rate of return (MARR) is the smallest interest rate or rate of return at which one is willing to invest money.

Example 15.18

Compute the rate of return for the investment represented by the following cash flow table.

Year:	0	1	2	3	4	5
Cash Flow:	–$595	+250	+200	+150	+100	+50

Solution

This declining uniform gradient series may be separated into two cash flows for which compound interest factors are available (Exhibit 3).

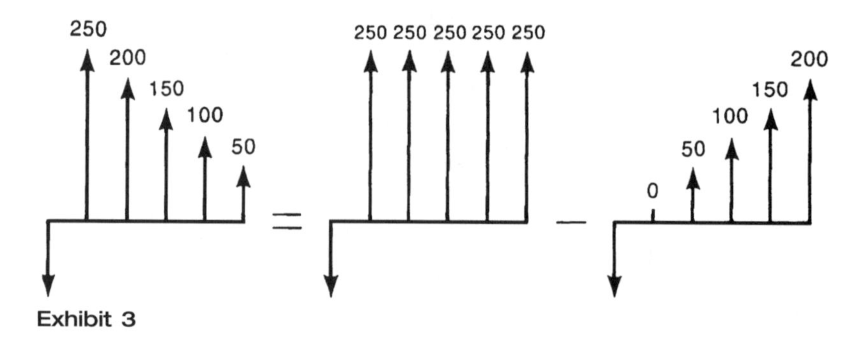

Exhibit 3

Note that the gradient series factors are based on an *increasing* gradient. Here the declining cash flow is resolved by subtracting an increasing uniform gradient, as indicated in Exhibit 3.

$$\text{PW of cost} - \text{PW of benefits} = 0$$
$$595 - [250(P/A,i,5) - 50\,(P/G,i,5)] = 0$$

Try $i = 10\%$:

$$595 - [250(3.791) - 50(6.862)] = -9.65.$$

Try $i = 12\%$:

$$595 - [250(3.605) - 50(6.397)] = +13.60$$

The rate of return is between 10% and 12%. It may be computed more accurately by linear interpolation:

$$\text{Rate of return} = 10\% + (2\%)\left(\frac{9.65 - 0}{13.60 + 9.65}\right) = 10.83\%$$

Two Alternatives

To solve an engineering economics problem in which two alternatives are presented, compute the incremental rate of return on the cash flow representing the difference between the two alternatives. Since we want to look at increments of *investment*, the cash flow for the difference between the alternatives is computed by taking the higher-initial-cost alternative minus the lower-initial-cost alternative. If the incremental rate of return is greater than or equal to the predetermined minimum attractive rate of return, choose the higher-cost alternative; otherwise, choose the lower-cost alternative.

Example **15.19**

Two alternatives have the following cash flows:

	Alternative	
Year	**A**	**B**
0	–$2000	–$2800
1	+800	+1100
2	+800	+1100
3	+880	+1100

If 4% is considered the minimum attractive rate of return, which alternative should be selected?

Solution

These two alternatives were previously examined in Examples 15.14 and 15.16 by present-worth and future-worth analysis. This time, the alternatives will be resolved using a rate-of-return analysis.

Note that the problem statement specifies a 4% minimum attractive rate of return, whereas Examples 15.14 and 15.16 referred to a 4% interest rate. These are really two different ways of saying the same thing: the minimum acceptable time value of money is 4%.

First, tabulate the cash flow that represents the increment of investment between the alternatives. This is done by taking the higher-initial-cost alternative minus the lower-initial-cost alternative:

Year	Alternative A	Alternative B	Difference Between Alternatives B − A
0	−$2000	−$2800	−$800
1	+800	+1100	+300
2	+800	+1100	+300
3	+800	+1100	+300

Then compute the rate of return on the increment of investment represented by the difference between the alternatives:

$$\text{PW of cost} = \text{PW of benefits}$$
$$800 = 300(P/A,i,3)$$
$$(P/A,i,3) = 800/300 = 2.67$$
$$i = 6.1\%$$

Since the incremental rate of return exceeds the 4% MARR, the increment of investment is desirable. Choose the higher-cost alternative B.

Before leaving this example, one should note something that relates to the rates of return on alternative A and on alternative B. These rates of return, if calculated, are:

	Rate of Return
Alternative A	9.7%
Alternative B	8.7%

The correct answer to this problem has been shown to be alternative B, even though alternative A has a higher rate of return. The higher-cost alternative may be thought of as the sum of the lower-cost alternative and the increment of investment between the two alternatives.

The important conclusion is that computing the rate of return for each alternative does *not* provide the basis for choosing between alternatives. Instead, incremental analysis is required.

Example 15.20

Consider the following:

Year	Alternative A	B
0	−$200.0	−$131.0
1	+77.6	+48.1
2	+77.6	+48.1
3	+77.6	+48.1

If the minimum attractive rate of return is 10%, which alternative should be selected?

Solution

To examine the increment of investment between the alternatives, we will examine the higher-initial-cost alternative minus the lower-initial-cost alternative, or A – B.

	Alternative		**Increment**
Year	**A**	**B**	**A – B**
0	–$200.0	–$131.0	–$69.0
1	+77.6	+48.1	+29.5
2	+77.6	+48.1	+29.5
3	+77.6	+48.1	+29.5

Solve for the incremental rate of return:

$$\text{PW of cost} = \text{PW of benefits}$$
$$69.0 = 29.5(P/A,i,3)$$
$$(P/A,i,3) = 69.0/29.5 = 2.339$$

From compound interest tables, the incremental rate of return is between 12% and 18%. This is a desirable increment of investment, hence we select the higher-initial-cost alternative A.

Benefit-Cost Analysis

Generally, in public works and governmental economic analyses, the dominant method of analysis is the **benefit-cost ratio (B/C)**. It is simply the value of benefits divided by costs, taking into account the time value of money.

$$\text{B/C} = \frac{\text{PW of benefits}}{\text{PW of cost}} = \frac{\text{equivalent uniform annual benefits}}{\text{equivalent uniform annual cost}}$$

For a given interest rate, a B/C ratio ≥ 1 reflects an acceptable project. The B/C analysis method is parallel to that of rate-of-return analysis. The same kind of incremental analysis is required.

Example 15.21

Solve Example 15.20 by benefit-cost analysis.

Solution

	Alternative		**Increment**
Year	**A**	**B**	**A – B**
0	–$200.0	–$131.0	–$69.0
1	+77.6	+48.1	+29.5
2	+77.6	+48.1	+29.5
3	+77.6	+48.1	+29.5

The benefit-cost ratio for the A – B increment is

$$\text{B/C} = \frac{\text{PW of benefits}}{\text{PW of cost}} = \frac{29.5(P/A,10\%,3)}{69.0} = \frac{73.37}{69.0} = 1.06$$

Since the B/C ratio exceeds 1, the increment of investment is desirable. Select the higher-cost alternative A.

Break-Even Analysis

In business, "break-even" is often defined as the percentage of capacity of a manufacturing operation for which income just covers costs. In engineering economics, the break-even point is defined as the point at which two alternatives are equivalent.

Example 15.22

A city is considering a new \$50,000 snowplow. The new machine will operate at a savings of \$600 per day compared to the present equipment. Assume the minimum attractive rate of return is 12%, and the machine's life is 10 years with zero resale value at that time. How many days per year must the machine be used to justify the investment?

Solution

This break-even problem may be readily solved by annual cost computations. We will set the equivalent uniform annual cost (EUAC) of the snowplow equal to its annual benefit and solve for the required annual utilization. Let X = break-even point = days of operation per year.

$$\text{EUAC} = \text{EUAB}$$
$$50{,}000(A/P,12\%,10) = 600X$$
$$X = 50{,}000(0.1770)/600 = 14.8 \text{ days/year}$$

BONDS

A **bond** is a form of debt represented by a certificate. The bond will specify the amount of the debt, the interest rate of the bond, how often the interest is paid, and when the debt will be repaid.

Bond Value

Bond value is the present worth of all the future interest payments plus the future repayment of the debt, computed at some selected interest rate.

Example 15.23

A \$5000 bond is being offered for sale. It has a stated interest rate of 7%, paid annually. At the end of 8 years the \$5000 debt will be repaid along with the last interest payment. It you want an 8% rate of return on this investment (bond yield), how much would you be willing to pay for the bond (bond value)?

Solution

The bond pays 7% × \$5000 = \$350 at the end of every year and will repay the \$5000 debt at the end of 8 years.

$$\begin{aligned}\text{Bond value} &= \text{PW of all future benefits}\\ &= 350(P/A,8\%,8) + 5000(P/F,8\%,8)\\ &= 350(5.747) + 5000(0.5403) = \$4712.95\end{aligned}$$

Bond Yield

Bond yield is the interest rate at which the benefits of owning the bond are equivalent to the cost of the bond.

Example 15.24

If the bond in Example 15.23 can actually be purchased for $4200, what is the bond yield?

Solution

Set the cost of the bond equal to the PW of the bond benefits and solve for the unknown interest rate. The resulting i^* is the bond yield.

$$\$4200 = 350(P/A,i,8) + 5000(P/F,i,8)$$

The equation must be solved by trial and error. Try $i = 10\%$.

$$\$4200 \stackrel{?}{=} 350(5.335) + 5000(0.467) = 4202.25$$

We see that i^* is very close to 10%. No further computations are required. The bond yield is very close to 10%.

PAYBACK PERIOD

Payback period is the period of time required for the profit or other benefits of an investment to equal the cost of the investment.

Example 15.25

A project has the following costs and benefits.

Year	Costs	Benefits
0	$1400	
1	500	$200
2	300	100
3–10		400/year

What is the payback period?

Solution

The total cost is $2200. At the end of Year 6 the total benefits will be 200 + 100 + 400 + 400 + 400 + 400 = $1900. And at the end of Year 7 benefits will be 1900 + 400 = $2300. The payback period is where benefits equal cost. Since a cash flow table is normally based on the end-of-year convention (See Example 15.1), the payback period is at the end of Year 7 and the answer is 7 years.

If, on the other hand, the problem had been stated in words something like ". and the benefits from the third year on are $400/year," I would assume the benefits occur uniformly throughout the year. In this situation, the correct answer would be 6.75 years.

VALUATION AND DEPRECIATION

Depreciation of capital equipment is an important component of many after-tax economic analyses. For this reason, one must understand the fundamentals of depreciation accounting.

Notation

BV = book value

C = cost of the property (basis)

D_j = depreciation in year j

S_n = salvage value in year n

Depreciation is the systematic allocation of the cost of a capital asset over its useful life. **Book value** is the original cost of an asset minus the accumulated depreciation of the asset.

$$\text{Book Value (BV)} = C - \Sigma(D_j)$$

In computing a schedule of depreciation charges four items are considered:

1. Cost of the property, C (called the *basis* in tax law).
2. Type of property. Property is classified either as **tangible** (such as machinery) or **intangible** (such as a franchise or a copyright) and as either **real property** (real estate) or **personal property** (everything not real property).
3. Depreciable life in years, n.
4. Salvage value of the property at the end of its depreciable (usable) life, S_n.

Straight-Line Depreciation

Depreciation charge in any year is given by

$$D_j = \frac{C - S_n}{n}$$

An alternate computation of the depreciation charge in year j is

$$D_j = \frac{\text{C} - \text{depreciation taken to beginning of year } j - S_n}{\text{Remaining useful life at beginning of year } j}$$

Double-Declining-Balance Depreciation

DDB depreciation in any year, $D_j = \frac{2}{n}$ (C – depreciation in years prior to j)
For 150% declining balance depreciation, replace the 2 in the equation with 1.5.

Modified Accelerated Cost Recovery System Depreciation

The modified accelerated cost recovery system (MACRS) depreciation method generally applies to property placed in service after 1986. To compute the MACRS depreciation for an item one must know the following:

1. Cost (basis) of the item.
2. Property class. All tangible property is classified in one of six classes (3, 5, 7, 10, 15, and 20 years), based on the life over which it is depreciated (see Table 15.3). Residential real estate and nonresidential real estate are in two separate real property classes of 27.5 years and 39 years, respectively.
3. Depreciation computation.
   - 3, 5, 7, and 10-year property classes use double-declining-balance depreciation with conversion to straight-line depreciation in the year that increases the deduction
   - 15 and 20-year property classes use 150%-declining-balance depreciation with conversion to straight-line depreciation in the year that increases the deduction
   - in MACRS the salvage value is assumed to be zero

Half-Year Convention

Except for real property, a half-year convention is used. Under this convention all property is considered to be placed in service in the middle of the tax year, and a half year of depreciation is allowed in the first year. For each of the remaining years, one is allowed a full year of depreciation. If the property is disposed of prior to the end of the recovery period (property class life), a half year of depreciation is allowed in that year. If the property is held for the entire recovery period, a half year of depreciation is allowed for the year following the end of the recovery period (see Table 15.3).

Table 15.3 Modified ACRS (MACRS) depreciation for personal property—half-year convention

	Applicable Percentage for the Class of Property			
Recovery Year	**3-year Recovery**	**5-year Recovery**	**7-year Recovery**	**10-year Recovery**
1	33.33	20.00	14.29	10.00
2	44.45	32.00	24.49	18.00
3	14.81†	19.20	17.49	14.40
4	7.41	11.52†	12.49	11.52
5		11.52	8.93†	9.22
6		5.76	8.92	7.37
7			8.93	6.55†
8			4.46	6.55
9				6.56
10				6.55
11				3.28

†Use straight-line depreciation for the year marked and all subsequent years.

Example 15.26

A $5000 computer has an anticipated $500 salvage value at the end of its five-year depreciable life. Compute the depreciation schedule by MACRS depreciation. Do the MACRS computation by hand, and then compare the results with the values from Table 15.3.

Solution

The depreciation method is double declining balance with conversion to straight line for the computer's five-year property class and the half-year convention is used. Salvage value S_n is assumed to be zero for MACRS. Using the equation for DDB depreciation in any year:

Year

$$1\left(\frac{1}{2}\text{year}\right)\quad D_1=\frac{1}{2}\times\frac{2}{5}(5000-0)=\$1000$$

$$2\quad D_2=\frac{2}{5}(5000-1000)=1600$$

$$3\quad D_3=\frac{2}{5}(5000-2600)=960$$

$$4\quad D_4=\frac{2}{5}(5000-3560)=346$$

$$5\quad D_5=\frac{2}{5}(5000-4136)=346$$

$$6\left(\frac{1}{2}\text{year}\right)\quad D_6=\frac{1}{2}\times\frac{2}{5}(5000-4482)=104$$

$$\$4586$$

The computation must now be modified to convert to straight-line depreciation at the point where the straight-line depreciation will be larger. Using the alternate straight-line computation,

$$D_5=\frac{5000-4136-0}{1.5\text{ years remaining}}=\$576$$

This is more than the \$346 computed using DDB, hence switch to the straight-line method for year 5 and beyond.

$$D_6\left(\frac{1}{2}\text{year}\right)=\frac{1}{2}(576)=\$288$$

Answers:

Year	Depreciation (MACRS)
1	\$1000
2	1600
3	960
4	576
5	576
6	288
	\$5000

The computed MACRS depreciation is identical with that obtained from Table 15.3.

INFLATION

Inflation is characterized by rising prices for goods and services, while deflation produces a decrease in prices. An inflationary trend makes future dollars have less purchasing power than present dollars. This benefits long-term borrowers of money because they may repay a loan of present dollars in the future with dollars of reduced buying power. The help to borrowers is at the expense of lenders. Deflation has the opposite effect. Money borrowed at a point in time followed by a deflationary period subjects the borrower to loan repayment with dollars of greater purchasing power than those he borrowed. This is to the lenders' advantage at the expense of borrowers.

Price changes occur in a variety of ways. One method of stating a price change is a uniform rate of price change per year.

$$f = \text{general inflation rate per interest period}$$

The following example problem will illustrate the computations.

Example 15.27

A mortgage will be repaid in three equal payments of $5000 at the end of years 1, 2, and 3. If the annual inflation rate, f, is 8% during this period, and the investor wishes a 12% annual interest rate (i), what is the maximum amount he would be willing to pay for the mortgage?

Solution

The computation is a two-step process. First, the three future payments must be converted into dollars with the same purchasing power as today's (Year 0) dollars.

Year	Actual Cash Flow		Multiplied by		Cash Flow Adjusted to Today's (Year 0) Dollars
0	—		—		—
1	+5000	×	$(1 + 0.08)^{-1}$	=	+4630
2	+5000	×	$(1 + 0.08)^{-2}$	=	+4286
3	+5000	×	$(1 + 0.08)^{-3}$	=	+3969

The general form of the adjusting multiplier is

$$(1 + f)^{-n} \quad \text{or} \quad (P/F, f, n)$$

Now that the problem has been converted to dollars of the same purchasing power (today's dollars in this example), we can proceed to compute the present worth of the future payments.

Year	Adjusted Cash Flow		Multiplied by		Present Worth
0	—		—		—
1	+4630	×	$(1 + 0.12)^{-1}$	=	+4134
2	+4286	×	$(1 + 0.12)^{-2}$	=	+3417
3	+3969	×	$(1 + 0.12)^{-3}$	=	+2825
					$10,376

The investor would pay $10,376.

Alternate Solution

Instead of doing the inflation and interest rate computations separately, one can compute a combined equivalent interest rate per interest period, d.

$$d = (1 + f)(1 + i) - 1 = i + f + (i \times f)$$

For this cash flow, $d = 0.12 + 0.08 + 0.12(0.08) = 0.2096$. Since we do not have 20.96% interest tables, the problem must be calculated using present-worth equations.

$$\begin{aligned} \text{PW} &= 5000(1 + 0.2096)^{-1} + 5000(1 + 0.2096)^{-2} + 5000(1 + 0.2096)^{-3} \\ &= 4134 + 3417 + 2825 = \$10{,}376 \end{aligned}$$

Example 15.28

One economist has predicted that there will be a 7% per year inflation of prices during the next 10 years. If this proves to be correct, an item that presently sells for $10 would sell for what price 10 years hence?

Solution

$f = 7\%, \quad P = \$10$

$F = ?, \quad n = 10 \text{ years}$

Here the computation is to find the future worth F, rather than the present worth, P.

$$F = P(1 + f)^{10} = 10(1 + 0.07)^{10} = \$19.67$$

Effect of Inflation on a Rate of Return

The effect of inflation on the computed rate of return for an investment depends on how future benefits respond to the inflation. If benefits produce constant dollars, which are not increased by inflation, the effect of inflation is to reduce the before-tax rate of return on the investment. If, on the other hand, the dollar benefits increase to keep up with the inflation, the before-tax rate of return will not be adversely affected by the inflation. This is not true when an after-tax analysis is made. Even if the future benefits increase to match the inflation rate, the allowable depreciation schedule does not increase. The result will be increased taxable income and income tax payments. This reduces the available after-tax benefits and, therefore, the after-tax rate of return.

Example 15.29

A man bought a 5% tax-free municipal bond. It cost $1000 and will pay $50 interest each year for 20 years. The bond will mature at the end of 20 years and return the original $1000. If there is 2% annual inflation during this period, what rate of return will the investor receive after considering the effect of inflation?

Solution

$$d = 0.05, \quad i = \text{unknown}, \quad f = 0.02$$

Combined effective interest rate/interest period,

$$d = i + f + (i \times f)$$
$$0.05 = i + 0.02 + 0.02i$$
$$1.02i = 0.03, \quad i = 0.294 = 2.94\%$$

REFERENCE

Newnan, D. G. et al., *Engineering Economic Analysis*, 6th ed. Engineering Press, San Jose, CA, 2000.

PROBLEMS

15.1 A retirement fund earns 8% interest, compounded quarterly. If $400 is deposited every three months for 25 years, the amount in the fund at the end of 25 years is nearest to

a. $50,000
b. $75,000
c. $100,000
d. $125,000

15.2 The repair costs for some handheld equipment are estimated to be $120 the first year, increasing by $30 per year in subsequent years. The amount a person needs to deposit into a bank account paying 4% interest to provide for the repair costs for the next five years is nearest to

a. $500
b. $600
c. $700
d. $800

15.3 One thousand dollars is borrowed for one year at an interest rate of 1% per month. If this same sum of money were borrowed for the same period at an interest rate of 12% per year, the saving in interest charges would be closest to

a. $0
b. $3
c. $5
d. $7

15.4 How much should a person invest in a fund that will pay 9%, compounded continuously, if he wishes to have $10,000 in the fund at the end of 10 years?

a. $4000
b. $5000
c. $6000
d. $7000

15.5 A store charges 1.5% interest per month on credit purchases. This is equivalent to a nominal annual interest rate of

a. 1.5%
b. 15.0%
c. 18.0%
d. 19.6%

15.6 A small company borrowed $10,000 to expand its business. The entire principal of $10,000 will be repaid in two years, but quarterly interest of $330 must be paid every three months. The nominal annual interest rate the company is paying is closest to

a. 3.3%
b. 5.0%
c. 6.6%
d. 13.2%

15.7 A store's policy is to charge 3% interest every two months on the unpaid balance in charge accounts. The effective interest rate is closest to

a. 6%
b. 12%
c. 15%
d. 19%

15.8 The effective interest rate on a loan is 19.56%. If there are 12 compounding periods per year, the nominal interest rate is closest to

a. 1.5%
b. 4.5%
c. 9.0%
d. 18.0%

15.9 A deposit of $300 was made one year ago into an account paying monthly interest. If the account now has $320.52, the effective annual interest rate is closest to

a. 7% c. 12%
b. 10% d. 15%

15.10 If the effective interest rate per year is 12%, based on monthly compounding, the nominal interest rate per year is closest to

a. 8.5% c. 10.0%
b. 9.3% d. 11.4%

15.11 If 10% nominal annual interest is compounded daily, the effective annual interest rate is nearest to

a. 10.00% c. 10.50%
b. 10.38% d. 10.75%

15.12 An individual wishes to deposit a certain quantity of money now so that he will have $500 at the end of five years. With interest at 4% per year, compounded semiannually, the amount of the deposit is nearest to

a. $340 c. $410
b. $400 d. $416

15.13 A steam boiler is purchased on the basis of guaranteed performance. A test indicates that the operating cost will be $300 more per year than the manufacturer guaranteed. If the expected life of the boiler is 20 years, and the time value of money is 8%, the amount the purchaser should deduct from the purchase price to compensate for the extra operating cost is nearest to

a. $2950 c. $4100
b. $3320 d. $5520

15.14 A consulting engineer bought a fax machine with one year's free maintenance. In the second year the maintenance cost is estimated at $20. In subsequent years the maintenance cost will increase $20 per year (that is, third year maintenance will be $40, fourth year maintenance will be $60, and so forth). The amount that must be set aside now at 6% interest to pay the maintenance costs on the fax machine for the first six years of ownership is nearest to

a. $101 c. $229
b. $164 d. $284

15.15 An investor is considering buying a 20-year corporate bond. The bond has a face value of $1000 and pays 6% interest per year in two semiannual payments. Thus the purchaser of the bond will receive $30 every six months, and in addition he will receive $1000 at the end of 20 years, along with the last $30 interest payment. If the investor believes he should receive 8% annual interest, compounded semiannually, the amount he is willing to pay for the bond (bond value) is closest to

a. $500 c. $700
b. $600 d. $800

15.16 Annual maintenance costs for a particular section of highway pavement are \$2000. The placement of a new surface would reduce the annual maintenance cost to \$500 per year for the first five years and to \$1000 per year for the next five years. The annual maintenance after 10 years would again be \$2000. If maintenance costs are the only saving, the maximum investment that can be justified for the new surface, with interest at 4%, is closest to

a. \$5500 c. \$10,000
b. \$7170 d. \$10,340

15.17 A project has an initial cost of \$10,000, uniform annual benefits of \$2400, and a salvage value of \$3000 at the end of its 10-year useful life. At 12% interest the net present worth (NPW) of the project is closest to

a. \$2500 c. \$4500
b. \$3500 d. \$5500

15.18 A person borrows \$5000 at an interest rate of 18%, compounded monthly. Monthly payments of \$167.10 are agreed upon. The length of the loan is closest to

a. 12 months c. 24 months
b. 20 months d. 40 months

15.19 A machine costing \$2000 to buy and \$300 per year to operate will save labor expenses of \$650 per year for eight years. The machine will be purchased if its salvage value at the end of eight years is sufficiently large to make the investment economically attractive. If an interest rate of 10% is used, the minimum salvage value must be closest to

a. \$100 c. \$300
b. \$200 d. \$400

15.20 The amount of money deposited 50 years ago at 8% interest that would now provide a perpetual payment of \$10,000 per year is nearest to

a. \$3000 c. \$50,000
b. \$8000 d. \$70,000

15.21 An industrial firm must pay a local jurisdiction the cost to expand its sewage treatment plant. In addition, the firm must pay \$12,000 annually toward the plant operating costs. The industrial firm will pay sufficient money into a fund that earns 5% per year to pay its share of the plant operating costs forever. The amount to be paid to the fund is nearest to

a. \$15,000 c. \$60,000
b. \$30,000 d. \$240,000

15.22 At an interest rate of 2% per month, money will double in value in how many months?

a. 20 months c. 24 months
b. 22 months d. 35 months

15.23 A woman deposited \$10,000 into an account at her credit union. The money was left on deposit for 80 months. During the first 50 months the woman earned 12% interest, compounded monthly. The credit union then changed its interest policy so that the woman earned 8% interest compounded quarterly during the next 30 months. The amount of money in the account at the end of 80 months is nearest to

a. \$10,000 c. \$15,000
b. \$12,500 d. \$20,000

15.24 An engineer deposited \$200 quarterly in her savings account for three years at 6% interest, compounded quarterly. Then for five years she made no deposits or withdrawals. The amount in the account after eight years is closest to

a. \$1200 c. \$2400
b. \$1800 d. \$3600

15.25 A sum of money, Q, will be received six years from now. At 6% annual interest the present worth of Q is \$60. At this same interest rate the value of Q 10 years from now is closest to

a. \$60 c. \$90
b. \$77 d. \$107

15.26 If \$200 is deposited in a savings account at the beginning of each year for 15 years and the account earns interest at 6%, compounded annually, the value of the account at the end of 15 years will be most nearly

a. \$4500 c. \$4900
b. \$4700 d. \$5100

15.27 The maintenance expense on a piece of machinery is estimated as follows:

Year	1	2	3	4
Maintenance	\$150	\$300	\$450	\$600

If interest is 8%, the equivalent uniform annual maintenance cost is closest to

a. \$250 c. \$350
b. \$300 d. \$400

15.28 A payment of \$12,000 six years from now is equivalent, at 10% interest, to an annual payment for eight years starting at the end of this year. The annual payment is closest to

a. \$1000 c. \$1400
b. \$1200 d. \$1600

15.29 A manufacturer purchased \$15,000 worth of equipment with a useful life of six years and a \$2000 salvage value at the end of the six years. Assuming a 12% interest rate, the equivalent uniform annual cost (EUAC) is nearest to

a. \$1500 c. \$3500
b. \$2500 d. \$4500

15.30 Consider a machine as follows:

Initial cost: $80,000

End-of-useful-life salvage value: $20,000

Annual operating cost: $18,000

Useful life: 20 years

Based on 10% interest, the equivalent uniform annual cost for the machine is closest to

a. $21,000 c. $25,000
b. $23,000 d. $27,000

15.31 Consider a machine as follows:

Initial cost: $80,000

Annual operating cost: $18,000

Useful life: 20 years

What must the salvage value of the machine at the end of 20 years be for the machine to have an equivalent uniform annual cost of $27,000? Assume a 10% interest rate. The salvage value S_{20} is closest to

a. $10,000 c. $30,000
b. $20,000 d. $40,000

15.32 Twenty-five thousand dollars is deposited in a savings account that pays 5% interest, compounded semiannually. Equal annual withdrawals are to be made from the account beginning one year from now and continuing forever. The maximum amount of the equal annual withdrawals is closest to

a. $625 c. $1250
b. $1000 d. $1265

15.33 An investor is considering the investment of $10,000 in a piece of land. The property taxes are $100 per year. The lowest selling price the investor must receive if she wishes to earn a 10% interest rate after keeping the land for 10 years is

a. $20,000 c. $23,000
b. $21,000 d. $27,000

15.34 The rate of return for a $10,000 investment that will yield $1000 per year for 20 years is closest to

a. 1% c. 8%
b. 4% d. 12%

15.35 An engineer invested $10,000 in a company. In return he received $600 per year for six years and his $10,000 investment back at the end of the six years. His rate of return on the investment was closes to

a. 6% c. 12%
b. 10% d. 15%

15.36 An engineer made 10 annual end-of-year purchases of \$1000 of common stock. At the end of the tenth year, just after the last purchase, the engineer sold all the stock for \$12,000. The rate of return received on the investment is closest to

a. 2% c. 8%
b. 4% d. 10%

15.37 A company is considering buying a new piece of machinery.

Initial cost: \$80,000

End-of-useful-life salvage value: \$20,000

Annual operating cost: \$18,000

Useful life: 20 years

The machine will produce an annual saving in material of \$25,700. What is the before-tax rate of return if the machine is installed? The rate of return is closest to

a. 6% c. 10%
b. 8% d. 15%

15.38 Consider the following situation: Invest \$100 now and receive two payments of \$102.15—one at the end of year 3, and one at the end of year 6. The rate of return is nearest to

a. 6% c. 10%
b. 8% d. 18%

15.39 Two mutually exclusive alternatives are being considered:

Year	A	B
0	–\$2500	–\$6000
1	+746	+1664
2	+746	+1664
3	+746	+1664
4	+746	+1664
5	+746	+1664

The rate of return on the difference between the alternatives is closest to

a. 6% c. 10%
b. 8% d. 12%

15.40 A project will cost \$50,000. The benefits at the end of the first year are estimated to be \$10,000, increasing \$1000 per year in subsequent years. Assuming a 12% interest rate, no salvage value, and an eight-year analysis period, the benefit-cost ratio is closest to

a. 0.78 c. 1.28
b. 1.00 d. 1.45

15.41 Two alternatives are being considered.

	A	B
Initial cost:	$500	$800
Uniform annual benefit:	$140	$200
Useful life, years:	8	8

The benefit-cost ratio of the difference between the alternatives, based on a 12% interest rate, is closest to

a. 0.60
b. 0.80
c. 1.00
d. 1.20

15.42 An engineer will invest in a mining project if the benefit-cost ratio is greater than one, based on an 18% interest rate. The project cost is $57,000. The net annual return is estimated at $14,000 for each of the next eight years. At the end of eight years the mining project will be worthless. The benefit-cost ratio is closest to

a. 1.00
b. 1.05
c. 1.21
d. 1.57

15.43 A city has retained your firm to do a benefit-cost analysis of the following project:

Project cost: $60,000,000

Gross income: $20,000,000 per year

Operating costs: $5,500,000 per year

Salvage value after 10 years: None

The project life is ten years. Use 8% interest in the analysis. The computed benefit-cost ratio is closest to

a. 0.80
b. 1.00
c. 1.20
d. 1.60

15.44 A piece of property is purchased for $10,000 and yields a $1000 yearly profit. If the property is sold after five years, the minimum price to break even, with interest at 6%, is closest to

a. $5000
b. $6500
c. $7700
d. $8300

15.45 Given two machines:

	A	B
Initial cost	$55,000	$75,000
Total annual costs	$16,200	$12,450

With interest at 10% per year, at what service life do these two machines have the same equivalent uniform annual cost? The service life is closest to

a. 4 years c. 6 years
b. 5 years d. 8 years

15.46 A machine part that is operating in a corrosive atmosphere is made of low-carbon steel. It costs $350 installed, and lasts six years. If the part is treated for corrosion resistance it will cost $700 installed. How long must the treated part last to be as economic as the untreated part, if money is worth 6%?

a. 8 years c. 15 years
b. 11 years d. 17 years

15.47 A firm has determined that the two best paints for its machinery are Tuff-Coat at $45 per gallon and Quick at $22 per gallon. The Quick paint is expected to prevent rust for five years. Both paints take $40 of labor per gallon to apply, and both cover the same area. If a 12% interest rate is used, how long must the Tuff-Coat paint prevent rust to justify its use?

a. 5 years c. 7 years
b. 6 years d. 8 years

15.48 Two alternatives are being considered:

	A	B
Cost:	$1000	$2000
Useful life in years:	10	10
End-of-useful-life salvage value:	$100	$400

The net annual benefit of alternative A is $150. If interest is 8%, what must be the net annual benefit of alternative B for the two alternatives to be equally desirable?

a. $150 c. $225
b. $200 d. $275

15.49 A $5000 municipal bond is offered for sale. It will provide 8% annual interest by paying $200 to the bond holder every six months. At the end of 10 years, the $5000 will be paid to the bond holder along with the final $200 interest payment. If you consider 12% nominal annual interest, compounded semiannually, an appropriate bond yield, the amount you would be willing to pay for the bond is closest to

a. $2750 c. $5000
b. $3850 d. $7400

15.50 A municipal bond is being offered for sale for $10,000. It is a zero-coupon bond, that is, the bond pays no interest during its 15-year life. At the end of 15 years the owner of the bond will receive a single payment of $26,639. The bond yield is closest to

a. 4% c. 6%
b. 5% d. 7%

15.51 A firm is considering purchasing $8000 of small hand tools for use on a production line. It is estimated that the tools will reduce the amount of required overtime work by $2000 the first year, with this amount increasing by $1000 per year thereafter. The payback period for the hand tools is closest to

a. 2.00 years
b. 2.50 years
c. 2.75 years
d. 3.00 years

15.52 Special tools for the manufacture of finished plastic products cost $15,000 and have an estimated $1000 salvage value at the end of an estimated three-year useful life and recovery period. The third-year straight-line depreciation is closest to

a. $3000
b. $3500
c. $4000
d. $4500

15.53 Refer to the facts of Problem 15-52. The first-year MACRS depreciation is closest to

a. $3000
b. $3500
c. $4000
d. $5000

15.54 An engineer is considering the purchase of an annuity that will pay $1000 per year for 10 years. The engineer feels he should obtain a 5% rate of return on the annuity after considering the effect of an estimated 6% inflation per year. The amount he would be willing to pay to purchase the annuity is closest to

a. $1500
b. $3000
c. $4500
d. $6000

15.55 An automobile costs $20,000 today. You can earn 12% tax-free on an "auto purchase account." If you expect the cost of the auto to increase by 10% per year, the amount you would need to deposit in the account to provide for the purchase of the auto five years from now is closest to

a. $12,000
b. $14,000
c. $16,000
d. $18,000

15.56 An engineer purchases a building lot for $40,000 cash and plans to sell it after five years. If he wants an 18% before-tax rate of return, after taking the 6% annual inflation rate into account, the selling price must be nearest to

a. $55,000
b. $65,000
c. $75,000
d. $125,000

15.57 A piece of equipment with a list price of $450 can actually be purchased for either $400 cash or $50 immediately plus four additional annual payments of $115.25. All values are in dollars of current purchasing power. If the typical customer considered a 5% interest rate appropriate, the inflation rate at which the two purchase alternatives are equivalent is nearest to

a. 5%
b. 6%
c. 8%
d. 10%

SOLUTIONS

15.1 **d.**

$$F = A(F/A,i,n) = 400(F/A,2\%,100)$$
$$= 400(312.23) = \$124,890$$

15.2 **d.**

$$P = A(P/A,i,n) + G(P/G,i,n)$$
$$= 120(P/A,4\%,5) + 30(P/G,4\%,5)$$
$$= 120(4.452) + 30(8.555) = \$791$$

15.3 **d.**

At i = 1%/month: $F = 1000(1 + 0.01)^{12} = \$1126.83$

At i = 12%/year: $F = 1000(1 + 0.12)^{1} = 1120.00$

Saving in interesting charges = 1126.83 – 1120.00 = \$6.83

15.4 **a.**

$$P = Fe^{-rn} = 10{,}000e^{-0.09(10)} = 4066$$

15.5 **c.** The nominal interest rate is the annual interest rate ignoring the effect of any compounding. Nominal interest rate = 1.5% × 12 = 18%.

15.6 **d.** The interest paid per year = 330 × 4 = 1320. The nominal annual interest rate = 1320/10,000 = 0.132 = 13.2%.

15.7 **d.**

$$i_e = (1 + r/m)^m - 1 = (1 + 0.03)^6 - 1 = 0.194 = 19.4\%$$

15.8 **d.**

$$i_e = (1 + r/m)^m - 1$$
$$r/m = (1 + i_e)^{1/m} - 1 = (1 + 0.1956)^{1/12} - 1 = 0.015$$
$$r = 0.015(m) = 0.015 \times 12 = 0.18 = 18\%$$

15.9 **a.**

$$i_e = 20.52/300 = 0.0684 = 6.84\%$$

15.10 **d.**

$$i_e = (1 + r/m)^m - 1$$
$$0.12 = (1 + r/12)^{12} - 1$$
$$(1.12)^{1/12} = (1 + r/12)$$
$$1.00949 = (1 + r/12)$$
$$r = 0.00949 \times 12 = 0.1138 = 11.38\%$$

15.11 **c.**

$$i_e = (1 + r/m)^m - 1 = (1 + 0.10/365)^{365} - 1 = 0.1052 = 10.52\%$$

15.12 c.

$$P = F(P/F,i,n) = 500(P/F,2\%,10) = 500(0.8203) = \$410$$

15.13 a.

$$P = 300(P/A,8\%,20) = 300(9.818) = \$2945$$

15.14 c. Using single payment present worth factors:

$$P = 20(P/F,6\%,2) + 40(P/F,6\%,3) + 60(P/F,6\%,4) + 80(P/F,6\%,5) + 100(P/F,6\%,6) = \$229$$

Alternate solution using the gradient present worth factor:

$$P = 20(P/G,6\%,6) = 20(11.459) = \$229$$

15.15 d.

$$\text{PW} = 30\ (P/A,4\%,40) + 1000(P/F,4\%,40) = 30(19.793) + 1000(0.2083) = \$802$$

15.16 d. Benefits are \$1500 per year for the first five years and \$1000 per year for the subsequent five years.

As Exhibit 15.16 indicates, the benefits may be considered as \$1000 per year for 10 years, plus an additional \$500 benefit in each of the first five years.

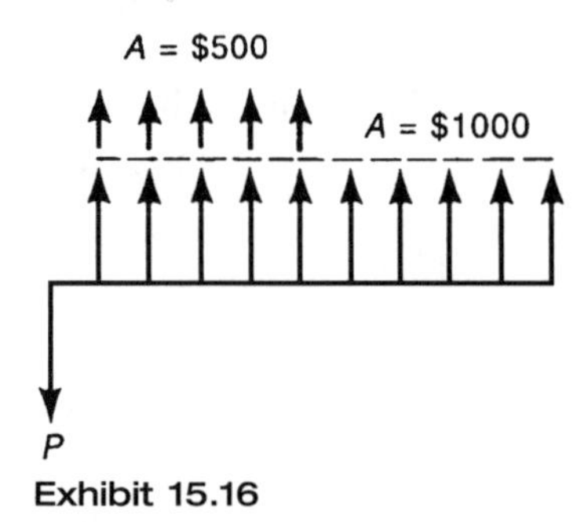

Exhibit 15.16

$$\begin{aligned}\text{maximum investment} &= \text{present worth of benefits}\\ &= 1000(P/A,4\%,10) + 500(P/A,4\%,5)\\ &= 1000(8.111) + 500(4.452) = \$10{,}337\end{aligned}$$

15.17 c.

$$\begin{aligned}\text{NPW} &= \text{PW of benefits} - \text{PW of cost}\\ &= 2400(P/A,12\%,10) + 3000(P/F,12\%,10) - 10{,}000 = \$4526\end{aligned}$$

15.18 d.

$$\begin{aligned}\text{PW of benefits} &= \text{PW of cost}\\ 5000 &= 167.10(P/A,1.5\%,n)\\ (P/A,1.5\%,n) &= 5000/167.10 = 29.92\end{aligned}$$

From the $1\frac{1}{2}\%$ interest table, $n = 40$.

15.19 **c.**

$$\begin{aligned} \text{NPW} &= \text{PW of benefits} - \text{PW of cost} = 0 \\ &= (650 - 300)(P/A,10\%,8) + S_8\,(P/F,10\%,8) - 2000 = 0 \\ &= 350(5.335) + S_8(0.4665) - 2000 = 0 \\ S_8 &= 132.75/0.4665 = \$285 \end{aligned}$$

15.20 **a.** The amount of money needed now to begin the perpetual payments is $P' = A/i = 10{,}000/0.08 = 125{,}000$. From this we can compute the amount of money, P, that would need to have been deposited 50 years ago:

$$P = 125{,}000(P/F,8\%,50) = 125{,}000(0.0213) = \$2663$$

15.21 **d.**

$$P = A/i = 12{,}000/0.5 = \$240{,}000$$

15.22 **d.**

$$\begin{aligned} 2 &= 1(F/P,i,n) \\ (F/P,2\%,n) &= 2 \end{aligned}$$

From the 2% interest table, n = about 35 months.

15.23 **d.** At the end of 50 months

$$F = 10{,}000(F/P,1\%,50) = 10{,}000(1.645) = \$16{,}450$$

At the end of 80 months

$$F = 16{,}450(F/P,2\%,10) = 16{,}450(1.219) = \$20{,}053$$

15.24 **d.**

$$\begin{aligned} \text{FW} &= 200(F/A,1.5\%,12)(F/P,\ 1.5\%,20) \\ &= 200(13.041)(1.347) = \$3513 \end{aligned}$$

15.25 **d.** The present amount $P = 60$ is equivalent to Q six years hence at 6% interest. The future sum F may be calculated by either of two methods:

$$F = Q(F/P,6\%,4) \quad \text{and} \quad Q = 60(F/P,6\%,6)$$

or

$$F = P(F/P,6\%,10)$$

Since P is known, the second equation may be solved directly.

$$F = P(F/P,6\%,10) = 60(1.791) = \$107$$

15.26 **c.**

$$\begin{aligned} F' &= A(F/A,i,n) = 200(F/A,6\%,15) = 200(23.276) = \$4655.20 \\ F &= F'(F/P,i,n) = 4655.20(F/P,6\%,1) = 4655.20(1.06) = \$4935 \end{aligned}$$

15.27 **c.**

$$\text{EUAC} = 150 + 150(A/G,8\%,4) = 150 + 150(1.404) = \$361$$

15.28 **b.**

$$\begin{aligned} \text{Annual payment} &= 12{,}000(P/F,10\%,6)(A/P,10\%,8) \\ &= 12{,}000(0.5645)(0.1874) = \$1269 \end{aligned}$$

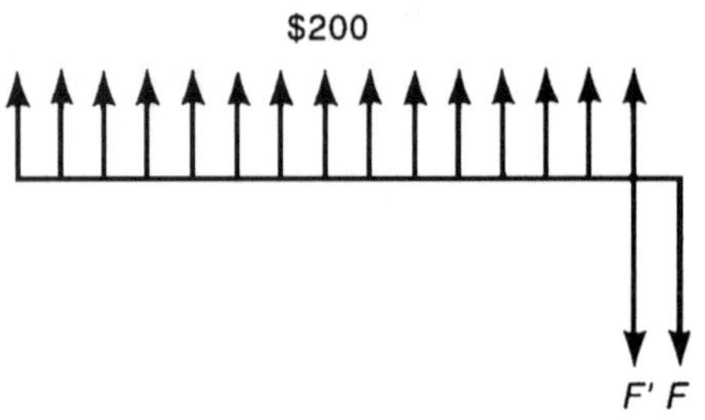

$200

F' F

Exhibit 15.26

15.29 **c.**

$$\begin{aligned} \text{EUAC} &= 15{,}000(A/P,12\%,6) - 2000(A/F,12\%,6) \\ &= 15{,}000(0.2432) - 2000(0.1232) = \$3402 \end{aligned}$$

15.30 **d.**

$$\begin{aligned} \text{EUAC} &= 80{,}000(A/P,10\%,20) - 20{,}000(A/F,10\%,20) \\ &\quad + \text{annual operating cost} \\ &= 80{,}000(0.1175) - 20{,}000(0.0175) + 18{,}000 \\ &= 9400 - 350 + 18{,}000 = \$27{,}050 \end{aligned}$$

15.31 **b.**

$$\begin{aligned} \text{EUAC} &= \text{EUAB} \\ 27{,}000 &= 80{,}000(A/P,10\%,20) + 18{,}000 - S_{20}(A/F,10\%,20) \\ &= 80{,}000(0.1175) + 18{,}000 - S_{20}(0.0175) \\ S_{20} &= (27{,}400 - 27{,}000)/0.0175 = \$22{,}857 \end{aligned}$$

15.32 **d.** The general equation for an infinite life, $P = A/i$, must be used to solve the problem.

$$i_e = (1 + 0.025)^2 - 1 = 0.050625$$

The maximum annual withdrawal will be $A = Pi = 25{,}000(0.050625)$ $= \$1266$.

15.33 **d.**

$$\begin{aligned} \text{Minimum sale price} &= 10{,}000(F/P,10\%,10) + 100(F/A,10\%,10) \\ &= 10{,}000(2.594) + 100(15.937) = \$27{,}530 \end{aligned}$$

15.34 **c.**

$$\begin{aligned} \text{NPW} &= 1000(P/A,i,20) - 10{,}000 = 0 \\ (P/A,i,20) &= 10{,}000/1000 = 10 \end{aligned}$$

From interest tables: $6\% < i < 8\%$.

15.35 **a.** The rate of return was 600/10,000 = 0.06 = 6%.

15.36 **b.**

$$\begin{aligned} F &= A(F/A,i,n) \\ 12{,}000 &= 1000(F/A,i,10) \\ (F/A,i,10) &= 12{,}000/1000 = 12 \end{aligned}$$

In the 4% interest table: $(F/A,4\%,10) = 12.006$, so $i = $%.

15.37 **b.**

PW of cost = PW of benefits

$$80{,}000 = (25{,}700 - 18{,}000)(P/A,i,20) + 20{,}000(P/F,i,20)$$

Try $i = 8\%$.

$$80{,}000 \stackrel{?}{=} 7700(9.818) + 20{,}000(0.2145) = 79{,}889$$

Therefore, the rate of return is very close to 8%.

15.38 **d.**

PW of cost = PW of benefits

$$100 = 102.15(P/F,i,3) + 102.15(P/F,i,6)$$

Solve by trial and error. Try $i = 12\%$.

$$100 \stackrel{?}{=} 102.15(0.7118) + 102.15(0.5066) = 124.46$$

The PW of benefits exceeds the PW of cost. This indicates that the interest rate i is too low. Try $i = 18\%$.

$$100 \stackrel{?}{=} 102.15(0.6086) + 102.15(0.3704) = 100.00$$

Therefore, the rate of return is 18%.

15.39 **c.** The difference between the alternatives:

$$\text{Incremental cost} = 6000 - 2500 = \$3500$$
$$\text{Incremental annual benefit} = 1664 - 746 = \$918$$
$$\text{PW of cost} = \text{PW of benefits}$$
$$3500 = 918(P/A,i,5)$$
$$(P/A,i,5) = 3500/918 = 3.81$$

From the interest tables, i is very close to 10%.

15.40 **c.**

$$B/C = \frac{\text{PW of benefits}}{\text{PW of cost}} = \frac{10{,}000\ (P/A,\ 12\%,\ 8) + 1000\ (P/G,\ 12\%,\ 8)}{50{,}000}$$

$$= \frac{10{,}000\ (4.968) + 1000\ (14.471)}{50{,}000} = 1.28$$

15.41 **c.**

$$B/C = \frac{\text{PW of benefits}}{\text{PW of cost}} = \frac{60(P/A,12\%,8)}{300} = \frac{60(4.968)}{300} = 0.99$$

Alternate Solution:

$$B/C = \frac{\text{EUAB}}{\text{EUAC}} = \frac{60}{300(A/P,12\%,8)} = \frac{60}{300(0.2013)} = 0.99$$

15.42 **a.**

$$B/C = \frac{\text{PW of benefits}}{\text{PW of cost}} = \frac{14{,}000(P/A,18\%,8)}{57{,}000} = \frac{14{,}000(4.078)}{57{,}000} = 1.00$$

15.43 **d.**

$$\text{B/C} = \frac{\text{EUAB}}{\text{EUAC}} = \frac{20{,}000{,}000 - 5{,}500{,}000}{60{,}000{,}000(A/P, 8\%, 10)} = 1.62$$

15.44 **c.**

$$\begin{aligned} F &= 10{,}000(F/P,6\%,5) - 1000(F/A,6\%,5) \\ &= 10{,}000(1.338) - 1000(5.637) = \$7743 \end{aligned}$$

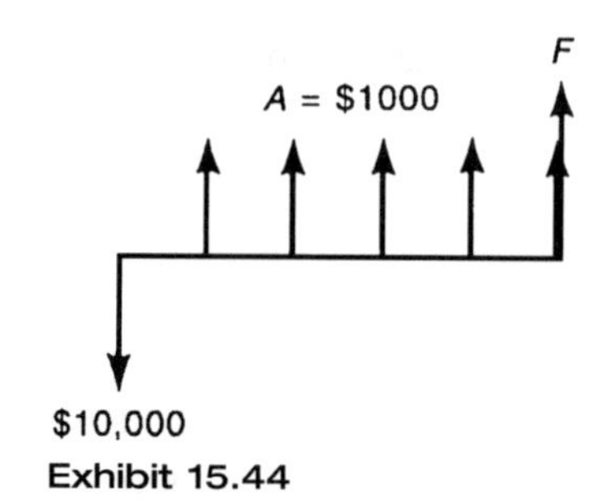

Exhibit 15.44

15.45 **d.**

$$\begin{aligned} \text{PW of cost}_A &= \text{PW of cost}_B \\ 55{,}000 + 16{,}200(P/A,10\%,n) &= 75{,}000 + 12{,}450(P/A,10\%,n) \\ (P/A,10\%,n) &= (75{,}000 - 55{,}000)/(16{,}200 - 12{,}450) \\ &= 5.33 \end{aligned}$$

From the 10% interest tables, $n = 8$ years.

15.46 **c.**

$$\begin{aligned} \text{EUAC}_{\text{untreated}} &= \text{EUAC}_{\text{treated}} \\ 350(A/P,6\%,6) &= 700(A/P,6\%,n) \\ 350(0.2034) &= 700(A/P,6\%,n) \\ (A/P,6\%,n) &= 71.19/700 = 0.1017 \end{aligned}$$

From the 6% interest table, $n =$ 15+ years.

15.47 **d.**

$$\begin{aligned} \text{EUAC}_{\text{T-C}} &= \text{EUAC}_{\text{Quick}} \\ (45 + 40)(A/P,12\%,n) &= (22 + 40)(A/P,12\%,5) \\ (A/P,12\%,n) &= 17.20/85 = 0.202 \end{aligned}$$

From the 12% interest table, $n = 8$.

15.48 **d.** At break-even,

$$\begin{aligned} \text{NPW}_A &= \text{NPW}_B \\ 150(P/A,8\%,10) + 100(P/F,8\%,10) - 1000 &= \text{NAB}(P/A,8\%,10) \\ &\quad + 400(P/F,8\%,10) - 2000 \\ 52.82 &= 6.71(\text{NAB}) - 1814.72 \end{aligned}$$

Net annual benefit (NAB) = (1814.72 + 52.82)/6.71 = \$278.

15.49 **b.** The number of six-month compounding periods in this problem is 20. So $n = 20$ and $12\%/2 = 6\%$ is the interest rate for the six-month interest period.

$$\begin{aligned} \text{Bond value} &= \text{PW of all future benefits} \\ &= 200(P/A,6\%,20) + 5000(P/F,6\%,20) \\ &= 200(11.470) + 5000(0.3118) = \$3853 \end{aligned}$$

15.50 **d.** We know $P = 10{,}000$, $F = 26{,}639$, $n = 15$, and i = bond yield. Using the equation for the single payment compound amount:

$$\begin{aligned} F &= P(1+i)^n \\ 26{,}639 &= 10{,}000(1+i)^{15} \\ 2.6639^{1/15} &= (1+i) \\ 1.0675 &= 1+i \\ i &= 0.0675 = 6.75\% \end{aligned}$$

15.51 **c.** The annual benefits are \$2000, \$3000, \$4000, \$5000, and so on. The payback period is the time when \$8000 of benefits are received. This will occur in 2.75 years.

15.52 **d.**

$$D_3 = (C - S)/n = (15{,}000 - 1000)/3 = \$4666$$

15.53 **d.** From the modified ACRS table (Table 15.3) read for the first recovery year and 3-year recovery the MACRS depreciation is $33.33\% \times 15{,}000 = \$5000$.

15.54 **d.**

$$d = i + f + (i \times f) = 0.05 + 0.06 + 0.05(0.06) = 0.113 = 11.3\%$$

$$P = A(P/A,11.3\%,10) = 1000\left[\frac{(1+0.113)^{10}-1}{0.113(1+0.113)^{10}}\right]$$

$$= 1000\left[\frac{1.9171}{0.3296}\right] = \$5816$$

15.55 **d.**

$$\begin{aligned} \text{Cost of auto five years hence } (F) &= P(1 + \text{inflation rate})^n \\ &= 20{,}000(1+0.10)^5 = 32{,}210 \end{aligned}$$

Amount to deposit now to have \$32,210 available five years hence:

$$P = F(P/F,i,n) = 32{,}210\ (P/F,12\%,5) = 32{,}210(0.5674) = \$18{,}276$$

15.56 **d.**

$$\begin{aligned} \text{Selling price } (F) &= 40{,}000(F/P,18\%,5)(F/P,6\%,5) \\ &= 40{,}000(2.288)(1.338) = \$122{,}500 \end{aligned}$$

15.57 **b.**

$$\text{PW of cash purchase} = \text{PW of installment purchase}$$
$$400 = 50 + 115.25(P/A,d,4)$$
$$(P/A,d,4) = 350/11.25 = 3.037$$

From the interest tables, $d = 12\%$.

$$d = i + f + i(f)$$
$$0.12 = 0.05 + f + 0.05f$$
$$f = 0.07/1.05 = 0.0667 = 6.67\%$$

APPENDIX A

Sample Exam

OUTLINE

The semiannual FE/EIT examination that is created and administered by the National Council of Examiners for Engineering and Surveying is simulated on the following pages. Individuals who are preparing to take the FE/EIT Examination will find it helpful to become familiar with its format and general level of difficulty.

The topics included in the morning and afternoon sessions of the NCEES exam are listed in Chapter 1. Also included there, with a list of subtopics, is the percentage of exam problems allotted to each topic. In the exam itself, the topics may not appear in the same order, as there are three versions of each exam.

There are 120 multiple-choice problems in the morning session and 60 in the afternoon session. The latter carry twice the grading value of the former; thus the possible score for each session is 120, or a total of 240 points for a perfect score in both sessions.

The only reference material that is allowed at the NCEES exam is the *Fundamentals of Engineering Supplied-Reference Handbook*, a copy of which is furnished to each examinee at each session. (Applicants are furnished a copy of the handbook as part of the application approval, but neither this particular copy nor any other is allowed to be carried into the exam.)

The problems in this simulated examination are grouped by topic. Following the problems is a section containing detailed solutions.

We recommend that you find four hours when you will not be interrupted. Use only the *Reference Handbook*, pencil, and calculator, and time yourself. Work in your strong areas first but complete all 120 AM and 60 PM problems in 4-hour sessions.

FUNDAMENTALS OF ENGINEERING EXAM

MORNING SESSION

ⒶⒷⒸⒹ Fill in the circle that matches your exam booklet

1 ⒶⒷⒸⒹ
2 ⒶⒷⒸⒹ
3 ⒶⒷⒸⒹ
4 ⒶⒷⒸⒹ
5 ⒶⒷⒸⒹ
6 ⒶⒷⒸⒹ
7 ⒶⒷⒸⒹ
8 ⒶⒷⒸⒹ
9 ⒶⒷⒸⒹ
10 ⒶⒷⒸⒹ
11 ⒶⒷⒸⒹ
12 ⒶⒷⒸⒹ
13 ⒶⒷⒸⒹ
14 ⒶⒷⒸⒹ
15 ⒶⒷⒸⒹ
16 ⒶⒷⒸⒹ
17 ⒶⒷⒸⒹ
18 ⒶⒷⒸⒹ
19 ⒶⒷⒸⒹ
20 ⒶⒷⒸⒹ
21 ⒶⒷⒸⒹ
22 ⒶⒷⒸⒹ
23 ⒶⒷⒸⒹ
24 ⒶⒷⒸⒹ
25 ⒶⒷⒸⒹ
26 ⒶⒷⒸⒹ
27 ⒶⒷⒸⒹ
28 ⒶⒷⒸⒹ
29 ⒶⒷⒸⒹ
30 ⒶⒷⒸⒹ
31 ⒶⒷⒸⒹ
32 ⒶⒷⒸⒹ
33 ⒶⒷⒸⒹ
34 ⒶⒷⒸⒹ
35 ⒶⒷⒸⒹ
36 ⒶⒷⒸⒹ
37 ⒶⒷⒸⒹ
38 ⒶⒷⒸⒹ
39 ⒶⒷⒸⒹ
40 ⒶⒷⒸⒹ
41 ⒶⒷⒸⒹ
42 ⒶⒷⒸⒹ
43 ⒶⒷⒸⒹ
44 ⒶⒷⒸⒹ
45 ⒶⒷⒸⒹ
46 ⒶⒷⒸⒹ
47 ⒶⒷⒸⒹ
48 ⒶⒷⒸⒹ
49 ⒶⒷⒸⒹ
50 ⒶⒷⒸⒹ
51 ⒶⒷⒸⒹ
52 ⒶⒷⒸⒹ
53 ⒶⒷⒸⒹ
54 ⒶⒷⒸⒹ
55 ⒶⒷⒸⒹ
56 ⒶⒷⒸⒹ
57 ⒶⒷⒸⒹ
58 ⒶⒷⒸⒹ
59 ⒶⒷⒸⒹ
60 ⒶⒷⒸⒹ
61 ⒶⒷⒸⒹ
62 ⒶⒷⒸⒹ
63 ⒶⒷⒸⒹ
64 ⒶⒷⒸⒹ
65 ⒶⒷⒸⒹ
66 ⒶⒷⒸⒹ
67 ⒶⒷⒸⒹ
68 ⒶⒷⒸⒹ
69 ⒶⒷⒸⒹ
70 ⒶⒷⒸⒹ
71 ⒶⒷⒸⒹ
72 ⒶⒷⒸⒹ
73 ⒶⒷⒸⒹ
74 ⒶⒷⒸⒹ
75 ⒶⒷⒸⒹ
76 ⒶⒷⒸⒹ
77 ⒶⒷⒸⒹ
78 ⒶⒷⒸⒹ
79 ⒶⒷⒸⒹ
80 ⒶⒷⒸⒹ
81 ⒶⒷⒸⒹ
82 ⒶⒷⒸⒹ
83 ⒶⒷⒸⒹ
84 ⒶⒷⒸⒹ
85 ⒶⒷⒸⒹ
86 ⒶⒷⒸⒹ
87 ⒶⒷⒸⒹ
88 ⒶⒷⒸⒹ
89 ⒶⒷⒸⒹ
90 ⒶⒷⒸⒹ
91 ⒶⒷⒸⒹ
92 ⒶⒷⒸⒹ
93 ⒶⒷⒸⒹ
94 ⒶⒷⒸⒹ
95 ⒶⒷⒸⒹ
96 ⒶⒷⒸⒹ
97 ⒶⒷⒸⒹ
98 ⒶⒷⒸⒹ
99 ⒶⒷⒸⒹ
100 ⒶⒷⒸⒹ
101 ⒶⒷⒸⒹ
102 ⒶⒷⒸⒹ
103 ⒶⒷⒸⒹ
104 ⒶⒷⒸⒹ
105 ⒶⒷⒸⒹ
106 ⒶⒷⒸⒹ
107 ⒶⒷⒸⒹ
108 ⒶⒷⒸⒹ
109 ⒶⒷⒸⒹ
110 ⒶⒷⒸⒹ
111 ⒶⒷⒸⒹ
112 ⒶⒷⒸⒹ
113 ⒶⒷⒸⒹ
114 ⒶⒷⒸⒹ
115 ⒶⒷⒸⒹ
116 ⒶⒷⒸⒹ
117 ⒶⒷⒸⒹ
118 ⒶⒷⒸⒹ
119 ⒶⒷⒸⒹ
120 ⒶⒷⒸⒹ

DO NOT WRITE IN BLANK AREAS

INSTRUCTIONS FOR MORNING SESSION

1. You have four hours to work on the morning session. You may use the *Fundamentals of Engineering Supplied-Reference Handbook* as your *only* reference. Do not write in this handbook.
2. Answer every question. There is no penalty for guessing.
3. Work rapidly and use your time effectively. If you do not know the correct answer, skip it and return to it later.
4. Some problems are presented in both metric and English units. Solve either problem.
5. Mark your answer sheet carefully. Fill in the answer space completely. No marks on the workbook will be evaluated. Multiple answers receive no credit. If you make a mistake, erase completely.

Work 120 morning problems in four hours.

Sample Examination Problems/AM—Mathematics

1. In finding the distance between two points $P_1(x_1, y_1)$ and $P_2(x_2, y_2)$, the most direct procedure is to use

a. The translation of the axes
b. The Pythagorean theorem
c. The slope of the line
d. The derivative

2. An angle between 90° and 180° has

a. A positive sine and cosine
b. A negative cotangent and cosecant
c. A negative secant and tangent
d. All of its trigonometric functions negative

3. The csc of 960° is equal to

a. $-\dfrac{2\sqrt{3}}{3}$
b. 1
c. 1/2
d. −2

4. If $i = \sqrt{-1}$, the quantity i^{27} is equal to

a. 0
b. i
c. $-i$
d. 1

5. The integral of $y = x^3 - x + 1$ is

a. $3x^2 - 1 + c$

b. $\frac{x^3}{3} - \frac{x^2}{2} + x$

c. $\frac{x^4}{3} - \frac{x^2}{2} + 1 + c$

d. $\frac{x^4}{4} - \frac{x^2}{2} + x + c$

6. The point of inflection on the curve representing the equation $y = x^3 + x^2 - 3$ is at x equals

a. −2/3

b. −1/3

c. 0

d. 1/3

7. If the second derivative of the equation of a curve is equal to the negative of the equation of that same curve, the curve is

a. an exponential

b. a tangent

c. a conic section

d. a sinusoid

8. For a given curve $f(x, y) = 0$, it is found that $f(x, y) = f(x, -y)$. This means that the curve is

a. unsymmetrical

b. a conic section

c. symmetric to both axes

d. symmetric to the x-axis

9. A bicycle rider rides away from home along a highway and back along the same road in such a way that her distance from home at time t is given by

$$x(t) = t^4 - 8t^3 + 16t^2$$

where t is in hours and x is in kilometers.

When does she get home?

a. After 1 hour

b. After 2 hours

c. After 3 hours

d. After 4 hours

10. How far from home does the rider in Problem 9 go?

a. 20 kilometers

b. 16 kilometers

c. 10 kilometers

d. 8 kilometers

11. What is her average speed?

a. 2 kph
b. 4 kph
c. 8 kph
d. 10 kph

12. When does her maximum speed occur?

a. $2 \pm \frac{2}{\sqrt{3}}$ hours out
b. 1 and 3 hours out
c. $2 \pm \frac{1}{2}$ hours out
d. 2 hours out

13. Solve $y'' + 4y = 8 \sin x$.

a. $y = \text{A}e^{2x} + \text{B}e^{-2x}$
b. $y = \text{A} \sin 2x + \text{B} \cos 2x$
c. $y = \text{A} \sin 2x + \text{B} \cos 2x + \sin x$
d. $y = \text{A} \sin 2x + \text{B} \cos 2x + (8/3) \sin x$

14. The solution to $y'' - 5y' + 6y = 0$ is

a. $y = 5x + \text{c}$
b. $y = e^{5x} + \text{c} + \text{d}e^{6x}$
c. $y = \text{A}e^{5x} + \text{B}e^{6x}$
d. $y = \text{A}e^{2x} + \text{B}e^{3x}$

15. The solution to $x^2y'' - 3xy' + 4y = 0$ is

a. $y = \text{A}e^{2x} + \text{B}xe^{2x}$
b. $y = \text{A}e^{x} + \text{B}e^{2x}$
c. $y = \text{A}x^2 + \text{B}x^2 \ln |x|$
d. $y = \text{A} \sin 3x + \text{B} \cos 3x$

16. The Laplace transform of the solution of the initial value problem $y'' - 3y' - 4y = e^t$ $[y(0) = 0, y'(0) = 1]$ is

a. $Y(s) = \frac{1}{s-1} + \frac{1}{s+1} + \frac{1}{s-4}$
b. $Y(s) = -\frac{1}{s-1} + \frac{1}{s+1} - \frac{2}{s-4}$
c. $Y(s) = \frac{\frac{-1}{6}}{s-1} - \frac{\frac{1}{10}}{s+1} + \frac{\frac{4}{15}}{s-4}$
d. $Y(s) = \frac{1}{(s-1)(s+1)(s-4)}$

17. What is the probability of drawing a royal flush (the five highest cards of one suit) from a standard deck of 52 playing cards?

a. 0.000 259 8
b. 0.000 011 54
c. 0.000 001 539
d. 0.000 003 078

18. Using the frequency table below, what is the probability of throwing two dice totaling less than six?

Frequency table

Sum	Frequency	Probability Density
2	1	1/36
3	2	2/36
4	3	3/36
5	4	4/36
6	5	5/36
7	6	6/36
8	5	5/36
9	4	4/36
10	3	3/36
11	2	2/36
12	1	1/36

a. 2/36
b. 4/36
c. 6/36
d. 10/36

19. How many ways can five guests sit around a circular table, assuming there is no head?
a. 24
b. 60
c. 72
d. 120

20. What is the probability of rolling a total of four given two six-sided dice?
a. 1/36
b. 2/36
c. 3/36
d. 4/36

21. Given

$$\begin{bmatrix} X & 5 \\ 3 & X+Y \end{bmatrix} = \begin{bmatrix} 0 & 5 \\ 3 & 2 \end{bmatrix}$$

the value of Y is nearest to
a. 0
b. 2
c. 3
d. 4

22. Let $\vec{a} = (1,2,2)$ and $\vec{b} = (3,0,4)$. The standard dot product $\mathbf{a} \bullet \mathbf{b}$ is closest to
a. 0
b. 2
c. 5
d. 11

23. Find the value of the determinant

$$\begin{vmatrix} 1 & 1 & 0 & 0 \\ 1 & 1 & 1 & 0 \\ 0 & 1 & 1 & 1 \\ 0 & 0 & 1 & 1 \end{vmatrix}$$

a. +1
b. 0
c. −2
d. −1

24. The following equation involves two determinants.

$$\begin{vmatrix} 3 & X \\ 2 & 2 \end{vmatrix} = \begin{vmatrix} 2 & -1 \\ X & -3 \end{vmatrix}$$

The value of X is nearest to
a. −1
b. 1
c. 2
d. 4

Sample Examination Problems/AM—Statics

25. The moment of the force F about point O has the value

a. $\frac{12}{5}(3\mathbf{e}_x - \mathbf{e}_y + 4\mathbf{e}_z)\text{m}F$

b. $\frac{12}{5}(-3\mathbf{e}_x + \mathbf{e}_y + 4\mathbf{e}_z)\text{m}F$

c. $\frac{1}{5}(-4\mathbf{e}_x + 3\mathbf{e}_y)\text{m}F$

d. $\frac{1}{13}(3\mathbf{e}_x + 12\mathbf{e}_y + 3\mathbf{e}_z)\text{m}F$

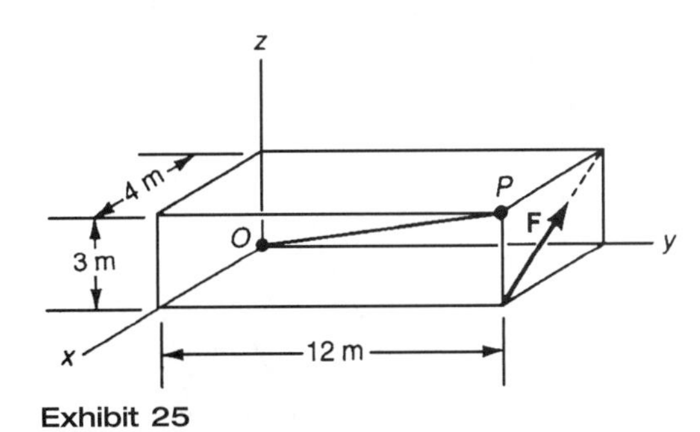

Exhibit 25

26. The moment of the force F about the axis OP of the previous problem has the value

a. $\left(\frac{108}{65}\text{ m}\right)F$

b. $\left(\frac{24}{65}\text{ m}\right)F$

c. $\left(\frac{19}{65}\text{ m}\right)F$

d. $\left(\frac{144}{65}\text{ m}\right)F$

27. The force in the member CD has the value

a. zero
b. 600 N
c. 849 N
d. 1109 N

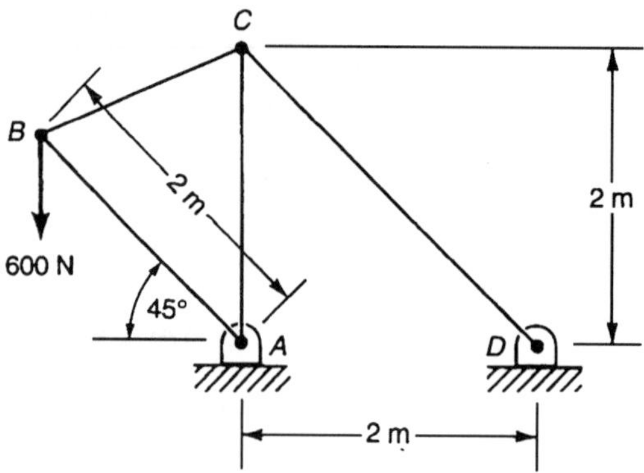

Exhibit 27

28. Which of the diagrams shown below are correct free-body diagrams for analyzing the truss?

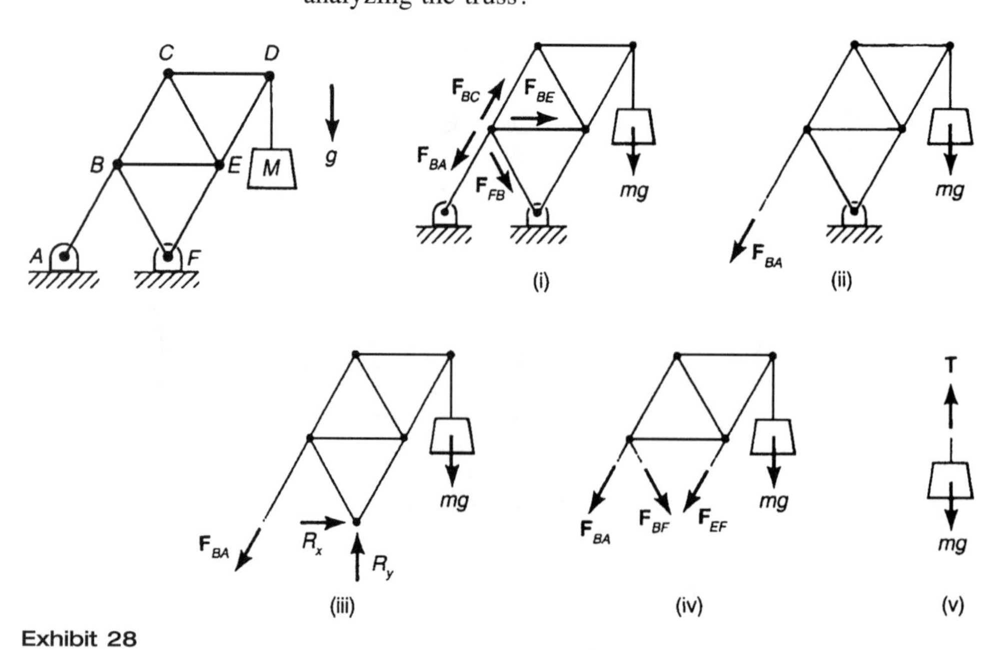

Exhibit 28

a. (i) only
b. (iv) only
c. all except (i) and (ii)
d. none of them

29. The force in the member *EH* of the truss has the value
a. –4 kN (Compression)
b. –5 kN (Compression)
c. –10 kN (Compression)
d. 10 kN (Tension)

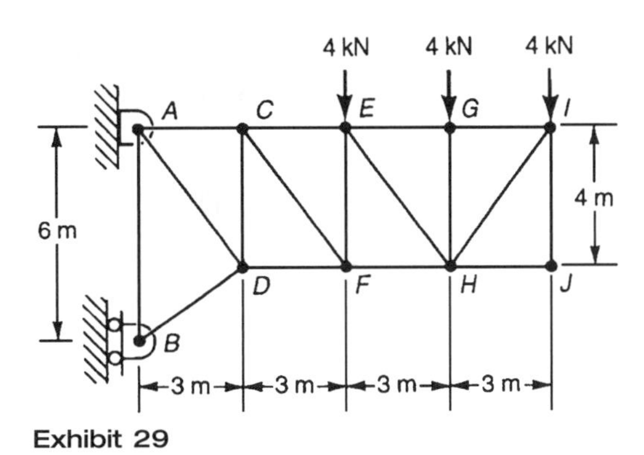

Exhibit 29

Exhibit 30

30. If the small ring is in equilibrium under the action of the three forces, the magnitude *R* must have the value
a. 58.3 kN
b. 20.0 kN
c. 80.0 kN
d. 74.4 kN

31. The value of the angle θ of the previous problem is
a. 28.4°
b. 22.5°
c. 30.0°
d. 61.6°

32. As shown in Exhibit 32 the twisting and bending moment components of the reaction transmitted by the support to the pipe where it is welded at *A* have the magnitudes

a. $M_t = bP, \quad M_b = aP$

b. $M_t = aP, \quad M_b = bP$

c. $M_t = \sqrt{a^2 + b^2}\,P, \quad M_b = \sqrt{a^2 - b^2}\,P$

d. $M_t = \dfrac{b^2}{a}P, \quad M_b = \dfrac{a^2}{b}P$

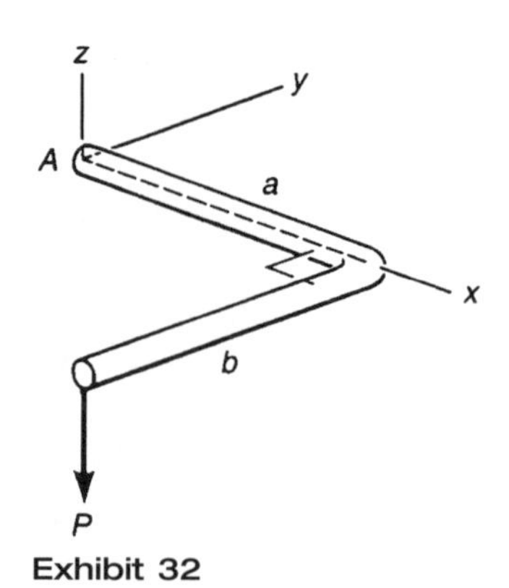

Exhibit 32

33. If the weights of the members of the linkage are negligible, equilibrium requires that the force P have the magnitude

a. $\sqrt{3}Q$

b. $4\sqrt{3}Q/3$

c. $3\sqrt{3}Q/4$

d. $\sqrt{3}Q/12$

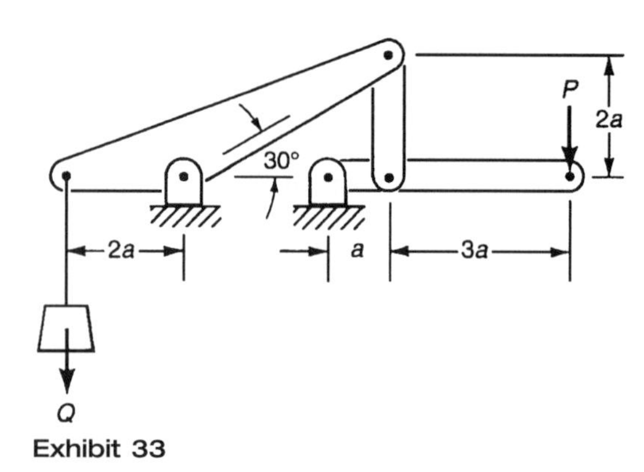

Exhibit 33

34. The mass of the bar is negligible compared with that of the block. The minimum coefficient of friction necessary to prevent the bar from rotating about O is

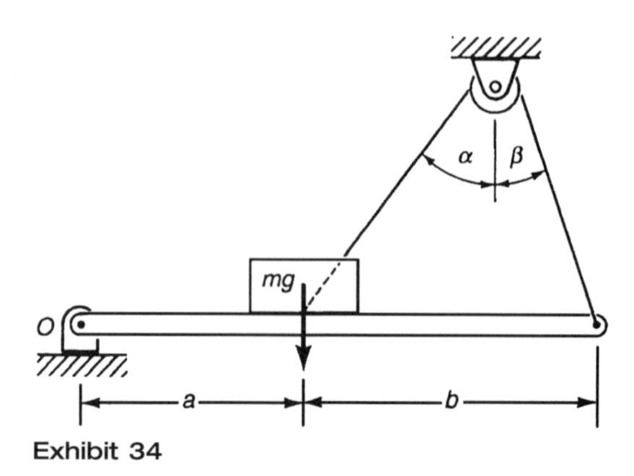

Exhibit 34

a. $\dfrac{a \sin \alpha}{(a+b)\cos\beta}$ o

b. $\dfrac{a \sin \beta}{(a+b)\cos\alpha}$

c. $\dfrac{a}{b}\tan(\alpha+\beta)$

d. $\dfrac{a \sin \beta}{b \cos\alpha}$

35. The moment about the axis *BC* of the gravity force *mg* has magnitude

a. (2.5 m)*mg*
b. (2.9 m)*mg*
c. (1.7 m)*mg*
d. (1.265 m)*mg*

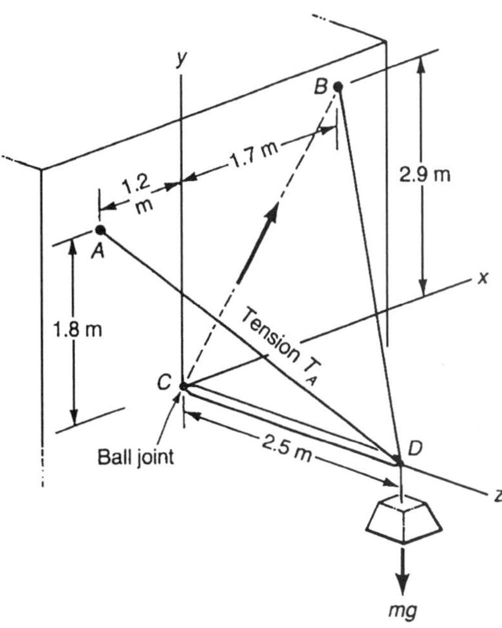

Exhibit 35

36. The tensile force in the cable *AD* of the previous two problems is

a. 0.860 *mg*
b. 1.164 *mg*
c. 0.500 *mg*
d. 1.417 *mg*

Sample Examination Problems/AM—Dynamics

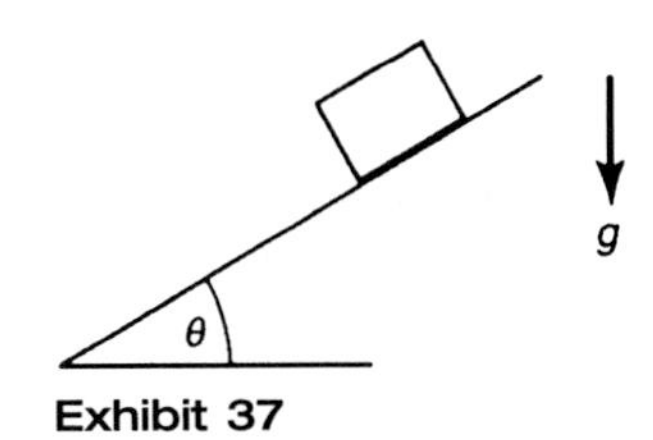

Exhibit 37

37. In Exhibit 37, the block is released from rest on a frictionless incline. Its acceleration down the incline is

a. g
b. $g \sec\theta$
c. $g \sin\theta$
d. $g \cos\theta$

38. A small object P moves along a straight path. The position coordinate of P is known to be $x = (t^2 - 2t + 10)$ meters, where t is time in seconds. The total distance traveled by P in 10 seconds is

a. 82 m
b. 90 m
c. 80 m
d. 100 m

39. The vehicle shown in Exhibit 39 is rolling forward on a straight horizontal path without its tires slipping. The outer radius of the front tires is two-thirds that of the rear tires. The vehicle's speedometer is attached to one of the front wheels and indicates a constant speed of 40 km/h. The ratio of the angular speed of the front wheel to that of the rear wheel is

a. 1
b. 1/3
c. 3/2
d. 3

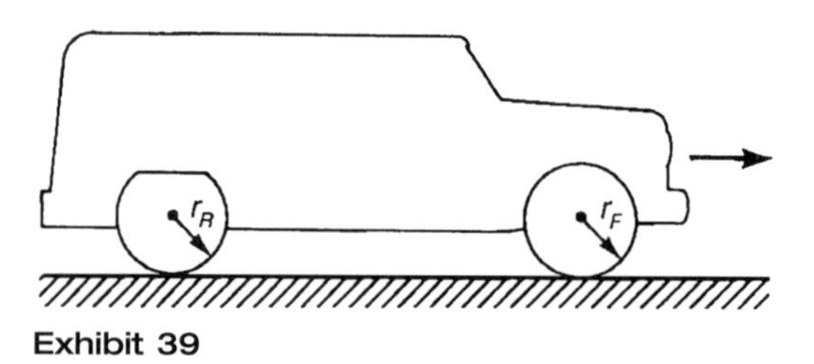

Exhibit 39

40. A projectile is launched at an angle of 60° to the vertical (Exhibit 40) and covers a horizontal distance of 150 m in 10 seconds. What is the launch speed? Neglect air resistance.

a. 30 m/s
b. 15 m/s
c. 507 m/s
d. 17 m/s

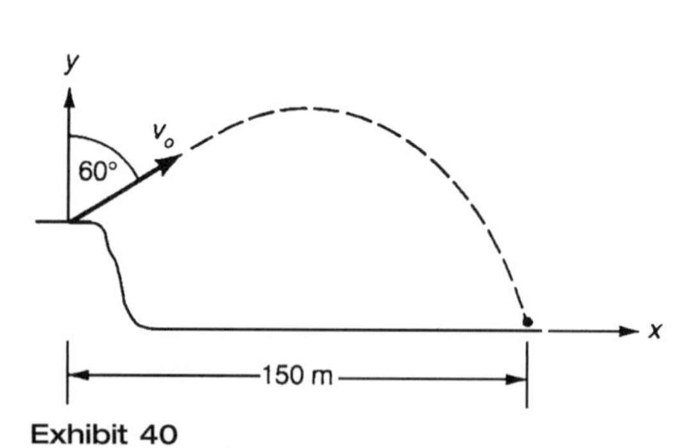

Exhibit 40

41. The fire truck in Exhibit 41 is stationary while its ladder is being raised and extended. The angle θ is increasing at the constant rate of 1 rad/s, and the ladder's length is being increased at a constant rate of 3 m/s. Determine the magnitude of the acceleration of the end A of the ladder at the instant when $b = 2$ meters.

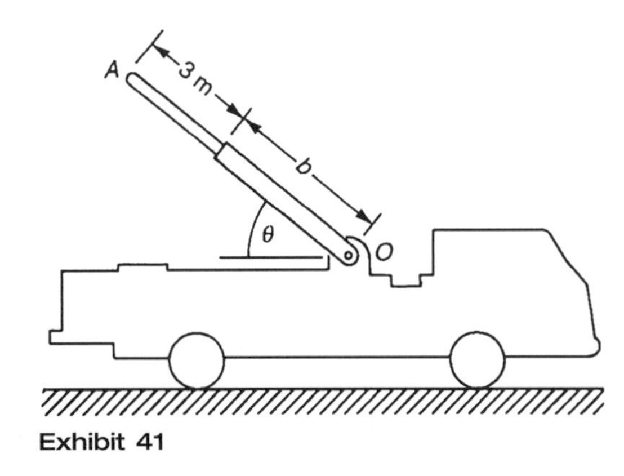

Exhibit 41

a. 5.0 m/s^2
b. 0
c. 6.0 m/s^2
d. 7.8 m/s^2

42. Two uniform slender rods of the same mass density are rigidly connected together as shown in Exhibit 42 to form the letter "*L*." The length of one of the rods is twice that of the other. The mass center, *G*, of the assemblage is located *x* meters above the lower rod, where *x* is

a. 1/3
b. 2/3
c. 1
d. 0

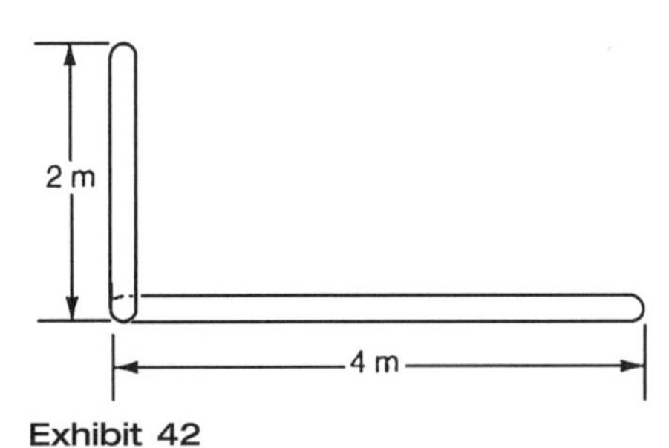

Exhibit 42

43. An airplane that is in the process of landing touches the ground with a velocity of 250 km/h. If the total mass of the airplane and its cargo is 6000 kg, determine the average retarding force to which the plane must be subjected in order for it to stop after traveling 150 m.

a. 2.9 kN
b. 193.0 kN
c. 210.0 kN
d. 96.5 kN

44. A rod of mass 2 kg and length 1.5 m is released from rest in the horizontal position where the torsional spring is compressed $\pi/4$ radian. The spring has a stiffness of 5 N-m/rad and is further compressed when the rod rotates from the horizontal position to the vertical position (shown as dotted lines in Exhibit 44). What is the work done by the spring forces during this motion?

a. −15.4 N•m
b. 15.4 N•m
c. −13.9 N•m
d. −12.3 N•m

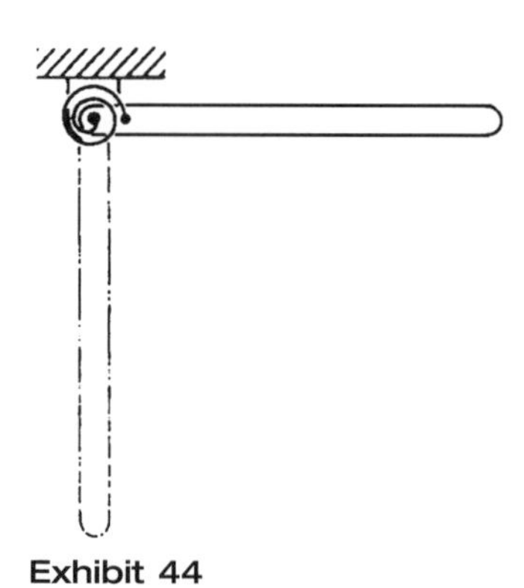

Exhibit 44

45. A constant torque $T = 10$ N • m is applied to a disk of mass 5 kg and radius 1 m that is rotating about a fixed axis through its center. If the effects of bearing friction can be approximated by a constant resistive torque of 4 N • m, find the angular velocity of the disk 7 seconds after it starts from rest.
a. 8.4 rad/s
b. 33.6 rad/s
c. 68.6 rad/s
d. 16.8 rad/s

Sample Examination Problems/AM—Mechanics of Materials

46. A uniform member carries three axial loads. The member has a cross-sectional area of 500 mm^2. The largest stress is most nearly
a. 16 MPa
b. 22 MPa
c. 33 MPa
d. 45 MPa

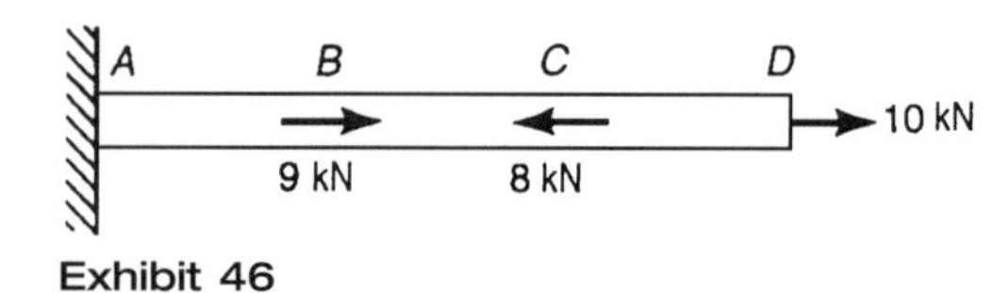

Exhibit 46

47. An axially loaded member is to have a maximum stress of 100 MPa and a maximum deflection of 0.25 mm. Assume $E = 210$ GPa. The minimum area to meet both of these criteria is most nearly
a. 340 mm^2
b. 400 mm^2
c. 460 mm^2
d. 520 mm^2

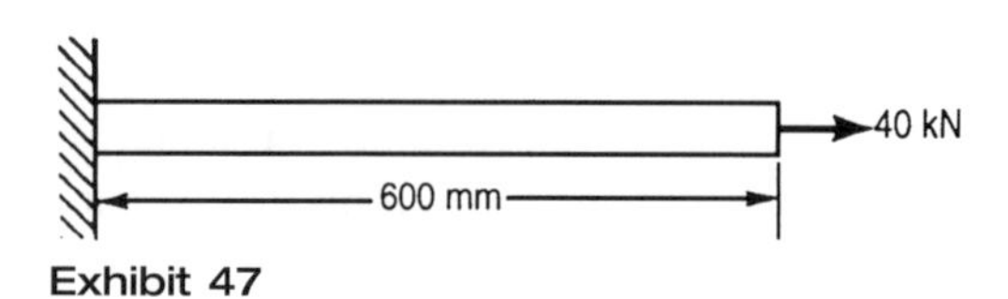

Exhibit 47

48. The hollow circular shaft is to carry a torque of 1 kN • m. It is to have a maximum shear stress of 100 MPa. The largest inner diameter which this shaft can have is most nearly

a. 10 mm
b. 24 mm
c. 36 mm
d. 44 mm

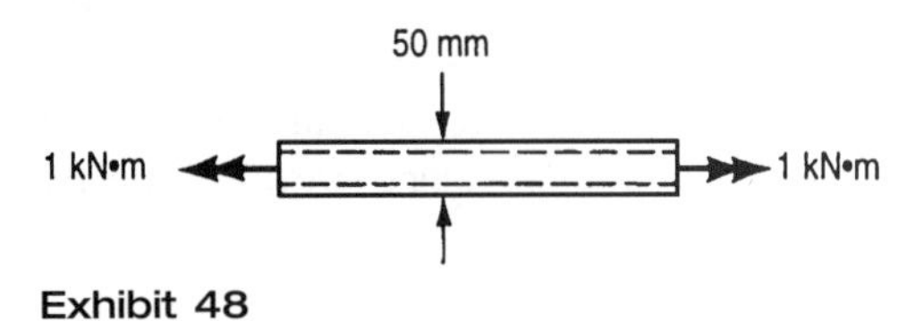

Exhibit 48

49. A beam has a concentrated load P at its third point and a concentrated moment at its other third point. The shear diagram is most nearly

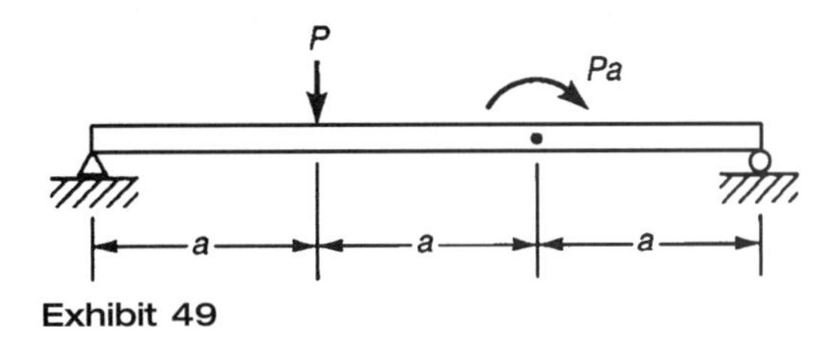

Exhibit 49

a. 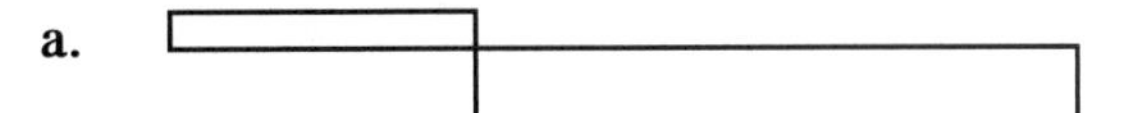

b. 

c. 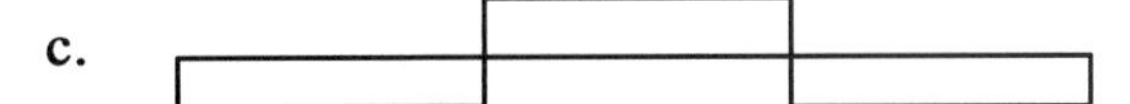

d. 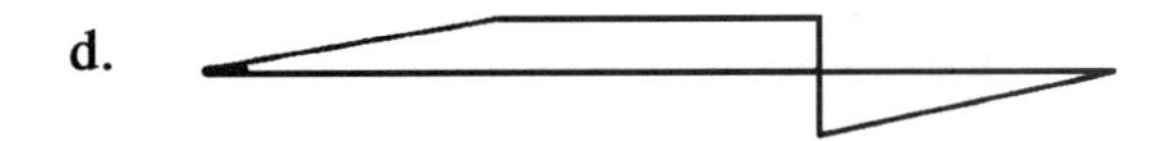

50. For the same beam as in Problem 49 the moment diagram is most nearly

a. 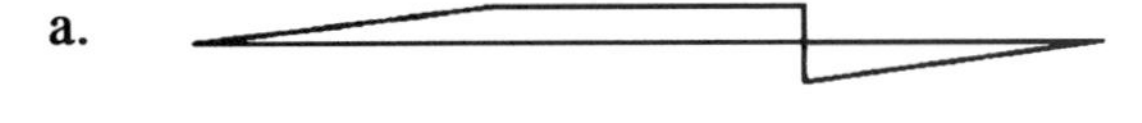

b. 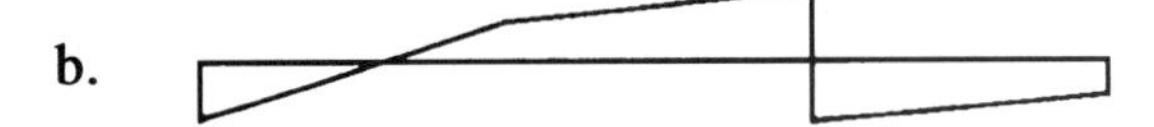

c. 

d. 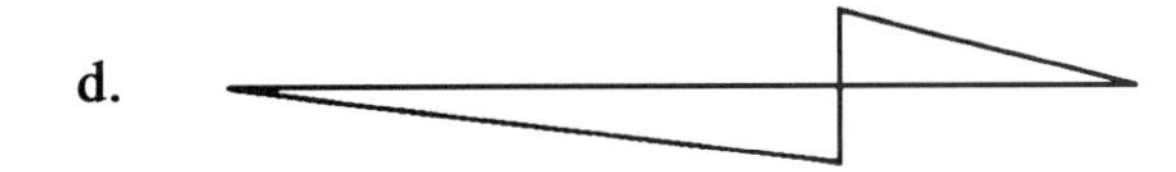

51. In a state of plane stress the stresses are $\sigma_x = 60$ MPa, $\sigma_y = 20$ MPa and $\tau_{xy} = 20$ MPa. The maximum principal stress is most nearly

a. 62 MPa
b. 64 MPa
c. 66 MPa
d. 68 MPa

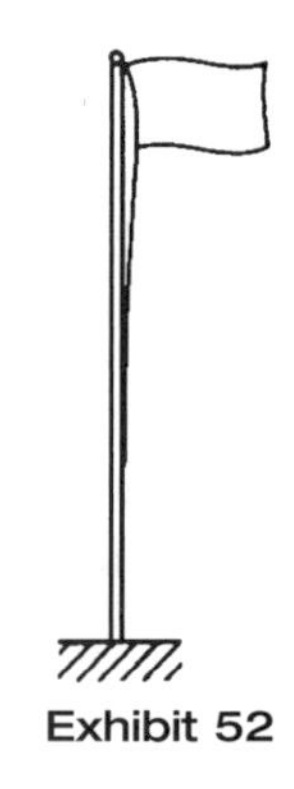

Exhibit 52

52. A wooden flagpole is 15 m high. It has a diameter of 100 mm and a modulus of elasticity of 12.6 GPa. A group of students decide to climb the pole to remove an objectionable flag. The maximum weight of a student who could successfully climb the pole would be most nearly

a. 350 N
b. 500 N
c. 680 N
d. 810 N

53. A square plate *ABCD* is loaded with uniform shear forces applied along its edges. After loading point *A* and *B* are unmoved and points *C* and *D* have moved downward by 0.50 mm. The shear strain in the plate is most nearly

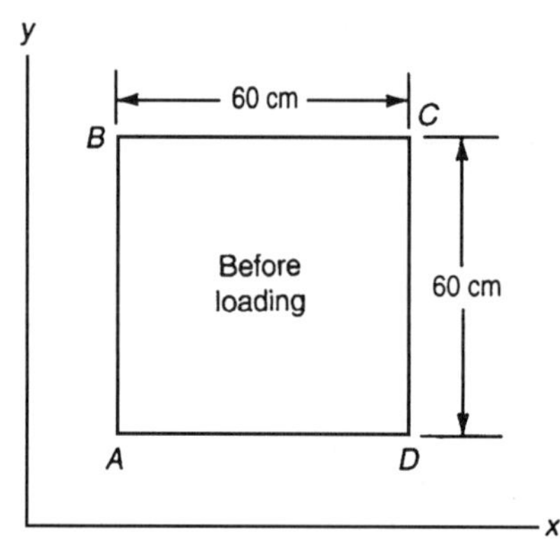

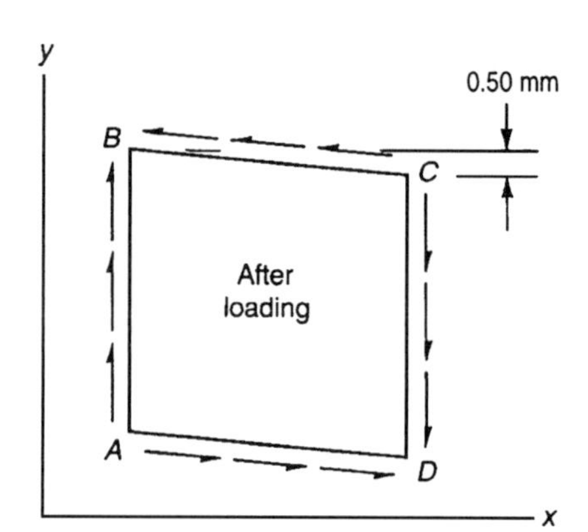

Exhibit 53

a. −0.0008
b. −0.0005
c. +0.0005
d. +0.0008

Sample Examination Problems/AM—Fluid Mechanics

54. The pressure at a point 100 feet (30.48 meters) below the surface of the ocean with a specific gravity of 1.03 is most nearly

a. 44.6 psig (308 kPa)
b. 0.715 psig (4.93 kPa)
c. 535 psig (3690 kPa)
d. 22.3 psig (154 kPa)

55. For stability to occur for a floating vessel
a. the vessel must be rectangular in shape
b. the center of gravity is below the metacenter
c. the center of gravity is at the metacenter
d. the center of gravity is above the metacenter

56. A rectangular tank has dimensions 9 feet (3 meters) long by 3 feet (1 meter) wide by 6 feet (2 meters) high. If the tank is filled with water to the top which is open to the atmosphere, the force on the 3 feet (1 meter) by 6 feet (2 meters) side is most nearly
a. 3370 lbf (19.6 kN)
b. 187 lbf (1.09 kN)
c. 6740 lbf (39.2 kN)
d. 187 psig (1288 kPa)

57. Which of the following describes the change in dynamic viscosity of a liquid and a gas if temperature is *increased*?
a. Viscosity of a gas increases, viscosity of a liquid decreases.
b. Viscosity of a gas decreases, viscosity of a liquid increases.
c. Viscosities of both decrease.
d. Viscosities of both are independent of temperature.

58. The resultant force created by a fluid on a submerged flat surface always acts
a. below the centroid of the flat surface area
b. at the centroid of the flat surface area
c. above the centroid of the flat surface area
d. halfway between the top and bottom of the flat surface area

59. The flow rate through an orifice meter is
a. proportional to the pressure drop in the meter
b. inversely proportional to the pressure drop in the meter
c. proportional to the square root of the pressure drop in the meter
d. inversely proportional to the square root of the pressure drop in the meter

60. The relationship which relates energy levels at two points in a steady stream according to

$$\frac{p}{\gamma} + \frac{V^2}{2g} + Z = \text{constant}$$

is known as
a. the Hagen-Poiseuille equation
b. the continuity equation
c. Bernoulli's equation
d. the Cauchy equation

61. According to the impulse-momentum relationship, the magnitude and direction of the resultant force on a stationary vane subject to steady flow of a liquid is a function of
   a. a change in flow direction only
   b. a velocity magnitude and change in flow direction
   c. the mass rate of flow, velocity magnitude, and change in flow direction
   d. the mass rate of flow and change in flow direction

Sample Examination Problems/AM—Thermodynamics

62. All of the following are thermodynamic properties of a system *except*
   a. enthalpy
   b. entropy
   c. pressure
   d. work

63. A cylinder fitted with a frictionless piston contains an ideal gas at 50°C and 10 atmospheres. If the gas expands reversibly and adiabatically until the pressure is 5 atmospheres, the work done by the gas is equal to
   a. the internal energy decrease of the gas
   b. the enthalpy decrease of the gas
   c. the heat absorbed by the gas
   d. 5 atmospheres times the volume change of the gas

64. For a process with a closed stationary thermodynamic system, the work done by the system is 10 kJ while the change in the energy of the system is −5 kJ. During this process the heat transfer is
   a. 5 kJ
   b. −5 kJ
   c. 10 kJ
   d. 15 kJ

65. Which of the following statements is true for the constant-pressure and constant-volume lines on a temperature-entropy (T-s) diagram for an ideal gas? (Temperature is plotted on the vertical axis.)
   a. Pressure lines at higher temperatures represent lower pressures.
   b. Volume lines at higher temperatures represent higher volumes.
   c. The pressure and volume lines are parallel.
   d. The pressure lines have a smaller slope than the volume lines.

66. The ideal Otto cycle consists of which of the following processes?
   a. two constant-volume and two isentropic processes
   b. two constant-pressure and two isentropic processes
   c. two constant-volume and two isothermal processes
   d. two constant-pressure and two isothermal processes

67. Which of the following is correct for the entropy of an isolated system?
   a. It always tends to decrease with time.
   b. It always remains constant with time.
   c. It either remains constant or decreases with time.
   d. It either remains constant or increases with time.

68. A Carnot engine operates between two reservoirs at temperatures of 900°C and 100°C, respectively. The thermal efficiency is most nearly

a. 0%
b. 68%
c. 79%
d. 92%

69. A refrigeration system operates as a closed cycle. Heat is added to the refrigerant at the rate of 60,000 watts and the net work input is 4 kW. The amount of heat that must be rejected per hour is most nearly

a. 50,000 W
b. 60,000 W
c. 64,000 W
d. 100,000 W

70. Saturated water vapor undergoes an isothermal volume increase. If the mass of the system is constant, the pressure

a. decreases
b. increases
c. remains constant
d. is double-valued in this region

71. A Carnot cycle refrigerator operates between –5°C and 40°C. The coefficient of performance of the cycle is most nearly

a. 0.88
b. 0.20
c. 1.13
d. 6.0

72. Heat transfer by conduction across a solid slab of area A and thickness d is dependent on all of the following *except*

a. the temperature difference between the two sides of the slab
b. the material composing the slab
c. the pressure difference between the two sides of the slab
d. area A

Sample Examination Problems/AM—Electrical Circuits

73. Determine R in the Exhibit 73.

a. 40 Ω
b. 20 Ω
c. 10 Ω
d. 8 Ω

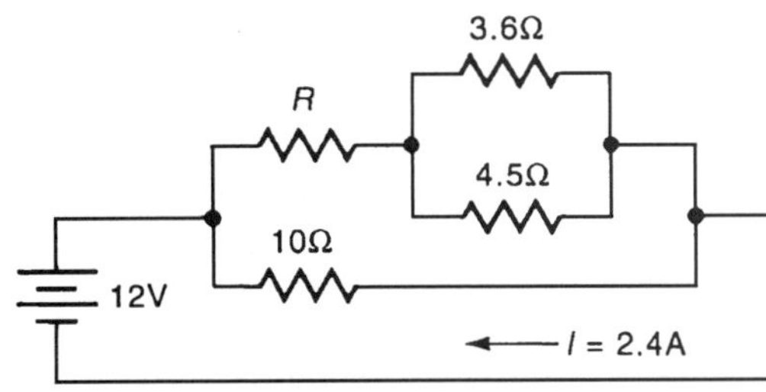

Exhibit 73 Circuit with unknown R

74. Determine the magnitude of the source voltage for the circuit shown if the magnitude of the individual voltages measured across each element is as given.

a. 50 V
b. 150 V
c. 71 V
d. 112 V

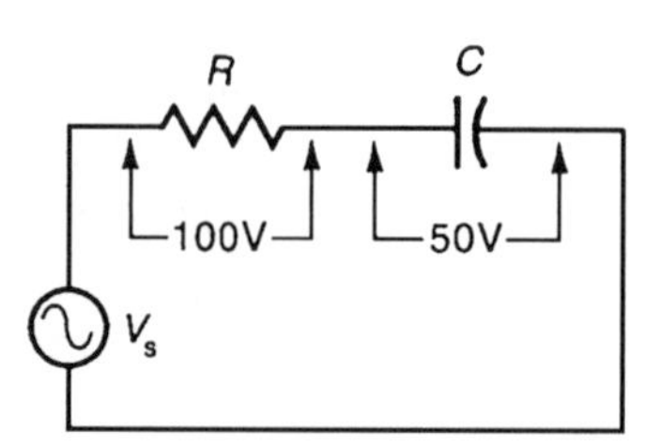

Exhibit 74 A series ac circuit

75. In Exhibit 75, the bridge circuit has a certain R_x whose value causes the ammeter to read zero. Determine R_x.

a. 100 Ω
b. 80 Ω
c. 20 Ω
d. 10 Ω

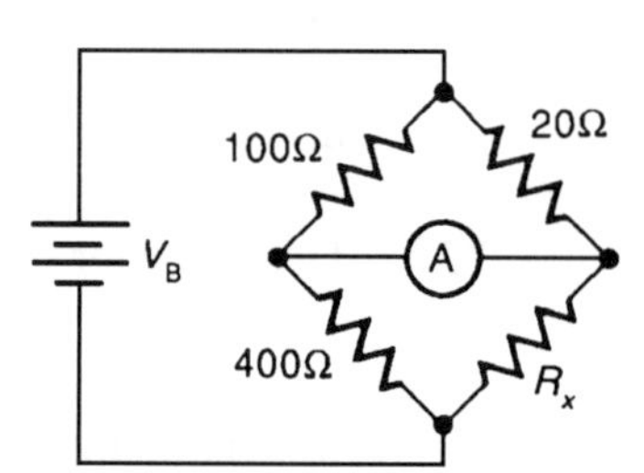

Exhibit 75 A resistive bridge circuit

76. An ac current is given in its polar form as $25\angle{-45°}$ amperes; what is the rectangular equivalent form?

a. $25 - j45$ A
b. $25 - j25$ A
c. $17.7 - j17.7$ A
d. $25 + j25$ A

77. Two capacitors, C_1 and C_2, are connected in series. If the series combination is 7.3 μF and it is known that C_1 is 9.6 μF, what is C_2 (given in microfarads)?

a. 2.3
b. 30.5
c. 35.0
d. 49.3

78. Two charges, A and B, are placed one meter apart. The charge on A is +5 μC and on B is −10 μC. Determine the electric field, **E**, at a point half way between them.

a. 0.54×10^6 toward B
b. 0.54×10^6 toward A
c. 0.18×10^6 toward B
d. 0.18×10^6 toward A

79. A summing op-amp circuit as shown in Exhibit 79 has an output voltage of

a. −2 V
b. +2 V
c. −5 V
d. +0.5 V

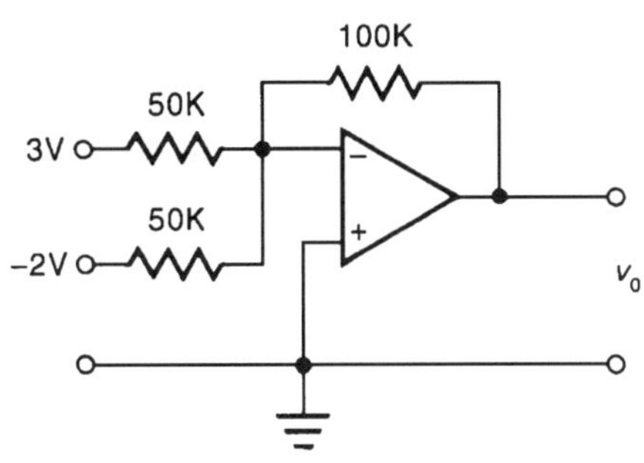

Exhibit 79 An op-amp circuit

80. For an ac source of 100 volts connected to a series R–C circuit of $10 - j10$ ohms, the real power dissipated by the load is

a. 250 watts
b. 400 watts
c. 500 watts
d. 750 watts

81. What is Thevenin's voltage as seen by R_L in Exhibit 81?

a. 10 V
b. 7.5 V
c. 5 V
d. 2.5 V

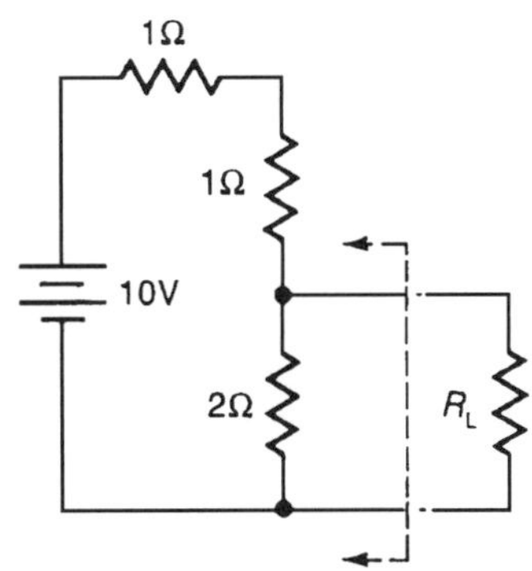

Exhibit 81 A dc circuit

82. For an *RLC* series ac circuit connected to a 100-volt source that has a (radian) frequency of 10 Hz, determine the magnitude of the current if $R = 10\ \Omega$, $L = 1$ H, and $C = 0.01$ F.

a. 3.3 A
b. 6.6 A
c. 9.1 A
d. 10 A

83. An ac voltage source of 100 volts supplies a load that is a parallel *R–C* combination and is given by $Y = 0.1 + j0.1$ siemens. What amount of power is dissipated in the load?

a. 250 watts
b. 400 watts
c. 500 watts
d. 1000 watts

84. A 5:1 transformer interfaces an amplifier and a speaker where the low side goes to the 8-ohm speaker. How many ohms does the speaker appear to have as viewed from the amplifier side?

a. 200 Ω
b. 64 Ω
c. 40 Ω
d. 32 Ω

Sample Examination Problems/AM—Materials Engineering

85. Which of the following crystalline imperfections will increase the strength of a metal most significantly?

a. interstitials
b. vacancies
c. dislocations
d. stoichiometry

86. Other factors being equal, which of the following variables will *not* affect the diffusion flux in a solid?

a. mass of the solid
b. concentration gradient of the solute
c. atomic packing factor of the solvent
d. temperature of diffusion

87. A structural variable in a multiphase microstructure that is *not* present in a single-phase microstructure is

a. numbers of dislocations
b. phase or grain shape
c. phase or grain size
d. amounts of phases or grains

88. Two steels have identical compositions (99.2Fe–0.8C) but different microstructures (pearlite and spheroidite). Which of the following statements is correct?

a. The pearlite was formed by rapid cooling (quenching) of the steel.
b. The phases in each are ferrite and austenite.
c. The pearlitic steel is denser.
d. The pearlitic steel will have less ductility.

89. In which one of the following processes does the material undergo a significant (>3%) change in volume?

a. extrusion
b. precipitation hardening
c. solution hardening
d. sintering

90. In which one of the following processes does the material undergo a noticeable change in shape?

a. extrusion
b. precipitation hardening
c. solution hardening
d. sintering

91. The value for which one of the following is commonly reported with results of fatigue testing?

a. fracture toughness
b. stress intensity factor
c. endurance limit
d. percent per hour

92. Which of the following is NOT associated with radiation damage?

a. vacancies
b. increased hardness
c. higher thermal conductivity
d. interstitial atoms

Sample Examination Problems/AM—Chemistry

93. A student is titrating 50 ml of 0.2 N HCI solution with a solution of 0.2 N KOH. He accidentally adds one ml too much titrant. What is the pH of the resulting solution?

a. 10.3
b. 11.3
c. 2.7
d. 7.3

94. $H_2(g)$, $I_2(g)$, and HI(g) are in a chamber at 25°C. It is found that the partial pressures are: $H_2(g) = 0.10$ atm, $I_2(g) = 0.066$ atm, and $HI(g) = 2.3$ atm. Calculate K_p and K_c for the reaction: $H_2(g) + I_2(g) = 2\ HI(g)$.

a. $K_p = 8.0 \times 10^1$, $K_c = 3.3$
b. $K_p = 1.2 \times 10^{-3}$, $K_c = 1.2 \times 10^{-3}$
c. $K_p = 8.0 \times 10^2$, $K_c = 8.0 \times 10^2$
d. $K_p = 3.5 \times 10^2\ \text{atm}^{-1}$, $K_c = 1.4 \times 10^1\ (\text{mol/L})^{-1}$

95. A balanced equation for the titration of ferrous iron in solution is

$$6\ FeCl_2 + K_2Cr_2O_7 + 14HCl = 6\ FeCl_3 + 2\ CrCl_3 + 2\ KCl + 7\ H_2O$$

In this reaction:

a. H^+ is reduced to the H^0 oxidation state
b. K^+ ions effect the oxidation of the ferrous ions
c. Cl^- ions catalyze the oxidation
d. Cr is reduced from a +6 oxidation state to a +3 oxidation state

96. An electrochemical cell has an Sn(s) anode and a Cu(s) cathode. The electrodes are in standard solutions of Sn^{2+} ions and Cu^{2+} ions, respectively. Calculate the standard cell potential (25°C) from the following standard reduction potentials for the half-cells:

$$Sn^{2+} + 2e^- = Sn(s);\ E^0 = -0.14\,V$$
$$Cu^{2+} + 2e^+ = Cu(s);\ E^0 = +0.34\,V$$

Also indicate whether the reaction is spontaneous electron flow in an external circuit from anode to cathode or cathode to anode.

a. +0.24 V (anode to cathode)
b. +0.20 V (anode to cathode)
c. −0.20 V (cathode to anode)
d. +0.48 V (anode to cathode)

97. Silver is electroplated from a solution of the cyanide complex according to the reduction reaction:

$$[Ag(CN)_2]^- + e^- = Ag(s) + (CN)^-$$

What current in amperes is needed to deposit 1.00 g silver on the workpiece in 0.50 hr? (a.w. Ag = 107.0 g/mol, F = 96,485 coulombs (C)/mol)

a. 0.50 A
b. 0.25 A
c. 1.00 A
d. 2.00 A

98. The gaseous reaction

$$F_2 + 2\ NO_2 \rightarrow 2\ NO_2F$$

is known to be first order in $[F_2]$. In an experiment, the initial concentration $[F_2]$ is held constant and the concentration $[NO_2]$ is doubled. The rate of reaction (R) is found to double. What is the rate law, and what is the overall order of the reaction?

a. $R = k\ [F_2]\ [NO_2]^2$, third order
b. $R = k\ [NO_2]$, first order
c. $R = k\ [F_2]\ [NO_2]$, second order
d. $R = k\ [F_2]^2\ [NO_2]$, third order

99. A 10-g sample of a hydrocarbon is analyzed by combustion and found to contain 9.37 g carbon and 0.63 g hydrogen. The molecular weight of the hydrocarbon sample is estimated from a freezing-point-lowering experiment to be 128 g/mol (a.w. C = 12.01, H = 1.01). The molecular formula of the sample is

a. C_8H_{18}
b. $C_{10}H_8$
c. C_9H_4O
d. C_5H_4

100. An iron-containing sample of 1.00 g is dissolved in acid solution and reduced. The reduced iron, Fe^{2+}, is titrated with 0.100 N (normal) $KMnO_4$ solution, and the titre to reach the endpoint is 100.0 ml. The balanced equation for the reaction is

$$10\ FeSO_4 + 2\ KMnO_4 + 8\ H_2SO_4$$
$$= 5\ Fe_2(SO_4)_3 + K_2SO_4 + 2\ MnSO_4 + 8\ H_2O$$

(a.w. Fe = 55.85, m.w. $FeSO_4$ = 151.92). Based on the weight of the original sample, what was the percent of iron in the sample?

a. 79.8
b. 27.9
c. 15.2
d. 55.8

101. Two gas storage bulbs are at the same temperature. The 1.00-L bulb contains 0.50 atm of helium (He) and the 2.00-L bulb contains 1.00 atm nitrogen (N_2). The valves connecting the two bulbs are opened and they are connected through a negligible volume. The pressure in the bulbs after they have pressure equilibrated and are at the initial temperature is

a. 1.50 atm
b. 0.67 atm
c. 0.50 atm
d. 0.84 atm

102. A solution of 32.00% sulfuric acid (H_2SO_4) (32.00 g solute/100.0 g solution) has a density of 1.2353 kg/L (m.w. H_2SO_4 = 98.08, H_2O = 18.02). What is the solute concentration in g/L?

a. 320.0
b. 395.3
c. 259.0
d. 403

103. What is the molar concentration of the acid solution of Problem 102 in g • mol/L?

a. 3.953
b. 3.263
c. 3.877
d. 4.030

Sample Examination Problems/AM—Engineering Economics

104. If a loan of $15,000 for four months yields $975 interest, the annual interest rate is most nearly:
a. 6.50%
b. 9.75%
c. 15.38%
d. 19.50%

105. If you had $1000 now and invested it at 6%, how much would it be worth 12 years from now?
a. $1720
b. $1820
c. $1920
d. $2012

106. What present sum is equivalent to a series of $1000 annual end-of-year payments, if a total of 10 payments are made and interest is 6%? The present sum is most nearly
a. $6260
b. $7350
c. $9400
d. $10,000

107. The present worth of an obligation of $10,000 due in 10 years, if money is worth 8%, is nearest to
a. $4500
b. $6000
c. $7500
d. $9000

108. A firm charges $1\frac{1}{2}$% interest per month on credit purchases. The equivalent effective annual interest rate is nearest to
a. 1.50%
b. 18.00%
c. 18.50%
d. 19.56%

Sample Examination Problems/AM—Computers

109. The binary representation 10101 corresponds to which base-10 number?
a. 3
b. 16
c. 21
d. 10,101

110. For the base-10 number of 30, the equivalent binary number is
a. 1110
b. 0111
c. 1111
d. 11110

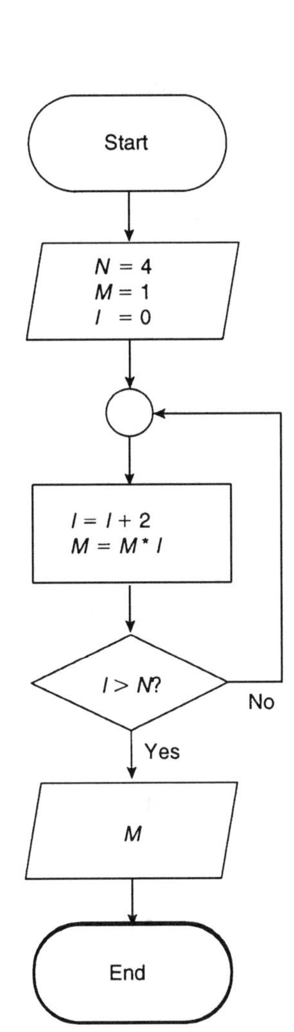

Exhibit 112

111. On personal computers the data storage device with the largest storage capacity is most likely to be

a. $3\frac{1}{2}$" diskette
b. hard disk
c. random access memory
d. $3\frac{1}{2}$" HD diskette

112. The output value of M in Exhibit 112 is closest to

a. 1
b. 2
c. 8
d. 48

113. Pseudocode can best be described as

a. a simple letter-substitution method of encryption.
b. an English-like language representation of computer programming.
c. a relational operator in a database.
d. the way data are stored on a diskette.

114. The pseudocode that best represents Exhibit 114 is

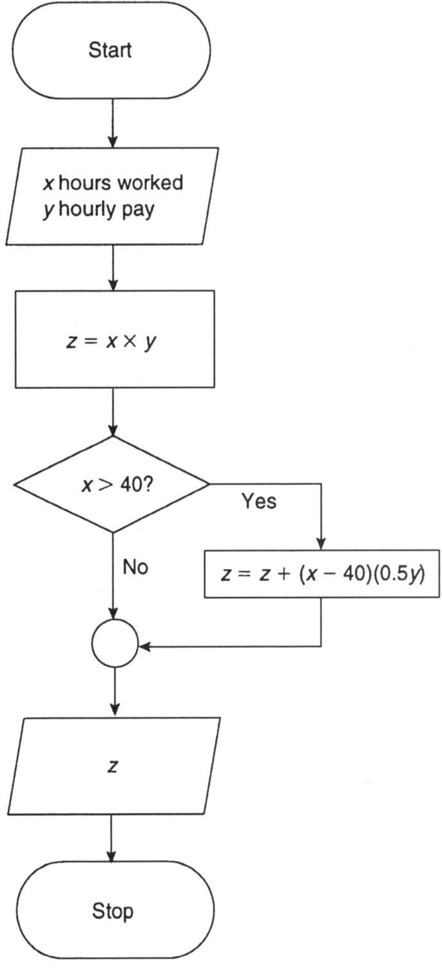

Exhibit 114

a. INPUT X, Y
WAGE = X × Y
IF X less than 40 THEN
WAGE = WAGE + OVERTIME PAY
END IF
OUTPUT WAGE

b. INPUT hours worked and hourly pay
WAGE = hours worked × hourly pay
IF hours worked greater than 40 THEN
WAGE = WAGE + (hours worked – 40)
(overtime wage supplement)
END IF
OUTPUT WAGE

c. INPUT hours worked and hourly pay
WAGE = (hours worked – 40)(overtime rate) + 40(hourly pay)
OUTPUT WAGE

d. PRINT hours worked and hourly pay
WAGE = hours worked × hourly pay
IF hours worked greater than 40 THEN
WAGE = hours worked × hourly pay + (hours worked – 40)
× overtime pay
END IF
OUTPUT WAGE

115. All of the following are true of spreadsheets **except**

a. A cell may contain label, value, formula, or function.
b. A cell reference is made by a numbered column and lettered row reference.
c. A cell content that is displayed is the result of a formula entered in that cell.
d. Line graphs, bar graphs, stacked bar graphs, and pie charts are typical graphs created from spreadsheets.

Sample Examination Problems/AM—Ethics

116. The *NCEES Model Rules of Professional Conduct* do not describe which one of the following?

a. Obligations to safeguard the welfare of the public.
b. Obligations to other college students.
c. Obligations to other engineers.
d. Obligations to an employer and/or clients.

117. The *NCEES Model Rules of Professional Conduct* do not allow an engineer to improve his/her competence in a specific area in which one of the following ways?

a. Studying the work of other engineers.
b. Study at a university outside of the United States.
c. Performing services in an area where the engineer is not yet competent.
d. Work in a specific area under the direction of a competent but unregistered engineer.

118. The *NCEES Model Rules of Professional Conduct* require the registered engineer to do all but one of the following actions. Which one should not be done?

a. Make no statements on technical matters unless an interested party for whom the engineer is acting is disclosed.
b. Approve documents only if they safeguard the life, health, property, and welfare of the public.
c. Reveal facts and data obtained in a professional capacity only after obtaining consent of the client or employer, unless required to reveal the facts and data by law.
d. Comment on professional matters publicly even though the engineer has not yet had an opportunity to make a careful evaluation of the facts and subject matter.

119. The *NCEES Model Rules of Professional Conduct* prohibit the registered engineer from doing all but one of the following. Which one can the engineer do?

a. Indiscriminately criticize the work of another registered engineer.
b. Giving gifts to people in order to promote a business and obtain work.
c. Affix a signature or seal to any plans or documents not prepared under the engineer's control and supervision.
d. Accept compensation from more than one party for services on the same project, even though the circumstances are fully disclosed and agreed to by all parties.

120. Jim, a young civil engineer, was assigned by his oil company employer to oversee the construction of a 40-mile portion of a pipeline to carry petroleum products. After the construction had begun he found that the pipeline contractor had a practice of having a big party and dinner every Friday night in the only restaurant in town. The contractor, his foremen, and other invited guests attended. The contractor told Jim that since he was alone on Friday evenings, he would like Jim to come to the weekly party and dinner as a guest of the contractor. Jim would like to attend but must decide what to do.

a. Jim decides to attend the weekly party and dinner. Jim feels as a professional, his professional judgement would not be distorted by these weekly gatherings.
b. Jim decides he would enjoy the weekly party and dinner but decides to attend only every other week.
c. Jim tells the contractor that his employer's rules would prevent him from being the contractor's guest, but he could attend if he paid his share of the cost.
d. Jim rejects the invitation saying it would not be appropriate for an oil company professional engineer to attend.

Sample Examination Solutions/AM—Mathematics

1. b. The Pythagorean theorem: $d^2 = (x_2 - x_1)^2 + (y_2 - y_1)^2$, Exhibit 1.

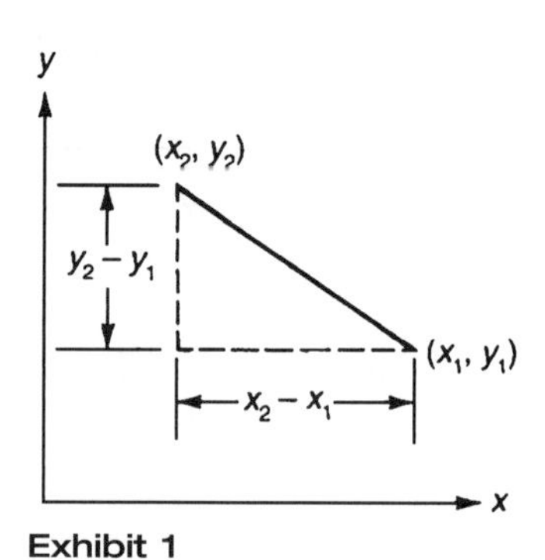

Exhibit 1

2. c. In the second quadrant (Exhibit 2) all functions are negative except the sine and cosecant.

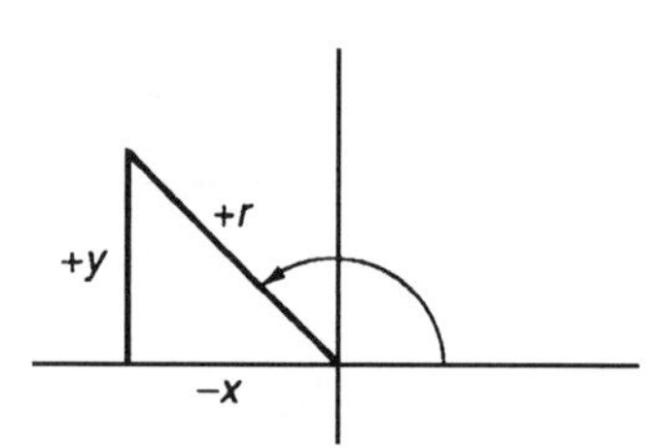

Exhibit 2

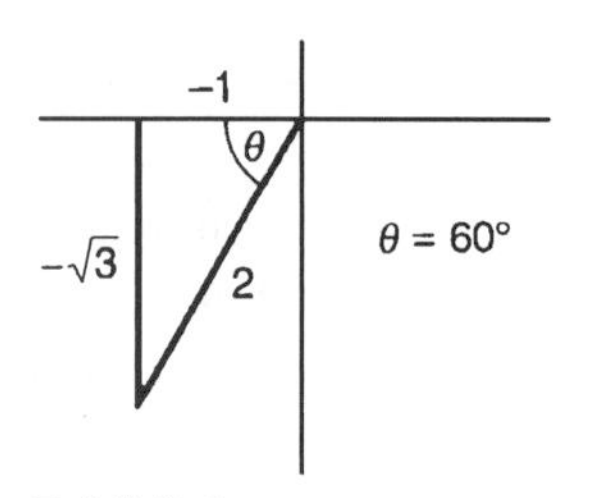

Exhibit 3

3. a.

$$\csc 960^\circ = \csc(5\pi + 60)$$
$$= \csc\theta = \frac{r}{y} = \frac{2}{-\sqrt{3}} = -\frac{2\sqrt{3}}{3}$$

4. c.

$$i^{27} = \left(\sqrt{-1}\right)^{27} = \sqrt{(-1)^{27}} = \sqrt{(-1)^{26}(-1)} = (-1)^{13}\sqrt{-1} = -1\sqrt{-1}$$

Now substituting back i for $\sqrt{-1}$, $i^{27} = -1(i) = -i$.

5. d.

$$\int y\,dx = \int (x^3 - x + 1)dx = \frac{x^4}{4} - \frac{x^2}{2} + x + c$$

6. b. To find maxima, minima, and points of inflection: Take the first derivative, equate it to zero, and solve for its roots; take the second derivative, and insert the above roots to determine whether each is a maximum or a minimum.

Equate the second derivative to zero and solve for its root(s) or point(s) of inflection.

Now apply these steps to the problem: $y = x^3 + x^2 - 3$

	Max.	Min.	Inflection
$\frac{dy}{dx}$	0	0	anything
$\frac{d^2y}{dx^2}$	–	+	0

$$\frac{dy}{dx} = 3x^2 + 2x = 0 \quad x(3x+2) = 0 \quad \therefore x = 0, -2/3, \text{ maximum and minimum exist}$$

$$\frac{d^2y}{dx^2} = 6x + 2 \quad \text{At } x = 0, \frac{d^2y}{dx^2} = +2, \text{ a minimum exists}$$

$$\text{At } x = -2/3, \quad \frac{d^2y}{dx^2} = -4 + 2 = -2, \text{ a maximum exists.}$$

$$\frac{d^2y}{dx^2} = 6x + 2 = 0$$

At $x = -1/3$, an inflection point exists.

7. d. Taking the sine curve,

$$f(x) = \sin x \quad f'(x) = \cos x \quad f''(x) = -\sin x$$

Thus a function whose second derivative is equal to the negative of the equation of that same function is a sinusoid.

8. d.

9. d. Find the roots of $x(t) = 0$. Since $t = 0$ is when she leaves, solve $t^2 - 8t + 16 = 0$. Thus $t = 4$ hours.

10. b. The maximum value of $x(t)$ occurs when $x'(t) = 4t^3 - 24t^2 + 32t = 0$ when $t = 0$ (wrong) or when $4t^2 - 24t + 32 = 0$, or $t^2 - 6t + 8 = (t - 2)(t - 4) = 0$. When $t = 4$, she is home (Problem 9). Thus $t = 2$, $x(2) = 16 - 64 + 64 = 16$ miles.

11. c. Since she travels 16 kilometers away and 16 kilometers back in 4 hours, her average speed is 8 kph.

12. a. Differentiate the velocity found in Problem 10:

$$x''(t) = 12t^2 - 48t + 32 = 0 \text{ or } 3t^2 - 12t + 8 = 0,$$

$$t = \frac{12 \pm \sqrt{144 - 96}}{6} = 2 \pm \frac{\sqrt{48}}{6} = 2 \pm \frac{2}{\sqrt{3}}$$

13. d. The associated homogeneous equation, $y'' + 4y = 0$, has the solution $y_h = \text{A} \sin 2x + \text{B} \cos 2x$. Using the method of undetermined coefficients,

$$y_p = \text{a} \sin x + \text{b} \cos x$$
$$y_p'' = -\text{a} \sin x - \text{b} \cos x,$$
$$y_p'' + 4y_p = (-\text{a} + 4\text{a}) \sin x + (-\text{b} + 4\text{b}) \cos x = 8 \sin x.$$

Thus b = 0 and a = 8/3.

14. d. The characteristic equation is $r^2 - 5r + 6 = 0$, $(r - 3)(r - 2) = 0$, $r = 2, 3$.

15. c. This is an Euler equation. Try $y = x^m$ to obtain $m(m-1)x^m - 3mx^m + 4x^m = 0$ or $m^2 - 4m + 4 = 0$. Thus $m = 2$ (twice).

16. c.

$$(p^2 - 3p - 4)Y(p) = \frac{1}{p-1} + 1$$

$$Y(p) = \frac{1 + p - 1}{(p-1)(p-4)(p+1)} = \frac{p}{(p-1)(p+1)(p-4)}$$

By partial fractions,

$$Y(p) = \frac{-\frac{1}{6}}{p-1} - \frac{\frac{1}{10}}{p+1} + \frac{\frac{4}{15}}{p-4}$$

The solution of the differential equation is:

$$Y(t) = -\frac{1}{6}e^{t} - \frac{1}{10}e^{-t} + \frac{4}{15}e^{4t}.$$

17. c. The number of ways of drawing five cards from the deck is given by

$$n = C(52,5) = \frac{52 \times 51 \times 50 \times 49 \times 48}{5 \times 4 \times 3 \times 2 \times 1} = 2{,}598{,}960$$

The number of ways, n, that a hand can be a royal flush in four, since there are four suits in a deck. Hence the probability of getting a royal flush is $P\{\text{royal flush}\} = n/m = 4/2{,}598{,}960 = 0.000\,001\,539$.

18. d. This is the cumulative probability and represents the area under the probability curve, including 2 through 5.

$$P(<6) = P(2) + P(3) + P(4) + P(5) = (1 + 2 + 3 + 4)/36 = 10/36$$

19. a. The ring permutation formula applies. So

$$P_{\text{ring}} = P(n,r)/r = P(5,5)/5 = 5!/(5 \times (5-5)!) = 5 \times 4 \times 3 \times 2 \times \frac{1}{5} = 24$$

20. c. Out of 36 possible outcomes, "four" occurs three times: (1 and 3), (2 and 2—the "hard way"), and (3 and 1).

21. b. If two matrices are equal, then their corresponding elements must be equal. Therefore $X = 0$, and $X + Y = 2$. Since $X = 0$, Y must equal 2.

22. d. $\vec{\mathbf{a}} \bullet \vec{\mathbf{b}} = 1(3) + 2(0) + 2(4) = 11$.

23. d. Expanding by minors along the first column,

$$\begin{vmatrix} 1 & 1 & 0 & 0 \\ 1 & 1 & 1 & 0 \\ 0 & 1 & 1 & 1 \\ 0 & 0 & 1 & 1 \end{vmatrix} = 1\begin{vmatrix} 1 & 1 & 0 \\ 1 & 1 & 1 \\ 0 & 1 & 1 \end{vmatrix} - 1\begin{vmatrix} 1 & 0 & 0 \\ 1 & 1 & 1 \\ 0 & 1 & 1 \end{vmatrix}$$

$$= 1\begin{vmatrix} 1 & 1 \\ 1 & 1 \end{vmatrix} - 1\begin{vmatrix} 1 & 0 \\ 1 & 1 \end{vmatrix} - 1\begin{vmatrix} 1 & 1 \\ 1 & 1 \end{vmatrix} + 1\begin{vmatrix} 0 & 0 \\ 1 & 1 \end{vmatrix}$$

$$= -1(1) + 1(0) = -1$$

24. d.

$$6 - 2X = -6 - (-X)$$
$$= -6 + X$$

Collecting like terms gives $12 = 3X$, so $4 = X$.

Sample Examination Solutions/AM—Statics

25. a.

$$\mathbf{M}_o = [(4\text{ m})\mathbf{e}_x + (12\text{ m})\mathbf{e}_y] \times F\left(-\frac{4}{5}\mathbf{e}_x + \frac{3}{5}\mathbf{e}_z\right) = \frac{12}{5}(3\mathbf{e}_x - \mathbf{e}_y + 4\mathbf{e}_z)\text{m}F$$

26. d.

$$\mathbf{e}_{OP} = \frac{4}{13}\mathbf{e}_x + \frac{12}{13}\mathbf{e}_y + \frac{3}{13}\mathbf{e}_z$$

$$M_{OP} = \boldsymbol{e}_{OP} \bullet \boldsymbol{M}_O = \frac{12\text{m}F}{5}\left[\frac{4}{13}(3) + \frac{12}{13}(-1) + \frac{3}{13}(4)\right] = \frac{144}{65}\text{m}F.$$

27. b.

$$\Sigma M_A = (2\text{ m})\cos 45°(600\text{ N}) - (2\text{ m})\cos 45°\ T_{CD} = 0$$
$$T_{CD} = 600\text{ N}$$

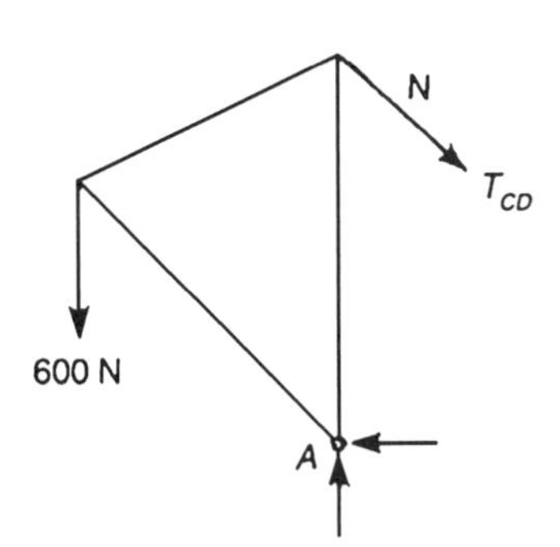

Exhibit 27a

28. c. Diagram (i) shows forces internal to the system, and the system is not completely isolated. The system depicted in diagram (ii) is not completely isolated. Diagrams (iii), (iv), and (v) are all correct.

29. d.

$$\sum F_{\text{vert}} = \tfrac{4}{5}T_{EH} - 4\,\text{kN} - 4\,\text{kN} = 0$$
$$T_{EH} = 10 \text{ kN (tension)}$$

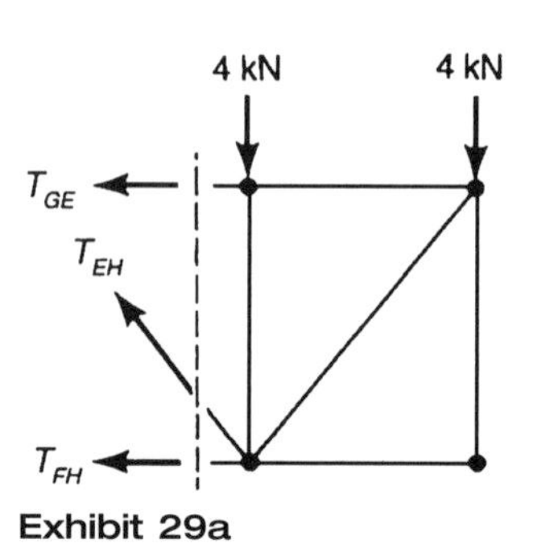

Exhibit 29a

30. d.

$$\mathbf{R} = -[(50 \text{ kN})(\sin 45° \,\mathbf{e}_x + \cos 45° \,\mathbf{e}_y) + 30 \text{ kN } \mathbf{e}_y]$$
$$= -(35.4 \text{ kN } \mathbf{e}_x + 65.4 \text{ kN } \mathbf{e}_y)$$
$$R = \sqrt{R_x^{\,2} + R_y^{\,2}} = 74.4\,\text{kN}$$

31. a. Look at the definition of θ in Exhibit 30.

$$\theta = \tan^{-1}\left(\frac{35.4}{65.4}\right) = 28.4°$$

32. a. $M_b = aP$; $M_t = bP$.

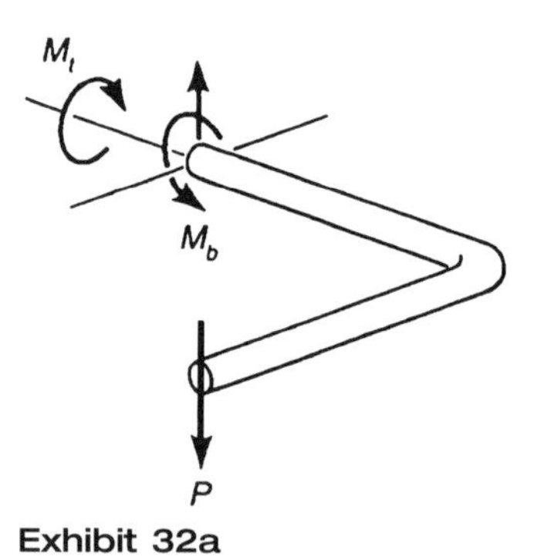

Exhibit 32a

33. d.

$$\Sigma M_A = 2aQ - (2a \text{ ctn } 30°)T = 0$$
$$T = \frac{Q}{\sqrt{3}}$$
$$\Sigma M_B = aT - 4aP = 0$$
$$P = \frac{T}{4} = \frac{Q/\sqrt{3}}{4} = \sqrt{3}Q/12$$

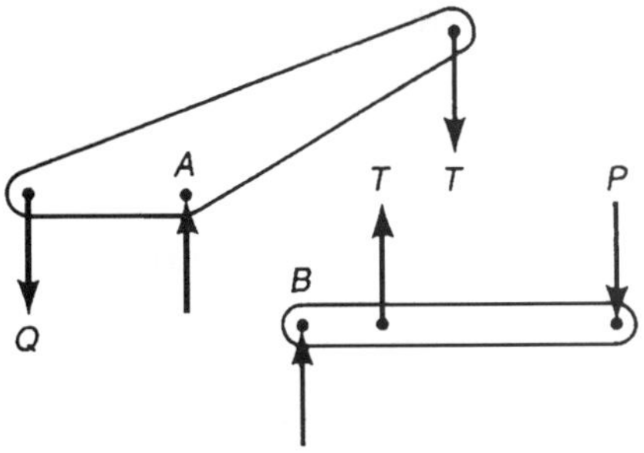

Exhibit 33a

34. a.

$$\sum F_x = 0, T\sin\alpha = \mu f_n$$

$$\sum M_O = 0, (a+b)T\cos\beta = af_n$$

$$\mu = \frac{T\sin\alpha}{f_n} = \frac{a\sin\alpha}{(a+b)\cos\beta}$$

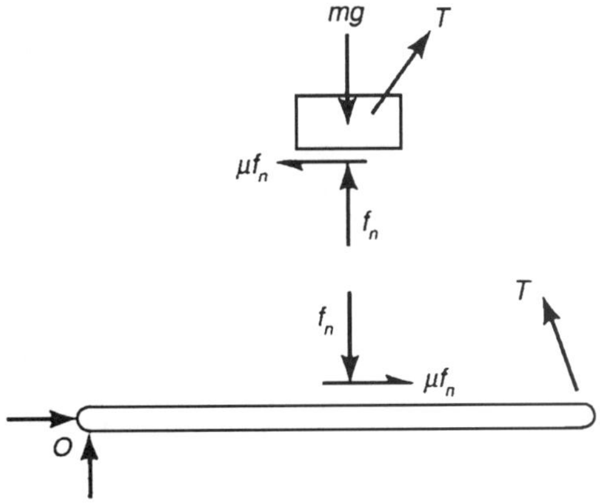

Exhibit 34a

35. d.

$$\mathbf{e}_{CB} = \frac{1.7\mathbf{e}_x + 2.9\mathbf{e}_y}{\sqrt{(1.7)^2 + (2.9)^2}} = 0.506\mathbf{e}_x + 0.863\mathbf{e}_y$$

$$\mathbf{M}_C = (2.5 \text{ m})mg\ \mathbf{e}_x$$

$$M_{CB} = \mathbf{e}_{CB} \bullet \mathbf{M}_C = (1.265 \text{ m})\ mg$$

36. a. $\Sigma M_{CB} = (1.265 \text{ m})mg - (1.471 \text{ m})T_A = 0;\ T_A = 0.860\ mg.$

Sample Examination Solutions/AM—Dynamics

37. c. In Exhibit 37a, $\Sigma F_x = ma_x$ implies $mg \sin \theta = ma_x$ and $a_x = g \sin \theta$.

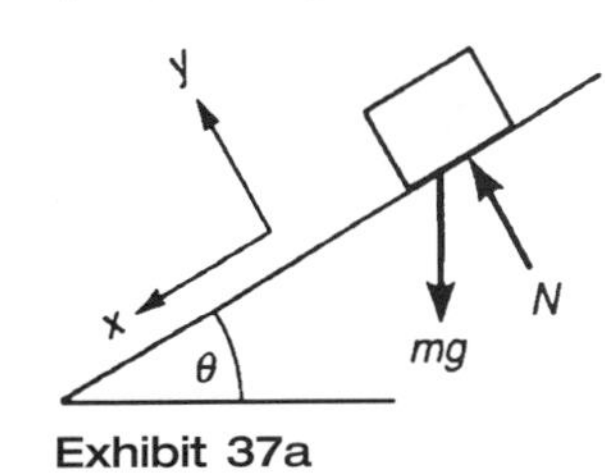

Exhibit 37a

38. a. Since $x = (t^2 - 2t + 10)$ m is given, direct differentiation gives $v = (2t - 2)$ m/s. Observe that $v = 0$ when $2t - 2 = 0$ or $t = 1$ s.

The total distance traveled is

$$
\begin{aligned}
D &= |x(0) - x(1)| + |x(1) - x(10)| \\
&= |10 - (1 - 2 + 10)| + |(1 - 2 + 10) - (100 - 20 + 10)| = 82\,\text{m}
\end{aligned}
$$

39. c. Let ω_F be the angular speed of the front wheel, and r_F its radius. The velocity of its center is $v_F = \omega_F r_F$. Since the vehicle is in translation, the velocity of the center of the rear wheel must be the same:

$$v_R = \omega_R r_R = \omega_F r_F$$

Thus,

$$\frac{\omega_F}{\omega_R} = \frac{r_R}{r_F} = \frac{r}{\frac{2}{3}r} = \frac{3}{2}$$

40. d. Horizontal motion:

$$x = x_0 + v_{x0}t + (a_x t^2)/2$$

Here $150 = 0 + v_0 \sin 60° (10) + 0$, so $v_0 = 17.3$ m/s.

41. d. Using the coordinate system shown in Exhibit 41a,

$$
\begin{aligned}
a_r &= \ddot{r} - r\dot{\theta}^2 = 0 - (3+2)(1)^2 = -5\,\text{m/s}^2 \\
a_\theta &= r\ddot{\theta} + 2\dot{r}\dot{\theta} = 0 + 2(3)(1) = 6\ \text{m/s}^2
\end{aligned}
$$

Hence,

$$a = \sqrt{a_r^2 + a_\theta^2} = \sqrt{(-5)^2 + 6^2} = 7.8\ \text{m/s}^2$$

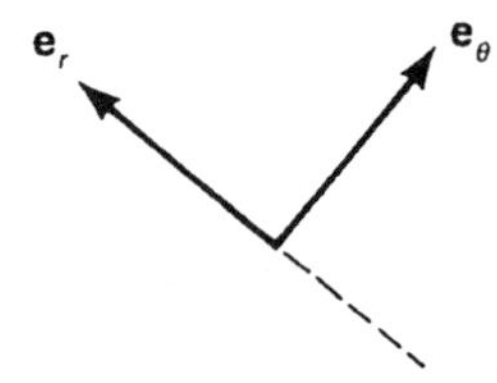

Exhibit 41a

42. a. Refer to Exhibit 42a. The y-coordinate of G is given by the relation

$$y_G = \frac{m_A y_A + m_B y_B}{m_A + m_B} = \frac{m(1) + (2m)(0)}{m + 2m} = \frac{1}{3}\ \text{m}$$

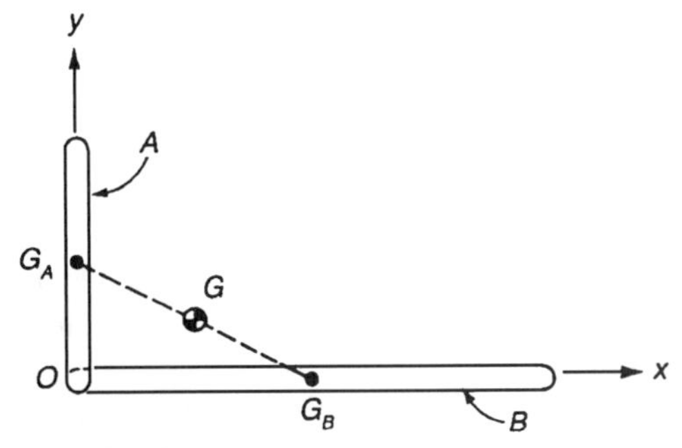

Exhibit 42a

43. d. With the aid of Exhibit 43a, apply the work-energy principle:

$$W_{1-2} = T_2 - T_1$$

$$-Rx = 0 - \frac{1}{2}mv_1^2 \quad \text{or} \quad R = \frac{mv_1^2}{2x}$$

Substituting values, we obtain $R = 96.5$ kN.

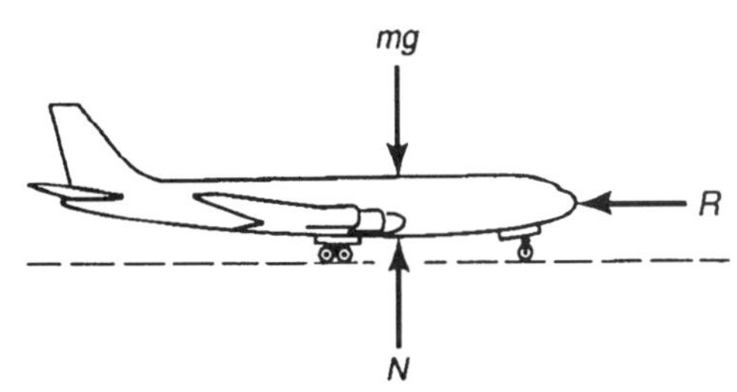

Exhibit 43a

44. d. Using subscripts H and V to indicate the horizontal and vertical positions respectively,

$$W_{H-v} = \frac{1}{2}k\theta_H^2 - \frac{1}{2}k\theta_v^2 = \frac{1}{2}(5)\left[\left(\frac{\pi}{4}\right)^2 - \left(\frac{\pi}{4} + \frac{\pi}{2}\right)^2\right] = -12.3\ \text{N}\bullet\text{m}$$

45. d. The net driving torque is $M = 10 - 4 = 6$ N • m and is constant. The angular momentum at time $t = 0$ is $H_0 = 0$, and the angular momentum at $t = 7$ s is $H = I\omega = 0.5(mr^2\omega) = (0.5)(5)(1)^2\omega = 2.5\omega$. The impulse momentum principle gives

$$H_0 + \int_0^7 Mdt = H \quad \text{or} \quad 0 + M\Delta t = H$$

Hence, $6(7) = 2.5\omega$, so $\omega = 16.8$ rad/s.

Sample Examination Solutions/AM—Mechanics of Materials

46. b. Draw the free-body diagrams. The maximum force is $F_{AB} = 11$ kN. The maximum stress is, therefore,

$$\sigma = \frac{P}{A} = \frac{11\,\text{kN}}{500\,\text{mm}^2} = 22\,\text{MPa}$$

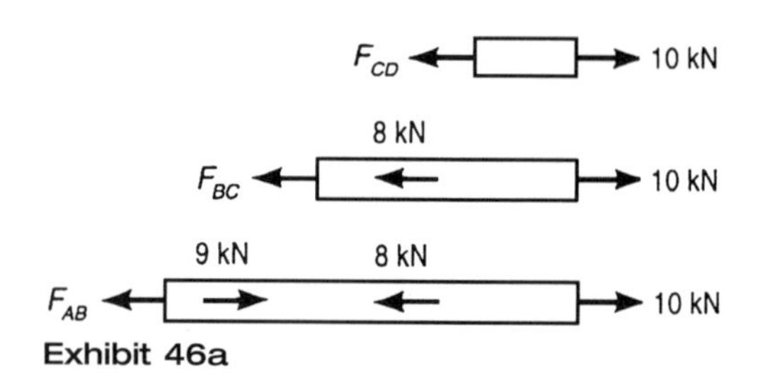

Exhibit 46a

47. c. For the stress criteria,

$$\sigma = \frac{P}{A} \qquad A = \frac{P}{\sigma} = \frac{40\,\text{kN}}{100\ \text{MPa}} = 400\,\text{mm}^2$$

For the displacement criteria,

$$\delta = \frac{PL}{AE} \qquad A = \frac{PL}{\delta E} = \frac{(40\ \text{kN})(600\ \text{mm})}{(0.25\ \text{mm})(210\ \text{GPa})} = 457\,\text{mm}^2$$

The larger of the two is the minimum area to satisfy both criteria.

48. d. The maximum shear stress is

$$\tau_{\max} = \frac{Tr_O}{J} = \frac{(1\ \text{kN} \bullet \text{m})(25\ \text{mm})}{0.5\pi\left(25^4 - r_i^4\right)} = 100\ \text{MPa}$$

Solving for r_i gives

$$r_i^4 = (25\ \text{mm})^4 - \frac{(1\ \text{kN}\bullet\text{m})(25\ \text{mm})}{0.5\pi(100\ \text{MPa})} = 231 \times 10^3\ \text{mm}^4$$

$$r_i = 21.9\ \text{mm} \qquad D_i = 43.9\ \text{mm}$$

49. a. This problem can be solved by doing the necessary steps to plot the shear diagram or by the process of elimination. Using the process of elimination the shear must be constant between loads which rules out (B) and (D). The jump at the concentrated load must be downward which rules out (C).

50. c. This problem can be solved by doing the necessary steps to plot the moment diagram or by the process of elimination. Using the process of elimination, the moment must be zero at the ends which rules out (B). The jump at the concentrated moment must be upward which rules out (A). There must be a discontinuity in the slope at the concentrated load which rules out (D).

51. d. Mohr's circle gives the following relations:

$$R = \sqrt{\left(\frac{\sigma_x - \sigma_y}{2}\right)^2 + \tau_{xy}{}^2} = \sqrt{\left(\frac{60 - 20}{2}\right)^2 + 20^2} = 28.3\ \text{MPa}$$

$$C = \frac{\sigma_x + \sigma_y}{2} = \frac{60 + 20}{2} = 40\ \text{MPa}$$

$$\sigma_1 = C + R = 68.3\ \text{ksi}$$

52. c. If the pole is not to buckle, the critical buckling load cannot be exceeded. The critical buckling load for a cantilever is

$$P_{cr} = \frac{\pi^2 EI}{4L^2} = \frac{\pi^2 (12.6 \text{ GPa})\left[\frac{\pi (50 \text{ mm})^4}{4}\right]}{4(15 \text{ m})^2} = 678 \text{N}$$

53. a. The definition of shear strain is the decrease in angle of lines originally coincident with the x and y axes. Since the angle is small and increasing, the shear strain is

$$\gamma_{xy} = -\frac{0.50 \text{ mm}}{600 \text{ mm}} = -0.000833$$

Sample Examination Solutions/AM—Fluid Mechanics

54. a.

(U.S.) $$p = \gamma h = \text{SG}\left(62.4\frac{\text{lbf}}{\text{ft}^3}\right)h = 1.03\left(62.4\frac{\text{lbf}}{\text{ft}^3}\right)100 \text{ ft} \bullet \frac{1 \text{ ft}^2}{144 \text{ in.}^2} = 44.6 \text{ psig}$$

(S.I.) $$P = \gamma h = \text{SG}\left(9.81\frac{\text{kN}}{\text{m}^3}\right)h = 1.03\left(9.81\frac{\text{kN}}{\text{m}^3}\right)30.48 \text{ m} = 308 \text{ kPa}$$

55. b. The center of gravity is below the metacenter.

56. a.

$$F = \gamma h_e A. \text{ Assume } \gamma = 62.4\frac{\text{lbf}}{\text{ft}^3} = 9.81\frac{\text{kN}}{\text{m}^3}.$$

(U.S.) $$F = 62.4\frac{\text{lbf}}{\text{ft}^3} \bullet 3\text{ft}(3\text{ft} \bullet 6 \text{ ft}) = 3370 \text{ lbf}$$

(S.I.) $$F = 9.81\frac{\text{kN}}{\text{m}^3} \bullet 1\text{m} \bullet (1 \text{ m} \bullet 2 \text{ m}) = 19.6 \text{kN}$$

57. a. The viscosity of a gas increases and the viscosity of a liquid decreases with increases in temperature.

58. a.

$$h_p = h_c + \frac{I_c \sin^2 \alpha}{h_c A}$$

h_p = vertical distance from the fluid surface to the point of force application

h_c = vertical distance from the fluid surface to the centroid of the area

Since all quantities in the second term are positive, then $h_p > h_c$. Therefore the force always acts below the centroid of the flat surface area.

59. c.

$$Q=\sqrt{2g\left(\frac{p_1-p_2}{\gamma}\right)+z_1-z_2}$$

For a horizontal meter (where $z_1 = z_2$), Q is proportional to the square root of the pressure differential, $p_1 - p_2$.

60. c. This is Bernoulli's equation.

61. c. $F = \rho QV(\cos \alpha_1 - \cos \alpha_2) = \dot{m}V(\cos \alpha_1 - \cos \alpha_2)$. Thus the force is a function of the mass flow rate, $\dot{m}$, the velocity magnitude, V, and the change in flow direction.

Sample Examination Solutions/AM—Thermodynamics

62. d. Work is a form of energy which crosses the boundary of a system and is a path function.

63. a. For this closed system $u_1 + q = u_2 + w$, adiabatic, $q = 0$. Hence $w = u_1 - u_2$.

64. a. In a closed system $u_1 + q = u_2 + w$. Here, $w = +10$ kJ, and $u_2 - u_1 = -5$ kJ, so one finds $q = u_2 - u_1 + w = -5 + 10 = 5$ kJ.

65. d. Pressure lines have less slope than volume lines because $C_p > C_v$. ($C_p - R = C_v$). So, for a given Q, the ΔT for a constant pressure heat addition is less than for a constant volume heat addition.

66. a. The Otto cycle, in sequence, consists of an isentropic compression, a constant volume heat addition, an isentropic expansion, and finally a constant volume heat rejection.

67. d. An isolated system has neither energy nor mass crossing its boundary. Entropy can either remain constant, if nothing happens in the system, or increase if disorder increases.

68. b.

$$\eta_{TH}=\frac{T_H-T_C}{T_H}=\frac{(900-100)}{900+273}=0.68$$

69. c. Conservation of energy in this case is

$$\begin{aligned}q_{\text{out}} &= q_{\text{in}}+w\\ &=60{,}000+(4\times 1000)\\ &=64{,}000\ \text{W}\end{aligned}$$

70. a. Constant temperature lines are as shown on a p-v diagram. The behavior could also be deduced from the ideal gas laws

$$\frac{p_1 v_1}{RT_1} = \frac{p_2 v_2}{RT_2}$$

Since $T_1 = T_2$, $p_1(v_1/v_2) = p_2 < p_1$ owing to the volume increase.

71. d.

$$\text{COP}_{\text{REFR}} = \frac{T_C}{T_H - T_C} = \frac{-5+273}{40-(-5)} = 5.96$$

72. c. One dimensional, steady state, conductive heat transfer is described by

$$q = \frac{kA(T_H - T_C)}{d}$$

where k is the conductivity of the material. It is not a function of the pressures.

Sample Examination Solutions/AM—Electrical Circuits

73. d. The top parallel combination reduces to 2.0 Ω; the top leg is then (R + 2.0) Ω; the bottom leg takes 12 V/10 Ω = 1.2 A. Therefore, I_{top} is 2.4 – 1.2 = 1.2 A. This requires a voltage drop across the top parallel resistance of 1.2 A × 2.0 Ω = 2.4 volts, leaving a drop across R of 12 – 2.4 = 9.6 volts, which yields $R = V/I$ = 9.6/1.2 = 8 Ω.

74. d. The voltage across R is in phase with current, while that of the capacitor lags by 90°; thus

$$\mathbf{V} = R\mathbf{I} - jX_C\mathbf{I} = 100 - j50 = 112\angle -26.6°$$

75. b. Since zero current flows through an ammeter, the voltages at each end must be the same. The voltage at the left side is found to be V_{Left} = [400/(400 + 100)]V_B = (4/5)V_B. Therefore (4/5)V_B = [R_x/(R_x + 20)]V_B, solving for R_x = 80 Ω.

76. c.

$$25\angle -45° = 25(\cos 45° - j\sin 45°) = 17.7 - j17.7.$$

77. b.

$$1/C_{eq} = 1/C_1 + 1/C_2, \qquad 1/7.3 = 1/9.6 + 1/C_2, \qquad C_2 = 30.4\ \mu\text{F}$$

78. a.

$\mathbf{E}_{mid} = (q_A\mathbf{a}_r + q_B\mathbf{a}_r)/(4\pi r^2\varepsilon_0) = (+5\mathbf{a}_x - 10\mathbf{a}_{-x}) \times 10^{-6}/[4\pi(0.5)^2 \times 8.85 \times 10^{-12}]$
$= 0.54 \times 10^6$ V/m

79. a.

$v_0 = -[(R_f/R_1)v_1 + (R_f/R_1)v_2] = -[(100/50)3 + (100/50)(-2)] = -2$ volts

80. c. In polar form, $Z = 10 - j10 = 10\sqrt{2}\angle -45°$. The current is

$$\mathbf{I} = \mathbf{V}/\mathbf{Z} = 100\angle 0°(10\sqrt{2}\angle -45°) = 7.07\angle 45° \text{ A}$$

$P = I^2R = (7.07)^2 10 = 500$ watts. (Check: $P = VI\cos\theta = 100 \times 7.07 \cos 45°$.)

81. c. With R_L disconnected, $V_{out} = V_{oc} = [2/(1 + 1 + 2)]V_B = 5$ V.

82. d.

$X_L = 2\pi fL = \omega L = 10 \times 1 = 10\Omega;\qquad X_C = 1/(\omega C) = 1/(10 \times 0.01) = 10\ \Omega$
$\mathbf{Z} = 10 + j10 - j10 = 10, I = V/Z = 100/10 = 10\text{ A}$

83. d. The only real power that may be dissipated in the circuit is from the resistance. The expression for Y is $\mathbf{Y} = G + jB = 0.1 + j0.1$ siemans, $R = 1/G = 1/0.1 = 10\ \Omega$, $I = V/R = 100/10 = 10$A, $P_R = I^2R = 10^2 10 =$ 1000 watts.

84. a. The impedance of the load reflected through the transformer is the turns ratio squared times the load: $Z'_L(\text{reflected}) = a^2R_L = 5^2 8 = 25 \times 8 = 200\Omega$.

Sample Examination Solutions/AM—Materials Engineering

85. c. Although dislocations are required for slip, increased numbers arising from plastic deformation produce dislocation entanglements that arrest further slip and increase the strength.

86. a. Flux is proportional to the concentration gradient; atoms move more readily in loosely packed solids; higher temperatures provide more activation energy; and grain boundaries provide a diffusion path. Mass, as such, is not a factor.

87. d. A single-phase material contains 100% of that phase.

88. d. Pearlite lamellae provide a brittle fracture path. The crack encounters very little of the ductile ferrite.

89. d. A powder is compacted into a porous shape. Sintering shrinks the product by eliminating the pores.

90. a. Extrusion forces the material through a shaped die.

91. c. The endurance limit is the stress level that withstands unlimited stress cycling.

92. c. Thermal conductivity is decreased because disorder is introduced into the material.

Sample Examination Solutions/AM—Chemistry

93. b. Since both solutions are the same normality, the equivalence point will be at 50 ml KOH solution. If one ml excess is added, there will be 51 ml of KOH but all will be neutralized by the HCI except for one ml. The resulting solution will have an excess of one ml of 0.2 N KOH in 101 ml solution. Since we are limited to one significant figure by the data, 100 ml is a valid approximation. Then

$$\frac{0.2 \text{ equivalents}}{\text{liter}} \times \frac{1 \text{ liter}}{1000 \text{ ml}} = 0.2 \times 10^{-3} \text{ equivalents in ml excess}$$

$$\frac{0.2 \times 10^{-3} \text{ equivalents}}{0.1 \text{ L solution liter}} = \frac{0.2 \times 10^{-4} \text{equivalents}}{1 \text{ L solution}}$$

Since equivalent KOH = moles KOH,

$$[\text{OH}-] = 0.2 \times 10^{-4} = 2 \times 10^{-3} \text{ mol/L}$$
$$\text{pOH} = -\log (2 \times 10^{-3}) = -(-3 + 0.30) = 2.7$$
$$\text{pH} = 14 - \text{pOH} = 11.3$$

94. c. The equation for K_p is

$$K_p = \frac{(P_{HI})^2}{(P_{H2})(P_{12})} = \frac{(2.3 \text{ atm})^2}{(0.10 \text{ atm})(0.066 \text{ atm})} = 8.0 \times 10^2$$

$K_c = 8.0 \times 10^2$. *Note*: The data justify only two significant figures. The units cancel, so K_c is dimensionless. There is no volume change in the reaction, so $\Delta n = 0$, $RT\Delta n = 0$, and $K_c = K_p$.

95. d. The equation represents an oxidation-reduction reaction used for the determination of iron. The H^+ is not reduced as no H_2 is formed. The K^+ and CI^- ions do not enter into the reaction or affect its rate. Cr in $K_2Cr_2O_7$ is in a +6 oxidation state, and Cr in $CrCI_3$ is in a +3 oxidation state.

96. d. Anode reaction: $Sn(s) = Sn^{2+} + 2e^-$; $E^0 = +0.14$ (reverse standard reduction half-cell reaction and change sign of E^0)

Cathode reaction: $Cu^{2+} + 2e^- = Cu(s)$; $E^0 = +0.34$ (same as standard reduction half-cell reaction)

Cell reaction: $Sn(s) + Cu^{2+} + 2e^- = Sn^{2+} + 2e^- + Cu(s)$; $E^0 = +0.14 + 0.34 = +0.48$ V (add anode half-reaction and cathode half-reaction and add standard potentials for each reaction as written) $Sn(s) + Cu^{2+} = Cu(s) + Sn^{2+}$; $E^0 = +0.48$ V

Positive cell potential shows that cell reaction proceeds spontaneously in direction to the right of an equation. Since electrons are given up in anode half-reaction and acquired in cathode half-reaction, the spontaneous electron flow is anode to cathode.

97. a.

$$1.00 \text{ g Ag} \times \frac{1 \text{ mol Ag}}{107.9 \text{ g Ag}} = 9.27 \times 10^{-3} \text{ mol Ag}$$

Reaction shows 1 mol Ag = 1 mol electron.

$$9.27 \times 10^{-3} \text{ mol e}^- \times \frac{96{,}485 \text{ C}}{1 \text{ mol e}^-} = 8.94 \times 10^2 \text{ C}$$

1 hr = 60 × 60 = 3600 s; 0.50 hr = 0.50 × 3600 s = 1.8×10^3 s;
C = A × s

$$\text{A} = \frac{\text{C}}{\text{s}} = \frac{8.94 \times 10^2 \text{C}}{1.8 \times 10^3 \text{s}} = 5.0 \times 10^{-1} \text{A} = 0.50 \text{A}$$

98. c. The rate equation is: $R = \text{k} [F_2]^m [NO_2]^n$, where m and n must be experimentally determined. However, since it is given that the reaction is first order in $[F_2]$, $m = 1$. So: $R = \text{k} [F_2] [NO_2]^n$. In the experiment, $[F_2]$ is held constant, and a doubling in $[NO_2]$ gives a doubling in R. Then

$$\frac{R_1}{R_2} = \left(\frac{[NO_2]_1}{[NO_2]_2}\right)^n$$

Since $R_2 = 2 \times R_1$ and $[NO_2]_2 = 2 \times [NO_2]_1$,

$$\frac{1}{2} = \left(\frac{1}{2}\right)^n$$

and $n = 1$. The rate law is $R = [F_2][NO_2]$. Since $m = 1$ and $n = 1$, the overall order of the reaction is $n + m = 2$, or second order.

99. b. The number of moles of carbon and hydrogen in the sample are:

$$\text{moles C} = \frac{9.37 \text{ g}}{\text{sample}} \times \frac{\text{moles}}{12.01 \text{ g}} = 0.780 \text{ mol}$$

$$\text{moles H} = \frac{0.63\text{g}}{\text{sample}} \times \frac{\text{moles}}{1.01 \text{ g}} = 0.624 \text{ mol}$$

The molar ratio of the sample is

$$\frac{0.780 \text{ mol C}}{0.624 \text{ mol H}} = \frac{1.25 \text{ mol C}}{1.00 \text{ mol H}}$$

Multiply by 4 to obtain the empirical formula C_5H_4. The molecular weight of $C_5H_4 = (5 \times 12.01) \times (4 \times 1.01) = 64.09$

The molecular weight was estimated to be 128/64.09 = 2.00. Multiply the empirical formula by 2 to get $C_{10}H_8$ (naphthalene).
Note: Significant figures were limited by the molecular weight of 128 obtained from the freezing point experiment, and would be 128.18 on the basis of the atomic weights given here. The molecular formula must be in whole numbers.

100. d. Since the normality and the volume of the $KMnO_4$ titre are given, the amount of equivalents used in the titration can be calculated from: N(normality) × L(volume) = equivalents 0.100 N × 0.1000 L = 0.0100 equivalents. The equivalents of Fe must be the same. Since Fe in the reaction is oxidized from Fe^{2+} to Fe^{3+}, there is only one mole of electrons transferred per mole of Fe, and the equivalent weight is the same as the formula weight. Then

$$\text{g Fe} = 0.01 \times 55.8 = .558; \qquad \text{g sample} = 1.00$$

$$\% \text{ FE} = \frac{0.558}{1.00} \times 100 = 55.8\%$$

101. d. Since the system is at constant temperature, Boyle's law applies, or $p_1v_1 = p_2v_2$. Letting the final pressure of the gas $= p_2$, the equation becomes

$$\text{for He, } p_2 = \frac{p_1v_1}{v_2} = \frac{0.50 \text{ atm} \times 1.00 \text{ L}}{1.00 \text{ L} + 2.00 \text{ L}} = \frac{0.5 \text{ atm l}}{3.00 \text{ L}} = 0.17 \text{ atm}$$

$$\text{for } N_2\text{, } p_2 = \frac{p_1v_1}{v_2} = \frac{1.00 \text{ atm} \times 2.00 \text{ L}}{2.00 \text{ L} + 1.00 \text{ L}} = \frac{2.00 \text{ atm l}}{3.00 \text{ L}} = 0.67 \text{ atm}$$

Dalton's law states that the total gas pressure in a vessel is the sum of the partial pressures, so the pressure in the connected system will be 0.17 atm + 0.67 atm = 0.84 atm.

102. b. The density in kg/L multiplied by the fractional weight of the solute gives the g/L of solute.

$$\frac{1.2353 \text{ kg}}{\text{L}} = \frac{32.00 \text{ g solute}}{100 \text{ g solution}} = \frac{0.3953 \text{kg solute}}{\text{L solution}} \times \frac{1000 \text{ g}}{\text{kg}} = \frac{395.3 \text{ g solute}}{\text{L solution}}$$

103. d. The g of solute (H_2SO_4) per L solution divided by the molecular weight.

$$\text{g solute/mol} = \text{the g moles/L or molar concentration}$$

$$\frac{395.3 \text{ g solute}}{\text{L}} \times \frac{\text{mol}}{98.08 \text{ g}} = \frac{4.030 \text{ mol}}{\text{L}}$$

Sample Examination Solutions/AM—Engineering Economics

104. d. The interest for the four month loan is 975/15,000 = 0.065 = 6 1/2%. For 12 months, the total interest would be three times as much.

$$3 \times 0.065 = 0.195 = 19\frac{1}{2}\%$$

105. d.

$$\begin{aligned} F &= P(F/P, i\%, n) \\ &= 1000(F/P, 6\%, 12) \\ &= 1000(2.012) = 2012 \end{aligned}$$

106. b.

$$\begin{aligned} P &= A(P/A, i\%, n) \\ &= 1000(P/A, 6\%, 10) \\ &= 1000(7.360) = 7360 \end{aligned}$$

107. a.

$$\begin{aligned} P &= F(P/F, i\%, n) \\ &= 10,000(P/F, 8\%, 10) \\ &= 10,000(0.4632) = 4632 \end{aligned}$$

108. d.

$$i_{\text{eff}} = [1 + (0.015 \times 12)/12]^{12} - 1 = (1 + 0.015)^{12} - 1 = 0.1956 = 19.56\%.$$

Sample Examination Solutions/AM—Computers

109. c. The binary (base-2) number system representation is

	2^4	2^3	2^2	2^1	2^0	
	16	8	4	2	1	
Binary number	1	0	1	0	1	
Base-10 number	16	0	4	0	1	= 21

110. d.

2^5	2^4	2^3	2^2	2^1	2^0	
32	16	8	4	2	1	
	1	1	1	1	0	= 30

111. b.

112. d.

113. b.

114. b.

115. b. A cell reference is made by a lettered column and a numbered row, for example C3.

Sample Examination Solutions/AM—Ethics

116. b. The *Model Rules* are divided into the three broad categories listed in (A), (C), and (D). Nothing is said about obligations to other college students. Note, however, that the *Model Rules* set the tone of one's relationships, so one could easily conclude that a professional is expected to deal with all people (including college students) in an ethical manner.

117. c. The answer (A), (B), and (D) represent practical and appropriate ways of improving one's competence. Answer (C) however, may subject the client or employer to work by an engineer who is not fully competent in the specific area. This is clearly a less appropriate way to gain competency in a technical area.

118. d. Answers (A), (B), and (C) are prescribed in the *Model Rules.* Section 240.15(A)(5) states that engineers shall express a professional opinion publicly when it is based on adequate knowledge and a competent evaluation. Answer (D) is commenting publicly *before* a careful evaluation.

119. d. Professional conduct precludes an engineer from doing the things in (A), (B), and (C). Section 240.15(B)(6) says the engineer shall not accept compensation from more than one party for services on the same project unless the circumstances are fully disclosed and agreed to by all parties. Thus the conduct in (D) is acceptable.

120. c. If as indicated in (C), Jim's employer has a company policy telling its engineers they must not accept significant gifts from contractors, then neither (A) or (B) are acceptable. And the *Model Rules* say the same thing. Most companies have a policy limiting gifts (or valuable considerations) to less than $25. Under $25 is acceptable; over $25 is unacceptable. So the value of the gift (a party and dinner each week) must be considered in reaching a conclusion. Answer (D) of rejecting the invitation seems unrealistically harsh and highhanded and hardly the conduct of a professional. Thus while Jim might decide to reject the offer, there certainly would be a better way to do it. Since he would like to go, (C) is an appropriate professional response.

INSTRUCTIONS FOR AFTERNOON SESSION

1. You have four hours to work on the afternoon session. You may use the *Fundamentals of Engineering Supplied-Reference Handbook* as your *only* reference. Do not write in this handbook.
2. Answer every question. There is no penalty for guessing.
3. Work rapidly and use your time effectively. If you do not know the correct answer, skip it and return to it later.
4. Some problems are presented in both metric and English units. Solve either problem.
5. Mark your answer sheet carefully. Fill in the answer space completely. No marks on the workbook will be evaluated. Multiple answers receive no credit. If you make a mistake, erase completely.

Work 60 afternoon problems in four hours.

FUNDAMENTALS OF ENGINEERING EXAM

AFTERNOON SESSION

ⒶⒷⒸⒹ Fill in the circle that matches your exam booklet

1 ⒶⒷⒸⒹ	16 ⒶⒷⒸⒹ	31 ⒶⒷⒸⒹ	46 ⒶⒷⒸⒹ
2 ⒶⒷⒸⒹ	17 ⒶⒷⒸⒹ	32 ⒶⒷⒸⒹ	47 ⒶⒷⒸⒹ
3 ⒶⒷⒸⒹ	18 ⒶⒷⒸⒹ	33 ⒶⒷⒸⒹ	48 ⒶⒷⒸⒹ
4 ⒶⒷⒸⒹ	19 ⒶⒷⒸⒹ	34 ⒶⒷⒸⒹ	49 ⒶⒷⒸⒹ
5 ⒶⒷⒸⒹ	20 ⒶⒷⒸⒹ	35 ⒶⒷⒸⒹ	50 ⒶⒷⒸⒹ
6 ⒶⒷⒸⒹ	21 ⒶⒷⒸⒹ	36 ⒶⒷⒸⒹ	51 ⒶⒷⒸⒹ
7 ⒶⒷⒸⒹ	22 ⒶⒷⒸⒹ	37 ⒶⒷⒸⒹ	52 ⒶⒷⒸⒹ
8 ⒶⒷⒸⒹ	23 ⒶⒷⒸⒹ	38 ⒶⒷⒸⒹ	53 ⒶⒷⒸⒹ
9 ⒶⒷⒸⒹ	24 ⒶⒷⒸⒹ	39 ⒶⒷⒸⒹ	54 ⒶⒷⒸⒹ
10 ⒶⒷⒸⒹ	25 ⒶⒷⒸⒹ	40 ⒶⒷⒸⒹ	55 ⒶⒷⒸⒹ
11 ⒶⒷⒸⒹ	26 ⒶⒷⒸⒹ	41 ⒶⒷⒸⒹ	56 ⒶⒷⒸⒹ
12 ⒶⒷⒸⒹ	27 ⒶⒷⒸⒹ	42 ⒶⒷⒸⒹ	57 ⒶⒷⒸⒹ
13 ⒶⒷⒸⒹ	28 ⒶⒷⒸⒹ	43 ⒶⒷⒸⒹ	58 ⒶⒷⒸⒹ
14 ⒶⒷⒸⒹ	29 ⒶⒷⒸⒹ	44 ⒶⒷⒸⒹ	59 ⒶⒷⒸⒹ
15 ⒶⒷⒸⒹ	30 ⒶⒷⒸⒹ	45 ⒶⒷⒸⒹ	60 ⒶⒷⒸⒹ

DO NOT WRITE IN BLANK AREAS

Sample Examination Problems/PM—Mathematics

1. A carpenter is making an octagonal table for a customer who wants each side (all equal) to be *about* 1 meter (Exhibit 1). *About* how large a square should he start with?

a. $S = 1.6$ m
b. $S = 2$ m
c. $S = 2.4$ m
d. $S = 2.8$ m

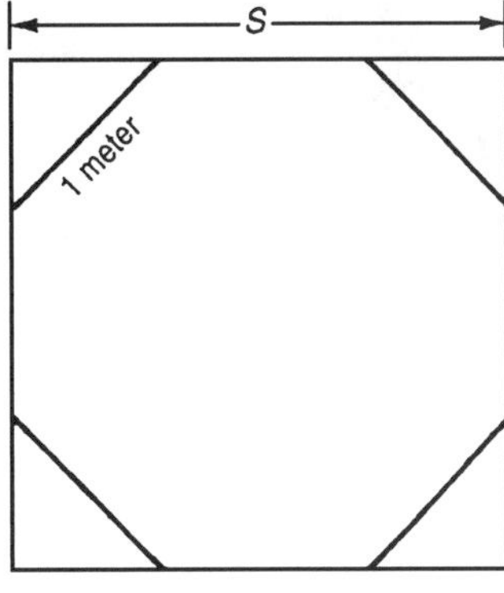

Exhibit 1

2. A fanbelt connects a motor with a pulley of radius 5 centimeters to a blower with a radius of 30 cm. If the motor is turning 1770 rpm, the blower will turn at about how many rpm?

a. 10,620
b. 3440
c. 1770
d. 295

3. The area of the octagon in Problem 1 on the previous page is approximately

a. 4.8 m^2
b. 5.0 m^2
c. 5.4 m^2
d. 6.0 m^2

4. A surveyor is measuring the area of a piece of land (Exhibit 4) for a client. He takes the following measurements: side $AB = 380$ m, side $BC = 500$ m, and angle $ABC = 111°$. The area, in hectares, is nearest

a. 5
b. 4
c. 3
d. 2

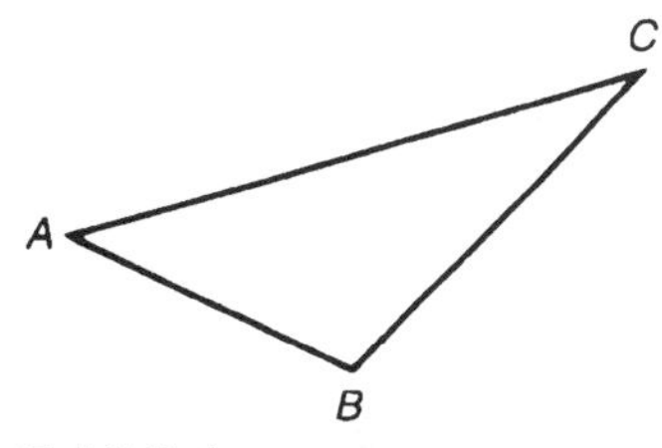

Exhibit 4

Questions 5-7 refer to the physical system shown in Exhibit 5. The block B of weight W is hanging on a spring/dashpot system with a spring constant k and a damping constant c. With no driving force, the motion of B is described by the differential equation

$$\frac{W}{g}\ddot{x} + c\dot{x} + kx = 0$$

where g is acceleration of gravity, and the position coordinate x is measured from the system's static equilibrium position. For this system, assume that, in appropriate units, $W/g = \frac{9}{16}$, $k = \frac{27}{8}$, and $c = \frac{9}{4}$.

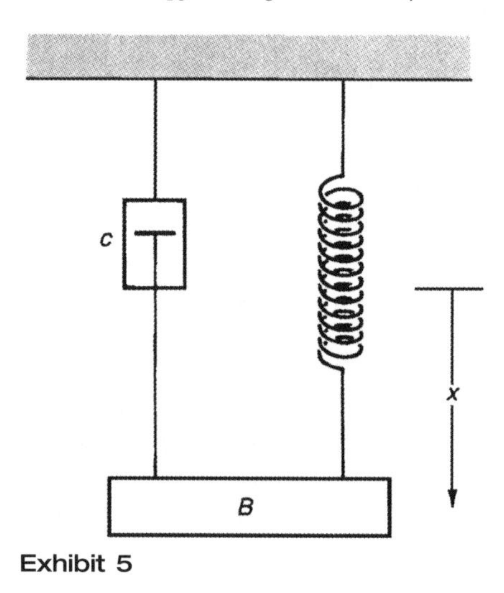

Exhibit 5

5. What is the general expression for the position of B as a function of time?

a. $x = A\sin\sqrt{2}t + B\cos\sqrt{2}t$

b. $x = Ae^{\sqrt{2}t} + Be^{-\sqrt{2}t}$

c. $x = e^{-2t}\left(Ae^{\sqrt{2}t} + Be^{-\sqrt{2}t}\right)$

d. $x = e^{-2t}\left(A\sin\sqrt{2}t + B\cos\sqrt{2}t\right)$

6. What is the quasiperiod?

a. $\frac{\pi}{\sqrt{2}}$

b. 2π

c. $\sqrt{2}\pi$

d. π

7. The weight will be first moving upward through the equilibrium position when t equals

a. $\frac{\pi}{\sqrt{2}}$

b. $\sqrt{2}\pi$

c. π

d. $2\sqrt{2}\pi$

8. How many three-letter symbols can be made with the five letters a, b, c, d, and e?

a. 15
b. 25
c. 60
d. 120

9. What is the probability of rolling an even number on two dice?

a. 1/2
b. 1/3
c. 1/6
d. 3/36

Problems 10 through 12 are based on the equations **AX** = **B** in which

$$\mathbf{A} = \begin{bmatrix} 1 & 3 & 2 \\ 2 & 4 & 6 \\ 3 & 5 & 7 \end{bmatrix} \qquad \mathbf{X} = \begin{bmatrix} x \\ y \\ z \end{bmatrix} \qquad \mathbf{B} = \begin{bmatrix} 1 \\ 8 \\ 12 \end{bmatrix}$$

10. The solution of the equations will give a value of z that is closest to

a. –2
b. 1
c. 3
d. 5

11. The determinant of the cofactor of a_{11} is nearest to

a. –2
b. 1
c. 3
d. 5

12. The inverse of the coefficient matrix **A** is $\mathbf{A}^{-1}$. The correct matrix for $\mathbf{A}^{-1}$ is

a. $\begin{bmatrix} -2 & 4 & -2 \\ -11 & 1 & 4 \\ 10 & -2 & -2 \end{bmatrix}$

b. $\begin{bmatrix} -2 & -11 & 10 \\ 4 & 1 & -2 \\ -2 & 4 & -2 \end{bmatrix}$

c. $\frac{1}{6}\begin{bmatrix} -2 & 4 & -2 \\ -11 & 1 & 4 \\ 10 & -2 & -2 \end{bmatrix}$

d. $\frac{1}{6}\begin{bmatrix} -2 & -11 & 10 \\ 4 & 1 & -2 \\ -2 & 4 & -2 \end{bmatrix}$

Sample Examination Problems/PM—Statics

13. In terms of the axial force P, the coefficient of friction, μ, and the inside and outside radii a and b, what is the moment M necessary to cause slipping of the clamp in Exhibit 13? Assume that the normal pressure is uniform over the surface of contact.

a. $\mu \dfrac{b^2 - a^2}{b - a} P$

b. $\mu \dfrac{b^3 - a^3}{b^2 - a^2} P$

c. $\dfrac{2\mu(b^3 - a^3)}{3(b^2 - a^2)} P$

d. $\mu(b - a)P$

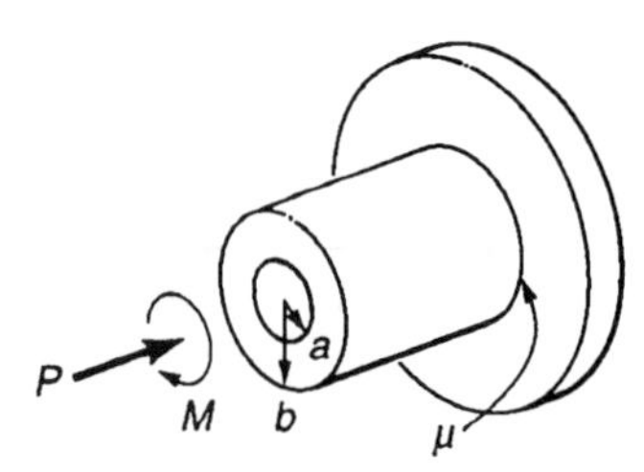

Exhibit 13

14. At what angle α will the plate in Exhibit 14 hang, suspended at rest?

a. 22.5°
b. 25.4°
c. 18.4°
d. 15.9°

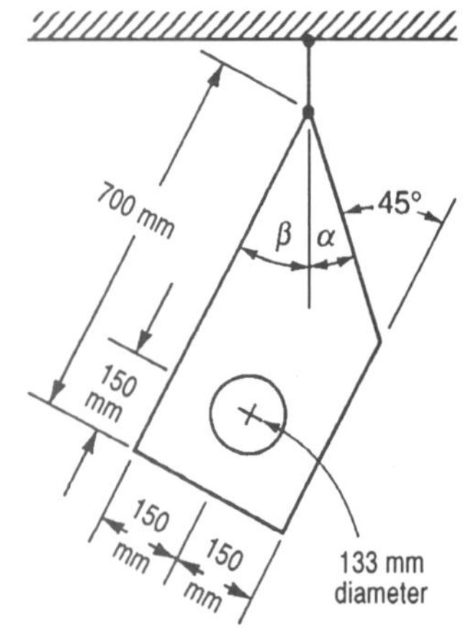

Exhibit 14

15. The trough in Exhibit 15 is constructed of 2 mm thick steel plate and has a semicircular cross section. The distance from the top of the trough to its center of mass is

a. 125 mm
b. 150 mm
c. 137.5 mm
d. 112.5 mm

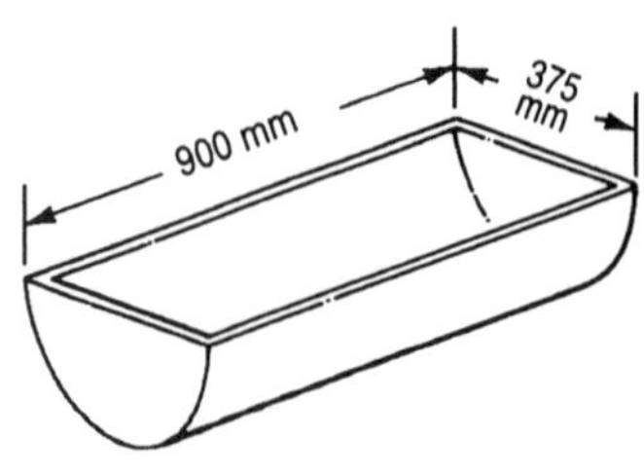

Exhibit 15

16. The thickness of the wall of the conical shell in Exhibit 16 is uniform and much smaller than the radius a. The distance from the apex to the center of mass of the shell is

a. $\dfrac{2h}{3}\sqrt{1+\left(\dfrac{2a}{\pi h}\right)^2}$

b. $\dfrac{2h}{3}\sqrt{1+\left(\dfrac{a}{\pi h}\right)^2}$

c. $\dfrac{3h}{4}\sqrt{1+\left(\dfrac{2a}{\pi h}\right)^2}$

d. $\dfrac{3h}{4}\sqrt{1+\left(\dfrac{a}{\pi h}\right)^2}$

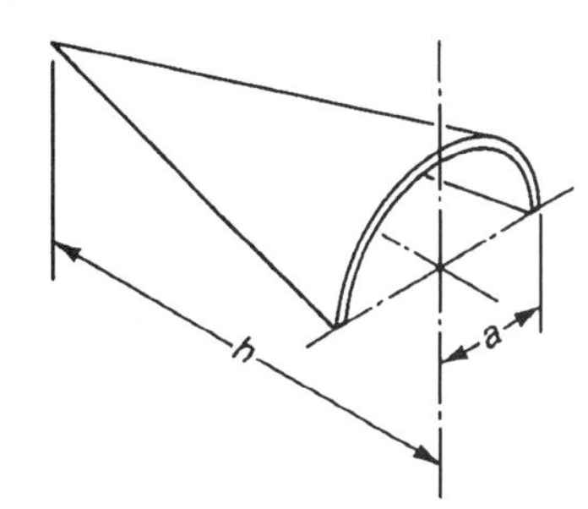

Exhibit 16

17. In Exhibit 17 the magnitude of the bending moment at a section 1.625 m from the left support is

a. 1641 N•m
b. 1250 N•m
c. 1500 N•m
d. 1750 N•m

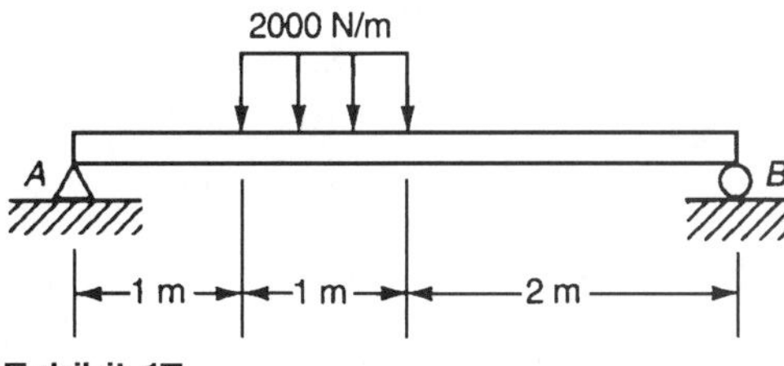

Exhibit 17

18. The reaction at the support at the center of the beam in Exhibit 18 has magnitude

a. $w_oL/2$
b. $2w_oL/3$
c. $5w_oL/6$
d. $3w_oL/4$

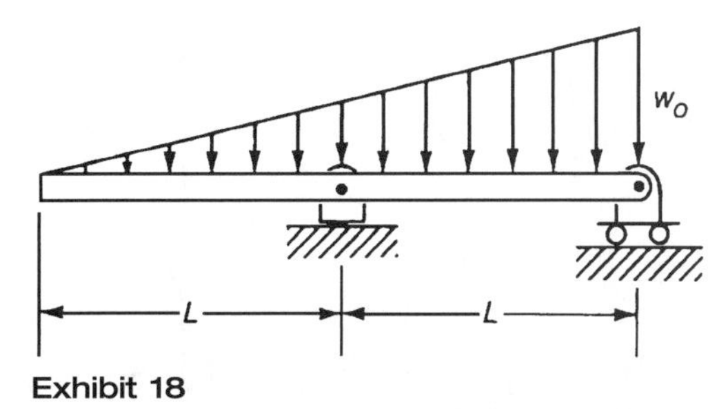

Exhibit 18

Sample Examination Problems/PM—Dynamics

19. An airplane maintains an acceleration of 5 m/s^2 on a straight path (Exhibit 19). As the plane moves along this path, it rolls about its longitudinal axis at a constant rate $\dot{\phi} = 20$ rpm. Simultaneously, an antenna is being extended from the plane at a constant rate of 1 m/s. What is the magnitude of the acceleration of the tip of the antenna when the antenna is 1.5 m long?

a. 6.52 m/s^2
b. 7.80 m/s^2
c. 8.37 m/s^2
d. 9.27 m/s^2

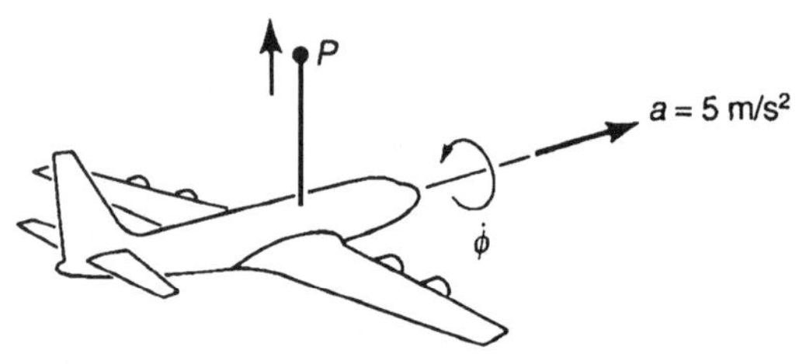

Exhibit 19

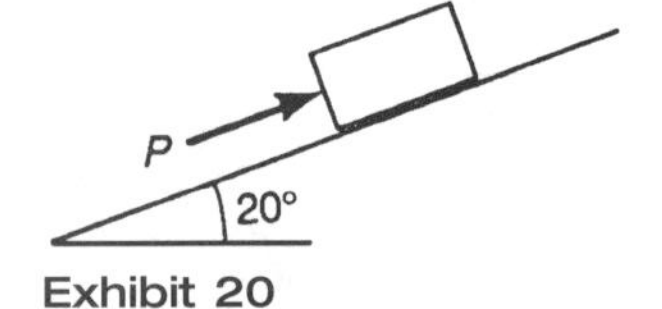

Exhibit 20

20. What value of the constant force, *P*, is required to bring the 4-kg block in Exhibit 20 from rest to a velocity of 9 m/s in 10 seconds? The coefficient of kinetic friction between the block and the inclined plane is 0.1.

a. 7.4 N
b. 20 N
c. 28 N
d. 17 N

21. The pendulum shown in Exhibit 21 consists of a uniform, slender, 2-kg rod rigidly attached to a 4-kg disk. The system is released from rest with the rod in the horizontal position. What is the angular acceleration of the rod immediately after the system is released?

a. 8.7 rad/s^2
b. 76.2 rad/s^2
c. 230.4 rad/s^2
d. 4.3 rad/s^2

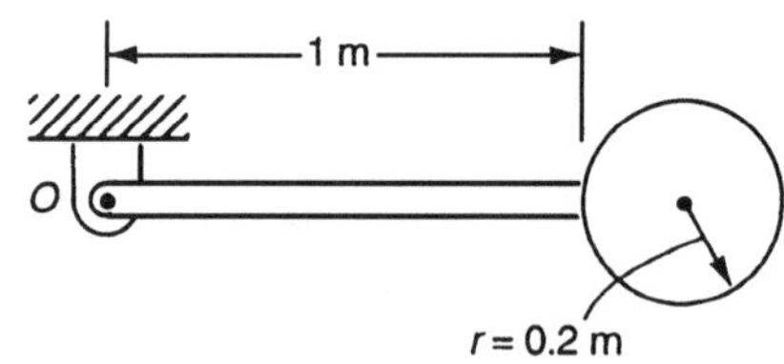

Exhibit 21

The following statement for Questions 22 and 23 applies to Exhibit 22: A projectile of mass 2 kg is launched from point *A*. As an approximation, air

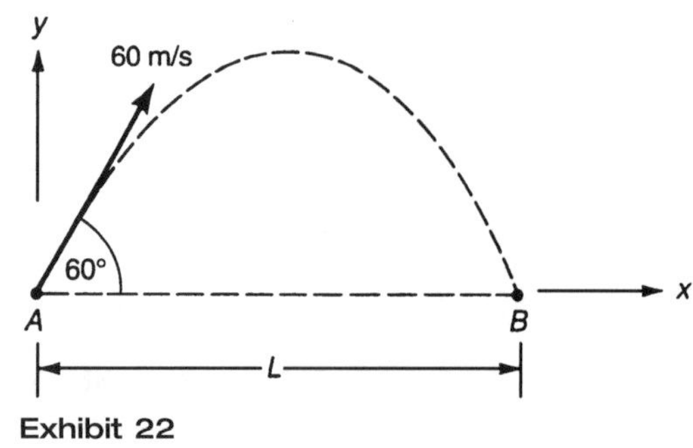

Exhibit 22

resistance creates a constant horizontal resistive force of 4 N during projectile travel.

22. What is the speed of the projectile after 10 seconds?
a. 47 m/s
b. 10 m/s
c. 46 m/s
d. 55 m/s

23. The projectile attains a maximum horizontal range, L, of
a. 318 m
b. 159 m
c. 206 m
d. 266 m

Sample Examination Problems/PM—Mechanics of Materials

Problem 24 through 28 relate to a rectangular beam 1.2 m long supported by a pin at A and by a vertical member at B. The vertical member BD, which is square, is pinned at both ends. A uniformly distributed load acts over the span BC. Assume the modulus of elasticity is $E = 210$ Gpa.

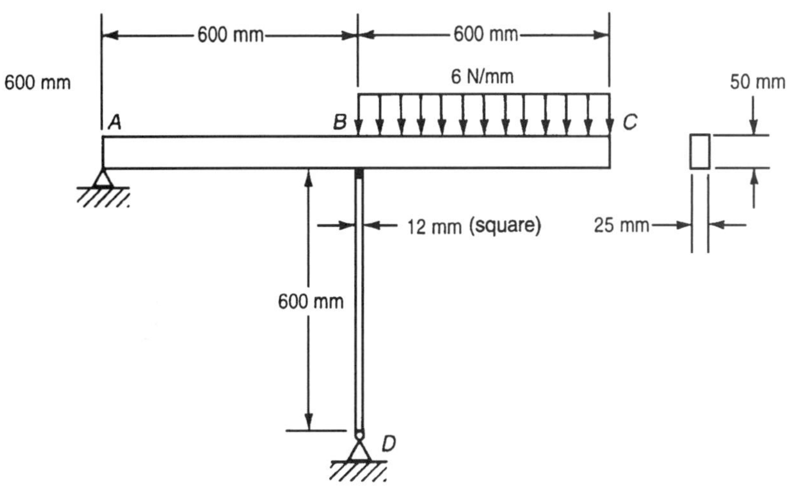

Exhibit 24

24 The vertical reaction at point B in Exhibit 24 is most nearly
a. 3600 N
b. 4200 N
c. 5000 N
d. 5400 N

25. The stress in the vertical member BD is most nearly
a. +56.3 MPa
b. –37.5 MPa
c. –21.2 MPa
d. +21.2 MPa

26. The maximum vertical shear in the member *AC* is most nearly
a. 1800 N
b. 2400 N
c. 3000 N
d. 3600 N

27. The maximum absolute value of the bending moment in the member *AC* is most nearly
a. 0.8 kN•m
b. 0.9 kN•m
c. 1.0 kN•m
d. 1.1 kN•m

28. The maximum bending stress in the member *AC* is most nearly
a. 50 MPa
b. 70 MPa
c. 85 MPa
d. 100 MPa

Sample Examination Problems/PM—Fluid Mechanics

Questions 29 and 30 relate to the tank system shown in Exhibit 29.

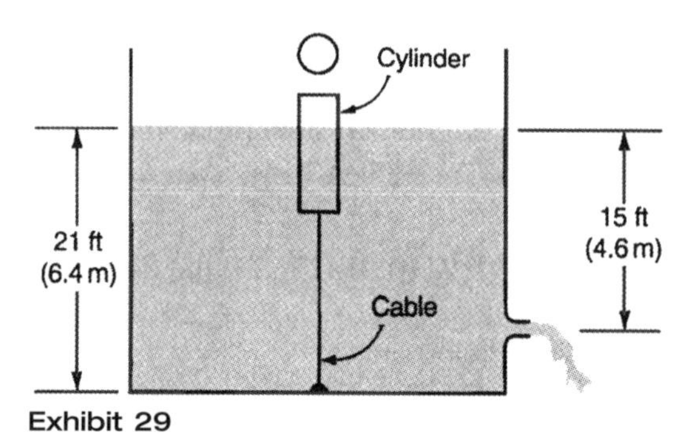

Exhibit 29

The tank contains water, $\gamma = 62.4$ lbf/ft^3 (9.81 kN/m^3).

The solid cylinder is wood, $\gamma = 25$ lbf/ft^3 (3.93 kN/m^3).

Cylinder dimensions: diameter = 1 foot (0.30 meter); height = 3 feet (0.91 meter).

Nozzle diameter: 2 inches (50 mm).

29. For the cylinder to float upright with half its volume submerged, the tension in the cable is most nearly
a. 4.9 lbf (21 N)
b. 14.6 lbf (63 N)
c. 58.4 lbf (252 N)
d. 0 (no tension)

30. The thrust on the tank created by the water discharging through the nozzle is most nearly

a. 0 lbf (0 kN)
b. 20.4 lbf (88.5 N)
c. 2940 lbf (12.7 kN)
d. 40.8 lbf (177 N)

Questions 31 through 33 relate to the system shown in Exhibit 31.

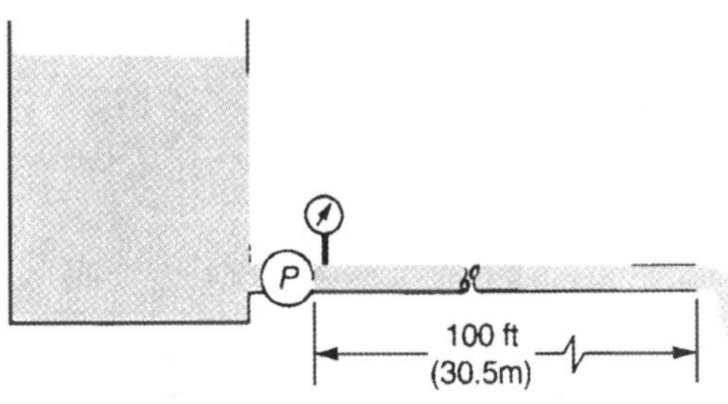

Exhibit 31

The tank contains crude oil with a specific gravity of 0.93 and a dynamic viscosity of 7×10^{-4} lbf•s/ft^2(3.35×10^{-2}N• s/m^2).

The pipe is 4-inch Schedule 40 commercial steel pipe with an inside diameter of 4.026 inches (102.3 mm) and a length of 100 ft (30.5 m).

31. For a flow rate of 0.5 ft^3/s (0.0142 m^3/s) the Reynolds number in the pipe is most nearly

a. 2.0×10^3
b. 3.2×10^4
c. 4.9×10^3
d. 4.9×10^{-3}

32. For a flow rate at 1.0 ft^3/s (0.0284 m^3/s) the head loss from friction in the 100-foot (30.5 m) length of pipe is most nearly

a. 18.35 ft (5.64 m)
b. 6.12 ft (1.88 m)
c. 12.36 ft (3.80 m)
d. 100 ft (30.5m)

33. If the pressure gage at the discharge of the pump reads 10 psig (69 kPa) when 1.0 ft^3/s (0.0284 m^3/s) are flowing, the power delivered to the fluid is most nearly

a. 16.8 HP (12.53 kW)
b. 0 HP (0 kW)
c. 5.3 HP (3.95 kW)
d. 1.77 HP (1.32 kW)

Sample Examination Problems/PM—Thermodynamics

Refrigeration questions 34–38 are about a vapor compression refrigeration system that uses a refrigerant as the working medium. The data are given directly in Exhibit 34. Assume that there is adiabatic compression, steady flow, and no loss of pressure in the lines. Neglect kinetic and potential energy effects. One ton of refrigeration equals 200 BTU/min (3.52 kW). Use the tables of the thermodynamic properties of the refrigerant.

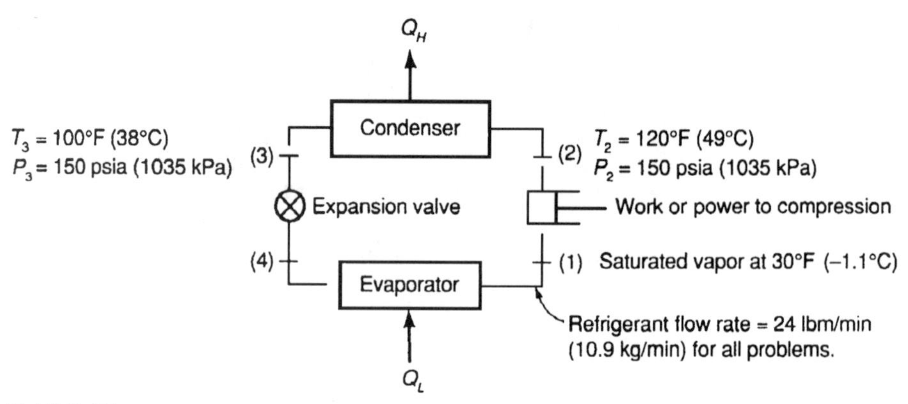

Exhibit 34

Properties of Saturated Refrigerant (SI units): Pressure Table

		v, cm³/g Specific Volume		u, kJ/kg Internal Energy		h, kJ/kg Enthalpy			s, kJ/(kg)(°K) Entropy	
Press., bars P	Temp., °C T	Sat. liquid v_f	Sat. vapor v_g	Sat. liquid u_f	Sat. vapor u_g	Sat. liquid h_f	Evap. h_{fg}	Sat. vapor h_g	Sat. liquid s_f	Sat. vapor s_g
10.0	41.64	0.8023	17.44	75.46	186.32	76.26	127.50	203.76	0.2770	0.6820

Properties of Superheated Refrigerant (SI units)

v, cm³/g; u and h, kJ/kg; s kJ/(kg)(°K)

Temp, C	v	u	h	s	v	u	h	s
	10.0 bars (1.0 MPa)(T_{sat} 41.64 C)				12.0 bars (1.2 MPa)(T_{sat} = 49.31 °C)			
50	18.37	191.95	210.32	0.7026	14.41	188.95	206.24	0.6799
60	19.41	198.56	217.97	0.7259	14.48	189.43	206.81	0.6816

Properties of Saturated Refrigerant (SI units): Temperature Table

v, cm³/g; u kJ/kg; h, kJ/kg; s, kJ/(kg)(°K)

		Specific Volume		Internal Energy		Enthalpy			Entropy	
Temp. °C T	Press., bars P	Sat. liquid v_f	Sat. vapor v_g	Sat. liquid u_f	Sat. vapor u_g	Sat. liquid h_f	Evap. h_{fg}	Sat. vapor h_g	Sat. liquid s_f	Sat. vapor s_g
–5	2.6096	0.7078	64.96	31.27	168.42	31.45	153.93	185.37	0.1251	0.6991
0	3.0861	0.7159	55.39	35.83	170.44	36.05	151.48	187.53	0.1420	0.6965

34. When the system operates as a cooler, the capacity of refrigeration is most nearly

a. 0.35 kW
b. 1.65 kW
c. 19.9 kW
d. 69.3 kW

35. The power required for the compressor is most nearly

a. 0.003 kW
b. 0.009 kW
c. 0.116 kW
d. 4.13 kW

36. The actual coefficient of performance as a refrigerator is most nearly

a. 0.186
b. 3.40
c. 4.85
d. 6.40

37. When the system operates as a heat pump, the rate at which heat is supplied is most nearly

a. 0.0172 kW
b. 0.413 kW
c. 1.03 kW
d. 24.0 kW

38. The actual coefficient of performance as a heat pump is most nearly

a. 2.89
b. 3.44
c. 4.40
d. 5.85

Sample Examination Problems/PM—Electrical Circuits

39. Determine the voltage drop across the 2-ohm resistance at $t = 1$ second. Assume the switch closes at $t = 0$.

a. 0.75 V
b. 1.25 V
c. 1.84 V
d. 3.68 V

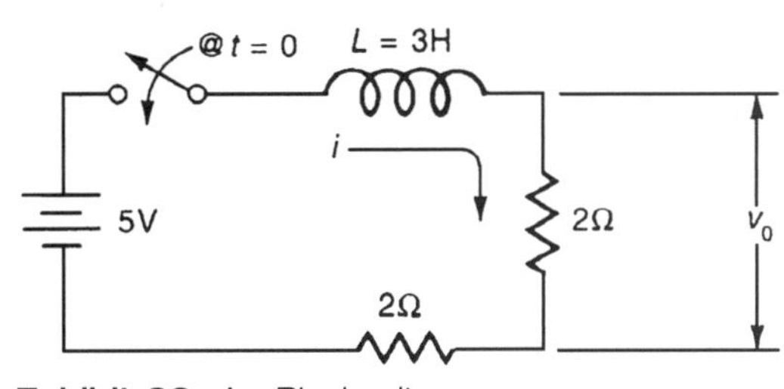

Exhibit 39 An *RL* circuit

40. Find the current through the 12-volt battery in Exhibit 40 and indicate the actual direction of current (up or down).

a. 0.86 A (up)
b. 0.86 A (down)
c. 1.43 A (up)
d. 2.00 A (down)

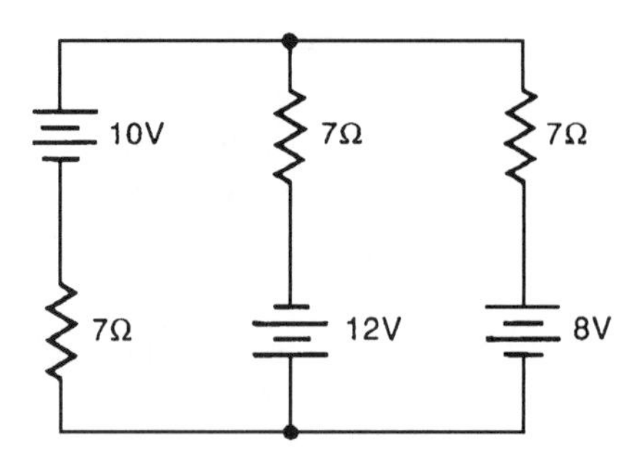

Exhibit 40 A two-loop circuit

41. Refer to Exhibit 41; find the power dissipated in the circuit.

a. 0.5 watt
b. 1.0 watt
c. $0.5\sqrt{2}$ watt
d. $1.0\sqrt{2}$ watt

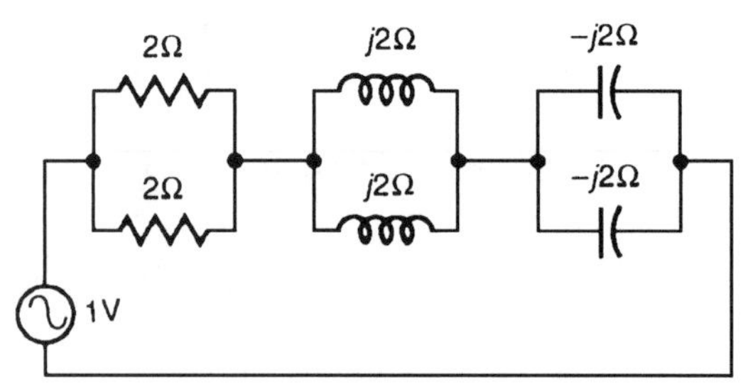

Exhibit 41 An ac circuit

42. The circuit shown in Exhibit 42 has both a voltage and a current source. Find the magnitude of the voltage across the current source.

a. 5 V
b. 25 V
c. 50 V
d. 72 V

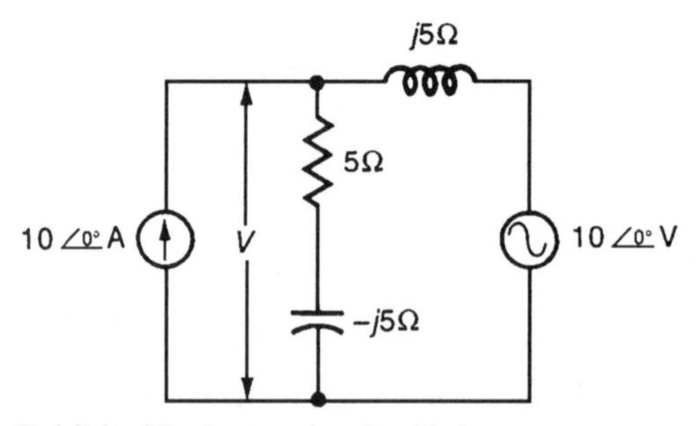

Exhibit 42 An ac circuit with two sources

43. For the operational amplifier shown in Exhibit 43, the configuration is a current source for the load resistance. For the parameters given, determine i_L.

a. 0.67 mA
b. 0.80 mA
c. 1.0 mA
d. 2.0 mA

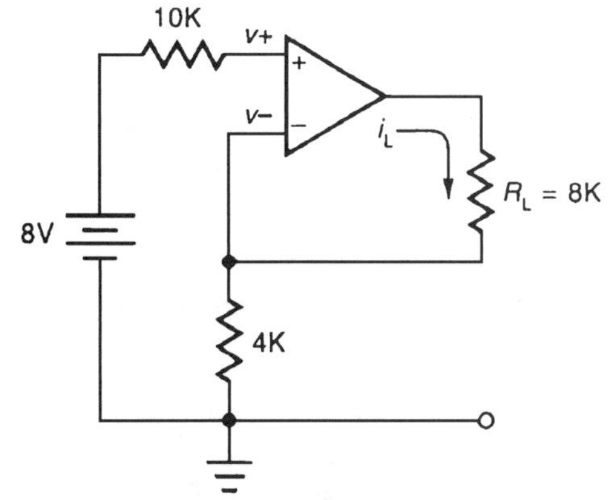

Exhibit 43 An op-amp circuit

44. For an initially uncharged 10 μF capacitor that will start charging at $t = 0$ from a constant current source of 2 amperes, determine the energy stored at the end of 0.001 second.

a. 20 J
b. 10 J
c. 5 J
d. 0.2 J

Sample Examination Problems/PM—Material Science

45. Natural copper contains Cu^{63} and Cu^{65}, with atomic masses of 62.930 and 64.928, respectively. The atomic weight of natural copper is 63.546/mole. What percent of the Cu^{65} isotope is present?

a. 75
b. 31
c. 69
d. 25

46. In aluminum, deformation readily occurs by slip along the (111) plane. A possible slip direction on that plane is

a. [111]
b. [110]
c. [0 01]
d. $[1\bar{1}0]$

47. Refer to the lead-tin diagram in Exhibit 47. Which statement is correct?

a. At 250°C, the solubility limit of tin in liquid is 38% Sn.
b. There is ~10% β in a (80Pb–20Sn) alloy at 150°C.
c. 240°C is the lowest temperature at which a (60Pb–40Sn) alloy can contain liquid.
d. A (96Pb–4Sn) alloy equilibrated at 200°C contains ~18% Sn in α.

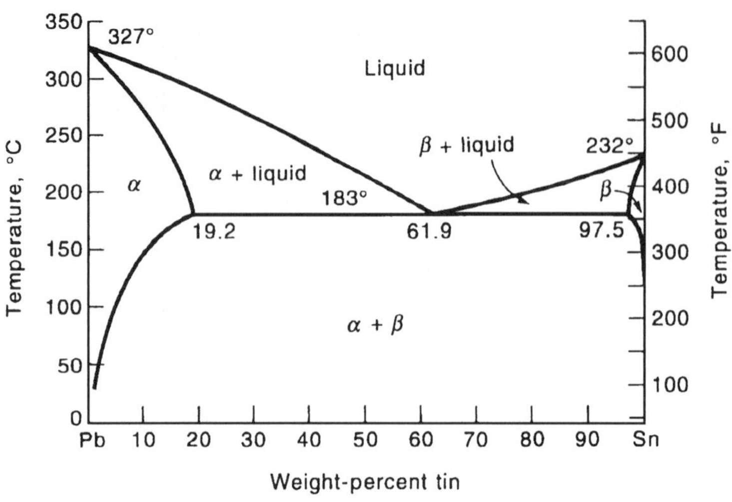

Exhibit 47

Sample Examination Problems/PM—Chemistry

48.

a. The covalent radius of elements in the periodic table:

a. is unaffected by the number of electron shells in the atom
b. decreases going down a group of elements
c. stays the same when the elements compared are isoelectronic
d. usually decreases from left to right across a period of elements

b. The ionic radius of elements in the periodic table:

a. is usually less than that of the preceding element in going down a group of elements with positive ions
b. is usually less than that of the preceding element in going down a group of elements with negative ions
c. is less than the covalent radius of the parent atom when the ion is positive
d. is more than the covalent radius of the parent atom when the ion is positive

49. Which of the following refers to intrinsic semiconductors?

a. There are equal numbers of p-type carriers in the conduction band and n-type carriers in the valence band.
b. The valence band is half filled, allowing limited conductivity.
c. The valence band is full. There is a large energy gap.
d. The valence band is full. There is a small energy gap.

50. Which of the following statements does NOT refer to an extrinsic silicon semiconductor?

a. There are equal numbers of n-type carriers in the conduction band and p-type carriers in the valence band.
b. Group III elements can accept electrons from the valence band.
c. Group V elements can donate electrons to the conduction band.
d. Group III elements lower the Fermi energy.

51. When metals react with water, they form
a. oxygen gas
b. basic hydroxides
c. soluble covalent species
d. acidic hydroxyl compounds

Sample Examination Problems/PM—Engineering Economics

52. What present sum would need to be put in a savings account now to provide a $1000 annual withdrawal for the next 50 years if interest is 6%? The present sum is nearest to
a. $1000
b. $10,000
c. $25,000
d. $37,500

53. What interest rate, compounded quarterly, is equivalent to a 9.31 percent effective rate? The interest rate is nearest to
a. 2.25%
b. 2.33%
c. 4.66%
d. 9.00%

54. A company deposits $1000 every year for ten years in a bank. The company makes no deposits during the subsequent five years. If the bank pays 8% interest, the amount in the account at the end of 15 years is most nearly
a. $10,800
b. $15,000
c. $16,200
d. $21,200

Sample Examination Problems/PM—Computers

55. The following polynomial obviously has three roots:

$$P(x) = x^3 + 4x^2 + 6x + 4$$

Is $x = -1$ one of the exact roots?
a. No, because it's less than the last coefficient.
b. Yes, because there is a remainder when using synthetic division.
c. No, because there is a remainder when using synthetic division.
d. Yes, there is no remainder.

56. For problem 55 if a second trial root is to be found using Newton's method what is this next trial division?
a. 1
b. 2
c. –1
d. –2

57. For a differential equation $dx/dt + 5x = 1$, $x(0) = 0$, what is the value of x suggested by the Euler's method after first iteration after initial condition?

a. 0.1
b. 0.2
c. 0.3
d. 0.4

Sample Examination Problems/PM—Ethics

58. Jane Doe is a self-employed consulting engineer in the engineering department of ABC company where you work as a full time staff process engineer. One of your expertises is relief system design. Things are slow, and even though you are well recognized, you are facing a potential layoff. She approached you for part time help, which may be summarized by the following conversation:

Jane: "Hey Phil, how about earning some extra money, say forty dollars per hour, from XYZ Company, for helping in my calculations to check the size of relief devices on weekends? I will do all the leg work, collect all the pertinent data, all you have to do is to check whether the device is OK, or not OK, and show the basis of your calculations."

Phil: "Is it not illegal to have a part-time job when I am a full-time employee?"

Jane: "I don't think so. I know John, your project manager, has a restaurant business. Roger, your own colleague, helps in income tax preparation on weekends."

What would you do?

a. Check the employment agreement, and make sure nothing was signed against accepting a part-time job, and accept the offer and work at home without using any company time or resources, and follow "don't-ask-don't-tell" principle.
b. Accept the offer, and work at the office outside working hours, and use the company's methods and copier.
c. Read the company ethics book and guide lines to check whether there is any clear definition of Conflict of Interest. If the definition is clear, reject the offer. If unclear, ask the Human Resources Department for permission by a short memo, and then decide.
d. Invite Jane and your boss for a dinner, and casually refer to the situation at an appropriate moment, and have his permission well heard so that Jane remains a witness, and then accept the offer.

59. Suppose you work for a large electrical designs firm that designs electrical control systems for industrial manufacturing sites. Assume your design requirements include specifying "short circuit and ground fault protection" for motor circuits. These device specifications should comply with the National Electric Code. Your design, started in early 1991, involved the selection of a non-time delay fuse for a three-phase, 230V, 50 hp, 115A (its name plate values) induction motor for a molded rubber parts plant. The code book states that the full rated current for this motor is to be 130 amperes. The book also states that for this type of non-time delay fuse the current should not be more than 300% of rated. However, you find that commercially available uses come only in increments of 50 amperes near that value. You choose the 400 amp fuses since the code implies that it is permissible (and is a widely accepted practice) to "round up" to this higher value. Your design was not implemented as the project was tentatively canceled due to lack of funding. In 1995 the exact same project was funded and the project went ahead. Assume you have gone on to work on other projects. You are aware that in the 1993 edition of this handbook a number of revisions were made, one requiring fusing and circuit breakers to use the next lower, or "rounding down," of the fuse and circuit breaker sizes. What action, if any, should you take?

a. Ignore the project since you are working in another area now.
b. Call the electrical engineer you know is working on the project reminding him that the prior work was based on the code in force in 1991 and not the latest code.
c. Write a memo to the project supervisor of the change in the code and the need to revise the fuse rating.
d. The client for the project is that firm that produces molded rubber parts. They have a project manager overseeing the plant design. Call this project manager and tell him of the needed change in fuse rating to match the present electrical code.

60. A certain "hi-tech" company, located in California, produces a very popular electronic device that emits an amount of radiation that is within established limits of radiation. The company sells all of the product that it produces, both domestically and abroad. The company is able to increase its production by buying some of the required components overseas. You are the supervisor of the company's test laboratory, which samples and tests for radiation levels. It is a standard and acceptable practice to sample a certain percentage of the devices produced to check for compliance. After a large production run, your sampling procedure shows that the limits of radiation have been exceeded. You advise your company's managers of this higher radiation level due to the variation of overseas components (that is, their quality control is poor). A meeting is called consisting of yourself, two other engineers, and two plant managers. The cause of the problem, of course, is the variation in the overseas components. The possible courses of action discussed are:

a. The tolerances of a few of the sampled components are usually detectable. To create a distorted testing record, selectively choose the samples to test. Because the test records will show the percent sampled components to be within limits, ignore the problem (there are not that many devices too far out of specs).
b. Curtail the sales in the United States and sell to the overseas countries that don't have the radiation requirements.
c. Stop all sales until a new supplier is found for higher quality components; and, attempt to modify existing devices to meet samples specifications. This action may delay an impeding cost of living salary increase for most of the plant's employees.
d. Remove the label on each of the devices that states that the device meets all radiation requirements, reduce the price to remove the inventory, then phase in newer devices made with higher quality components.

Each member of the committee is highly respected for his judgement; the management wants honest opinions. Four of the members have voted, one for each of the possible courses of action, your vote then will be the deciding one. Which will you vote for?

Sample Examination Solutions/PM—Mathematics

1. **c.** Each corner's "missing triangle" is a 45° isoceles right triangle, so its leg is 1 cos 45° = .7m. Hence the original square must be 1 m + 2(.7 m) = 2.4 m. The answer is (C).

2. **d.** Since the ratio $\frac{\text{driving}}{\text{driven}} = \frac{2}{12} = \frac{1}{6}$, the blower will turn at 1770/6 = 295 rpm.

3. **a.** Each corner has area $(0.7)^2/2 \cong .25\ \text{m}^2$, so, multiplying by 4 (number of triangles) we see 1 m^2 missing from a square of side 2.4 m. (Solution 1). Area is $(2.4)^2 - 1\ \text{m}^2 = 4.8\ \text{m}^2$.

4. **d.** The area is $\overline{AB} \bullet \overline{BC} \bullet \sin B \bullet 0.5 = 17{,}738\ \text{m}^2$. Since a hectare contains 10,000 m^2, the area is 1.77 hectares.

5. **d.** The general solution of the given differential equation is needed. This is found by setting $x = e^{rt}$. Thus $(r^2 + 4r + 6)e^{rt} = 0$, or $r = \frac{-4 \pm \sqrt{16-24}}{2} = -2 \pm \sqrt{2}t$. Hence $x = e^{-2t}\left(A \sin \sqrt{2}t + B\cos \sqrt{2}t\right)$.

6. **c.** The quasiperiod is determined by setting the argument of the sine function equal to 2π: $\sqrt{2}t = 2\pi,\quad t = \frac{2\pi}{\sqrt{2}} = \sqrt{2}\pi$.

7. **a.** Set $x(t) = 0$: $2e^{-2t} \sin \sqrt{2}t = 0$. This first occurs at $\sqrt{2}t = \pi$, that is, $t = \frac{\pi}{\sqrt{2}}$.

8. **c.** $P(n,r) = P(5,3) = 5!/(5-3)! = 5 \times 4 \times 3 \times 2 \times 1/(2 \times 1) = 60.$

9. **a.** Since the outcomes are mutually exclusive, the probabilities are additive, so

$$\begin{aligned} P\{\text{even}\} &= P\{2\} + P\{4\} + P\{6\} + P\{8\} + P\{10\} + P\{12\} \\ &= 1/36 + 3/36 + 5/36 + 5/36 + 3/36 + 1/36 = 18/36 = 1/2 \end{aligned}$$

10. **b.** The value of z, using determinants, is given by z = $\mathbf{D_3/D}$. In this instance $\mathbf{D_3}$ is the determinant

$$D_3 = \begin{vmatrix} 1 & 3 & 1 \\ 2 & 4 & 8 \\ 3 & 5 & 12 \end{vmatrix} = 1\begin{vmatrix} 4 & 8 \\ 5 & 12 \end{vmatrix} - 2\begin{vmatrix} 3 & 1 \\ 5 & 12 \end{vmatrix} + 3\begin{vmatrix} 3 & 1 \\ 4 & 8 \end{vmatrix}$$
$$= 1(48-40) - 2(36-5) + 3(24-4) = 6$$

Hence, z = $\mathbf{D_3/D}$ = 6/6 = 1.

11. **a.** The cofactor of $\mathbf{A}_{11}$ is found by deleting Row 1 and Column 1 from the matrix **A,** multiplied by $(-1)^{m+n}$ where m and n are the row and column numbers, respectively. Thus the cofactor is

$$A_{11} = \begin{bmatrix} 4 & 6 \\ 5 & 7 \end{bmatrix}$$

Evaluating the determinant of $\mathbf{A}_{11}$ yields

$$\begin{bmatrix} 4 & 6 \\ 5 & 7 \end{bmatrix} = 28 - 30 = -2$$

12. d. The inverse matrix can be found by the method described in Chapter 6, Example 6.4, or by the method described here. It is well known that

$$\mathbf{A}^{-1} = \frac{\text{Adj A}}{\text{Det A}}$$

The matrix Adj A is called the adjoint of **A**. Each entry is a cofactor of the transpose of **A**; that is, each entry in the adjoint is computed from the transpose of the original **A** matrix via the method described in the solution immediately preceding this solution. Thus

$$\text{Adj }\mathbf{A} = \begin{bmatrix} \begin{vmatrix} 4 & 5 \\ 6 & 7 \end{vmatrix} & -\begin{vmatrix} 3 & 5 \\ 2 & 7 \end{vmatrix} & \begin{vmatrix} 3 & 4 \\ 2 & 6 \end{vmatrix} \\ -\begin{vmatrix} 2 & 3 \\ 6 & 7 \end{vmatrix} & \begin{vmatrix} 1 & 3 \\ 2 & 7 \end{vmatrix} & -\begin{vmatrix} 1 & 2 \\ 2 & 6 \end{vmatrix} \\ \begin{vmatrix} 2 & 3 \\ 4 & 5 \end{vmatrix} & -\begin{vmatrix} 1 & 3 \\ 3 & 5 \end{vmatrix} & \begin{vmatrix} 1 & 2 \\ 3 & 4 \end{vmatrix} \end{bmatrix} = \begin{bmatrix} -2 & -11 & 10 \\ 4 & 1 & -2 \\ -2 & 4 & -2 \end{bmatrix}$$

Since Det **A** = 6 (from Problem 10), the inverse is

$$\mathbf{A}^{-1} = \frac{1}{6}\begin{bmatrix} -2 & -11 & 10 \\ 4 & 1 & -2 \\ -2 & 4 & -2 \end{bmatrix}$$

Sample Examination Solutions/PM—Statics

13. c.

$$M = \int_A r\,df = \int_a^b \int_0^{2\pi} r\frac{\mu P\, r\, dr\, d\theta}{\pi(b^2 - a^2)} = \frac{2\mu P}{b^2 - a^2}\int_a^b r^2 dr = \frac{2\mu(b^3 - a^3)}{3(b^2 - a^2)}P$$

$dA = rdrd\theta$

$df = \frac{\mu PdA}{\pi(b^2 - a^2)}$

Exhibit 13a

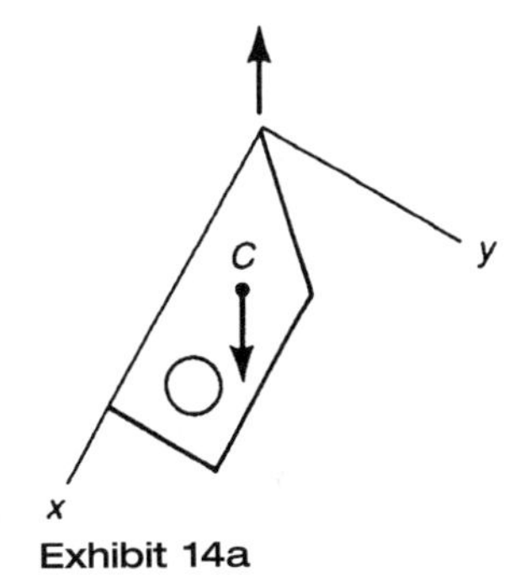

Exhibit 14a

14. b. The area of the plate is $A = 1.359\ \text{ft}^2$.

$$Ac_x = \left[(0.3)(0.7)(0.35) - \frac{1}{2}(0.3)^2(.15) - \frac{\pi}{4}(.133)^2(.55)\right]\text{m}^3$$

$$c_x = .0638\ \text{m}$$

$$Ac_y = \left[(0.3)(0.7)(0.15) - \frac{1}{2}(0.3)^2(0.15) - \frac{\pi}{4}(.133)^2(.15)\right]\text{m}^3$$

$$c_y = 0.2727\ \text{m}$$

$$\beta = \tan^{-1}\frac{c_y}{c_x} = 19.6°;\ \alpha + \beta = 45°,\ \alpha = 25.4$$

15. **d.** Let c_z be the distance from top of tank down to its center of mass.

$$c_z = \frac{\pi(187.5)(900)\frac{2(187.5)}{\pi} + \pi(187.5)^2 \frac{4(187.5)}{3\pi}}{\pi(187.5)(900) + \pi(187.5)^2} = 112.5 \text{ mm}$$

16. **a.**

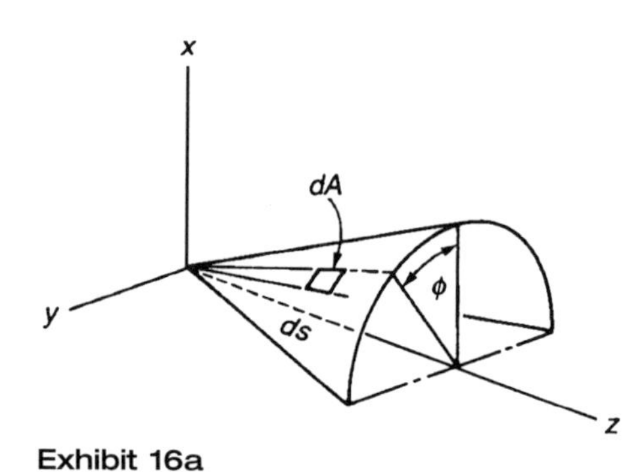

Exhibit 16a

$$dA = ds\frac{z}{h}ad\phi = \frac{1}{h}dz\frac{z}{h}ad\phi = \frac{al}{h^2}zdzd\phi$$

where $l = \sqrt{a^2 + h^2}$.

$$A = \frac{al}{h^2}\int_{-\pi/2}^{\pi/2}\int_0^h z\,d\,zd\phi = \frac{\pi al}{2}$$

$$c_z = \frac{al}{h^2 A}\int_{-\pi/2}^{\pi/2}\int_0^h z^2 dzd\phi = \frac{2}{\pi h^2}\frac{h^3}{3}\pi = \frac{2h}{3}$$

$$c_x = \frac{al}{h^2 A}\int_{-\pi/2}^{\pi/2}\int_0^h \frac{az}{h}\cos\phi\, zdz\, d\phi = \frac{2a}{\pi h^3}\int_{-\pi/2}^{\pi/2}\int_0^h z^2 \cos\phi\, dz\, d\phi = \frac{4a}{3\pi}$$

$$\sqrt{c_x{}^2 + c_z{}^2} = \sqrt{\left(\frac{2h}{3}\right)^2 + \left(\frac{4a}{3\pi}\right)^2} = \frac{2h}{3}\sqrt{1 + \left(\frac{2a}{\pi h}\right)^2}$$

17. **a.**

$$\sum M_B = (4 \text{ m})R_A - (2.5 \text{ m})(2000 \text{ N}) = 0$$

$$R_A = 1250 \text{ N}$$

$$M = (1.625 \text{ m})(1250 \text{ N}) - \left(\frac{0.625}{2}\text{ m}\right)(1250 \text{ N}) = 1641 \text{ N}\bullet\text{m}$$

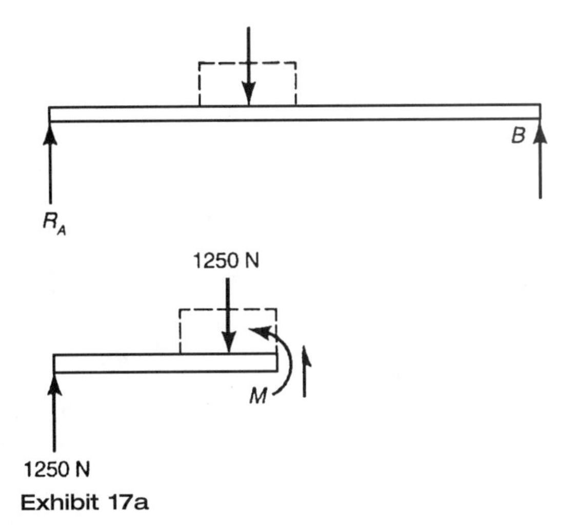

Exhibit 17a

18. b.

$$\sum M_B = \left(\frac{2}{3}L\right)w_0 L - LR = 0$$

$$R = \frac{2w_0}{3}L$$

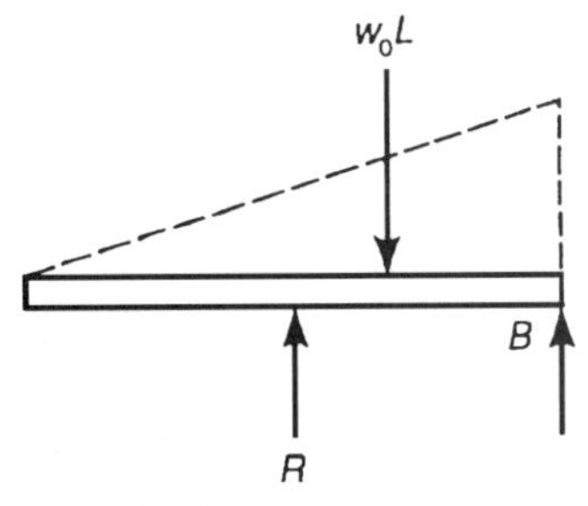

Exhibit 18a

Sample Examination Solutions/PM—Dynamics

19. d. See Exhibit 19a. The accelerations are

$$a_z = 5 \text{ m/s}^2$$

$$a_r = \ddot{r} - r\dot{\phi}^2 = 0 - 1.5\left[20\frac{2\pi}{60}\right]^2 = -6.58 \text{ m/s}^2$$

$$a_\phi = r\ddot{\phi} + 2\dot{r}\dot{\phi} = 0 + 2(1)\left[20\frac{2\pi}{60}\right] = 4.19 \text{ m/s}^2$$

Thus,

$$a = \sqrt{a_r^{\,2} + a_\phi^{\,2} + a_z^{\,2}} = \sqrt{5^2 + (-6.58)^2 + (4.19)^2} = 9.27 \text{ m/s}^2$$

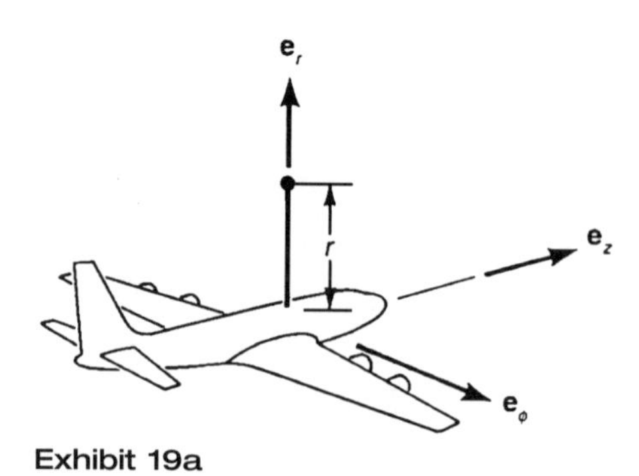

Exhibit 19a

20. b. In Exhibit 20a, the impulse-momentum principle applied along the x-direction gives $mv_1 + \int \sum F_x dt = mv_2$.

$$0 + (P - \mu N - \text{mg} \sin 20°)\Delta\text{t} = \text{m}v_2 \tag{1}$$

and $\Sigma F_y = 0$ produces

$$N = \text{mg} \cos 20° \tag{2}$$

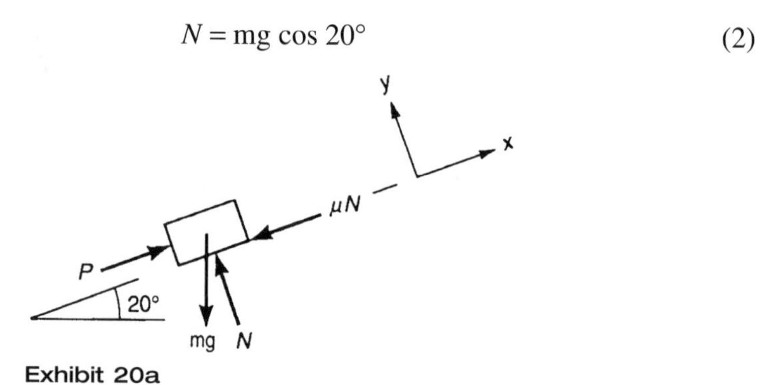

Exhibit 20a

Combining Eqs. (1) and (2),

$$(P - \mu \text{ mg} \cos 20° - \text{mg} \sin 20°)\Delta t = \text{m}v_2$$

so that

$$P = \frac{\text{m}v_2}{\Delta t} + \text{mg}(\mu \cos 20° + \sin 20°)$$

Substituting values, P = 20 N.

21. a. Referring to Exhibit 21a,

$$\sum M_o = I_o\alpha \text{ leads to } m_1 g\frac{1}{2} + m_2 g(l + r) = I_o\alpha$$

Thus,

$$\alpha = \frac{g}{I_o}[0.5m_1 l + m_2(l + r)]$$

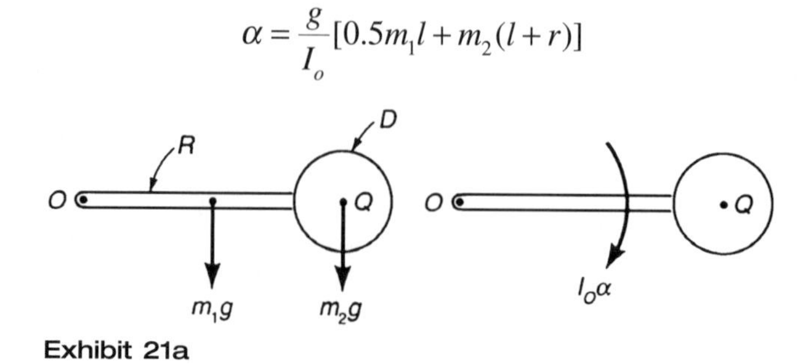

Exhibit 21a

Now, $I_O = I_{R/O} + I_{D/O} = I_{R/O} + I_{D/Q} + m_2(l+r)^2 = (1/3)m_1 l^2 + 0.5 m_2 r^2 + m_2(l+r)^2$. Substituting values,

$$I_O = 6.51 \text{ kg} \bullet \text{m}^2 \quad \text{and} \quad \alpha = 8.74 \text{ rad/s}^2$$

F = 4N

mg

Exhibit 22a

22. a. According to Exhibit 22a, $\Sigma F_x = ma_x$ leads to $-F = ma_x$. Hence, $a_x = -F/m$ $-4/2 = -2$ m/s^2 (a constant). Thus,

$$v_x = v_{x0} + a_x t = 60 \cos 60° - 2(10) = 10 \text{ m/s}.$$

Similarly $\Sigma F_y = ma_y$ gives $-mg = ma_y$, or $a_y = -g = -9.8$ m/s^2. Hence, $v_y = v_{y0} + a_y t = 60 \sin 60° - 9.8(10) = -46.0$ m/s, and

$$v = \sqrt{v_x^2 + v_y^2} = \sqrt{10^2 + (-46.0)^2} = 47.1 \text{ m/s}$$

23. c. For the vertical motion between A and B, $y = y_0 + v_{y0}t + a_y t^2/2$. At B, the y coordinate is again zero. Hence,

$$0 = 0 + 60 \sin 60° t - 9.8t^2/2$$

or

$$t = \frac{2(60)\sin 60°}{9.8} = 10.6 \text{sec}.$$

The horizontal distance covered in this time is

$$x = x_0 + v_{x0}t + a_x t^2/2$$
$$= 0 + 60 \cos 60° (10.6) - 2(10.6)^2/2 = 206 \text{ m}.$$

Sample Examination Solutions/PM—Mechanics of Materials

24. d. Draw the free-body diagram of the beam AC, replacing the uniform load with its static equivalent. Summation of moments about A gives

$$R_B(600 \text{ mm}) = (3{,}600 \text{ N})(900 \text{ mm}); \qquad R_B = \frac{(3600)(900)}{(600)} = 5400 \text{ N}$$

Summation of forces in the vertical direction gives $R_A = R_B - 3600 = 1800$ N.

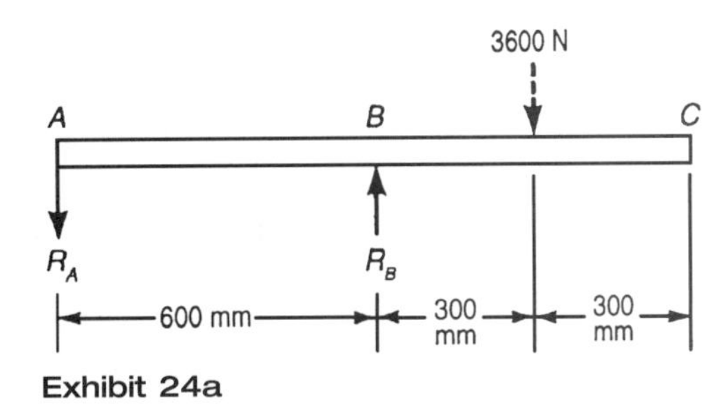

Exhibit 24a

25. b. The member *BD* has a compressive force equal to the reaction at *B*; therefore,

$$\sigma = \frac{P}{A} = \frac{-5400\ \text{N}}{(12\ \text{mm})^2} = -37.5\ \text{MPa}$$

26. d. Draw the shear diagram for the beam *AC*. The maximum shear is 840 lb just to the right of point *B*.

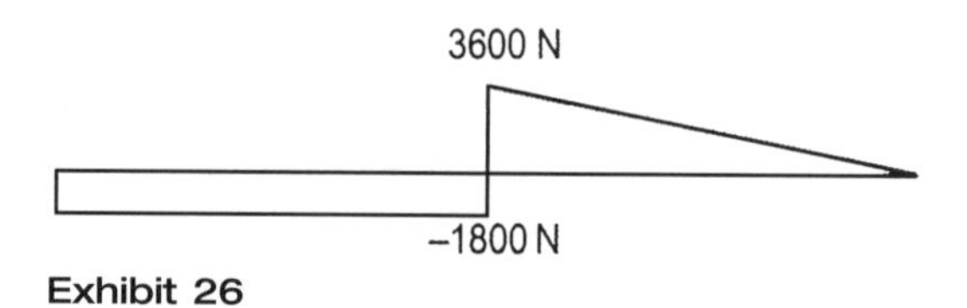

Exhibit 26

27. d. From the previous shear diagram, the moment at *B* is the area *A* shown in Exhibit 27.

$$M_B = A = (-1800\ \text{N})(600\ \text{mm}) = -1.080\ \ \text{kN} \bullet \text{m}$$

The moment diagram can then be drawn as shown in Exhibit 27a.

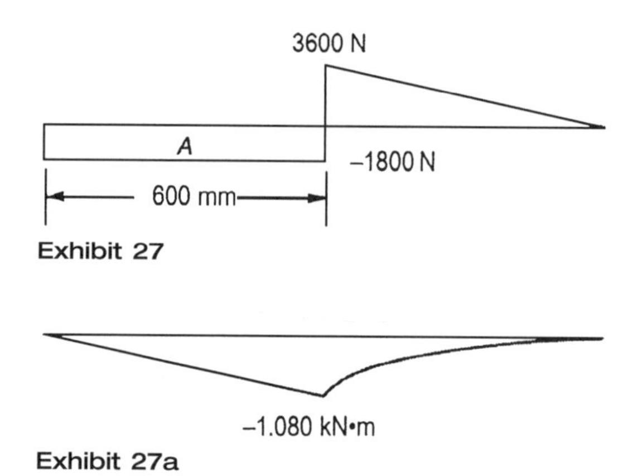

Exhibit 27

Exhibit 27a

28. d. The maximum bending stress is

$$\sigma_{\text{max}} = \frac{M_{\text{max}} C}{I}$$

where for the rectangular cross-section

$$c = \frac{h}{2} = 25\ \text{mm} \quad \text{and} \quad I = \frac{bh^3}{12} = \frac{(25\ \text{mm})(50\ \text{mm})^3}{12} = 260 \times 10^3\ \text{mm}^4$$

The maximum bending stress is

$$\sigma_{\text{max}} = \frac{M_{\text{max}} C}{I} = \frac{(1.080\ \text{kN} \bullet \text{m})(25\ \text{mm})}{260 \times 10^3\ \text{mm}^4} = 103.7\ \text{MPa}$$

Sample Examination Solutions/PM—Fluid Mechanics

29. b.

$$W + T = F_B = \gamma_W V_D$$

$$T = \gamma_w V_D - W = \gamma_w \bullet \frac{V_c}{2} - \gamma_c V_c = V_c\left(\frac{\gamma_w}{2} - \gamma_c\right) \quad V = \frac{Q}{A} = \frac{Q}{\pi d^2/4} = \frac{4Q}{\pi d^2}$$

(U.S.) $$T = \frac{\pi(1\text{ ft})^2}{4} \bullet 3\text{ ft}\left(\frac{62.4\frac{\text{lbf}}{\text{ft}^3}}{2} - 25\frac{\text{lbf}}{\text{ft}^3}\right) = 14.6\text{ lbf}$$

(S.I.) $$T = \frac{\pi(0.3\text{ m})^2}{4} \bullet 0.91\text{ m}\left(\frac{9.81\frac{\text{kN}}{\text{m}^3}}{2} - 3.93\frac{\text{kN}}{\text{m}^3}\right) = 63\text{ N}$$

30. d.

$$F = pQV_2 = pA_2V_2^2 = 2\gamma_w A_2 h, \quad \text{since } V_2^{\,2} = 2gh \text{ and } \rho g = \gamma.$$

(U.S.) $$F = 2 \bullet 62.4\frac{\text{lbf}}{\text{ft}^3} \bullet \frac{\pi(2\text{ in.})^2}{4} \bullet \frac{1\text{ ft}^2}{144\text{ in.}} \bullet 15\frac{\text{ft}}{\text{s}} = 40.8\text{ lbf}$$

(S.I.) $$F = 2 \bullet 9.81\frac{\text{kN}}{\text{m}^3} \bullet \frac{\pi(0.05\text{ m})^2}{4} \bullet 4.6\frac{\text{m}}{\text{s}} = 177\text{ N}$$

31. c.

(U.S.) $$\rho = \text{SG} \bullet 1.94\frac{\text{slug}}{\text{ft}^3} = 0.93 \bullet 1.94\frac{\text{slug}}{\text{ft}^3} = 1.80\frac{\text{slug}}{\text{ft}^3}$$

$$V = \frac{Q}{A} = \frac{Q}{\frac{\pi d^2}{4}} = \frac{4Q}{\pi d^2} = \frac{4 \bullet 0.5\frac{\text{ft}^3}{\text{s}}}{\pi(4.026\text{ in.})^2} \bullet \frac{144\text{ in.}^2}{1\text{ ft}^2} = 5.66\frac{\text{ft}}{\text{s}}$$

$$\text{Re} = \frac{\rho \text{Vd}}{\mu} = \frac{1.80\frac{\text{slug}}{\text{ft}^3} \bullet 5.66\frac{\text{ft}}{\text{s}} \bullet 4.026\text{ in.} \bullet \frac{1\text{ ft}}{12\text{ in.}}}{7\times10^{-4}\frac{\text{lbf}-\text{s}}{\text{ft}^2}} = 4.9\times10^3$$

(S.I.) $$\rho = \text{SG}\left(1000\frac{\text{kg}}{\text{m}^3}\right) = 0.93 \bullet 1000\frac{\text{kg}}{\text{m}^3} = 930\frac{\text{kg}}{\text{m}^3}$$

$$V = \frac{4Q}{\pi d^2} = \frac{4 \bullet 0.0142\frac{\text{m}^3}{\text{s}}}{\pi(0.1023\text{ m})^2} = 1.73\frac{\text{m}}{\text{s}}$$

$$\text{Re} = \frac{\rho \text{Vd}}{\mu} = \frac{930\frac{\text{kg}}{\text{m}^3} \bullet 1.73\frac{\text{m}}{\text{s}} \bullet 0.1023\text{ m}}{3.35\times10^{-2}\frac{\text{N}\bullet\text{S}}{\text{m}^2}} = 4.9\times10^3$$

32. a. (U.S.) For steel,

$$\varepsilon = 1.5\times10^{-4}\text{ft},\ \frac{\varepsilon}{d} = \frac{1.5\times10^{-4}\text{ft}}{4.026\text{ in.}} \bullet \frac{12\text{ in.}}{\text{ft}} = 0.00045$$

$$V = \frac{Q}{A} = \frac{4Q}{\pi d^2} = \frac{4\bullet 1.0\frac{\text{ft}^3}{\text{s}}}{\pi(4.026\text{ in.})^2} \bullet \frac{144\text{ in.}^2}{\text{ft}^2} = 11.31\frac{\text{ft}}{\text{s}}$$

When Q and V are twice that in the previous problem, Re = 9.8×10^3. From the Moody diagram with

$\frac{\varepsilon}{d} = 0.00045, f = 0.031.$ From Darcy's equation

$$h_f = f\frac{L}{d}\frac{V^2}{2g} = 0.031\bullet\frac{100\text{ ft}}{4.026\text{ in.}}\bullet\frac{12\text{ in.}}{\text{ft}}\bullet\frac{\left(11.32\frac{\text{ft}}{\text{s}}\right)^2}{2\bullet 32.2\frac{\text{ft}}{\text{s}^2}} = 18.35\text{ ft}$$

(S.I.) For steel,

$$\varepsilon = 4.6\times10^{-5}\text{m},\ \frac{\varepsilon}{d} = \frac{4.6\times10^{-5}\text{m}}{0.1023\text{ m}} = 0.00045$$

$$V = \frac{Q}{a} = \frac{4Q}{\pi d^2} = \frac{4\bullet 0.0284\frac{\text{m}^3}{\text{s}}}{\pi(0.1023\text{m})^2} = 3.46\frac{\text{m}}{\text{s}}$$

When Q and V are twice that in the previous problem, Re = 9.8×10^3. From the Moody diagram with $\frac{\varepsilon}{d} = 0.00045, f = 0.031$. From Darcy's equation,

$$h_f = f\frac{L}{d}\frac{V^2}{2g} = 0.031\bullet\frac{30.5\text{ m}}{0.1023\text{ m}}\bullet\frac{\left(3.46\frac{\text{m}}{\text{s}}\right)^2}{2\bullet 9.81\frac{\text{m}}{\text{s}^2}} = 5.64\text{ m}$$

33. d.

$$h_a = \frac{p_2 - p_1}{\gamma} + \frac{V_2^2 - V_1^2}{2g} + Z_2 - Z_1$$

for point (1) at the tank free surface and point (2) at the pump discharge; here $p_1 = V_1 = Z_2 = 0$.

(U.S.) $\gamma = \text{SG}\bullet 62.4\frac{\text{lbf}}{\text{ft}^3} = 0.93\bullet 62.4\frac{\text{lbf}}{\text{ft}^3} = 58.0\frac{\text{lbf}}{\text{ft}^3}$

$$V_2 = 11.31\frac{\text{ft}}{\text{s}},\ Z_1 = 10\text{ ft},\ Q = 1.0\frac{\text{ft}^3}{\text{s}},\ p_2 = 10\text{ psig}$$

$$h_A = \frac{10\frac{\text{lbf}}{\text{in.}^2}\bullet\frac{144\text{ in.}^2}{\text{ft}^2}}{58\frac{\text{lbf}}{\text{ft}^3}} + \frac{\left(11.31\frac{\text{ft}}{\text{s}}\right)^2}{2\bullet 32.2\frac{\text{ft}}{\text{s}^2}} - 10\text{ ft} = 16.8\text{ ft}$$

$$P = \gamma Q h_A = 58.0\frac{\text{lbf}}{\text{ft}^3}\bullet 1.0\frac{\text{ft}^3}{\text{s}}\bullet 16.8\text{ ft}\bullet\frac{1\text{ HP}}{550\frac{\text{ft - lbf}}{\text{s}}} = 1.77\text{ HP}$$

$$\text{(S.I.)} \quad \gamma = \text{SG} \bullet 9.81\frac{\text{kN}}{\text{m}^3} = 0.93 \bullet 9.81\frac{\text{kN}}{\text{m}^3} = 9.12\frac{\text{kN}}{\text{m}^3}$$

$$V_2 = 3.46\frac{\text{m}}{\text{s}}, \quad Z_1 = 3.05\text{ m}, \quad Q = 0.0284\frac{\text{m}^3}{\text{s}}, \quad p_2 = 69\text{ kPa}$$

$$h_A = \frac{69\frac{\text{kN}}{\text{m}^2}}{9.12\frac{\text{kN}}{\text{m}^3}} + \frac{\left(3.46\frac{\text{m}}{\text{s}}\right)^2}{2 \bullet 9.81\frac{\text{m}}{\text{s}^2}} - 3.05\text{ m} = 5.12\text{ m}$$

$$P = \gamma Q h_A = 9.12\frac{\text{kN}}{\text{m}^3} \bullet 0.0284\frac{\text{m}^3}{\text{s}} \bullet 5.12\text{ m} = 1.32\frac{\text{kN} \bullet \text{m}}{\text{s}} = 1.32\text{ kW}$$

Sample Examination Solutions/PM—Thermodynamics

34. c.

$$\dot{Q}_{\text{REFR}} = \dot{m}\Delta h_{\text{EVAP}} = \dot{m}(h_1 - h_4) = \dot{m}(h_1 - h_3)$$
$$h_3 = h_4$$

From the table at saturated vapor h_1 = 186.9 kJ/kg (at –1.1°C). From the table at saturated liquid h_3 = 77.6 kJ/kg (at 1035 KPa).

35. d. $\dot{W}_{\text{COMP}} = \dot{m}\Delta h_{\text{COMP}} = \dot{m}(h_2 - h_1)$
h_2 is found at p_2 = 1035 kPa and 49°C; h_2 = 209.5 kJ/kg.

$$\dot{W}_{\text{COMP}} = \frac{10.9}{60} \times (209.5 - 186.9) = 4.13\text{ k}W$$

36. c.

$$\text{COP}_{\text{REFR}} = \frac{\dot{W}_{\text{REFR}}}{\dot{W}_{\text{COMP}}} = \frac{19.9}{4.1} = 4.85$$

37. d.

$$\dot{Q}_{\text{COND}} = \dot{m}\Delta h_{\text{COND}} = \dot{m}(h_2 - h_3)$$
$$= \frac{10.9}{60} \times (209.5 - 77.6) = 24\text{ kW}$$

38. d.

$$\text{COP}_{\text{MP}} = \dot{Q}_{\text{COND}}/\dot{W}_{\text{COMP}} = \text{COP}_{\text{REFR}} + 1.0$$
$$= 4.85 + 1 = 5.85$$

Sample Examination Solutions/PM—Electrical Circuits

39. c. For $t > 0$, the loop equation is $V = R\text{i} + L(di/dt)$. The standard form for the solution of this differential equation is

$$i(t) = I_{\text{fv}}(1 - e^{-t/\tau}), \quad \tau = \text{L/R} = 3/4 = 0.75\text{ s}, \quad I_{\text{fv}} = \text{V}/\sum R = 5/4 = 1.25\text{ A}$$
$$i(t = 1) = 1.25(1 - e^{-1/0.75}) = 0.92\text{ A}, \quad v_{2\Omega} = iR_{2\Omega} = 0.92 \times 2 = 1.84\text{ A}$$

40. d. Assume two loop currents. The one in the left block is I_1 (assume clockwise), and the one in the right block is I_2 (assume counterclockwise). The loop voltage equations are

$$\text{Loop 1: } 10 = 7(I_1 + I_2) + (-12) + 7I_1;\ 22 = 14I_1 + 7I_2$$

$$\text{Loop 2: } 8 = 7I_2 + 7(I_1 + I_2) + (-12);\ 20 = 7I_1 + 14I_2$$

$$I_1 = \frac{\begin{vmatrix} 22 & 7 \\ 20 & 14 \end{vmatrix}}{\begin{vmatrix} 14 & 7 \\ 7 & 14 \end{vmatrix}} = \frac{308-140}{196-49} = 1.14 \text{ A}$$

$$I_2 = \frac{\begin{vmatrix} 14 & 22 \\ 7 & 20 \end{vmatrix}}{\begin{vmatrix} 14 & 7 \\ 7 & 14 \end{vmatrix}} = \frac{280-154}{147} = 0.86 \text{ A}$$

$$I_{\text{total}} = I_1 + I_2 = 1.14 + 0.86 = 2.00 \text{ A}$$

41. b. Convert each parallel branch to its series equivalent.

$$R_{\text{eq}} = 1\ \Omega;\ X_{\text{Leq}} = 1\ \Omega;\ X_{\text{C}} = 1\ \Omega;\ \mathbf{Z} = 1 + j1 - j1 = 1\ \angle 0°$$

$$\mathbf{I}_{\text{total}} = \mathbf{V}/\mathbf{Z} = 1\text{V}/1\ \Omega = 1 \text{ A};\ P = I^2R_{\text{eq}} = 1 \text{ watt}$$

42. d. Summing currents at V gives

$$-10\angle 0° + \mathbf{V}/(5 - j5) + (\mathbf{V} - 10\angle 0°)/(j5) = 0$$

$$-10 + \mathbf{V}/(5\sqrt{2}\ \angle{-45°}) + (\mathbf{V} - 10)/(5\angle 90°) = 0$$

Now multiply the equation by 10: $100 = (\sqrt{2}\angle 45° + 2\angle{-90°})\mathbf{V} - 20\angle 90°$.
$100 - j20 = [(1 + j1) - j2]\mathbf{V}$; $\mathbf{V} = 20[(5 - j1)/(1 - j1)] = 60 + j40 = 72.1\angle 33.7°$ V

43. d. Make the following two assumptions for an ideal op-amp: $v(+)$ input = $v(-)$ input, $i(+)$ input = 0 = $i(-)$ input, then $v(+)$ input = 8 V = $v(-)$ input. The current through the 4 kΩ resistor is 8V/4 kΩ = 2 mA; this must be the same as through R_{L}.

44. d. Since $i = dq/dt$, $q = \int_0^{0.001} i\,dt$, and $v = q/C$; $q = \int_0^{0.001} 2\,dt = 0.002$ coulombs, and $v = 0.002/(10 \times 10^{-6}) = 200$ V. Then the energy is $w = Cv^2/2 = (10 \times 10^{-6})(200)^2/2 = 0.2$ J.

Sample Examination Solutions/PM—Materials Engineering

45. b. $x(64.928) + (1 - x)(62.930) = 63.546$; $x = 0.31$.

46. d. Make a sketch. Also, the dot product of (111) and $[1\bar{1}0]$ is zero.

47. b. By interpolation at 150°C, an 80Pb–20Sn alloy is 10% of the distance from the α composition limit to the β composition limit.

Sample Examination Solutions/PM—Chemistry

48.

a. **d.** In going across the periodic table from left to right in a given period, the covalent radius of the elements usually decreases as the electrons are held by increasing nuclear charge. The transition metals series and some adjacent metals deviate from this trend. Going down a group of elements, succeeding elements have a greater number of electron shells, leading to increased covalent radii. Series of isoelectronic elements have both ionic and covalent radii and the trend will be to smaller radius for the element with higher nuclear charge.

b. **c.** Ionic radii are larger than covalent radii for negative ions and smaller for positive ions. Going down a group both positive and negative ions tend to have larger radii because of the larger electron shells. In an isolectronic series the ions with higher atomic number (greater nuclear charge) tend to have smaller ionic radii.

49. d. The valence band is full, but the gap is small enough so that a useable number of electrons jump the gap. (Answer (A) has the carrier types interchanged.)

50. a. This question pertains to extrinsic semiconductors and NOT to intrinsic semiconductors.

51. b. The more active metals, including all of Group IA, most of Group IIA, and some other metals, react with water to yield the hydroxyl compound and hydrogen. No oxygen is evolved. The soluble species is ionic, with OH^- acting as a base. The nonmetal hydroxyl compounds formed with water are acidic. Superoxide forms from the reaction of Group IA metals with oxygen.

Sample Examination Solutions/PM—Engineering Economics

52. b.

$$P = A(P/A,i\%,n) = 1000(P/A,6\%,50) = 1000(15.762) = 15{,}762$$

53. d.

$$i_{\text{eff}} = \left[1+(r/k)\right]^k - 1$$
$$0.0931 = \left[1+(r/k)\right]^4 - 1$$
$$1 + r/4 = 1.0931^{1/4} = 1.0225$$
$$r = 0.09 = 9\% \text{ per year}$$

54. d.

$$F = 1000(F/A,8\%,10)(F/P,8\%,5) = 1000(14.487)(1.469) = 21{,}281$$

55. c. Synthetic division

1	4	6	4	−1	trial root
	−1	−3	−3		a remainder of 1
1	3	3	1		

56. d.

$$x_{n+1} = x_n - P_n(x_n)/P_n'(x_n)$$
$$\text{for } x_1 = -1,\ P_n'(x) = 3x^2 + 8x + 6\,|_{x=-1} = 3 - 8 + 6 = 1$$
$$P_n(x)\,|_{x=-1} = -1 + 4 - 6 + 4 = +1$$
$$\therefore x_{n+1} = (-1) - (1/1) = -2$$

57. c. Using Eq. (4.9a),

$$(x_{k+1}) = x_k - Tax_k + Tf_k \qquad \text{for fixed increments}$$
$$x_1 = x_0 - Ta_{x_0} + TF = 0 - 0 + 0.3 \times 1 = 0.3$$

Sample Examination Solutions/PM—Ethics

58. c. The best answer is (C). If (A) is chosen and discovered, you may jeopardize your employment. Working at the office, using company resources, tools, or proprietary methods, would be in conflict of interest, so answer (B) must be excluded. Answer (D) might be acceptable but relies on involved parties as witnesses to an agreement.

59. c. As a professional engineer you have a responsibility to safeguard life, health, and property. Thus you should take appropriate action to advise the proper person or people of the need to review the fuse selection. The question becomes who should one contact. While alternative (B) might be all that is needed, Alternative (C) is a more formal action (memo instead of a call), is addressed to the supervisor level, and is therefore preferred. Alternative (D) is premature since you do not know the current state of the design (the fuse size may already have been changed). Almost certainly (D) is not acceptable to your employer or in accordance with his policies on how to handle matters like this.

60. c. Although, in a legal sense, only action (A) is unacceptable; items (B) and (D) may possibly not be in the best interest of everyone because of unknown human effects; therefore item (C) is in the best interest of public safety.

INDEX

T

U